DESIGNING FOR AN aging population

Ten Years of Human Factors/Ergonomics Research

EDITED BY WENDY A. ROGERS

Human Factors and Ergonomics Society

ISBN 0-945289-08-1

When citing work from this book, please refer to the **original** proceedings publication, as in the following example:

Ellis, S. R., & Menges, B. M. (1996). Effects of age on the judged distance to virtual objects in the near visual field. In *Proceedings of the Human Factors and Ergonomics Society 40th Annual Meeting* (pp. 1197–1201). Santa Monica, CA: Human Factors and Ergonomics Society.

(Note that in 1993, the Society's name changed to the Human Factors and Ergonomics Society.)

The following HFES proceedings are represented in this book:

1996, 40th Annual Meeting	1991, 35th Annual Meeting
1995, 39th Annual Meeting	1990, 34th Annual Meeting
1994, 38th Annual Meeting	1989, 33rd Annual Meeting
1993, 37th Annual Meeting	1988, 32nd Annual Meeting
1992, 36th Annual Meeting	1987, 31st Annual Meeting

The HFES annual meeting proceedings are indexed or abstracted in the following publications or services: *Applied Mechanics Reviews, Engineering Index Annual, EI Monthly, Cambridge Scientific Abstracts, EI Bioengineering Abstracts, EI Energy Abstracts, Ergonomics Abstracts, ISI Index to Scientific & Technical Proceedings,* and *International Aerospace Abstracts.* This publication is also available on microfilm from University Microfilms International, 300 N. Zeeb Road, Department P.R., Ann Arbor, MI 48106; 18 Bedford Row, Department P.R., London WC1R 4EJ, England.

To obtain copies of papers not included in this book, readers may:

1. purchase back volumes (see back page for ordering information);
2. access documents available through the above-listed indexing/abstracting services;
3. obtain a microfilm/microfiche copy through UMI (see above), or
4. order photocopies from HFES at the cost of $7.50 per paper ($17.50/paper for rush orders).

Additional copies of this book may be purchased from the Human Factors and Ergonomics Society at $42 per copy for HFES members and $58 for nonmembers. Add California sales tax for deliveries in California. Discounts apply on purchases of 5 or more copies. In addition, a 10% discount applies when ordering this book as part of a four-book set of HFES proceedings collections. The other three books are:

Human Factors Perspectives on Warnings: Selections from Human Factors and Ergonomics Society Annual Meetings, 1980–1993 (ISBN 0-945289-02-2, 1994, $35 HFES members, $50 nonmembers)

Human Factors Perspectives on Human-Computer Interaction: Selections from Proceedings of Human Factors and Ergonomics Society Annual Meetings, 1983–1994 (ISBN 0-945289-05-7, 1995, $42 HFES members, $58 nonmembers)

Ergonomics and Musculoskeletal Disorders: Research on Manual Materials Handling, 1982–1996 (ISBN 0-945289-09-X, 1997, $35 HFES members, $50 nonmembers)

To order the four-book set, send $180 plus $15 shipping/handling and sales tax if applicable to HFES at the address above. Orders are accepted with prepayment by check (payable to Human Factors and Ergonomics Society), MasterCard, or VISA.

Library of Congress Cataloging-in-Publication Data

Designing for an aging population : ten years of human factors/ergonomics research / edited by Wendy A. Rogers.
p. cm.
Includes indexes.
ISBN 0-945289-08 1 (alk. paper)
1. Human engineering. 2. Aged. I. Rogers, Wendy A. II. Human Factors and Ergonomics Society.
TA166.D48 1997
620.8'2'0846--dc21 97-35926
CIP

Contents

Note: The original proceedings year and page numbers are given for each paper in parentheses.

Note: The original proceedings year and page numbers are given for each paper in parentheses.

Part III: Technology and Computers

Part IV: Workplace

Note: The original proceedings year and page numbers are given for each paper in parentheses.

Part V: Health, Warnings, and Safety

Part VI: Driving

Note: The original proceedings year and page numbers are given for each paper in parentheses.

Preface

The purpose of this collection is to provide easy access to important research about the ways in which systems, tools, and environments can be improved for optimum use by older adults. The discipline of human factors/ergonomics provides a unique technology for understanding the needs and capabilities of humans as they interact with systems; the present focus is on the issues that must be considered as those humans grow older.

The papers included here provide an in-depth look at the empirical studies of human factors and aging that have been conducted in the last decade. They should appeal to human factors practitioners, researchers, and students. For practitioners both within the Human Factors and Ergonomics Society and in the general community, the collection provides an overview of the recent research and issues to be considered in the development of products and the design of systems whose user population will include older adults. The book could also provide a solid basis for a graduate or advanced undergraduate seminar in aging and human factors.

Background and Content

The older population within developed countries is expanding. Some predict that by the year 2000, there will be 35 million people in the United States over the age of 65. There is a good deal of evidence to suggest that individuals over age 65 experience declines in sensory, perceptual, motor, and cognitive abilities that may interfere with their ability to interact with systems ranging from doorknobs to microwave ovens to computers. Thus there is—and will continue to be—a large population of individuals who can benefit from human factors solutions that are tailored to their needs and capabilities.

According to the report, *Human Factors Needs for an Aging Population* (Czaja, 1990), one of the primary ways to address the problems of an aging population is to "get existing . . . human factors knowledge into effective use" (p. 68). This collection is intended to help fill that need.

The 84 selections are organized into six parts. "Perception, Movement Control, and Biomechanics" contains papers that focus on the basic changes that accompany aging of which human factors practitioners should be aware. In "Cognition," the papers cover a variety of cognitive processes, ranging from simple reaction-time tasks to more complex dual- and multiple-tasks. These papers provide insight as to how age-related changes in cognition influence human factors issues.

The third section is on "Technology and Computers." Not surprisingly, most of the contributions in this section are fairly recent and focus on making technology accessible to people of all ages. Technologies covered in this part include mouse input devices, synthetic speech, remote controls, automatic teller machines, on-line library catalogs, and text editors. In the fourth part, "Workplace," the focus is on issues relevant to the older worker, encompassing concerns such as subjective workload, accidents, shiftwork, and retirement issues.

The fifth section is a combination of papers with the common theme of "Health, Warnings, and Safety." Topics include medication instructions and adherence, warning labels, and general safety issues. Finally, in "Driving," the authors cover the subjects of highway signs, individual differences in capabilities relevant to driving, general issues of vehicle design, and more.

Paper Selection

Human Factors and Ergonomics Society annual meeting proceedings from 1987 to 1996 were carefully reviewed to select papers with any relevance to older adults or the aging process. Brian Jamieson, Nina Lamson, Amber Robinson, and Gabe Rousseau assisted me in selecting a total of 184 papers. Each paper was then reviewed by two people. Reviewers were asked to base their judgments on the quality of the paper and its relevance to the goals of the collection. They rated the candidate papers on a scale of 1 (must be included) to 7 (must not be included). Given the page limitations for the collection, 84 papers (46%) were chosen. Their mean rating was 2.25 (*SD* = .72).

My sincere thanks go to the reviewers, each of whom reviewed 10–12 papers:

Nancy Anderson
Karlene Ball
Karel Brookhuis
Neil Charness
Regina Colonia-Willner
Brian Cooper
Sara Czaja
Beth Davis
Cindy Dulaney
Jerry Duncan
Katharina Echt
Darin Ellis
Kristen Gilbert
José Guerrier
Darryl Humphrey
Don Lassiter
Lila Laux
Rob Mahan
Tom Malone
Beth Meyer
Roger Morrell
Dan Morrow
Erik Olsen
Brian Peacock
Gabe Rousseau
Ben Somberg
Pamela Tsang
Marilyn Turner
Frank Winn
Mike Wogalter

I also thank Barry Beith and Mike Wogalter of the HFES Communications Subcouncil for their support, and Lois Smith, Dan Fisk, and Lynn Strother for their advice and guidance.

The process of identifying, reviewing, and selecting the papers for this collection has yielded two primary benefits. First is the obvious product of a well-organized collection that represents the cutting-edge, high-quality research that has recently been conducted in the realm of aging and human factors. The second, perhaps less apparent, product is the identification of areas of human factors in which there is a dearth of knowledge about the influence of the aging process. For example, we found very few papers about decision making, communication, consumer products, environmental design, industrial ergonomics, macroergonomics, and virtual environments. Thus, researchers and practitioners in these areas should consider addressing the importance of the age variable in their future projects. Perhaps their work will be highlighted in a collection of articles from the next decade.

Wendy A. Rogers
Athens, Georgia

Additional Resources

Birren, J. E., & Schaie, K. W. (1996). *Handbook of the psychology of aging.* San Diego, CA: Academic.

Bouma, H., & Graafmans, J. A. M. (1991). *Gerontechnology.* Amsterdam: IOS Press.

Craik, F. I. M., & Salthouse, T. A. (1991). *The handbook of aging and cognition.* Hillsdale, NJ: Erlbaum.

Czaja, S. J. (1990). *Human factors research needs for an aging population.* Washington, DC: National Academy Press.

Fisk, A. D., & Rogers, W. A. (1997). *Handbook of human factors and the older adult.* San Diego, CA: Academic.

INFLUENCE OF AGE ON THE ABILITY TO HEAR TELEPHONE RINGERS OF DIFFERENT SPECTRAL CONTENT

J. P. Berkowitz and S. P. Casali
Auditory Systems Laboratory
Virginia Polytechnic Institute and State University
Blacksburg, VA 24061-0118

The failure to detect a telephone ringer signal can prove frustrating or even hazardous in certain situations, especially for older individuals who rely heavily on telephone access. This study was conducted to investigate the detectability of telephone ringer signals with individuals having elevated hearing levels. Specifically, the study investigated the detectability of three acoustically different telephone ringer signals under two masking noise conditions (quiet and 65 dBA pink noise) for two subject age groups : 20-30 years of age and over 70 years of age. Common residential telephone ringers were sampled, with three acoustically different ringers selected for study. To determine hearing ability, pure tone audiograms were administered to all subjects. Subjects' threshold levels for each ringer were then determined. Significant differences were found between the two age groups, both across telephone ringers and across noise conditions. For the older group, an advantage was found for the ringer signal which contained prominent low-to-mid range frequency components. In addition, the threshold level in noise of one ringer (a high frequency "beeper" type ringer) proved to be approximately equal to the naturally occurring decibel level of that ringer. Thus, the beeper ringer in moderate level noise (65 dBA) was effectively inaudible. The results suggest that certain electronic ringers which are currently in vogue may be unsuitable for use by the elderly or by any individual with significant high-frequency hearing loss.

INTRODUCTION

Hearing Impairment in the Elderly

A substantial proportion of individuals 65 years and older suffer from functional hearing loss, with some noticeable loss occurring as early as the age of 40 (Bess, Lichtenstein, Logan, and Burger, 1989; Stelmachowicz, Beauchaine, Kalberer, and Jesteadt, 1989; Bingea, Raffin, Aune, Baye, and Shae, 1982). In terms of strict demographics, the greatest incidence of hearing loss in the United States occurs among males; however, the disparity in hearing loss between the sexes diminishes with increasing age, especially above 85 years. Hotchkiss (1989) reported that approximately 30 percent of persons 65 years and older consider themselves functionally hearing-impaired with the percentage increasing to almost 50 percent for the over 85 population. Thus 7.9 million people over the age of 65 years reported having a functional hearing loss in 1985. Projections indicate that this number will increase to 12.6 million people by the year 2015 (Hotchkiss, 1989). Because these results were based on self-reports of functional hearing difficulties, and not audiometric measures, they likely underestimate the actual magnitude of hearing impairment among the aged.

Within industrialized societies, hearing loss among the aging population, whether in the form of sensory, conductive or neural deficits, can be linked to number of different causal conditions (Mader, 1984; Zarnoch, 1982). *Presbycusis*, or hearing loss directly associated with aging, is typified by age-related structural changes within the mechanical and sensory portions of the ear. A second form of hearing deficit, *noise-induced hearing loss*, is particularly prevalent among older adults. Having lived in an industrialized society, they now suffer from the cumulative effect of lifelong exposure to harmful noise. A third category of hearing loss, *nosoacusis*, is associated with traumatic events, disease, or hereditary conditions. In addition, ototoxicity from prescribed medications can also contribute to significant hearing loss among the elderly population.

Age related hearing loss is not constant for all sound stimuli, but rather is a function of the frequency of the sound. Figures 1 and 2 show typical hearing threshold levels as a function of frequency and age for both males and females not exposed to high workplace noise levels. As illustrated, threshold levels at low frequencies increase only minimally with increasing age. However, as the frequency increases, the degree of hearing loss with advancing age increases dramatically. Recent work with very high frequencies (between 8000 and 20000 Hz) has revealed a similar frequency by age interaction (Stelmachowicz, et. al., 1989).

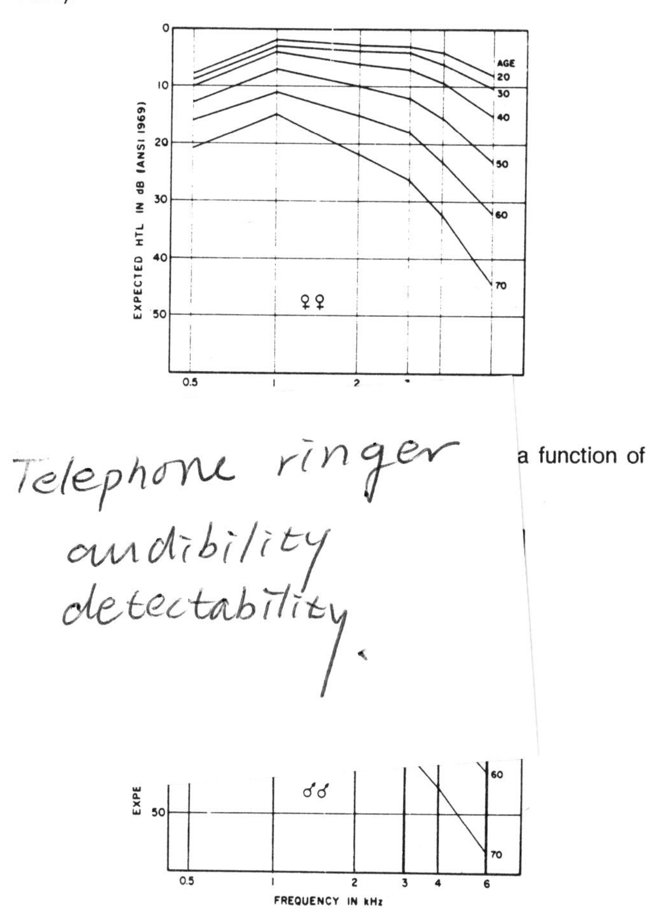

a function of

Figure 2. Hearing threshold levels for males as a function of age (Reproduced with permission from Ward, 1986).

Telephone Ringer Audibility

For the older individual, access to a telephone is critical to the maintenance of independent living. Loss of such access may create feelings of social isolation and pose safety problems (Infante-Rivard, Krieger, Petitclerc, and Baumgarten, 1988). In a study of 500 new hearing aid applicants, Stephens (1980) noted that 20% of those polled listed difficulty in detecting telephone ringers as a primary problem. However, limited human factors research attention has been applied to the design of telephone ringers, especially for the elderly. Hunt (1970) examined the acoustics of a tone ringer and reported that effective ringers consist of a signal with at least two spectral components between 500 and 4500 Hz with the prominent components remaining below 2000 Hz, and that the telephone ringer signal should maintain a minimum level of 77 dB for adequate detectability.

Despite the existence of such guidelines, some recently-produced consumer telephones appear to exhibit problems in ringer signal detectability. Various forms of electronically-generated tone signals, electronically-synthesized bells, and mechanical bells are currently available, and most phones incorporate one of these technologies for the ringer. The older-style mechanical bells appear to be dwindling in use compared to the electronic ringers.

Research Objectives

With the typical hearing impairments of the elderly in mind, the fundamental objective of this study was to experimentally compare the detectability of three common but very acoustically-different, telephone ringers for both young and older listeners. Given that telephones must be heard in a variety of ambient noise conditions, detection thresholds for each ringer were obtained in quiet and continuous broadband noise.

METHOD

Subjects

Eighteen volunteer subjects participated in the study. Nine subjects (group 1) were between the ages of 20 and 30 years with a mean age of 25 years. The remaining nine subjects (group 2) were between the ages of 70 and 95 years with a mean age of 83 years. Each group consisted of 4 females and 5 males. All ringer detection testing was done binaurally with unassisted (no hearing aids) ears.

Experimental Design

The mixed three-factor design included within-subject variables of telephone ringer and background noise and the between-subjects variable of age. A variety of commercially available telephones were obtained and considered for testing. The three selected incorporated widely diverse ringer designs, representing three perceptually distinctive categories of available ringers. The AT&T Trimline model 221 includes an "electronic bell," the Cobra model WP-142 utilizes an "electronic beeper" ringer, and the Unisonic model 6462 incorporates a "mechanical bell" ringer. The measured sound pressure levels at 18 inches for the three telephones were 80 dBA, 67 dBA, and 77 dBA, respectively. All ringers were measured with the volume control set to maximum, if controllable. These ringer descriptors are used herein to designate the particular phones, but are not intended to imply that the three ringer spectra are global standards across the market. Each telephone simply represents one example of that particular ringer type. Two ambient sound conditions were also investigated: quiet (24 dBA ambient in the booth) and 65 dBA pink noise (representing noise levels in home recreational areas, such as in television rooms).

The presentation order of the 3 (phone) by 2 (noise) treatment conditions was balanced using a Latin Square design. The measure obtained for each trial was ringer threshold sound pressure level, measured in dBA at head center.

Equipment

All data were collected in the Auditory Systems Laboratory at Virginia Tech. The sound spectrum of each telephone ringer (Figures 3, 4, and 5) was measured in an Eckel anechoic chamber, using a Larson-Davis 800-B 1/3 octave analyzer with ACO 7013 microphone positioned 18 inches from the phone which was suspended in the center of the free-field space. Stimuli tapes for the telephone signals were recorded inside the anechoic chamber using an AKG C414-B microphone and a TEAC 124 tape deck. All audiometric screening (using a Beltone 114 pure-tone audiometer), as well as ringer detection tests were conducted in a semi-reverberant, sound-treated booth. A Sony TC-W7R tape deck was used for playback of the recorded ringer signals, which were spectrally-shaped to match the originally measured output spectra of Figures 3-5 using a Yamaha GE-60 equalizer. Signal amplification was provided by an NAD-1020B preamplifier and NAD 2200 power amplifier system, powering an Infinity RS-9B loudspeaker located 18 inches behiind the seated subject's head. Calibrated signal level control was accomplished by passing the ringer signal through the attenuator network of the Beltone 114 audiometer. Signal levels were calibrated with the Larson-Davis analyzer.

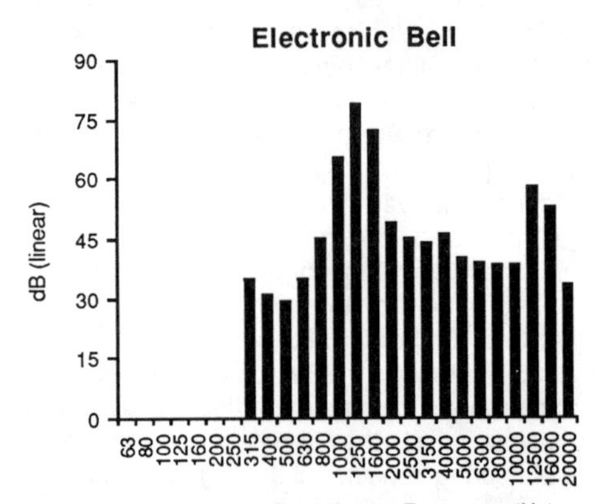

Figure 3. Sound level spectrum of the electronic bell.

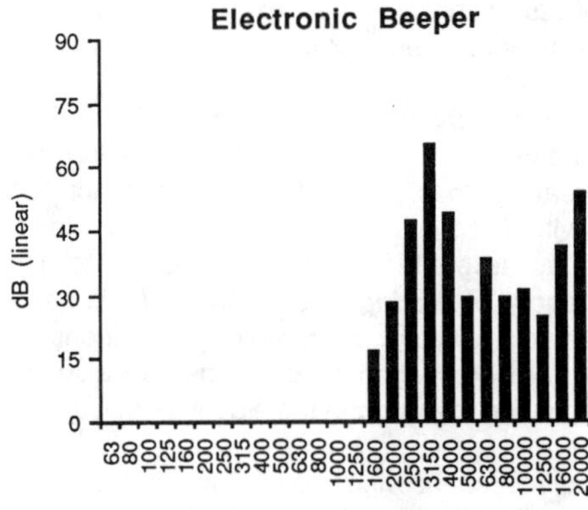

Figure 4. Sound level spectrum of the electronic beeper.

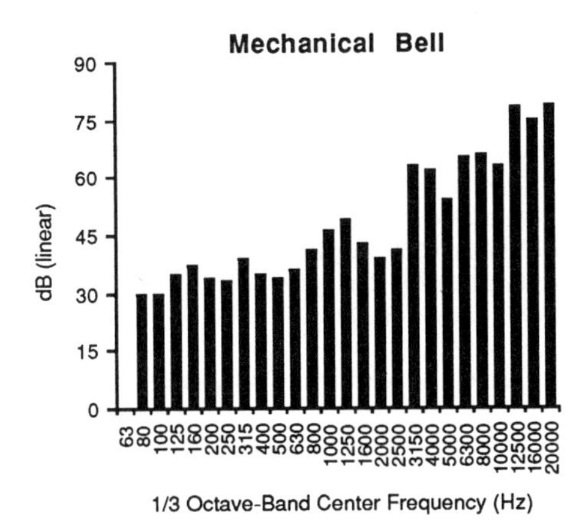

Figure 5. Sound level spectrum of the mechanical bell.

The background pink noise spectrum, the energy in which was flat (+/-2 dB) by octaves between 100 and 10000 Hz, was originated from tape on a TEAC 124 deck, equalized through a Ross R31M spectrum shaper, and amplified with a Realistic SCR 3010 receiver. The noise was presented through four Infinity RS-6B loudspeakers located equidistant at 39 inches from the subject's head and positioned in the room's corners. Noise was presented at a constant 65 dBA.

Experimental Protocol

After practice on the method of limits threshold determination procedures, each subject was given a standard audiogram at frequencies of 125, 250, 500, 1000, 2000, 4000, and 8000 Hz. Following instruction on the experimental task, the subject underwent a ringer detection threshold test for each of the six conditions. A bracketing procedure, using 2.5 dB step increments of ringer level, was followed to determine threshold. At each ringer level step, three rings were allowed to occur before proceeding to the next level. The subject's task was to indicate audibility of the ringer signal by pressing a hand-held push button. For each experimental condition, five estimates of threshold were obtained.

RESULTS

Mean threshold levels of the seven audiogram frequencies for each age group appear in Table 1. These mean thresholds coincide closely with what would be predicted for each age group based on survey data for non-noise exposed populations (Ward, 1986).

Table 1. Mean pure-tone audiogram hearing thresholds, in dBA, for the two age groups.

Hz	Older Subjects	Young Subjects
125	25.25	5.55
250	23.9	5.25
500	23.3	2.8
1000	21.95	3.35
2000	36.15	3.65
4000	54.45	4.45
8000	79.45	9.75

A single estimate of ringer threshold, in dBA, for each subject in each condition was determined by computing the arithmetic mean of all estimates which lay within +/- 5 dBA of the mode. It was necessary to discard only one data point of the 540 collected; this provided clear evidence of the stability of the within-subject trials. An analysis of variance was performed on the resulting threshold measure using ringer, noise, and age as the main effects and all possible interactions thereof. (Complete ANOVA summary tables are available from the authors.) A post-hoc Newman-Keuls test ($p < 0.05$) was performed for pairwise means comparisons within each significant main effect and interaction.

The three-way interaction of ringer, noise, and age was significant, $F(2,32) = 6.61$, p 0.004, as were the embedded two-way effects of ringer by age, $F(2,32) = 9.58$, $p < 0.0005$, and noise by age, $F(1,16) = 25.64$, $p < 0.0001$. Ringer by noise was nonsignificant, thus the further breakdown of the complex three-way interaction was performed only on the ringer by age and noise by age effects.

Mean detection thresholds for each ringer for both age groups are plotted in Figure 6. The rank ordering of detection thresholds for the three ringers was the same for both age groups, with better detectability for the electronic ringer, the mechanical bell, and the electronic beeper, in that order. The electronic beeper and mechanical bell ringers resulted in significantly higher threshold values than the electronic bell for the older group, while the electronic beeper ringer resulted in a higher threshold value than the other ringers for the younger group. Reviewing the frequency spectra of the three ringers (Figures 3, 4, and 5) aids in explaining these effects. For frequencies between 315 and 2000 Hz, the mechanical bell and electronic bell are largely similar except for the prominent energy peak of the electronic bell between 1000 and 1600 Hz. This energy peak is in the sensitive range of both age groups, leading to easy detection of the electronic bell by both young and older listeners. The mechanical bell, on the other hand, has significantly greater power above 3150 Hz than the electronic bell. This high-frequency content is more useful to the younger rather than the older group because, as evidenced in the audiograms, the older group has greatly reduced sensitivity in the upper frequency range. This apparently results in the frequency spectrum of the mechanical bell being equivalent in detectability to the electronic bell for the younger group, but less detectable for the older. The electronic beeper has essentially no output below 1600 Hz, and only a very limited amount below 2000 Hz, although a peak does exist at 3150 Hz, where normal hearing is quite sensitive. However, the lack of low and midrange energy content for this ringer probably put it at a disadvantage, especially for the older listener whose audiogram thresholds above 2000 Hz were elevated to 35 dBHL (hearing level) and above.

The significant ANOVA interaction between age and noise was further analyzed and results plotted in Figure 7. For both age groups, mean threshold in noise was significantly greater than under quiet conditions, as would be expected due to masking effects. On average across the three ringers, the young group required only 7.1 dBA for detection in quiet, as compared to 42.3 dBA for the older group. Though still statistically significant, the young group's advantage was reduced by the presence of the 65 dBA noise. Noise raised the threshold of the young group by 38.3 dBA and that of the older group by 15.0 dBA.

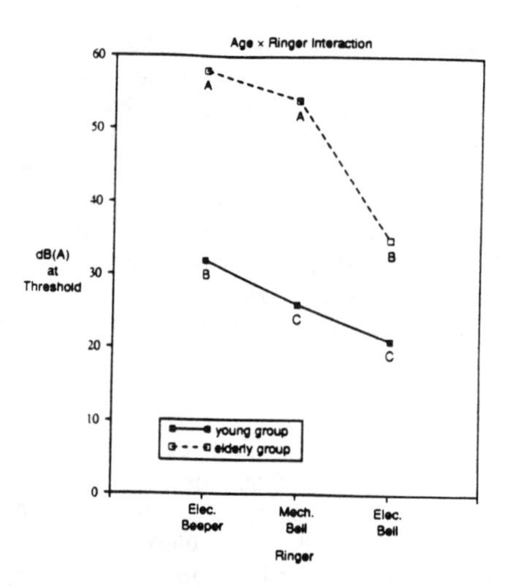

Figure 6. Mean detection thresholds as a function of ringer and age conditions (means with different letters are significantly different across the interaction according to the Newman-Keuls test at $p < 0.05$).

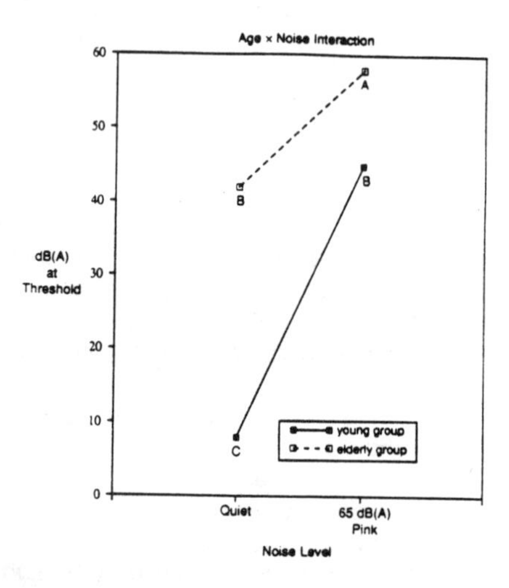

Figure 7. Mean detection thresholds as a function of age and noise conditions (means with different letters are significantly different across the interaction according to the Newman-Keuls test at $p < 0.05$).

Though restricted to their inclusion in the previously-discussed interactions, all three main effects (ringer, noise, and age) were found statistically significant in the ANOVA ($F(2,32) = 45.50$, $p<0.0001$; $F(1,16) = 133.39$, $p<0.0001$; $F(1,16) = 41.91$, $p< 0.0001$, respectively.) The post-hoc analysis revealed that all three ringers were significantly different ($p<0.05$) from one another in detectability, with the electronic bell resulting in the lowest mean threshold dBA value (28.7), the electronic beeper resulting in the highest (45.3), and the mechanical bell requiring 40.1. The overall superiority of the electronic bell ringer is probably due to the peak sound energy from 1000-1600 Hz, a sensitive range of hearing for both the young and old subjects as verified by their audiograms. In comparison, the mechanical bell has only moderate energy throughout the low to midrange frequencies of 80-2500 Hz, with slightly higher concentrations above 3150 Hz. The electronic beeper, which was found to require the highest level for detection, has negligible energy below 1600 Hz and only moderate energy elsewhere (in comparison to the other ringers), with the exception of moderate energy peaks at 3150 and 20000 Hz.

As expected, the quiet condition produced lower threshold values than the noise condition (24.7 dBA and 51.4 dBA, respectively), indicating that the noise produced masking effects on the detection of all of the telephone ringers. The younger group of subjects had a mean detection threshold much lower than that of the older subjects (26.3 dBA and 49.8 dBA, respectively). This result, again, was anticipated based on the well-documented hearing loss associated with age as well as the audiograms of these participants.

DISCUSSION

The results of this study provide evidence that current telephone ringer sound spectra differ widely in their aural detectability and that the older person, in particular, may encounter considerable difficulty in hearing some telephones ring. Furthermore, there is a significant masking effect from even moderate level broadband background noise which inhibits the detection of the telephone ringer.

The question remains as to whether the revealed detection threshold differences are actually important for the selection or design of a telephone ringer for a particular user group or noise environment. As measured in the anechoic chamber, the sound pressure levels at 18 inches for the three phones when allowed to ring normally were: electronic bell, 80 dBA; electronic beeper, 67 dBA; and mechanical bell, 77 dBA. In the quiet condition, the average required sound pressure level for detection by older subjects was, in the same ringer order, 26.9, 51.7, and 48.3 dBA. Each average detection level was well below the natural occurring sound pressure level for the respective telephone. However, under the noise condition, the detection thresholds increased to 45.1, 66.3, and 60.3 dBA respectively. Thus, the normal ringer output (67 dBA) for the beeper-type ringer was just above threshold (66.3 dBA). Considering that the distance from the ringer source to the subject's ear was only 18 inches, and that the subject was instructed to be attentive and dedicated to the singular task of detecting the ring, the experimental conditions were probably more favorable to ringer detection than typical household listening conditions which have longer distances and numerous distractions. Therefore, the selection of the beeper-type ringer spectrum (Figure 4) for an older hearing-impaired person is probably inappropriate, especially if reliable detection in noise is required. And it is often the case that older individuals construct a noisy home environment due to the need for high television and radio volume levels to offset hearing impairment.

A second important consideration lies in the applicability of Hunt's (1970) design recommendations for tone ringers. The most poorly detected ringer, which utilizes a beep with a tonal quality, apparently violates two of Hunt's central design principles. The overall output sound pressure level of 67 dBA is below Hunt's 77 dB target. Furthermore, the central spike in the frequency distribution at 3150 Hz, although still below the upper bound of 4500 Hz, does violate the recommendation for central components to remain below 2000 Hz. This appears to be important for two psychoacoustic reasons: 1) age-induced hearing loss is generally less prominent in the low to midrange frequencies, so the elderly have a better chance of being accommodated by these signal frequency components, and 2) noise masking effects primarily gravitate upwards, interfering with frequency components near to and higher than the masking sound. The results of this study are in general agreement with Hunt's recommendations about the proper frequency content and sound pressure level for a tone telephone ringer signal.

A significant noise by ringer interaction effect was not found. Therefore, the data cannot be interpreted as support for the selection of a specific telephone ringer type for differing noise situations. However, the background noise utilized in this study was broadband (pink), providing the opportunity for direct and upward masking by many noise frequency components which overlapped the components of all three ringers. The elevation of detection threshold (from that in quiet) due to noise masking was of similar magnitude for all three ringers, these elevation differences being 25.6 dBA for the electronic ringer, 27.4 dBA for the beeper, and 27.0 dBA for the mechanical bell. This evidences that the lack of a significant noise by ringer interaction is possibly accounted for by the similar masking effects of pink noise across the three ringer types. Furthermore, the masking noise was presented at a moderate level, 65 dBA, and in many realistic circumstances, a ringing telephone must be detected under higher noise levels which might necessitate more careful ringer selection. In any case, the need for further investigation of different, naturally occurring masking noise effects on ringer spectra audibility is clearly indicated. It is important to note, however, from these results that the electronic bell ringer spectrum required a considerably lower measured dBA output than either of the other ringers for both the quiet and noise conditions.

The capabilities and limitations of a user population must be considered in the design and implementation of any new technology which relies on human perception for successful system performance. In many cases, the perceptual changes associated with age are not regarded, making product usage difficult, if not impossible, for older subjects. The results of the present study suggest that some telephone ringers create greater detection problems than others for older adults, and that high frequency, beeper-type ringer signals in particular may be a poor selection for hearing-impaired listeners. Further research is need to precisely define the signal characteristics for telephone ringers in order to ensure reliable detection for listeners of all ages and under various ambient noise conditions.

ACKNOWLEDGEMENTS

The authors wish to thank the administration and residents of Warm Hearth Retirement Village and Ann Marie Connor for their assistance.

REFERENCES

Bess, F.H., Lichtenstein, M.J., Logan, S.A., and Burger, M.C. (1989). Comparing criteria of hearing impairment in the elderly: A functional approach. *Journal of Speech and Hearing Research, 32*, 795-802.

Bess, F.H., Lichtenstein, M.J., Logan, S.A., Burger, M.C., and Nelson, E. (1989). Hearing impairment as a determinant of function in the elderly. *Journal of the American Geriatrics Society, 37*, 123-128.

Bingea, R.L., Raffin, M.J.M., Aune, K.J., Baye, L., and Shea, S. (1982). Incidence of hearing loss among geriatric nursing home residents. *Journal of Auditory Research, 22*, 275-283.

Brown, S.C., Hotchkiss, D.R., Allen, T.E., Schein, J.D., and Adams, D.L. (1989). *Current and Future Needs of the Hearing Impaired Elderly Population, GRI Monograph Series B, No. 1.* Washington, DC: Gallaudet Research Institute.

Hinchcliffe, R. (1958). The pattern of the threshold of perception for hearing and other special senses as a function of age. *Gerontologica, 2*, 311.

Hotchkiss, D.R. (1989), *The Hearing Impaired Elderly Population: Estimation, Projection, and Assessment, GRI Monograph Series A, No. 1.* Washington, DC: Gallaudet Research Institute.

Hunt, R.M. (1970). Determination of an effective tone ringer signal. *Paper presented at the 38th Convention of the Audio Engineering Society.* New York, NY: Audio Engineering Society.

Infante-Rivard, C., Krieger, M., Petitclerc, M, and Baumgarten, M. (1988). A telephone support service to reduce medical care use among the elderly. *Journal of the American Geriatrics Society, 36*, 306-311.

Keith, R.W. (1982). Central auditory tests. In N.J. Lass, L.V. MacReynolds, J.L. Northern, and D.E. Yoder (Eds.), *Speech, Language and Hearing, Volume III: Hearing Disorders*. Philadelphia, PA: W.B. Saunders Company.

Mader, S. (1984). Hearing impairment in elderly persons. *Journal of the American Geriatrics Society*, 32, 548-553.

Stelmachowicz, P.G., Beauchaine, K.A., Kalberer, A., and Jesteadt, J. (1989). Normative thresholds in the 8- to 20-kHz range as a function of age. *Journal of the Acoustical Society of America, 86*(4), 1384-1391.

Stevens, S. (1980). Evaluating the problems of the hearing impaired. *Audiology, 19*, 205-220.

Ward, W.D. (1986). Anatomy and physiology of the ear: normal and damaged hearing. In E.H. Berger, W.D. Ward, J.C. Morrill, and L.H. Royster (Eds.), *Noise and Hearing Conservation Manual.* Akron, OH: American Industrial Hygiene Association.

Zarnoch, J.M. (1982). Hearing disorders: Audiologic manifestations. In N.J. Lass, L.V. MacReynolds, J.L. Northern, and D.E. Yoder (Eds.), *Speech, Language and Hearing, Volume III: Hearing Disorders.* Philadelphia, PA: W.B. Saunders Company.

AUDIBLE PERFORMANCE OF SMOKE ALARM SOUNDS

Richard W. Huey, Dawn S. Buckley and Neil D. Lerner
COMSIS Corporation
Silver Spring, Maryland

This paper concerns a study aimed at selection of alarm sounds with improved audible performance characteristics for older listeners over current conventional residential smoke detectors. Many current residential smoke detectors possess alarms that have their primary frequency peak in the 4000 Hz region of the audible spectrum. Additionally, many of these alarms are constant instead of providing temporal modulation of the signal. This study analyzed a variety of alternative sounds for selection as a better choice for an "age sensitive" smoke alarm signal. The study presented a battery of candidate sounds to pairs of subjects aged 65 and older with varying levels of hearing impairment (0 to 45 dB) in their own homes to see which sounds performed best in terms of detection, localization, and perceived attention-getting value. Subjects were placed in various location- and masking-based conditions within their homes during listening periods and subjected to sounds played at a constant level. A computerized system collected response data as the battery of stimuli was presented. The data showed a fairly predictable positive trend in detection and localization performance level as the frequency of the stimuli decreased from 4000 Hz to 500 Hz. The data also showed that pulsed signals were more detectable than steady alarms.

INTRODUCTION

Most older Americans live in private residences and are primarily responsible for their own fire safety. Fire death rates are substantially higher for the elderly (Karter, 1986). This population is particularly in need of the warnings provided by smoke alarms; unfortunately, current alarm features are poorly suited for them. This study is aimed at selection of alarm sounds with improved audible performance characteristics over current conventional residential smoke detector alarms, which typically have their primary frequency peak in the 4000 Hz region of the audible spectrum. The UL standards are set up at 85 dB, 10 ft. from the source. No particular frequency range has been outlined; however, technology has gravitated toward the primary frequency of 4000 Hz due to the resonance frequency of the most commonly used sound output devices. Unfortunately, this region and those higher in frequency tend to be prone to hearing loss in the elderly caused by presbycusis. Hotchkiss (1989) reported that approximately 30 percent of persons 65 years and older consider themselves functionally hearing-impaired, while persons over 85 raise this percentage to almost 50. These results were based on self-reports and are likely an underestimate of the actual magnitude of hearing impairment among these populations. This age-related, hearing impairment is not constant for all alarm frequencies. Berkowitz and Casali (1990) state that threshold levels at low frequencies (500 & 1000 Hz) increase slightly with age, while higher frequencies show dramatic increases in hearing loss with age. Additionally, many of these alarms have constant (flat) temporal waveforms instead of providing any modulation of the signal. Lower frequency signals should also provide better acoustic transmission through the physical structures of a house. In an attempt to develop a smoke detector alarm signal that promotes safety and addresses the unique needs of aging ears, this study was undertaken via a grant from the National Institutes of Health's National Institute on Aging.

METHOD

Thirty seven subjects over the age of 65 participated in the study. The sample was divided into three subgroups, based on their level of hearing loss. Potential subjects were given a preliminary audiometric screen and were classified according to the average pure tone threshold at frequencies of 500 through 8000 Hz. Those subjects showing a mean threshold elevation of 15 dB or less were classified as "normal hearing"; those with between 15 and 30 dB, as "mild loss"; and those with between 30 and 45 dB, as "moderate loss". People suffering more than a 45 dB hearing loss are considered "severely" or "profoundly" impaired, and were not included in this study. Subjects were not allowed to use hearing aids during audiometric testing nor during the actual study sessions, since such devices may be adjusted to different levels, invalidating the hearing level measurements.

The data collection system was comprised of sound generating equipment, sound playback equipment, and a data collection system. Sound generation was accomplished using a function generator to create pure and square wave signals at frequencies between 500 and 4000 Hz. An oscilloscope was used for confirmation of the wave form, frequency, and temporal characteristics of the signal, while a sound level meter provided signal level calibrations during recording. Signals were recorded using a digital audio tape

(DAT) deck to ensure minimal signal loss. A PC sound card allowed sounds to be edited and then re-recorded onto the DAT in various modulated forms. This device was used to modulate all of the sounds created by the function generator, as well as to record actual smoke detector output and less conventional sounds from an electric guitar.

The sound playback system consisted of the same DAT deck, an amplifier, a stereo frequency equalizer, and two high fidelity speakers. The DAT was used to play back the stimuli during the experiment. One speaker was placed on the main floor and one in an upstairs corridor near the bedrooms. To maintain realism, the speakers were situated in a location consistent with common placement of existing smoke alarms at approximately eye level.

The data collection system was computerized using a BASIC program on a notebook PC to provide event randomization, timing and data collection during the experiment. The subjects entered their responses by pressing buttons on custom consoles when they heard the stimuli. These button presses were input via RS-232 through a conversion box. This input was collected and interpreted to provide feedback to the experimenter and create a database of subject responses. The subject consoles were connected to the PC by long (75 foot) umbilicals to allow adequate separation from the centrally located sound playback and data collection system location.

The eight listening conditions were derived from the combination of two alarm locations (first floor and second floor) and four listener situations (first floor quiet living room; a second floor quiet bedroom with an open door; a second floor bedroom with a closed door and a radio or TV playing; and a basement laundry room). These listening conditions in combination with the sound stimuli represented the primary independent variables. Fourteen sounds were recorded onto the DAT player separately on the right and left channels for a total of 28 stimuli. The sounds recorded on the left and right channels were presented using the downstairs and upstairs speakers, respectively. All stimuli were recorded at a constant sound input level. The stimulus sounds were presented in the following forms:

- 500 Hz ISO & fast pulsed (about 10-12 Hz)
- 1,000 Hz ISO & fast pulsed
- 2,000 Hz ISO & fast pulsed
- Combined 1,000 and 4,000 Hz ISO, continuous & fast pulsed
- 4,000 Hz ISO & fast pulsed
- 0 to 4,000 Hz swept (up in .025 and down in .475 second)
- "Elephant" (electric guitar simulation modulated with ISO spacing)
- Modulated 900-1100 Hz skewed sine wave (up in .025 and down in .475 second)

NOTE: The ISO standard recommends that the sound be presented in half second bursts followed by a half second of silence for three cycles with each three-cycle segment separated by an additional one second of silence.

After visiting the subjects' houses to confirm acceptable architecture, define the listening locations, measure the ambient sound pressure levels for characterization at the listening locations, and to document the floor plan of the home including current smoke detector locations, the actual testing began. Setup included placing speakers in typical smoke alarm mounting locations, connecting and powering up the audio and data collection equipment and carrying subject response consoles to the initial listening locations. After setting up the equipment, calibration of the sound output level was performed using a pink noise source played from both speakers simultaneously. The sound pressure level of the pink noise source from one of the speakers was adjusted with the amplifier volume control until it reached 85 dB at 10 feet from the speaker. This reading was normally off axis because of space constraints within the subjects' homes. Then the other speaker, using the noise source and preset amplifier level was adjusted to the same sound pressure level using the balance control on the amplifier. After calibration, octave band measurements were taken at each of the listening locations for ambient and pink noise levels from each of the sound output locations. All of these measurements were taken before the session began except for the listening location characterized by the door being closed and the radio playing. In this case, the subject was asked to choose either a music or talk program and to adjust the volume of the radio to a comfortable level before the noise levels were measured.

Two subjects (often married couples) were tested simultaneously to facilitate the data collection for this task. Test sessions were performed in the subjects' two-story, single-family homes (with basements). Subjects were located in different rooms of their home during testing to ensure that they would not influence each other's responses. Before retreating to their respective listening locations for the first battery of sounds, a standardized set of instructions was read to each subject. Subjects were allowed to read or work on a hobby simulating a realistic "at home" situation as opposed to merely listening attentively and responding to sounds. Subjects were asked to remain in each location for the entire thirty minutes though measures were taken to accommodate various household interruptions without affecting the integrity of the data.

Listening location, stimulus order and output location, and inter-stimulus timing were randomized by the data collection PC and then activated manually by an experimenter. For each alarm activation, the particular alarm was active for no more than 30 seconds, providing a standard maximum signal length to ensure equal opportunity to hear and respond to the stimulus. When subjects heard an alarm, they were to immediately press a large button located on the input console. Their response time and all relevant parameters were recorded on the PC. The subject then pressed another button to indicate perceived signal location (i.e., upstairs or downstairs). Finally, a judgment of attention-getting value was keyed in, using a 7-point scale, with 1 being "barely noticeable" and 7 being "impossible to ignore". If a subject did not respond to a stimulus, that response was marked as a "miss" by the experimenter. The experimenter also kept an "unusual event" log throughout the experiment. Various items were recorded, such as dogs barking and clocks chiming during a particular stimulus presentation. These logs were used to explain strange subject responses and/or unusual data records. Exposure to the full battery of signal conditions was possible in approximately 45 minutes before subjects were moved to the next listening location, with the entire experiment taking about 4.5 hours.

A structured questionnaire was administered to the subjects upon completion of the testing. This questionnaire tapped subjects' overall impressions of the signals, suggested improvements, smoke alarm feature desires, willingness to purchase such enhancements, etc.

RESULTS

The data from this study indicated that the lower frequency tones modulated at a fast rate performed the best in every perspective. The results of the experimentation were compared by way of frequency counts which were used to calculate success rates for various conditions of the study parameters. Measurements of reaction time were dismissed since speed of reaction was not stressed as the most critical factor when subjects were instructed in the test procedure. Also, there were some instances of subjects leaving their assigned listening positions, potentially adding some unreliability to the response timing data.

Figure 1 presents the overall detection success rates of the candidate sounds, showing that sounds with their major peak in the 500 Hz frequency band were most likely to be heard. Low frequencies are more easily heard by younger as well as older residents. Zepelin (1984) showed that among subjects with normal hearing, there was similar threshold responses, but with some loss of sensitivity, as expected for the older subjects. The average hearing levels at 750 Hz were 13, 14 and 20 dB for age groups: 18 to 25; 40 to 48 and 52 to 71, respectively. Figure 2 shows the relationship between detection success rate and the modulation pattern of the sounds, with faster modulation leading to slightly greater detection success rates. The differences in the detection success rates were most pronounced in the worst case scenario (subject located upstairs in a noisy bedroom with the door closed while the stimulus was downstairs). This is illustrated in Figure 3. Figure 4 shows the relative localization success for the various sounds. None of the sounds were substantially better than any of the others, but again the 500 Hz peak and fast modulation performed at least slightly better in this measure. Figure 5 depicts the overall average ratings of attention-getting value for the sounds. This subjective measure held true to the objective performance ratings collected. Again the 500 Hz sounds were near the top of the group, and the sounds employing fast modulation were rated higher. The "elephant" and 0-4000 Hz Swept sounds were also rated high, though possibly for novelty reasons rather than purely attention-getting qualities.

An error in the calibration of the 2000 Hz sounds occurred during the recording process. Unfortunately, these sounds were calibrated using the wrong weighting scale (i.e., dB(A) instead of dB(C)). Thus, when the weighting scale problem was detected, a 3 dB difference was noted. This magnitude of difference may well have been audibly detectable and skewed the data. It is interesting to note that the same modulation speed phenomenon described earlier was reflected for this group as well as the appropriately calibrated samples. It did, however, fare worse from a detection standpoint than the other groups of similarly constructed sounds due to the intensity difference.

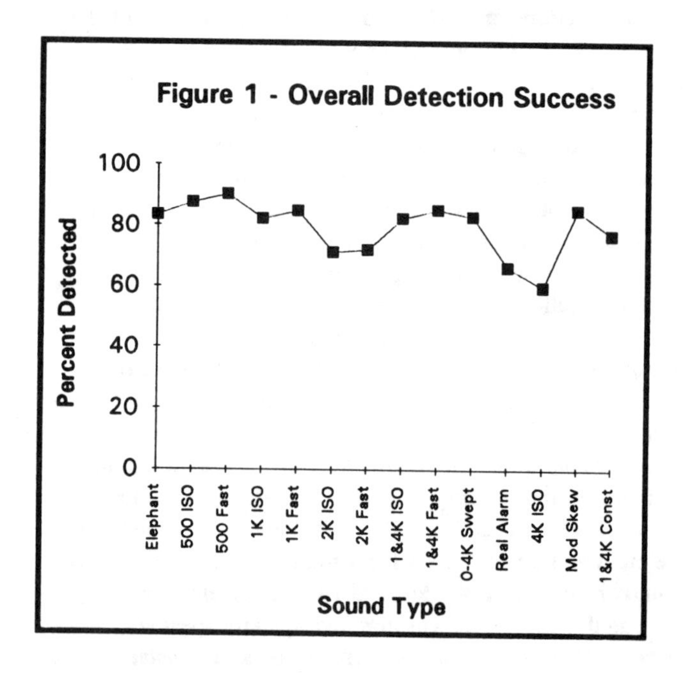

DISCUSSION

Based on the results, several features are suggested for the design of an improved smoke alarm for elderly and younger users. These results point to sounds with a primary peak at 500 Hz with a fast modulation rate to improve detection, perceived attention-getting value, and even localization effectiveness to some degree. Because of a calibration error, a contiguous curve cannot be drawn between the similar sound samples at 500, 1000, 2000, and 4000 Hz. However, if 2000 Hz is excluded from this grouping, the general trend shows improved performance at the lower frequency peaks. It appears that a new alarm design incorporating lower frequency sound could help to improve detection among elderly users and the magnitude of results depicted in the figures suggests that the improvement may well be a substantial one.

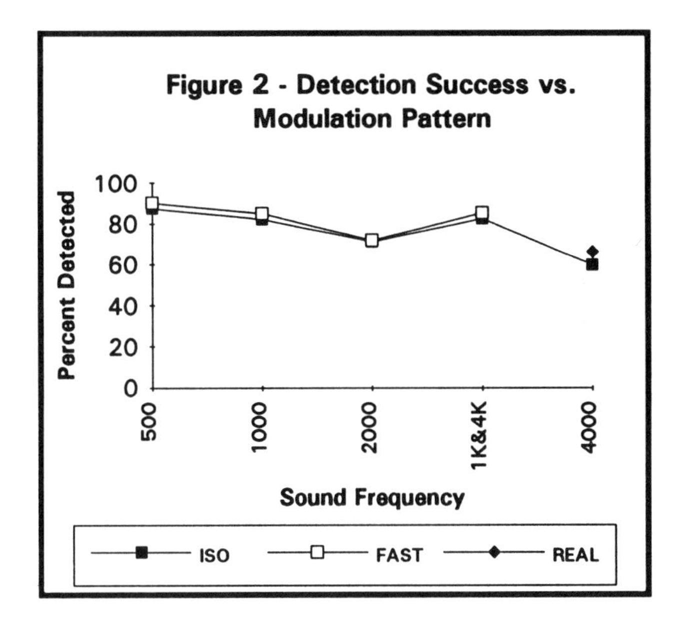
Figure 2 - Detection Success vs. Modulation Pattern
Percent Detected
100
80
60
40
20
0
500
1000
2000
1K&4K
4000
Sound Frequency
ISO
FAST
REAL

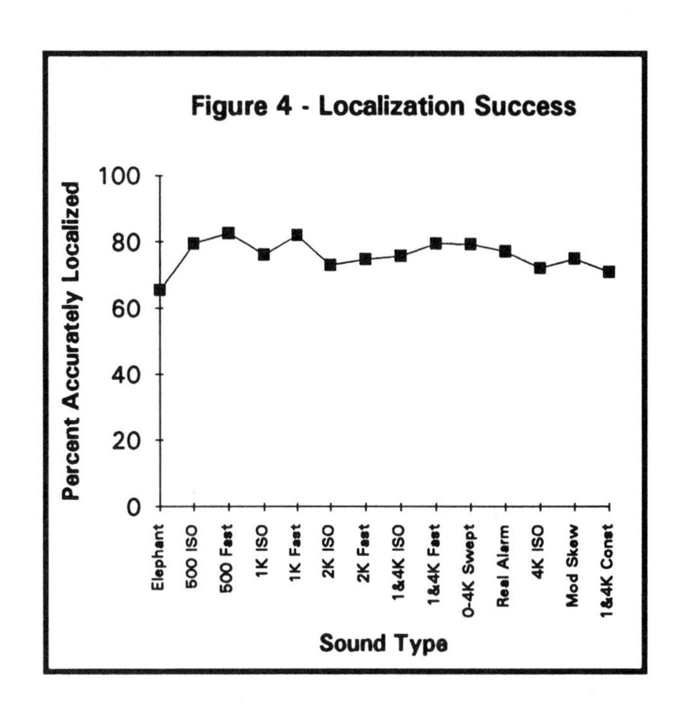
Figure 4 - Localization Success
Percent Accurately Localized
100
80
60
40
20
0
Elephant
500 ISO
500 Fast
1K ISO
1K Fast
2K ISO
2K Fast
1&4K ISO
1&4K Fast
0-4K Swept
Real Alarm
4K ISO
Mod Skew
1&4K Const
Sound Type

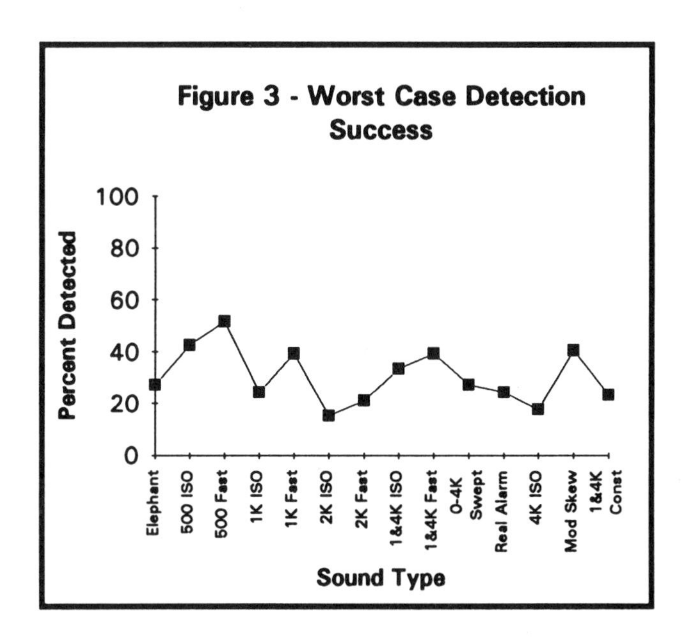
Figure 3 - Worst Case Detection Success
Percent Detected
100
80
60
40
20
0
Elephant
500 ISO
500 Fast
1K ISO
1K Fast
2K ISO
2K Fast
1&4K ISO
1&4K Fast
0-4K Swept
Real Alarm
4K ISO
Mod Skew
1&4K Const
Sound Type

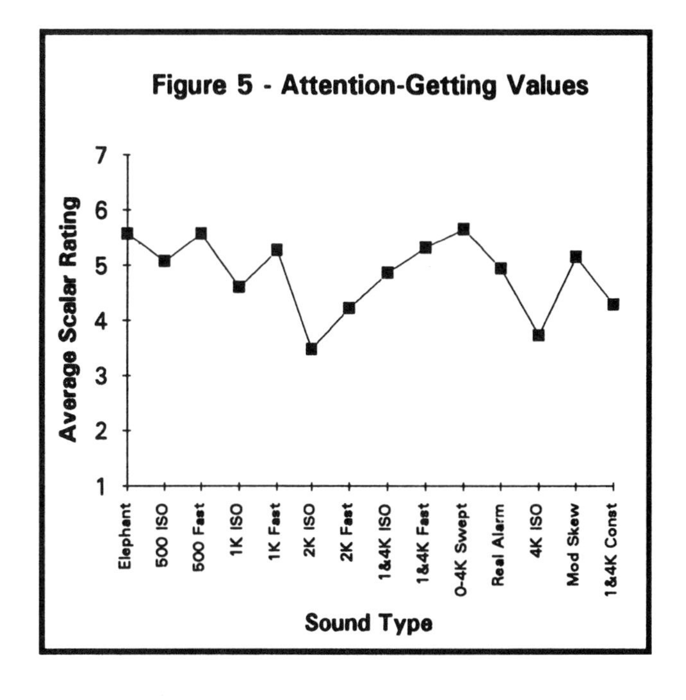
Figure 5 - Attention-Getting Values
Average Scalar Rating
7
6
5
4
3
2
1
Elephant
500 ISO
500 Fast
1K ISO
1K Fast
2K ISO
2K Fast
1&4K ISO
1&4K Fast
0-4K Swept
Real Alarm
4K ISO
Mod Skew
1&4K Const
Sound Type

REFERENCES

Berkowitz, J. and Casali, S. (1990). Influence of age on the ability to hear telephone ringers of different spectral content. Proceedings of the Human Factors Society 34th Annual Meeting, 132-136.

Hotchkiss, D. (1989). The hearing impaired elderly population: Estimation, projection and assessment. ***In Berkowitz, J. and Casali, S. (1990). Influence of age on the ability to hear telephone ringers of different spectral content. Proceedings of the Human Factors Society 34th Annual Meeting, 132-136.***

Karter, M. (1986). Patterns of fire deaths among the elderly and children in the home. Fire Journal, 80(2), 19-22.

Zepelin, H., McDonald, C. and Zammit, G. (1984). Effects of age on auditory awakening thresholds. Journal of Gerontology, 39(3), 294-300.

AGE AND GLARE RECOVERY TIME FOR LOW-CONTRAST STIMULI

Frank Schieber
Department of Psychology
University of South Dakota
Vermillion, SD 57069
Email: schieber@charlie.usd.edu

The purpose of this study was to obtain a rigorous experimental estimate of the time required to recover from the deleterious effects of glare. Low contrast test stimuli were employed to increase the potential sensitivity of the procedure. Multiple age groups were sampled since susceptibility to glare effects is known to increase with advancing years. Glare recovery time assessments were collected from 12 young, 12 middle-aged and 16 older adults. Subjects were presented with 10 sec exposures to an intense glare source under highly controlled experimental conditions. Upon the offset of the glare exposure period, the time required to regain sensitivity for low contrast test stimuli was measured. Relative to their younger counterparts, older subjects required 3-times longer to recover from glare exposure. These findings suggest that the dynamic components of glare effects must be considered when designing environments - especially where older observers are involved.

INTRODUCTION

Disability glare occurs when the introduction of a stray light source reduces one's ability to resolve spatial detail. Advancing adult age is known to be associated with significant increases in the susceptibility to the deleterious effects of glare. Most models of disability glare attribute these effects to intraocular light scatter (Schieber, et al., 1991; Schieber, 1992). Such off-axis scattering of light within the eye covers the retina with a "veiling luminance" which effectively reduces the contrast of stimulus images formed upon it. For this reason, Schieber (1988) has proposed that low contrast optotypes are better suited for the quantification of disability glare effects than high contrast optotypes which are traditionally employed in glare assessment procedures.

One aspect of disability glare which remains poorly understood is the time required to recover visual sensitivity following a transient exposure to a brief glare source (e.g., exposure to oncoming headlights while driving at night). There is ample clinical and anecdotal evidence that the time needed to recover from glare increases with age. However, little systematic work has been done to substantiate this claim (see Olson and Sivak, 1984). The purpose of this study was to obtain rigorous estimates of the time needed to recover from the deleterious effects of glare and to examine how this index differs as a function of advancing adult age.

METHOD

Subjects. Letter contrast sensitivity and glare recovery times were collected from 12 young (ages 18-24), 12 middle-aged (ages 40-55) and 16 older (ages 65-74) adult volunteers. All participants were in reported good health and demonstrated visual acuities of 20/25 or better.

Apparatus and stimuli. Stimulus generation and control was accomplished with an IBM PC/AT computer equipped with a Metrabyte CTM-5 programmable clock module which was used to record response times with millisecond precision. The system also contained a modified Matrox PIP-

1024 image processing board which enabled stimuli to be presented upon an Electrohome high-resolution monitor (white phosphor) with a gray-scale resolution of 12 bits (i.e., over 4000 unique gray levels). Test stimuli for both the contrast sensitivity and glare recovery assessment procedures consisted of the 26 letters of the alphabet which subtended 2-degrees of visual angle at the 6 m viewing distance. Test letters were presented in gray-on-white format against a constant background of 22 cd/m2. The glare source consisted of a pair of 50W incandescent flood lamps which were mounted 5 degrees to the left and the right of the center of the display monitor. The glare lamps were mounted in hooded enclosures which prevented the extraocular mixing of light from the stimulus monitor and glare sources. Illuminance of the glare sources - measured at the entrance pupil to the eye - was 78 lux.

Procedure. Letter contrast sensitivity was assessed first so that a "challenging" - yet visible - low contrast target level could be established for each observer in the glare recovery time procedure. Following an 8 min dark adaptation period, the assessment was begun. Subjects were presented with a random pair of letters and asked to identify them. The first set of letters was shown at 50% contrast. Letter contrast was reduced in 0.1 log unit steps and a new randomized pair of letters was presented until the observer made an error reporting one or both of the stimuli. At this point, stimulus contrast was incremented in 0.03 log unit steps until the observer could correctly report both items in the randomized stimulus letter pair. The contrast at this point was recorded as the *letter contrast threshold* (1/threshold = sensitivity). The odds of correctly reporting a randomized letter pair by chance alone were 1 in 380.

Glare recovery time was measured next. Subjects experienced a 10 sec exposure to the dual glare sources. During the glare exposure phase, subjects were required to shadow (i.e., name aloud as quickly as possible) high-contrast letters which periodically appeared in the center of the stimulus display monitor. This assured that all subjects were looking directly between the dual glare sources. At the end of the 10 sec exposure, the glare sources were extinguished and a randomized pair of low contrast test letters appeared simultaneously on the display screen (The letter stimuli used to assess glare recovery time were presented at a luminance contrast which was 0.1 log units higher than each observer's previously determined letter contrast threshold). The subject was instructed to press a button as soon as the letters could be recognized. When the button was pressed, the test letter stimuli disappeared and the subject was required to report which letters had appeared on the screen. Following the opportunity for a few practice trials, the time between the offset of the glare source and the button press signifying stimulus recognition served as a highly reliable index of glare recovery time. Three glare recovery time trials were collected and averaged. If a recognition error was made the trial was repeated.

RESULTS

Reference to Figure 1 reveals that advanced adult age was associated with an increase in the amount of contrast required to identify large stimulus letters. An analysis of variance revealed this age-related elevation in letter contrast threshold to be statistically significant ($F(2,37) = 5.62$, $p < 0.01$). Glare recovery time for suprathreshold low contrast letters was found to slow with increasing age ($F(2,37) = 12.66$, $p < 0.0001$). Reference to Figure 2 indicates that glare recovery time slowed progressively for both the middle-aged as well as the oldest group of observers. Statistical analysis support this impression: middle-aged observers required significantly more time to recover from glare than young observers ($F(1,22) = 7.06$, $p < 0.01$) but significantly less time than old observers ($F(1,26) = 8.52$, $p < 0.01$).

DISCUSSION

The observed age-difference in letter contrast threshold was consistent with previous observations of late-life declines in contrast sensitivity (Adams, et al., 1988; Schieber, et al., 1992). The more interesting findings involved the demonstrated age-differences in the time required to recover contrast sensitivity for large objects following exposure to

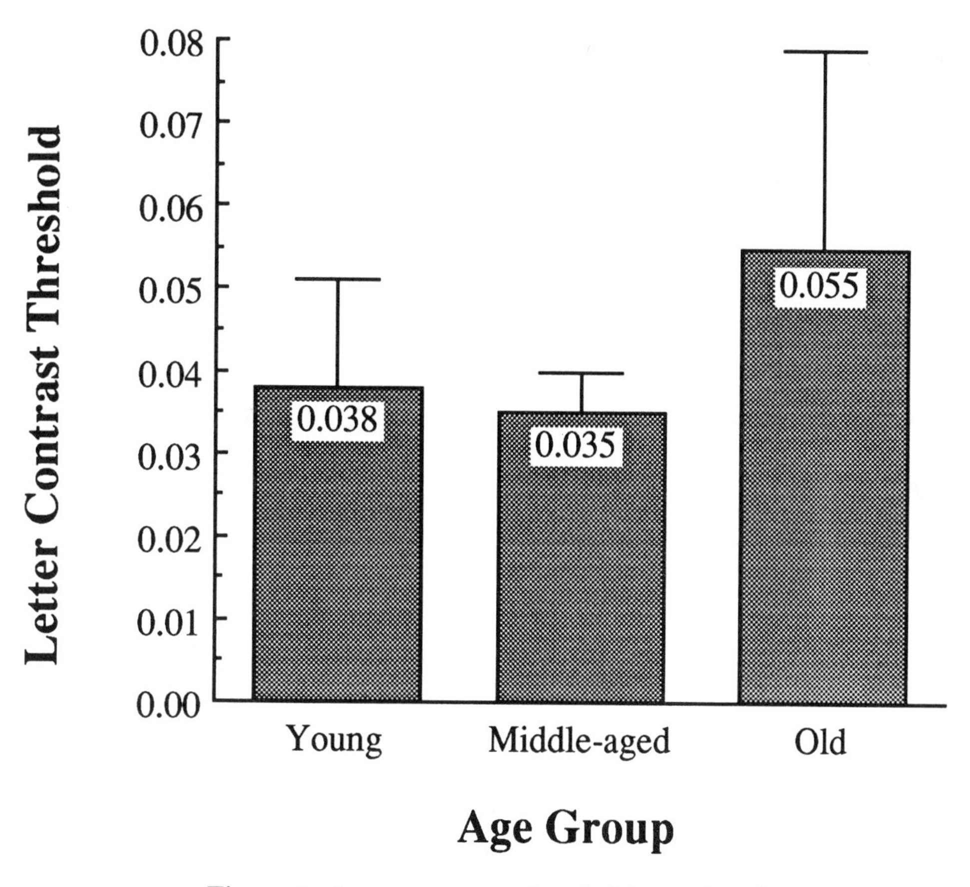

Figure 1. Letter contrast threshold as a function of age.

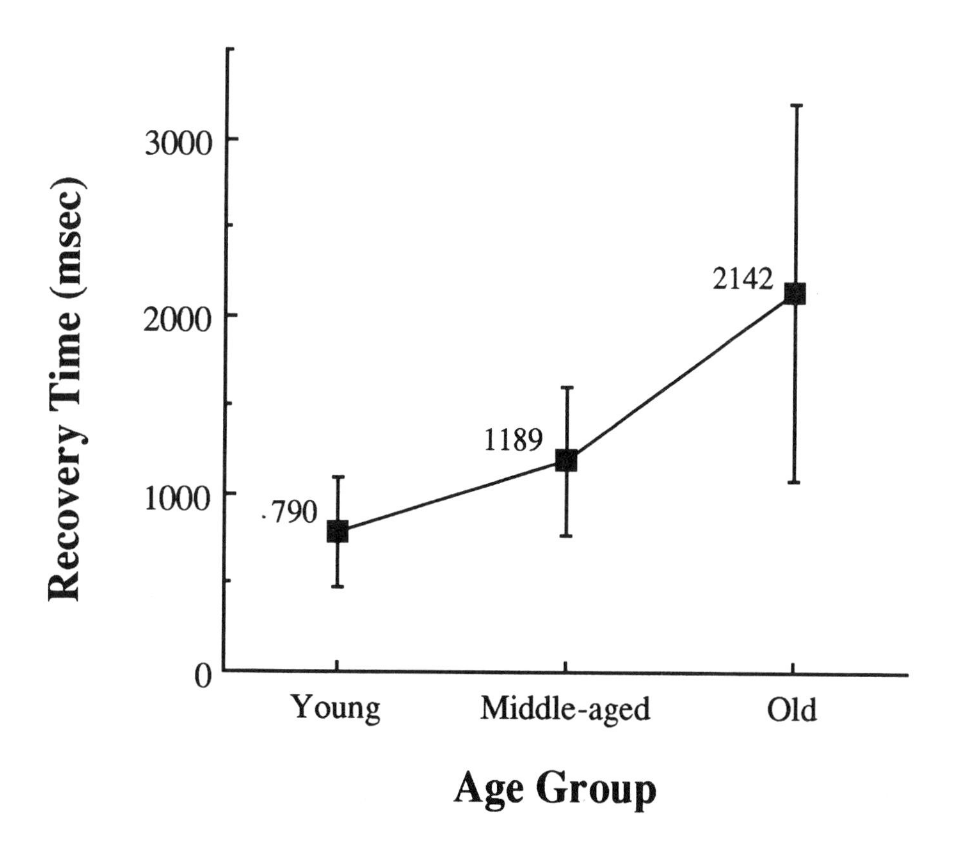

Figure 2. Glare recovery time as a function of age.

an intense glare source. Older subjects required nearly 3-times longer to recover than their young counterparts (i.e., 2142 vs 790 msec). This difference occurred despite the fact that older observers - on average - were tested using stimuli presented at higher levels of luminance contrast (viz., 0.1 log units above their "elevated" letter recognition thresholds). Age-differences in simple motor response time cannot begin to account for the sizable magnitude of the observed increase in glare recovery time. That is, past research suggests that age-related differences in simple reaction time paradigms like the one employed here would be on the order of 50-100 msec (see Kausler, 1991) - far smaller than the 1352 msec age-difference observed. Perhaps the most surprising result was the significant increase in glare recovery time required by middle-aged observers (i.e., 1189 msec) - despite the fact that they demonstrated letter contrast thresholds equivalent to those of the youngest subjects.

The observed age-related increases in glare recovery time for low contrast stimuli have design implications for nighttime driving environments. The target luminance level (22 cd/m2) was chosen to represent the adaptation state experienced by a nighttime driver facing oncoming low-beam headlight traffic at a distance of approximately 150 ft (Olson and Aoki, 1989). The intensity of the glare source (79 lux) represented the challenge offered by an approaching (or closely following) vehicle using high-beam headlamps (Olson and Sivak, 1984; Olson and Aoki, 1989). Under these conditions, older drivers would lose visual contact with targets having contrasts in the 0-10% range for a period of over 2 sec following exposure to a challenging glare source. Most rural and secondary roads without delineation treatments have effective contrasts which fall within this range of transient invisibility. Perhaps this is one of the reasons why older persons universally report problems with nighttime driving (e.g., Schieber, et al., 1992). On the other hand, it should be noted that the glare recovery time for high contrast targets approached zero for healthy adult observers regardless of age. That is, all of our observers could identify the high contrast letter targets which were presented while the glare source was exposed (Although this would not have been the case for some older adults had we not screened subjects for ocular pathology such as advanced cataract). This clearly indicates the potential benefits which may be realized by older drivers following the broad application and maintenance of high-contrast roadway delineation treatments such as retroreflective roadway edge lines.

REFERENCES

Adams, A.J., Wong, L.S., Wong, L. and Gould, B. (1988). Visual acuity changes with age: Some new perspectives. American Journal of Optometry and Physiological Optics, 65, 403-406.

Kausler, D.H. (1991). Experimental psychology, cognition and human aging. NY: Springer-Verlag.

Olson, P.L. and Sivak, M. (1984). Glare from automobile rear-vision mirrors. Human Factors, 26, 269-282.

Olson, P.L. and Aoki, T. (1989). The measurement of dark adaptation level in the presence of glare. Ann Arbor, MI: Transportation Research Institute, University of Michigan. [Report UMTRI-89-34].

Schieber, F. (1988). Vision assessment technology and screening older drivers. In Transportation in an aging society: Improving mobility and safety for older persons. Volume 2. Washington, D.C.: National Research Council. pp. 325-378.

Schieber, F., Fozard, J.L., Gordon-Salant, S. & Weiffenbach, J.(1991). Optimizing the sensory-perceptual environment of older adults. International J. of Industrial Ergonomics, 7, 133-162.

Schieber, F. (1992). Aging and the senses. In J.E. Birren, R.B. Sloan and G. Cohen (Eds.), Handbook of mental health and aging. New York: Academic Press. pp. 251-306.

Schieber, F., Kline, D.W., Kline, T.J.B. and Fozard, J.L. (1992). Contrast sensitivity and the visual problems of older drivers. Warrendale, PA: Society of Automotive Engineers. [SAE No. 920613].

EFFECTS OF AGE ON THE JUDGED DISTANCE TO VIRTUAL OBJECTS IN THE NEAR VISUAL FIELD

Stephen R. Ellis
NASA Ames Research Center
Moffett Field, CA

Brian M. Menges
San José State University
San José, CA

The following experiment examines effects of observers' age on their judged depth of nearby virtual objects displayed with see-through, Helmet Mounted Displays (HMDs). Monocular, biocular or stereoscopic viewing conditions were used. Two previous finding were investigated: 1) the effect of accommodative demand on judged target depth (Ellis & Menges, 1995) and 2) the change in judged depth of a virtual object due to the introduction of a physical surface at its previously judged depth (Ellis, Bucher & Menges, 1995). Observed effects were consistent with the older subjects' loss of accommodative response and reflexive accommodative vergence. In the present study it was found that only subjects younger than 38 yrs were able to benefit from correctly presented accommodative demand in the monocular viewing situations. Older subjects were especially likely to make greater depth judgment errors in the monocular viewing situations and would especially benefit from compensatory design of the stimulus to correct their larger judgment errors. Biocular and stereo viewing conditions were approximately equivalent with respect to observed judgment bias and produced roughly comparable accuracy for young and old subjects.

INTRODUCTION

The increasingly complete computerization of mechanical design and manufacturing has led industrial laboratories during the past four years to investigate the communication of mechanical assembly sequences directly from the design databases to the workers performing the fabrication. Lightweight, low power, head-mounted see-through displays are one economical display format considered for this purpose. Assemblers using such a display, may be instructed or guided in assembly sequences by computer graphics generated icons and images that may be optically superimposed over their work surface. These icons may or may not be geometrically conformal to the physical surfaces over which they are projected, depending upon the specific application. But in all see-through formats the overlaid graphics appear as virtual images against a background of physical surfaces. In a biocular or stereo viewing mode this kind of composite image adds oculomotor conflict to the classic conflict between vergence and accommodation in binocular displays. Since concerns about binocular rivalry and accurate depth rendering may incline designers to utilize such binocular viewing modes, interest in the role of accommodative demand in such viewing conditions has developed.

Accommodation itself is generally not thought to be a potent influence on perceived depth (Cuiffreda, 1992). However, through the accommodation-vergence reflex, it can indirectly, but significantly influence perceived depth through binocular-vergence-rescaling of perceptual space. Previous results, for example, have suggested that vergence changes when viewing virtual objects presented via head-mounted see-through displays are associated with changes in the judged distance of these objects (Ellis, Bucher & Menges, 1995). Since subjects' ability to respond to changes in accommodative demand varies with age, it was anticipated that the changes in depth judgments presented with monocular, biocular, and stereoscopic conditions would also interact with age.

The following experiment examines age effects on the two principle findings previously reported concerning depth judgments with different viewing conditions: 1) The differential effects of accommodative demand on the judged depth of targets seen with monocular, biocular and stereo viewing conditions and 2) the change in judged depth of a virtual object due to the introduction of a physical surface at its judged position (Ellis, Bucher & Menges, 1995). In general, it was expected that only the younger subjects (< 38 yrs) should be able to benefit from correctly presented accommodative demand. But in any case it was anticipated that information regarding the impact of age on the accuracy of the judged depth in see-through displays of nearby virtual objects would be useful for industrial designers.

METHODS

The Apparatus

The entire display system, called an electronic haploscope, is built around a rigid headband weighing 1.26 kg. in the configuration used. It is the similar apparatus to that described in previous reports (Ellis & Menges,

1995). In the present experiment the band was fitted to each subject's head and then supported by a special pivoted mount at the end of a 1.8 m table. Subjects sat at this end during the course of each experiment. The mount and chair were adjusted so that the virtual objects could be presented at eye level. Head movement was restricted but a residual pitch of about ±10° was allowed for subject comfort.

Electronic display

The display system used two vertically mounted Citizen 1.5 ' 1000 line Miniature CRTs in NTSC mode which were driven by an SGI graphics computer (4D/210GTXB) through custom video conditioning circuits. This video signal conditioning allowed lateral adjustment of the video frame.

Optical features

The CRT images were infinity collimated by standard telescope eyepieces with minimal optical abberation (pairs Ploessl 42 mm) mounted directly under the CRTs. Rotateable prisms in front of each eye allowed fine optical alignment that could be coordinated with electronic lateral displacements. Consequently, the center of each graphics viewport could be individually adjusted for alignment with monocular bore-sight references. After the signal transformation from the RGB to NTSC, individual pixels which corresponded to at least 5 arcmin horizontal resolution measured from subjects' eyes were easily discriminated. The left and right viewing channels could be mechanically adjusted between 55 mm and 71 mm separations for different subject's interpupillary distances. The system was used at 100% overlap and 20° field of view.

Stimuli

A monocular, biocular or stereoscopic virtual image of an upside down, axially rotating, (approximately 2 rpm) wire frame tetrahedron was presented at a distance of 58 cm away from the subjects' eyes. This display was operated under moderate indoor artificial illumination (approximately 50 lux). The virtual image was presented with either 2 diopters accommodative relief or at optical infinity (0 diopters). The monocular display was simply the stereo channel of the subject's dominant eye. The biocular display was produced by positioning the graphical eye point midway between the subject's eyes. The left and right images were identical copies of this view but were shifted laterally so that when the subject's eyes converged to 58 cm the centers of the images would have 0 disparity relative to the convergence point. This technique was used in general for all biocular stimuli at different depths which were occasionally interjected as described below.

The depicted size of the virtual object was randomly scaled from 70 to 130% of its nominal size for each trial preventing use of angular size as a depth cue. The tetrahedron had a nominal 10 cm base and 5 cm height. The lines of the tetrahedron had a luminance of about 65 cd/m^2. and were seen against a 2.9 cd/m^2 gray cloth background placed 2.4 m from the subject. Under half of the conditions with 2 D accommodative relief a physical surface was introduced along the line of sight to the tetrahedron (as illustrated in Figure 1.) and was a slowly, irregularly rotating checkerboard (~2 rpm) made of Xeroxed paper glued on foamcore. The checkerboard was a disk 29 cm in diameter with 5 cm black and white checks having 1.3 cd/m^2 or 17.8 cd/m^2 luminance respectively. It was positioned such that the virtual image of the tetrahedron was seen near the lower rim of the disk in order to allow the subjects to adjust the physical cursor to the apparent distance of the virtual image. Subjects viewed the stimuli in each condition with either monocular, biocular or stereoscopic view conditions and made judgments of the distance to the tetrahedron.

Stereo Calibration

The haploscope display system was calibrated as previously described (Ellis & Menges, 1995) by monocular superimposition of virtual reference images upon corresponding physical diagrams presented at a distance of 2.4 m.

Design

Viewing Conditions (monocular, biocular, and stereoscopic) were crossed with Accommodation (0 or 2 diopters) and nested within Age. The experiment used a blocked design in which blocks of 5 replications of a given condition were presented for each of the three viewing conditions producing uninterrupted 15 judgment sequences. The sequence of viewing conditions was randomly assigned to each subject and thereafter systematically permuted after each set of 3 viewing conditions were presented. Four out of every 30 judgments were based on unanalyzed random variations in the depicted depth of the tetrahedron. In general, it was possible to randomly vary the monocular, biocular, or stereo viewing conditions and the virtual target distance solely through software. Thus, the subjects and experimenters were generally unaware which specific experimental condition was presented. Since the perceptual variation in apparent depth caused by variation in viewing condition appeared to the subjects as variation in depicted depth and the occasional true variation in depth provided added variation, the subjects were unable to notice that many depicted depths were in fact repeatedly presented. The viewing conditions were blocked for a given

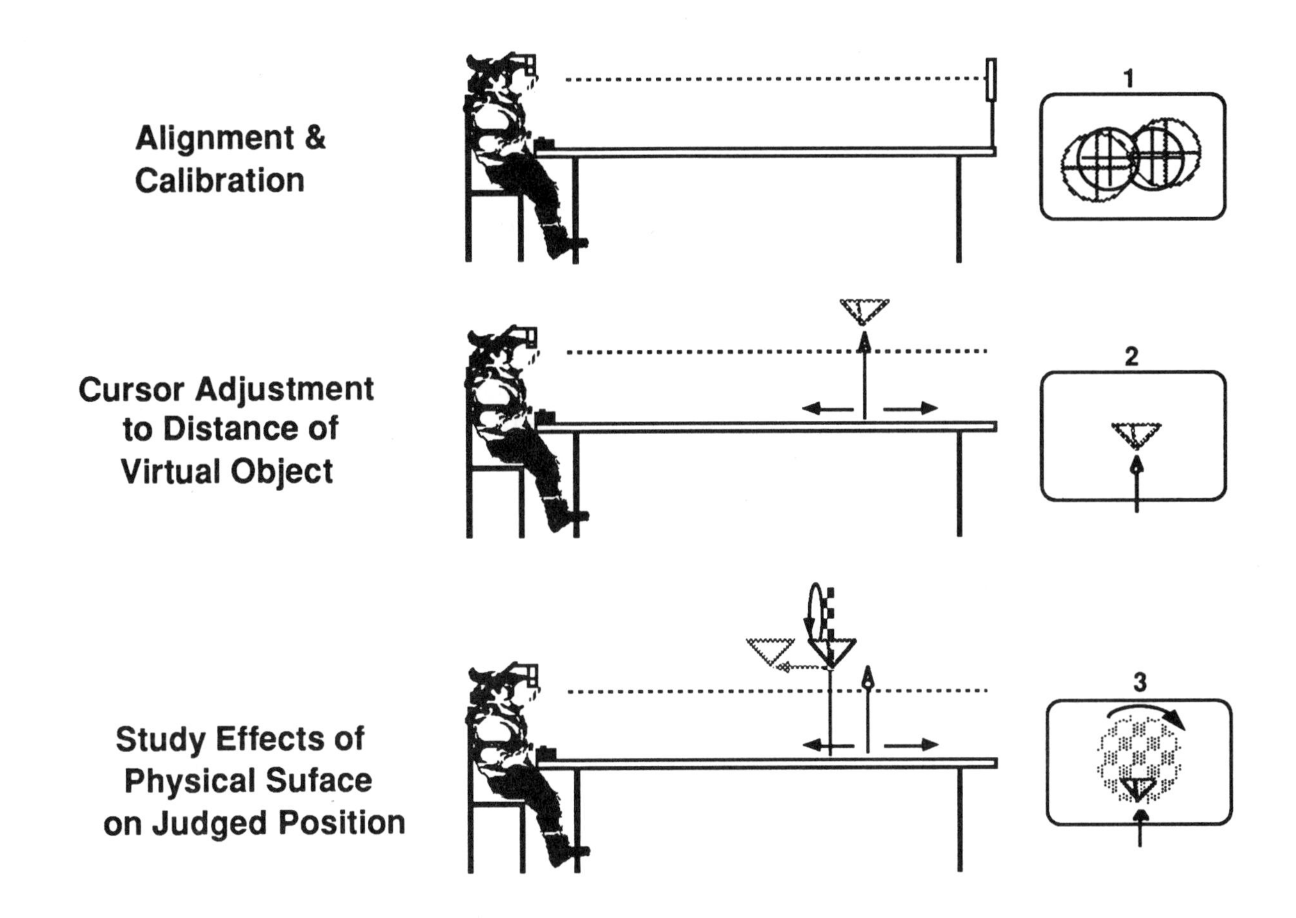

Figure 1. Top: alignment, magnification, and interpupillary adjustment, Middle: Judgment of distance to virtual image Bottom: Judgment of distance to virtual image in presence of physical checkerboard surface.

accommodative demand which was switched by interrupting the experiment after every 15 trials to change the viewing lenses.

Task

The first part of the subject's task was to place the yellow-green LED pointer under the nadir of the slowly rotating virtual tetrahedron. The physical pointer was moved on a track by a chain and gear mechanism attached to a shaft encoder providing position input to the computer. The second part of the task involved the introduction of the opaque checkerboard at the judged depth from the first part and a second adjustment of the pointer to the tetrahedron's depth. The physical surface was only introduced for those conditions in which 2D of accommodation was introduced, an accommodative demand appropriate to the depicted depth of the tetrahedron. The depicted depth of the tetrahedron remained the same as in the first part. In this new configuration the planned variations of depicted size and viewing condition, generally concealed the constant depicted depth from the subjects so that the naive subjects believed each trial, with or without the checkerboard, involved a slightly different depicted depth.

Subjects

Ten subjects, 5 young (<38, mean age =19.2, SD=2.17 range 17-22 yrs) and 5 older (≥38, mean age 43.6, SD=4.27, 38-47 range in yrs.) participated in the first part of the experiment. All but one young and one older subject (the authors) were naive with respect to the purpose of the experiment. The others were either lab personnel or recruited through the Ames Bionetics contractor. All subjects were tested on the Bauch & Lomb Orthorater for stereo acuity better than 1 arcmin. Subjects who normally wore prescription spectacles were allowed to wear them during screening and the experiment. An additional 2 young and 3 older subjects were tested in the second part of the experiment in which a physical surface was introducted. The resulting groups were for the < 38 group: mean age =21.1, SD=4.77 range 15-29 yrs; and for the ≥38 group : mean age = 41.5 SD=2.67, 38-47 range yrs. The selected age ranges were based on observations that the progressive loss of accommodative ability with age, presbyopia, have generally stopped by the age of 40 so that all subjects in our older group could be presumed to be presbyopic (Cuiffreda, 1991). Because of the age distribution in our subject pool we found 38 to be a natural breakpoint since we did not have access to many subjects in the 29-38 age range and it was not sensible to include 38 year olds with a much younger group.

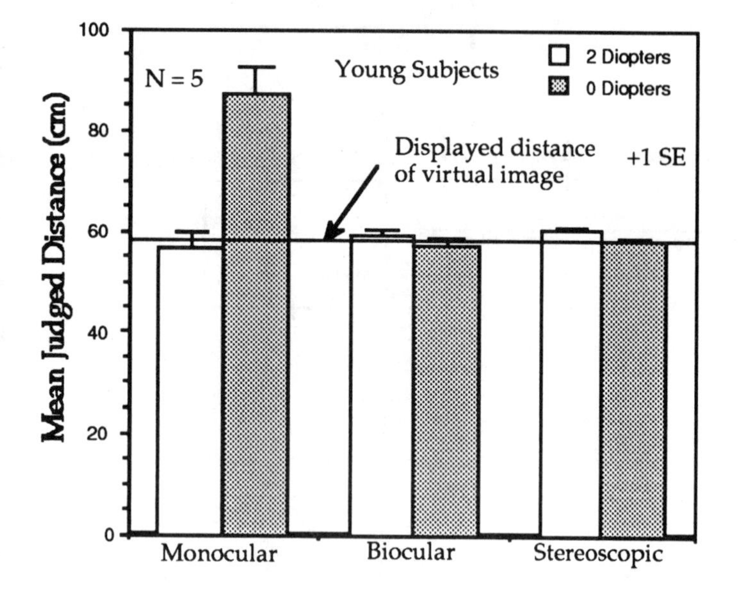

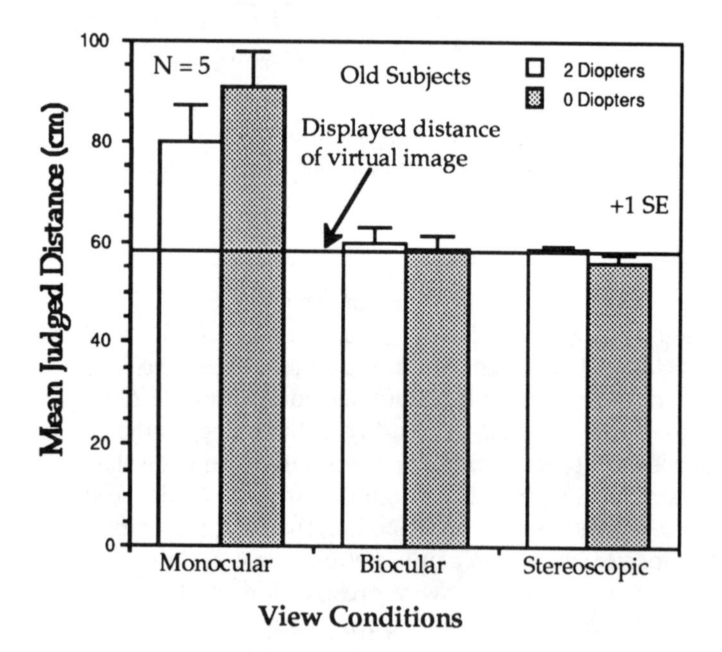

Figure 2

RESULTS

For the first part of the experiment, the analysis of variance (ANOVA) showed that the various viewing conditions strongly interacted with age as expected. The mean stereoscopic and biocular distance judgments were almost completely correct, but a judgment bias appeared as an overestimate when the stereo depth cues associated with the virtual object were removed by monocular viewing. This effect of viewing condition interacted with accommodative demand and age, as indicated in Figure 2. ($F(2,16) = 7.760$ $p < 0.004$). All other significant effects were indirect consequences of this three way interaction.

Analysis of the change of the mean judged distance to the virtual object associated with the introduction of the physical surface showed a main effect of viewing condition ($F(2,26) = 91.340$ $p < 0.001$) and a significant interaction between viewing condition and age ($F(2,26) = 21.921$ $p < 0.001$) (Figure 3). (Note: the additional subjects run increased the Ns , N(young) = 7 and N(old) = 8.) This interaction is plotted both in terms of cm of offset and the corresponding change in meter-angles to give a sense of the magnitude of convergence demand in the viewing conditions used. These interaction effects are modulations of the overall significant change across all viewing conditions ($F(1,13) = 90.623$ $p < 0.001$) of the judged distance to the virtual object closer to the viewer. This main effect appeared as an overall change in the judgment of -4.88 cm towards the viewer.

DISCUSSION

The different initial depth judgments of the younger and older subjects for the differing viewing conditions and accommodative demand shows that when accommodative demand is the only cue to the depth of the virtual object (monocular condition), only the younger subjects are able to fully use it. The older subjects who may be presumed to have impaired accommodative response show a minimal difference in depth judgment when accommodative demand is varied. In the biocular and stereo conditions in which disparity cues to correct depth are present both groups of subjects are able to judge the virtual object depth substantially correctly. The present data do not, of course, demonstrate that these two viewing conditions are indistinguishable but any undetected significant difference would have to be on the order of only a few centimeters or less than about 5% of the depicted distance.

The two age groups also respond differently to the introduction of the physical surface in a manner reflecting their differing abilities to use accommodative information. While superposition of the virtual object on the physical surface causes the object to appear to be closer to the subject for all viewing conditions, the effect is strongest for the monocular condition. This difference would clearly be expected since the monocular condition provides a strong vergence stimulus, bringing convergence closer to the subject by the action of the introduced surface, and thereby providing a changed convergence cue during observation of the monocular stimulus. It is, in fact, quite remarkable that the monocular stimulus seems to appear at a specific depth at all!. In particular, when subjects were informally queried con-

target, they generally were quite certain they were pointing accurately. The problem they had was that the target appeared to move around and follow their adjustment of the pointer. Since in the monocular conditions the subjects may reasonably be presumed to have converged onto the physical pointer, the efference and afference associated with the convergence is the likely cause of the virtual target's apparent distance.

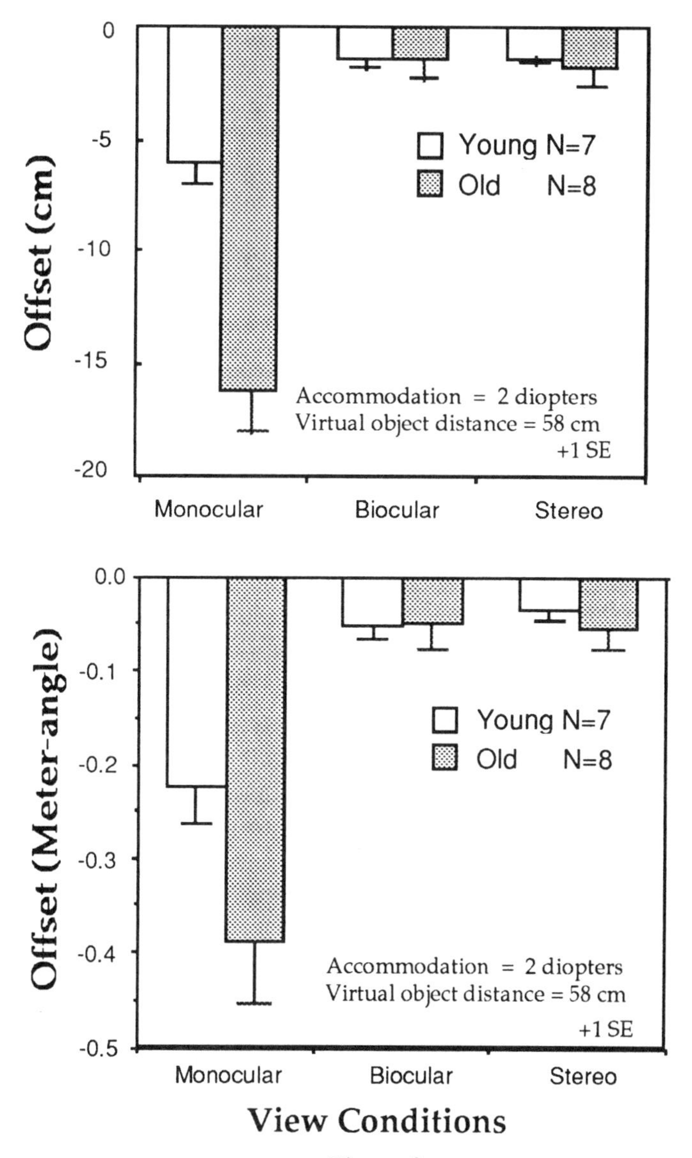

Figure 3.

Interposition of a physical surface caused the distance of virtual targets to be judged closer

The closer judgment of the virtual object distance in the biocular and stereo viewing conditions may have a separate cause from the effect in the monocular situation. In these conditions the closer judgment caused by the physical surface is much smaller than for the monocular case and does not differ for the two age groups. Thus, the loss of accommodative sensitivity and consequent accommodative vergence is unlikely to be its cause. Previous reports have shown that under conditions like these in which disparity information about the virtual object is available, the introduction of the physical surface can produce small changes in static convergence (Ellis, Bucher & Menges, 1995). But these changes could themselves be a proximal convergence response to the interpretation that the optically overlaid virtual object is occluding the physical surface and is therefore closer. Consequently, convergence itself may not be the cause in the change of judged distance.

Design Implications

The present results for the monocular case, which are generally a consequence of the near-response (Semlow & Hung, 1983), indicate the displays of virtual objects for near work should be adjustably focused at the depicted depth of the objects. Results also show that the age of the operator of such a display should be considered since presbyopic operators may be especially subject to certain depth illusions during use in a variety of proposed applications (e.g. Janin, et al, 1993; Rolland, et al 1995). Current work is underway to determine how well other depth cues such as motion parallax can remove the depth errors for the older subjects reported here.

See Ellis, S. R., Menges B. M., Jacoby, R.H., Adelstein, B. D. & McCandless, J.W., (1997) Influence of head motion on the judged distance of monocularly presented virtual objects. Proc. of the Human Factors and Ergonomics Soc for report of the effect of freer head movement on the monocular judgment bias.

REFERENCES

Cuiffreda, K. J. (1992) Components of clinical near vergence testing. *Journal of Behavl. Optom., 3,1,3*-13.

Cuiffreda, K. J. (1991) Accommodation and its anomalies in *Visual Optics and Instrumentation, vol 1,Vision and visual dysfunction*, W. N.Charman, ed., MacMillan Press 231-279.

Ellis, S. R. Bucher, U. J. & Menges, B. M. (1995) The relationship of binocular convergence to error in the judged distance of virtual objects. Proc. of the Intern. Fed. for Automatic Control, Boston, June 26-27, 1995.

Ellis, S. R. & Menges, B.M.(1995) Judged distance to virtual objects in the near visual field. Proceedings of the Human Factors and Ergonomics Soc., 39th Ann. Meeting. San Diego, CA, 1400-1404.

Janin, A.L.; Mizell, D.W.; & Caudell, T.P.(1993) Calibration of head-mounted displays for augmented reality applications Proceedings of IEEE VRAIS '93, Seattle, WA.

Koenderink, J.J., Kappers, A.M.L., Todd, J.T., Norman, J.F., and Phillips, F. (1996) Surface range and attitude probing in stereoscopically dynamic scenes. Journal of Exp. Psych (in the press).

Rolland, J. P., A., D. & Gibson, W. (1995) Towards quantifying depth and size perception in 3D virtual environments. *Presence 4* , 1, 24-49

Semlow J. L & Hung G. K (1983) The near response: theories of control In Schor CM and Ciuffreda, K. J. eds. *Vergence eye movements: basic and clinic aspects.* Boston, Butterworths.

PERCEPTUAL ORGANIZATION AND GROUPING FACTORS: AGE RELATED EFFECTS

Darryl G. Humphrey[1], Arthur F. Kramer[2], Sheryl S. Gore[3]

[1]Wichita State University, Wichita, Kansas [2]Beckman Institute, University of Illinois, Urbana Illinois
[3]State Univeristy of New York, Albany

Older adults have evidenced a poorer ability to use grouping factors in such tasks as Embedded Figures, Incomplete Figures, and partial report. Difficulties in disambiguating the findings of these studies has left unanswered the cause of this age-related difference. By taking into account age-related differences in visual short-term memory, the results of the current study suggest that older adults maintain the ability to capitalize on the perceptual organization of the visual environment as a means of facilitating recall performance. These results have implications for the design of information displays, product labels, codes, and instructions.

We are all familiar with displays and input devices that make use of proximity and similarity to increase ease of use and facilitate operator performance. A review of the aging literature suggests that while grouping by proximity and similarity benefits the performance of younger adults, older adults fail to capitalize on these methods of organizing the perceptual environment. However, the choice of tasks may have masked the ability of the older adults to use proximity and similarity to facilitate task performance. The results reported here suggest that older adults are capable of utilizing grouping factors to achieve a perceptual organization of the visual environment that facilitates recall performance.

The embedded figure task requires subjects to indicate if a complex probe stimulus contains a simpler target stimulus. Basowitz and Korchin (1957) reported that older adults were unable to resist closure as evidenced by a decreased ability to identify a simple stimulus embedded in a more complex stimulus. Cohen and Axelrod (1962) report a similar finding for an embedded figure task. Basowitz and Korchin (1957) also employed the Gestalt completion task which involves trying to identify an object from a picture in which portions of the object have been removed. The older adults were unable to effect closure as evidenced by poor recognition of incomplete objects in the Gestalt completion task. These two studies suggest that age results in a deficiency in the use of aspects of the visual environment to facilitate task performance.

Danziger and Salthouse (1978) extended this line of research through a series of studies. In considering why aging was associated with decreased efficiency in identifying incomplete figures, Danziger and Salthouse suggest four possibilities: criterion differences, stimulus familiarity differences, differences in the knowledge of the information quality of the figure segments, and differences in the utilization of partial information. All three of their studies used the same basic stimuli: pictures of objects in which from 4.4 to 25% of the object had been removed.

On the basis of their results, Danziger and Salthouse (1978) concluded that age differences in identification accuracy for incomplete figures are due to deficiencies in cognitive inference processes. This is an interesting but vague suggestion in that it does not link the performance deficits with specific mechanisms or processes but to a general class of processes. The use of the incomplete figure task incorporates Gestalt principles such as closure and good figure. However, the manner and extent to which these principles are used, or not used, by older adults is left unspecified.

Fryklund (1975) and Merikle (1980) have both demonstrated that young adults benefit from grouping by similarity and proximity in a partial report paradigm based on Sperling's methodology (Sperling, 1960). Building on this work, several researches have reported impaired abilities on the part of older adults to use grouping principles to facilitate stimulus recall (Coyne et al, 1987; Gilmore, Allen, & Royer, 1986; Salthouse, 1976). The studies contrasting the performance of young and older adults in partial report tasks call into question the visual short term memory abilities of older adults and their ability to capitalize on grouping factors. However, it is important to note that the partial report paradigm as used by these researches can not distinguish between deficits in visual short term memory and deficits in the ability to use grouping principles to facilitate the processing of the visual array.

To summarize the studies just reviewed, there is some evidence that older adults experience difficulty in utilizing aspects of the visual environment to facilitate the processing of task relevant information. Unfortunately, these studies have left unspecified the mechanisms that are responsible for age-related performance decrements. One possible reason for this lack of resolution is the complexity of the tasks used. Successful performance of the Gestalt completion task, the embedded figure task, and the incomplete figure task surely involves a variety of processing mechanisms and strategies since several different grouping principles may be employed when achieving a perceptual organization of the stimuli in the visual array. Tasks employing simpler stimuli and simpler task requirements may provide more definitive conclusions regarding age-related differences in the use of aspects of the visual environment for the processing of task relevant stimuli. A second issue is separating age-related decline in visual short-term memory from changes in the ability to benefit from grouping factors. A modification of a paradigm developed by Kahneman and Henik (1977) allows both of these issues to be addressed.

At the core of Kahneman and Henik's (1977) research are three assumptions subsequently found in many models of visual selective attention: (a) a limited capacity attentional system, (b) a queuing rule (serial processing of different objects), and (c) a hierarchical organization in that objects are defined first, then their attributes are processed. Kahneman and Henik's (1977) basic methodology had stimuli organized into groups based on similarity (color) or proximity. The task was to recall the display string, serial order of the elements being important. Subjects were run in groups, stimuli were presented on a projector screen, and responses were written on a form provided by the experimenter. Group boundaries were formed based on the proximity or similarity of elements. The first group in the display was processed to the detriment of the second. Recall for elements within a group was uniform. The size of the recall effects suggests that this paradigm may provide an alternative methodology by which to examine the generalizability of age-related differences in the ability to use display structure to direct attention. The experiment reported below adds a baseline condition to the basic methodology of Kahneman and Henik (1977) as means of separating the effects of visual short term memory changes from changes in the use of grouping principles to increase the percent of the display string that is recalled.

Subjects. Thirty young and 34 older adults participated in this study (mean age of 19.58 and 69.7 years, respectively). The young subjects had less education than the older adults (13.4 years versus 15.3 years; Cochrane's $T(56)=-3.22$, $p<0.01$). However, the younger adults had larger forward digit spans and total digit spans as measured by the WAIS digit span test (10.3 versus 8.4 with $T(54)=-3.22, p<.01$ and 18.2 versus 15.8 with $T(52)=2.24, p<.05$ respectively). The younger adults also possessed higher visual acuity than the older adults for both near and far vision (21.7 versus 25.5 with Fischer's Exact Test $F(1)=7.2, p<.05$ and 21.9 versus 25.3 with $F(1)=6.77, p<.05$ respectively). Debriefing indicated that neither age group experienced difficulty in resolving the stimuli. The two age groups were equivalent in terms of IQ as measured by the K-Bit (115.6 and 114.2 for young and older adults respectively, $T(53)=0.57, p<.6$).

An effort was made to screen older adults for visual health problems with 18 of the 34 older adults returning a completed opthomological questionnaire (details available upon request). The near and far visual acuity of those subjects who did not return a questionnaire was no different from the acuity of those subjects who did return a completed questionnaire ($F(1)=3.85, p<.2$ and $F(1)=1.93$, $p<.5$ for near and far visual acuity, respectively). In terms of performance in the monochrome condition, the two groups of older adults did not differ ($F(1,32)=0.84, p<.4$). The results of these two analyses supports retaining all 34 of the older adults.

Design. Two display conditions were used to examine the effects of two different grouping principles, proximity and similarity of color, on serial recall. A third display condition, monochrome, was used to establish a baseline for each subject's recall performance. Examples of each of the conditions can be found in Figure 1. In the monochrome condition all display elements were one color (red or green) and the elements were contiguous. That is, the display string in the monochrome condition did not have an extra space inserted at any point in the string. In the Proximity condition the seven display letters were separated into two groups by inserting an extra space into the string. In the Similarity condition the two groups were formed on the basis of color (red and green). The second factor was change point: where in the string the extra space was inserted or the change in color occurred. The grouping principle condition (2) was completely crossed with change point (6).

Proximity	Similarity
1 234567	1**234567**
12 34567	12**34567**
123 4567	123**4567**
1234 567	1234**567**
12345 67	12345**67**
123456 7	123456**7**

Figure 1. An example of the stimuli from the visual segmentation task. For the similarity condition the bold font represents red stimuli, the normal font represents green stimuli.

Task. The subject's task was to recall a string of briefly presented letters, serial position being important. The dependent measure was percent of letters correctly recalled for each position in the display string. Each trial began with a fixation rectangle on the screen. The fixation rectangle encompassed the area of the screen where the character string was displayed. A trial was initiated by the subject pressing the space bar. Three hundred msec after the space bar the fixation cross was removed giving a blank screen for 100 msec followed by the 300 msec stimulus display. Subjects gave their responses verbally and the experimenter recorded the responses on a response sheet. The subject then initiated the next trial subject to a minimum ITI of 5200 msec. This ITI was designed to allow the experimenter sufficient time to record the subject's response.

Subjects completed 15 blocks of 24 trials including three practice blocks. The data from the practice blocks were not analyzed. This design provided 16 trials for each combination of Grouping principle and change point. Grouping principle was blocked and the order randomized for each subject. Change point was randomized within each block of trials. No feedback was provided. Subjects were encouraged to take short breaks between each block. The session lasted approximately 90 minutes.

Apparatus and stimuli. Stimulus presentation was accomplished with 386 class computers equipped with a VGA color monitor and a standard keyboard. Subjects were seated in a small comfortably lighted room, given instructions, and then performed the task. The stimuli consisted of letters of the English alphabet (excluding the letter 'U') with no letters repeating within a display string. With a viewing distance of 110 cm the entire display string subtended 4.5° of visual angle in the proximity condition and 4.0° in the similarity and monochrome condition. Each individual letter was 0.43° by 0.48° of visual angle, being greater in height than in width. The separation of adjacent letters was 0.38°. The inserted space was equivalent in size to a single letter. The screen background was white, the default stimulus color red, the altered stimulus color green, and the fixation rectangle and space bar prompt were blue.

Predictions. The data of most interest in this study is the percent recall for each element of the string in the proximity and similarity conditions, once the baseline performance of the monochrome condition is subtracted. Subtracting the percent recall of the monochrome condition standardizes each subject to their own baseline ability to recall the elements of the display string. This subtraction also has the benefit of removing individual and age-related differences in information processing rate that may contribute to performance differences in tasks that rely on visual short term memory (Coyne, Burger, Berry, & Botwinick, 1987; Di Lollo, Arnett, & Kruk, 1982) (Coyne, Burger, Berry, and Botwinick, 1987) (Di Lollo, Arnett, and Kruk, 1982). The resulting difference scores should provide an uncontaminated measure of the ability to use grouping by proximity or similarity to facilitate the recall of the display string.

Results and Discussion. The results for the baseline (monochrome) condition can be found in Table 1. These data were analyzed in a two-way mixed factors repeated measures analysis of variance with age as the between subjects factor and position as a within subject factor. As expected, percent recall declined with serial position in the display string ($F(6,372)=980.21$ $p<.01$). Older subjects demonstrated poorer recall overall ($F(1,62)=9.99$, $p<.01$) due to a greater deficit at later positions relative to the younger subjects ($F(6,372)=7.63, p>.01$).

Table 1. Percent recall in the baseline (monochrome) condition.

	Serial Position						
	1	2	3	4	5	6	7
Young	94	90	87	68	24	8	11
Old	93	90	86	49	13	4	5

To assess the effects of proximity and similarity on the percent of recall, each subject's monochrome performance was subtracted from their performance in the proximity and similarity condition. The resulting difference score provides a measure of the costs and benefits of the two grouping principles. The complete results of the four-way mixed factors repeated measures analysis of variance is contained in Table 2. Age was a between subjects factor while condition (proximity or similarity), change point (where in the string the break or change in color occurred) and position (the serial position in the display string) were within subject factors.

Table 2. Analysis of the mean percent recall data from Experiment 1.

Factor	F	Df	p<
Age	0.35	1,62	.5
Condition	0.45	1,62	.5
Change Point	31.50	5,310	.01
Position	7.21	6,372	.05
Age X Condition	1.23	1,62	.3
Age X Change Point	2.03	5,310	.07
Age X Position	2.11	6,372	.05
Condition X Change Point	7.79	5,310	.01
Condition X Position	13.75	6,372	.01
Change Point X Position	40.23	30,1860	.01
Age X Condition X Change Point	0.76	5,310	.6
Age X Condition X Position	4.14	6,372	.01
Condition X Change Point X Position	9.17	30,1860	.01
Age X Condition X Change Point X Position	2.24	30,1860	.01

Of primary interest is the four-way interaction of condition X change point X position X age. This four-way interaction is due to the older adults being more sensitive to the costs and benefits of grouping, especially at serial position 3 and 4 when the change point manipulation places these serial positions in the first versus the second group, and to a greater degree when grouping by proximity than when grouping by similarity. The older adults also show a greater cost at position three when this position is not included in the first group. A summary of these effects is depicted in Figure 1. The top panel shows the change in percent recall when the element position is the last position in the first group whereas the bottom panel shows the change in percent recall when the element position is the first position of the second group.

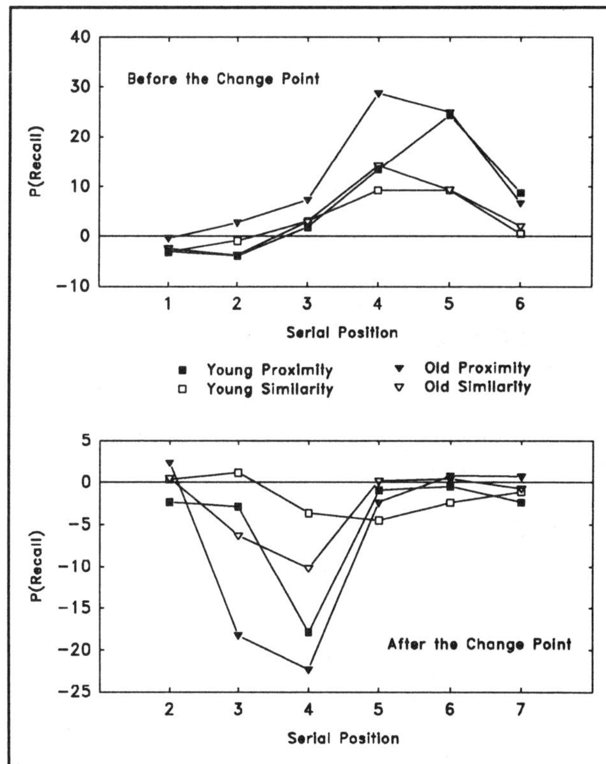

Figure 1. Change in percent recall for each position as a function of belonging to the group before or after the change point.

An important outcome of these data is the relatively cost-free benefit of having five or six elements in the first group of the display string. For both young and older adults, with the change point at position 5 or 6, recall performance is increased for the positions included in the first group without the recall performance of the second group suffering appreciably (all t-tests gave $p>.2$). This effect indicates that segmenting the display string at the fifth or sixth position can afford a performance benefit without a concomitant cost for later positions. This is in contrast to conditions where the change point comes earlier in the string (ie. third position) where performance benefits for the positions of the first group come at the expense of recall for later positions.

An alternative explanation for the present results is that the pattern of recall performance represents the von Restorff effect (von Restorff, 1933). Briefly, the von Restorff effect is the increased probability of recall for an item that has been made salient or distinct from other items. For instance, the positions on either side of the change point (position three and four for change point 3_4) could be considered more salient than the other positions. According to the logic of the von Restorff effect, we would expect an increase in percent recall for each of the positions for change point 3_4. What we do find is an increase in percent recall for position three and a decrease in percent recall for position four. This pattern of effects is present for both young and old adults in both the proximity and similarity conditions. This consistent pattern of results is incompatible with an explanation based on the von Restorff effect, being more easily accounted for by models such as the queuing model of Kahneman and Henik (1977).

Conclusions. Taken together, these results suggest that older adult's recall performance is as at least as sensitive as that of younger adults to the effects of grouping factors in organizing the visual environment. Previous research suggesting that older adults suffered from a reduced ability to utilize grouping factors to facilitate display recall may have lacked an adequate baseline from which to measure recall performance. Manipulating the size of the two groups of letters within the display string afforded both costs and benefits to recall performance. The recall of the middle letters of the string was more sensitive to the effects of proximity and similarity grouping. All of the effects were more pronounced for older adults. These results suggest that grouping character strings by proximity or similarity of color can enhance recall performance. However, the pattern of the present results suggests that the change point needs to be selected with care. Choosing a change point too early in the string will result in decreased recall for later positions in the string. Selecting a change point between the fifth and sixth position should yield the greatest performance benefit. It is important to note that older adults benefited to a greater extent than did younger adults but they also experienced larger costs when the change point was early in the string. This finding has implications for product labels and codes, safety codes, and usage codes for products or devices that are targeted primarily for use by older adults. Grouping the units of the codes or messages into groups of five will increase the likelihood that all the units of the group are recalled. Most importantly, this benefit in recall will not be at the expense of elements that come later in the string. Grouping by proximity had the larger effect on recall performance but grouping by similarity did provide some benefit. Combining the two factors should not harm performance but it remains to be demonstrated if the redundancy will produce increased benefits in recall performance. Further research is needed to determine if the benefits in recall performance demonstrated in the current paradigm can be elicited with stimuli such as icons and if the performance benefits will persist over a longer time scale.

References

Cohen, L. D., & Axelrod, S. (1962). Performance of young and elderly persons on embedded-figure tasks in two sensory modalities. In C. Tibbits & W. Donahue (Eds.), Aging around the world: Social and psychological aspects of aging. (pp. 740-750). New York: Columbia University Press.

Coyne, A., Burger, M., Berry, J., & Botwinick, J. (1987). Adult age, information processing, and partial report performance. Journal of Genetic Psychology, 128(2), 219-224.

Danziger, W. L., & Salthouse, T. A. (1978). Age and the perception of incomplete figures. Experimental Aging Research., 4, 67-80.

Di Lollo, V., Arnett, J., & Kruk, R. (1982). Age-related changes in rate of visual information processing. Journal of Experimental Psychology: Human Perception and Performance, 8(2), 225-237.

Fryklund, I. (1975). Effects of cued-set spatial arrangement and target-background similarity in the partial-report paradigm. Perception & Psychophysics, 17, 375-386.

Gilmore, G., Allan, T., & Royer, F. (1986). Iconic memory and aging. Journal of Gerontology, 41(2), 183-190.

Kahneman, D., & Henik, A. (1977). Effects of visual grouping on immediate recall and selective attention. In S. Dornic (Ed.), Attention and Performance VI (pp. 307-332). Hillsdale, NJ: Erlbaum.

Merikle, P. M. (1980). Selection from visual persistence by perceptual groups and category membership. Journal of Experimental Psychology: General, 109, 279-295.

Salthouse, T. (1976). Age and tachistoscopic perception. Experimental Aging Research, 2, 91-103.

Sperling, G. (1960). The information available in brief visual presentations. Psychological Monographs, 74, 1-29.

von Restorff, H. (1933). Uber die Wirkung von Bereichsbildungen im Spurenfeld. Psychologische Forschung, 18, 299-342.

Acknowledgements

This research was supported by grant # AG12203 from the National Institute on Aging.

EFFECTS OF AGE AND SEX ON SPEED AND ACCURACY OF HAND MOVEMENTS: AND THE REFINEMENTS THEY SUGGEST FOR FITTS' LAW

George Erich Brogmus
Human Factors Program, ISSM; Lab of Attention and Motor Performance, Gerontology
University of Southern California, Los Angeles, CA

Based on modifications of Fitts' Law suggested in the literature, 121 unique formulas were tested against reciprocal tapping data from 1,318 subjects (1,047 males and 271 females) who participated in the Baltimore Longitudinal Study of Aging from 1960 to 1981 in order to determine the best formula (based on the standard error of estimate) and to examine age and sex differences using this formula. The best formula for males differed from that found for females, resulting in a set of new formulas which take into consideration age and sex and which fit the experimental data better than past formulations. While females were faster than males and young were faster than the old, a substantial portion of age and sex differences might be explained by a speed-accuracy tradeoff.

INTRODUCTION

The field of Human Factors can claim only two fundamental principles which have risen to the status of "laws": Hick-Hyman Law and Fitts' Law. The latter is the subject of this paper. Long before Fitts (1954) described the relationship between movement time, movement distance, and accuracy (target width), the qualitative nature of the relationship between these elements were well known and were documented by Woodworth (1899). Fitts, however, has been credited with consolidating the concepts into an information theory law, now known as Fitts' Law, the most frequently cited form of which is:

$$MT = a + b\text{Log}_2(2A/W) \quad (1)$$

where "a" and "b" are the empirically derived y-intercept and slope respectively, "A" is the center-to-center distance between targets, and "W" is the constructed target width. (The portion of the formula "$\text{Log}_2(2A/W)$" is usually referred to as the index of difficulty.) Since Fitts (1954) proposed this relationship between movement time, amplitude, and target width, there have been numerous attempts to develop a formula that would better fit his original data as well as data from subsequent investigations (Bullock & Grossberg 1988; Crossman & Goodeve 1983; Howarth & Beggs 1981; Howarth, Beggs, & Bowden, 1971; MacKenzie, Martiniuk, Dugas, Liske, & Eickmeier, 1987; MacKenzie, 1989; Meyer, Abrams, Kornblum, Wright, & Smith, 1988; Meyer, Smith, & Wright, 1982; Schmidt, 1988; Schmidt, Zelazink, Hawkins, Frank, & Quinn, 1979; Wallace, Hawkins, & Mood, 1983; Welford, 1958, 1960, 1968; Welford, Norris, & Shock, 1969; Wright, & Meyer, 1983). With these modifications have come additional theoretical considerations for the information processing aspects of movement as well as methodological considerations for the design of tapping tasks. Nevertheless, Fitts' Law has been repeatedly shown to be quite robust.

Surprisingly, although much work has been done in substantiating (or attempting to refute) Fitts' Law under various conditions and for many different applications, and even though there has been some work done to improve the formula itself and to advance the psychomotor theory, there has been little work examining age and sex differences (Crossman & Goodeve, 1983; Murrell & Entwisle, 1960; Welford, 1958, 1960, 1968; Welford, Norris, & Shock, 1969). (See Brogmus, 1991, for a review of recent work on aging and gender as it relates to Fitts' Law.) Recently, with the intent to examine possible gender-related speed-accuracy tradeoffs that may take place with age, York and Biederman (1990) administered the Fitts' tapping task to 62 males and 84 females, 20 to 89 years of age. They found that women appear to perform better than men on the Fitts tapping task. They also found that the young men seem to favor speed over accuracy. The slopes for the lines fitted to the data using Fitts's original formula showed that women perform better on the more difficult tasks than the men and that, with age, women show less slowing. This work by York and Biederman is the first to closely examine the factors of speed, accuracy, age, and gender using the Fitts' tapping task. Most recently, Welford (1990) reviewed the current literature, including York and Biederman (1990), and concluded that the best formula for Fitts' Law is given by either $\text{Log}_2(A/W + 0.5)$ or $\text{Log}_2(A/W + 1)$. Welford also evaluated the apparent speed-accuracy tradeoff between the young and old and outlined an explanation for a shift from an emphasis on speed for the young to an emphasis on accuracy for the old.

Based on past research it is safe to say that a large sample size would probably be needed to clearly identify sex differences. This, combined with the desire for longitudinal research on the subject, makes it economically difficult to conduct such research. Fortunately, however, the data gathering for such a project had actually already been done. The Baltimore longitudinal study of aging (BLSA), a federally-sponsored research project, (Shock, 1984) is considered by many to be the "gold standard" of longitudinal research. The main purpose of the BLSA is to study "normal" aging. This extensive research project has been ongoing since February, 1958 and continues to conduct a multitude of tests including a battery of psychomotor performance measures on its subjects.

METHOD

Welford (Welford, Norris & Shock, 1969; Shock, 1984) designed a reciprocal tapping task (similar to Fitts's original tapping task) that was administered to BLSA participants on each visit from 1960 to 1981. Only the results of analysis of the first visit data from this task are presented here. Welford's design allowed recording of each "hit" by using a pencil and paper to record the actual hits made. Initial testing of 325 males who were participating in this study was reported by Welford, Norris, and Shock (1969). In January, 1978 women began to be tested as part of the BLSA. "As of June 30, 1981, more than 300 [women] had been examined and tested at least once, 150 two or more times." (Shock, 1984, p.1) The data for the present study includes usable data from 1,047 male subjects and 271 female subjects (aged 17 to 100, **M**=50.2; **SD**=16.2). Study participants were self-recruited volunteers from the Baltimore area, and therefore were neither a random sample nor a representative sample of the Baltimore population, who differ from the general population in that they are happier, healthier, and more likely to be married than the general population (Andres, 1978). Over 70 percent of the subjects were college graduates, and over 40 percent held advanced degrees.

Each target configuration consisted of four parallel lines - two lines per target. Each line was approximately 4 1/2 inches (115 millimeters) long. Three different target widths were used (4, 11, and 32 millimeters) in conjunction with three different movement requirements (50, 142, and 402 millimeters, measured from the inside edge of one target to the center of the other target) for a total of nine target configurations. Subjects were seated at a desk and given the target configuration to be used. They were allowed to position the paper in front of them so that they were comfortable with its orientation, holding the paper with the non-dominant hand and tapping back and forth from one target to the other using the dominant hand. The instructions read to the subjects included "be accurate in hitting the target and at the same time maintain maximum speed". The time in minutes for a total of 100 hits, each of the widths of both the right and left scatters of hits as well as the distance between the leftmost hit and the rightmost hit, excluding any wild deviant hits, were recorded. The nine target combinations were presented in different orders to different subjects in such a way that the serial positions both of the conditions and of the transitions from any condition to any other were appropriately balanced. One practice trial using one target configuration was given prior to the recorded trials.

In order to fairly evaluate age and sex differences, it is critically important to have a formula that provides a good fit to the data for each age group by sex. Therefore the pursuit of the "best" equation was of primary initial concern. The treatment of data can be outlined as follows: 1) first, "suitable" data was selected from the raw data, 2) next, formulas were tested using Se, the standard error of estimate, to see which one provided the best fit to the first visit data, 3) this formula was used to evaluate age and sex effects based on first visit data, 4) based on this evaluation, a new formula was created which incorporated age and sex, 5) this new formula was then tested similar to step 2 above to evaluate how well this formula fit the first visit data. One hundred and twenty one (121) unique formulas were applied to several groupings of the data to determine which formula was the best representation of the data. The 121 formulas were derived from all possible combinations of six (6) basic logarithmic expressions, five (5) different measures of movement amplitude, four (4) different measures of target width, and a formula derived by Welford (Welford, Norris, and Shock, 1969). For each of the 121 formulas, r^2 (the squared coefficient of correlation, also known as the coefficient of determination) and Se were calculated for first visit data. The calculations were carried out in two ways: first, for all test conditions; second, omitting the "easiest" condition (i.e. small movement amplitude and wide target - SW) on the grounds that it does not fit the major trend of the data and instead represents a lower limit to movement time of about 200 msec (Welford, Norris and Shock, 1969, and Welford, 1990). The primary justification for the omission of this data point is that movement times of less than about 200 msec are too fast to be part of an information processing loop and thus are not indicative of the process modeled by Fitts' Law.

RESULTS AND DISCUSSION

The best formula for males differed from the best formula for females. The best formula for males turns out to be:

$$MT = a + bLog_2(D'/W + 1) \quad (2)$$

which is of the form of Shannon's Theorem 17 (Shannon, 1948; which is presumedly where Fitts derived his original formula from) with the primary distance of measure being D' - the distance between the *far edges* of the *scatters* of hits instead of "A" - the center-to-center distance between targets. Formula 2 yields an Se= 15.43 and an r^2=0.9935 for males as compared to Formula 1 which resulted in an Se=23.66 and an r^2=0.98472. The best formula for females is the same as Formula 2 except D' is replaced by D - the *constructed* distance between the *far edges* of the targets:

$$MT = a + bLog_2(D/W + 1) \quad (3)$$

Formula 3 resulted in an Se=21.38 and an r^2=0.98531 for females as compared to Formula 1 which resulted in an Se=27.50 and an r^2=0.97571 for females. Figure 1 shows Formula 2 fitted to the data for all subjects combined for each target configuration.

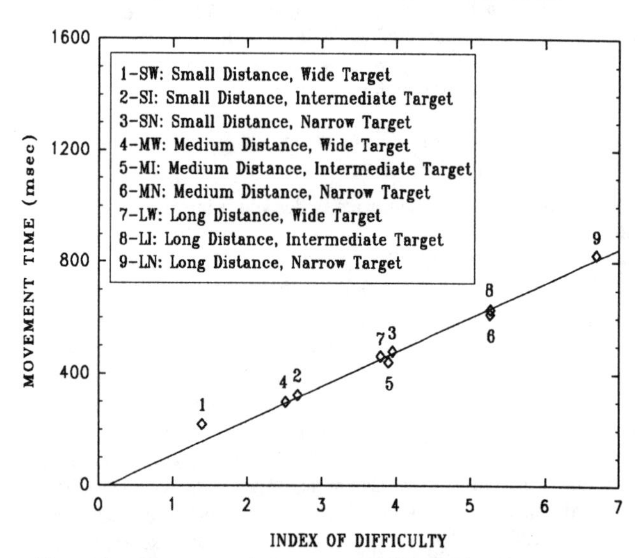

Figure 1. Movement time versus the index of difficulty using Formula 2 for all subjects combined. Standard error bars are smaller than the symbols and therefore are not observable. Target configuration SW (the point with the lowest index of difficulty) is not included in the regression line drawn.

While it could be argued that the mean of all subjects for each target configuration is not the best baseline since it may emphasize the largest population age group, a nearly identical plot is achieved if the mean of the means of each age group are used instead.

Females showed lower movement times than men for each target configuration and for every age range except those with very few subjects, but the main effects of sex were not significant (F=2.58, p=0.1083). However, there was a marginally significant interaction between sex and target configuration (F=1.85, p=0.0625), such that females were disproportionately faster than males on the targets with a higher index of difficulty.

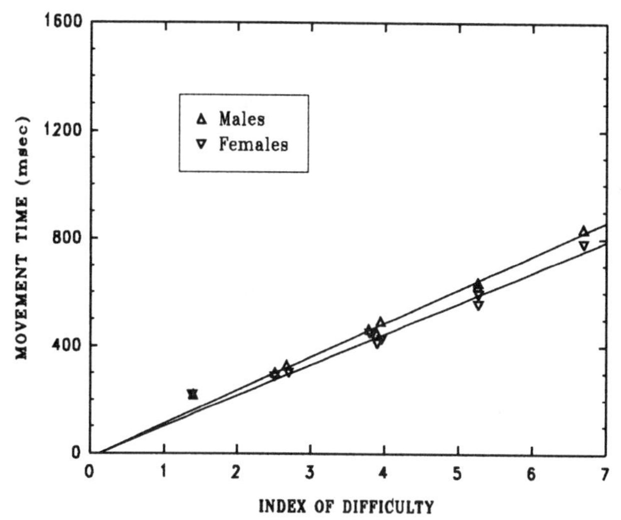

Figure 2. Formula 2 fitted to the mean movement times for males and females separately. Standard error bars are smaller than the symbols and therefore are not observable.

The consistent trend of females being faster than males for each age group and for nearly every target within each age group suggests that women may have a slight advantage in rapid hand movements that require precision. Whether this is the result of a biological advantage, performance strategy, or gender difference due to work and hobby stereotypes is open to question. Part of this difference may also be explained in terms of a speed-accuracy tradeoff. Another explanation may be related to the smaller average mass of female forearms, which might be allowing for fewer corrections to be made to the arm trajectory in order to overcome the inertia of the arm during deceleration.

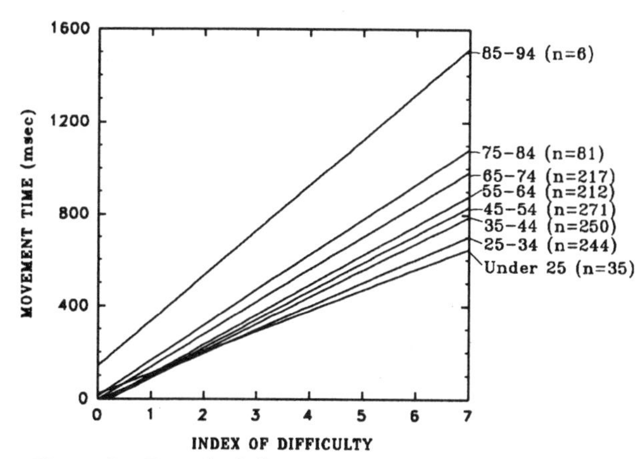

Figure 3. Formula 2 fitted to each age decade's movement times for all subjects.

Age differences in movement times were significant between every age group ($F=38.03$, $p<0.0001$). As expected, older subjects experience a disproportionate slowing on more difficult targets (i.e. targets with a high index of difficulty). Nevertheless, even for the target with the lowest index of difficulty (SW), movement time was significantly higher for the older subjects than the young. Figure 3 illustrates the age differences in movement time. (Sample size for each group is given in parenthesis.) A steady increase in the slopes is evidence of a disproportionate slowing of performance on the "difficult" tasks.

With the exception of the groups with very small sample sizes, all y-intercepts are near zero and most are negative. Differences in y-intercepts between age groups were not statistically significant. Nevertheless, a trend is indicated: Young subjects tend to have a positive y-intercept. As they age this drops below zero and bottoms out around -30 ms around the 50's. It then begins to rise in the latter years. The y-intercepts for the females appear to decrease more linearly with age.

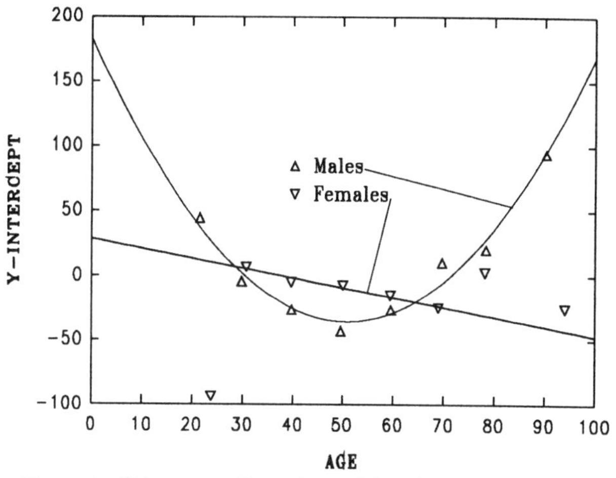

Figure 4. Y-intercepts for males and females by age using the means of the y-intercepts calculated for each subject. Formula 2 was used for males and Formula 3 was used for females.

The more prominent trend is for the slope. The differences in slopes with age was significant at $p<0.0001$ for both Formula 2 and 3. The difference between slopes by sex was not statistically significant at the $p=0.05$ level of significance for both formulas, however, trends between males and females can be seen. With the exception of the youngest females, the slopes for both formulas have a strong linear increase with age for all groups. The slopes for females do not change quite as much as those for males. Figure 5 breaks out males and females separately. A very slight interaction of age and sex can be seen in the slopes.

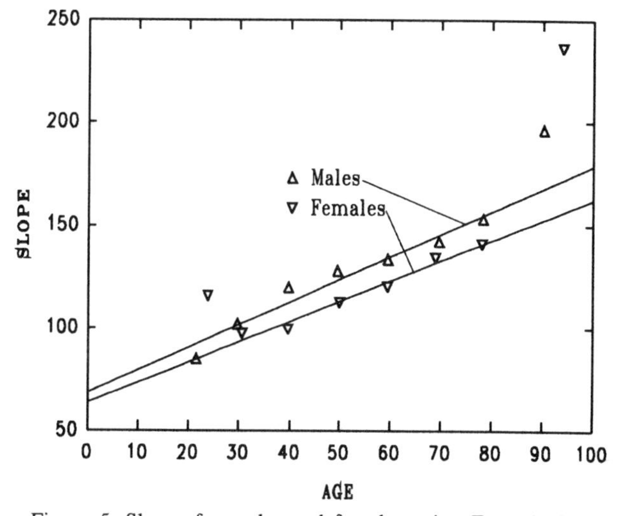

Figure 5. Slopes for males and females using Formula 2 for males and Formula 3 for females. Slopes are based on the mean of slopes of each subject.

The resulting "best" formula for males is given by:

$$MT= -40+[-15+0.3(AGE)]2 + [69+1.1(AGE)]Log_2(D'/W+1) \quad (4)$$

The resulting "best" formula for females is given by:

$$MT= 28-0.75(AGE) + (64+AGE)Log_2(D/W+1) \quad (5)$$

A further (yet slight) improvement in the fit of the above formulas can be achieved if, instead of D' and D, the length of the arc between the far edges of the scatters of hits and the length of the arc between the far edges of the constructed targets is used where the radius of the arc is given by the average length of the distance between the elbow and the tip of the thumb.

The data indicate that for the older age ranges the average scatter width (W') is much smaller for old subjects than for young. Consistent with the work of York and Biederman (1990) *it appears that older subjects are emphasizing accuracy more than the young subjects.* This may account for some of the decreased speed with the old subjects. It may also account for some of the sex differences as well since the males have smaller scatters of hits than the females except for the target configurations with wide targets, where the females have the (slightly) smaller scatters. Exactly how much of the variance in MT can be accounted for by the differences in W' is uncertain. For the most difficult target configuration (LN - long amplitude, narrow target) the scatter width (W') for the decade8 subjects is 42% of the W' for the decade2 subjects. In other words, the scatter of hits for the older subjects for the most difficult target width is less than half the scatter of hits made by the young subjects on the same target!

The slight, but consistent, difference between the fit of the two formulas for males and females presents some interesting possibilities for explaining the difference. First, the basic form of these formulas suggests again that Fitts's original inspiration from Shannon's Theorem 17 may, in fact, for whatever real reason, have been the best form of the formula. This supports the findings of MacKenzie (1989), and more recently of Welford (1990) using Fitts's original data and some of the first visit data from the present study (approximately 300 visits). Second, the basic form of Formulas 2 and 3, with their inclusion of the measures of D and D' provide a much more intuitive measure of distance than A. In order for the brain to utilize the measure of A, an estimation must be made between the centers of the targets or else recognition of the outside edge of one target and the inside edge of the other target must be made. It seems far simpler to input the readily available outside edges of the targets.

Nevertheless Formulas 2 and 3 present some interpretive difficulties. Intuitively, what does $D/W + 1$ (= $(D + W)/W$) represent? What is $D + W$? While it could be that subjects are constrained not by the amplitude of movement but by the outer edges of the targets, it seems difficult to apply to this formula. A more appropriate formula might have been $Log_2(A/W + 1)$ which is equivalent to $Log_2(D/W)$. The following is an attempt to explain this finding: Perhaps as one taps back and forth between two targets, motor programing takes into consideration the total possible distance to be covered (D) and *a basic parameter of the constraint of that distance* (W). If this is the case, the W in the numerator becomes a constraining parameter - the bigger W is, the more control must be exercised so as not to miss the opportunity to hit just inside the close edge of the target being hit. In the denominator, W would then have to take on a different meaning - it would have the traditional interpretation that more control is needed to hit a small target, and therefore more time-consuming motor programs to be executed.

The difference between the males' formula (2) and the females' formula (3) may suggest a difference in perception of the task as it proceeds. It may be that males tend to use existing hits (as they are being made) for the parameter of the "outside edges", while the females may be focusing on the constructed edges thought the tapping task. If this is true, it may be that the "strategy" of the females results in faster movement, albeit less accurate. Perhaps more information is processed/programmed by the males to account for a variable "hit edge".

CONCLUSIONS

The implications for this research are several. First, a new formulation of the Fitts' Law relationship has been indicated as superior over past formulations. Unfortunately, this new formula, while mathematically superior, has produced more questions than answers in the effort to provide an explanation of the workings of psychomotor performance. As Welford (Welford, 1990) has pointed out, a choice of formulas must consider the theoretical implications. Some researchers may feel more comfortable with the traditional formulas since they may provide for easier interpretations; nevertheless, these new formulas have some interpretive benefits. Furthermore, the superior fit of the formulas described in this study indicate that a re-thinking of the theory underlying rapid aimed hand movements should be considered. The present study has contributed to the overall understanding of psychomotor behavior in at least three ways:

1. It has demonstrated that there is a linear increase with age in the slope of the Fitts' Law equation (modified - Formulas 2 and 3), implying that as we age we tend to take a disproportionately greater amount of time on difficult rapid hand movement tasks.

2. Different formulas for males and females provide a better representation of the data.

3. The decrease in performance with age appears to be partially, if not substantially due to an emphasis on greater accuracy.

A possible fourth contribution is in disturbing the common assumption that males have faster movement times than women. These contributions can be applied, in a more quantitative way, to the design of products, systems, and jobs, so that the capabilities of the people using those products or systems or performing those jobs are adequately considered.

Since no statistical tests were applied to the formula-testing step, this should be included in future tests to determine if differences in the Se's for different formulas are statistically different. A more thorough screening of outliers should also precede the data analysis. Consideration should also be given to designing the task so that visual clues of the hits is minimized so that the theoretical implications of Formula 2 can be tested. A more detailed investigation should be made into the contribution of the speed-accuracy tradeoff that was observed between the young and old and between the females and males. An analysis of covariance using 1/W' is recommended. It must be pointed out that even though the formulas proposed in this paper do appear to improve the fit of the formulas to the data, there is still the hint of something missing from the formulas. The graphs of the data seem to indicate times for the most difficult task which are disproportionately slower than the rest and thus these points end up off of the regression line. A second order spline of all points fits remarkably well and should be researched more thoroughly.

ACKNOWLEDGMENTS

This paper was based on the author's masters thesis (Brogmus, 1991). Thesis committee members were Max Vercruyssen, David B. D. Smith, Alan T. Welford, James E. Birren, James L. Fozard, and Michelle Robertson. Data were collected as part of the Baltimore Longitudinal Study of Aging (BLSA) conducted by the National Institute on Aging (NIA) at the Nathan W. Shock Laboratories of the Gerontology Research Center (GRC), Baltimore, MD 21224. Programing and preparation of computer-based analyses used for this research were supported in part by a contract to the University of Southern California (Max Vercruyssen, Principal Investigator) by the NIA (James L. Fozard, Technical Contract Officer). Computer assistance was provided by Carolyn Eames and Olukayode Olofinboba.

REFERENCES

Andres, R. O. (1978). The normality of aging: The Baltimore Longitudinal Study. Part of the National Institute on Aging Science Writer Seminar Series. Washington D.C.: U.S. Department of Health, Education, and Welfare.

Brogmus, G. E. (1991). Effects of age and gender on speed and accuracy of hand movements: and the refinements they suggest for Fitts' Law. Unpublished masters thesis, University of Southern California, Los Angeles, CA.

Bullock, D. & Grossberg, S. (1988). Neural dynamics of planned arm movements: Emergent invariants and speed-accuracy properties during trajectory formation. In S. Grossberg (Ed.), Neural networks and natural intelligence. Cambridge, MA: MIT .

Crossman, E. R. & Goodeve, P. J. (1963 and 1983). Feedback control of hand-movement and Fitts' Law. Quarterly Journal of Experimental Psychology - Human Experimental Psychology, 35A(2), 251-278, (originally presented at the Oxford meeting of the Experimental Psychology Society in July, 1963).

Fitts, P. M. (1954). The information capacity of the human motor system in controlling the amplitude of movement. Journal of Experimental Psychology, 47(6), 381-391.

Howarth, C. I., & Beggs, W. D. A. (1981). Discreet movements. In D. Holding (Ed.), Human Skills, (pp. 91-117).Chichester: Wiley.

Howarth, C. I., Beggs, W. D. A., & Bowden, J. M. (1971). The relationship between speed and accuracy of movement aimed at a target. Acta Psychologica, 35, 207-218.

MacKenzie, C. L., Martiniuk, R. G., Dugas, C., Liske, D., & Eickmeier, B. (1987). Three-dimensional movement trajectories in Fitts' task: Implications for control. The Quarterly Journal of Experimental Psychology, 39A, 629-647.

MacKenzie, I. S. (1989). A note on the information-theoretic basis for Fitts' Law. Journal of Motor Behavior, 21(3), 323-330.

Meyer, D. E., Abrams, R. A., Kornblum, S., Wright, C. E., & Smith, J. E. K. (1988). Optimality in human motor performance: Ideal control of rapid aimed movements. Psychological Review, 95(3), 340-370.

Meyer, D. E., Smith, J. E. K., & Wright, C. E. (1982). Models for the speed and accuracy of aimed movements. Psychological Review, 89, 449-482.

Murrell, K. F. H., & Entwisle, D. G. (1960). Age differences in movement pattern. Nature, 185 , p. 948.

Schmidt, R. A. (1988). Laws of simple movement. Motor control and learning: A behavioral emphasis. Chapter 9, Champaign, IL: Human Kinetics Publishers.

Schmidt, R. A., Zelazink, H., Hawkins, B., Frank, J. S., & Quinn, J. T. (1979). Motor-output variability: A theory for the accuracy of rapid motor acts. Psychological Review, 86(5), 415-451.

Shannon, C. (1948). A mathematical theory of communication. Bell Systems Technical Journal, 27, 379-423, 623-656.

Shock, N. W. (1984). Normal human aging: The Baltimore Longitudinal Study of Aging. (NIH publication no. 84-2450) U.S. Department of Health and Human Services.

Wallace, S. A., Hawkins, B., & Mood, D. P. (1983). Can Fitts' Law be improved?: Predicting movement time based on more than one dimension. (Technical Report No. 126). Boulder, CO: University of Colorado, Institute of Cognitive Science.

Welford, A. T. (1958). Ageing and human skill. London: Oxford University Press.

Welford, A. T. (1960). The measurement of sensory-motor performance: Survey and re-appraisal of twelve years' progress. Ergonomics, 3(3), 189-230.

Welford, A. T. (1968). Fundamentals of Skill. pp.137-160, London: Methuen and Company.

Welford, A. T. (1990). Fitts' Law revisited, with a note on changes associated with age. Private communication, December, 1990.

Welford, A. T., Norris, A. H., & Shock, N. W. (1969). Speed and accuracy and their changes with age. Acta Psychologica, 30, 3-15.

Woodworth, R. S. (1899). The accuracy of voluntary movement. Psychological Monographs, 3, (3, Whole No. 13), 1-114.

Wright, C. E., & Meyer, D. E. (1983). Conditions for a linear speed-accuracy trade-off in aimed movements. Quarterly Journal of Experimental Psychology, 34A, 279-296.

York, J. L., & Biederman, I. (1990). Effects of age and sex on reciprocal tapping performance. Perceptual and Motor Skills, 71, 675-684.

MANUAL PERFORMANCE OF OLDER AND YOUNGER ADULTS WITH SUPPLEMENTARY AUDITORY CUES

Richard J. Jagacinski, Neil Greenberg, and Min-Ju Liao
Ohio State University
Columbus, Ohio

Jian Wang
Hangzhou University
Hangzhou, Zhejiang, People's Republic of China

ABSTRACT

Subjects attempted to perform the same manual movement pattern on repeated trials using a visual display of error. Additionally, some subjects heard a tone that was proportional to either the position or velocity of the ideal movement pattern. With the tone, both older and younger adults demonstrated increased anticipation in the form of an increased correlation of their movement pattern with the ideal velocity pattern. However, males exhibited this effect most with the tone that was proportional to ideal velocity, and females, with the tone that was proportional to ideal position. The benefit of the auditory displays did not carry over after they were withdrawn. These results demonstrate one technique for improving perceptual/motor performance. Although older adults exhibited a longer effective time delay, the older and younger adults benefitted from the additional cues to comparable degrees

INTRODUCTION

One conception of perceptual-motor skill development is that performers have a template or internal model of ideal performance. Skill learning involves gradually refining this internal model, and correcting discrepancies between actual performance and the model (Keele, 1977). Supplementary cues to enhance skill learning may either facilitate establishing the internal model, making comparisons between actual performance and the model, or both (e.g., Stenius, 1976, cited in Keele, 1977).

Another conception of movement skill development, the Progression Hypothesis (Fitts, Bahrick, Noble, & Briggs, 1959), is that with practice performers emphasize higher derivatives of movement trajectories. This hypothesis has received support from the modeling of visual-manual tracking performance (Fuchs, 1962; Garvey, 1960; Jagacinski & Hah, 1988) and a movement production task without concurrent visual feedback (Marteniuk & Romanow, 1983). With practice, subjects' movement patterns are increasingly correlated with the velocity and/or acceleration of displayed signals.

An additional hypothesis is that supplementary cues that emphasize higher derivatives will accelerate training, lead to higher levels of skill development, and/or permit better performance (e.g., Birmingham & Taylor, 1954; Marteniuk & Romanow, 1983). This hypothesis was tested in the present experiment by providing supplementary auditory cues in a tracking task that required subjects to repeatedly perform a back and forth movement pattern. In previous research, supplementary auditory displays have been found to improve performance over levels achieved with visual displays alone (e.g., Benepe, Narasimhan, & Ellson, 1954, cited in Poulton, 1974; McGee & Christ, 1971; Mirchandani, 1972; Pitkin & Vinje, 1973; Janiga & Mayne, 1977). The present experiment tested whether the effectiveness of a supplementary auditory display would be greater if it emphasized velocity rather than position information, and whether such effects would differ between older and younger adults and between males and females.

METHOD

Subjects

Fifteen male and 15 female older adults aged 60 to 69, and 15 male and 15 female younger adults aged 18 to 26 served as subjects. All subjects were right-handed, had generally good health, had no arthritis in their hands and arms, exercised at least twice per week, and did not take any of a list of drugs that might affect performance. Additionally, all subjects had to demonstrate 20/20 vision and be able to detect 50 Hz increases or decreases in tones in the range from 200 to 900 Hz. Subjects' mean scores on the Digit Symbol Substitution Test from the WAIS-R corresponded to about the 75th percentile for each age group.

Apparatus

Subjects sat in a chair and manipulated a one-dimensional position control stick (Measurement Systems 525) with their right hand. Movement of the control stick controlled the position of a green dot that was displayed on a Tektronix 604 oscilloscope with P31 phosphor. One degree of movement of the control stick resulted in 0.14 degrees of visual angle displacement of the green dot. The oscilloscope screen was 50 cm from the subjects' eyes. In the center of the screen was a 1-mm square yellow marker that was a zero error reference. Subjects wore Sennheiser HD 520 earphones through which they heard tones that were generated by a Wavetek 193 modulation generator.

Procedure

Subjects were introduced to the tracking task and were shown a graph of the ideal movement pattern that they would be required to produce (a sinusoidal pattern with both amplitude and phase modulation). Subjects performed a total of 12 blocks of trials over a period of three days. Each block consisted of one practice trial followed by ten data trials. Each trial lasted 14 s. There was a 12 s rest period between trials followed by a 2 s warning interval for the next trial.

On Block 1, all subjects used only the visual display of error. On Blocks 2-8, one-third of the subjects received no supplementary auditory display, one-third of the subjects heard a tone with pitch proportional to the ideal control stick position, and one-third of the subjects heard a tone with pitch proportional to the absolute magnitude of the ideal control stick velocity. After Block 8 the supplementary tone stimuli were withdrawn.

RESULTS

Block 1

The mean absolute error was calculated for the middle 10 s of each trial. In addition, the ensemble average movement trajectory was calculated for blocks of ten trials by averaging the control stick position across trials at 50 ms intervals. The resulting ensemble average movement trajectory was approximated with the regression equation:

$$m(t) = B_0 + B_1 I(t-0.150) + B_2 \dot{I}(t-0.150) \quad \text{(Eq. 1)}$$

where $\underline{m}(\underline{t})$ is the ensemble average movement trajectory, $\underline{I}(\underline{t}-0.150)$ is the input position delayed by a reaction time of 0.150 s, $\dot{\underline{I}}(t-0.150)$ is the input velocity similarly delayed, and $\underline{B}_0$, $\underline{B}_1$, and $\underline{B}_2$ are constants. An equation of this form has been used by previous investigators (Elkind, 1956; Ware, 1971; Jagacinski & Hah, 1988). The velocity term provides some local anticipation of the input to compensate for the subjects' 0.150 s time delay. The higher the value of $\underline{B}_2$, the more heavily does this velocity term contribute to local anticipation and shorten the "effective time delay," 0.150-($\underline{B}_2/\underline{B}_1$) s. Given the mathematical structure of the model, if one were to increase the value of the fixed delay above 0.150 s, there would be a corresponding increase in the fitted values of $\underline{B}_2$, such that the effective time delay would stay approximately constant. Choice of the fixed delay parameter is therefore not crucial, and for convenience a value of 0.150 s was used for all subjects.

For Block 1, analyses of variance were conducted on absolute error, $\underline{B}_1$, and $\underline{B}_2$. Each analysis was a between subjects design with three experimental factors, age group, sex, and display condition. The analysis of the mean absolute error revealed significant main effects of age group and sex. Younger subjects performed better than older subjects (1.22 vs. 1.65 degrees of visual angle), and males performed better than females (1.25 vs. 1.62 degrees of visual angle). Similar effects were found in an analysis of $\underline{B}_1$. In contrast, there were no significant effects for $\underline{B}_2$ or for the effective time delay. The values of $\underline{B}_2$ were close to zero, which is consistent with an initial effective time delay near 0.150 s.

Block 8

On Block 8, the last block with the auditory cues present, the analyses of absolute error and $\underline{B}_1$ again revealed significantly better performance by younger adults and by males. There were no significant effects of the auditory display and no significant interactions in either of these analyses.

The pattern of results for $\underline{B}_2$ for Block 8 were more complex. The average values of $\underline{B}_2$ were significantly higher for younger adults (0.083 vs. 0.038 s) and for male subjects (0.071 vs. 0.050 s). Additionally, there was a significant age group by sex interaction, which corresponded to a larger difference between younger males and females than among older males and females. Finally, there was a significant sex by display group interaction. In order to clarify this interaction, separate two-way analyses of variance were conducted for the male and female subjects.

The analysis of Block 8 for female subjects revealed significantly higher values of $\underline{B}_2$ for the younger subjects (0.065 vs. 0.035 s). There was also a significant main effect of the display condition. A one-tailed Dunnett's $\underline{t}$-test indicated that the position display was superior to the control (0.066 vs. 0.044 s); however, the velocity display was not superior to the control (0.040 vs. 0.044 s). The older females with the position display had a value of $\underline{B}_2$ of 0.061 s, which is much closer to the value of the younger female control subjects (0.066 s) than the older female control subjects (0.022 s).

The analysis of Block 8 for male subjects revealed higher values of $\underline{B}_2$ for the younger subjects (0.101 vs. 0.041 s). There was also a significant main effect of the display condition. A one-tailed Dunnett's $\underline{t}$-test indicated that the velocity display was superior to the control condition (0.091 vs. 0.058 s); however, the position display was not superior to the control (0.064 vs. 0.058 s). The older males with the velocity display had a value of $\underline{B}_2$ of 0.064 s, which was roughly midway between the older male control subjects (0.027 s) and the younger male control subjects (0.089 s).

Analyses of the effective time delay, $0.150-(\underline{B}_2/\underline{B}_1)$, revealed significantly shorter delays for younger subjects (0.044 vs. 0.090 s) and a sex by display interaction. Three older female subjects were excluded from this analysis as outliers. A separate analysis of variance for the males revealed significant differences across displays that mimicked the results for the $\underline{B}_2$ analysis. A separate analysis for the females did not reveal a significant display effect.

Block 10

Block 10 was the second block of trials after the auditory displays were discontinued. Analysis of the absolute error for Block 10 indicated that younger subjects performed significantly better than older subjects and that male subjects performed significantly better than female subjects. Analyses of $\underline{B}_1$, $\underline{B}_2$, and the effective time delay showed similar effects. The effective delays were 0.096 s for older subjects and 0.049 s for younger subjects. Two older females were excluded from this latter analysis as outliers. In none of these analyses was there a significant main effect or interaction involving auditory display condition.

DISCUSSION

The Progression Hypothesis (Fitts et al. 1959) suggests that higher derivatives of movement will become increasingly important with practice. Consistent with this hypothesis, all groups showed increases in the velocity weighting with practice. Additionally, both older and younger adults benefitted from supplementary auditory displays as exhibited in increased weighting of the velocity cues. However, it was expected that the supplementary velocity display would be superior in promoting this change with practice. While males did exhibit a higher velocity weighting with the supplementary velocity display, females exhibited a higher velocity weighting with the position display.

The pattern of influence of the displays on B_1 and B_2 suggests that Equation 1 is not a fundamental process model for the present task. While it may be possible for a position display to increase the salience of velocity information (Pitkin & Vinje, 1973), one would also expect such a display to increase the weighting on position, B_1. This latter effect did not occur. Also, the velocity display only increased B_2 for the male subjects. The lack of a simple relation between display changes and performance changes suggests the need for a different performance model. Equation 1 has nevertheless proven useful in providing more sensitive measures of performance than average absolute error.

With the present experimental design, one cannot attribute the main effect of age group solely to age differences. One particularly marked cohort difference was that all of the younger subjects had some experience with video games, whereas few of the older subjects had similar experience. Consistent with previous evidence of slower perceptual-motor performance by older adults (Salthouse, 1985), the older subjects in the present experiment had longer effective time delays. Of greater interest, however, was the lack of a significant age by display interaction, indicating that older and younger adults benefitted from the tones to comparable degrees. While some research has indicated that older adults have greater difficulty attending to multiple, simultaneous sources of information (McDowd & Birren, 1990), such an effect was not apparent in the present data. Supplementary auditory cues for older adults may therefore provide one methodology bringing their performance levels closer to those of their younger counterparts. One limitation of this benefit is that it occurred only while the tones were present, and did not continue when they were withdrawn. A second limitation is that some older subjects would have benefitted much more from increasing their position weighting, B_1, and the supplementary displays did not produce this effect. Over the years the suggested applications of supplementary auditory displays have included vehicular control (e.g., Janiga & Mayne, 1977), sports skills (Keele, 1977), and home appliances (Koncelik, 1982). While generalizations across tasks and display formats are uncertain, the present results provide some evidence that older adults can benefit from supplementary auditory displays.

ACKNOWLEDGMENTS

This research was supported by National Institute on Aging Grant AG09179 to The Ohio State University. The project monitor was Deborah Claman.

The authors wish to thank Mark Acker, Shelly Mullen, Fara McGrady, Suzanne Beason-Hazen, Jeanette Lee, Craig Oshima, David Lozano, Larry Campbell, Tom Lydon, and Pat Neidhart for their help.

REFERENCES

Benepe, O. J., Narasimhan, R., & Ellson, D. G. (1954). An experimental evaluation of the application of harmonic analysis to the tracking behavior of the human operator. USAF Wright Air Development Center, Technical Report 53-384. Wright-Patterson Air Force Base, Ohio. Cited in Poulton, (1974).

Birmingham, H. P., & Taylor, F. V. (1954). A design philosophy for man-machine control systems. Proceedings of the IEEE, 54, 1748-1758.

Elkind, J. I. (1956). Characteristics of simple manual control systems. Technical Report 111, M.I.T. Lincoln Laboratory, Lexington, Massachusetts.

Fitts, P. M., Bahrick, H. P., Noble, M. E., & Briggs, G. E. (1959). Skilled performance. Ohio State University Final Report for Contract AF 41(657)-70. Wright Air Development Center, Wright-Patterson Air Force Base, Dayton, Ohio.

Fuchs, A. H. (1962). The progression-regression hypothesis in perceptual-motor skill learning. Journal of Experimental Psychology, 63, 177-182.

Garvey, W. D. (1960). A comparison of the effects of training and secondary tasks on tracking behavior. Journal of Applied Psychology, 44, 370-375.

Jagacinski, R. J., & Hah, S. (1988). Progression-regression effects in tracking repeated patterns. Journal of Experimental Psychology: Human Perception and Performance, 14, 77-88.

Janiga, D. V. & Mayne, R. W. (1977). Use of a nonvisual display for improving the manual control of an unstable system. IEEE Transactions on Systems, Man and Cybernetics, SMC-7, 530-537.

Keele, S. W. (1977). Current status of the motor program concept. In R. W. Christina & D. M. Landers (Eds.), Psychology of motor behavior and sport, volume 2. Champaign, Illinois: Human Kinetics.

Koncelik, J. A. (1982). Aging and the product environment. Stroudsburg, Pennsylvania: Hutchinson Ross.

Marteniuk, R. G., & Romanow, S. K. E. (1983). Human movement organization and learning as revealed by variability of movement, use of kinematic information, and Fourier analysis. In R. A. Magill (Ed.), Memory and control of action (167-197). Amsterdam: North Holland.

McDowd, J. M., & Birren, J. E. (1990). Aging and attentional processes. In J. E. Birren & K. W. Schaie, (Eds.), Handbook of the psychology of aging (222-233). New York: Academic.

McGee, D. H., & Christ, R. E. (1971). Effects of bimodal stimulus presentation on tracking performance. Journal of Experimental Psychology, 91, 110-114.

Mirchandani, P. B. (1972). An auditory display in a dual-axis tracking task. IEEE Transactions on Systems, Man, and Cybernetics, SMC-2, 375-380.

Pitkin, E. T., & Vinje, E. W. (1973). Comparison of human operator critical tracking task performance with aural and visual displays. IEEE Transactions on Systems, Man, and Cybernetics, SMC-3, 184-187.

Poulton, E. C. (1974). Tracking skill and manual control. New York: Academic.

Salthouse, T. A. (1985). Speed of behavior and its implications for cognition. In J. E. Birren & K. W. Schaie, (Eds.), Handbook of the psychology of aging. New York: Van Nostrand Reinhold, 400-426.

Stenius, R. K. (1976). Qualitative auditory action feedback in acquisition of a novel motor act. Unpublished doctoral dissertation, University of Oregon, 1976. Cited in Keele, (1977).

Ware, J. R. (1971). An input adaptive, pursuit tracking model of the human operator. Proceedings of the Seventh Annual Conference on Manual Control (33-45). Washington, D.C.: NASA.

THERMOSTATS FOR INDIVIDUALS WITH MOVEMENT DISABILITIES: DESIGN OPTIONS AND MANIPULATION STRATEGIES

Stephen Metz and Brian Isle
Honeywell Sensor and System Development Center
3660 Technology Drive
Minneapolis, MN 55418

Sandra Denno
Center for Rehabilitation Technology
University of Wisconsin–Stout
Menomonie, WI 54751

James Odom
Honeywell Home and Building Control Division
1885 Douglas Drive N.
Golden Valley, MN 55422

ABSTRACT

Using common household products is often difficult for people with neuromuscular disorders, spinal cord injury, or arthritis. We need to better understand their capabilities when designing and adapting products that are easier for them to use. In this study, individuals with movement impairments used two experimental home control thermostats with features that allowed easier positioning and viewing. The participants employed a variety of grasping and manipulation strategies, including some that were not anticipated by the designers. Participants' preferences indicated that the appearance of the product, not just effective control design, was an important factor in their judgments. We discuss the implications of the study results for universal design and adaptation of traditional products for the elderly and those with disabilities.

INTRODUCTION

Adapting products for people with disabilities is a research and design activity that has acquired new urgency with the passage of the Americans with Disabilities Act. A central precept of good human factors design is to match the design of the machine to the person rather than the person to the machine. To advance consumer product human factors design for these people, we need to learn about their capabilities, not just their disabilities (Ellner & Bender, 1981; Wylde, 1991).

In recent years, several researchers have reported capabilities for grasping, pinching, and similar movements by individuals with movement disabilities such as arthritis, multiple sclerosis, spinal cord injury, and cerebral palsy (Kanis, 1988; Metz, Isle, Denno, & Li, 1990; Steinfeld & Mullick, 1990) and without a specific disability (Berns, 1981; Kohl, 1983; Pheasant & O'Neill, 1975; Voelz & Hunt, 1987). Some of these studies have used standardized laboratory equipment, while others have used typical consumer products for testing. Both approaches provide researchers and designers with better descriptions of these movements. They assist in understanding the capabilities of individuals with these disabilities to perform relevant everyday tasks at home or the office.

In this study, we tested two thermostat designs adapted for use by individuals with movement disabilities. These prototypes were designed to be adjusted without a precise grasp of a typical adjustment knob. We expected that a large handle, similar to the levers on some door handles, would be a good solution, because it could be moved without grasping. We also developed an alternative design that provided a large adjustment ring or dial. Each of these solutions was developed as an accessory that could be added to an existing thermostat.

METHOD

Participants

Participants included individuals who had decreased hand function because of a specific condition: rheumatoid or osteoarthritis (n = 23) (mean age = 66.7), neuromuscular disorders, including multiple sclerosis and cerebral palsy (n = 48) (mean age = 45.9), and spinal cord injury (n = 14) (mean age = 29.7). Thirty-four elderly individuals with no known disability of the hand or wrist also participated (mean age = 68.9).

Equipment Description

Three thermostat designs were tested, two experimental designs and one control. A Honeywell Round T87F thermostat, an adapted design for the visually impaired, was used as the control. This product has large numerals to show the temperature settings and modified detents to provide tactile feedback of control position (Metz & Isle, 1991). Figure 1 shows a front view of this product. Figure 2 shows the basic T87 thermostat for comparison. Note the smaller numerals for setting the temperature on the basic model.

Figure 1. T87F Round Thermostat for the Visually Impaired. This model served as the control thermostat design in the present study.

Figure 2. Standard Honeywell Round Thermostat

The first experimental prototype used a plastic part that fit in front of the T87F thermostat. The principal feature was a vertical lever for control. The thermostat setting could be changed by moving the lever approximately 45 degrees to the left or right. Figure 3 is a drawing of this prototype. We will refer to it as the "lever model" in this report.

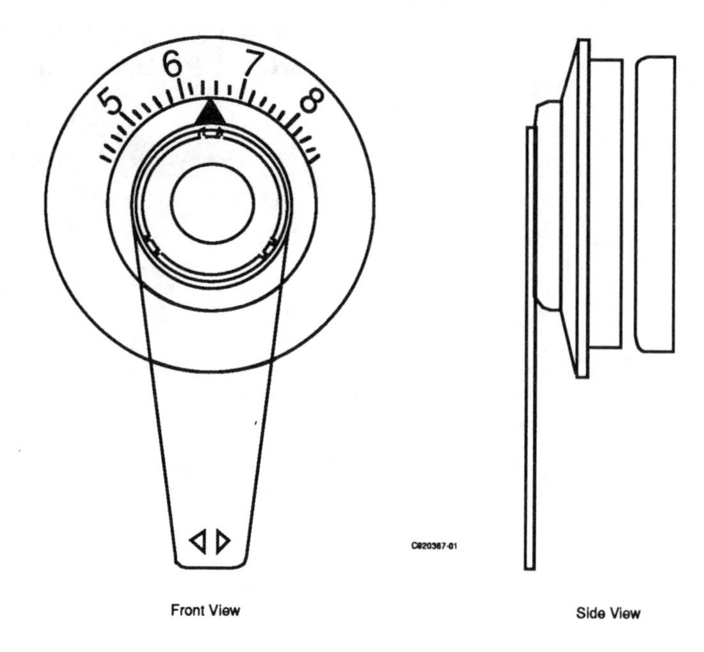

Figure 3. Drawing of "Lever" Adapter for T87F (Center portion has been left blank for clarity.)

The second experimental design was plastic ring with a coarse knurled edge that covered the T87F thermostat. The thermostat setting could be changed by rotating the ring clockwise or counterclockwise. Figure 4 shows this design. We will refer to it as the "ring model" for the balance of this paper.

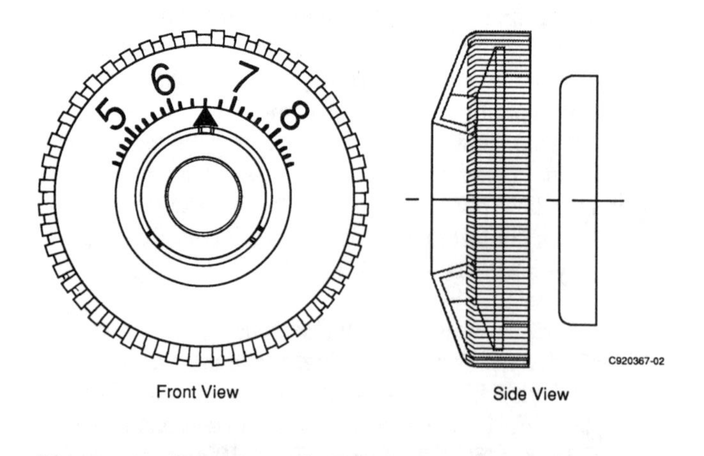

Figure 4. Drawing of "Ring" Adapter for T87F (Center portion has been left blank for clarity.)

The three thermostats were each mounted approximately 0.4 m apart on a white plywood board. The board, in turn, was mounted vertically on an easel that placed the thermostats at either 1.1 m or 1.5 m above the surface of the floor. (A mounting height of 1.5 m is an industry standard for home thermostats; the lower 1.1 m height was judged to be suitable for those seated in wheelchairs.) A fourth model, the basic T87, was also on the board for visual comparison.

Procedure

The participants were asked to use each of the three thermostats. Individuals seated in wheelchairs were presented with the thermostats at both mounting heights. Those who were not in a wheelchair used the higher mounting position.

We provided a setting where the participants could use the thermostats in a typical manner, setting the control to a value on the display, allowing us to observe their behavior and provide them with the experience necessary for later comparison. We did not offer any verbal prompts or modeling for the "correct" way to use the thermostats.

The specific task of setting the thermostat was adapted to the capabilities of each participant. In general, individuals with arthritis were asked to set the thermostat to a particular value, such as 60°, while individuals with neuromuscular disorders such as multiple sclerosis were simply asked to use the thermostat and set it to any value. (For some of the participants with multiple sclerosis, introducing a particular set-point value presented cognitive difficulties, and they could not complete the original task.)

For all participants, we noted their relative success and behavioral strategies as they used the different models, though we did not record any specific dependent measure of performance.

After using the thermostats, each participant was asked to select their preferred model and justify their choice.

RESULTS

The data collected during testing included both observations of how the participants manipulated the thermostats and their preferences for the different models.

Manipulation Strategies

Our observations focused on participants' hand position and movement during use of the thermostats. We did not determine the precision of the movement, although this was considered satisfactory.

In general, participants used one hand to loosely grasp the lever model, either cradling the lever between the thumb on top of the lever and four fingers underneath, or pinching the edges of the lever between the thumb and forefinger. The ring model was also operated with one hand, although occasionally users grasped it with both hands. Most chose to push alongside the edge of the device instead of grasping it with the fingertips.

The control thermostat with the smaller adjustment ring presented some difficulties for many of the participants with movement disabilities. Its stiffer action and shorter moment arm made turning more difficult for each of the groups with a disability. The group with no known hand limitations did not have difficulty using the control model.

While the users with spinal cord injury, who had very limited hand strength, were able to use both experimental thermostats, the ring model appeared to provide better stability for their hand than the lever model. These users could easily move the ring model by brushing or rolling the side or distal edge of the hand along the knurled edge. Some individuals had difficulty getting their hand to the device accurately, but once the hand was on the device, they appeared to move it easily.

Individuals with arthritis generally used an open grasp of the ring model, cupping their hand to fit the external circumference of the ring. The coarse knurled edges appeared to provide enough friction so that they did not have to grip the device, as such. One individual with arthritis attempted to grasp the ring thermostat in a more conventional manner, experienced some discomfort in doing so, and stopped using that device.

The thermostat mounted at the higher 1.5 m level was not particularly difficult to use for those in wheelchairs, with the exception of those participants with weak upper arm strength. The larger numerals on the bezel appeared to simplify reading the settings; however, the current temperature display in the center of the device was harder to read, and many commented on this problem. Reading that temperature was not required to complete our task, but it would be necessary in a normal home.

Preferences

Table 1 presents the preference data for individuals who were seated in wheelchairs during testing. (Thirteen individuals with neuromuscular disorder could not reach the thermostats when mounted 1.5 m high.) As a group, those with neuromuscular disorders preferred the lever model when it was mounted at 1.5 m but did not indicate a preference when the thermostats were mounted at a lower height. Most individuals with spinal cord injury preferred the ring model when mounted at the lower height, but did not indicate a strong preference when it was mounted at 1.5 m.

In Table 1, the distribution of the preferences for each group was significantly different from chance (Chi-square test at the .05, .01, and .01 level, respectively, for the first three groups in the table) and not significant for the last group (spinal cord injury—thermostat mounted at 1.5 m).

Table 2 presents the preference results for those standing at the device mounted at 1.5 m above the floor. (Approximately 25% of the individuals with neuromuscular

Table 1. Preference Measurements for Individuals in Wheelchairs

	Thermostat Mounted 1.1 m High		Thermostat Mounted 1.5 m High	
Design	Neuro-muscular Disorder (n = 41)	Spinal Cord Injury (n = 14)	Neuro-muscular Disorder (n = 28)	Spinal Cord Injury (n = 14)
Ring	17 (41%)	10 (70%)	7 (25%)	6 (43%)
Lever	18 (45%)	4 (30%)	18 (64%)	7 (50%)
Standard	6 (15%)	0 (0%)	3 (11%)	1 (7%)

disorders could stand and use the apparatus.) All groups using the thermostats in this configuration preferred the ring model.

In Table 2, the distribution of the preferences for each group was significantly different from chance (Chi-square test at the .10, .10, and .01 levels, respectively, for the three groups).

Table 2. Preference Data for Those Standing (thermostats mounted at 1.5 m)

Design	Neuro-muscular Disorder (n = 13)	Arthritis (n = 26)	Elderly n = 34)	Total (n = 73)
Ring	8 (62%)	15 (58%)	20 (57%)	43 (59%)
Lever	3 (23%)	6 (23%)	8 (24%)	17 (23%)
Standard	2 (15%)	5 (19%)	6 (16%)	13 (18%)

When asked to explain their preferences, users listed differences between the experimental models and the control model in reaching, turning, and adjusting, but many users also told us the lever model looked like a special "adapted" device. They liked the ring model because it looked more like a standard thermostat.

DISCUSSION

The expectation of the research and design team prior to testing was that users would prefer the lever model because it offered better leverage and a larger grasping area. However, the results did not support this expectation, although both designs were reasonably effective.

Both the individuals with arthritis and the elderly participants preferred the ring model. We believe they preferred it for several reasons.

Based on our observations, the ring model offered better stability. The lever model was somewhat harder to use when the hand was braced near the device or on the wall. Moreover, the large dial allowed the device to be "rolled" into position, and those with difficulties with grasping could use it relatively easily.

The preference for the ring model was also due to dissatisfaction with the lever model as an "adapted" device. This information reminds us that the physical appearance of products often communicates a message. In this case, it was a message that many users consider personal and would prefer not to send.

Admittedly, the lever model was larger, so some of the negative comments may have been motivated by the physical size of the adaptor rather than its shape. Nevertheless, the ring model was a more attractive form and was similar to the standard design. Certainly individuals have expectations about what looks "normal" in their home. In this respect, users with disabilities are no different from people without disabilities.

Those who were in wheelchairs showed a preference for the lever model when mounted above them at 1.5 m, which may have been because it was difficult to reach the ring model at that height. They did not indicate a preference when the devices were mounted lower. However, compared to the standard thermostat model, they clearly preferred the experimental designs.

We were pleased with the success of these simple adaptive accessories to improve the usability of this product. While we human factors specialists and industrial designers need to take more opportunities to develop universal designs that accommodate both able and disabled users, many products already installed in homes and offices are not easily or economically replaceable with modern electronic devices. In these cases, adaptive accessories that improve usability for individuals with disabilities and do not restrict use by those without disabilities are needed and help bridge the usability gap.

Testing provided us with direct feedback on our prototype designs and also increased our understanding of how individuals with these disabilities can better use consumer products and will assist us in designing future adaptive solutions.

ACKNOWLEDGMENTS

The authors wish to thank the following organizations for their support for this study: Honeywell Foundation; Courage Center, Golden Valley, Minnesota; MS Achievement Center, St. Paul, Minnesota.

REFERENCES

Berns, T. (1981). The handling of consumer packaging. *Applied Ergonomics*, *12*, 153–161.

Ellner, J.R., and Bender, H.E. (1981). Hiring the handicapped: The role of the human factors practitioner. *Proceedings of the 25th Annual Meeting of the Human Factors Society*, 759–761.

Kanis, H. (1988). Design for all? The use of consumer products by the physically disabled. *Proceedings of the 32nd Annual Meeting of the Human Factors Society,* 416–420.

Kohl, G.A. (1983). Effects of shape and size of knobs on maximal hand-turning forces applied by females. *The Bell System Technical Journal, 62,* 1705–1712.

Metz, S.V., and Isle, B. (1991). Products for special people: A case study of thermostat development. *Insight, 2,* National Association of Senior Living Industries, 9–16.

Metz, S.V., Isle, B., Denno, S., and Li, W. (1990). Small rotary controls: Limitations for people with arthritis. *Proceedings of the 34th Annual Meeting of the Human Factors Society,* 137–140.

Pheasant, S.T., and O'Neill, D. (1975). Performance in gripping and turning: A study of hand/handle effectiveness. *Applied Ergonomics, 6,* 205–208.

Steinfeld, E., and Mullick, A. (1990). Universal design: The case of the hand. *Innovation, 9,* 27–30.

Voelz, S.L., and Hunt, F.E. (1987). Measurement of hand strength in arthritic women and design of appliance control knobs. *Home Economics Research Journal, 16,* 65–69.

Wylde, M.A. (1991). Toilets, tubs and automobiles: The mature consumer's perspective of consumer products. *Insight, 2,* National Association of Senior Living Industries, 17–24.

SHAPE AND PLACEMENT OF FAUCET HANDLES FOR THE ELDERLY

Beverly A. Meindl
Andris Freivalds
The Department of Industrial and Management Systems Engineering
The Pennsylvania State University
University Park, PA 16802

The 'modern' bathroom is an area which poses considerable barriers to the elderly, specifically a mismatch between the shape and the placement of faucet handles and their physiological capabilities. Fifteen residents of a retirement facility, with a mean age of 79.9 years, exerted their maximum turning torques in a randomized fully-crossed design incorporating the following factors: two positions (low-21 inches and high-42 inches), two angles (45° and 90°) and three types of handles (acrylic, star, and lever). Analysis of covariance (ANCOVA) of the data indicated that the type of handle and the position, along with the covariates: age, gender and subject height, were significant at $p<.05$. Angle and the covariates: weight and arthritis were not significant, most likely because weight is correlated with height and arthritis with age. The second order interactions of handle-angle and position-angle were also significant at $p<.05$.

The lever handle was clearly superior with average torques produced being 50% greater than those from the acrylic and star handles. Torque levels on the acrylic handle and the star handle were very similar, with the acrylic handle slightly superior in the 45°-angled position and the star handle being slightly better in the 90°-angled position. Overall, the 45°-low position resulted in the lowest torques, while the 45° high position resulted in the highest torques. Age had a profound effect, with torque values decreasing an average of 10% over the 15 year age span of the subjects. Based on the study, it is recommended that plumbing systems with both high (42 inches) and low (21 inches) faucets with lever handles be installed where ever possible.

INTRODUCTION

The make-up of the United States population is in a clear state of flux, with the total and proportion of the older adults increasing significantly. From 1900 to 1980, the U.S. population doubled in size, while the proportion of those over 65 increased nearly seven times. At present, there are almost 24 million persons over 65, of which 13 million are women and 10.5 million are men. The median age has increased from 23 in 1900 to 30 in 1980 and is will increase to over 37 in the early 2000's. In 1900, one of every 25 persons was over 65, now it is one of every 9, and by 2020, it will be one of every 6 or possibly 5 (Small, 1986).

Unfortunately, this aging population will have to cope with the associated progressive loss of physiological capabilities that can dramatically reduce the individual's ability to participate in everyday activities. This decrease in capabilities and mobility tends to confine them and most of their activities to their homes. However, functioning in a poorly designed home environment can become a demanding task and an additional burden. The 'modern' bathroom is one such area which poses almost insurmountable barriers to the elderly with the frequent daily activities occurring there. Ever more common shower stalls are not popular with the elderly, because they dislike standing for extended periods of time and have been used to bathing in a tub (Kira, 1976). In fact, falls in the bathtub are one of the most common causes of accidental injury in the home for the elderly (Czaja, 1990). The shape and position of bathtub/shower faucets can cause difficulty in reaching or turning the faucet. For example, a scalding accident could occur if the elderly person could not react with sufficient force to turn the faucet off when the water suddenly increases in temperature (Czaja, 1990). An additional barrier is the psychological one that the bathroom has always been considered a place where a person can achieve complete privacy. This, coupled with the fact that what takes place in bathroom is of a very private nature, makes assistance in the bathroom a touchy subject (Kira, 1976).

Operating a faucet handle is a complex biomechanical process requiring the action of several bones, joints and muscle groups and depends on many factors. The way in which a person grasps a handle is a factor determining the amount of torque that can be generated, with the power grip being the strongest (Kroemer, 1986). Knurling (Swain et al., 1970; Pheasant and O'Neil, 1975) and larger diameters (Pheasant and O'Neil, 1975; Cochran and Riley, 1986) increase the amount of torque generated. Shape is another important factor because it dictates the size of the available moment arm as well as the type of grip used (Cochran and Riley, 1986), with the lever type handle being clearly superior to other types of handles (Bordett et al., 1988). The position of the handle relative to the operator's body is another factor, with standing postures and shorter reaches generating larger torques than sitting postures and long reaches (Huston et al., 1984; Bordett et al., 1988).

Physical disabilities and the natural consequences of aging also affect torque production. Hand strength decreases by more than 50% (Fisher and Birren, 1947). Also, upwards of 50% of the elderly population is afflicted with osteoarthritis, the gradual wearing away of the joint cartilage. With pain in the joints, the elderly will refrain from the using those joints, further decreasing the muscle strength and the range of motion. Thus, the objective of this study was to determine the optimum shape and placement of bathroom faucet fixtures in order to best conform to the needs and desires of the elderly.

METHODS

Apparatus - A testing apparatus to hold two sets of three faucet handles, referred to as acrylic, star and lever because of their overall appearance (Fig. 1), were constructed from Unistrut (King of Prussia, PA). These handles, selected from the commercial plumbing fixture line offered by Kohler, Inc., were chosen because they are commonly found in retirement facilities and were similar to the handles used in Bordett et al. (1988). The faucet handles were mounted on two levels, the first being designated the low position, represented the standard bathtub fixture height of 21 inches, while the second, being designated the high position, represented a height of 42 inches, which could be reached without bending. The upright position of the frame was the 90° angle (Fig. 2a), while tipping the entire frame backward produced a 45° angled position (Fig. 2b). A Sears Craftsman Microtork torque wrench, slipped over a bolt fused to each valve stem, provided the torque readings.

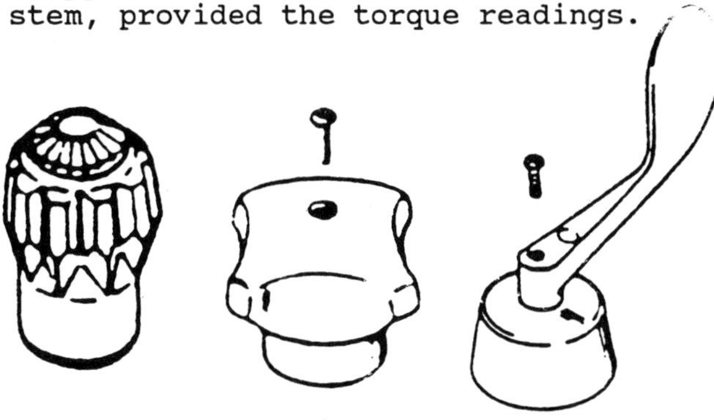

Fig. 1 - Faucet Handles Tested

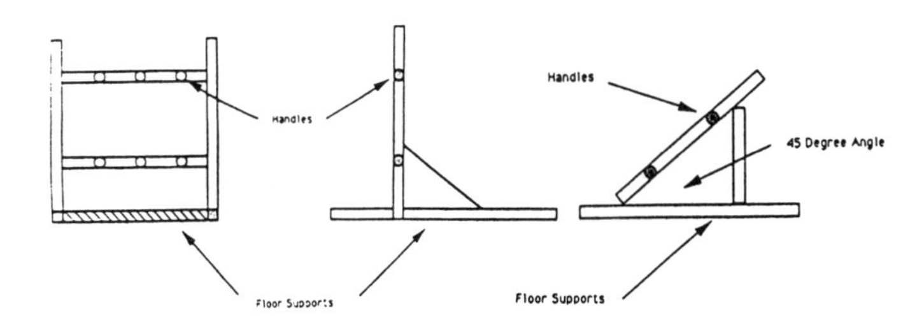

a) 90° Position b) 45° Position

Fig. 2 - Testing Apparatus

Subjects - Fifteen residents of a retirement facility volunteered for the study, with their attributes shown in Table 1. Results from a questionnaire provided valuable insight into the subject population. The average age was 79.9 years, and 13 out of 15 subjects (86.7%) were female. Only two of the 15 subjects required assistance in using the bathtub/shower. However, both of these subjects required assistance specifically in getting in and out of the tub. One subject expressed difficulties in using the faucet handles. Specifically cited were problems with handle shape, handle placement, and turning of the handles. Seven had been diagnosed with arthritis. Three subjects mentioned ailments that could limit the range of motion of the arms, wrists, hands, or fingers. One subject had been stricken with polio and confined to a wheel chair since the age of 30. One subject had had a major stroke that affected the left side. All signed the informed consent form.

Table 1 - Subject Attributes

Subject	Sex (M/F)	Age (yrs.)	Height (in.)	Weight (lbs.)	Arthritis (Y/N)
1	F	75	67.5	147	N
2	M	76	70.0	185	N
3	F	78	66.0	138	N
4	F	80	64.0	137	Y
5	F	83	63.5	127	Y
6	F	79	65.0	145	Y
7	M	80	69.0	175	N
8	F	90	59.0	115	Y
9	F	78	66.0	127	N
10	F	79	64.0	123	Y
11	F	78	67.5	139	N
12	F	79	63.5	137	N
13	F	85	61.0	117	N
14	F	83	65.0	135	Y
15	F	75	61.0	132	Y
Mean		79.867	64.80	138.60	
St.Dev.		3.998	3.04	19.25	

Procedures - To accommodate their personal schedules and to provide as little disruption to their daily lives as possible, subject were allowed to select their individual testing times. The subjects were instructed to stand with both feet parallel, facing the testing apparatus, as close as desired (most chose about 15 inches). The testing apparatus simulated a right handed bathtub meaning that the subjects would have to bend to the right and use their right hand in turning the faucets. The foot placement was integrated so as to approximate the bending that would be required to operate the faucet handles in an actual bathtub. They were asked to exert their maximum force against each faucet handle. This exertion lasted 2-5 seconds. Approximately seven exertions per handle were required in order to the best three exertions. Subjects were allotted one minute rest between exertions.

Experimental Design - A randomized fully-crossed design incorporated the following factors: two positions (low-21 inches and high-42 inches), two angles (45° and 90°) and three handles (star, acrylic, lever). With 15 subjects and three trials per subject, a total of 36 observations were recorded.

RESULTS

Average torques generated on the lever handle were 50% greater than those from the acrylic and star handles (shown in Fig. 3). Torque levels on the acrylic handle and the star handle were very similar, with the acrylic handle slightly superior in the 45°-angled position and the star handle being slightly better in the 90°-angled position. Overall, the 45°-low position resulted in the lowest torques, while the 45° high position resulted in the highest torques.

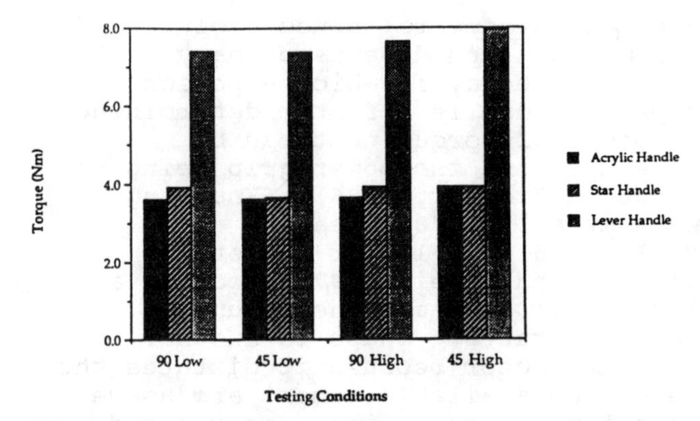

Fig. 3 - Torques Averaged over Subjects

Separate analysis of the lever handle produced similar results (Fig. 4); 11 of 15 subjects were able to produce higher torque values at 45° than at 90°. However, the two angled conditions differed in torque values by only 5%. In the low position, the torque values for the lever handles were more equal for the two angled conditions, with five subjects producing higher torque values at 45° and six subjects producing higher torque values at 90°. The remainder exhibited approximately the same torque values for each condition. Overall, it appears that the high position was a better selection for torque production regardless of the angled condition used.

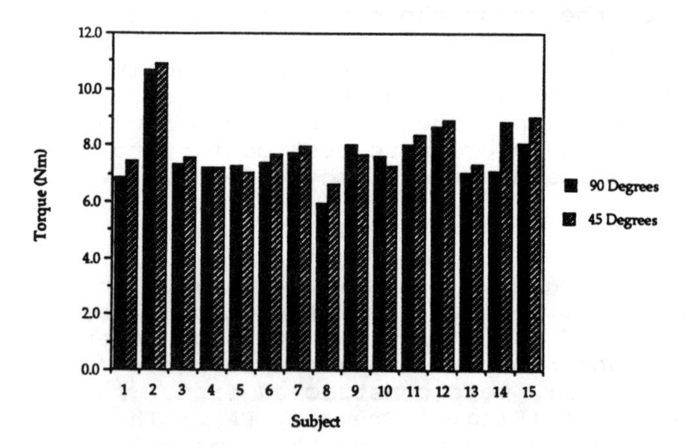

Fig. 4 - Torques by Subject

To test for the significance of the these results, an analysis of covariance (ANCOVA) was performed on the raw data (Table 2). The type of handle and the position, along with the covariates: age, gender and subject height, were significant at $p<.05$. Angle and the covariates: weight and arthritis were not significant, most likely because weight is correlated with height and arthritis with age. The second order interactions of handle-angle and position-angle were also significant at $p<.05$. The best fit regression model accounted for 91.5% of the variance and was significant at $p<.05$:

$$Torque = 18.7 - 3.94*Star - 3.77*Acrylic + .251*Position - .017*Age - .994*Gender - .0331*Height$$

Table 2 - ANOVA of Torques

Source	DF	*F*	*p*
Age	1	117.72	0.000
Gender	1	23.91	0.000
Weight	1	0.49	0.483
Height	1	4.07	0.044
Arthritis	1	0.24	0.623
Handle	2	2692.21	0.000
Position	1	25.68	0.000
Angle	1	0.70	0.402
Handle*Position	2	2.63	0.073
Handle*Angle	2	3.62	0.028
Position*Angle	1	10.22	0.001
Handle*Position*Angle	2	0.11	0.894
Error	523		
Total	539		

The age effect is more clearly shown in Fig. 5 with torque values decreasing an average of 10% over the 15 year age span of the subjects. However, for any particular age, there is considerable variation in torque capability, most probably due to musculoskeletal and joint disorders such as arthritis.

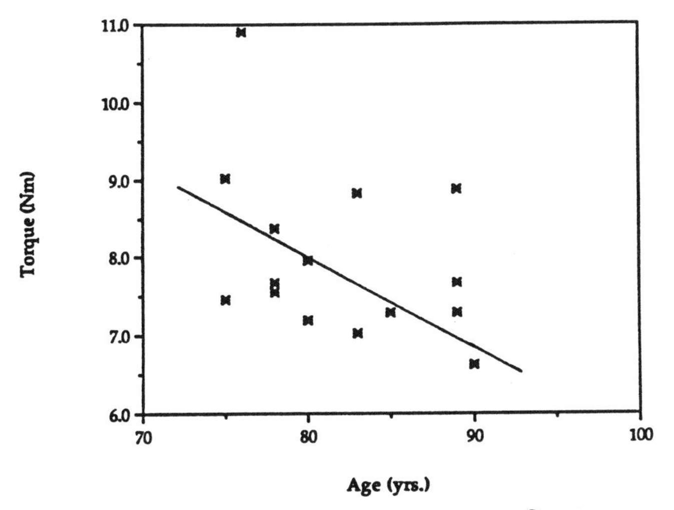

Fig. 5 - Torque vs. Age (45°-high)

DISCUSSION

The statistical analyses and subjective responses indicated that handle type was the most significant factor in torque generation capability with the lever handle being the best. The subjects claimed it was easier to operate, allowing greater flexibility in exerting force, and didn't require them to wrap their fingers around the handle in order to operate it. Force could easily be exerted against the handle by using the heel of the hand or even by using an elbow. A common complaint with the star handle was its small size and a pinching of the skin that occurred when the fingers were wrapped around the handle. The unfinished metal edges of the handle provided an uncomfortable grasp. To remedy this situation, subjects either used their fingers in a flexed position (usually three fingers) or used primarily the heel of their hand with the fingers bent and perched around the circumference of the handle. These results compare favorably with Bordett et al. (1988) who also found a 50% increase in torque production using the lever handles.

These results also support the need for a larger handle, like the lever handle, for generating torque. Swain et al. (1970), Cochran and Riley (1986) and Pheasant and O'Neill (1975) all found that a larger handle produces a larger moment arm and thus greater torque. The lever handle had a 6-inch moment arm, while the acrylic and star handles had only 1-inch and 11/16-inch moment arms, respectively.

While the high-low position was significant, angle was not. Although, the 45° high position yielded slightly larger torques, these values were not significant ($p>.05$). Subjective opinions were also mixed. Some felt that exerting a component of the force down helped, while others felt that the angle did not really make a difference. However, the subjects clearly preferred a handle position that did not require them to assume awkward or uncomfortable bending postures. The 45° low position was definitely a disliked condition for this reason.

While the high position is a better choice for torque production, it does not conform to the standard set forth by Kira (1976). Forty-two inches is too high to be easily reached from a sitting position in the tub. A compromise would be a position between 42 and 21 inches that emphasizes the positive aspects of the high position but yet mirrors the standard developed by Kira (1976). It should also be noted that the 45°-angled position presents some interesting plumbing problems. Currently, there are few bathtub designs that deviate from the common 90°-angled walls. Changing the wall of the bathtub would require a redesign of the bathtub/shower space, and plumbing. While this redesign is not an impossibility, its implementation in an existing facility would be expensive for the administration and disruptive to the residents.

Age was shown to have a significant effect on torque generation, declining 10% over the 15 year age span. This decrease would then be an additional decrease over the 17% up to age 65 observed by Fisher and Birren (1947). However, the wide range of torques within a particular age group also indicated that the torque generation capabilities of the elderly are quite varied. Some subjects appeared to be very strong while other seemed weaker. The subjects who expressed having the greatest difficulties with the bathtub/ shower were the subjects who suffered from physical ailments that affected the arms and hands. But, it should be noted that the subject with the highest torques was the one confined to the wheelchair for 46 years. Obviously, this subject's grip strength had not decreased, and perhaps even increased considerably.

The existing bathroom faucet in the retirement facility was a one-piece unit. The subjects expressed problems in adjusting the unit to achieve the desired water temperature and pressure. The handles would stick and would be especially difficult to manipulate with wet hands. Furthermore, the elderly residents had become accustomed, through the years, of using two faucets, left for hot and right for cold water. Switching to an unfamiliar unit only compounded their problems in using the faucet.

RECOMMENDATIONS

Based on the experimental data and the results of the questionnaire, the following changes are recommended for retirement facilities:

1) Use lever handles where ever possible in the living areas, especially the bathroom.

2) Adhere to the convention of one handle for hot water and one handle for cold water. Use handles that clearly designate the hot and cold temperatures with tradition read and blue markings.

3) Install a plumbing system that has both a high (42 inches) and a low (21 inches) set of faucets. This would eliminate problems associated with bending while permitting easy reach from a seated position within a tub. If space and financial constraints do not allow this, introduce a new standard height of 32 inches for the faucets. This would allow for an acceptable reach from both standing and sitting positions.

REFERENCES

Bordett, H.M., Koppa, R.J., Congleton, J.J. (1988) Torque required from elderly females to operate faucet handles of various shapes, Human Factors, 30:339-346.

Cochran, D.J. and Riley, M.W. (1986) The effects of handle shape and size on exerted forces, Human Factors, 28:253-265.

Czaja, S.J. (1990) Human Factors Research Needs for an Aging Population, Washington, D.C: National Academy Press.

Fisher, M.B. and Birren, J.E. (1947) Age and strength, J. Applied Physiology, 31:490-497.

Huston, T.R., Sanghavi, N., Mital, A. (1984) Human torque exertion capabilities on a fastener device with wrenches and screwdrivers, in A. Mital (ed) Trends in Ergonomics/Human Factors I, Amsterdam: Elsevier, pp. 51-64.

Kira, A. (1976) The Bathroom. New York: Viking Press.

Kroemer, K. (1986) Coupling the hand with the handle: An improved notation of touch, grip and grasp, Human Factors, 28:337-339.

Pheasant, S. and O'Neill, D. (1975) Performance in gripping and turning - a study in hand/handle effectiveness, Applied Ergonomics, 6:205-208.

Small, A.M. (1986) Design for older people, in G. Salvendy (ed) Handbook of Human Factors, Wiley-Interscience, N.Y. pp. 495-504.

Swain, A.D., Shelton, G.C. and Rigby, L.V. (1970) Maximum torque for small knobs operated with and without gloves, Ergonomics, 13:201-208.

MODELLING AGE DIFFERENCES IN ISOMETRIC ELBOW FLEXION USING HILL'S THREE-ELEMENT VISCO-ELASTIC MODEL.

R. Darin Ellis
Kentaro Kotani
The Pennsylvania State University
University Park, PA, 16802

A visco-elastic model of the mechanical properties of muscle was used to describe age-differences in the buildup of force in isometric elbow flexion. Given information from the literature on age-related physiological changes, such as decreasing connective-tissue elasticity, one would expect changes in the mechanical properties of skeletal muscle and their related model parameters. Force vs. time curves were obtained for 7 young (aged 21-27) and 7 old (aged 69-83) female subject. There were significant age group differences in steady-state force level and the best fitting model parameters. In particular, the viscous damping element of the model plays a large role in describing the increased time to reach steady-state force levels in the older subject group. Implications of this research include incorporating parameter differences into more complex models, such as crash impact models.

INTRODUCTION

Muscular strength is an essential component in the successful completion of everyday activities (eg. walking or carrying groceries) for the aging adult (Kovar and LaCroix, 1987). Additionally, muscular strength plays a large role in the ability to avoid accidents, such as falls. In the home of the older individual, falls are the leading cause of accidental fatality (Czaja, 1990). The continued independence of the elderly adult is maintained in parallel with the preservation of skeletal muscle. In order for appropriate interventions to be designed, we must understand the underlying mechanisms of age-related change in muscle function.

Many investigations have shown that the peak values for maximal strength are achieved in the second and third decade of life (Fisher and Birren, 1947; Larsson, 1978), and that the time-course of contraction is slowed in older adults (Gutmann, 1977; McDonagh *et al.*, 1984). The objective of this paper is to integrate the research on these topics using a well-documented model, Hill's three-element visco-elastic model (Hill, 1922), to account for age-related differences in the parameters of the force vs. time curve.

Hill's three element model is composed of an elastic element (spring) in series with a unit comprised of a contractile element, damping element (dashpot), and spring all in parallel with each other (see Figure 1).

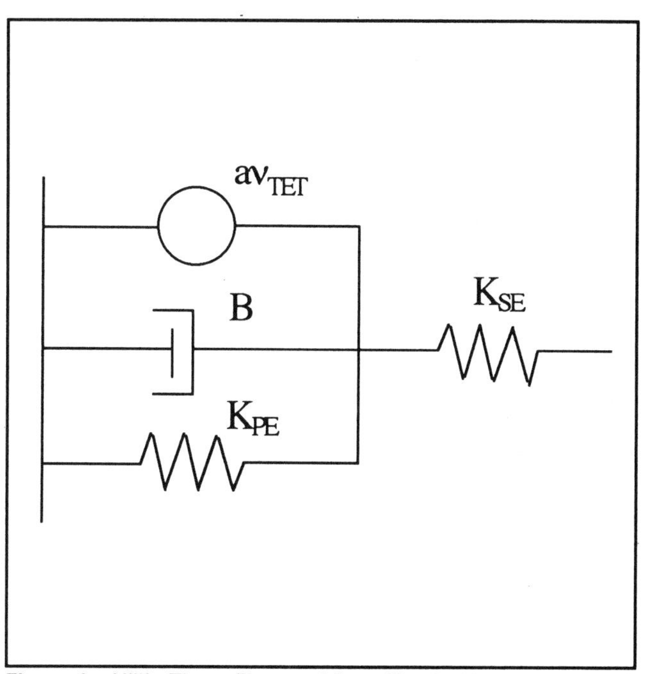

Figure 1. Hill's Three-Element Visco-Elastic Model of Muscle.

The elastic elements are representative of tendons (**K_{SE}**) and intramuscular connective tissue (**K_{PE}**). The contractile element (**av_{TET}**) is symbolic of sarcomere activity (sliding filaments of actin and myosin) on a lumped macroscopic level. The damping element, **B**, represents the viscous properties of muscle that oppose rapid changes in inertia. It can be shown that for active contraction, the force-time relationship derived from the model takes the following form:

$$F(t) = \frac{av_{TET}K_{SE}}{K_{SE}+K_{PE}}\left[1 - e^{-\left(\frac{K_{SE}+K_{PE}}{B}\right)*t}\right] \qquad (1)$$

Further, in isometric contraction, **K_{PE}** does not play a significant role, and the equation simplifies to the following (McMahon, 1984, p.22):

$$F(t) = F_{\substack{steady\\state}}*\left[1 - e^{-\left(\frac{K_{SE}}{B}\right)*t}\right] \qquad (2)$$

There are age-related changes in all of these model components. First, connective tissue (modelled with **K_{SE}** and **K_{PE}**) undergoes collagen fiber cross-linking and a loss of active elastin with increasing age. This is compounded by the fact that the collagen in connective tissues undergoes relatively little repair and replacement after maturity. The result is increased stiffness, as well as other transformations such as increased chemical stabilization, and greater insolubility in various reagents (Arking, 1991, p.141). Testing the mechanical properties of tendon, Yamada (1970) found that 70-79 year old tissue had only 60% of the ultimate tensile strength and 91% of the ultimate percentage elongation of 20-29 year-old tissue. These results would predict increases in **K_{SE}** and **K_{PE}** with age.

The contractile element has also been shown to undergo change with age. Lean body mass declines throughout the adult life-span, and a substantial portion of this loss takes place in skeletal muscle. McCarter (1978) presented evidence for both a loss in the number of skeletal muscle fibers and a decrease in the size of the remaining fibers. This would be reflected by a decrease in **av_{TET}** for older adults.

Finally, there are also changes in the viscous properties of muscle. The dashpot, **B**, provides a mechanism to explain Hill's (1922) observation that the force produced by active-state muscle is a function its shortening velocity. Intra-cellular water was originally proposed as the damping mechanism. Arking (1991) noted a loss of intra-cellular water and a corresponding increase in lipofuscin (cellular garbage) in non-dividing cells with age. This could cause an increase in viscosity (and thus increases in **B**), however, this contention has not been investigated. Although the simple mechanical dashpot is inadequate to account for some muscle properties, such as sensitivity to changes in temperature, it remains an integral part of models which describe purely mechanical properties of muscle acting against a load (McMahon, 1984). All other things being equal, increases in **B** would increase the time to peak contraction in older muscle; there is empirical evidence which shows age-related increases in the time to peak contraction (Gutmann, 1977, p.458; McDonagh *et al.*, 1984).

The hypothesis of this investigation is that, in addition to the decrease in the overall level of isometric elbow flexion force, there will be age-group differences in the time-course of contraction, and thus differences in the parameters that best fit equation 2. Although there is evidence in the literature to support the existence of age-related change in all of the model parameters, the experimental evidence of Gutman (1977) and McDonagh *et al.* (1984) which showed longer time to peak contraction in older adults would be explained by a lower value for the ratio K_{SE}/B. We therefore predict that the older age group will have a lower average parameter ratio.

METHODS

Subjects

Fourteen unpaid volunteers participated in the investigation. Subjects were divided into two age groups: Young (21 to 27) and old (69 to 83). The young subject group was recruited from the campus population. The old group was recruited from a local senior citizen activity center. The study was limited to female subjects due to the lower accessibility of older men. All subjects reported being in good health, using the PAR-Q questionnaire (Chisholm, et al., 1975). Subject characteristics are shown in Table 1.

Table 1. Subject characteristics.

Characteristic	Young Group	Old Group
Age mean (yrs)	23	77
Age range (yrs)	21-27	69-83
Exercise mean (min/wk)	60	38
Exercise range (min/wk)	0-180	0-120

Procedure and Apparatus

The Caldwell Regimen (Caldwell et al., 1974) was followed. Subjects were seated in a chair with a specially fitted adjustable armrest. The armrest allowed the upper arm to remain vertical with an included elbow angle of 90 degrees. The subject's arm was extended directly forward and a non-elastic cuff was placed around the subject's wrist. The cuff was connected to an Interface model SM500 load cell directly underneath. After ensuring that the apparatus was secure, the investigator began the countdown to the start of the muscle contraction. The countdown for each trial was as follows: "3, 2, 1, start, 1, 2, 3, 4, 5, stop." The subjects began muscle contraction on the word "start" and achieved maximum voluntary contraction as quickly as possible without jerking the apparatus. The force levels were automatically recorded using a DAS-16 analog to digital converter mounted in a Daewoo 286 PC. Each subject performed 3 trials with a minimum of 5 minutes rest between trials. The dependent variable was maximal voluntary force as a function of time, and the independent variable was age group.

RESULTS

There were significant differences between the two groups in the steady-state level of force attained ($F_{1, 40}$ = 88.73, $p < 0.001$). On average, the older group attained 41% of the steady-state force level attained by the younger group; the young group averaged 11 kg, while the older group averaged 4.5 kg (see Figure 2).

Optimal values of K_{SE}/B were determined by minimizing the χ^2 statistic for each subject's data using equation 2; all of the individual parameters produced fits which not could be distinguished from the expected function ($\chi^2_{100, .05}$ = 124.3). Figures 3 and 4 exhibit the normalized experimental data and best fitting curves for Young Subject 1 and Old Subject 7, respectively. Comparing these two subjects, it is quite apparent that the older subject required more time to reach the steady state force level, indicating a lower value of K_{SE}/B. The individual and group average parameter ratios are presented in Table 2, where it can be seen that the older group had much lower optimal values of K_{SE}/B. A t-test established group differences in the value of K_{SE}/B ($t_{11, .05}$ = 2.62, $p < 0.05$). The correlation between age and K_{SE}/B was analyzed post-hoc within the older subject group. The regression of age on K_{SE}/B was marginally significant ($p < 0.10$, R^2 = 56.0%), showing that the older-old subjects had somewhat lower values of K_{SE}/B than their younger-old counterparts.

Table 2. Model Parameters by Age Group.

	Young		Old	
Subject	K_{SE}/B	χ^2	K_{SE}/B	χ^2
1	5.1	1.8	1.5	2.8
2	3.0	0.4	2.4	1.6
3	7.2	2.2	1.2	0.6
4	5.5	1.6	1.5	1.6
5	6.4	2.7	5.5	2.9
6	2.5	0.1	2.6	2.3
7	3.2	1.1	2.0	1.9
Mean	4.7		2.4	
s.d.	1.8		1.5	

DISCUSSION

The overall steady-state force level differences obtained were somewhat greater than most previous studies. There is some consensus that strength decreases by about 20% by the mid-60s (Larsson, 1982; Montoye and Lamphiear, 1977). For example, McDonagh et al (1984) found that elderly males had 76% of the elbow flexion force of young males. Fisher and Birren (1947) noted only a 17% decrease in handgrip strength. The only study found in the literature that paralleled the differences found in the present results

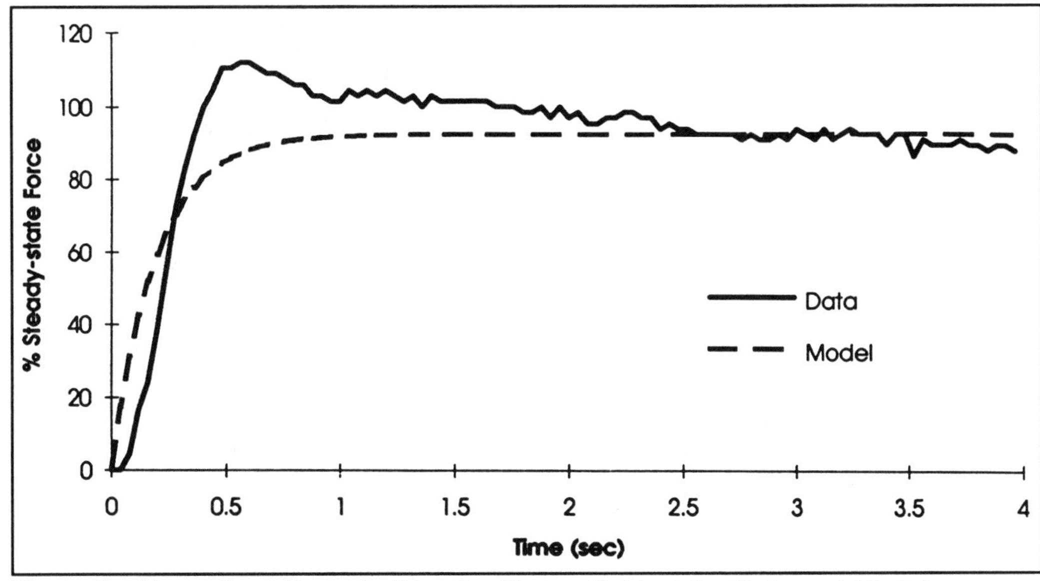

Figure 2. Normalized Force-buildup for Young Subject 1 - Data and Model

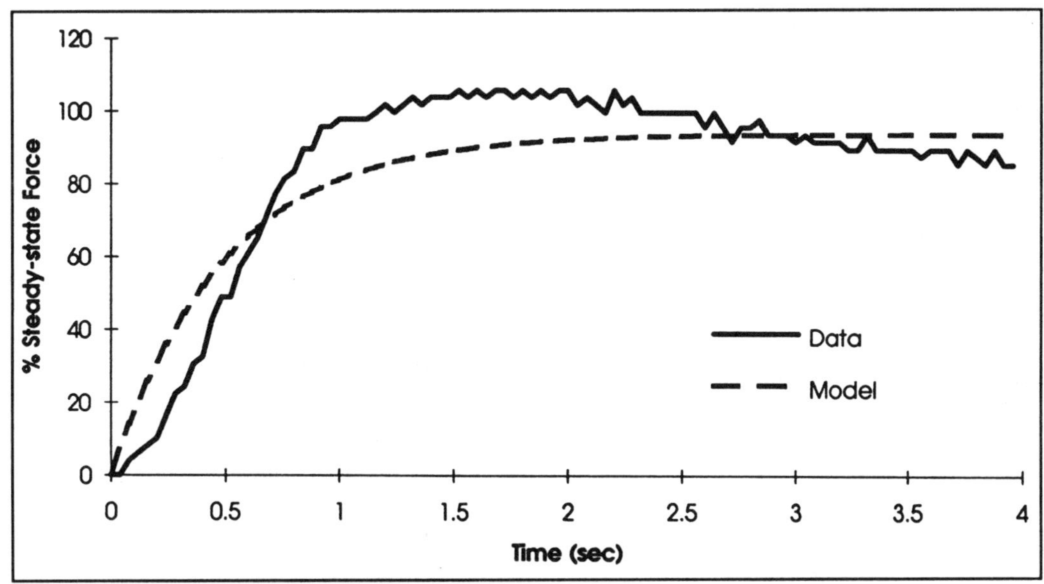

Figure 3. Normalized Force-buildup for Old Subject 7 - Data and Model

was by Mathiowetz, et al.(1985). They found that in an older sample of women (aged 75-94) grip strength was 40% of the younger group. These findings are much more consistent with the data presented here. The lack of consistent data in the literature show the need for more research with broader and better defined subject populations.

All in all, Hill's model provided a good description of age-differences in the active state force buildup curve, given the *a priori* knowledge of age-related physiological changes associated with model parameters. Despite the fact that the present results do not provide a unique solution of the values of K_{SE} and B, the following can be said: Given the indirect evidence that the values of K_{SE} and B both increase and the older subject group's lower values of K_{SE}/B in the present results, the value of B must increase faster than K_{SE} with age. Direct evidence for age-related change in B was provided by a longitudinal study by Aniansson, et al. (1986). They reported force-velocity curves which differed in slope with age; the older subjects' average curve had a 20% steeper slope in the initial part of the curve. This is a strong finding, due to the fact that the ages-at-test were only 5 years apart (70 and 75 years old).

The 3-element model developed by Hill provided some insight into the operation of basic physiological changes as mechanisms of declining strength and slowing contractile times with increasing age. While there are age-related changes that the model is not equipped to handle (such as changes in innervation and proportion of type I and II fibers), the effects are strong enough that the use of Hill's model can not be ruled out for further research.

Further research using Hill's model to describe age-differences in muscle response could be integrated into more complex models such as crash impact models (eg. Fleck and Butler, 1978; Freivalds, 1985). Older drivers have the highest crash per mile rate of all age groups and a higher risk of morbidity and mortality due to a crash (Pike, 1988; Barr, 1991). Waller (1991) noted that current occupant restraint systems are designed primarily for younger occupants in a frontal collision. Incorporating age-dependent mechanical properties of muscle into crash impact models would allow more realistic prediction of the effect of design changes on the risk of injury for older adults.

Acknowledgements.

This work was performed while the first author was supported by a NIA Research Traineeship (National Institute on Aging Grant T32-AG00048 to The Pennsylvania State University).

REFERENCES

Aniansson, A., Sperling, L., Rundgren, A. and Lehnberg, E., (1983). Muscle function in 75 year old men and women, a longitudinal study. Scandinavian Journal of Rehabilitative Medicine (Suppl.), 193: 92-102.

Arking, R. (1991). The Biology of Aging. Englewood Cliffs, NJ: Prentice-Hall.

Barr, R.A. (1991). Recent changes in driving in older adults. Human Factors, 33(5): 597-600.

Caldwell, L.S., Chaffin, D.B., Dukes-Dobos, F.N., Kroemer, K.H.E., Laubach, L.L., Snook, S.H. and Wasserman, D.E. (1974). A proposed standard procedure for static muscle strength testing. American Industrial Hygiene Association Journal, 35: 201-206.

Chisholm, D.M., Collins, M.I., Davenport, W., Gruber, N. and Kulak, L.L. (1975). PAR-Q Validation Report. British Columbia Medical Journal, 17.

Czaja, S.J. (1990). Human Factors Research Needs for an Aging Population. Washington, DC: National Academy Press.

Fisher, M.B. and Birren, J.E. (1947). Age and strength. Journal of Applied Physiology, 31: 490-497.

Fleck, J.T. and Butler, F.E. (1978). Development of an improved computer model of the human body and extremity dynamics. AMRL-TR-75-14. Dayton, OH: Wright Patterson AFB.

Freivalds, A. (1985). Incorporation of active elements into the Articulated Total Body model. AAMRL-TR-85-061. Dayton, OH: Wright Patterson AFB

Gutmann, E. (1977). Muscle. In C.E. Finch and L. Hayflick (eds.), Handbook of the Biology of Aging. New York: Van Nostrand Reinhold.

Hill, A.V. (1922). The maximum work and mechanical efficiency of human muscles, and their most economical speed. Journal of Physiology, 56: 19-41.

Kovar, M.G. and LaCroix, A.Z. (1987). Aging in the eighties: Ability to perform work related activities. Vital and Health Statistics, No. 136, Supplement on Aging to the National Health Interview Survey. Hyattsville, MD: Public Health Service, National Center for Health Statistics.

Larsson, L. (1982). Physical training effects on muscle morphology in sedentary males at different areas. Medicine and Science in Sports and Exercise, 14: 203-206.

Larsson, L. (1978). Morphological and functional characteristics of the ageing of skeletal muscle in man. Acta Physiologica Scandinavica (Suppl.), 457: 1-29.

Mathiowetz, V., Kashman, N., Volland, G., Weber, K., Dowe, M., and Rogers, S. (1985). Grip and pinch strength: normative data for adults. Arch Phys Med Rehabil, 66: 69-72.

McCarter, R. (1978). Effects of age on contraction of mammalian skeletal muscle. In G. Kaldor and W.J. DiBattista (eds.) Aging in Muscle, NY: Raven Press.

McDonagh, M.J.N., White, M.J., and Davies, C.T.M. (1984). Different effects of ageing on the mechanical properties of human arm and leg muscles. Gerontology, 30: 49-54.

McMahon, T.A. (1984). Muscles, Reflexes and Locomotion. Princeton, NJ: Princeton University Press.

Montoye, H.J., and Lamphiear, D.E. (1977). Grip and arm strength in males and females, age 10 to 69. Research Quarterly, 48: 109-120.

Pike, J.A. (1988). The elderly driver and vehicle-related injury. In Effects of Aging on Driver Performance (pp. 61-71). Warren-dale, PA: SAE.

Waller, P.F. (1991). The older driver. Human Factors, 33(5): 499-505.

Yamada, H. (1970). Strength of Biological Materials. Baltimore, MD: Williams and Wilkins.

PERCEIVED EXERTION IN ISOMETRIC MUSCULAR CONTRACTIONS RELATED TO AGE, MUSCLE, FORCE LEVEL AND DURATION

Joseph M. Deeb
Department of Industrial Engineering
North Carolina A&T State University
and
Colin G. Drury
Department of Industrial Engineering
State University of New York at Buffalo

ABSTRACT

This research was concerned with studying the development and growth of perceived effort of long-term isometric contractions as a function of muscle group (biceps vs quadriceps), of subjects with different age groups (20-29 vs. 50-59 years old) on long-term muscular isometric contractions (5 minutes) at different levels of %MVC (20,40,60,80 and 100 %MVC). An experiment testing 20 subjects each performing 10 conditions (two muscle groups x five levels of %MVC) showed that the older age group reported Significantly higher perceived exertion at higher levels of %MVC and across time. Furthermore, subjects experienced a higher and faster increase in their perceived exertion when the level of %MVC and time increased.

1. INTRODUCTION

Physical workload is one of the leading causes of physical stress and dissatisfaction of the workplace. Many of the psychological problems are caused by a decrease in the physical work capacity of the individual. Objective measurements of physical stress (i,e. measured by the physical laboratory tests) do not present the whole picture of what is really going on inside the person and that person's perception of the physical stress. Since people react to the world as they perceive it and not as pure reality, it becomes extremely important to study and establish some kind of relationship between subjective (as reported by the person) and objective (as measured physically) measurements of physical stress (Borg, 1970).

A number of studies (Caldwell and Smith, 1966; Lloyd et al., 1970; Barbonis, 1979) have examined the subjective and objective aspects of fatigue. These studies required the subjects to maintain the load, applied force, as long as possible and to report the pain intensity that they experienced as it developed during the endurance task on a five-point scale. The results showed that the rate of pain increases in a curvilinear fashion. These studies and others have examined pain intensity during the endurance time of a given load using young subjects.

The objective of this study was to investigate the development and growth of perceived effort as reflected by subjective ratings using Borg's scale (1982). Another goal was to explore the effects of muscle group, age group, relative force level, and duration of task on the perceived effort of long-term isometric contractions.

2. METHOD

2.1 Subjects

Two age groups of male subjects (20-29; 50-59 years old) with ten subjects in each group were recruited. Subjects were not involved in any regular exercise program.

2.2 Apparatus

The equipment consisted of a specially built chair where subjects rested their arms on the armrest, with the elbow angle at 90° for the elbow flexion (biceps muscle group) strength test. For the knee extension (quadriceps muscle group) strength test, the thigh was resting on the seat pan, the lower leg hanging down, with no foot support provided, giving a knee angle of 90°. Two nonelastic cuffs (around the wrist and around the ankle) were connected to two load cells transmitting the force exerted to an IBM-PC. Borg's scale (1982) was used to collect the subjective ratings.

2.3 Procedure

After reading the instructions and signing a consent form, each subject was tested for his maximum voluntary exertion (100 %MVC) following the standard procedure developed by Caldwell et al. (1974). Subjects were instructed to exert a force and reach a designated level (%MVC) and maintain that level for as long as possible until asked by the experimenter to stop. The testing time was for five minutes. Only one session, with one level of %MVC for both elbow flexion and knee extension, was scheduled per day with at least 24 hours between sessions. All conditions were assigned randomly and in different random order for each subject. Throughout the exertion time, subjects were asked to rate verbally their perceived exertion (RPE) when a beep sounded. The first rating was obtained at the time the force level was reached (starting time of each trial) and once every 15 seconds afterwards giving a total of 21 ratings for the five minutes test.

2.4 Experimental Design

The design of the experiment was a five-factor repeated measures with subjects nested under age group (20-29; 50-59) and crossed with muscle group (biceps vs quadriceps), level of exertion (20,40,60,80 and 100 %MVC) and time (0, 15, 30,....,300 seconds). Ten subjects participated in each age group and treated as a random variable. The other independent variables, namely muscle group, level of exertion and time were treated as fixed variables. The dependent variable was the rated perceived exertion.

3. RESULTS

The analyses of variance demonstrated that all main effects of age group (G), muscle group (M), level of exertion (P) and time (T) were significant. Furthermore, the first order interactions of GxP, GxT, MxT, and PxT were also significant. Only the first order interactions will be discussed.

The GxP interaction (figure 1) revealed that both age groups reported higher RPE values at higher levels of %MVC. Furthermore, a Newman-Keuls test ($p<0.05$) showed significantly higher RPE values for the older age group. The GxT interaction (figure 2) demonstrated that the reported perceived effort followed a negatively accelerating increase with the RPE for the older age group continuing to increase with increasing time while it appeared to reach a steady state for the younger age group. The MxT interaction (figure 3) showed that the increase of RPE for both muscle groups was negatively accelerated with a higher and faster increase for the quadriceps muscle group. The PxT interaction (figure 4) revealed that the subjects experienced higher and faster increase in their RPE when the level of %MVC and time increased. Furthermore, the indication is that the RPE followed a negatively accelerating increase with time except for the 20 %MVC which seemed to increase linearly with time.

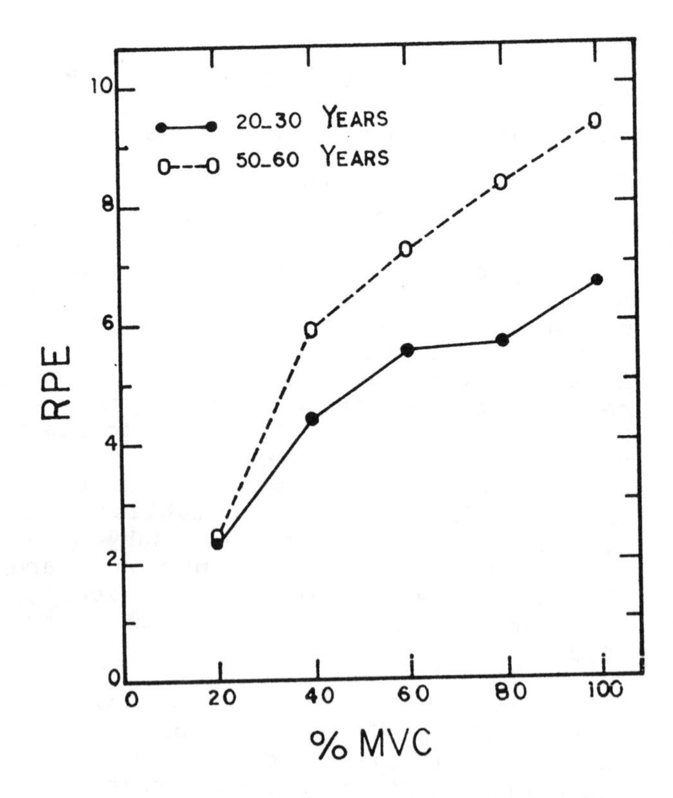

Figure 1. Means of RPE for age group and %MVC interaction, G X P.

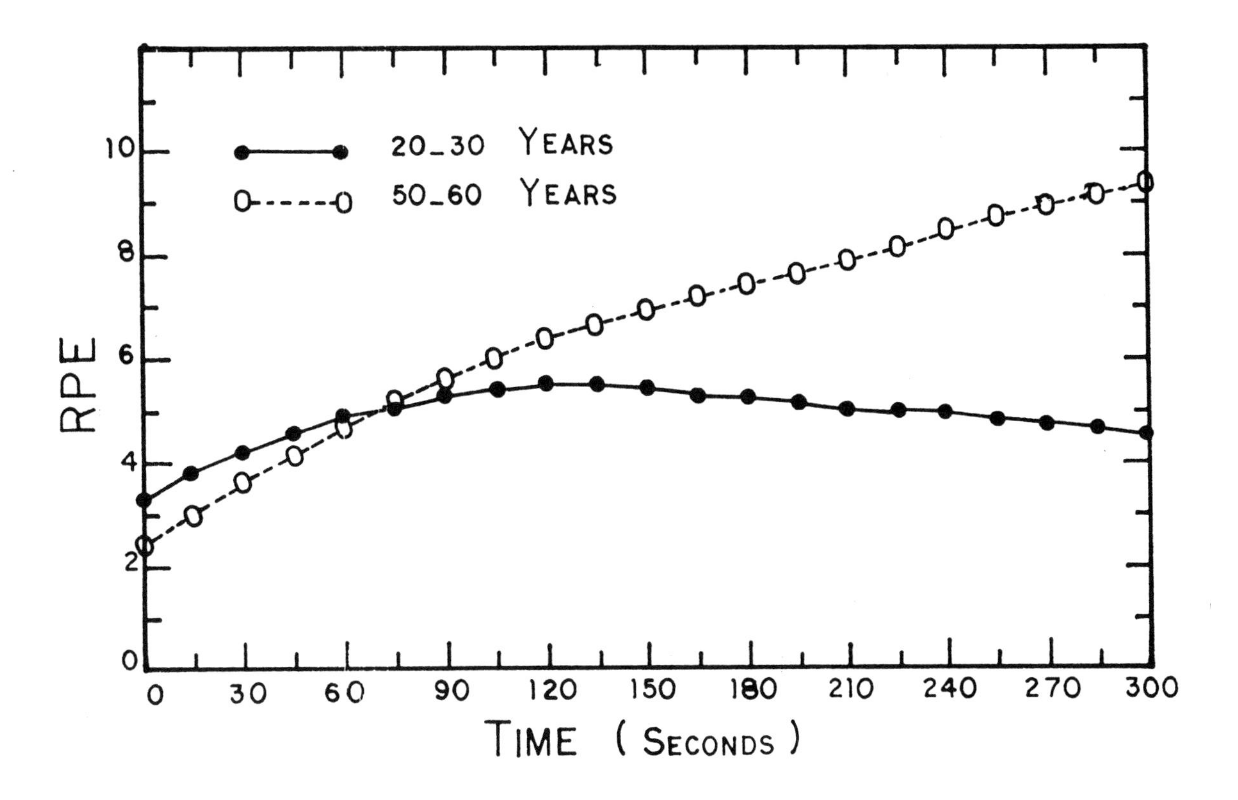

Figure 2. Means of RPE for age group and time interaction, G X T.

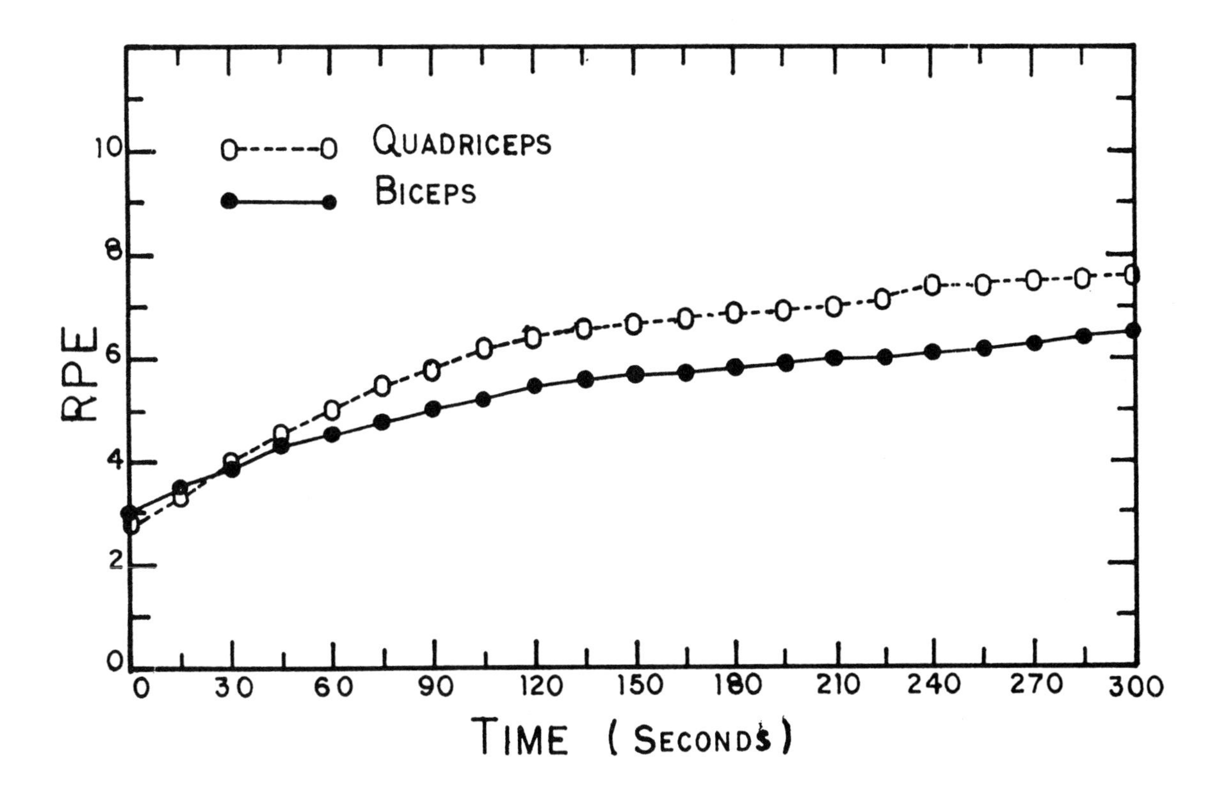

Figure 3. Means of RPE for muscle group and time interaction, M X T.

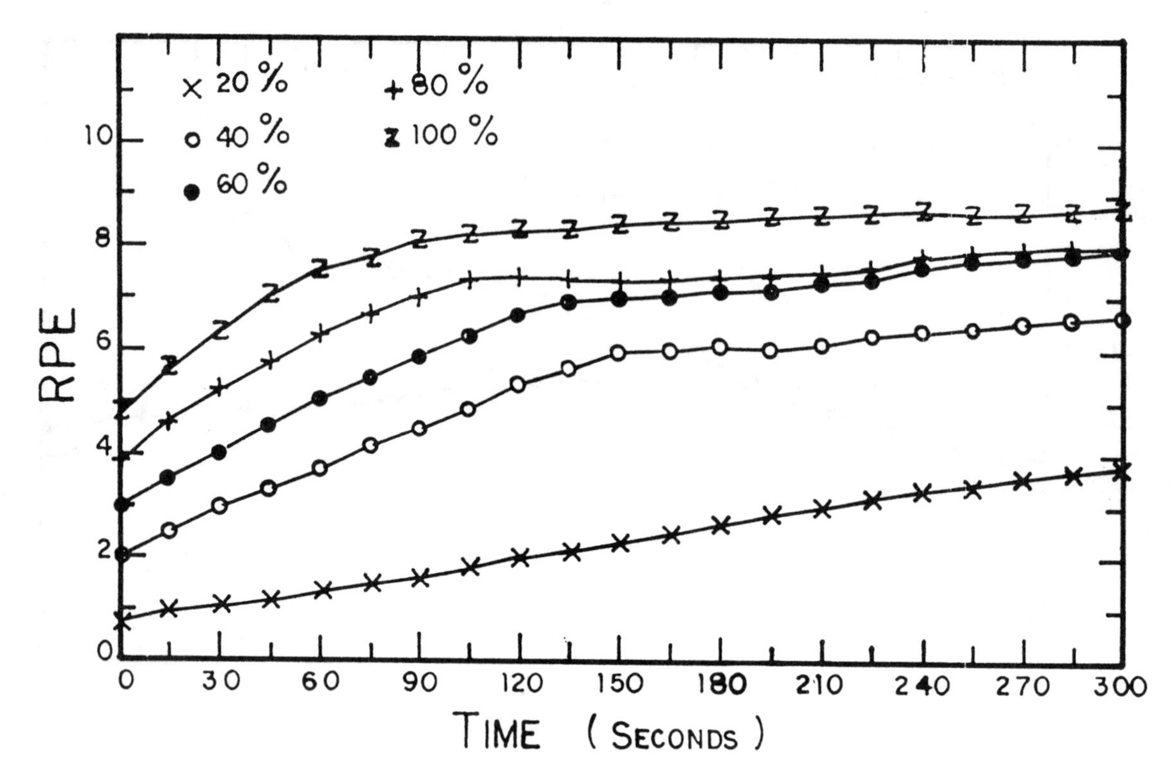

Figure 4. Means of RPE for % MVC and time interaction, P X T.

4. DISCUSSION

The effects of GxP and GxT interaction on the perceived effort of isometric contraction revealed that the reported PRE showed a faster increase for the older group. This experience of higher perceived effort by the older group is consistent with the literature (Borg and Linderholm, 1967; BAR-OR, 1977). However, those authors utilized the bicycle ergometer and treadmill to study the whole body perception of effort with age in dynamic tasks. No other studies were located that had investigated the effects of GxT interaction on the perceived effort in an isometric contraction task of a long duration. It can be concluded that the increase in perceived effort with increasing age holds for both dynamic and static tasks.

The subjects in this study appeared to experience faster and higher increase in the perceived effort with their quadriceps than their biceps muscle groups (figure 3). Banister (1979) studied the adductor polices and quadriceps muscles by eliciting perception of effort at different levels of exertion, and reported that the perception of effort grew much faster for the adductor polices muscle than for the quadriceps and concluded that this perception depends on the muscle group involved. The present study tends to agree with the conclusion of Banister's study.

The perceived exertion of constant effort isometric contraction (PxT interaction) revealed a faster increase with the duration of the exertion as the level of %MVC increased. Studies (Stevens and Cain, 1970; McClosky et al., 1974) have shown that fatigue caused by sustained contraction increased the subject's perception of force and this increase was nonlinear. Further, Stevens and Cain (1970) requesting subjects to maintain a constant level of effort for a duration of 140 seconds, demonstrated that sustainment of a greater force increases the perceived intensity more rapidly than sustainment of a smaller force for the same duration. The results from the present study agree with the findings of the literature. The perceived effort increased linearly at 20% MVC but followed a faster negatively accelerating increase as force level increases.

In summary, this study has provided an insight into the perception of effort with only two age groups and muscle groups tested. It would be of interest to test a wide range of age groups and different muscle groups, that are known to have different composition of fast twitch and slow twitch fibers, to assess the perception of effort more clearly.

REFERENCES

Banister, E.W, 1979. The Perception of Effort: An Inductive Approach. Europ. J. Appl. Physiol. and Occup. Physiol., 41, 141-150.

Barbonis, P., 1979. Work Design and Ergonomics. Unpublished Dissertation. Birmingham University, Department of Engineering Production.

BAR-OR, O., 1977. Age Related Changes In Exercise Perception. In: Physical Work and Effort, G. Borg (Ed) New York: Pergamon Press, 255-266.

Borg, G.A.V., 1970. Perceived Exertion As An Indicator of Somatic Stress. Scand. J. Rehab. Med., 2-3: 92-98.

Borg, G.A.V., 1982. Psychophysical Bases of Perceived Exertion. Medicine and Science In Sports and Exercise, 4(5), 377-381.

Borg, G.A.V., and Linderholm, H., 1967. Perceived Exertion and Pulse Rate During Graded Exercise In Various Age Groups. Acta Med. Scand., 472, 194-206.

Caldwell, L.S., Chaffin, D.B., Dukes-Dobos,, F.N., Kroemer, K.H.E., Laubach, L.L., Snook, S.K., and Wasserman, D.E., 1974. A Proposed Standard Procedure for Static Muscle Strength Testing. Am. Industrial Hygiene Association J., 35, 201-206.

Caldwell, L.S. and Smith, R.P., 1966. Pain and Endurance of Isometric Muscle Contraction. J. of Engineering Psychology, 5(1), 25-32.

Lloyd, A.J., Voor, J.H. and Thurman, T.J., 1970. Subjective and Electromyographic Assessment of Isometric Muscle Contractions Ergonomics, Vol. 13, No. 6, 685-691.

McClosky, D.I., Eberling, P., and Goodwin, G.M., 1974. Estimation of Weights and Tensions and Apparent Involvement of A "Sense of Effort". Experimental Neurology, 42, 220-232.

Stevens, J.C. and Cain, W.S., 1970. Effort In Isometric Muscular Contractions related to Force Level and Duration. Perception and Psychophysics, 8, 4, 240-244.

MULTIVARIATE MODEL FOR DEFINING CHANGES IN MAXIMAL PHYSICAL WORKING CAPACITY OF MEN, AGES 25 TO 70 YEARS

Andrew S. Jackson, Earl F. Beard, Larry T. Wier, and J. E. Stuteville
Cardiopulmonary Laboratory, NASA/Johnson Space Center
and Department of HHP,
University of Houston
Houston, Texas 77204-5331

ABSTRACT

The purpose of this study was to develop a multivariate model with cross-sectional data that defined the decline in VO2max over time, and cross-validate the model with longitudinal data. The cross-sectional sample consisted of 1,608 healthy men who ranged in age from 25 to 70 years. VO2max was directly measured during a maximum Bruce treadmill stress test . Regression analysis showed that the cross-sectional age and VO2max relationship was linear, r = 0.45 and the age decline in VO2max was 0.48 ml/kg/min/year. Multiple regression developed the multivariate model from age, percent body fat (%fat), self-report physical activity (SR-PA), and the interaction of SR-PA and %fat (R = 0.793). Accounting for the variance in percent body fat and exercise habits decreased the influence of age on the decline of VO2max to just -0.27 ml/kg/min/year. This showed that much of decline in maximal physical working capacity was due to physical activity level and percent body fat, not aging. The multivariate equation was applied to the data of the longitudinal sample of 156 men who had been tested twice (Mean AgeΔ = 3.1 ± 1.2 years). The correlation between the measured and estimated change in VO2max over time (ΔVO2max) was 0.75. The results of the study showed that changes in body composition and exercise habits had more of an influence on changes in maximal physical working capacity than aging. The developed model provides a useful way to quantify the changes in physical working capacity with aging.

INTRODUCTION

Maximal physical working capacity (VO2max) of adult men declines with age (Åstrand, Åstrand, Hallback, & Kilbom, 1973; Dill, Robinson, & Ross, 1967; Kasch, Boyer, VanCamp, Verity, & Wallace, 1990; Mitchell, Sproule, & Chapman, 1958; Pollock, Foster, Knapp, Rod, & Schmidt, 1987; Robinson, 1938; Robinson, Dill, Robinson, Tzankoff, & Wagner, 1976). Since VO2max is the product of maximum heart rate, stroke volume and arteriovenous O2 difference, the age related decline in VO2max is partially due to the decline in maximum heart rate with age (Åstrand et al., 1973; Hagberg, Allen, Seals, Hurley, Ehsani, & Holloszy, 1985; Jones, Makrides, Hitchcock, Chypcar, & McCartney, 1985; Pollock et al., 1987). Cross-sectional data of men exhibit a uniform decline in VO2max over time (ΔVO2max) ranging from 0.40 to 0.50 ml/kg/min/year. In contrast, the longitudinal decline in VO2max was much more variable, ranging from 0.04 to 1.43 ml/kg/min/year (Buskirk & Hodgson, 1987). The difference reported in cross-sectional and longitudinal studies may be partly due to the method used to compute ΔVO2max. The simple linear regression beta weight defines the cross-section ΔVO2max; whereas, the longitudinal ΔVO2max is computed from VO2max measured at two points in time. In a cross-sectional study only age defines ΔVO2max, but changes in body composition and exercise habits over time influence ΔVO2max.

The purpose of this study was to quantify the age decline in VO2max by developing cross-sectional multivariate models that not only considered age, but also exercise habits and body composition, and then find if the models were sensitive to detect longitudinal changes. The goal of this study was not only to define the cross-sectional ΔVO2max, but also identify the factors other than age that affect it.

METHODS

The subjects were healthy male employees at NASA/ Johnson Space Center, Houston, Texas who choose to have a graded exercise stress test as part of their annual health examination. The subjects ranged in age from 25 to 70 years. The data came from retrospective cross-sectional and longitudinal databases. A total of 1,604 men comprised the cross-sectional sample. The 156 men of longitudinal sample were former participants in a health-related fitness program and each was tested twice within seven years.

An open-circuit, computer-controlled spirometry system continuously measured VO2 during the Bruce treadmill protocol. VO2max was the highest oxygen uptake observed during the final minute of the test. The databases included only those who: 1) reached voluntary exhaustion, 2) exceeded a respiratory exchange ratio of 1.0; and 3) exceeded 90% of their age-predicted maximum heart rate. Percent body fat was estimated from published equations that used the sum of chest, abdominal and thigh skinfolds estimated %fat. The NASA self-report physical activity scale (SR-PA) surveyed the subjects' exercise habits. The SR-PA scale provides a global estimate of physical activity during the previous 30 days.

Table 1. Means and standard deviations (M ± SD) for the cross-sectional sample and by age groups.

VARIABLE	TOTAL N=1,604	AGE GROUPS 25-35 N=150	35-44 N=752	45-54 N=835	≥ 55 N=167
Age (yrs)	45.9 ± 7.7	30.0 ± 2.8	40.8 ± 2.7	49.1 ± 2.8	58.1 ± 3.0
Weight (kg.)	80.1 ± 11.0	78.6 ± 11.1	79.7 ± 10.7	80.8 ± 11.3	79.0 ± 10.4
Percent Fat (%)	20.5 ± 5.9	16.4 ± 6.6	19.2 ± 5.9	21.4 ± 5.4	22.9 ± 5.7
Self-Report Physical Activity	4.1 ± 2.1	4.7 ± 1.9	4.4 ± 2.1	3.9 ± 2.1	3.6 ± 2.0
Max Heart Rate (b/min)	176.2 ± 11.4	187.0 ± 8.1.	179.2 ± 9.1	174.5 ± 11.2	167.4 ± 10.8
VO2max (ml/kg/min)	38.5 ± 8.2	46.1 ± 7.9	41.3 ± 8.4	36.8 ± 7.1	33.2 ± 6.0

Multiple regression analyzed the cross-sectional data. The dependent variable was VO2max (ml/kg/min), and the independent variables were age, SR-PA, skinfold %fat, and the interaction of %fat and SR-PA. The multivariate models developed on the cross-sectional data were then applied to the longitudinal data and compared with metabolically measured VO2max. The true rate of VO2max change per year, ΔVO2max (ml/kg/min/year), was computed from the metabolic data and compared to the change estimated with the multivariate model. The product-moment correlation between the measured and estimated ΔVO2max quantified the accuracy of the multivariate model to assess the true change VO2max over time.

RESULTS

Table 1 describes the cross-sectional sample. Furnished are the means, and standard deviations of the total sample, and then contrasted by age groups. The range for the variables of primary interest is: age, 25 to 70 years; VO2max, 20 to 72 ml/kg/min; %fat, 5 to 39%; and SR-PA, 0 to 7, the entire range of the self-report scale. The means of the age groups reflect a general aging trend. The significant correlations with age were: VO2max (-0.45); maximum heart rate (-0.47); %fat (0.32); and SR-PA (-0.15). Figure 1 graphically shows the cross-sectional decline in VO2max associated with age. Simple linear regression defined the rate of decline to be 0.48 ml/kg/min/year. The significant age correlations with %fat and SR-PA showed that older men tended to have higher levels of body fat and were less active then younger men. The correlations between VO2max and %fat (-0.62) and SR-PA (0.58) were both higher than the correlation between age and VO2max of -0.45. The correlation between resting heart rate and VO2max was lower, -0.38.

Stepdown multiple regression analyses showed that in addition to age, percent body fat, SR-PA, and their interaction accounted additional VO2max variance. Figure 2 shows the relationship between measured VO2max and estimated from the multiple regression equation. The effect of exercise and body composition is shown by comparing the age regression weights of the simple linear models with those obtained with the multivariate analyses. Adding %fat and SR-PA to the regression models reduced the influence of age on the decline in VO2max from 0.48 less than 0.27 ml/kg/min/year. Age accounted for about 20% of VO2max variance, while the %fat, SR-PA, and their interaction accounted for an additional 42% of VO2max variation.

Figure 3 shows the %fat and SR-PA interaction and illustrates importance of physical activity and body compo-

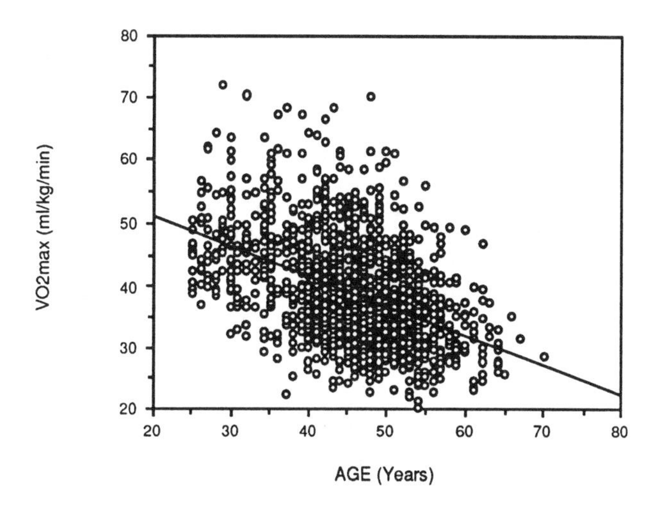

Figure 1. Scatterplot showing the cross-sectional relationship between age and VO2max. Provided for reference is the linear regression line that defined the linear trend as: VO2max = 60.45 -0.48(Age). (r = 0.45, SEE = 7.4 ml/kg/min)

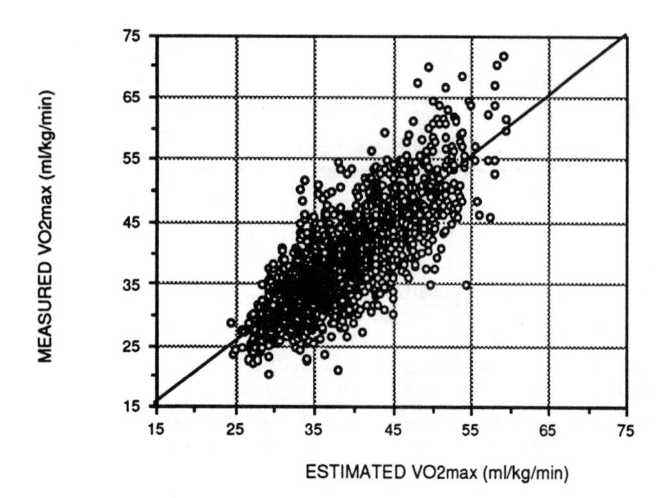

Figure 2. Scatterplot showing the cross-sectional relationship between VO2max and the multivariate relationship of age, SR-PA, and %fat (Model 4). The multiple regression model provides a better fit of the data than just age (Figure 1) reducing the standard error of estimate from 7.35 to 5.01 ml/kg/min. Provided for reference is the line of identity. VO2max = 47.90 - 0.27(Age) + 3.41(SR-PA) - 0.20(%fat) - 0.09(%fat x SR-PA) (R = 0.79, SEE = 5.0 ml/kg/min)

sition on VO2max. The graph shows that level of physical activity has more of an influence on VO2max than age. When percent body fat is ≤ 25 %, the average VO2max of an active 60-year old man is higher than a 30-year old man who is sedentary and the same body composition.

The multivariate model was applied to data of the longitudinal sample. These 156 men were tested twice. The time between tests 1 and 2 was not the same for all subjects (ΔMean = 4.1 ±1.2 years). To control for this bias, VO2max was standardized to represent the average yearly change (ml/kg/min/year). The zero-order correlation between measured and estimated changes in VO2max was 0.75 (Figure 4). These supports the multivariate model's validity for estimating longitudinal changes in VO2max with cross-sectional data.

DISCUSSION

The results of this study showed that changes in VO2max over time were more of a function body composition and exercise changes than aging. Classic data with champion athletes (Dill et al., 1967) support the findings of this study that ΔVO2max is more of a function %fat than aging. In a longitudinal study of 16 champion runners followed for more than 20 years, the mean longitudinal ΔVO2max was -1.04 ml/kg/min/year, but individual changes ranged from -1.94 to -0.41 ml/kg/min/year. At their initial test, the trained runners were fit (mean VO2max = 66.9 ±9.5 ml/kg/min, and mean %fat = 8.6 ± 2.9%). As they aged (mean Δ = 24.2 ±3.5 years) their mean VO2max decreased to 41.7 (±9.9) ml/kg/min, and their %fat increased to 19.4% (±5.4). Since Dill and Robinson published each runner's data, it was possible to relate the ΔVO2max with %fat changes and aging. The correlation between the ΔVO2max, and Δ%fat was -0.77 ($p < 0.01$), while the correlation with the change in age was lower and not statistically significant (0.29, $p > 0.05$).

We believe that the significant SR-PA and %fat interaction samples a habitual adherence to exercise. We have shown (Wier & Jackson, 1989) that exercise adherence and VO2max were related, those most physically active were more likely to continue exercise. Additionally, the data reported by Kasch and associates (Kasch, Wallace, & VanCamp, 1985) showed that the mean %fat of individuals who remained physically active was stable. The runners studied by Dill and Robinson (Dill et al., 1967) decreased their exercise between tests. This suggests that part of the runners' accumulation of body fat (mean Δ 10.8 %fat) were likely due to changes in physical activity.

The results of this study showed that the age decline in VO2max is due to several factors. This complicates the process of defining the true age decline in VO2max, but the multiple regression weights suggest that the decline in VO2max due to aging is much less than the published range of 0.4-0.5 ml/kg/min/year. The independent contribution of aging defined by the multivariate model is a decline of 0.27 ml/kg/min/year. The longitudinal data support the multivariate estimate. Kasch and associates (Kasch et al., 1990) followed a sample of 15 men who maintained their exercise and body composition levels over the 25 years of the study. The mean ΔVO2max of this active group was -0.25 ml/kg/min/year. The data of the active men were compared to a random sample of 15 inactive men who dropped from the exercise program.

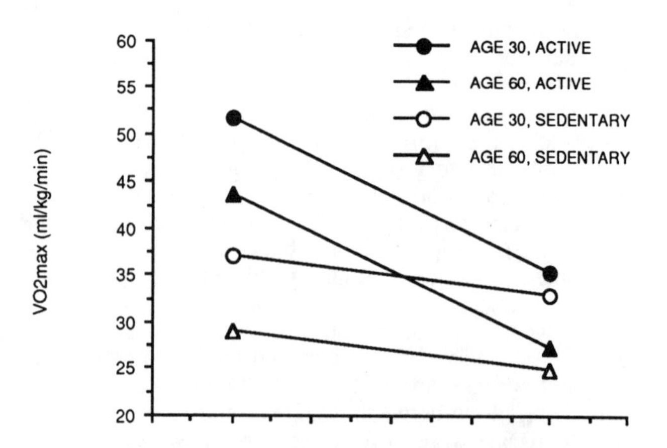

Figure 3. Graphic representation of the VO2max for 30- and 60-year old activeand sedentary menfor levels of body fatness. When %fat is less than 25%, the VO2max of active 60-year men can be expected to exceed that of sedentary men 25 years younger.

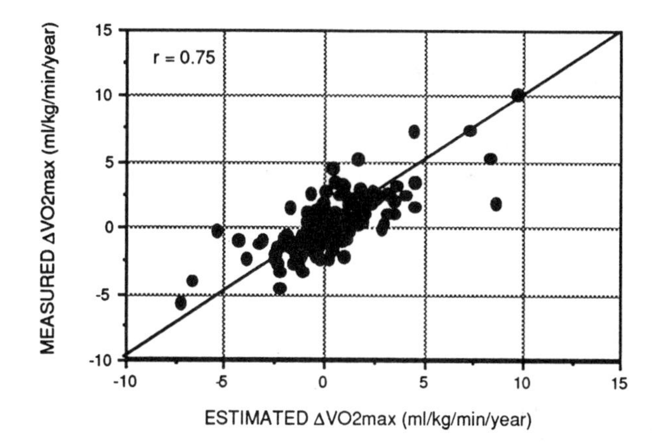

Figure 4. The scattergram between changes in measured and estimated VO2max. This shows that changes in exercise and %fat are sensitive to longitudinal changes in VO2max.

The mean ΔVO2max of the inactive men was over 3 times higher, -0.77 ml/kg/min/year. During the study, the active subjects loss 3.4 kg. of weight while the inactive subjects gained 3.2 kg. Assuming the VO2max loss of the active subjects represented the "true aging effect," the authors estimated that 33% of the decline in VO2max of the inactive subjects was due to age while 67% was due to inactivity.

The results of this study showed that body composition and exercise habits had more of an influence on the longitudinal decline of VO2max than aging. The discrepancy in the age-VO2max cross-sectional and longitudinal slopes reported in the literature (Buskirk & Hodgson, 1987) was due largely to the failure to include changes in body composition and exercise habits. Longitudinal studies, like cross-sectional, need to account for changes in body composition and exercise habits when attempting to define the age-decline in VO2max. With body composition and activity statistically controlled, the effect of age on the decline of VO2max went from -0.48 to under -0.27 ml/kg/min/year. The data of the longitudinal sample support this. The multivariate model provides a valuable tool for not only assessing age-related changes in physical working capacity, but the effects when level of physical activity and percent body fat is changed.

REFERENCES

Åstrand, I.,Åstrand, P.-O.,Hallback, I., & Kilbom, Å. (1973). Reduction in maximal oxygen uptake with age. *Journal of Applied Physiology*, 35(5), 649-654.

Buskirk, E. R., & Hodgson, J. L. (1987). Age and aerobic power: the rate of change in men and women. *Federation Proceedings,* 46, 1824-1829.

Dill, D. B.,Robinson, S., & Ross, J. C. (1967). A longitudinal study of 16 champion runners. *Journal of Sports Medicine and Physical Fitness,* 7, 4-27.

Hagberg, J. M.,Allen, W. K.,Seals, D. R.,Hurley, B. F.,Ehsani, A. A., & Holloszy, J. O. (1985). A hemodynamic comparison of young and older endurance athletes during exercise. *Journal of Applied Physiology,* 58(6), 2041-2046.

Jones, N. L.,Makrides, L.,Hitchcock, C.,Chypcar, T., & McCartney, N. (1985). Normal standard for an incremental progressive cycle ergometer test. *American Review of Respiratory Diseases,* 131, 700-708.

Kasch, F. W.,Boyer, J. L.,VanCamp, S. P.,Verity, L. S., & Wallace, J. (1990). The effect of physical activity and inactivity on aerobic power in older men (a longitudinal study). *The Physician and Sportsmedicine,* 18, 73-81.

Kasch, F. W.,Wallace, J. P., & VanCamp, S. P. (1985). Effects of 18 years of endurance exercise on the physical work capacity of older men. Journal of Cardiopulmonary Rehabilitation, 5, 308-312.

Mitchell, J. H.,Sproule, B. J., & Chapman, C. B. (1958). The physiological meaning of the maximal oxygen intake test. *Journal of Clinical Investigation,* 37, 538-547.

Pollock, M. L.,Foster, C.,Knapp, D.,Rod, J. L., & Schmidt, D. H. (1987). Effect of age and training on aerobic capacity and body composition of master athletes. *Journal of Applied Physiology,* 62(2), 725-731.

Robinson, S. (1938). Experimental studies of physical fitness in relation to age. *Arbeitsphysiologie*, 10, 251-323.

Robinson, S. D.,Dill, D. B.,Robinson, R. D.,Tzankoff, S. P., & Wagner, J. A. (1976). Physiological aging of champion runners. *Journal of Applied Physiology,* 41, 46-51.

Wier, L. T., & A.S.Jackson (1989). Factors affecting compliance in the NASA/JSC fitness program. *Sports Medicine,* 8, 9-14.

MAXIMAL POWER GRASP FORCE AS A FUNCTION OF WRIST POSITION, AGE AND GLOVE TYPE: A PILOT STUDY

D. L. McMullin and M. S. Hallbeck
Industrial and Management Systems Engineering Department
University of Nebraska -- Lincoln
Lincoln, Nebraska

In many industries, environmental and/or safety considerations require the use of gloves. In an effort to quantify the effects of gloves on physical capabilities; wrist position, glove type, age, gender, and dominant/non-dominant hand differences were examined. Power grasp force was used as the dependent measure of physical capabilities. Task design that requires less grasp muscle effort and neutral wrist postures are assumed to be less likely to cause CTDs. Six glove types: bare hand, thermal, knit, reinforced knit, a layered combination of thermal and knit, and a layered combination of thermal and reinforced knit were used as the independent variables. Subjects were selected from three age categories (20-25, 40-45, and 60-65). For each force exertion condition, one of four wrist positions were employed: neutral, 45° extension, 45° flexion, and 65° flexion. Five subjects within each age-gender category were tested giving a total of 30 subjects. Each subject was asked to build up to his or her maximal voluntary contraction using the Caldwell regimen. Results were analyzed using analysis of variance (ANOVA) with significant effects (gender, handedness, glove type, wrist position, and various interactions) tested using post hoc analysis.

INTRODUCTION

Many industries today have high incidence rates of cumulative trauma disorders (CTD) including tendonitis, tenosynovitis, carpal tunnel syndrome, and others which affect the wrist, hand, and arm of the worker (Tanaka and McGlothlin, 1989). Possible causes of CTDs include repetitions, deviated postures, and high force. Repeated application of high forces in deviated postures inflame the tendons through the carpal tunnel (Smith, Sontegard, and Anderson, 1977). If the tendons are not allowed to recover from this inflammation, they will eventually swell and put pressure on the median nerve, causing pain and numbness (Drury, Begbie, Ulate, and Deeb, 1985). In spite of the frequent occurrence of CTDs, surprisingly little is known about the combination of factors responsible for its development. A reasonable rationale is that a task design that requires less grasp muscle effort and neutral wrist postures will be less likely to cause CTD. The current study was conducted to confirm and quantify power grasp force magnitude differences for six glove conditions in four wrist positions for both hands. Both male and female subjects were selected from three age categories.

REVIEW OF LITERATURE

Age

After approximately age 30, muscle mass decreases due to a loss of muscle protein. The decreased mass leads to a reduction in muscle strength (Astrand and Rodahl, 1986; McArdle, Katch and Katch, 1986; Rodahl, 1989). Many researchers have found that muscle strength peaks between ages 20 and 30 and declines thereafter (Astrand and Rodahl, 1986; Bailey, 1982; Grandjean, 1982; McArdle et al., 1986; Rodahl, 1989). By age 40, grip strength has declined to approximately 95% of peak force at age 20 (Mathiowetz, Kashman, Volland, Weber, Dowe, and Rogers, 1985) and by age 60 to 70, grip force is 65 - 74% of the grip force exerted by 20 year olds for men and between 42 - 78% of grip force exerted by 20 year olds for women (Kellor, Frost, Silberberg, Iverson, and Cummings, 1971; Mathiowetz et al., 1985). However, Anderson (1965) found little difference between the grip strength of 18 years olds and that of 65 year olds.

Gender

Human skeletal muscle can generate approximately 3-4 kg of force per square centimeter of cross-sectional muscle. However, men contain a greater amount of muscle area than females therefore, they can generate a greater amount of force (McArdle et al., 1986). Female grip strength has been found to be approximately 50 - 60% of male grip strength (An, Chao, and Askew, 1983; Kellor et al., 1971; Mathiowetz et al., 1985).

Handedness

Anderson (1965) reported that the dominant hand can be up to 30% stronger than the non-dominant hand. Results from separate studies on grip strength show that on the average the non-dominant hand has about 85-95% of the strength of the dominant hand (Anderson, 1965; Kellor et al., 1971; Mathiowetz et al., 1985).

Glove Type

Greater forces can be generated by a bare hand as compared to a gloved hand (Cochran, Albin, Bishu, and Riley, 1986; Sudhakar, Schoenmarklin, Lavender, and Marras, 1988). This has been attributed to glove interference with the hand-handle interface (Cochran et al., 1986).

Wrist Position

For grasp strength, a neutral wrist angle has been shown to be superior to a bent wrist position (Anderson, 1965; Kraft and Detels, 1972; Pryce, 1980; Putz-Anderson, 1988; Skovly, 1967). Putz-Anderson (1988) reports that 45° extension was 75%, 45° flexion was 60%, and 65° flexion was 45% of the force that can be generated in a neutral wrist angle.

METHOD

Subjects

Thirty subjects (15 females and 15 males) participated in this study. They were selected from 3 age ranges: 20-25 years, 40-45 years, and 60-65 years of age. All subjects reported no previous wrist injuries, upper limb nerve damage, and good health.

Apparatus

A Vital Signs hand dynamometer (model 68800) was used to test the power grasp force. Three gloves were used: a 100% cotton inspection thermal glove (thermal protection), a Polar Bear® Plus (Spectra® knit glove) mesh glove, and a Polar Bear® Supreme (three-strand steel reinforced knit glove) reinforced glove.

Procedure

Upon arrival, the subjects received instructions for the study. A standardized body posture, standing with a 90° elbow angle and a relaxed shoulder angle, was employed for all trials. After putting on the gloves and positioning his or her wrist in the selected position, the subject was instructed to build up to maximum voluntary contraction (MVC) using the modified Caldwell regimen (1974). Following this regimen, they held the MVC for 4 seconds. The peak MVC exerted was recorded. This procedure was repeated with a different glove-wrist position combination with the opposite hand. The subject alternated between hands until all combinations were completed on both hands. Due to the time consumed during wrist positioning and glove changing, more than 1 minute elapsed between trials; therefore, no additional time was needed for fatigue recovery. Short breaks were taken if requested by the subject.

Experimental Design

This study evaluated the effects of glove type, wrist position, age, gender, and handedness on force exertion. Glove type had 6 levels: bare hand, thermal, knit, reinforced knit, a combination of thermal and knit, and a combination of thermal and reinforced knit. The order of gloves within the experiment was randomized as was wrist position. The wrist position variable replicated the positions discussed by Putz-Anderson (1988) (neutral, 45° extension, 45° flexion and 65° flexion). Subjects were nested within age-gender interaction. The age variable consisted of 3 levels: 20-25, 40-45, and 60-65 years of age. The 2 levels within the variable handedness referred to dominant or non-dominant. The resultant design was a 6 (glove) by 4 (wrist position) by 3 (age) by 2 (handedness) by 2 (gender) mixed factor model.

RESULTS

An ANOVA was conducted using peak force exertion as the dependent variable and glove type, wrist position, age, handedness, gender, and their interactions as the independent variables. Significant effects were found to be gender, glove type, handedness, wrist position, and the glove type-wrist position interaction, glove type-age-gender interaction, and glove type-wrist position-gender interaction as shown in Table 1.

A Tukey-studentized range test was conducted for the significant main effects. It showed that the addition of gloves significantly reduces the amount of force that can be exerted, also the bulkier and less form fitting the glove, the less amount of force that can be generated. Extreme wrist position also significantly reduces the amount of force exerted. Means and percentages are reported in Tables 2 - 5.

DISCUSSION

A surprising result of the study was that there was no significant age effect. This matches the results found by Anderson (1965). Since all subjects were healthy and were screened for nerve, arm, and wrist problems, this could have possibly excluded typical aging effects.

The gender difference found in the current study is not as extreme as those reported in previous studies (An et al., 1983; Kellor et al., 1971).

The glove type effect illustrates that the bare hand is more efficient than a gloved hand. This corresponds to the findings of Cochran et al., 1986 and Sudhakar et al., 1988. If gloves are necessary, the post-hoc tests demonstrate that they should be form-fitting rather than loose and that a single layer is preferable to multiple layers. This suggests that the strength decrement may be due to bunching of material at the joints causing hand-handle interference.

The results of the wrist position variable show the same general trend as found by Putz-Anderson (1988); however, the decrements in force are not as extreme (Table 5).

The results of this study demonstrate the problems associated with gloves and extreme wrist postures. These problems occur frequently in industry. A reasonable assumption is that a task design that requires less grasp muscle effort will be less likely to contribute to cumulative trauma disorders. Thus, wherever possible extreme wrist postures and multiple layers of gloves should be avoided.

REFERENCES

An, K.N., Chao, E.Y. and Askew, L.J. (1983). Functional assessment of upper extremity joints. IEEE Frontiers of Engineering and Computing in Health Care, 136-139.

Anderson, C.T. (1965). Wrist joint position influençes normal hand function. Unpublished masters thesis, University of Iowa, Iowa City, IA.

Astrand, P.O. and Rodahl, K. (1986). Textbook of Work Physiology: Physiological bases of exercise. New York: McGraw-Hill.

Bailey, R.W. (1982). Human Performance Engineering: A guide for systems designers. Englewood Cliffs, NJ: Prentice Hall.

Caldwell, L.S., Chaffin, D.B., Dukes-Dobos, F.N., Kroemer, K.H.E., Laubach, L.L., Snook, S.H.,and Wasserman, D.E. (1974). A proposed standard procedure for static muscle strength testing. American Industrial Hygiene Association Journal, 35(4), 201-206.

Cochran, D.J., Albin, T.J., Bishu, R.R., and Riley, M.W. (1986). An analysis of grasp degradation with commercially available gloves. In Proceedings of the Human Factors Society 30th Annual Meeting (pp. 852-855). Santa Monica, CA: Human Factors Society.

Drury, C.G., Begbie, K., Ulate, C., and Deeb, J.B.(1985). Experiments on wrist deviation in manual materials handling. Ergonomics, 28(4), 577-589.

Grandjean, E. (1982). Fitting the task to the Man: An Ergonomic Approach. London: Taylor and Francis.

Kellor, M., Frost, J., Silberberg, N., Iverson, I., and Cummings, R. (1971). Hand strength and dexterity. American Journal of Occupational Therapy, 25 (2), 77-83.

Kraft, G.H. and Detels, P.E. (1972). Position of function of the wrist. Archives of Physical Medicine and Rehabilitation, 53, 272-275.

Mathiowetz, V., Kashman, N., Volland, G., Weber, K., Dowe, M., and Rogers, S. (1985). Grip and pinch strength: Normative data for adults. Archives of Physical Medicine and Rehabilitation, 66, 16-21.

McArdle, W.D., Katch, F.I., and Katch, V.L. (1986). Exercise Physiology: Energy, Nutrition, and Human Performance (2nd ed.). Philadelphia: Lea and Febiger.

Pryce, J.C. (1980). The wrist position between neutral and ulnar deviation that facilitates the maximum power grip strength. Journal of Biomechanics, 13, 505-511.

Putz-Anderson, V. (1988). Cumulative Trauma Disorder - A Manual for Musculo-skeletal Disease of the Upper Limbs. London: Taylor & Francis.

Rodahl, K. (1989). The Physiology of Work. London: Taylor and Francis.

Skovly, R.C. (1967). A study of power grip strength and how it is influenced by wrist joint position. Unpublished masters thesis, University of Iowa, Iowa City, IA.

Smith, E.M., Sontegard, D.A., and Anderson, W.H.(1977). Carpal tunnel syndrome: contribution of flexor tendons. Archives of Physical Medicine and Rehabilitation, 58, 379-385.

Sudhakar, L.R., Schoenmarklin, R.W., Lavender, S.A., and Marras, W.S. (1988). The effects of gloves on grip strength and muscle activity. In Proceedings of Human Factors Society 32nd Annual Meeting (pp. 647-650). Santa Monica, CA: Human Factors Society.

Tanaka, S. and McGlothlin, J.D. (1989). A conceptual model to assess musculo-skeletal stress of manual work for establishment of quantitative guidelines to prevent hand and wrist cumulative trauma disorders (CTDs). Advances in Industrial Ergonomics and Safety I (pp. 419-426). London: Taylor & Francis.

AGE EFFECTS IN BIOMECHANICAL MODELING OF STATIC LIFTING STRENGTHS

Don B. Chaffin, Charles B. Woolley, Trina Buhr, Lois Verbrugge
The University of Michigan

ABSTRACT

There is growing awareness that age results in reduced strengths in the population, and that significant decreases start in the 5th decade. The magnitude of the decrease in strength depends on the specific muscle function being tested. Because of differential effects it is not clear how various decreases could alter whole-body strength performance. This paper describes how specific strength decreases measured in an older population of men and women could affect their whole-body exertion capabilities in selected scenarios. A computerized strength prediction program is used to both predict the whole-body strength changes with age, and to study how older populations can alter their postures to achieve maximum exertion capability. The results indicate that different muscle group strengths decline by 5% to 70% with age, depending on which muscle group is tested. These changes have profound effects on whole-body exertion capabilities, which also are shown to depend on specific postures used to perform the exertions.

INTRODUCTION

There is a general recognition in the literature that age changes in muscular strengths are profound. A recent report of the National Research Council (Czaja and Guion, 1990) summarizes strength performance declines with age as follows:

"There is, on average, a decrease in muscle mass with age, which results from a decrease both in the number and the size of muscle fibers. Translated into performance, there are rough estimates that by age 40 average muscle strength is about 95 percent of an earlier maximum in the late 20s; by age 50 it drops to about 85 percent; and by age 65 only 75 percent of the earlier output is still available, with further declines thereafter. However, these are population mean differences and there is a great deal of variability in different muscle groups, in types of muscular performance and individuals."

Viitasalo et al. (1985) compared the static strengths of healthy Finnish men in cohorts of 31-44 years and 71-75 years. They showed average declines of about 30% to 50%, depending on specific muscle groups.

It is the intent of this paper to describe how age affects a selected set of muscle strengths, and how the resulting decreased strengths may affect whole-body lifting capabilities.

METHODS

This study involves two phases. The first phase required the development of a set of static strength data on specific muscle functions for an older population of healthy men and women. Fortunately, data of this type were available from a study of 98 men and women of ages 40 to 80+ years living in the Ann Arbor area, conducted by these investigators (Verbrugge et al. 1991). Table 1 provides some descriptive statistics on these people. In this study 12 different bilateral isometric strengths were measured using a specially designed strength testing chair (Verbrugge et al. 1994). For the purpose of this paper the data were expressed as strength moments for each muscle functions. This provided a set of strength norms for this older population which could then be compared to the population norms now used in the University of Michigan's 2D Static Strength Prediction Program™ (Chaffin and Andersson, 1991). The latter values were collected from several different population studies of younger men and women (18-48 years) employed in manual labor.

In phase two of this investigation the new static strength values were inserted into the existing 2DSSPP™ program. The program was then used to simulate a set of whole-body lifting exertions with both the younger and older aged population norms used for reference. The lifting postures chosen were those available from previous studies (see Chaffin and Andersson, 1991 for summary). Incremental changes in the postures were then made to evaluate how postural compensation may affect the predicted lifting strengths.

RESULTS

Differences in the mean strength values between the older population and the existing norms are shown in Figure 1.

Table 1 Older Population Statistics (Verbrugge et al. 1991)

		Men	Women
	Mean Age (yrs)	71.8	74.3
	Mean Stature (m)	1.77	1.60
	Mean Body Wt. (Kg.)	80.5	62.0
Total 98 =	Sample Size	18	80

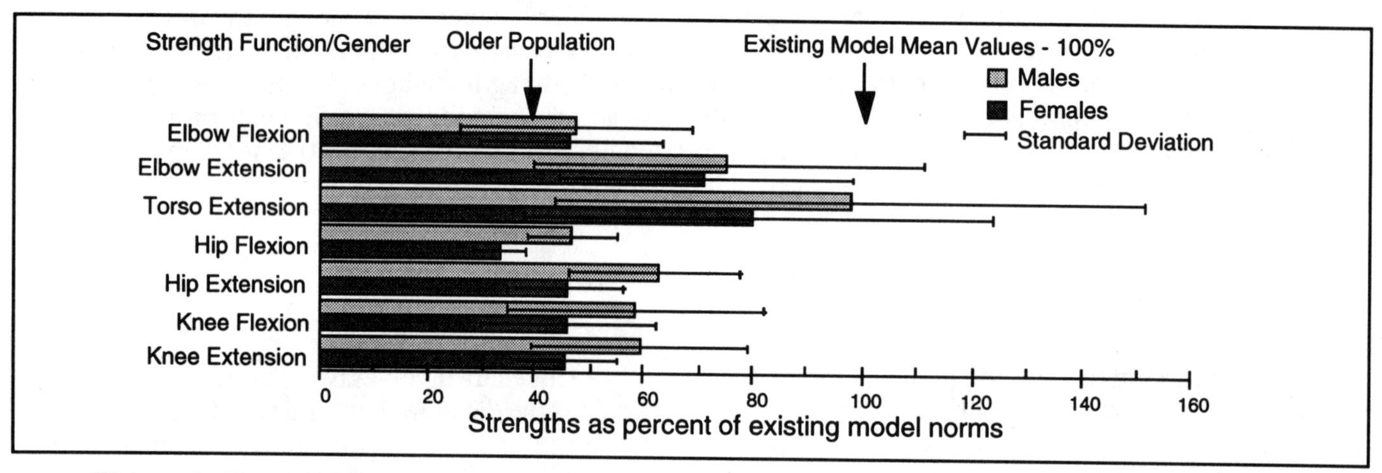

Figure 1: Comparison of older population strengths with existing norms in UM-2DSSPP™

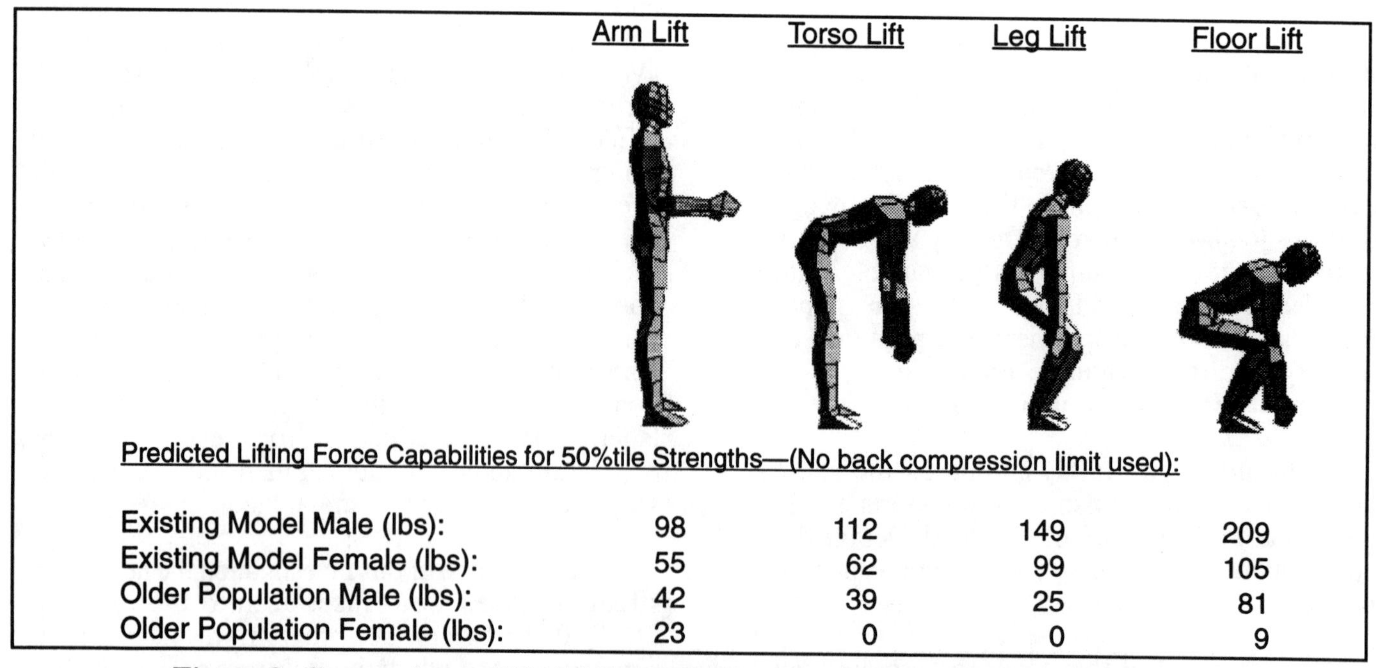

	Arm Lift	Torso Lift	Leg Lift	Floor Lift
Existing Model Male (lbs):	98	112	149	209
Existing Model Female (lbs):	55	62	99	105
Older Population Male (lbs):	42	39	25	81
Older Population Female (lbs):	23	0	0	9

Figure 2: Comparison of Model Predicted 50%tile Lifting Strengths for Existing Norms and Older Population Norms

Four arbitrary but different lifting postures were simulated. The postures and 50%tile lifting capability predictions for the original chosen postures are shown in Figure 2. From inspection it is clear that the lifting capabilities of older men and women is severely compromised by decreases in specific muscle strengths. In this latter regard, knee and hip extension strength declines had the greatest effects for the postures used in the simulations, especially for older females. In fact, it was predicted that 50% of older females would not be able to lift any weight in the torso and leg lift postures.

Clearly the exact postures chosen by individuals attempting a whole-body maximum exertion are critical, and will depend to a large extent on individual muscle strengths at various joints. The reason the older female population was predicted to have no useful lifting capability in the initial torso and leg lift postures illustrated in Figure 2 is because of hip and knee extension strength limits, respectively. By modifying the initial postures to both reduce the joint moment requirements at those joints and increase the moment loads at the joints having higher strength capabilities, increased lifting capabilities were predicted. This is illustrated in Figure 3, wherein lifting capabilities associated with four slightly modified postures are compared to the initial, arbitrary postures for the older female populations.

Though an exhaustive study of strength and postural effects has not been undertaken in this study, the initial results certainly indicate why it is so important to use a model of whole-body exertions to study the complex interactions. This is particularly true when interest in special populations, wherein muscle strengths will not consistently or proportionally vary from one muscle group to the next. Thus a posture which produces high exertion capabilities in one population may not be appropriate for another population.

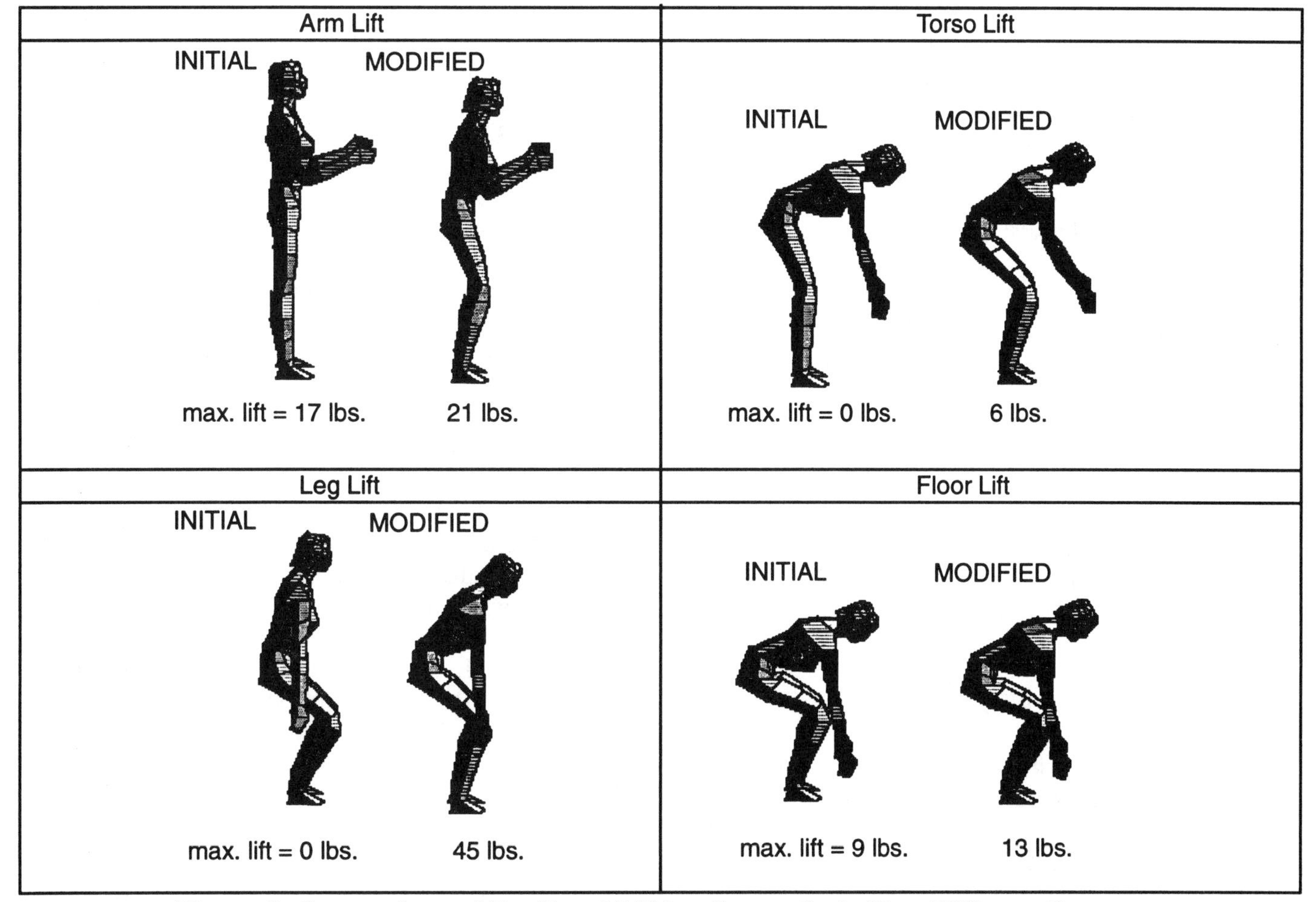

Figure 3. Comparison of Predicted Lifting Strengths in Two Different Postures for Older Female Population.

DISCUSSION

This investigation demonstrates how important specific muscle strengths are in providing whole-body exertion capabilities. Though limited by the availability of normative strength data on older populations, (e.g., shoulder strength declines with age could drastically affect lifting to high shelves) the initial results certainly demonstrate the magnitude of the problem of aging in such physical tasks. Additional tasks are being simulated as part of this investigation.

As reported by several investigators, small changes in postures can have a large effect on predicted, whole-body strength performance (Chaffin and Erig, 1991). Since the declines in muscle strengths with age are not proportional for different muscle groups, it is not clear which types of exertions are most sensitive to postural compensations. Further simulations are planned to explore this aspect of the problem.

Lastly, it is well documented that low-back pain and injury prevalence rates increase in older populations (Biergin-Sorenson, 1982). In the lifting simulations used in this investigation, the higher lifting strengths in the younger populations resulted in much higher spinal disc compression forces than was the case for the older populations. The high spinal compression forces in the young male population may explain why the incident rate of low back pain is high for young men in heavy manual labor (Chaffin and Page, 1993). One could conjecture that reduced back and hip muscle strengths in older populations are protective of the spinal column, even with its reduced compression tolerance in later years (Chaffin and Ashton-Miller, 1991). This conjecture is to be studied more fully, but one must wonder what specific lifting tasks may be particularly hazardous to the low-back of older individuals.

ACKNOWLEDGMENTS

The authors with to acknowledge partial support from the Multipurpose Arthritis Center NIH Grant No. AR20557, and the Whitaker Foundation Predoctoral Fellowship Program.

REFERENCES

Biergin-Sorenson, F. (1982) Low back trouble in a general population of 30-, 40-, 50-, and 60-year old men and women. Study design representativeness, and basic results, *Danish Medical Bulletin,* 29(6), 289-299.

Chaffin, D.B. and Andersson, G.B.J. (1991) *Occupational Biomechanics,* J. Wiley, New York.

Chaffin, D.B. and Ashton-Miller, J.A. (1991) Biomechanical aspects of low-back pain in the older worker. *Experimental Aging Research,* 17(3), 177-187.

Chaffin, D.B. and Erig, M. (1991) Three-dimensional biomechanical static strength prediction model sensitivity to postural and anthropometric inaccuracies, *IIE Transactions*, 23(3):215-227.

Chaffin, D.B. and Page G. (1993) Postural Effects on Biomechanical and Psychophysical Weight-lifting Limits. *Ergonomics,* 37(4):663-676.

Czaja, S.J. and Guion, R.M. (1990) Human factors research needs for an aging population, Panel on Human Factors Research Issues for an aging population,

Verbrugge, L.M., Reoma, J.R., Woolley, C.B., Chaffin, D.B. and Wery, S.D. (1991) The results of the field testing of quantitative musculoskeletal function in older workers. Report to the National Institute of Occupational Institute of Occupational Safety and Health. NIOSH Contract RFQ 90-40,

Verbrugge, L.M., Chaffin, D.B., Woolley, C.B., Sekulski, R.A. (1994) Musculoskeletal functioning in older adults. To be presented at Symposium on Ergonomics and Aging Research (IEA 1994)

Viitasalo, J.T., Era, P., Leskinen, P.A. and Heikkinen, E. (1985) Muscular strength profiles and anthropometry in random samples of men aged 31-35, 51-55 and 71-75 years, *Ergonomics,* 28(1), 1563-1574.

REACH DESIGN DATA FOR THE ELDERLY

Ursula Wright, G. Major Kumar, Anil Mital
Industrial Engineering
University of Cincinnati, Cincinnati, Ohio 45221-0116

The increased life expectancy of the elderly may require substantial redesigning of environments in order to accommodate age related body changes. One of the most important aspects allowing the elderly to function independently is the ability to reach for items comfortably during daily activities. Designing for an independent elder requires knowledge of reach measurements that determines the optimal design of working/living environments. This paper provides reach capability design data of elderly males and females between 65 and 89 years of age. Results show that direct and indirect reach indicators decrease substantially above the age of 80. The elderly participating in this study were compared with younger cohorts, showing significantly lower reach capabilities. This indicates the necessity of considering anthropometric data, such as reach, to design for the elderly. The need is particularly dire for those above 80 years of age.

INTRODUCTION

The elderly population that lives independently is expected to increase tremendously (Cjaza, 1990). To accommodate age-related decline in capabilities, redesigning environments for the elderly is essential.

An important aspect of living independently is the ability to perform activities of daily living (ADL) which require a substantial amount of reaching to grasp objects or operate controls (Clark et al., 1990). Only a few studies are available that have considered the anthropometric functional capabilities of the elderly in the design process (Kelly and Kroemer, 1990). This is particularly true of the reach envelope. The designs have focused primarily on the reach of younger populations while ignoring the needs of the elderly user.

The purpose of this study was to collect and report anthropometric reach design data on the elderly (age 65 and above). It was also important to know at what age, age-related changes in the reach capability become significant.

METHOD

Participants

The subjects in this study consisted of elderly men and women aged 65 and above. Each subject was in good general health and living independently. Volunteers with walkers, in wheel chairs, or with canes were not considered. A cross-sectional design method was used such that a cohort of age groups were measured at one point in time.

Some of the reach measurements recorded dealt with the extension of the dominant arm in the vertical and horizontal planes. Information on the dexterity of the individual was also recorded, along with work status, physical ailments present, or major surgery having drastic effects on the measurement of the extended arm. A total of 67 females and 66 males have participated in the study thus far.

Apparatus

The body dimensions were measured using an anthropometer, a goniometer for recording joint angles, and a tape measure. All measurements were taken in millimeters. The measurements for the reach envelope in the vertical plane required angular dimensions on a radial scale drawn on the floor. The radial scale was marked at the center with markings representing degrees. The 180° and 0° were on a radial line parallel to the wall.

Procedure

The techniques used to measure each of the fourteen measurements are explained here briefly. Standard procedures were used to measure body dimensions such as stature, acromial standing height, etc. (Roebuck et al., 1975).

Vertical reach measurements were taken on the dominant arm while subjects stood with feet flat and arm raised over the shoulder as straight and steady as possible. The hand was extended to reach the maximum possible height. The anthropometer was positioned vertically sliding the pointer to touch the tip of the third finger to take the measurement.

The vertical reach angle was determined while measuring the vertical reach. A goniometer was used on the subject's raised arm. One lever of the goniometer was aligned parallel to the trunk of the body, while the other lever was adjusted parallel to the arm. The angle read from the goniometer was subtracted from 180° to obtain the vertical reach angle.

Each of the next measurements is a part of the reach envelope. These measurements were the primary focus of this study. Horizontal reaches (grip and tip of finger) at 0°

were measured while seated. The measurement was taken on the dominant arm extended as straight as possible at the side of the body, parallel to the floor. Two measurements were taken at this point - first with the hand in a fist position (gripped) and second with the fingers extended out (tip).

To measure the horizontal reach with the hand extended at 0°, the subjects were asked to sit with their backs comfortably upright and against the chair. The chair was then adjusted so that the shoulder joint was over the center of the radial scale drawn on the floor through visual inspection. The subject extended the arm sideways along the radial line coinciding with the 0° mark on the floor.

A vertical rod was placed on the radial line and moved until it touched the tip of the subject's stretched arm and hand. There was no set distance above the leveled floor that the arm was required for measure. The distance between the vertical rod at the tip of the finger and the shoulder ball joint determined the horizontal reach of the subject at the 0° radial position.

The same method was used to measure arm extensions at 90°. For the horizontal reach with the hand gripped, similar procedures were used. The grip measurement was made by having the subject grip the vertical rod with the third finger exactly over the rod. This allowed the measurements to be taken from the rod position on the floor to the ball of the shoulder.

The maximum extension of the arm in the horizontal plane was measured using similar methods. The vertical rod was placed at the top of the third finger and positioned along the radial markings on the floor. This method not only determined the horizontal length of the arm but the angle that the arm could be extended. The angle was measured from the 0° marking on the floor.

Measurements for biacromial breadth, and arm and hand lengths were also taken. Due to incompatibilities (lack of control in measurement due to several experimenters) in measuring techniques, these dimensions were eliminated from the study.

DATA PROCESSING AND ANALYSIS

The collected data were divided into five age categories of 4-year intervals. This was done to reduce the variation in measurements among subjects. The 4-year intervals were: 65 - 69 yrs, 70 - 74 yrs, 75 - 79 yrs, 80 - 84 yrs, and 85 yrs and above. The mean, standard deviation, and range for each of the measurements recorded were determined and are shown in Tables 1 and 2.

Using an anthropometrical database of younger adults, (NASA, 1978), comparisons were made with the means of the data collected in this study. Assuming normal distribution, statistical tests were carried out to determine any significant differences between the cohorts. The t - test was used to analyze the five age groups of the elderly to determine which groups differed significantly in reach capabilities from others. This also provided information about the maximum age-related decline in reach capabilities. Reach capabilities of the young adult population were also compared with those of the elderly population. Significant differences between the two populations (younger adults vs. older adults) would demonstrate a need to consider using design data for the elderly when designing for them.

The direct reach indicators (vertical and horizontal reaches with the hand gripped and extended) were analyzed for the age effect using the Analysis of Variance (ANOVA). The seven reach measurements were analyzed for both genders to determine the existence of significant differences among the five elderly age categories. For those reach measurements showing significance, the Duncan's multiple range test was carried out to determine which age categories significantly differed from others.

RESULTS

The statistical analysis revealed that males, as expected, had larger reach capabilities than females ($p < 0.05$). However, for both men and women only vertical reach capability of those 85 years or above in age was significantly lower than their cohorts in other age groups ($p < 0.05$); all other reach capabilities showed no age-related effect ($p \geq 0.10$).

For both men and women of age 85 or above, several body size measurements were found to be significantly smaller than the respective body size measurements of their cohorts ($p < 0.05$). The measurements included were stature, acromial standing height, bideltoid breath, and sitting height. The acromial sitting height of men 85 and above was also different from their cohorts ($p < 0.05$). Also, the bideltoid breadth and sitting height of females between 80 years and 84 years of age were also different from their cohorts ($p < 0.05$).

DISCUSSION

Living independently with the ability to perform ADLs influences the continued lifestyle of the elderly without the need for an institutionalized home environment. Successful performance of ADLs depends on the capabilities of the elderly and the task at hand. Enabling the elderly to function in their environments with limited assistance, necessitates the importance for the environment to be designed to accommodate age-related changes. It is important for designers to identify physical capabilities that change with age and create and revise the guidance for designers. Designs should satisfy the needs of younger populations as well as the elderly. It is advantageous to consider designing for declines in capacity that occur throughout life not solely at retirement age.

To alleviate design problems, anthropometrical data for reach measurements should be available for those 65 and above. These measurements should consider the

TABLE 1: Collected statistics on reach body dimensions of male elderly subjects

MALE*	ELDERLY AGE COHORTS (YEARS)														
	65 - 69			70 - 74			75 - 79			80 - 84			85 +		
	AVG	SD	R	AVG	SD	R	AVG	SD	R	AVG	SD	R	AVG	SD	R
Stature	1722	67.6	1584-1855	1747	50.5	1629-1825	1735	37.3	1636-1795	1742	70.0	1627-1851	1678	42.8	1623-1761
Acromial standing hgt.	1441	59.2	1313-1527	1471	55.4	1341-1585	1458	37.3	1366-1519	1463	77.8	1348-1599	1408	36.8	1340-1463
Sitting height	854	31.8	780 - 896	863	39.2	756 - 911	873	29.6	804 - 905	853	25.2	820 - 889	828	20.3	805 - 863
Acromial sitting hgt.	584	31.8	540 - 630	595	32.6	512 - 634	596	24.9	541 - 620	569	20.4	535 - 600	563	14.8	544 - 589
Bideltoid breadth	463	35.3	411 - 557	450	23.7	399 - 485	449	14.6	420 - 471	446	28.9	408 - 495	437	18.5	410 - 470
Vertical reach	2145	111.4	1874-2350	2177	81.5	2040-2330	2157	70.8	1984-2290	2155	117.1	2070-2421	2027	59.7	1940-2146
Vertical reach angle	13	8.23	2 - 28	24	13.3	3 - 50	21	13.8	2 - 60	22	8.3	8 - 33	31	6.53	21 - 41
Horiz. grip reach at 0°	593	42.1	541 - 701	601	45.3	435 - 644	612	28.4	561 - 671	622	40.4	560 - 671	600	25.6	561 - 640
Horiz. grip reach at 90°	606	40.8	543 - 715	635	24.1	591 - 678	625	37.8	555 - 700	626	22.7	575 - 646	610	26.4	560 - 665
Maximum grip reach	613	38.5	558 - 708	635	33.7	584 - 726	624	35.7	550 - 680	629	20.8	590 - 651	609	16.6	582 - 645
Horiz. tip reach at 0°	709	45.9	632 - 805	719	37.3	604 - 765	727	25.2	681 - 778	731	53.6	648 - 794	731	32.2	684 - 784
Horiz. tip reach at 90°	726	45.3	648 - 826	743	24.7	690 - 785	744	31.6	684 - 800	759	34.3	698 - 801	738	28.3	691 - 786
Maximum tip reach	727	39.7	606 - 808	738	33.6	647 - 798	742	29.7	692 - 798	751	34.2	690 - 790	730	37.2	634 - 784
Maximum reach angle	120	10.2	106 - 145	113	7.84	95 - 129	116	37.5	115 - 165	114	10.9	95 - 130	110	10.1	102 - 132

* Linear dimensions in millimeters; Angular dimensions in degrees; Averages and Ranges rounded to nearest millimeter.

TABLE 2: Collected statistics on reach body dimensions of female elderly subjects

FEMALE*	ELDERLY AGE COHORTS (YEARS)														
	65 - 69			70 - 74			75 - 79			80 - 84			85 +		
	AVG	SD	R	AVG	SD	R	AVG	SD	R	AVG	SD	R	AVG	SD	R
Stature	1609	55.6	1515-1729	1585	66.7	1459-1775	1592	83.9	1490-1755	1571	66.8	1471-1687	1541	75.9	1420-1669
Acromial standing hgt.	1343	48.7	1260-1456	1334	60.0	1223-1495	1328	72.3	1256-1489	1327	62.2	1221-1443	1292	69.4	1187-1411
Sitting height	797	33.8	714 - 854	800	36.5	739 - 860	790	63.4	666 - 896	758	61.2	594 - 832	739	44.1	664 - 797
Acromial sitting hgt.	531	42.3	414 - 589	527	42.6	392 - 583	536	55.7	448 - 644	522	35.9	467 - 588	514	77.7	435 - 726
Bideltoid breadth	411	30.4	360 - 500	389	17.2	358 - 430	397	26.5	346 - 425	386	26.2	315 - 416	376	22.7	335 - 401
Vertical reach	2022	85.0	1817-2185	1991	83.1	1810-2221	1987	114.8	1794-2220	1949	89.1	1815-2130	1878	139.3	1592-2064
Vertical reach angle	10	8.0	0 - 26	16	8.9	2 - 42	19	10.4	5 - 40	27	11.0	13 - 50	20	9.6	0 - 36
Horiz. grip reach at 0°	558	31.3	491 - 632	554	31.3	498 - 629	533	37.6	495 - 598	559	32.1	508 - 621	548	34.5	484 - 601
Horiz. grip reach at 90°	578	33.3	529 - 642	564	36.9	472 - 642	555	34.4	500 - 608	570	29.9	530 - 624	572	25.3	531 - 614
Maximum grip reach	575	35.5	509 - 630	564	40.7	498 - 640	559	38.7	504 - 610	569	24.6	538 - 605	580	34.5	520 - 639
Horiz. tip reach at 0°	668	32.1	615 - 732	666	32.2	614 - 732	652	38.0	601 - 714	670	38.6	615 - 734	665	47.8	560 - 748
Horiz. tip reach at 90°	689	34.6	635 - 754	674	44.5	565 - 776	669	39.6	618 - 726	678	28.6	648 - 720	687	31.3	630 - 724
Maximum tip reach	686	37.7	614 - 745	671	42.1	601 - 765	673	45.9	610 - 742	667	30.0	631 - 730	678	36.6	610 - 736
Maximum reach angle	124	11.7	95 - 142	118	10.3	101 - 145	119	10.9	105 - 145	119	14.7	100 - 147	112	7.4	101 - 126

* Linear dimensions in millimeters; Angular dimensions in degrees; Averages and Ranges rounded to nearest millimeter.

heterogeneousness of the elderly population by using samples in small age intervals for each gender. This will give a more accurate view of when and to what degree the bodies reach capabilities become limited.

As the above discussion indicates, reaching capability is needed in tasks of daily living for the elderly. And yet, there are no studies that report such data. This study provides horizontal and vertical reach data on men and women who are 65 years in age or older.

While the results show that age tends to affect only the vertical reach, and that too at age 85 and above, this conclusion can be misleading. Reach capability is one of those physical capabilities where a lack of statistical significance is less relevant than the actual difference. For instance, the difference in a specific reach capability between two age groups may be only a few millimeters, and statistically insignificant, from a practical standpoint this could mean ability or inability to grasp an object. With this logic in mind, the reach capabilities of the elderly, as a uniform gender group, was compared with the reach capabilities of the younger population (NASA, 1978). It should be noted that only limited comparison was possible since the study dealing with the younger population included only limited reach data (vertical reach for males only).

The comparison indicated that elderly men (between the ages of 65 and 89 years) have a significantly lower vertical reach capability than their younger colleagues (age 25 to 40 years). It is very likely that even though other reaches, such as the horizontal reach, did not show any age effect for the ages included in this study (between 65 and 89 years), a significant difference will be observed when comparisons are made with the younger population; with the younger population having a larger reach capability.

The logical inference drawn above is largely based on the following findings: (1) the body size measurements (such as stature, sitting height, and bideltoid breadth) of the younger population (25 to 40 years age) were significantly larger in comparison to the older population studied in this work and (2) both vertical and horizontal reach angles changed substantially with age (the vertical reach angle increased with age and the horizontal reach angle decreased (details are not reported in ths paper)), particularly so for the age group 85 and above.

CONCLUSION

Based on the results and limited discussion, it is possible to conclude that age has a profound effect on reach capabilities of individuals. This is particularly true when individuals reach the age of 85 years. The discussion also leads us to conclude that elderly, as a whole, form a distinct population when compared with younger individuals who are in their 20s, 30s or 40s. Therefore, it is critical that designers consider the reaching capabilities of the elderly when designing working or living environments for them.

Given the dearth of studies in this area and the potential growth in the elderly population, it is recommended that large scale data collection studies be carried out in the near future.

REFERENCES

National Aerospace and Space Administration (1978). **Anthropometric Source Book,** Volumes I and II, Houston, Texas.

Clark, M. C., Czaja, S. J. and Weber, R. A. (1990). Older Adults and Daily Living Task Profiles. **Human Factors, 32,** 537 - 549.

Cjaza, S. J. (1990). Human Factors Research Needs for an Aging Population. Washington, DC: National Academy Press.

Kelly, P. and Kroemer, K. (1990). Anthropometry of the Elderly: Status and Recommendations. **Human Factors, 32,** 571 - 595.

Roebuck, J.A., Kroemer, K.H.E., and Thomson, W.G. (1975). **Engineering Anthropometry Methods,** Wiley Interscience, New York.

Editor's Note: For an updated report of this research, see Ursula Wright, "Reach Profiles of Men and Women 65-89 Years of Age," *Experimental Aging Research, 23*(4), 367-393.

REPEATED MEASURES BATTERY FOR THE AGED*

Richard H. Shannon, Ph.D.
General Physics Corporation
Columbia, Maryland

ABSTRACT

A battery of 29 reliable, valid and repeatable cognitive and psychomotor paper-and-pencil tests, with each test measuring a specific construct, was used to assess the performance of 48 older males and females. These subjects were divided into three separate age groups: 55-60, 65-70, and 75-80 years. In addition, a group of 16 men and women aged 25-35 served as a control group. This battery is divided into three sub-batteries (A, B and C) which were given on three separate weeks. The emphasis of this paper will be to describe the results of the nine tests contained in sub-battery C. Each test of a basic ability was analyzed separately across a total of five days and fifteen trials, with three trials being given each day. Total test time for each trial was approximately 35 minutes.

INTRODUCTION

Although there is a growing gerontological literature that demonstrates age-related differences across a wide variety of critical behavioral processes, any given study typically has assessed the impact of age on a very limited set of variables. Relatedly, most of these studies have utilizied experimental laboratory paradigms that have little utility for behavioral screening or assessment in everyday and/or clinical settings. The implications of this state of the geronotological literature are clear. The rich mosaic of age-related differences in inter-process relationships has gone largely unexplored, and with this comes the opportunity to develop effective but factor-efficient and easy-to-administer screening/evaluation tests. The goal of this project was to systematically examine in the same individuals age differences in the factor structure of a comprehensive range of behavioral processes.

One of the major shortcomings in the gerontological literature is that studies have not covered the full range of basic cognitive and perceptual abilities. The number of variables studied are narrow in scope. For this reason, the conclusions from most of these studies are limited as to the comparisons between age groups, their factor structure, and the possible explanations for changes in the factor structure. Cunningham (1980) feels that a more diverse and representative sample of basic abilities would be desirable in future studies of aging. The present project met this requirement by using a wide variety of tests that have construct validity. Another fundamental problem with the geronotological literature is that it has rarely been concerned with behavioral variability. For example, Willis and Baltes (1980) stated that up to a few years ago, the aging literature on intelligence did not separate or identify important components of variability which would assist in the identification of development changes and differences across age groups. Only a repeated measures design using numerous trials or equivalent forms, over a wide sampling of basic abilities, such as the design in this paper, can address this shortcoming.

METHOD

The research design used an existing battery of 29 paper-and-pencil tests that measured 23 cognitive and 4 psychomotor abilities. A repeated measures paradigm of 15 trials or forms was used to measure performance during periods prior to stability (time to establish equal means and correlations) and during stability (time to establish reliable test scores). Evaluation of problem-solving strategies, practice and learning effects, and reliable and stable performance for each construct for older individuals was then capable using this design.

In the initial design of the battery, construct validity was considered critical to test selection and development. The Kit of Factor-Referenced Cognitive Tests (Ekstrom, French, Harman & Dermen, 1976) outlines three tests, which were selected by experts for each of the 23 cognitive factors or constructs isolated by prior psychological research. Each of the tests in this Kit has two alternative forms. Grammatical Reasoning (Baddeley, 1968) and Digit Letter (Weschler, 1958) are not contained within the Kit but were selected for use in this battery because they have been observed to be sensitive to human behavioral changes and individual differences. In addition, Theologus and Fleishman (1971) isolated another nine psychomotor factors or abilities with appropriate stimulus items and apparatus equipment. Four of these psychomotor factors can be measured in a paper-and-pencil format.

In order to perform repeated measurements on a set of basic abilities, it is necessary that different and parallel forms of tests be

*Work was conducted under NIA Grant 7R43AG06516-01 while the author was employed by Human Systems Technology, Columbia, Maryland

developed. Each test within this battery was expanded to 15 alternate forms, with each form being developed to contain the same number of items with similar content, range, level of difficulty and format. This research required that 15 trials of each test be given over a five-day period with three different forms being given each day. Performance in general followed a pattern of having more change initially and of more consistent behavior following practice. In addition, subjects responded differently during the early stages of practice and became more uniform with repeated measurements. After a reasonable amount of practice, statistical stability among the subjects and tests could be demonstrated by analysis of the means and correlations across trials and days.

The research plan divided the battery into three parts or sub-batteries, which were given on three separate weeks. Each sub-battery contained either nine or eleven tests and was qualitatively balanced as to reasoning, perception, fluency, and psychomotor tests. This paper describes only the results using Battery C. The following nine tests which compose Battery C, are separated into groups that used the same measurement parameter (construct being measured and trial times are in parentheses): (1) number correct - symbols (figural fluency, 5 minutes), digit letter (perceptual speed, 2 min.), hidden words (verbal closure, 2-1/2 min.), word beginnings (word fluency, 3 min.); (2) number correct minus number missed/incorrect - form board (visualization, 4 min.), pursuit aiming (aiming, 1 min.); (3) number correct minus (.25) number incorrect - following directions (integrative processes, 2-1/2 min), mathematics aptitude (general reasoning, 6 min.), building memory (visual memory, 4 min.)

Forty-eight men and women subjects were divided equally into three age groups of 55-60, 65-70, and 75-80. In addition, sixteen men and women from a younger age sample (25-35) participated in the study as a control group. Each age by sex group contained eight subjects. Each test of a basic ability was analyzed separately across five days and fifteen trials. In all, there were nine independent analyses using the same design. Each test was randomly given in a different sequence for each trial in order to control for order effects. Test time for each trial was approximately 35 minutes. With all three trials being given, plus rest periods, a daily session took approximately two hours to administer. The total participatory time for each individual was approximately 30 hours (two hours per day for three weeks).

Subjects were paid volunteers who were recruited from the greater Baltimore, Maryland area by advertisements at local senior citizen centers and in newpapers. Subjects possessing the following characteristics were selected: U.S. Citizenship, spoke English well, good health, corrected vision of 20/30 or better, a minimum of a high school education and maximum of two-four years of college.

RESULTS

Analysis of Variance

Nine analysis of variance (ANOVA) tests were performed on two nested variables (two levels of sex, and four levels of age) in a repeated measures design across 15 trials and five days. The dependent variables were the performance scores on each of the paper-and-pencil tests. There were 64 subjects divided equally into eight sex/age groups containing eight subjects each. Analyses were conducted using this ANOVA design (Winer, 1970) in order to determine significant main and interactive effects. Pooling procedures were used in the ANOVA. All sources of interaction, that were not of interest to the experimenter or did not contribute significantly to the total variation, became part of the error term. Multiple comparisons of the means among the levels of the significant main and interactive effects were also performed using the Tukey test (Kirk, 1968). Care was taken during these analyses to avoid Type I errors (rejection of the null hypothesis when it is true). This error was partially controlled by varying the levels of significance with the number of observations tested.

Days. The three trials within a day were pooled and the resultant scores were compared for equality at the .01 level across days. Multiple comparisons of the daily means were performed in order to determine stability and the asymptotic level of the learning curve. Table 1 contains the daily means, daily average standard deviations of the pooled trials, and the stabilized days on each of the nine tests of Battery C. In addition, Figures 1 - 9 depict curves of the means of the four age groups across the five days on each of the nine tests. Most of these curves clearly show the periods prior to and during stability within each of these four groups, as well as the performance differences between the four groups. Effects within the days - sex and days - age interactions are not discussed in this paper due to space limitations. However, the stable periods of the age and sex groups within the nine tests are very similar to the stable periods of the total sample.

Trials. Since analysis performed on this data prior to stable performance could be misleading, the determination of alternate form equality should be made from a comparison of trial means across and within stabilized days. In general the tests, that statistically indicate the possibility of having equal forms at the trial level, were usually those developed using either a randomization process of equal stimulus items or the same form on each trial. There are four tests (digit letter, hidden words, pursuit aiming and building memory) considered to have equal alternate forms at the trial level.

Sex. When the results of the multiple comparison tests are compared across stabilized days, only mathematics aptitude has a significant difference (performance scores: males, 3.7; females, 2.4). The same trend is evident when the results of Batteries A and B are studied. Therefore, the sex variable is considered to have had little affect upon performance testing in this study.

Age. Table 1 contains the means and average standard deviations (of the pooled trials) within stabilized days of each test across the four age samples. This Table also depicts the results of the multiple comparison tests of these means using a .01 level of significance. In general, three clusters appear to emerge from these results: (1) those tests that have differences across most aged groups (55-80) and the younger control group (25-35); (2) those tests that have differences only with the older sample (75-80); and (3) those tests that show no differences across the four age groups. The nine tests of Battery C can be categorized as follows: Cluster 1 (form board), Cluster 2 (digit letter, following directions, pursuit aiming, building memory) and Cluster 3 (symbols, hidden words, mathematics aptitude, word beginnings).

These clusters of tests appear, in general, to follow a Fluid and Crystallized Intelligence model, as outlined by Horn and Cattell (1967). Fluid Intelligence tests tend to favor performance by younger subjects and represent abilities that contain novel information that is not based on culture. Tests that measure reasoning, short-term memory and spatial relations can be categorized under this heading. A related group to fluid abilities are those representing speed and motor coordination, such as aiming, tapping, and perceptual speed. Crystallized Intelligence tests tend not to discriminate among age groups and represent abilities that are influenced by prior learning and culture. Tests that observe performance on numerical facility, verbal comprehension and fluency, and long-term memory can be placed in this category.

Effects of the sex-age interaction are not discussed in-depth in the paper due to space limitations. However, in general, the analysis of the eight sex-age groups at a .01 level across stabilized days within the nine tests of Battery C indicate that (1) sex is not a major variable while age accounts for most of the variance, and (2) the results are similar to those at the main effects levels of sex and age.

Multivariate Analyses.

Of the two parameters (means and correlations) used in this study to determine stability, correlational stability is considered to be the more important for the following reasons: (1) the relative rank order among the subjects is established, (2) the reliabilities of the individual test scores are calculated, (3) all of the correlations within the stable period are considered to be equal or stable, and (4) the analyses of the means within a test are reduced to the established period of equality. Correlational analyses on each test were performed across the 64 subjects on the fifteen trials and the five days (by pooling the three trial scores by subject within a day). In addition, the performance scores of each subject on each of the 29 tests across stabilized trials and days were pooled. The resulting stable scores were then factor analyzed and rotated in order to produce a factor-analytic model for the data in this study.

Correlational analyses. The average correlation between the five days, the average correlation between stabilized days and the reliability estimate of stabilized days are the three reliability estimates of the data collected on the 64 subjects and outlined sequentially below in parentheses for each of the nine tests: pursuit aiming (.851, .894. .944), digit letter (.845, .867, .951), hidden words (.772, .820, .901), symbols (.779, .779, .946), mathematics aptitude (.625, .694, .872), building memory (.634, .683, .866), following directions (.538, .654, .791), form board (.507, .556, .883), and word beginnings (.338, .480, .649). The average correlation across the five days (unstabilized and stabilized) is a reliability index of the correlational stability within the matrix at the days level over time. The average correlation between stabilized days (unstabilized days are excluded) is utilized as the quoted reliability of one day's performance on a particular test. An estimated statistical reliability using the Spearman-Brown formula across stabilized days is determined from the reliability of one day's performance on a stabilized day. Reliability estimation increases as the number of pooled stabilized days increase on a particular test. For example, if the average correlation between stabilized days is .650, the Spearman-Brown estimate for two pooled days is .788, for three days is .848, for four days is .881, and for five days is .903.

Factor analysis. Factor analysis was performed on the stable scores for each aged subject (N=48) on each of the 29 tests. As seen from the data box, the abilities or tests are treated as variables, and subjects are cases. In this approach, the resultant factors are clusters of abilities as they covary over people. A Principal Components Analysis was conducted which tended to maximize the amount of variance shared commonly among the factors. Factoring was halted when the Eigenvalue slipped below 1.0. Accordingly, six factors explaining 76.8% of the variance emerged. Varimax rotation was performed so that each variable loaded mainly on only one factor. In this way, factorial interpretation is as simple as possible. The six factor solution with Battery C tests in parentheses are interpreted as follows: Factor 1 - Spatial Orientation (form board, building memory), Factor 2 - Verbal Content (hidden words, following directions, mathematical aptitude, beginning words, Factor 3 - psychomotor Speed and Precision (pursuit aiming), Factor 4 - Cognitive Speed (digit letter, hidden words), Factor 5 - Numerical Memory and Facility (none), and Factor 6 - Figural Fluency (symbols). In general, the test loadings on Factors 1, 3, and 4 appear to relate more to Fluid Intelligence, while the loadings on Factors 2, 5, and 6 seem to describe Crytallized Intelligence abilities.

DISCUSSION

Nine tests, which measure performance on a diverse and representative sample of basic abilities in a repeated measures design, evaluated individual differences across the independent variables of age, sex and time (trials and days). A repeated measures design in this research assisted in the determination on each test of: (1) the effects of practice and learning upon performance, (2) the equality of the alternate forms, (3) the stable period of performance across trials and days, (4) a stable and highly reliable score for each subject to be used in factor analysis, and (5) an average reliability estimate across stabilized days. This information was then used to select those tests that best discriminate performance differences across age groups and fit into the resultant factor structure. The sex variable was considered to have little influence upon performance testing in this study.

A distinction is made in this paper between a test having equal forms at the trials or days level. Equality at the trials level indicates the possibility of performance comparisons among the 15 trials of a test. Equivalence at the days level means that the three trials within a day have to be pooled prior to the comparisons of performance among the five days of a test. The significance of this information is that it lends support to future longitudinal studies on aged samples using either five or fifteen equal alternative forms.

The results of the factor analysis indicate that there are six factors that appear to follow a Fluid and Crystallized Intelligence model. It is recommended that two performance batteries be developed based upon these findings: One to evaluate individual levels of proficiency (crystallized) for job placement, and the other to determine the effects of age (fluid) upon performance. Using the criteria of reasonable loadings on the six factors, acceptable times to stability, and moderate average correlations (reliabilities) between stabilized days, six of the nine tests of Battery C are selected for inclusion in one of these two batteries as follows: Fluid battery containing tests having age differences (building memory, pursuit aiming, digit letter) and Crystallized battery not containing tests having age differences (hidden words, mathematics aptitude, symbols).

REFERENCES

Baddeley, A.D. (1968). A three-minute reasoning test based on grammatical transformation. Psychonomic Science, 10, 341-342.

Cunningham, W.R. (1980). Speed, age and qualitative differences in cognitive functioning. In L.W. Poon (Ed.), Aging in the 1980's: Psychological issues (pp. 327-331). Washington, D.C.: American Psychological Association.

Ekstrom, R.B., French, J.W., Harman, H.H., & Dermen, D. (1976). Manual for the Kit of Factor-Referenced Cognitive Tests, Princeton, N.J.: Educational Testing Services.

Horn, J.L., & Cattell, R.B. (1967). Age differences in fluid and crystallized intelligence. Acta Psychologica, 26, 107-179.

Kirk, R.E. (1968). Experimental design: Procedures for the behavioral sciences. Belmont, CA: Brooks/Cole.

Theologus, G., & Fleishman, E. (1971). Development of a taxonomy of human performance: Validation study of ability scales for classifying human tasks. Washington, D.C.: American Institute for Research.

Weschler, D. (1958). Measurement and appraisal of adult intelligence. Baltimore: Williams and Wilkins.

Willis, S.L., & Baltes, P.B. (1980). Intelligence in adulthood and aging: Contemporary issues. In L. W. Poon (Ed.), Aging in the 1980's: Psychological issues, (pp. 260-272). Washington, D.C: American Psychological Association.

Winer, B.J. (1971). Statistical principles in experimental design. New York: McGraw-Hill.

Table 1. Statistics for Day and Age Variables for Total Sample (N=64)

TESTS	DAILY MEANS AND AVERAGE STANDARD DEVIATIONS					STAB[(1)] DAYS	AGE GROUP MEANS AND AVERAGE STD. DEVIATIONS ON STABILIZED DAYS				AGE[(2)] COMP
	1	2	3	4	5		1 : 55-60	2 : 65-70	3 : 75-80	4 : 25-35	
Form Board	12.2 (9.8)	15.3 (12.9)	15.8 (12.1)	14.8 (11.3)	17.4 (11.9)	2-5	16.1 (10.3)	12.0 (7.9)	9.1 (7.7)	26.2 (11.2)	4-1, 4-2, 4-3
Symbols	11.5 (4.5)	11.5 (5.0)	11.3 (4.7)	11.6 (4.7)	11.3 (4.7)	1-5	12.6 (5.4)	12.3 (4.6)	10.2 (4.1)	10.9 (4.3)	NONE
Digit Letter	49.3 (15.7)	52.8 (14.1)	53.8 (14.7)	54.4 (15.2)	58.6 (16.0)	2-4	54.3 (11.7)	51.2 (11.8)	41.0 (10.0)	68.2 (10.8)	4-3
Hidden Words	23.0 (6.9)	27.9 (9.8)	30.1 (9.6)	32.5 (9.6)	33.6 (9.0)	4-5	33.8 (9.7)	33.7 (9.4)	26.7 (8.6)	38.4 (4.5)	NONE
Following Directions	2.2 (1.7)	2.0 (1.9)	1.9 (1.7)	2.5 (1.8)	2.3 (1.7)	4-5	2.8 (2.0)	2.1 (1.5)	1.3 (1.0)	3.6 (1.3)	4-3
Pursuit Aiming	88.2 (28.2)	98.4 (22.3)	100.5 (22.7)	102.2 (24.0)	103.8 (23.0)	4-5	102.7 (18.4)	99.4 (23.8)	86.1 (17.9)	123.8 (17.9)	4-3
Mathematics Aptitude	2.5 (2.0)	2.6 (2.5)	3.0 (2.2)	3.3 (2.5)	2.9 (2.5)	3-5	3.8 (2.9)	3.6 (2.4)	1.7 (1.4)	3.1 (1.7)	NONE
Building Memory	4.4 (3.0)	4.1 (2.8)	4.9 (3.2)	4.4 (2.9)	4.5 (3.1)	3-5	4.7 (3.3)	4.8 (2.3)	2.3 (2.2)	6.7 (2.6)	4-3
Word Beginnings	12.0 (5.8)	12.8 (5.7)	12.5 (4.6)	13.4 (4.7)	14.0 (4.3)	4-5	13.1 (4.5)	14.5 (4.7)	12.9 (4.0)	14.3 (3.7)	NONE

(1) Stabilized Days from Tukey Comparisons of Daily Means at .01 Level (2) Significant Age Comparisons of Means using Tukey Tests at .01 Level

Figures 1-9. Comparison of Age Across Days by Test

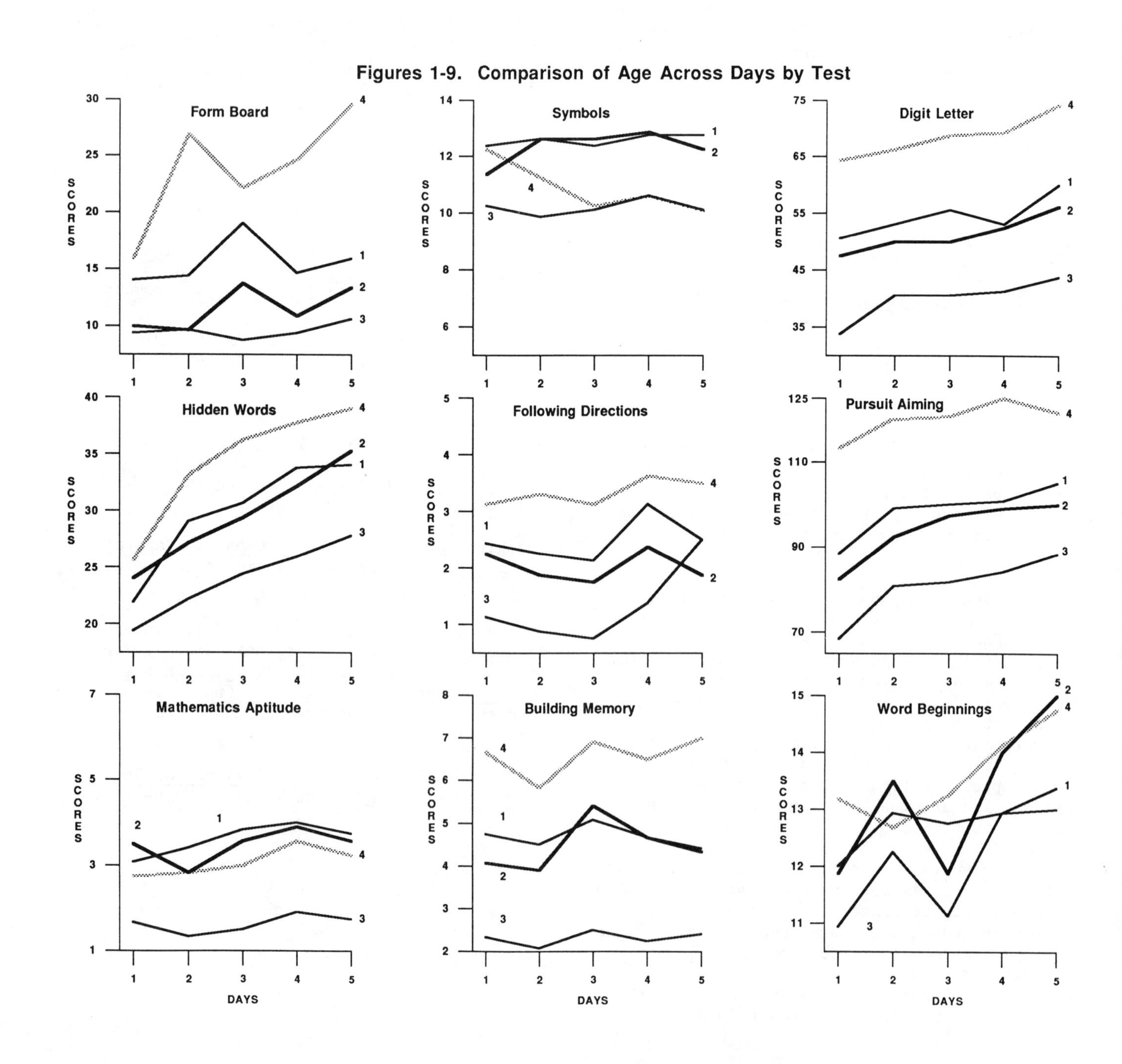

EFFECTS OF AGE, GENDER, ACTIVATION, STIMULUS DEGRADATION AND PRACTICE ON ATTENTION AND VISUAL CHOICE REACTION TIME

Max Vercruyssen[13], Michael T. Cann[13], Joan M. McDowd[23], James E. Birren[23], Barbara L. Carlton[123], Jane Burton[13], and P. A. Hancock[13]

[1] Human Factors Department, ISSM
[2] Department of Psychology
[3] Safety Science Department, ISSM
[4] Laboratory of Attention & Motor Performance, Gerontology
University of Southern California, Los Angeles, CA 90089-0021

ABSTRACT

This paper presents research conducted by the authors and others investigating the interaction of a variety of variables which are presumed to affect reaction time in hopes of obtaining much needed information on factors influencing age effects on attention and information processing. Reported is progress to date on an experiment which shows that the effects of age on central nervous system speed, as measured by visual choice reaction time, depends on many factors, including the gender, neural activation level, and skill of the subject as well as the stimulus quality and type of reaction task employed.

INTRODUCTION

While the effects of a variety of variables on speed of response and attentional capacity have been fairly well documented in previous research, the interaction of these variables has not been closely studied. Many studies may give incomplete and skewed pictures of the factors affecting information processing because they have failed to examine the 'big picture'. This paper, in addition to presenting a brief review of the literature concerned with speed of response and attention, will attempt to demonstrate that the analysis of a range of variables is crucial if one is to get an accurate picture of human information processing. Therefore, the purpose of this experiment was to determine the effects of age, gender, activation, degradation and practice on speed of response and attentional capacity.

BACKGROUND

Effects of Age. According to a compendium of 26 studies, from 20 to 60 years of age, simple reaction time (RT) slows by 20% (Birren, Woods, & Williams, 1980). Indeed, Birren (1965, 1974) has noted that as a consequence of their slowness of behavior, older adults may be living in a qualitatively and functionally different environment than younger people.

Effects of Gender. Botwinick (1984) has documented gender differences in the development and maintenance of set or expectancy in older individuals. It appears that aged men seem to demonstrate both a decreased ability to maintain set and a need for a longer period to recover from a poor set for short preparatory intervals with increasing age than do aged women.

With respect to RT, some investigators postulate that men are faster than women (e.g., Noble, Baker, & Jones, 1964) while others cannot find such effects (e.g., Botwinick & Thompson, 1967). Men are faster in midlife while women are faster at early and later ages. An alternative theory is that while men have been shown to have generally superior spatial skills (Waber, 1977), women tend to perform better on attention and vigilance tasks such as card sorting (Quinkert & Baker, 1984). This is an interesting issue because the effects of gender appear to interact with practice and age on RT tasks such that both old and young groups of women improve more than men with increasing amounts of practice on RT tasks using a variable foreperiod (Botwinick & Thompson, 1967).

Effects of Activation. The Ascending Reticular Activating System (ARAS) is responsible for maintaining a state of activation (Hebb 1955; Lindsley, 1951, 1958, 1970; Malmo, 1959) and certain attentional processes. Isaac (1960) found that stimulating the ARAS through increased sensory input or electrical impulses generally improved RT. deVries (1970) has also linked this stimulation to improved cognitive functioning. Of particular interest is a dissertation by Woods (1981) which examined the effects of activation as induced by postural changes (lying, sitting, and standing) on RT across age groups. She found a significant interaction between age and posture whereby older subjects performed significantly faster when standing than when lying, although such effects were not found for the young subjects.

Effects of Practice. Another variable which has been shown to influence the performance on information processing tasks is the level of training or practice given to the subjects. Brinley and Botwinick (1959) compared RT's of young and old men on simple and choice reaction tasks with four different PI intervals at differing levels of practice. They found that the additional practice sessions significantly reduced the mean age difference on the tasks. These results were replicated in a later study by Botwinick and Thompson (1967) as well as Falduto & Baron (1986).

In general, the large initial differences between age groups may reflect factors such as inexperience and test anxiety rather than true ability differences. Older adults do seem to exhibit the same qualitative changes as do young adults, and the qualitative relationship between the speed of behavior of old and young groups can be altered through practice. Indeed, some research shows that some of these initial patterns of age differences may disappear with extended practice (Falduto & Baron, 1986; Murrell, 1970).

Attention Capacity and Preparatory State. Preparatory set, as measured by RT has been used as a valid and sensitive measure of attention by studies in studies going as far back as Woodrow (1914) and Mowrer (1940). It seems that RT is affected not only by the immediate PI but also by the range and length of the other PI's used within a particular block of trials. Attempts have been made to identify the specific length of preparatory interval after which attentional capacity and expectancy are at their maximum levels. In an early experiment, Woodrow (1914) found that the optimal length of PI in a discrete simple reaction task was 2 sec in that it yielded the fastest RT in comparison with RT's produced by PI's ranging from 1 to 24 sec. These results were supported by studies of Woodrow (1930) and Woodworth (1938). Several researchers have found a steady increase in RT as the PI increases after the 'optimal PI' of 2 sec (Botwinick, Brinley, & Birren, 1957; Karlin, 1959; Klemmer, 1957; Woodworth, 1938; Woodrow, 1914, 1930).

It has been theorized by Woodrow (1914) and Karlin (1959) that this direct linear relation between RT and PI reflects progressive changes in the subject's state of readiness. Thus, as the PI increases, the subject becomes increasingly less prepared for the arrival of the forthcoming stimulus. This is consistent with Alegria's (1975) view that the state of preparation caused by the warning signal is dissipative in nature. Theoretically, by extending the preparatory period over a long enough period, corresponding delays in RT can be shown to be indicative of differential attentive states.

Research on expectancy has typically used simple RT. While this measure has been informative, even more insight might be attained using choice RT because of its sensitivity and utility with numerous models of cognitive functioning. Therefore, our research (like that of Green, Smith, & von Gierke, 1983) manipulates the response-stimulus interval (RSI) on a variety of choice reaction tasks (CRT). These measures include a variable CRT (VCRT) in which the length of the RSI is randomly varies between 0, 1, 2, 3, and 6 ms, and a serial choice reaction task in which the RSI is held constant at zero. It is hoped that by employing multiple measures, the most sensitive measure to the effects of each respective independent variable can be identified.

METHOD

Subjects. Four groups of healthy, non smoking volunteers, including four older men ($\underline{M}$ = 72 yrs), four older women ($\underline{M}$ = 73 yrs), eight younger men ($\underline{M}$ = 24 yrs) and eight younger women ($\underline{M}$ = 24 yrs), were recruited for this experiment. The older subjects were affiliated with the Andrus Gerontology Center at the University of Southern California (USC) and can be considered above average in level of mental and physical functioning for their age groups. The young subjects were recruited USC undergraduates and graduate students.

Apparatus. The apparatus consisted of a Commodore 64 computer together with a XETEC 20 megabyte hard drive controlled by a DCS software package (Wheeler, Vercruyssen, & Olofinboba, 1988) and a pair of subject response boxes consisting of two subminiature SPDT lever switches (Radio Shack catalog # 275-06) requiring 600 grams of pressure to activate. This resistance allowed for rapid depression of the switch by both age groups while preventing accidental activation. Each of the four microswitches corresponded directly to one of four possible stimuli appearing on the video monitor placed in front of the subject. The imperative stimuli consisted of a horizontal arrow which pointed either to the right or to the left while appearing in one of two adjacent boxes displayed on the monitor.

Independent Variable. The subjects performed the tasks over a two day period with the first day's data corresponding to the naive state and the second day's corresponding to the skilled state. Two age groups were studied with a young group mean of 24 years and an old group mean of 73.5 years for both males and females. Postural stance was used to induce changes in level of activation. The standing condition corresponding to the high activation while the sitting condition corresponded to the low level of activation. The actual stimulus was either degraded or intact, with the degraded image consisting of the intact image surrounded by and suffused within a tight array of dots. The array of dots served as a masking grid which would theoretically attenuate the stimulus' perceptibility.

Dependent Measures. The subjects performed a variety of four choice visual reaction tasks entailing either variable foreperiods of 0, 1, 2, 3, or 6 ms (VCRT) or discrete foreperiods of 0 ms (SCRT). In the VCRT task equal numbers of the different RSI's randomly preceded the stimulus. Heart rate measure was collected five times for each subject on every task, and was then averaged to obtain a mean heart rate for that task.

Procedure. The subjects were instructed to react as fast as possible while maintaining a ten percent error rate by rapidly depressing and releasing the microswitches upon the arrival of the stimulus. The subjects were also told to maintain a light touch on the microswitches to prevent unrealistically long RT's due to potentially long movement times. The maintenance of a ten percent error rate was particularly important due to the need to control for a potential speed accuracy trade off which could invalidate any observed reaction time differences.

Each subject performed a total of approximately 2400 reaction time trials which were spread across the various conditions. The subjects were asked to fill out both health questionnaires and consent forms on day before the day's testing began. Also, the subject was allowed 25 trials on each task (degraded and intact stimulus) before data collection on day to familiarize them with the nature of the task. In addition to the actual test trials each subject performed a series of practice blocks after the first day of testing and before the second day of testing.

Treatment of Data. The data were analyzed according to Age_2 x $Gender_2$ x $Degradation_2$ x $Days_2$ x $Activation_2$ x RSI_5 mixed ANOVA design with repeated measures for the last four factors. Descriptive statistics were produced by IBMDAS with inferential statistics done by BMDP and ANOVR on a Compaq 386/387. Post hoc analyses were according to a Tukey WSD procedure All statistical contrasts were made at the .05 level of significance.

RESULTS

SCRT. The main effects indicated that younger subjects ($\underline{M}$ = 462) were significantly faster than older subjects ($\underline{M}$ = 653; $\underline{F}_{1,20}$ = 67.3; $\underline{p}$ < .00005). The degraded stimulus ($\underline{M}$ = 582) produced significantly slower RT's than the intact condition ($\underline{M}$ = 470; $\underline{F}_{1,20}$ = 65.7; $\underline{p}$ < .00005). The ANOVA also showed a significant main effect for practice with skilled subjects ($\underline{M}$ = 490) performing significantly faster than naive subjects ($\underline{M}$ = 561; $\underline{F}_{1,20}$ = 61.1; $\underline{p}$ < .00005). SCRT also showed a main effect for activation with standing subjects ($\underline{M}$ = 520) faster than sitting subjects ($\underline{M}$ = 532; $\underline{F}_{1,20}$ = 4.5; $\underline{p}$ < .05).

An interaction between practice and age was obtained ($\underline{F}_{1,20}$ = 5.2; $\underline{p}$ < .03) such that the difference between the young ($\underline{M}$ = 492) and old ($\underline{M}$ = 700) on day 1 was greater than the was the difference between the young ($\underline{M}$ = 433)and old ($\underline{M}$ = 605) on day 2. This interaction takes the same form as the interaction between practice and age for VCRT.

Age also interacted with degradation ($\underline{F}_{1,20}$ = 5.15; $\underline{p}$ < .03) such that the difference between the young ($\underline{M}$ = 508) and old (M = 728) on the degraded task was greater than the was the difference between the young (M = 417) and old (M = 577) on the intact task. This interaction takes the same form as the interaction between degradation and age for VCRT (Figure 4).

SCRT also produced a marginally ($F_{1,20}$ = 3.21; $p < .08$) significant interaction between practice and activation. Standing (M = 550) produced faster RT's than did (M = 572) sitting on day 1 while on day 2 the differences between standing (M = 490) and sitting (M = 491) disappeared.

There was an interaction between gender and degradation ($F_{1,20}$ = 7.76; $p < .011$) such that during the degraded task females (M = 551) were faster than males (M = 613), while on the intact task there was little difference between the speed of females (M = 474) and males (M = 465). This interaction is shown in Figure 1.

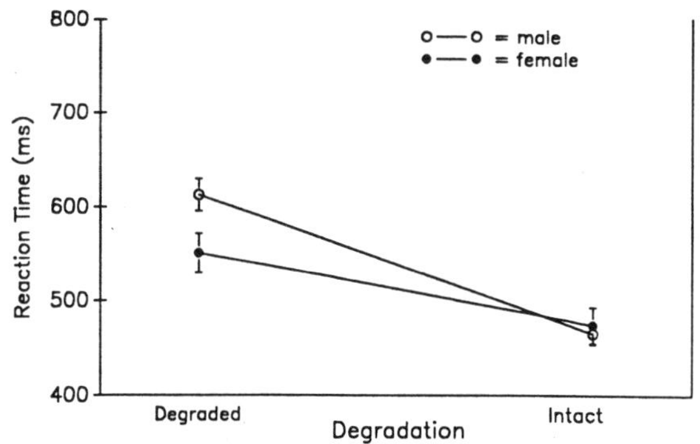

Figure 1. Mean RT and SEM for SCRT as a function of gender and degradation.

VCRT. The main effects indicated that younger subjects (M = 517) were significantly faster than older subjects (M = 679; $F_{1,20}$ = 40.3; $p < .00005$). The degraded stimulus (M = 632) produced significantly slower RT's than the intact condition (M = 510; $F_{1,20}$ = 85.36 $p < .00005$). The ANOVA also showed a significant main effect for practice with skilled subjects (M = 540) performing significantly faster than naive subjects (M = 601; $F_{1,20}$ = 62.24 $p < .00005$).

RSI produced a significant main effect with the RSIs of 0 seconds (M = 586) and 6 seconds (M = 574) being slower than the RSI of 2 seconds (M = 562) at p < .02 (F_4 = 4.4). (The 2ms RSI produced the fastest response times of any RSI.) This relationship is presented in Figure 2.

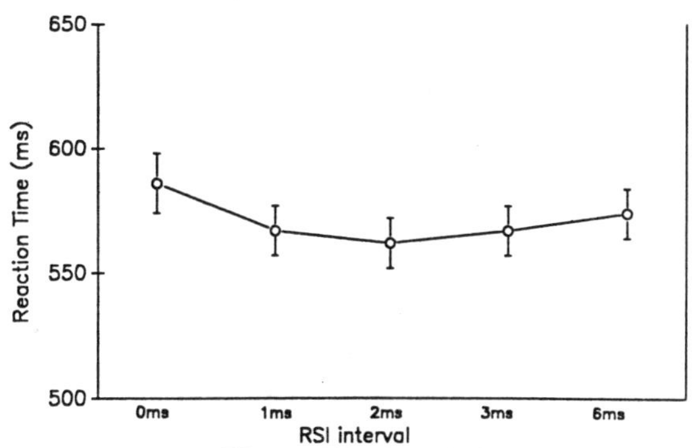

Figure 2. Mean RT and SEM for VCRT as a function of response–stimulus interval

An interaction between practice and age was obtained ($F_{1,20}$ = 5.2; $p < .03$) such that the difference between the young (M = 537) and old (M = 730) on day 1 was greater than the was the difference between the young (M = 497) and old (M = 627) on day 2 and is described by Figure 3.

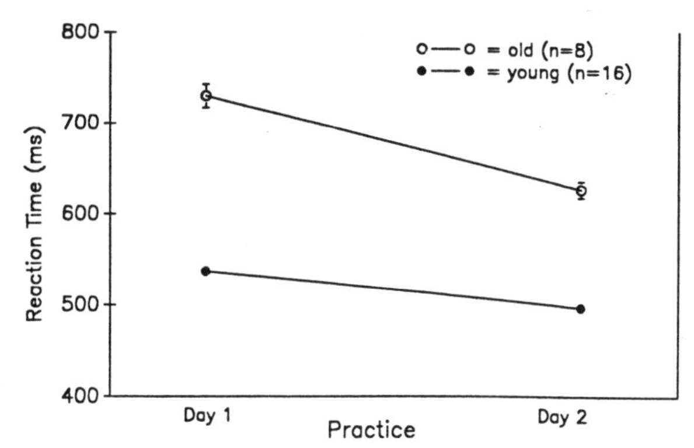

Figure 3. Mean RT and SEM for VCRT as a function age and practice.

Age also interacted with degradation ($F_{1,20}$ = 8.88; $p < .0074$) such that the difference between the young (M = 563) and old (M = 769) on the degraded task was greater than the was the difference between the young (M = 471) and old (M = 588) on the intact task and is displayed in Figure 4.

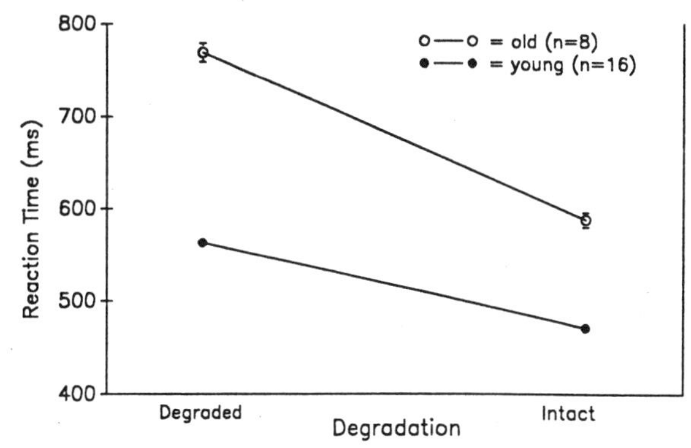

Figure 4. Mean RT and SEM for VCRT as a function of age and degradation.

The ANOVA also produced an interaction between degradation and practice for VCRT ($F_{1,20}$ = 16.65; $p < .0006$). There was a greater difference between the intact (M = 526) and degraded (M = 677) tasks on day 1 than there was between the intact (M = 493) and degraded (M = 587) tasks on day 2. This interaction is displayed in Figure 5.

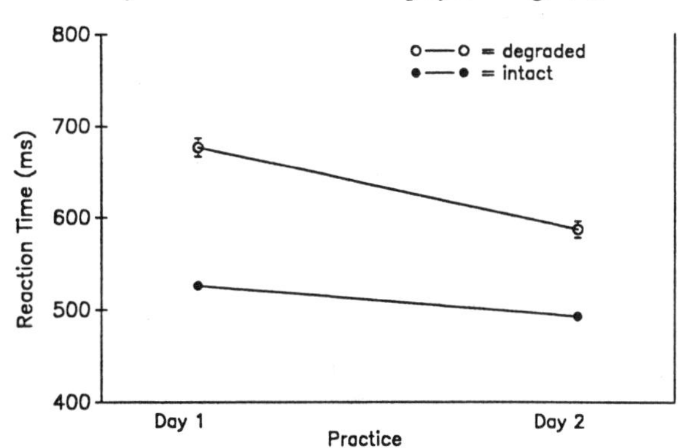

Figure 5. Mean RT and SEM for VCRT as a function of practice and degradation.

An interaction between gender and degradation was revealed ($F_{1,20}$ = 5.55; p < .028). During the degraded task the females (M = 551) were faster than the males (M = 613), while on the intact task there was little difference between the speed of the females (M = 474) and the males (M = 465). This interaction is displayed in Figure 6. A marginally significant three way interaction between RSI, age, and gender was obtained by VCRT ($F_{1,20}$ = 2.27; p < .0692). This result is shown in Figure 6.

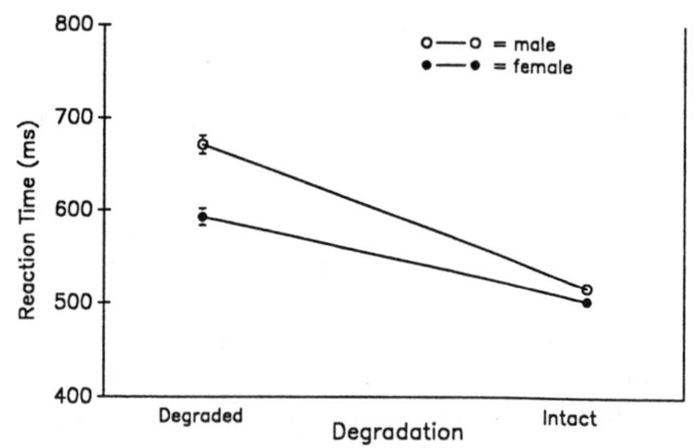

Figure 6. Mean RT and SEM for VCRT as a function of gender and degradation.

A marginally significant three way interaction between RSI, age, and gender was obtained by VCRT ($F_{1,20}$ = 2.27; p < .0692). This result is shown in Figure 7.

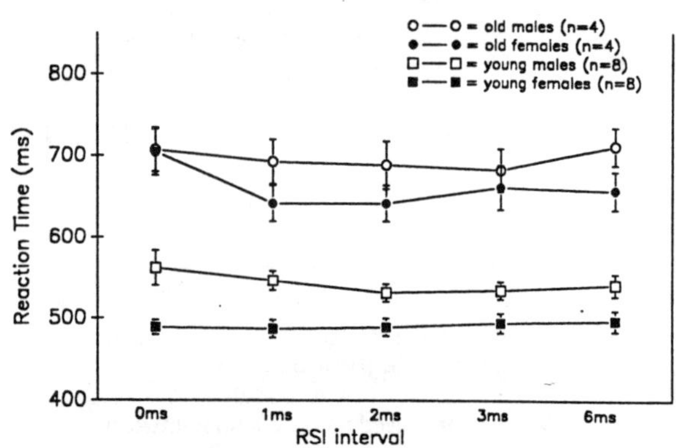

Figure 7. Mean RT and SEM for VCRT as a function of RSI, age, and gender.

DISCUSSION

Speed of Response. An interesting finding is that SCRT and VCRT, while revealing many of the same effects, appear to be sensitive to different effects. For instance, VCRT reveals an interaction between degradation and practice (Figure 5) which is not revealed by SCRT. Likewise, SCRT appears sensitive to the interaction between practice and activation which is not shown by VCRT. These results indicate that reaction time tasks may not be reflecting the same cognitive processes.

The main effects for both SCRT and VCRT support the results of past speed of response research. Age effects are extremely well defined, with younger subjects performing far faster than the older subjects. The results of the degradation by age interaction seems to indicate that the older subject have proportionally greater impairment of feature extraction/encoding processes in comparison with younger subjects.

Gender. A major result of the current study is the finding that gender interacts with stimulus degradation. While the interactions of degradation with age and with practice were expected and easily explained in light of past studies involving these variables, the interaction of degradation with gender revealed by both SCRT (Figure 1) and VCRT (Figure 6) seems to be a more novel finding. The interactions behave in the same general way, with men performing slower than females on the degraded task condition while performing with relatively equal speed on the intact condition. It seems that men have difficulty with early stages of processing (ie. stimulus encoding/feature extraction) in comparison with women. In neither measure were any interactions revealed between age groups. It seems that these effects are therefore not restricted to either young or old age groups but instead seem to indicate general gender characteristics.

Activation. Neural activation as facilitated by posture has received tentative support as a moderator of RT. The main effect of activation as measured by VCRT indicates faster RTs when standing than when sitting. While the interaction between posture and practice supports this finding, by day 2 these differences are attenuated. Neural activation may influence the extent to which an organism can access the attentional capacity already developed. Presumably, higher activation levels improve speed of response by allowing greater access to attentional resources.

Practice. Other practice effects include the reduction of degradation differences from day to day 2 on VCRT as indicated by Figure 5. The reduction of age related differences with practice as indicated by Figure 3 supports the notion that practice has the general effect of reducing initial differences between groups by differentially improving performance more in the slower group than in the faster group. It is important to note that practice exerts an extremely powerful influence and, as indicated in the introduction, can often attenuate or eliminate differences caused by virtually all other between group factors.

Attention. The main effects of RSI (Figure 2) suggest that the optimal attentional capacity as indicated by the swiftest reaction times occur at the RSI of 2 seconds. The slower RT witnessed at the 0 second RSI implicates a difficulty in the immediate generation of the preparatory state which is presumed to precede the response on choice reaction tasks. While at RSIs of this short length, the subject has still not fully recovered from the last response. However, at the two second interval, the subject is able to clear his or her processing system for the arrival of the next stimuli. The slow RT witnessed at the 6 second RSI is likely due to the fact that the subject has allowed his concentration to wander. It therefore seems that the differential performance following this interval is likely to implicate attentional capacity. With this in mind, the three way interaction between RSI, age, and gender, (Figure 7) seems to implicate gender differences in maintenance of preparatory state. Men seem to be less able to maintain set across the 6 second RSI than are the women, with more robust effects occurring in the older group. These results Botwinick's 1984 findings discussed in the introduction which indicate similar age by gender interactions in the maintenance of set.

Conclusions. The effects of aging on CNS functioning,

as measured by reaction time and preparatory "set" states, depends in part on the age, gender, level of neural activation, and skill of the performer. All of these variables have received support as influencing reactive capacity and research on the multiple interactions of experimental variables is absolutely essential. Until we understand the whole picture it seems naive to consider any one of these factors in isolation.

Attention reduces the likelihood of the misinterpretation or failure to process information. By understanding the factors affecting attentional capacity in humans, particularly the aging population, we can more effectively design new products, systems and technologies. This could allow for both increased productivity and the reduced risk of injury by controlling the conditions under which the neglect or misperception of important informational cues could occur. These findings could theoretically have the greatest impact on the growing older populations which seem to exibit the greatest variablility in attentional and reactive capacity.

REFERENCES

Alegria, J. (1975). Sequential effects of foreperiod duration: Some strategical factors involving time uncertainty. In P. Rabbitt & &S. Dornic (Eds.), Attention and Performance V. London: Academic Press.

Birren, J. (1965). Age changes in speed of response: Its central nature and physiological correlates. In A. Welford & J. Birren (Eds.) Behavior, Aging and the Nervous System, Springfield: Charles C. Thomas.

Birren, J. (1974). Translations in gerontology-From lab to life: Psychophysiology and the speed of response. American Psychologist, 29, 808-815.

Birren, J.E., Woods, A.M., Williams, M.V. (1980). Behavioral slowing with age: Causes, organization and consequences. In L.W. Poon, Aging in the 1980's: Psychological issues, Washington, D.C.: American Psychological Association, 293-308.

Botwinick, J. (1984). Aging and behavior: A Comprehensive Integration of Research Findings, NY: Springer.

Botwinick J., Brinley J., & Birren, J. (1957). Set in relation to age. Journal of Gerontology, 12, 300-305.

Botwinick, J., Brinley, J., & Robbin, J. (1959). Maintaining set in relation to motivation and age. American Journal of Psychology, 72, 140.

Botwinick, J. & Thompson, L. (1967). Practice of speeded response in relation to age, sex, and set. Journal of Gerontology, 22, 72-76.

Brinley, J. & Botwinick, J. (1959). Preparation time and choice in relation to age differences in reponse speed. Journal of Gerontology, 14, 226-228.

deVries, H.A. (1970). Physiological effects of an exercise training regimen upon men aged 52 to 88. Journal of Gerontology, 25, 325-336.

Falduto, L. & Baron, A. (1986). Age related effects of practice and task complexity on card sorting, Journal of Gerontology, 41(5), 659-661.

Green, D.M., Smith, A.F., & von Gierke, S.M. (1983). Choice reaction time with a random foreperiod. Perception and Psychophysics, 34(3), 195-208.

Hebb, D.O. (1955). Drives and the C.N.S. (Conceptual Nervous System). Psychological Review, 62, 243-254.

Isaac, W. (1960). Arousal and reaction time in cats. Journal of Comparative and Physiological Psychology, 53, 234-236.

Lindsley, D.B. (1951). Emotion. In S.S. Stevens (Ed.), Handbook of experimental psychology. New York: Wiley.

Lindsley, D.B. (1958). The reticular activating system and perceptual integration. In D.E. Sheer (Ed.), Electrical stimulation of the brain. Austin, Tx: University of Texas Press.

Lindsley, D.B. (1970). The role of reticulothalmocortical systems in emotion. In P. Black (Ed.), Physiological correlates of emotion. New York: Academic Press.

Malmo, R.B. (1959). Activation: A neurophysiological dimension. Psychological Review, 66, 367-386.

Mowrer, O.H. (1940). Preparatory state (expectancy)--Some methods of measurement. Psychological Monographs, 52, (2, whole no. 233).

Murrell, F. (1970). The effect of extensive practice on age differences in reaction time. Journal of Gerontology, 25, 268-274.

Noble, C.E., Baker, B.L., & Jones, T.A. (1964). Age and sex parameters in psychomotor learning. Perceptual and Motor Skills, 19, 935-945.

Quinkert, K. & Baker, M. (1984), Effects of gender, personality, and time of day on human performance. In A. Mital (Ed.), Trends in Ergonomics/Human Factors I. North Holland: Elsevier Science Publishers.

Salthouse, T.A. (1979). Adult age and the speed-accuracy tradeoff. Ergonomics, 22(7), 811-821.

Sherman, J. (1967). Problem of sex differences in space perception and aspects of intellectual functioning. Psychological Review, 74, 290-299.

Waber, D.P. (1977). Sex differences in mental abilities, hemispheric lateralization, and rate of physical growth at adolescence. Developmental Psychology, 13, 29-36.

Welford, A. T. (1958). Aging and Human Skill. London: Oxford University Press.

Welford, A.T. (1977). Motor Performance. In J.E. Birren and K.W. Schaie (Eds). Handbook of the psychology of aging. New York: Van Nostrand Reinhold Co.

Woodrow, H. (1914). The Measurement of Attention. Psychological Monographs, 17, (5, whole no. 76).

Woods, A.M. (1981). Age differences in the effects of physical activity and postural changes on information processing speed. Doctoral dissertation, Psychology Department, University of Southern California, L.A., CA 90089.

Woodworth, R.S. (1938). Experimental Psychology. New York: Holt.

AGING, REACTION TIME, AND STAGES OF INFORMATION PROCESSING

Max Vercruyssen[1,2], Barbara L. Carlton[2,3], and Virginia Diggles-Buckles[2]

[1] *Human Factors Department, ISSM, University of Southern California, Los Angeles, CA 90089-0021*
[2] *Lab of Attention and Motor Performance, Gerontology, University of Southern California, Los Angeles, CA 90089-0191*
[3] *Department of Defense, Polygraph Institute, P.O. Box 5310, Fort McClellan, AL 36205-5310*

ABSTRACT

Using Sternberg's (1969) Additive Factors Method (AFM), previous investigations in search of the locus of age-related slowing in reactive capacity have found conflicting results possibly due to inconsistencies in research methodologies. This experiment was conducted to examine age differences in the performance of AFM intratask manipulations of a reaction time task using both fixed and variable foreperiod conditions with subject testing at both naive and practiced skill levels. Twenty male subjects, ten young and ten old, performed a visual four-choice RT task with intratask manipulations of stimulus-degradation, stimulus-response compatibility, and response-stimulus intervals (RSIs were fixed at 0, 2, and 5 sec and variable with random presentations at 0, 2, and 5 sec), once when subjects were naive and again when practiced. The results varied by level of practice and RSI, but clearly the older subjects had difficulty with the intratask manipulations. The older subjects took twice as long, on the average, to respond. Interactions of age by compatibility suggest that, according to the AFM, with age comes inordinately long delays in the response selection stage of information processing. Conclusions are made with caution since this research points to limitations and methodological confounds which serve to explain many of the equivocal findings in previous studies.

INTRODUCTION

This paper provides a brief overview of three areas of research interest: age-related slowing of CNS functions as measured by reactive capacity, the effects of age on stages of information processing, and the influence of practice on the reaction time findings from the first two areas, in a fashion capable of addressing some of the equivocal findings in the literature. To date, little has been done to examine these areas simultaneously.

Age-related Slowing of Behavior. One of the most robust effects of aging is a general slowing of behavior (Birren, Riegal, & Morrison, 1962; Birren, Woods, & Williams, 1979, 1980), with the effects manifest in central rather than peripheral processes (Simon, 1968; Welford, 1977, 1980). While much of the early research was conducted using simple reaction time (e.g., Botwinick, Brinley, & Birren, 1957), the age effects are especially pronounced with choice reaction time (Botwinick, 1984). In general, the more complex the task, the greater the predicted age-related slowing.

Stages of Information Processing. Sternberg's (1969) additive factors method (AFM) has been extremely helpful as a tool for developing a model of information processing and speed of response. AFM assumes information is processed through serial stages, each requiring some time for processing, thereby making the overall RT a sum of the individual processing times for each stage. If a task manipulation affects one stage, the overall RT will be longer as a result. The strength of this model lies in the assumption of orthogonal stages. In a factorial experiment, task manipulations are additive if, and only if, no interactions are found between the intratask variables. Significant interactions between independent variables indicates that the task manipulations are not affecting individual component stages but are exhibiting an influence on RT at one, or more, common stages (Moraal, 1982ab; Sanders, 1977, 1980, 1983; Sternberg, 1969; Vercruyssen, 1984). The AFM has been successfully applied to determine the specific stages influenced by amphetamine and barbiturates (Frowein, 1981ab), sleep deprivation (Sanders, Wijnen, & von Arkel, 1982), alcohol (Huntley, 1972, 1974; Tharp, Rundell, Lester, & Williams, 1974), and toxic gases (Vercruyssen, 1984).

Several studies have used the AFM to examine age-related effects of behavioral slowing, however, the results have been equivocal. Simon and Pouraghabagher (1978) included two intratask manipulations using a 2-choice RT task, to ascertain the locus of slowing. The first intratask manipulation involved discriminability or stimulus quality and was designed to affect stimulus encoding. The second intratask factor manipulated the direction of a relevant/irrelevant cue and was designed to affect response selection. The results showed the usual main effects of age and stimulus quality, as well as, a significant interaction between stimulus quality and age. This interaction was in the direction of greater age differences between the young and old occurred when the stimulus was degraded (i.e., lower discriminability) than when the stimulus was presented in an intact form, thereby leading the authors to conclude that the locus of age-related behavioral slowing was central, not peripheral, specific in nature, rather than general, and impacted the earlier, stimulus encoding stage of information processing. The investigators went on to show that the age by stimulus quality interaction was not merely a sensory problem by demonstrating the absence of an interaction between age and stimulus quality when the task simply required stimulus detection. However, failing to find significant main effects of the irrelevant directional cue conditions questions the validity of this intratask manipulation to serve as a measure reflecting activity in a particular stage of information processing, and hence, may not applicable within the framework of the AFM.

Salthouse and Somberg (1982), using the AFM, included three intratask manipulations: stimulus degradation, response type, and comparison set size. The results found main effects of the intratask manipulations and age as well as interactions of each intratask manipulation with age leading them to the conclusion that age-related behavioral slowing of a response was general in nature and may reflect an overall speed reduction in central nervous system activity. However, this experiment also yielded a significant interaction between the intratask manipulations themselves which complicates the issue somewhat since failure to select intratask factors which are independent limits our ability to make inferences about effects on stages of information processing. At least in the area of stressor research, the intratask factors must be additive before administering the stressor (e.g., Sanders, 1980; Vercruyssen, 1984). A stronger statement might have been made if intratask manipulations were used which did not interact when testing young subjects.

Finally, Moraal (1982ab) used the AFM to examine the locus of age-related slowing with two intratask manipulations of stimulus degradation and S-R compatibility. Moraal found main effects of age, stimulus degradation and S-R compatibility with a significant interaction of age by stimulus degradation. Moraal concluded that the nature of age-related behavioral slowing was, indeed, general in nature, in spite of the significant stimulus degradation by age interaction. The conclusions based on this research were consistent with Salthouse and Somberg (1982). Moraal decided that although this interaction was significant it was still too small to indicate a specific locus of the slowing.

The effects of Practice on RT. Several investigations have shown that age differences in performance are affected by allowing the subject to practice a given task (Birren, Woods, & Williams, 1979, 1980). In fact, Birren (1964) states that the actual capability of an individual are not necessarily reflected in performance without the benefit of the learning that occurs with practice. Practice appears to be especially important when evaluating age differences in performance. Botwinick and Thompson (1967) demonstrated that under certain conditions, e.g., preparatory intervals near 500 msec, older subjects benefitted more from practice than did the younger group.

A second finding emerged from Moraal's experiment which was actually first questioned by Botwinick et al. (1957) which involves the use of a variable foreperiod in the reaction time task. Moraal included one condition in which successive presentations of the stimuli were separated by a variable response-stimulus-interval, in that the subjects could not predict the presentation of the next stimulus as they could in the fixed RSI conditions. Moraal found main effects for the variable RSI condition; subjects performed worse when the RSI was variable than when it was fixed. However, there was no interaction with age, the variable foreperiod affected both the young and the old subjects. Botwinick et al., however, demonstrated that the older groups' performance suffered more with the variable RSI condition than did the younger subjects' performance. Specifically, when a short RSI directly followed a long RSI, RT was higher than when the reverse was true. Botwinick, et al. speculated that the older subjects had a more difficult time recovering from the "set" of the first RSI, that subjects were not expecting a short interval and had a more difficult time recovering from this incorrect expectation.

Therefore, the purpose of the present experiment was to determine the effects of age and practice on choice reaction time while applying the AFM to examine stages of information processing and manipulating the response stimulus interval to monitor age differences in expectancy.

METHOD

Subjects. Twenty healthy male recruited volunteers, ten young (18-24 yrs) and ten old (69-82 yrs), served as subjects in this experiment. The young subjects were recruited from an Introductory Psychology class at the University of Southern California (USC) for extra credit. Subjects for the older group were recruited from volunteer organizations on the USC campus (i.e., Andrus Volunteers Organization and the USC Emeriti Center) and through a mailing process in which USC Alumni were encouraged to support ongoing research at USC. All of the subjects had normal or corrected to normal vision.

Stimulus and Apparatus. The visual display consisted of two adjacent boxes approximately 4.5 cm in height and 5 cm in width presented on a computer monitor. The imperative stimuli consisted of horizontal arrows presented within the borders of the boxes pointing either to the left or to the right. The display was presented at eye level (with subjects seated) approximately 60-75 cm away from the subjects, who sat facing the monitor. Directly in front of the subject were a pair of two-response units made up of small, blue plastic boxes, each containing two microswitches which were operated by the index and middle fingers of each hand. The presentation of the stimuli and the collection of the data were both performed by a Commodore 64 computer with a Xetec Lt. Kernal 20 megabyte hard disk system running DCS 1.05 software (Wheeler, Vercruyssen & Olofinboba, 1988). Data were then transferred to IBM compatible microcomputers for analyses and storage.

Stimulus Degradation. The stimulus was presented in each of two forms: intact and degraded. Stimulus degradation was achieved by superimposing a computer-generated random visual noise mask of the same color as the background over the arrow image. For each presentation of the stimulus, the mask would rotate so that the integrity of arrow diminished and the contours became difficult to distinguish while maintaining a constant amount of stimulus degradation and intensity over trials.

Stimulus-Response Compatibility. Compatibility mapping of the stimulus presentation and the required response was either 'high' or 'low'. During the 'high' compatibility condition the relationship between the stimulus presentation and the response was direct, spatial, one-to-one in correspondence. The subject response units were placed side by side and the right hand unit corresponded to the right stimulus box with the left hand unit to the left stimulus box. The direction of the arrow indicated which finger, either the index or middle finger, was to be used in making the response. If an arrow appeared in the left box, pointing to the right, the correct response would be depressing the key under the index finger of the left hand. In the 'low' compatibility condition, the subjects arms were crossed using an arm rest under the left arm to elevate the hand and arm. The relationship between the stimulus box, the direction of the arrow and the correct response was also changed. During this condition, the stimulus box indicated which finger would be used for the response and the direction of the arrow indicated which hand would be used. An arrow appearing in the left box was associated with a middle finger and the right indicated an index finger response. The direction of the arrow indicated which hand was to be used. With the arms crossed, an arrow pointing the right meant the farthest hand to the right which would be the left hand. For example, for an arrow appearing in the left box, pointing in the left direction, the correct response would be the middle finger of the right hand.

Response-Stimulus Interval. Along with stimulus degradation and S-R compatibility, four different RSI conditions were included. The following four RSI conditions were used: (a) **Fixed--RSI=0**, or serial choice RT, the completion of a response signalled the presentation of the next response, 40 trials; (b) **Fixed--RSI=2**, a two-sec foreperiod, 40 trials; (c) **Fixed--RSI=5**, a five-sec foreperiod, 40 trials; (d) **Variable--RSI=0,2,5**, the RSI consisted of a range of foreperiods, i.e. 0 sec, 2 sec and 5 sec. A total of 120 trials (40 of each) were presented in a random manner.

Procedure. The experiment was conducted over a two-day period. On the first day, subjects filled out medical health forms to determine the general level of health and completed informed consent forms. The subjects were then seated in a sound-attenuated chamber in front of the response units and the stimulus display monitor. The subjects were given two series of training conditions consisting of approximately 10 trials each (10 presentations of the arrow) in which to learn and become acquainted with the intratask manipulations. After the training conditions the subjects were tested on all 16 conditions. At the end of the experimental blocks, subjects were given an additional 15 mins of practice. The first session lasted approximately 90-120 mins. The second session was similar to the first with the exception that the first 15 mins were devoted to re-training of the task manipulations and practice. The subjects were then presented with the 16 testing conditions. The second session lasted approximately 75-90 mins. The order of the conditions was partially counterbalanced by RSI conditions. Subjects were given the same order for both days. The two groups, (young and old) were matched for order.

Treatment of Data. Data for RT, dwell time, percentage errors, overall and for each RSI were analyzed according to a 2 x 2 x 2 x 2 (age x degradation x compatibility x practice) mixed analysis of variance (ANOVA) design with repeated measures on the last three factors. Descriptive and inferential statistical analyses were performed on a Compaq 386/387 using DAS (Olofinboba, Vercruyssen, & Wheeler, 1989) and BMDP (version 1987) statistical packages. *Post-hoc* analyses were according to the Tukey WSD procedure (Vercruyssen & Hendrick, 1990). All contrasts were at the .05 level of significance.

RESULTS

To conserve space, only the most relevant and significant results on mean RTs for the second day will be presented. Analyses of percent errors and dwell times provided no evidence to suggest the possibility of speed-accuracy or reaction time-movement time trade-offs. Furthermore, analyses of median RTs provided near identical results as those obtained using means, suggesting the degree of skewness is fairly constant across groups and conditions. Since the variable RSI condition was used as an attentional measure and it is not considered suitable as a criterion measure for the AFM, results from the variable choice RT data analyses will not be presented here.

Practice. The strongest and most pervasive effect in this experiment was *skill level* as evidenced by a large number of interactions with practice, most of which were well below the .05 level of significance. The subjects' behavior changed dramatically from the first day to the second day. The older subjects improved more than the young from Day 1 to Day 2, particularly in the more difficult intratask loadings, i.e., when stimuli were degraded and incompatible with responses. Improvements were greater for the degraded than intact stimulus conditions and for the stimulus-response incompatible condition than the compatible. Since the AFM was designed for use with practiced subjects, **this paper will present the second day results** and simply caution the reader that results may differ depending on the subjects' skill level.

Age. Across criterion measures on the second day, the older subjects (1205 ms) took nearly twice as long, on the average, as the young (612 ms) to perform the RT tasks in this experiment ($p < .00005$).

Intratask Factors. With a similar pattern across criterion measures on the second day, performance on the intact stimulus conditions ($\underline{M}$ = 757ms) was significantly faster than the degraded conditions ($\underline{M}$ = 1082ms; $p < .00005$); performance on the compatible stimulus-response conditions ($\underline{M}$ = 635ms) was significantly faster than the incompatible condition ($\underline{M}$ = 1204ms; $p < .00005$).

Age x S-R Compatibility. On Day 2, each RSI condition revealed an age by compatibility interaction ($.0001 < p < .0005$) where the difference in overall mean RTs for the old from compatible (765ms) to incompatible (1645ms) was greater than those for the young (495ms, 730ms), i.e., differences of 879ms vs. 235ms. Figure 1 illustrates this interaction for a fixed RSI of 0 sec ($F_{1,18}$ = 24.37; p = .0001) which is representative of all four RSI conditions.

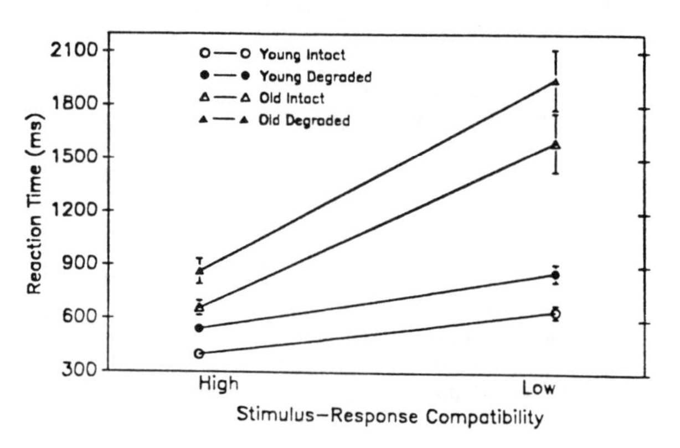

Figure 1. Mean serial four-choice reaction time and standard error of the mean as a function of age and stimulus-response compatibility.

Age x Degradation x Compatibility Interaction. On Day 2, there were no degradation by compatibility interactions in the fixed RSI measures administered to young subjects ($p > .05$). However, for the fixed RSI conditions of 2 and 5 sec presented to the old subjects, the interactions were significant (ps = .012 and .0294, respectively). Thus, the age by degradation by compatibility interactions were significant for the Fixed RSI=2 (p = .0015) and the Fixed RSI=5 (p = .0089). For the Fixed RSI=0 RT task, none of the groupings produced significant degradation by compatibility interactions.

DISCUSSION

This research project was undertaken to quantify age effects on speed of behavior when consideration is also given to the subject's skill on the RT task and with intratask manipulations designed to identify disruption in cognitive functioning through stages of information processing. Furthermore, the results of performance on RT tasks with both variable and fixed foreperiods was analyzed to determine sensitivities of the dependent measures to age differences while testing their utility for use with the AFM.

As expected, across RSI conditions, RT was slower for the old subjects than the young, for the degraded conditions than the intact, for the low compatibility condition than for the high compatibility condition, and for the first day of testing than on the second day.

Stages of Processing. Finding an interaction of age by compatibility points to the elderly having impairments in the response selection stage of information processing. This is in conflict with the findings of Simon and Pouraghabagher (1982) but in agreement with Hoyer and Plude (1980), at least with respect to the pattern of age-related decrement in performance being dependent on the degree to which effortful processing is involved. Findings from this experiment point to the possibility that age-related slowing is more specific than general and that the locus of this effect is associated with later stages of information processing. We are reluctant to conclude this without further evidence, however, since such a statement is not supported by the results of Moraal (1982ab) or Salthouse and Somberg (1982) as well as earlier studies by Birren et al. (1979, 1980). We are particularly interested in determining whether such results might be an artifact produced by the relative level of difficulty / complexity for the intratask manipulations.

AFM Limitations. It appears that the AFM is extremely sensitive to the skill of the subject. Analyses of the data which included the practice variable, produced a large number of interactions between the task manipulations and practice, as well as significant intratask manipulations which violates AFM assumptions of independent stages. This suggests that the AFM breaks down when used with naive or unpracticed subjects. In fact, Sternberg (1969) warned that subjects must be practiced, yet few of the aging studies which have employed the AFM have done so.

The most costly interactions in this experiment are those for degradation by compatibility for the elderly found in all of the RSI conditions except RSI=0. Such an interaction indicates that our task manipulations are not independent or that for the elderly degradation and compatibility affect a stage in common. This might suggest that the AFM is not a valid procedure for these data or that there are true fundamental differences between younger and older age groups in the way in which information is processed. Future research will need to take into consideration whether these differential age results may be eliminated with enough practice, or whether there are other factors involved in these manipulations.

Research conducted by McDowd and Craik (1988) show strong interactions between conditions differing by stimulus complexity on a divided attention task and age, with the older group showing larger RTs on the most difficult or complex conditions. Regarding the amount of practice, and even earlier study by McDowd (1986) found that differential responding on a divided attention task by age did not wash out, even after 6 hrs of testing (1 hr per week for 6 weeks).

The subject's skill level and the difficulty or intensity of intratask factors may very well account for the apparent equivocal findings obtained when using the AFM in aging research. Subsequent research needs to address the changing pattern of results produced when intratask manipulations are viewed as a function of difficulty intensity levels. The authors hope to see continued research which examines the robustness of the AFM for aging research relative to the known interactions of age with the rate of skill acquisition and different levels of intratask manipulations, while also considering individual differences such as gender, physical fitness, tonic level of arousal, and previous experience (e.g., Vercruyssen, Cann, Birren, McDowd, & Hancock, 1989; Vercruyssen, Cann, & Hancock, 1989).

In summary, the results obtained from testing 12 additional subjects to complete this experiment corroborated the conclusions made in our initial report (Carlton, Vercruyssen, McDowd, & Birren, 1988): (1) practice is often a major confound in reaction time research, (2) with age comes a pronounced slowing in speed of behavior, (3) the AFM may be a useful tool to assist in identifying the locus of aging effects on CNS functioning, (4) aging may slow information processing more in the response selection stage than an earlier encoding stage, and (5) the effects of aging, practice, and intratask factors depend on the response-stimulus interval characterizing the RT task.

Implications. This research has both practical and scientific implications. Understanding the nature of age-related slowing of behavior requires an examination of whether the 'slowing' is of a general nature, affecting all subprocesses, mechanisms or stages of information processing, or whether the slowing is of a selective nature, i.e., whether it can be 'localized' in one or more specific stages. The answer to this question has implications on human engineering and design for the elderly (Moraal, 1982ab). Continued multifactor research along the lines herein mentioned will yield much needed guidelines for laboratory testing methodologies.

ACKNOWLEDGMENTS

This research was funded in part by a USC Biomedical Research Support Grant to the first author, a post-doctoral stipend to the second author from a Research Initiate Grant to the Southern California Educational Resource Center of the National Institute of Occupational Safety and Health, a fellowship to the third author from a National Institute on Aging post-doctoral training grant in aging and attention (Grant No. 1 T32 MH18913-01), and an equipment grant from the Parsons Foundation. Programming and technical support were provided by O.B.A. Olofinboba and W.F. Wheeler. Valuable comments were provided by J.E. Birren, J.M. McDowd, and M.T. Cann. The authors wish to thank the subjects for participating in this study. This research is a continuation of the work by Carlton, Vercruyssen, McDowd, and Birren (1988).

REFERENCES

Birren, J.E. (1964). *The psychology of aging.* Englewood Cliffs, NJ: Prentice Hall.

Birren, J., Riegal, K., & Morrison, D. (1962). Age differences in response speeds as a function of controlled variations of stimulus conditions: Evidence of a general speed factor. *Gerontologia*, **6**, 1-18.

Birren, J.E., Woods, A.M., & Williams, M.V. (1979). Speed of behavior as an indicator of age changes and the integrity of the nervous system. In F. Hoffmeister & C. Mueller, *Bayer-Symposium VII: Brain Function in Old Age.* Berlin, W. Germany: Springer-Verlag, 10-44.

Birren, J.E., Woods, A.M., & Williams, M.V. (1980). Behavioral slowing with age: Causes, organization and consequences. In L.W. Poon (Ed.), *Aging in the 1980's: Psychological issues.* Washington D.C.: American Psychological Association, 293-308.

BMDP Statistical Software, Version 1987. 1440 Sepulveda Blvd., Los Angeles, CA 90025.

Botwinick, J. (1980). Contact with the environment: The senses. In *Aging and behavior.* New York: Springer, 185-206.

Botwinick, J., Brinley, & Birren, J.E. (1957). Set in relation to age. *Journal of Gerontology*, **12**, 300-305.

Botwinick, J., & Thompson, L.W. (1967). Practice of speeded response in relation to age, sex, and set. *Journal of Gerontology*, **22**, 72-76.

Carlton, B.L., Vercruyssen, M., McDowd, J.M., & Birren, J.E. (1988). Effects of age and practice on attention and stages of information processing using CRT with fixed and variable foreperiods. *Proceedings of the Human Factors Society--32nd Annual Meeting*. Santa Monica, CA: Human Factors Society, 208-212.

Frowein, H.W. (1981a). Selective effects of barbiturate and amphetamine on information processing and response execution. *Acta Psychologica*, **47**, 105-115.

Frowein, H.W. (1981b). Selective drug effects on information processing. *Thesis for doctor in de sociale wetenschappen* (Ph.D. in Social Sciences) at de Katholieke Hogeschool (Catholic High School, a university), Tilburg, The Netherlands. Druk: Sneldruck Boulevard Enschede, 4 September 1981.

Hoyer, W.J., & Plude, D.J. (1980). Attention and perceptual processes in the study of cognitive aging. In L.W. Poon (Ed.), *Aging in the 1980's: Psychological Issues*. Washington, D.C.: American Psychological Association, Chapter 16, 227-238.

Huntley, M.S., Jr. (1972). Influences of alcohol and S-R uncertainty upon spatial localization time. *Psychopharmocologia*, **27**, 131-140.

Huntley, M.S., Jr. (1974). Effects of alcohol, uncertainty and novelty upon response selection. *Psychopharmocologia*, **39**, 259-266.

Moraal, J. (1982a). Age and information processing. *Proceedings of the Human Factors Society--26th Annual Meeting*. Santa Monica, CA: Human Factors Society, 184-188.

Moraal, J. (1982b). *Age and information processing: An application of Sternberg's additive factors method*. Technical report No. IZF 1982-12. Soesterberg, Netherlands: TNO Institute for Perception, 1-19.

Olofinboba, O.B.A., Vercruyssen, M., & Wheeler, W.F. (1988). *DAS: Data analysis system for IBM compatible microcomputers, Version 2.02*. Computer software by Psy-Med Associates, 2329 Sonoma Street, Honolulu, Hi. 96822.

Salthouse, T.A., & Somberg, B.L. (1982). Isolating the age deficit in speeded processes. *Journal of Gerontology*, **37**(1), 59-63.

Sanders, A.F. (1977). Structural and functional aspects of the reaction process. In S. Dornic (Ed.), *Attention and Performance VI*. Hillsdale, NJ: Erlbaum, 3-25.

Sanders, A.F. (1980). Stage analysis of reaction processes. In G. Stelmach & J. Requin (Eds.), *Tutorials on motor behavior*. Amsterdam: North Holland, 331-354.

Sanders, A.F. (1983). Towards a model of stress and human performance. *Acta Psychologica*, **53**, 62-97.

Sanders, A.F., Wijnen, J.L.C., & von Arkel, A.E. (1982). An additive factors analysis of the effects of sleep loss on reaction processes. *Acta Psychologica*, **51**, 41-59.

Simon, J.R. (1968). Signal processing time as a function of aging. *Journal of Experimental Psychology*, **78**, 76-80.

Simon, J.R., & Pouraghabagher, A.R. (1982). The effect of aging on the stages of information processing in a choice reaction time task. *Journal of Gerontology*, **33**(4), 553-561.

Sternberg, S. (1969). The discovery of processing stages: Extensions of Donder's method. *Acta Psychologica*, **30**, 276-315.

Tharp, V.K., Rundell, D.H., Lester, B.K., & Williams, H.L. (1974). Alcohol and information processing. In, M.M. Gross (Ed.), *Alcohol intoxication and withdrawal, experimental studies: Advances in experimental medicine and biology*. New York: Plenum.

Vercruyssen, M. (1984). *Carbon dioxide inhalation and information processing: Effects of an environmental stressor on cognition*. Doctoral dissertation, The Pennsylvania State University, University Park, PA.

Vercruyssen, M., Cann, M.T., Birren, J.E., McDowd, J.M., & Hancock, P.A. (1989). Effects of aging, gender, physical fitness, neural activation on CNS speed of functioning. *Proceedings of the 1988 Symposium Osaka*. Osaka, Japan: International Council for Physical Fitness Research, in press.

Vercruyssen, M., Cann, M.T., & Hancock, P.A. (1989). Gender differences in posture effects on cognition. *Proceedings of the Human Factors Society--33rd Annual Meeting*. Santa Monica, CA: Human Factors Society, (elsewhere in these proceedings).

Vercruyssen, M. & Hendrick, H.W. (1990). *Behavioral research and analysis: Introduction to statistics within the context of experimental design, Edition 3*. Lawrence, KS: Ergosyst Associates, in press.

Welford, A.T. (1977). Motor performance. In J.E. Birren & K.W. Schaie (Eds.), *Handbook of the psychology of aging*. New York: Van Nostrand Reinhold, 450-496.

Welford, A.T. (1980). Relationships between reaction times and fatigue, stress and sex. In A.T. Welford, (Ed.), *Reaction times*. London: Academic Press, 321-354.

Wheeler, W.F., Vercruyssen, M., & Olofinboba, O.B.A. (1988). *DCS: Data collection system for the C-64/128, Version 1.05*. Computer software by Psy-Med Associates, 2329 Sonoma Street, Honolulu, HI 96822.

AGE-RELATED SLOWING, S-R COMPATIBILITY, AND STAGES OF INFORMATION PROCESSING

Virginia Diggles-Buckles[1] and Max Vercruyssen[12]

[1]Laboratory of Attention and Motor Performance
Andrus Gerontology Center
University of Southern California
Los Angeles, CA 90089-0191

[2]Human Factors Department,ISSM
University of Southern California
Los Angeles, CA 90089-0021

Previous work in this laboratory (Vercruyssen, Carlton & Diggles-Buckles, 1989) has found that older individuals are at a disproportional disadvantage when stimulus-response (S-R) compatibility relationships are made more difficult. When stimulus quality and S-R compatibility were manipulated, age interacted with the S-R manipulation, suggesting in an additive factors framework that the locus of age-related slowing was the response selection stage. In that study S-R compatibility was manipulated by changing the S-R spatial map as well as changing the environment (subjects were required to cross their arms). The present study attempted to tease apart factors that might be contributing to that age x S-R compatibility relationship by using S-R maps of simple, moderate, and high difficulty as one factor and the arm position (crossed or uncrossed, a test of the Simon effect, Simon, Sly & Vilapakkam, 1981) as a different factor. In addition, stimulus quality was manipulated as a factor in this 4 factor design: age x stimulus quality x S-R map x arm position. Results revealed that both factors, S-R compatibility and arm position interacted with age. The conclusion from an additive factors perspective is that the stages of decision making (S-R compatibility) and response preparation (arm position) show age-dependent slowing whereas the stimulus encoding stage (stimulus quality) does not.

INTRODUCTION

Age-related slowing has been the topic of much empirical study. Birren (1955, 1965, 1970, 1974) has postulated the Generalized Slowing Hypothesis which suggests that the time of fundamental neural events becomes slower with increasing age to an equal extent for all cognitive functions. An alternative perspective is that age-related slowing may be isolated in one or more stages of information processing. One approach in this endeavor has been the application of Sternberg's Additive Factors Method (AFM) in which attributions about the locus of a variable's effect are made based on the statistical outcome of an analysis of variance (Sternberg, 1969). The AFM adopts the assumptions of the information processing framework and adds interpretations regarding the statistical outcome of a factorial design. If the effects of the manipulated variable are additive, it is concluded that the variables are affecting separate stages of processing. If the variables interact, it is concluded that variables affect a stage or stages in common. In aging, a few attempts have been made to apply this logic to aging as a variable, the desired result being to identify particular stages of processing that are affected by aging.

Simon and Pouraghabagher (1978) used additive factors logic in manipulating stimulus quality and stimulus-response compatibility (by using an irrelevant directional cue). With two age groups, young and old, they manipulated these variables in a two choice reaction time task. They found that age interacted with stimulus quality and concluded that aging affected the stimulus encoding stage. Salthouse and Somberg (1982) used a target identification task to examine the effects of stimulus quality, response type, and set size on the speeded performance of young and old subjects. They found that all variables interacted with age and concluded that Birren's (1974) generalized slowing hypothesis was supported (i.e. all cognitive functions were affected). Moraal (1982) examined young and elderly performance on a four choice reaction time task in which he manipulated foreperiod duration, stimulus intensity, and S-R compatibility. No variables were found to interact with aging and Moraal concluded as did Salthouse and Somberg, that support was found for the generalized slowing hypothesis, although his results were opposite in nature to Salthouse and Somberg's.

Vercruyssen, Carlton and Diggles-Buckles (1989) used the same approach to examine the effects of stimulus quality, S-R compatibility and response-stimulus interval on aging performance. They found that age interacted with S-R compatibility and concluded that the locus of age-related slowing was the response selection stage. In that study S-R compatibility was manipulated by changing the S-R spatial map as well as changing the environment (subjects were required to cross their arms). It was thought that this condition may confound variables affecting two stages: S-R mapping affecting the response selection stage and arm position affecting the response preparation stage. It was also suggested that the difficulty of this condition (an arbitrary spatial map with no logical relationships) might have completely overwhelmed the elderly subjects.

The present study was conducted to tease apart these two variables in S-R compatibility, replicate parts of the earlier Vercruyssen et. al. (1989) study and examine a continuum of S-R map difficulty in two age groups.

METHOD

Subjects

Subjects were 12 USC students, (16-35, mean age = 25.8 yrs.) and 12 older adults (64-80, mean age = 70.8 yrs.) from the Andrus Volunteers. Equal numbers of each gender participated in each age group. Subjects filled out a health and activity questionnaire to assess current and past health problems and fitness levels. All subjects had normal or corrected to normal vision.

Task

The task was a four choice reaction time (RT) task in which two boxes (4.5 x 5 cm) appeared on side by side, on a monitor. The imperative stimulus was the presentation of a horizontal arrow within one of the boxes, pointing either left or right. The combination of two boxes with the two arrow directions yielded 4 alternatives. The display was presented at eye level (with subjects seated) approximately 60-75 cm away from the subjects, who sat facing the monitor. The response keys were four microswitches, two each mounted on two response boxes, placed on a platform in front of the subject. Subjects used index and middle fingers of each hand to respond. Stimuli were presented in 40 trial blocks for each of the 12 combinations of task factors. RTs and errors were recorded and stored for later analyses. Speed and accuracy tradeoffs were explained to subjects and they were encouraged to go as fast as possible to maintain a 5% error rate. Any block where the error rate exceeded 20% was repeated. The intertrial interval was one second.

Procedures

Subjects were tested on two occasions within the same week, the second a complete replication of the first day. Only the second day's data were analyzed. The presentation of the stimuli and the collection of the data were both performed by a Commodore 64 computer and later transferred to IBM compatible microcomputers for analyses and storage.

Stimulus Quality. The stimulus was presented in each of two forms: intact or degraded. Stimulus degradation was achieved by superimposing a computer-generated random visual noise mask of the same color as the background over the arrow image. For each presentation the mask rotates so that no consistent pattern could be discerned. This manipulation maintained the constant amount of stimulus degradation and stimulus intensity.

S-R Compatibility. The different S-R mappings were achieved by varying the information communicated by the boxes and arrows. In the easiest or most direct mapping (**Direct S-R Map**), the boxes indicated which hand to use, left or right, and the arrow pointed to which finger to use, left or right. In the map of intermediate difficulty, this relationship was reversed (**Reversed Map**), i.e. the arrow indicated which hand, left or right and the box indicated which finger, left or right. In the most difficult map, the arrow also indicated which hand, left or right, and the boxes indicated which finger but in an arbitrary fashion (**Arbitrary Map**). If the arrow occurred in the left box, it indicated that the middle finger was to be used, if it occurred in the right box, the index finger was to be used.

Arm Position. Crossed with the other factors was the position of the arms. Simon (Simon, Sly & Vilapakkam, 1981) has found that when that changing environmental S-R maps has a greater effect on RT than changing anatomical S-R maps. That is crossing the arms results in faster RTs than reversing a left right S-R map. In the earlier study this condition was confounded with the low compatibility map. Subjects performed all conditions both with arms uncrossed and crossed.

RESULTS

Errors. Errors (incorrect choices) were analyzed in a 2 (age) x 3 (S-R maps) x 2 (stimulus quality) x 2 (arm position) repeated measures ANOVA. There was no main effect of age in the number of errors committed. There were main effects of S-R mapping ($F_{2,40}$=19.14, p<.001), with the two more difficult maps being significantly different from the easiest (Direct Map M=4.3, Reversed Map M=6.7, and Arbitrary Map M=8.1). The degraded stimulus caused more errors ($F_{1,20}$=4.9, p<.039). Crossing the arms significantly increased errors ($F_{1,20}$=21.1, p<001). An age by stimulus quality interaction was found in which the older subjects made disproportionately more errors under the degraded stimulus condition ($F_{1,20}$=7.076, p<.015). Any interpretation of errors must be tempered by the fact that there was an upper limit on the number of errors that a subject could make. If the subject exceeded a 20% error rate, they repeated the condition.

RT Data. The RTs from were analyzed in a repeated measures ANOVA with the same design as that one used on errors. All main effects were significant in RTs. The older subjects were overall slower than the young (Old M=1618, Young M=909, $F_{1,20}$=13.25, p<.002). The two more difficult S-R maps were yielded slower RTs than the easiest map (Direct Map M=731, Reversed Map M=1220, Arbitrary Map M=1839, $F_{2,40}$=38.95, p<.001). The degraded stimulus slowed RT by 424 ms (intact M=1052, degraded M=1476, $F_{1,20}$=44.78, p<.001). Crossing the arms and altering the anatomic and environmental relationships slowed RT by 279 ms (uncrossed M=1124, crossed M=1403, $F_{1,20}$=11.68, p<.003). Three interactions were found to be significant, two involving age. An S-R compatibility x stimulus quality interaction was significant in which the effect of the degraded stimulus was more than doubled under the reversed S-R map relative to the other two maps ($F_{2,40}$=8.2, p<.001). Subject age interacted with S-R compatibility and with arm position. In the age x S-R compatibility interaction, older subjects were disproportionately slower in the most difficult S-R mapping ($F_{2,40}$=7.5, p<.002). Figure 1 illustrates this interaction. Where age interacted with arm position ($F_{1,20}$=6.69, p<.018), the older subjects had greatest difficulty with the crossed position. Crossing arms added 490 ms for the elderly but added only 68 ms for the young subjects.

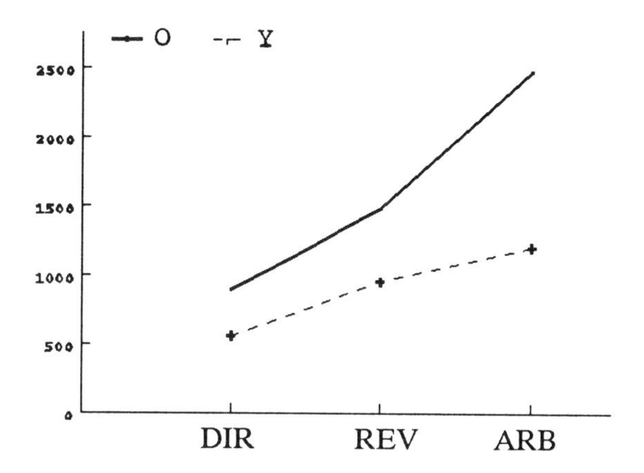

Figure 1. Age by S-R compatibility interaction

DISCUSSION

The two statistical interactions with age seem to support the notion that the stages of response selection and response generation manifest greater difficulty for speeded responses in the elderly than other stages of information processing. Age interacted with S-R compatibility and arm position on RT but did not interact with stimulus quality. From an additive factors framework, this finding would suggest that the stimulus encoding is not particularly affected by aging. These findings support previous research from this lab (Vercruyssen et. al., 1989) and the findings of Corpolongo and Salmon (1981) where age was found to interact with S-R compatibility.

When the arm position was tested as a separate factor, independent of S-R compatibility, S-R compatibility still interacted with age. The earlier notion that the combination of crossed arms and arbitrary S-R map may have overwhelmed the elderly does have some support. When an intermediate level of S-R compatibility was added, older subjects were only proportionally slower, the indication being that the difficulty of the S-R map is the determining factor. When the difficulty is only moderate, the older subjects process it in the same fashion as younger subjects, its difficulty being additive. This finding also supports the complexity hypothesis which maintains that aging interacts with the complexity of the task (Cerella, Poon & Williams, 1980).

The interaction between age and arm position suggests that the "Simon" effect presents a greater difficulty for the elderly than it does for the young. The "Simon" effect refers to the altering of S-R compatibility while maintaining environmental relationships, e.g. crossing the arms such that the hand on the left side (the anatomical right hand) still responds to the stimulus on the left. This condition is slower than if the hands weren't crossed but faster than a condition where arms are uncrossed and the S-R map is reversed (right hand on right side responds to left stimulus). It also suggests that the later stage of information processing, response programming, in this respect suffers with aging. Light and Spirduso (1990) also found evidence to support this contention by manipulating the complexity of the motor response. Age interacted with movement complexity, leading to the conclusion that movement planning and preparation were slowed in the elderly.

The purpose of this experiment was to investigate the locus of age-related slowing in the information processing system. Previous results in which S-R compatibility was found to interact with age were examined more carefully by using an intermediate level of S-R compatibility difficulty and by treating the arm position as a separate factor. Stimulus quality was also manipulated to assess earlier stages of information processing. The results indicated that the earlier stage of stimulus encoding was not differentially affected by age. However, response selection and response generation and programming were differentially affected by age, indicating these two stages as particularly affected by age-dependent changes.

REFERENCES

Birren, J.E. (1955) Age changes in speed of response and their significance for complex behavior. In *Old age and the modern world.* London: Livingstone.

Birren, J.E. (1965) Age changes in speed of response: Its central nature and physiological correlates. In A.T. Welford and J.E. Birren (Eds.), *Behavior, aging and the nervous system.* Springfield, IL: Thomas.

Birren, J.E. (1970) Toward an experimental psychology of aging. *American Psychologist*, **25**, 124-135.

Birren, J.E. (1974) Translations in gerontology - from lab to life: Psychophysiology and the speed of response. *American Physiologist*, **29**, 808-815.

Cerella, J., Poon, L.W., & Williams, D. M. (1980) Age and the complexity hypothesis. In L.W. Poon (Ed.), *Aging in the 1980s: Psychological Issues*. Washington, D.C.: American Psychological Association, 1980.

Corpolongo, M., & Salmon, P. (1981) Comparison of information-processing capacities in young and aged subjects using reaction times. *Perceptual and Motor Skills*, **52**, 987-994.

Light, K.E., & Spirduso, W.W. (1990) Effects of adult aging on the movement complexity factor of response programming. *Journal of Gerontology*, **45**, P107-109.

Moraal,J. (1982) Age and information processing: An application of Sternberg's additive factor method. *Report Nr. IZF 1982-18.* Soesterberg, the Netherlands: Institute for Perception, TNO.

Salthouse, T.A., & Somberg, B.L. (1982) Isolating the age deficit in speeded performance. *Journal of Gerontology*, **37**, 59-63.

Simon, J.R., Sly, P.E., & Vilapakkam, S. (1981) Effect of compatibility of S-R mapping on reactions toward the stimulus source. *Acta Psychologica*, **47**, 63-81.

Simon, J.R., & Pouraghabagher, A.R. (1978) The effect of aging on the stages of processing in a choice reaction time task. *Journal of Gerontology*, **33**, 553-561.

Sternberg, S. The discovery of processing stages: Extension of Donders' method. *Acta Psychologica*, **30**, 276-315.

Vercruyssen, M., Carlton, B.L., & Diggles-Buckles, V.A. (1989) Aging, reaction time, and stages of information processing. *Proceedings of the Human Factors Society, 33rd Annual Meeting* Santa Monica, CA: Human Factors Society.

Longitudinal Analysis of Age-Related Slowing: BLSA Reaction Time Data

James L. Fozard[1], Max Vercruyssen[2], Sara L. Reynolds[2], and P.A. Hancock[3]

[1] National Institutes of Health, National Institute on Aging
Gerontology Research Center, Rm 1E08, Baltimore, MD 21224

[2] Human Factors Department, ISSM
Lab of Attention and Motor Performance, Gerontology
University of Southern California, Los Angeles, CA 90089-0021

[3] Human Factors Research Laboratory, 164 Norris Hall
University of Minnesota, Minneapolis, MN 55455

Reported are preliminary findings from analyses of cross-sectional and longitudinal reaction time data collected on 865 male and 453 female volunteers who ranged in age from 20 to 96 years. Evident in both simple and disjunctive reaction time measures was a consistent slowing with age. In nearly all cases, males were faster than females but gender differences were negligible for the simple reaction time (SRT) compared to disjunctive reaction time (DRT). Repeated testing within subjects over 2-8 years also showed age-related slowing across decades. Cross-sectional studies have been criticized for overestimating the actual age-related slowing found in longitudinal analysis. However, this was not the case in the present research. Similar effects were observed in analyses of data from all subjects on their first visit (n = 1318 subjects) compared to data from all subjects over all of their visits (n = 3855 subject visits) compared to data from only those subjects across decades who were tested repeatedly over at least 8 years (n = 314 subjects X 5 visits = 1570 subject visits). Findings from this research have human factors implications for task design, personnel selection, performance prediction, accident analysis, human tests and measurements, and demographic norms, to mention a few.

INTRODUCTION

The Baltimore Longitudinal Study of Aging (BLSA), which has been gathering data since 1959, is the gold standard against which we compare all other longitudinal aging research. This federally sponsored research effort has extensively tested over 1300 adult volunteers (20 to 96 years of age) from the Baltimore area on a large battery of biographical, physiological, and psychological measures. Simple (SRT) and disjunctive (DRT) reaction time tasks were introduced in 1973 to male subjects and in 1978 to female subjects. Now there are sufficient data for describing longitudinal patterns in age-related slowing.

While there are disadvantages to using longitudinal analysis in the study of aging (*e.g.*, Damon, 1965; Shock, 1985), such as practice and period effects, recruitment of an "elite" subject pool, and subject drop-out, they are far outweighed by the advantages. Longitudinal research is capable of eliminating birth-cohort effects, and identifying changes in individuals' performance as they age. In addition, its reliability increases with increased duration and frequency of testing. Cross-sectional analysis, on the other hand, can only identify differences between age groups. And, while practice and period effects are not a problem, the performance differences observed may be partly a result of factors other than age, such as cohort effects or selective survival (Damon, 1965; Shock, 1985).

Most cross-sectional findings relative to age-related slowing of reaction time (RT) show increases in RT with age (Bleecker, Bolla-Wilson, Agner, & Meyers, 1987; Botwinick & Storandt, 1974; Era, Jokela, & Heikkinen, 1986; Goldfarb, 1941; Harkins, Nowlin, Ramm, & Schroeder, 1974; Lahtela, Niemi, & Kuusela, 1985; Pierson & Montoye, 1958; Simon, 1967; Spirduso, 1975; Spirduso & Clifford, 1978; Szafran, 1951; Vrtunski, Patterson, & Hill, 1984; Waugh, Fozard, Talland, & Erwin, 1973). Some studies also have found consistent gender differences favoring males (Bleecker, Bolla-Wilson, Agnew, & Meyers, 1987; Botwinick & Storandt, 1974; Harkins, Nowlin, Ramm, & Schroeder, 1974; Lahtela, Niemi, & Kuusela, 1985; Simon, 1967). In contrast, Noble, Baker, and Jones (1964) found women slightly faster than men in the 71-87 year age group, and Landauer (1981) and Landauer, Armstrong, and Digwood (1980) found no significant gender difference in total reaction time (*i.e.*, premotor time plus motor time).

Few longitudinal studies have published the results of their reaction time data analysis. Of those which have, namely the Duke Longitudinal Studies and the Bonn Longitudinal Study of Aging, findings are consistent with those of the cross-sectional studies. Reaction time increased with age, and men were faster than women (Mathey, 1976; Wilkie, Eisdorfer, & Siegler, 1975).

Never before has the reaction time data from the BLSA effort been analyzed and reported. Of particular interest is the comparison of cross-sectional versus longitudinal findings on age-related slowing and gender differences. Therefore, the purpose of this paper is to present findings from analyses of the reaction time data, both from cross-sectional and longitudinal approaches.

METHOD

Subjects. Data from the first visits of 865 male and 453 female volunteers were used to conduct the cross-sectional analysis by decades (20-90 $\pm$ 5 yrs). Longitudinal analysis was conducted using data from repeated testing of subjects according to two-year intervals between visits: 471 males and 145 females made three visits, and 284 males and 30 females made five visits.

The subjects comprised a self-recruited group and therefore were not a random or representative sample of the Baltimore population. Subjects recruited relatives, friends, and co-workers, who then similarly recruited new subjects. This has led to a somewhat homogeneous group of subjects from the upper-middle socioeconomic level who differ from the general population in that they are happier, healthier, and more likely to be married (Andres, 1978). Over 70 percent of the subjects were college graduates, over 40 percent with advanced degrees.

Auditory Reaction Time Tasks. Subjects responded to low and high auditory tones. In the simple reaction time (SRT) task, subjects responded to both high and low tones by depressing a hand-held response button as quickly as possible.

Subjects were instructed to respond only to the high tone in the disjunctive reaction time (DRT) task. A total of 122 RT trials were presented in four blocks of approximately 5 min each.

Procedure. Subjects were seated in a sound-proof audiometry booth with their backs to the experimenter's viewing window. The booth was well lit and ventilated with a fan. The experimenter presented both the low and high tones (250 and 1000 Hz, respectively) through a loudspeaker at a level of 56 DBA and asked the subject if he/she could hear the tones comfortably. If the tones were not easily audible, the experimenter increased the sound level to 62 dBA and tried the tones again. The test was discontinued if the 62-dBA tones were not easily heard by the subject. Tones were 3.0 sec duration during practice and 0.3 sec during testing. In addition, a single random order of variable foreperiods (6-13 sec) was used.

The first 5-min reaction time test began after the appropriate stimulus intensity was determined. After this 5-min practice test, the experimenter administered the test a second time, presenting 36 trials of which the last 20 were recorded as simple auditory RT raw data in ms. The third 5-min test was one of disjunctive reaction time. After practicing the DRT task, the experimenter administered the fourth block of trials, recording the last 13 high-tone trials as DRT raw data.

Each subject visited the Gerontology Research Center (GRC) every two years, annually if over 70 yrs old, for 2.5 days of extensive testing using a battery of physiological and psychological tests. From visit to visit across subjects, the 20-min reaction time testing took place at different times during the 2.5-day period. The RT tests were optional, so occasionally subjects elected only to do only one of the two tasks or to skip RT testing altogether for that particular visit.

Treatment of Data. Custom software was used to convert data from the BLSA storage format to easily manageable data files which were analyzed using microcomputer statistical packages. All descriptive statistics were conducted using DAS (Olofinboba & Vercruyssen, 1990) with univariate analysis of variance (ANOVA) conducted using BMDP (BMDP Statistical Software, 1988). All statistical contrasts were tested at the .05 level of significance with the more conservative Huynh-Feldt probabilities (H*p*) presented along with conventional values (*p*) for all repeated measures ANOVAs.

Reaction time summary data (mean of means, mean of medians, mean of variance about the means, and mean percent errors) for SRT, DRT, and both tasks combined were analyzed according to a variety of univariate ANOVA designs. First visit analyses used a 2 X 7 (gender X decade) between groups design. Analyses of visits 1-3 and 1-5 added a third factor of visits (repeated measures factor with levels of 3 and 5, respectively). When examining the interaction of SRT and DRT tasks, a fourth factor of task complexity (with two levels) was added. In order to get the most accurate representation of age-related changes within subjects and to overcome problems of irregular visitations across subjects, each subject's repeated measures data (means, medians, variances, and percent errors of SRT and DRT for each visit) were plotted and regressed according to time since first visit. Dependent measures were then extracted in two-year intervals (visits) from each subject's line of best fit. In this way, data were obtained from all subjects, regardless of how many years passed between visits. However, data points were not extrapolated beyond the last actual data point. This estimation procedure was unlikely to inflate the degrees of freedom (*df*) because there were nearly as many actual data points as estimated values, and, with data based on over 1300 subjects, it was doubtful that slight changes in *df* would alter the results.

Errors of omission (*i.e.,* failure to respond when appropriate within 800 ms) and commission (*i.e.,* incorrect responses -- those less than 150 ms or those to the low tones) were expressed as a percentage of the RT trials analyzed. Unusually high error rates (as high as 100%) occurred in some subjects on some visits; therefore, all visits in which errors were greater than 45% were excluded from analyses (this occurred on 2.12% of total SRT visits and 4.96% of total DRT visits).

RESULTS

Analyses were conducted on data from 3,855 simple reaction time and 3,781 disjunctive reaction time test sessions (*i.e.,* all data from multiple visits by 1318 subjects). Cross-sectional analyses used data from only the first visit of each subject (SRT n = 1318; DRT n = 1298) in scatterplot fashion and according to interval groupings by decade and gender. Longitudinal analyses were conducted by examining data from those subjects who returned for testing over four years (SRT n = 616, DRT n = 598) and eight years (SRT n = 314, DRT n = 311). Fewer DRT than SRT sessions were available due to subjects electing to discontinue testing after doing an SRT session, and there being a larger number of DRT visits excluded from analysis because error rates were greater than 45 percent. Simple and disjunctive reaction time data were analyzed separately except when discussing decision time, when SRT and DRT were treated as levels of a task type (complexity) factor in order to examine age (decade) by task interactions ($\underline{M}$ = mean; $\pm$ = plus or minus one standard deviation).

Cross-Sectional Analyses

First Visit Analysis by Decades

Simple Reaction Time. Analysis of variance on the means of median SRT performance revealed significant increases with age ($F_{7,1301}$ = 12.17; p < .00005) and gender (male $\underline{M}$ = 255.5 $\pm$ 57.0; female $\underline{M}$ = 269.5 $\pm$ 58.0; $F_{1,1301}$ = 6.17; p = .0131), but the interaction was not significant (p = .3664). Similar results were obtained using mean RT data. ANOVA on percentage errors also found significant increases with age ($F_{7,1301}$ = 6.29; p < .00005) and gender (male $\underline{M}$ = 3.0 $\pm$ 6.7 $\underline{SD}$; female $\underline{M}$ = 1.9 $\pm$ 4.7 $\underline{SD}$; $F_{1,1301}$ = 8.89; p = .0029), without a significant interaction (p = .2048). These results are depicted in Figure 1.

Disjunctive Reaction Time. Exactly like the results obtained for simple reaction time, analysis of variance on the means of median DRTs revealed significant increases with age ($F_{7,1282}$ = 29.62; p < .00005) and gender (male $\underline{M}$ = 402.8 $\pm$ 82.4; female $\underline{M}$ = 451.3 $\pm$ 87.5; $F_{1,1282}$ = 48.41; p < .00005), but the interaction (illustrated in Figure 1) was not significant (p = .3833). Results obtained using mean RT data were similar to those using medians. Percentage errors increased significantly with age ($F_{7,1282}$ = 12.09; p < .00005) and gender (male $\underline{M}$ = 3.8 $\pm$ 7.5; female $\underline{M}$ = 4.9 $\pm$ 8.5; $F_{1,1282}$ = 15.39; p = .0001), with a significant age by gender interaction ($F_{7,1282}$ = 2.57; p = .0124). The interaction showed similar error rates for each gender through the middle decades (< 5%) with a rise to 20% errors by the women and 10% by the men during the ninth decade.

Decision Time Analysis. Figure 1 illustrates the most important cross-sectional results for the means of median RTs as a function of age, gender, and task type (SRT and DRT).

Decision time analysis is concerned with changes in the difference between SRT and DRT as a function of age and gender. To examine age and gender interactions with task type, SRT and DRT data were combined and then analyzed as two levels of a single factor. Analyses of the mean of median SRT and DRT (combined) revealed significant differences for age ($F_{7,1278}$ = 24.64; p < .00005), gender ($F_{7,1278}$ = 39.27; p < .00005), task type ($F_{1,1278}$ = 3547.10; p < .00005, Hp < .00005), task type by age ($F_{7,1278}$ = 19.09; p < .00005), task type by gender ($F_{1,1278}$ = 31.98; p < .00005), and task type by age by gender ($F_{7,1278}$ = 2.61; p = .0113). No other contrasts were significant (p > .05). Similar results were obtained for percentage errors of the SRT-DRT data. Significant differences were obtained for age ($F_{7,1278}$ = 12.75; p < .00005), task type ($F_{1,1278}$ = 55.77; p < .00005), task type by age ($F_{7,1278}$ = 3.36; p = .0015), task type by gender ($F_{1,1278}$ = 30.65; p < .00005), and task type by age by gender ($F_{7,1278}$ = 4.12; p = .0002).

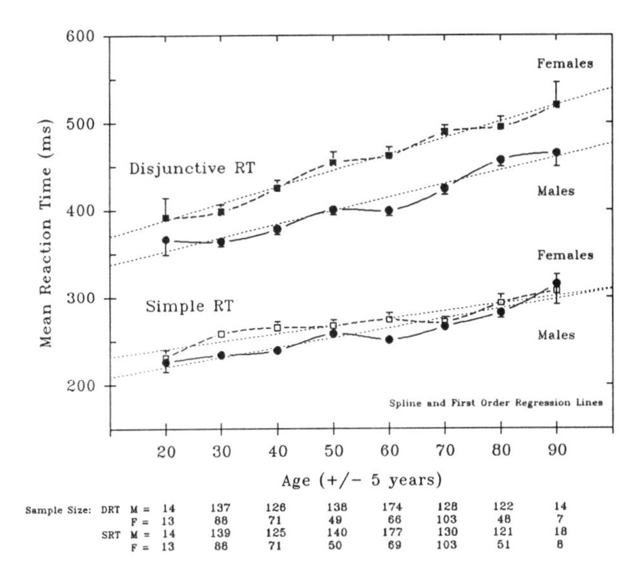

Figure 1. Mean of median auditory reaction times for each subject's first visit as a function of age, gender, and task type expressed with standard error bars, spline fitting, and first order regression lines.

Longitudinal Analyses

Three Visits over Four Years

Simple Reaction Time. After selecting only those subjects who completed testing at least four years following their first visit (n = 616), and sorting them into decades, ANOVAs revealed significant age-related slowing of the mean of median SRTs ($F_{6,602}$ = 3.50; p = .0021) where SRT went from 229.1 to 292.6 ms. Males (M = 253.6 ± 46.4 ms) were also significantly faster than females (M = 273.5 ± 54.6; $F_{1,602}$ = 8.35; p = .0040). There was a significant increase in SRT with visits (Visit 1 M = 256.1 ± 15.0 ms, Visit 2 M = 258.3 ± 18.5 ms, Visit 3 M = 260.5 ± 23.3 ms; $F_{2,1204}$ = 8.10; p = .0003, Hp = .0043). However, there was also a significant age by gender interaction ($F_{6,602}$ = 2.37; p = .0283). No other interactions were significant (p > .05). SRT errors increased significantly with age (20's M = 0.06 ± 0.28, 30's M = 1.26 ± 2.71, 40's M = 1.59 ± 3.63, 50's M = 3.15 ± 6.54, 60's M = 1.93 ± 3.58, 70's M = 3.59 ± 5.04, 80's M = 4.68 ± 6.05; $F_{6,602}$ = 3.47; p = .0022). No other main effects or interactions were significant (p > .05).

Disjunctive Reaction Time. Analysis of DRT data (n = 598) showed main effects for age ($F_{6,584}$ = 6.46; p < .00005; DRT increased per decade from 374.7 to 459.0 ms), gender (male M = 398.3 ± 70.4 ms, female M = 453.6 ± 71.0 ms; $F_{1,584}$ = 28.52; p < .00005), and visit (Visit 1 M = 406.1 ± 77.0 ms, Visit 2 M = 411.3 ± 71.8 ms, Visit 3 M = 416.4 ± 73.8 ms; $F_{2,1168}$ = 5.54; p = .0040, Hp = .0813). However, there was a visit by age interaction ($F_{12,1168}$ = 2.66; p = .0016, Hp = .0141). No other interactions were significant (p > .05). Concerning analysis of DRT errors, results were similar to those found in analysis of the SRT errors. There were significant increases with age (20's M = 1.6 ± 4.3, 30's M = 2.1 ± 4.2, 40's M = 2.9 ± 5.0, 50's M = 3.9 ± 6.3, 60's M = 3.4 ± 5.5, 70's M = 5.0 ± 6.8, 80's M = 4.9 ± 7.3; $F_{6,584}$ = 2.40; p = .0268). Gender differences were close to being significant (male M = 3.2 ± 5.3, female M = 4.6 ± 7.0; $F_{1,584}$ = 3.62; p =.0574). No other main effects or interactions were significant (p > .05).

Five Visits over Eight Years

[Descriptive statistics show all subjects in the database, but, in order to conduct the ANOVAs, the 80+ yrs group was removed (10 SRT and 9 DRT males) because there were no corresponding female subjects for that age group. Thus, the figures will show data in the 80+ age group (males only) but ANOVAs were conducted only on decades 2-7.]

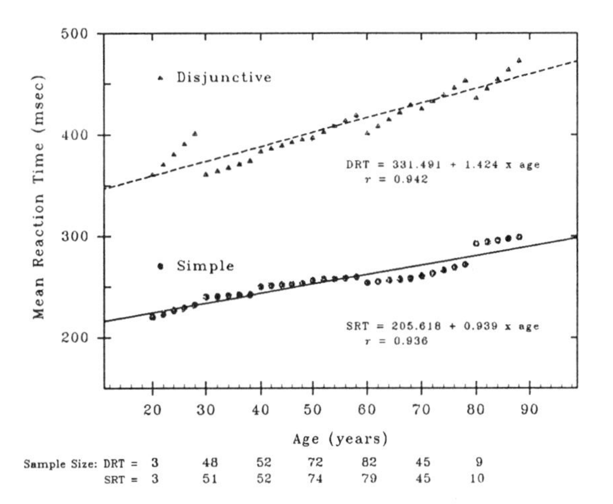

Figure 2. Mean of median reaction times for each of five visits (first visit plus 2, 4, 6, and 8 yrs) as a function of age and task. (Simple reaction time data were not collected on the eighth decade subjects.)

Simple Reaction Time. For ANOVA on mean of median SRTs (n = 314), age differences were not significant (M's ranged from 226.0 ± 21.0 ms to 266.1 ± 46.8; p = .3780), nor were gender differences (males M = 252.0 ± 43.3 ms, female M = 275.7 ± 57.5 ms; p = .0783). SRT slowed with consecutive visits (visit 1 M = 260.3 ± 50.3, visit 2 M = 261.8 ± 46.8, visit 3 M = 263.2 ± 45.6, visit 4 M = 264.7 ± 46.8, visit 5 M = 266.2 ± 50.3; $F_{4,1168}$ = 4.30; p = .0018, Hp = .0375). The age by visit interaction was marginally significant ($F_{20,1168}$ = 1.60; p = .0460, Hp = .1581). No other contrasts were significant.

While the trends in SRT errors appear significant (as shown in Figure 2), none are (age p = .0618; gender p = .8281; visit $F_{4,1168}$ = 3.52, p = .0073-.0593 (2.3-2.8); visit by age interaction $F_{20,1168}$ = 1.94, p = .0078, Hp = .0835).

Disjunctive Reaction Time. For DRT (n = 311), age differences were not significant (decade means ranged from 381.5 to 439.8; p=.2008). However, males (M = 400.9 ± 71.4) were significantly faster than females (M = 443.8 ± 67.5; $F_{1,290}$ = 7.19; p = .0078). The age by gender interaction was not significant (p = .2853). While the cross-sectional decade differences were not significant, the slowing of DRT within subjects across visits was (visit 1 M = 401.1 ± 65.9, visit 2 M = 406.4 ± 64.0, visit 3 M = 411.7 ± 64.6, visit 4 M = 417.0 ± 67.6, visit 5 M = 422.3 ± 72.8; $F_{4,1160}$ = 5.64; p = .0002, Hp = .0171; see Figure 2). The only other significant finding was the age by gender by visits interaction ($F_{20,1160}$ = 2.30; p = .0009, Hp = .0426). This interaction was the only case in all of the statistical analyses where decisions for the median were different from those for the means (M Hp = .0426; median Hp = .0847).

While differences in the errors made on the SRT task were not significant, most contrasts for DRT errors were significant, including those for age ($F_{5,290}$ = 5.31; p = .0001), gender (male M = 3.0 ± 4.7, female M = 5.9 ± 7.7; $F_{1,290}$ = 6.69; p = .0102), age by gender interaction ($F_{5,290}$ = 2.37; p = .0395), visit ($F_{4,1160}$ = 2.95; p = .0194, Hp = .0850), and visit by gender interaction ($F_{4,1160}$ = 3.18; p = .0131, Hp = .0735 (no change in males, large increase in females)).

DISCUSSION

Similar results were obtained whether data were analyzed cross-sectionally, based on all subjects' first visit, or longitudinally, according to repeated visits on each subject: (1) with age comes a slowing of RT, (2) across decades, males are faster than females, (3) DRT is always slower than SRT, (3) decision time (*i.e.*, DRT-SRT) increases with age, and (4) gender differences in SRT are small and unaffected by age, but gender differences in DRT are larger and increase with advancing age.

Cross-sectional Research. Several other cross-sectional studies obtained results consistent with the BLSA age results (*e.g.*, Bleecker, Bolla-Wilson, Agner, & Meyers, 1987; Botwinick & Storandt, 1974; Era, Jokela, & Heikkinen, 1986; Goldfarb, 1941; Harkins, Nowlin, Ramm, & Schroeder, 1974; Lahtela, Niemi, & Kuusela, 1985; Pierson & Montoye, 1958; Simon, 1967; Spirduso, 1975; Spirduso & Clifford, 1978; Szafran, 1951; Vrtunski, Patterson, & Hill, 1984; Waugh, Fozard, Talland, & Erwin, 1973).

Gender differences favoring males have been found in many studies (*e.g.*, Bleecker, Bolla-Wilson, Agnew, & Meyers, 1987; Botwinick & Storandt, 1974; Harkins, Nowlin, Ramm, & Schroeder, 1974; Lahtela, Niemi, & Kuusela, 1985; Simon, 1967). Not all cross-sectional studies, however, observed such uniform results. For example, Noble, Baker, and Jones (1964) found males faster than females, except in the 71-87 year age group where females responded faster. Landauer (1981) and Landauer, Armstrong, & Digwood (1980) found no significant gender differences in total reaction time. However, both studies observed significantly faster decision times for females, and significantly faster movement times for males. Results of Lahtela, Niemi, and Kuusela's (1985) CRT task showed that men generally had higher error rates than women. In the BLSA, on the other hand, women had lower error rates than men for DRT, but higher rates for SRT.

Longitudinal Research. Even though the Bonn Longitudinal Study of Aging employed a different type of CRT task, their findings were similar to the BLSAs (Mathey, 1976). In the Bonn study, subjects responded to a colored-light stimulus by pressing the button of the same color. The rate of stimulus presentation was gradually increased until the subject's error rate reached 50 percent. The total time taken to respond to a set number of stimuli was called the "circulation period."

Bonn researchers found that the mean and standard deviation of the circulation periods increased with age. These changes were more dramatic between the 60-70 year olds and the 70-80 year olds than between the 20-30 year olds and the 60-70 year olds (Mathey, 1976). This is consistent with the age-related slowing observed in the BLSA data.

Regarding gender differences, the BLSA was again consistent with the Bonn study. Though not as sizeable as the age-related differences, significant gender differences were found. Males were faster than females in all age groups, across all measurement points. The sex-related differences, like the age-related differences, became larger with age; gender differences were smallest in the 20-30 year old group, larger for the 60-70 year olds, and largest for the 70-80 year olds. In addition, women tended to have slightly higher standard deviations than men, but these differences were not observed in all samples at all measurement points (Mathey, 1976).

In discussing the Duke results, Shock (1985) states that "[t]he inference, drawn from averages based on cross-sectional observations, that functions gradually decline over the entire adult life span was contradicted by the longitudinal finding that a substantial number of subjects aged 65 and over showed no decline in health status or intellectual function, and that some actually showed improvement in health (Maddox & Douglass, 1973, 1974) over a number of years."

Longitudinal analysis of the reaction time data from the Duke Longitudinal Studies produced somewhat different results. Siegler (1977) found only small SRT and CRT changes across the first five examinations. "Cumulative frequency distributions developed for a subset of the subjects indicated that the major longitudinal change was in the increased variance of performance across time" (Busse & Maddox, 1985, p. 82).

Finally, a technical point deserves some attention. There is evidence that the high and low RT cutoffs were too restrictive, thereby ignoring valid data from the fast young and slow old subjects. The data collection software was designed with a response window of 150 to 800 ms. In retrospect, omitting data less than 150 ms is actually truncating the fast tails of simple reaction time distributions from many subjects, particularly young, physically fit males. This may slightly bias the data against younger or faster subjects whose best efforts would be scored as errors. On the other side of the RT distribution, data may be biased against the slow responders in cases where RT trials longer than 800 ms are ignored and labelled as errors. For example, some older subjects had average DRTs of greater than 700 ms with error rates over 50 percent suggesting that this paper is under-representing disjunctive reaction times of the elderly. Some five percent of the DRT data was not used in analysis because error rates were in excess of 45 percent. Several older subjects performed at 90-100 percent error rates because their responses were greater than 800 ms. Thus, a slightly larger sampling window (*e.g.*, 120-1200 ms) would have shown even greater differences between young SRTs and old DRTs (*i.e.*, the age x task interaction).

CONCLUSIONS

This research quantifies, via cross-sectional and

longitudinal experimental procedures, the degree to which reactive capacity declines with age as a function of gender and type of reaction task. Simple reaction time was less influenced by age and gender than was disjunctive reaction time, which supports the notion that, in addition to a generalized slowing of central nervous system functions, aging disrupts decision making processes and higher cortical functions. Longitudinal analysis revealed a constant rate of slowing across the life span comparable with that revealed by the cross-sectional analysis. The longitudinal analyses, therefore, did not produce a more optimistic picture of age-related slowing.

ACKNOWLEDGMENTS

Data collection was funded by the National Institutes of Health, National Institute on Aging (NIA). Data analyses were funded in part by the Baltimore Longitudinal Study of Aging (BLSA) and the National Institute of Occupational Safety and Health's (NIOSH) Southern California Educational Resource Center. This research was also supported by equipment grants from the Parsons Foundation and the University of Southern California Biomedical Research Support Fund. Data were collected at the GRC by J. Wood, J. Carre, and D. Baroc. The authors are particularly grateful for ideas developed from conversations with J. E. Birren, A. T. Welford, J. M. McDowd, V. D. Buckles, and coworkers both in the USC Laboratory of Attention and Motor Performance and the NIA's Gerontology Research Center; and computer assistance from J-Y Chang (USC), F. deBalogh (USC), C. Eames (BLSA), and O. Olofinboba (USC).

REFERENCES

Andres, R. (1978). *The normality of aging: The Baltimore Longitudinal Study.* Part of the National Institute on Aging Science Writer Seminar Series. Washington, D.C.: U.S. Department of Health, Education, and Welfare.

Birren, J. E., Vercruyssen, M., & Fisher, L. M. (1990). Aging and speed of behavior: Its scientific and practical significance. *The 1990 Sandoz lectures in gerontology*. New York: Academic Press, in press.

Birren, J. E., Woods, A. M., & Williams, M. V. (1980). Behavioral slowing with age: Causes, organization, and consequences. In L. W. Poon (Ed.), *Aging in the 1980s: Selected contemporary issues in the psychology of aging* (293-308). Washington, D.C.: American Psychological Association.

Bleecker, M. L., Bolla-Wilson, K., Agnew, J., & Meyers, D. A. (1987). Simple visual reaction time: Sex and age differences. *Developmental Neuropsychology*, **3**(2), 165-172.

BMDP Statistical Software, Inc. (1988). *BMDP [statistical] programs* and *BMDP Statistical Software Manual*. Statistical software, BMDP Statistical Software, Inc., Los Angeles, CA.

Botwinick, J., & Storandt, M. (1974). Cardiovascular status, depressive affect, and other factors in reaction time. *Journal of Gerontology*, **29**, 543-548.

Busse, E. W., & Maddox, G. L. (Eds.). (1985). The Duke Longitudinal Studies of normal aging, 1955-1980. New York: Springer.

Damon, A. (1965). Discrepancies between findings of longitudinal and cross-sectional studies in adult life: Physique and physiology. *Human Development*, **8**, 16-22.

Era, P., Jokela, J., & Hoikkinen, E. (1986). Reaction and movement times in men of different ages: A population study. *Perceptual and Motor Skills*, **63**, 111-130.

Finch, C. E., & Schneider, E. L. (Eds.). (1985). *Handbook of the biology of aging*. New York: Van Nostrand Reinhold.

Fozard, J. L. (1981). Speed of mental performance and aging: Costs of age and benefits of wisdom. In F. J. Pirozzolo and G. J. Maletta (Eds.), *Behavioral assessment and psychopharmacology*, (pp. 59-94). New York: Praeger Scientific.

Fozard, J. L., & Fisk, A. D. (1988). *Human factors and the aging population*. *Human Factors Bulletin*, **31**(11), 1-2.

Fozard, J. L., Thomas, J. C., Jr., & Waugh, N. C. (1976). Effects of age and frequency of stimulus repetitions on two-choice reaction time. *Journal of Gerontology*, **31**, 556-563.

Goldfarb, W. (1941). An investigation of reaction time in older adults. *Teachers College Contribution to Education*, 831. New York: Teachers College, Columbia University, Bureau of Publication.

Harkins, S. W., Nowlin, J. B., Ramm, D., & Schroeder, S. (1974). In Palmore, E. (Ed.), *Normal aging II: Reports from the Duke Longitudinal Studies, 1970-1973*. Durham, NC: Duke University Press.

Lahtela, K., Niemi, P., & Kuusela, V. (1985). Adult visual choice-reaction time, age, sex and preparedness. *Scandinavian Journal of Psychology*, **26**, 357-362.

Landauer, A. A., Armstrong, S., & Digwood, J. (1980). Sex difference in choice reaction time. *British Journal of Psychology*, **71**, 551-555.

Mathey, F. J. (1976). Psychomotor performance and reaction speed in old age. In H. Thomae (Ed.), *Contributions to human development, Volume 3: Patterns of aging -- Findings from the Bonn Longitudinal Study of Aging* (pp. 36-50). New York: Karger.

Munro, S. J. (1951). The retention of the increase in speed of movement transferred from a motivated simple response. *Research Quarterly*, **22**, 229-233.

Noble, C. E., Baker, B. L., & Jones, T. A. (1964). Age and sex parameters in psychomotor learning. *Perception and Motor Skills*, **19**, 935-945.

Olofinboba, O., & Vercruyssen, M. (1990). *DAS: Data analysis system for IBM compatible microcomputers, Version 2.05.* Statistical software by M. Vercruyssen, Human Factors Department, ISSM, University of Southern California, Los Angeles, CA 90089-0021.

Pierson, W. R., & Montoye, H. J. (1958). Movement time, reaction time, and age. *Journal of Gerontology*, **13**, 418-421.

Sage, G. H. (1977). *Introduction to motor behavior: A neuropsychological approach*. Menlo Park, CA: Addison Wesley.

Shock, N. W. (1985). Longitudinal studies of aging in humans. In C. E. Finch and E. L. Schneider, *Handbook of the biology of aging*, (pp. 721-743). New York: Van Nostrand Reinhold.

Siegler, I. C. (1977). Longitudinal reaction time patterns (Unpublished manuscript, Duke University).

Simon, J. R. (1967). Choice reaction time as a function of auditory SR correspondence, age and sex. *Ergonomics*, **10**, 659-664.

Spirduso, W. W. (1975). Reaction and movement time as a function of age and physical activity level. *Journal of Gerontology*, **30**, 435-440.

Spirduso, W. W., & Clifford, P. (1978). Replication of age and physical activity effects on reaction and movement time. *Journal of Gerontology*, **33**, 26-30.

Szafran, J. (1951). Changes with age and with exclusion of vision in performance at an aiming task. *Quarterly Journal of Experimental Psychology*, **3**, 111-118.

Thomae, H. (Ed.) (1976). *Patterns of aging -- Findings from the Bonn Longitudinal Study of Aging (Contributions to human development, Volume 3)*. New York: Karger.

Vrtunski, P. B., Patterson, M. B., & Hill, G.O. (1984). Factor analysis of choice reaction time in young and elderly subjects. *Perceptual and Motor Skills*, **59**, 659-676.

Waugh, N. C., Fozard, J. L., Talland, G. A., & Erwin, D. E. (1973). Effects of age and stimulus repetition on two-choice reaction time. *Journal of Gerontology*, **28**, 466-470.

Wilkie, F., Eisdorfer, C., & Siegler, I. (1975). Reaction time changes in the aged. In *Proceedings of 10th International Congress of Gerontology*, Vol. 2, 177. (Jerusalem, Israel).

AGE AND FITNESS DIFFERENCES IN THE EFFECTS OF POSTURE AND EXERCISE ON INFORMATION PROCESSING SPEED

Anita M. Woods[1], Max Vercruyssen[2,3], and James E. Birren[4]

[1] *Center on Aging, University of Texas, Houston;* [2] *Human Factors Research Laboratory, University of Minnesota, Minneapolis;* [3] *University of Hawai'i, Honolulu;* [4] *Borun Center for Gerontological Research, University of California, Los Angeles*

This experiment sought to determine if posture- and exercise-induced neuro-stimulation influences age differences in reaction (RT) and movement (MT) time, and whether obtained effects varied with physical fitness level. Thirty-six healthy male participants (18 young (19-29 yrs) and 18 old (60-69 yrs), with each group divided into the fit or unfit) performed both simple and two-choice visual reaction time tasks under six arousal/activation conditions: three postural changes (supine, sitting, standing) and three different relative workloads on a cycle ergometer (free pedaling, 20% HRR_{max}, 40% HRR_{max}). Consistently, RTs were slower for the older vs. young adults but the elderly performed fastest when Standing than when Sitting or Lying, whereas posture effects were negligible in the young. During exercise SRTs in the young and Old Fit were not greatly influenced by fitness level or arousal/activation condition, but the Old Unfit benefitted from moderate (20% HRRmax) exercise-induced neuromuscular activation thereby accounting for a portion of age-related cognitive slowing by providing evidence that the elderly function at a less activated (aroused) level than young adults and may benefit from circumstances which elevate these levels. An opposite pattern occurred in MTs for the Old Unfit -- for both posture and exercise: increases in arousal/activation caused increases in MT but in a fashion not supporting an RT-MT tradeoff in response strategy. Posture and exercise does affect speed of response, and may reduce age differences especially for those who possess already slowed response latencies.

INTRODUCTION

Slowing of response speed with advancing age is one of the most replicated findings in life span developmental studies (e.g., Birren, Vercruyssen, & Fisher, 1990; Birren, Woods, & Williams, 1980; Salthouse, 1985; Vercruyssen, in press), and this effect becomes more evident with increases in task complexity (e.g., Birren, Riegel, & Morrison, 1962). A major factor in behavioral slowing is a change in the nervous system at a subcortical level, which is reflected in processes mediated by the central nervous system (CNS). The ascending reticular activating system (ARAS) may be this subcortical mechanism because this complex system influences most behavioral processes by virtue of its anatomy as well as its integral function in attention and arousal. The reticular formation plays a major role in modulating postural reflexes and exerts a great deal of influence in maintaining an upright position against the forces of gravity. Arousal/activation theory in relation to aging suggests that the older CNS functions at a lower level of arousal/activation than the younger CNS and that because it is less activated, it is less sensitive to environmental input, takes longer to integrate information, and is slower in responding. Thus, neuromuscular activity, induced by postural changes or cycling exercise, might be used to create different levels of ARAS stimulation, and thereby activate or arouse the older CNS to assess whether a portion of age differences in speed of behavior can be attributed to sampling bias.

Numerous studies have shown psychomotor performance advantages for regular exercisers compared to less fit individuals (e.g., Botwinick & Thompson, 1968; Clement, 1966; Powell, 1974; Powell & Pohndorf, 1971; Spirduso, 1975, 1980, in press; Spirduso & Clifford, 1978; Tredway, 1978). de Vries (1970) proposed that improved cognitive performance exhibited by the more fit individuals resulted from "central stimulation" or arousal, rather than from physical conditioning. Arousal theory (e.g., Hebb, 1955; Lindsley, 1951, 1958, 1970; Malmo, 1959, 1972; Moruzzi & Magoun, 1949; Yerkes & Dodson, 1908) suggests that at low levels of activation individuals are less sensitive to environmental cues and integrate information less efficiently, resulting in slower reaction times (Woods, 1981). Such arousal is presumed to result in optimal performance at moderate levels--performance tends to be poor if the arousal is too high or too low.

Obrist (1965) proposed that the degree to which age differences in response speed are manifest relies upon the extent to which the aging cortex is capable of being activated by sensory stimulation via the reticular system. Arousal theory places emphasis on the ascending reticular activating system (ARAS) as the primary source of excitation to the cortex and responsible for maintaining arousal or activation (Kilner, McCullough, & Blum, 1969; Pribram & McGuinness, 1975). Issac (1960) reported that stimulation of the RAS, either electrically or by increased sensory input, shortened RT. Thus a major concern of this research was to look at age-related differences in speed of behavior within the framework of arousal and general activation theories. Specific questions were: (1) Do experimentally manipulated levels of arousal/activation affect speed of information processing and, if so, do they interact with age? (2) Do arousal/activation effects vary according to fitness level of the performer and task complexity?

METHOD

Participants

Thirty-six healthy male participants were tested: 18 young participants between the ages of 18 and 29 (nine Young Fit, $\underline{M}$ = 24.7 $\pm$ 2.7 yrs; nine Young Unfit, $\underline{M}$ = 23.1 $\pm$ 2.3 yrs), and 18 older participants between 60 and 69 years (nine Old Fit, $\underline{M}$ = 65.4 $\pm$ 4.3 yrs; nine Old Unfit, $\underline{M}$ = 66.2 $\pm$ 2.1 yrs). The younger participants were recruited from the University of Southern California (USC) Introductory Psychology and Physical Education courses. The older participants were USC faculty and staff, Senior Olympics participants, and volunteers at the Andrus Gerontology Center. Although the older fit volunteers had a few years more formal education than the other groups, every effort was made to insure educational background was similar across groups (young fit $\underline{M}$ = 16.8 $\pm$ 2.3 yrs; young unfit $\underline{M}$ = 16.0 $\pm$ 2.3 yrs; old fit $\underline{M}$ = 19.4 $\pm$ 2.0 yrs; old unfit $\underline{M}$ = 16.8 $\pm$ 1.9 yrs).

All volunteers were screened via an initial health status interview and questionnaire; participants not meeting the health criteria were excluded. Volunteers were excluded who showed a history of problems of the joints, back, or cardiovascular system. Also, volunteers were excluded who were found to be even mildly hypertensive by measurement of blood pressure, i.e., having a diastolic pressure reading greater than 90 mm.

All participants received the same experimental procedures and counterbalanced conditions, but were assigned on an *a priori* basis to two subgroups--fit or unfit--for purposes of analyses. The fitness criterion was based on the time spent pedaling a bicycle ergometer under a linearly increasing submaximal workload. The young adult volunteers capable of continuing beyond 200-225 watts during minute 9 were classified as Young Fit and those unable to continue were called Young Unfit. Three stopped at the cutoff point and were split to make an even nine in each group. A similar procedure was used for separating the older adults by fitness but the cutoff was 150-175 watts during minute seven. Five stopped at the cutoff point and were split to make an even nine in each of the Old Fit and Old Unfit groups.

Task and Apparatus

Simple and two-choice reaction times were measured from the onset of tachistoscopically presented visual stimuli consisting of either an "X" or a "Y" to a manual release of a microswitch. Stimuli were preceded by a variable foreperiod (500 ms average) visual signal to avoid anticipation and presented by a Gerbrands (model #G1175) tachistoscope with three Kodak Ektagraphic carrousels which projected stimuli onto a rear projection screen. An Apple II computer system controlled the random presentation of the stimuli for both RT tasks. The screen position, distance, and stimulus contrast were constant across conditions. Data for each RT trial was printed offline, including RT, MT, and correctness of each response.

Participants responded to each visual stimulus by releasing a microswitch "start" button ("S") with the middle fingers of their dominant hand and then quickly pressing either the "X" or "Y" microswitch. The microswitches were encased in a 5.1 by 7.6 cm metal box, or "RT response box" with raised buttons which made easier their depression. In the postural conditions, the RT box was fitted into a miniature slant board apparatus to allow participants to raise or lower the RT response box depending on their individual needs for comfort as the postural condition demanded. In the exercise (cycling) conditions, the RT response box was fastened to the middle point of the exercycle handlebars. Participants could comfortably use his dominant hand to respond while holding onto the handlebars with the other hand. The bicycle ergometer was connected to a Collins Pedal-Mode Ergometer which displayed information on workload (watts) and pedal revolutions per minute (rpm). A Morehouse-Collins cardiotachometer recorded true heart rate and could be programmed to maintain a pre-set heart rate level for the by modifying workload on the cycle ergometer. Participants were instructed to hold down the button marked "S" (for "starting") on the RT response box at the beginning of each RT trial. Following a variable foreperiod and visual warning signal, participants were instructed to, as quickly as possible, at stimulus onset remove their finger from the starting button and to depress one of the two response buttons, marked "X" or "Y," corresponding to the stimulus which was flashed on a viewing screen. Instructions implicitly emphasized accuracy over speed.

Procedure

Testing was conducted over two days in a manner which intentionally confounded posture and exercise: the posture manipulations always preceded the exercise conditions.

Test Day 1. Prior to the experiment, each participant was told of the nature of the procedures. He then completed a medical questionnaire and informed consent form. Resting heart rate (HR) was then measured and a 6-min submaximal stress test administered using a bicycle ergometer under a linearly increasing workload followed by a 2-min recovery. Exercise heart rate was measured continuously and recorded every minute.

Fifteen practice trials with feedback on the SRT and CRT tasks were performed in each of the three postural conditions each day. Data collection began following this practice period. For each postural condition, one block of 30 SRT and one block of 30 more complex (two-choice) RT trials were presented. The participant was informed whether the upcoming block of trials would consist of simple or choice RT trials. In the sitting postural condition, each participant sat facing the projection screen with his dominant hand resting on the RT box, which was situated in a slant bored adjusted to a comfortable height. In the standing position, each participant stood facing the projection screen with his hand positioned on the RT response box with the slant board again adjusted to a comfortable height such that it was not necessary to bend over to press the RT response buttons. In the lying position, participants were supine on a long table with a pillow propped under his head to facilitate full view of the projection screen directly in front of him. Visual angle was constant across posture and exercise conditions. A pre-determined counterbalanced order of the postural conditions (and the exercise conditions on Day 2) to ensure that all possible orders were used equally. The order of blocks of SRT and CRT trial blocks per condition were randomly determined.

Test Day 2. The same procedures were repeated on the second day; however, arousal/activation was manipulated via physical activity (pedaling) rather than postural changes. Upon entering the laboratory, the experimenter explained the treatment conditions and asked participants to sit on the bicycle ergometer. To familiarize each participant with the new positioning of the RT response box on the ergometer handlebars, 15 practice trials of both the SRT and CRT tasks were administered. The nondominant hand held the handlebar while the RT task was performed with the dominant hand. Also, practice was given in matching metronome beats and cycling rate. All participants across pedaling conditions maintained a pedal rate of 60 rpm. Heart rate was monitored via ear plethysmograph connected to an electric device which displayed average HR and pedal rpm while automatically regulating workload through pedal resistance to maintain each participant's target heart rate.

The experiment began after this practice period. If the first exercise condition was either 20% and 40% HRR_{max}, each participant pedaled for two min under a linearly increasing workload to attain his target HR. Then, during the third min, workload was set to allow HR to stabilize. During the first three min of the "free-pedal" condition, participants cycled under no resistance (i.e., 0.0 watts). Following this initial 3-min adaptation period, one block of 30 SRT and one block of 30 CRT trials were completed. Before the next condition began, heart rate was allowed to return to within 10 beats of resting level. Testing continued until RT data were recorded for each of the three exercise conditions. Testing took approximately ?? min per person per session.

Treatment of Data

Dependent measures recorded for each experimental condition were: median and mean RTs and MTs plus the percentage errors of commission for both simple and choice RT tasks. Data were analyzed using SAS and BMDP on an intel 80486 microcomputer according to 2 x 2 x 6 (age x fitness x activation) mixed ANOVA designs with repeated measures on the third factor (three postural conditions on Day 1 and three exercise intensities on Day 2) with degrees of freedom epsilon and Huynh-Feldt adjustments of the *p* values and Tukey *WSD post hoc* tests.

RESULTS

While the pattern of results were similar for CRT and SRT, allowing CRT, MT, and errors to vary simultaneously made interpretation of the CRT data difficult. A number of interactions suggest that there may have been RT-MT and speed-accuracy tradeoffs that require more careful interpretation. The SRT results were without complication and are herein reported. Details of the method and preliminary analyses are presented elsewhere (e.g., Woods, 1981). Analyses of median and mean SRT and MT produced nearly identical results so only average median values for correct responses will be reported.

In general, SRT of the younger participants and the Old Fit were not greatly influenced by either posture or exercise conditions. Speed of the Old Unfit participants, however, was improved by posture and moderate exercise. In addition to the obvious age [$F(1,32)=23.182$; $p<.0005$; $WSD=21.7$] and fitness [$F(1,32)=16.345$; $p<.0005$; $WSD=21.7$] differences, Figures 1 and 2 illustrate almost no effect of posture or exercise on SRT in the young and a slight improvement for the Old Fit from lying to sitting to standing. This posture effect is most pronounced in the Old Unfit with no further gain with increased arousal/activation from exercise, in fact, an increase RT with the highest exercise intensity tested (40% HRRmax). Across arousal conditions there was an age x fitness interaction [$F(1,32)=5.154$; $p=.030$; $WSD=40.8$]. Arousal significantly affected SRT [$F(5,160)=2.953$; $p=.014$, $Hp=.023$; $WSD=24.9$] but there was an interaction of arousal conditions with age [$F(5,160)=2.465$; $p=.035$, $Hp=.048$; $WSD=40.5$; all probabilities were lower for analyses of the means, e.g. here they were .004 and .016 respectively].

Percentage errors of commission are also shown in Figures 1 and 2. Not only were the unfit slower than the fit, but they also made more errors [$F(1,32)=7.830$; $p=.009$; $WSD=0.76$] and fitness interacted with arousal levels [$F(5,160)=3.084$; $p=.011$, $Hp=.019$].

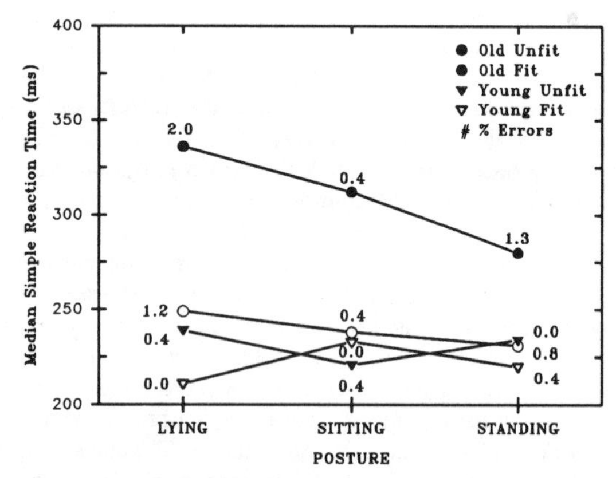

Figure 1. Mean of median simple reaction times and percentage errors as a function of age, fitness, and posture ($n=9$ per point).

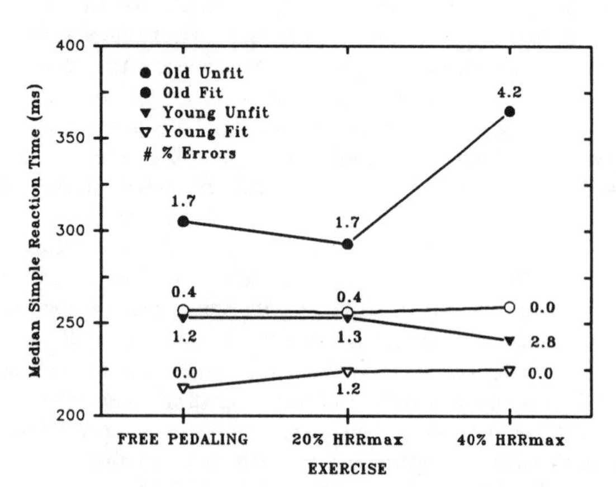

Figure 2. Mean of median simple reaction times and percent errors as a function of age, fitness, and exercise ($n=9$ per point).

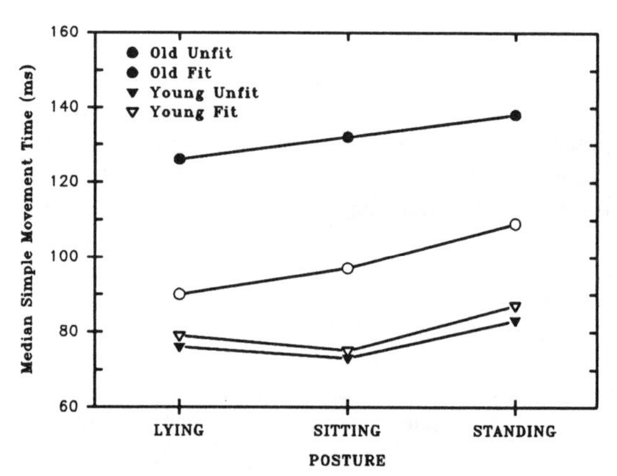

Figure 3. Mean of median simple movement times as a function of age, fitness, and posture ($n=9$ per point).

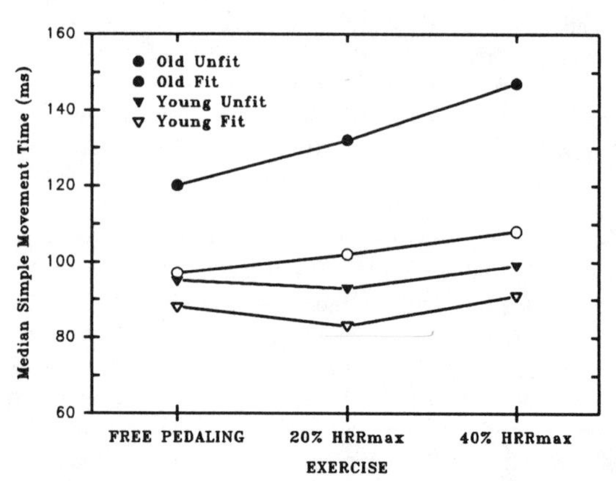

Figure 4. Mean of median simple movement times as a function of age, fitness, and exercise intensity ($n=9$ per point).

Median movement times (MT) for correct responses are shown across treatment in Figures 3 and 4. With progressive increments in arousal/activation, MT increases in older participants, particularly the Old Unfit. Overall, the older were slower than the young [$F(1,32)=27.399$; $p<.0005$; $WSD=12.3$] and the fit were faster than the unfit [$F(1,32)=8.231$; $p=.007$; $WSD=12.3$] except in Figure 3 where a reversal occurred in the young making a age x fitness interaction [$F(1,32)=5,883$; $p=.021$; $WSD=23.1$]. Increments in arousal/activation caused increases in MT [$F(5,160)=5.069$; $p<.0005$, $Hp=.003$; $WSD=12.5$]. The correlation of median SRT with median MT was .49 overall, .31 for the young, .47 for the old, .38 for the fit, and .54 for the unfit.

DISCUSSION

This study manipulated arousal (neuromuscular activation) via postural changes and increments of exercise intensity and found that varying levels of arousal/activation influenced psychomotor performance in older but not young participants. A significant main effect was found for age across all activation conditions and at both levels of task complexity--young participants were faster than the older participants in SRT, CRT, and corresponding MTs. The Old Fit were significantly faster than the Old Unfit across all treatments and measures. For the SRT tasks, the Young Unfit and the Old Fit were not significantly different. In general, RTs of the Old Fit group were more like those of the Young Unfit than those of their unfit peers. That is, differences in RT and MT between the Old Fit and the Young Unfit were smaller than those mean differences between the Old Fit and Old Unfit. Both older groups performed fastest in the Standing condition as compared to the Sitting and Supine conditions, whereas posture tended not to be a significant factor in the performance of the younger participants. Using the Sitting RTs for resting level measures, significant improvements in RT for the Old Unfit between their resting and 20% activation conditions, and significant improvements for the Old Fit between their resting and 40% activation conditions were found. In general, RTs and errors of commission were not greatly influenced by fitness level or activation condition in the young participants. There were significant fitness by exercise effects for the older participants, with the Old Fit group performing consistently faster and more accurately than the Old Unfit. Perhaps the older unfit CNS is more sensitive to subtle or relative changes in arousal/activation level than the Old Fit, and thus experiences the greater impact of those changes on performance.

Although merely speculation, three explanations may be advanced to explain faster RTs for the Old Fit than the Old Unfit: (1) frequent exercise causes adaptation to environmental demands (i.e., use of neural processes influences function by enhancing the manufacture and availability of neurotransmitter substances, preventing circulatory insufficiencies, and generally enhancing synaptic efficiency), (2) exercise may increase an older individual's attention or alertness and, in turn, may increase his sensitivity and responsiveness to environmental stimuli even when s/he is not engaged in physical activity (i.e., posture and exercise acts on the ARAS to elicit or "prime" the old CNS for optimal performance), and (3) the exercise effect may be a sampling artifact since fit and unfit groups are not alike in all ways except fitness (e.g., if the fit are better educated and more motivated, or compulsive, than the unfit, what drives them to exercise and maintain a healthier lifestyle may be responsible for the effect).

Practical implication for workstation design, personnel selection, and work-rest schedules are obvious (see also Vercruyssen & Simonton, in press). Less clear are the collateral or indirect benefits. For instance, central nervous system speed changes with posture also correspond to those found in strength. Isokinetic arm pull strength (e.g., Mital & Faard, 1990) and grip strength is greater when standing than when sitting and, at least for grip strength, this effect increases with duration in the posture (e.g., Clarke, Robertson, Gillies, & Ellis, 1991). Might normal declines in strength with age (e.g., Kallman, Plato, & Tobin, 1990) be lessened if older individuals were measured while standing?

CONCLUSIONS

Both simple and choice RT were found to vary as a function of arousal/activation level in the elderly but not in the young. For the Old Unfit a gradient effect of posture on RT held true -- RTs were progressively improved as arousal/activation state was increased (i.e., lying to standing to sitting). The Old Unfit seem to have a lower optimum "ceiling" of activation level--their performance peaked at 20% HRR_{max} and then dropped dramatically both in speed and accuracy in the 40% condition. The Old Fit, on the other hand, may not have even reached their optimal activation level--perhaps they could have gone up to 60% HRR_{max} and their performance may have continued to improve. Clearly, the older participants and most particularly the Old Unfit, benefitted more than the young from the induced neuromuscular activation of the postural changes and physical activity. These findings support the notion that the old, especially the Old Unfit individuals, are functioning at a less aroused/activated level than younger persons and a portion of age-related cognitive slowing can be removed by testing each group at their optimal level of activation. Slowing of behavior with age varies with physical fitness level and can be modified by induced neuromuscular stimulation. Age-related declines in CNS processing speed, and possibly many other cognitive functions, may be reduced immediately via postural stimulation and acute exercise bouts. Also, improvements in cardiovascular fitness from chronic exercise may mitigate cognitive slowing with increasing age.

ACKNOWLEDGMENTS

This paper is based on a reanalysis of data collected as part of a doctoral dissertation (Woods, 1981) and subsequent master's theses by Mihaly (1988), Cann (1990), Greatorex (1991), and Simonton (1992). Valuable comments on earlier drafts of this manuscript were provided by M.T. Cann, T. Mihaly, and anonymous Human Factors and Ergonomics Society reviewers. S.H. Zarit and R.A. Wiswell contributed to the design and methodology of this research.

REFERENCES

Birren, J.E., Riegel, K.F., & Morrison, D.F. (1962). Age differences in response to speed as function of controlled variations of stimulus conditions: Evidence of a general speed factor. *Gerontologia*, **6**, 1-18.

Birren, J.E., Vercruyssen, M., & Fisher, L.M. (1990). Aging and speed of behavior: It's scientific and practical significance. In M. Bergener, M. Ermini, & H.B. Stahelin (Eds.), *The Sandoz lectures in gerontology: Challenges in aging* (pp. 3-23). New York: Academic Press.

Birren, J.E., Woods, A.M., & Williams, M.V. (1980). Behavioral slowing with age: Causes, organization, and consequences. In L.W. Poon (Ed.), *Aging in the 1980s: Psychological issues* (pp. 293-308). Washington, D.C.: American Psychological Association.

Botwinick, J., & Thompson, L.W. (1968). Age differences in reaction time: An artifact? *The Gerontologist*, **8**, 25-28.

Cann, M.T. (1990). *Effects of age, gender, posture-induced arousal, and task loading on speed of behavior.* Unpublished master's thesis in Human Factors. University of Southern California, Los Angeles.

Clarke, M.M., Robertson, J.C., Gillies, J.H., & Ellis, R.M. (1991). Effect of body posture and time on grip strength in patients with cervical spondylosis. *Clinical Biomechanics*, **6**(2), 123-126.

Clement, F. (1966). Effects of physical activity on the maintenance of intellectual capacities. *The Gerontologist*, **6**(2), 91-102.

de Vries, H.A. (1970). Physiological effects of an exercise training regimen upon men aged 52 to 88. *Journal of Gerontology*, **25**, 325-366.

Greatorex, G.L. (1991). Aging and speed of behavior: CNS arousal and reaction time distribution analyses. Unpublished master's thesis in Human Factors. University of Southern California, Los Angeles.

Hebb, D.O. (1955). Drives and the C.N.S. (Conceptual nervous system). *Psychological Review*, **62**, 243-254.

Kallman, D.A., Plato, C.C., & Tobin, J.D. (1990). The role of muscle loss in the age-related decline of grip strength: Cross-sectional and longitudinal perspectives. *Journal of Gerontology: Medical Sciences*, **45**(3), M82-M88.

Lindsley, D.B. (1958). The reticular activating system and perceptual integration. In D.E. Sheer (Ed.), *Electrical stimulation of the brain.* Austin, TX: University of Texas Press.

Lindsley, D.B. (1970). The role of nonspecific reticulothalmocortical systems in emotion. In P. Black (Ed.), *Physiological correlates of emotion.* New York: Academic Press.

Malmo, R.B. (1959). Activation: A neurophysiological dimension. *Psychology Review*, **66**, 367-386.

Malmo, R.B. (1972). In N.S. Greenfield and R.A. Sternbach (Eds.), *Handbook of psychophysiology* (pp. 967-980). New York: Holt, Rinehart & Winston.

Mihaly, T. (1988). *Arousal and cognition: Effects of exercise-induced arousal (elevated heart rate), limb movement, and practice on four speed of response measures.* Unpublished master's thesis in Safety Science. University of Southern California, Los Angeles.

Mital, A., & Faard, H.F. (1990). Effects of sitting and standing, reach distance, and arm orientation on isokinetic pull strength in the horizontal plane. *International Journal of Industrial Ergonomics*, **6**(3), 241-248.

Moruzzi, G., & Magoun, H.W. (1949). Brain stem reticular formation and activation of the EEG. *Electroencephalography and Clinical Neurophysiology*, **1**, 455-473.

Obrist, W.D. (1965). Electroencephalographic approach to age changes in response speed. In A.T. Welford & J.E. Birren (Eds.), *Behavior, aging and the nervous system.* Springfield, IL: C.C. Thomas.

Powell, R.R. (1974). Psychological effects of exercise therapy upon institutionalized geriatric mental patients. *Journal of Gerontology*, **29**(2), 157-161.

Powell, R.R., & Pohndorf, R.H. (1971). Comparison of adult exercisers and non-exercisers on fluid intelligence and selected physiological variables. *Research Quarterly*, **23**, 70-77.

Pribram, K. H., & McGuinness, D. (1975). Arousal, activation and effort in the control of attention. *Psychological Review*, **82**, 116-149.

Salthouse, T.A. (1985). Speed of behavior and its implications for cognition. In J.E. Birren and K.W. Schaie (Eds.), *Handbook of the psychology of aging, 2nd ed.* (pp. 400-426). New York: Van Nostrand Reinhold.

Simonton, K. (1992). *Effects of posture on speed of behavior.* Unpublished master's thesis in Human Factors. University of Southern California, Los Angeles.

Spirduso, W.W. (1975). Reaction and movement time as a function of age and physical activity level. *Journal of Gerontology*, **30**, 435-440.

Spirduso, W.W. (1980). Physical fitness, aging, and psychomotor speed: A review. *Journal of Gerontology*, **35**, 850-865.

Spirduso, W.W. (book in press). *Physical dimensions of aging.* Champaign, IL: Human Kinetics.

Spirduso, W.W., & Clifford, P. (1978). Replication of age and physical activity effects on reaction and movement time. *Journal of Gerontology*, **33**(1), 26-30.

Tredway, V.A. (1978). *Mood effects of exercise program for older adults.* Unpublished doctoral dissertation. University of Southern California, Los Angeles.

Vercruyssen, M. (in press). Slowing of behavior with age. In R. Kastenbaum (Ed.), *Encyclopedia of adult development.* Phoenix, AZ: Oryx.

Vercruyssen, M., & Simonton, K. (in press). Effects of posture on mental performance: We think faster on our feet than on our seat. In R. Lueder & K. Noro (Eds.), *Science of seating.* London: Taylor & Francis.

Woods, A.M. (1981). *Age differences in the effect of physical activity and postural changes on information processing speed.* Unpublished doctoral dissertation. University of Southern California, Los Angeles.

Yerkes, R.M., & Dodson, J.D. (1908). The relation of strength of stimulus to rapidity of habit-formation. *Journal of Comparative and Neurological Psychology*, **18**, 459-482.

ARITHMETIC STROOP INTERFERENCE AS A FUNCTION OF AGE: MAINTENANCE AND MODIFICATION OF AUTOMATIC PROCESSES

Wendy A. Rogers and Arthur D. Fisk
Georgia Institute of Technology, Atlanta, GA

ABSTRACT

This experiment investigated whether well-learned "automatic" processes remain stable as a function of age, as well as whether the ability to modify automatic processes is disrupted for older adults. We used an arithmetic "Stroop" task. Nineteen young (mean 22) and 19 old adults (mean 75) participated in three sessions for a total of 450 trials. The young subjects had faster verification times, overall, than the old adults. Both young and old subjects showed significant Stroop interference. These results support the hypothesis that automatic processes, in this case access of addition and multiplication tables, are maintained for old adults. Furthermore, both groups reduced their RT with practice. For the young adults, there was a decrease in interference with practice suggesting that they were learning to inhibit the automatic process of performing the arithmetical operation. However, the old adults showed no significant decrease in interference, which implies that they were impaired in their ability to inhibit automatic processes, even when those processes interfered with performance. Theoretical and practical training implications are discussed.

INTRODUCTION

Given the ever-changing technology in today's work environment, understanding skills is critical, not only in terms of the maintenance of skills across time, but also the ease with which skills can be modified to adapt to changing situations. A related issue is whether the maintenance and modification of skills differs as a function of age. The purpose of this experiment was to investigate age-related changes (or lack thereof) in automatic processing, an important factor in skilled behavior. We investigated whether well-learned processes remain stable as a function of age, as well as whether the ability to modify automatic processes is disrupted for older adults.

The performance of most complex tasks involves the separate and interactive influences of two types of information processing: controlled and automatic processing (see Shiffrin and Dumais, 1981). Controlled processes govern the performance of novel tasks and are characteristically slow, serial in nature, and effortful. Automatic processes generally develop after extensive practice for the consistent components of tasks. The execution of an automatic process requires minimal resources, can occur without intention, and, once evoked, generally runs to completion. Logan (1985) has delineated the importance of automatic processing for most skilled tasks. He notes that, while skill and automaticity clearly are not synonymous terms, most skilled behaviors consist of various components which have been automatized.

Maintenance of Automatic Processing

Within the realm of aging research, it is generally assumed that automatic processes remain intact as a function of age.[1] In fact Hasher and Zacks (1979) include age-independence as a criterion for automaticity (although this is tautological, it does demonstrate the prevalence of the idea of age-independence of automatic processes developed prior to senescence).

Much of the evidence for the maintenance of automatic processes across the adult life span has come from investigations of the automaticity of semantic priming. For example, Chiarello, Church, and Hoyer, (1985) used a lexical access task and demonstrated that the automatic activation of words is not disrupted in older adults. Similarly, Cohn, Dustman, and Bradford (1984) demonstrated that automatic activation of words as measured by the Stroop color-word interference effect (see below) is not disrupted in older adults. Finally, the results of Light and Singh (1987) in the area of implicit memory, suggest that repetition priming effects occur equally for young and old adults.

Stroop Interference as a Measure of Automaticity

The traditional Stroop color-word test is frequently used to demonstrate the automaticity of lexical access. In the original experiment (Stroop, 1935) subjects were presented with squares of colored ink or color-words printed in an incongruent ink color (the word 'RED' printed in green ink). The subject's task was to name the ink color as quickly and accurately as possible. There was consistently a longer latency to name the ink color when it formed an incongruent color-word.

The interference effect has been attributed to perceptual conflict, response competition, or some combination thereof (see Dyer, 1973 for a review).

[1]This assumption is valid only for those processes developed prior to senescence. See Fisk, McGee, and Giambra (1988) for an example of differential age effects for the development of new automatic processes.

Perceptual conflict results from the necessity to ignore or inhibit one dimension of a stimulus (the word) while attending to another dimension of the same stimulus (the color). The response competition hypothesis is based on the contention that the processes of reading words and naming colors proceed in parallel to the point of response initiation but word reading is faster and thus occupies the response channel thereby interfering with the output of the color name. Both explanations assume that the presentation of the word results in its activation or as Shiffrin (1988) explains, "...the subject's intent to emit color names primes these responses and accounts for the interference they produce" (p. 23).

Dyer (1973) reviews various Stroop-like tasks which yield interference effects (for young adults). These tasks include: presenting north, east, south, and west in incongruent positions; combining direction names with the dimension of movement direction; and numerosity responses to counting a series of incongruent numerals. Many of these tasks produce smaller interference effects than the traditional Stroop with the exception of "The fairly high interference found...for enumeration tasks [which] suggests some comparability between color and numerosity dimensions" (p.110).

This "comparability" between semantic interference and numerical interference may be related to the proposal by Winkelman and Schmidt (1974) that there are associations in memory between pairs of digits and their sums and their products. They provide evidence to support their prediction that "...stimuli of the type 3+3=9 and 3x3=6 will produce a tendency to respond yes [in a verification task] because of associative interference..." (p. 374). They termed these occurrences of interference "associative confusions". Zbrodoff and Logan (1986) more precisely defined the conditions under which associative confusions occur. They required subjects to verify the correctness of equations which were: (1) Correct (3+4=7); (2) Associative - equations that are incorrect but become correct if the addition operation is substituted for the multiplication operation, or vice versa (3+4=12; 5x2=7); or (3) Nonassociative - equations that are incorrect (4+3=9). They concluded that the results "suggest that the process underlying simple arithmetic can begin without intention" (p. 129) which is a criterion for automaticity (see Shiffrin and Dumais, 1981).

To summarize, associative equations produce substantial slowing in verification time and are referred to as Stroop interference equations; essentially, the arithmetic operations are performed automatically and thus interfere with performance when the operators are reversed.

Our goal was to utilize this arithmetic Stroop task to determine if the associations of numbers with their products and sums (which result in their automatic activation upon presentation) are maintained for old adults. Based on the results of the arithmetic Stroop effects for young adults and color-word Stroop effects for old adults (e.g., Cohn, et al., 1984) we hypothesized that both young and old adults would show substantial Stroop-like interference on the associative equations and the degree of interference for the old adults would be at least as much, if not more, as the young adults.

Modification of Automatic Processing

Our second goal was to investigate the modification of the automatic process as evidenced by a reduction in interference with practice. There is evidence that interference resulting from automatic activation is reduced with practice. In the original experiments, Stroop (1935) found a 34% reduction in interference after eight days of practice. More recently, Harbeson, et al. (1982) provided 15 days of practice on the Stroop task reducing interference by 63%. Similarly, Ackerman and Schneider (1984) provided ten days of practice to reduce Stroop interference by 69%. They also tested a consistent Stroop condition in which the colors and words were distinct such that a specific color of ink was not used as a word representation. Thus the subjects could consistently ignore the words and selectively attend to the colors. There was a 77% reduction in this consistent Stroop condition.

The important question is why there is a reduction in the amount of interference on Stroop tasks as a function of practice. What is the subject learning that allows for this improvement? According to Shiffrin (1988) "...the subjects can, through attentive means, control the input to an automatic process, thereby exerting control over the process itself" (p. 26). Thus, while the encoding of the name presented by the printed word (or the product or sum of two numbers) may at first be automatically activated, the subject can learn to inhibit or selectively attend to the color, or in the arithmetic task, attend to the operation sign and inhibit the activation of the product or sum. Therefore, while it may be true that individuals at first cannot choose to avoid processing aspects of an input item that they desire to ignore, it may also be true that they can learn to ignore the information that is irrelevant, although this may take some time and practice.

If, in fact, the reduction of the interference effect is due to the learned inhibition of the word input then we predict that older adults will show less of a reduction in interference with practice than the young adults. Evidence for this prediction is threefold. First, as Ackerman and Schneider (1984) demonstrated, if there is an opportunity to `consistently` ignore the irrelevant information in the Stroop task (i.e., the word) there is a greater reduction in the interference effect with practice. Second, Fisk and Rogers (1989) showed that, relative to young adults, older adults have an attenuated ability to learn to ignore distracting or irrelevant information. Third, Cohn et al. (1984) suggest that their finding of greater Stroop interference for older adults may represent "deficits [relative to young adults] in inhibiting input of one stimulus dimension (words) while one is attending to a second stimulus characteristic (color)" (p. 1249).

METHOD

Subjects. Nineteen young adults (15 females) aged 19 to 24 (mean age 22) and 19 old adults (9 females) aged 70 to 89 (mean age 75) participated in the study. The young adults received course credit or cash for their participation. The old adults received $4.00 per hour. All subjects were screened for psychotropic drug use. Their corrected or uncorrected visual acuity was at least 20/40 for both far and near vision.

Apparatus and Stimuli. IBM PC XTs were programmed to present the appropriate stimuli, collect responses, and control timing of the display presentations. Stimuli were displayed on a standard IBM monochrome VDT. The stimuli were true and false addition and multiplication equations. All possible combinations of the numbers 1 through 5 were used to make up the left half of the equations. The equations were: (1) Correct (e.g., 3+4=7); (2) Associative - equations that are incorrect but become correct if the addition operation is substituted for the multiplication operation, or vice versa (e.g., 3+4=12; 5x2=7); or (3) Nonassociative - equations that are incorrect (e.g., 4+3=9). The false equations were balanced on each side such that the mean difference between the left and the right sides of the nonassociative equations were chosen to balance the mean difference between the left and the right sides of the associative equations (see Restle, 1970). There were 24 equations generated for each of the six conditions (addition and multiplication versions of the correct, nonassociative and associative type).

Procedure. There was an orientation plus three experimental sessions which were completed on two separate days within a 48 hour period. During orientation the subjects were given the eye and WAIS tests, written instructions on the procedure, and three 15-trial blocks of practice.

An individual trial consisted of the following sequence of events. Subjects pressed the '3' button on the IBM keyboard to initiate the trial. An equation would then appear in the center of the screen. The subject's task was to indicate whether the equation was true or false by pressing either the '9' (true) or '0' (false) button. The equation remained on the screen until the subject responded or for 5 seconds. Subjects were given an opportunity for a self-paced break after each block of 30 trials.

Following an incorrect response an error tone was sounded. Within each block, cumulative averages of reaction time and accuracy were displayed after each trial. The subjects were encouraged to respond as quickly and as accurately as possible.

Design. Fifty percent of the equations were true equations, half of which were multiplication (7 per block) and the other half were addition (7 per block). The remaining trials were 20% nonassociative addition (6 per block), 20% nonassociative multiplication (6 per block), 5% associative addition (2 per block), and 5% associative multiplication (2 per block) (see Zbrodoff and Logan, 1986). The order of trials was permuted with the stipulation that the same equation was not presented twice in a row. The subjects participated in three experimental sessions, each of which consisted of 15 blocks of 30 trials. The independent variable was Trial Type (true addition, true multiplication, nonassociative addition, nonassociative multiplication, associative addition, associative multiplication) which was a within-subject manipulation. The dependent variables were reaction time and accuracy. The nonexperimental variable was age.

RESULTS

Reaction Time. Mean reaction times (RT) for correct trials are presented in Table 1. These data were analyzed with an Age x Trial Type x Practice analysis of variance (ANOVA). The main effects of Age, $F(1, 36) = 57.65$, Trial Type, $F(5,180) = 111.85$, and Practice, $F(2,72) = 53.13$, were significant, as were the interactions of Age x Practice, $F(2,72) = 3.56$, and Age x Trial Type, $F(5,180) = 9.97$ (alpha level was set at $p < .05$ unless otherwise indicated). The young subjects had faster verification times overall than did the old subjects. The source of the Age x Trial Type interaction is the fact that for the old subjects, the difference between the associative equations and the other equations was greater than that for the young adults. The older subjects also showed a greater overall reduction in RT which accounts for the Age x Practice interaction (but, as reported below, the amount of interference was not reduced with practice for the old subjects). In fact, the planned test of Trial Type x Practice was significant for young adults, $F(10,180) = 2.85$, but not for old adults ($F < 1$).

Accuracy. An ANOVA conducted on performance accuracy showed that the main effects of Trial Type, $F(5,180) = 37.98$, and Practice, $F(2,72) = 8.45$, were significant, as were the interactions of Age x Trial Type, $F(5,180) = 2.39$, and Trial Type x Practice, $F(10,360) = 1.89$. The old subjects were less accurate on the associative trials, and for both age groups, the associative trials showed greater improvements in accuracy with practice.

Table 1. Mean RT for each equation type for young and old subjects across sessions.

YOUNG

	True		False			
			NonAssoc		Associative	
Sess	x	+	x	+	x	+
1	1011	1007	1011	974	1245	1226
2	913	919	890	883	1107	1080
3	890	894	871	860	1071	1055

OLD

	True		False			
			NonAssoc		Associative	
S	x	+	x	+	x	+
1	1808	1704	1825	1814	2206	2080
2	1632	1552	1665	1638	1966	1909
3	1570	1498	1578	1549	1926	1872

Associative Interference. The arithmetic Stroop effect was calculated by subtracting the RT for the associative trials from the RT for the nonassociative trials for each subject. This derived score provides an estimate of the associative confusion effect (Zbrodoff and Logan, 1986). The multiplication and addition interference effects are presented in Table 2. Both young and old adults were greatly slowed on the associative trials relative to the nonassociative trials. An Age x Trial Type (Addition or Multiplication) ANOVA on these data revealed a significant effect of Age, $F(1,36) = 8.68$, as well as Trial Type, $F(1,36) = 4.21$. These findings illustrate that the associative interference effect was greater for the old adults. The effect of Trial Type is significant because there is a greater interference effect for the multiplication than for the addition trials; this effect is stronger for the older adults as illustrated by the marginally significant Age x Trial Type, $F(1,36) = 3.72$, $p < .06$.

Table 2. Associative Interference

	Sess	Mult	Add
YOUNG	1	233	253
	2	217	197
	3	200	195
OLD	1	381	266
	2	301	270
	3	347	324

Practice Effects. For the young adults, there was a decrease in the interference effect with practice, $F(2,36) = 4.06$, suggesting that they were learning to inhibit the automatic process of performing the mathematical operation. However, the old adults showed no significant decrease in the interference effects observed for the associative equations ($F < 1$).

DISCUSSION

For both young and old subjects, response times for the Stroop equations were slower and less accurate than either the correct equations or the nonassociative false equations. There were also significant interference effects for both young and old subjects. These findings support the hypothesis that well-learned automatic processes, in this case access of addition and multiplication tables, are maintained for old adults.

In general, both young and old subjects reduced their RT with practice. For the young adults, there was a decrease in the interference effect with practice suggesting that they were learning to inhibit the automatic process of accessing both sums and products and perhaps simultaneously learning to attend more specifically to the mathematical operation. However, the old adults showed no significant decrease in the interference effects observed for the associative equations. This implies that the older adults were impaired in their ability to inhibit automatic processes, even when those processes interfered with performance.

The present findings have fundamental implications, both theoretically and from a practical training perspective. Theoretically, the results raise questions about a relationship between previous findings of an attenuated ability for old adults to develop new automatic processes and the present findings of age differences in the ability to modify or inhibit previously learned automatic processes. Fisk, McGee, and Giambra (1988) recently reported a series of experiments in which they have demonstrated (and replicated across task situations) an attenuated ability for older adults to develop new automatic processes. Fisk and Rogers (1989) propose a learning mechanism which may be disrupted as a function of age and accounts for the pattern of visual search performance in the literature. Briefly, the pattern shows similar improvements in performance for both young and old adults early in practice but divergent performance late in practice resulting in the development of automatic processes for young but an attenuation of this development for older adults. The learning mechanism is termed priority learning (Schneider and Detweiler, 1988) and allows for the development of automaticity by strengthening the associations of important (relevant) information and weakening information which is irrelevant for a particular situation.

Fisk and Rogers (1989) have proposed that an age-related disruption in this learning mechanism can explain the attenuated ability of old adults to develop new automatic processes because they are unable to sufficiently differentiate between important, target information and irrelevant, distracting information. This same mechanism, priority learning, may account for the present data. It is proposed that young adults show a reduction in Stroop interference due to their learning to inhibit the irrelevant information in the task, that is, inhibit the automatic access of the addition and multiplication tables and instead focus on the mathematical operator provided in the equation. The process of inhibiting the irrelevant information in the task is presumed to be a function of priority learning. Therefore, if the priority learning mechanism is disrupted with age, then we would expect that old adults would have difficulty inhibiting the distracting information and thus show greater and more persistent interference effects. This is the pattern of the present data.

Practically, the findings have implications for the development of training, and especially retraining, programs to either capitalize on processes which are already automatic or to allow sufficient time for the modification of automatic processes. The findings further support the contention that previously well-learned, automatic processes (i.e., those developed prior to senescence) will remain intact across the adult life span. This is an encouraging finding. However, the present results also suggest that if the need arises for the modification or inhibition of automatic processes, older adults may have difficulties. Thus, it is important to consider the degree to which changes in tasks or job requirements will require that individuals not only learn something new, but, in effect, have to unlearn previous methods. While the interference due to an automatic response to irrelevant task features was not eliminated for the young adults, the amount of interference did decrease with practice. It is possible that given sufficient practice, the interference could be eliminated. It is also possible that given more extensive practice, the amount of interference could be reduced for old adults. These possibilities remain to be tested and the results will provide important information for training applications.

ACKNOWLEDGMENT

This research was supported by grant No. 1R01AG07654) from the National Institutes of Health (National Institute of Aging). The authors would like to thank Nancy D. McGee for her assistance in the data collection and analyses.

REFERENCES

Ackerman, P.L., and Schneider, W. (1984). Practice effects and a model for Stroop interference. Paper presented at the Annual Meeting of the Psychonomic Society, San Antonio, TX.

Chiarello, C., Church, K.L., and Hoyer, W.J. (1985). Automatic and controlled semantic priming: Accuracy, response bias, and aging. Journal of Gerontology, 40, 595-600.

Cohn, N.B., Dustman, R.E., and Bradford, D.C. (1984). Age-related decrements in Stroop color test performance. Journal of Clinical Psychology, 40, 1244-1250.

Dyer, F.N. (1973). The Stroop phenomenon and its use in the study of perceptual, cognitive, and response processes. Memory and Cognition, 1, 106-120.

Fisk, A.D., McGee, N.D., and Giambra, L.M. (1988). The influence of age on consistent and varied semantic category search performance. Psychology and Aging, 3, 323-333.

Fisk, A.D., and Rogers, W.A. (1989). Toward an understanding of age-related memory and visual search effects: Why older adults are deficient in automatic process development. Submitted for publication.

Harbeson, M.M., Krause, M., Kennedy, R.S., and Bittner, A.C. (1982). The Stroop as a performance evaluation test for environmental research. Journal of Psychology, 111, 223-233.

Hasher, L., and Zacks, R.T. (1979) Automatic and effortful processing in memory. Journal of Experimental Psychology: General, 108, 356-388.

Light, L.L., and Singh, A. (1987). Implicit and Explicit memory in young and older adults. Journal of Experimental Psychology: Learning, Memory, and Cognition, 13, 531-541.

Logan, G.D. (1985). Skill and automaticity: Relations, implications, and future directions. Canadian Journal of Psychology, 39, 367-386.

Restle, F. (1970). Speed of adding and comparing numbers. Journal of Experimental Psychology, 83, 274-278.

Schneider, W., and Detweiler, M. (1988). The role of practice in dual-task performance: Toward workload modeling in a connectionist/control architecture. Human Factors, 30, 539-566.

Shiffrin, R.M. (1988). Attention. In R.C Atkinson, R.J. Herrnstein, and R.D. Luce (Eds.), Stevens' handbook of experimental psychology. NY: Wiley and Sons, Inc.

Shiffrin, R.M., and Dumais, S. (1981). The development of automatism. In J.A. Anderson (Ed.), Cognitive skills and their acquisition. Hillsdale, NJ: Erlbaum.

Stroop, J.R. (1935). Studies of interference in serial verbal reactions. Journal of Experimental Psychology, 18, 643-662.

Winkelman, J.H., and Schmidt, J. (1974). Associative confusions in mental arithmetic. Journal of Experimental Psychology, 102, 734-736.

Zbrodoff, N.J., and Logan, G.D. (1986). On the autonomy of mental processes: A case study of arithmetic. Journal of Experimental Psychology: General, 115, 118-130.

PAIRED ASSOCIATE LEARNING: AGE DIFFERENCES

Konstantinos V. Katsikopoulos, Donald L. Fisher, Michael T. Pullen
University of Massachusetts Amherst
Amherst, MA

The issue of age related differences in performance during the acqusition phase of a paired associate learning task is discussed within the framework of a precise mathematical tool. A two-stage, four-state Markov model is employed to analyze the data sets from two age groups consisting of 24 subjects each. The relative efficiencies of the acqusition processes of the younger and the older groups of adults are reflected in the different values of parameters. (These values were obtained by optimizing the fit of the model to the two data sets). The two major findings are: (i) the younger adults form associations (even temporary ones more easily) and (ii) these associations tend to decay less quickly, again in the younger adults. The results speak against the general decrement hypothesis, allthough further investigation is needed.

INTRODUCTION

This study was designed to generate evidence that could be used to understand better the differences in paired associate learning between younger (up to 30 years old) and older (above 60 years old) adults. Our goal was not only to investigate if such differences exist but also their nature as well. Specifically, the differences in performance can occur as a result of changes with age in any one of a number of different learning processes. These processes include the initial forming of an association between the stimulus and response in short term memory and the subsequent coding of an association first in what we call a critical state (an intermediate state between the short and long term memories) and then in long term memory. If the older adults are performing less efficiently, then for both theoretical and practical purposes it is important to know if all processes are degraded equally and, if not, which processes are degraded the most.

At this point it is important to understand *why* we study the paired associate task. We are motivated not only by the desire to make theoretical contributions, but because we hope to add to the practical store of benefits too. If the points where the elderly are especially weak are known, we can design more suitable training methods. We are indeed implying that the paired associate task is in essence a laboratory simulation of many real world tasks which the elderly often find difficult to perform, such as voice mail and the operation of ATMs (Fisher, in press). Learning the vocabulary of a foreign language is another example. Generally, we believe that the results can generalize to many everyday activities in a fairly straightforward fashion.

There has been considerable work on the matter. The first study to investigate age differences in paired associate learning was done by Ruch (1934). He concluded that age differences do exist. Most of the studies since have replicated this result (Korchin and Basowitz, 1957; Salthouse, Kausler and Saults, 1988). These differences though are not always of the same magnitude as demonstrated by Canestrari (1963) and Monge and Hultsch (1971) but depend on factors such as the time available to produce a response and the difficulty of the items. These findings actually imply that the psychological processes that underlie the task are *not* equally degraded. On the other hand, Fisher (in press) analyzed the results of a study by Kausler and Puckett (1980) and found the opposite pattern of results, that is all the processes were affected equally by the aging factor. Kausler (1992) calls this hypothesis the *general decrement hypothesis*. In this study, we want exactly to examine the validity of this hypothesis.

Arenberg and Tchabo (1994) have demonstrated that age differences in learning would probably be explained more adequately by integrating theoretical developments from many areas not currently considered. We believe that one of those areas that could prove useful is mathematical learning theory. During the 1950's and 1960's many models were developed that attempted to capture the nature of paired associate learning (see for example Estes, 1950; Bush and Mosteller, 1951; and Norman, 1963). But up to date we are not aware of any attempts to view age related differences in the context of such mathematical tools. In earlier studies, the emphasis was given to the quality of fits of the models to data generated mainly by using *younger* adults as subjects. Furthermore, the values of the parameters were not viewed as characterizing the core psychological processes of the task, and therefore would not have been seen as being able to provide a basis for comparing the efficiencies of the processes in the two age groups. Using a standard experimental paired associate paradigm (Pullen, 1995) but treating the data within the context of a Markov model proposed by Fisher (in press), we will view the problem in just this way. That is, we will argue that the parameters of the Markov model reflect the operation of the underlying processes and that, therefore, the effect of age on the efficiency of the core psychological processes can be determined by analyzing the values of the parameters for the younger and older adults.

METHOD

Twenty four younger and twenty four older adults were recruited from the University of Massachusetts and the local area. Each subject attended five consecutive hour long sessions approximately 24 hours apart.

The stimuli were presented on a monitor with a black background and amber or white characters using an IBM compatible computer. Responses were made on a standard computer keyboard.

Each stimulus in the paired associate was a CVC (consonant-vowel-consonant grouping) and it was always meaningless. Each response was one of four letters. The CVCs were taken from the Richardson and Erlebacher (1957) study. The four response sets were chosen so that the subject's index and middle fingers would be identically spaced for each session.

During each of four sessions, subjects were asked to learn a list of 24 pairs. Half of the stimuli in each list were classified as easy and half as difficult in terms of how familiar they were. Each paired associate was presented 16 times. Each list was divided in blocks with possible sizes of 24,12,8 or 4 pairs. Each block was repeated a fixed number of times before a new block was introduced and all blocks were randomized each time. The repetition level took the values of 16, 8 or 4. The repetition level was varied between subjects, the block size within subjects. Each subject was assigned to a repetition level and on each of four days was trained on a different list with 1 (of size 24), 2 (of size 12), 3 (of size 8) or 6 (of size 4) different blocks respectively. Eight subjects per age group were assigned to each repetition level. A given block size and a given repetition level defined a daily session (for example 4B4R defined a block size of 4 and a repetition level of 4). A unique set of CVCs was used in each training session. In all equivalent training sessions (fixed block size and repetition level) the presentation schedule remained the same. The dependent variable was accuracy.

Each subject was seated in front of the computer. At the beginning of each trial a '+' was displayed on the screen. Then the CVC would appear in the same position and the subject was prompted for a response. The subjects were given as much time as they wanted to provide the response. After responding, they would be informed of the correct response and would be given 2 seconds to study the pair.

RESULTS

Before presenting the data, we should talk about the Lag Sensitive Model that we shall use to analyze the results (Fisher, in press). Although more complex than many models of paired associate learning, this complexity is needed to explain results such as those we observed in the condition 4B4R (also see Young, 1971). It is a Markov model which assumes that the training process for each pair consists of two stages (the learning stage which occurs on a trial where the pair is trained and the forgetting stage which occurs on a trial where the the pair is not trained) and four states in each of the two stages. These states represent the *level of proficiency* of the subject in relation to a given pair or alternatively the *different memories* where the association the subject has formed resides. These different memories we use are much influenced by the short term memory theories that were developed in the 1960's; at the same time the mathematical learning models were developed too. So, if the association is learned (or equivalently it resides in the long term memory), the paired associate is assumed to be in the *Learned state L*. If the correct association between the stimulus and the response member of the pair is not learned, the paired associate is said to be in the *Unconditioned state U*. If the association that has been formed resides in short term memory (but has not yet been coded for storage in the long term memory), the paired associate is said to be in the *Short term state S*. For the cases where the association has been coded for storage in long term memory but is not yet completely learned, we introduce the *Critical state C*.

We need not feel uncomfortable if cognitive psychology theories of the past do not provide consistent support for the postulation of these states. When Estes (1960) proposed his famous two state model, he postulated the existence of only two states: *Unconditioned* and *Learned*. Of course there exist more possibilities for the status of an association and Estes was aware of that, but in an attempt to model a task mathematically one has to approximate the psychological theories which are usually much more descriptive. In our model it is mainly the existence of the *Critical state C* which can be questioned by theorists. Perhaps doubts could be overcome more easily if we focus first more on its mathematical than its psychological interpratation.

The transitions between the states are governed by two 4x4 (one for each stage) transition matrices. Each entry m_{ij} is the probability that after a trial (training or nontraining trial, depending on the stage the matrix represents) the pair will move from state i to j, where i,j take the values L,C,S and U. Thus, for example $p_T(C,L)$ would be the entry in the row C and the column L for the training matrix M_T and it would denote the probability that after a training trial a paired associate that resided in the *critical state C* would enter the *learned state L*. All these entries are probabilities, so they range from 0 to 1. A value of 0 means that the corresponding transition is impossible, while a value of 1 means that the corresponding transition is certain. Furthermore, all the entries of a given row

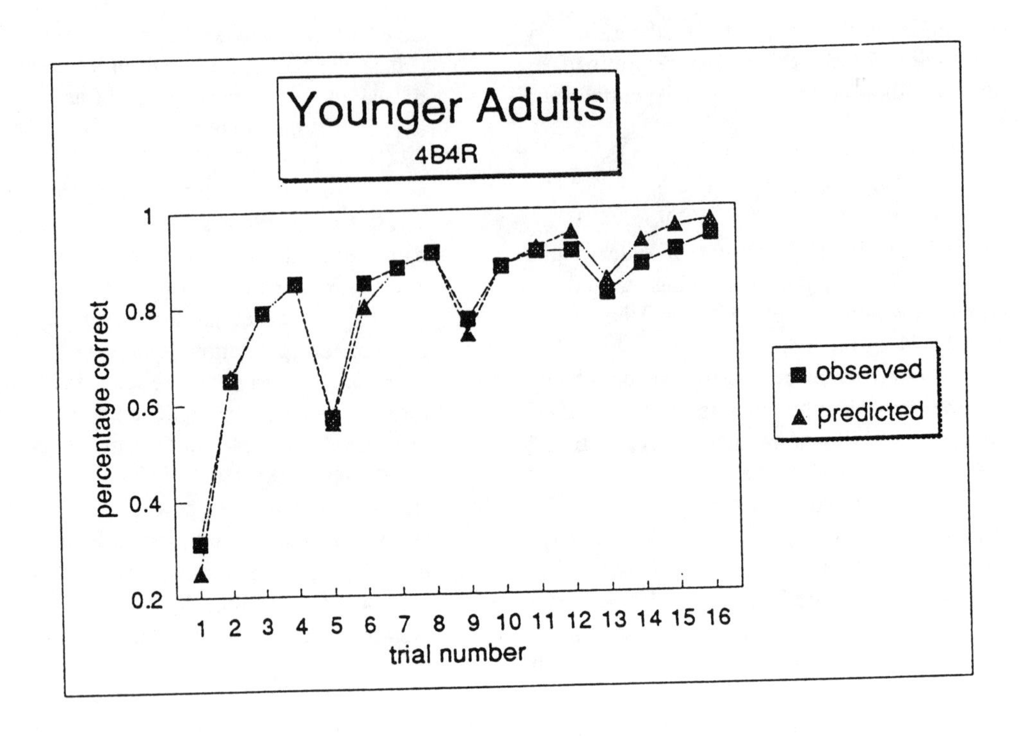

Figure 1. Predicted and Observed Learning Curve for Younger Adults

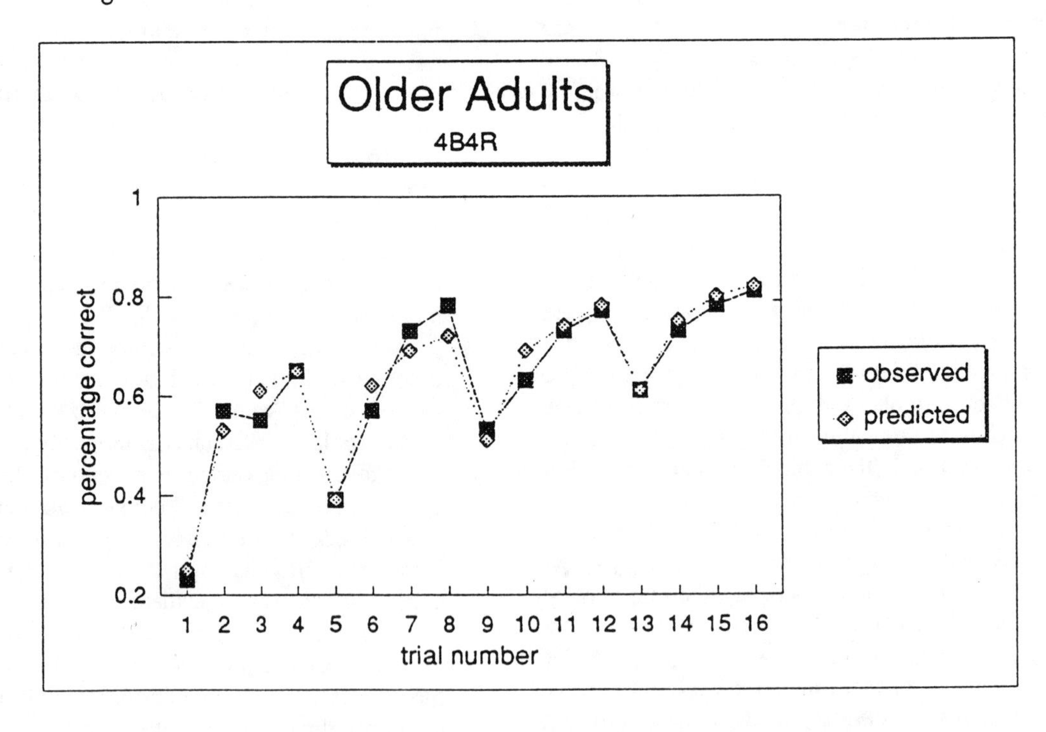

Figure 2. Predicted and Observed Learning Curve for Older Adults

should sum to 1, since it is impossible that none of the transitions occurs (including remaining in the same state). These constraints are actually very useful in our attempt at reducing the number of free parameters. Note that it seems we need 4 × 4=16 parameters for each matrix, so 32 for the model. But the fact that every four parameters have to sum to 1 and the fact that we have a consistent set of learning and forgetting axioms enable us to describe the process fully with six free parameters [which are $p_T(C,L)$, $p_T(C,C)$, $p_T(S,C)$, $p_T(U,S)$, $p_N(C,C)$ and $p_N(C,S)$].

The set of the axioms are: state L is *absorbing* in both stages, meaning that since it is entered it can not be left. We also assume that a pair which is not learned, can not be learned during a nontraining trial and that even during training a paired associate can not advance more than one state at a time. Finally, during nontraining a permanent association can not be formed and during training an association which already exists can not be lost (even if it is not permanent). We also assume that the subject guesses the correct response with probability g when in state U (*guessing axiom*).

The model is called the Lag Sensitive Model because it can exactly predict the lag effects on performance (see also Young, 1971). To do that we actually modify the chain of the forgetting stage to a non-markov model by letting $p_N(S_i,S_{i+1})$=exp(-*NII*/8) and $p_N(S_i,C_{i+1})=p_N(S_i,U_{i+1})$ where *NII* is the Number of Intervening Items.

Now, if we know the above 6 parameters and the training schedule we can predict the probability of a correct response on each of the 16 training trials in the 12 different conditions (4 block sizes × 3 repetition levels). In our analysis we will focus on the results for the 4B4R condition since other one stage models can not predict them (see Figures 1 and 2). This data has the form of 16 probabilities per age group and each probability is based on 24 (pairs) × 8 (subjects) = 192 observations. We can see a common (for both age groups) pattern in the data: accuracy increases during the first four trials, then it decreases in the fifth trial, then it increases again until the eighth trial, it drops in the ninth trial and so on. The most profound and logical explanation is the fact that the repetition level is 4, so each paired associate is trained 4 times with an average lag of 4 (block size), but then it will be trained again (for the fifth time) after *all* other paired associates that belong to other blocks are presented 4 times and similarly for the other circles until the last trial. Except for this common pattern, we can also see that the performance of the younger group is consistently superior.

Using an iterative search we determined the values of the parameters which yield the minimum chi-square (best fit). Both fits (for younger and older adults) were quite satisfactory and thus the Lag Sensitive Model can not be rejected out of hand (see Figures 1 and 2). Additionally, the observations did not differ form the predictions by more than an amount that one would expect by chance alone (the chi-square values were 5.50 and 5.79 for the younger and the older group respectively).

The values of the parameters were:

	Younger	Older
$p_T(C,L)$	0.6	0.9
$p_T(C,C)$	0.3	0.1
$p_T(S,C)$	1.0	1.0
$p_T(U,S)$	0.8	0.6
$p_N(C,C)$	0.7	0.1
$p_N(C,S)$	0.3	0.8

DISCUSSION

Now that we have a uniform framework to work within (our model explains the performance of both the younger and the older adults), we can examine the *general decrement hypothesis*. As stated before, this hypothesis argues that all the processes involved in the task are degraded equally in the case of the elderly. We will argue that this hypothesis must be rejected (since one of the processes is degraded in the case of the older adults and a second process operates more efficiently in older adults).

It is very important that we clarify some points: by *process* we understand all the transitions from one of the *memories* we described above, to one of the other *memories*. These transitions represent the rules that govern the processing of information in each state. And second we assume that a given transition probability reflects the succefullness of the corresponding transition or equivalently the percentage of material that is *on the average* transmitted between states. These two assumptions will lead us to our conclusions.

First suppose that a paired associate is not being trained. Thus we consider the matrix M_N. We have $p_N(C,C)$=.7 and $p_N(C,U)=1-p_N(C,C)-p_N(C,S)=0$ for the younger group. So, the younger adults retain 70% of the material that already existed in the critical state memory after a nontraining trial and it is impossible for them to lose completely an association. But for the older adults we have $p_N(C,C)$=.1 and $p_N(C,U)$=.1, so we can conclude that the processing of information in the critical state memory is degraded in the case of the older adults.

If we consider a paired associate that is trained, for the hypothesis to be valid we would want the transition from the critical state to the learned state to be more frequent for the younger group. But clearly that is not true: for the younger group $p_T(C,L)$=.6 and for the older group $p_T(C,L)$=.9

Much remains to be done. Some of the future directions could be: we should analyze in a similar fashion the results from the remaining conditions and ensure that our conclusions are not a product of chance factors. Regardless of the outcome of this more complete analysis, we believe that we can benefit considerably by viewing behavior in general and learning more specifically as a probabilistic process. The fact that uncertainty underlies the paired associate learning task need not make us uncomfortable; in fact this random element is a central feature of our approach and we believe it can help us explain age related differences.

ACKNOWLEDGEMENTS

Portions of this research were supported by a grant from the National Institute of Aging (R01-AG12461).

REFERENCES

Arenberg D. and Robertson-Tchabo E.A (1994). *Learning and Aging: Behavioral Pocesses. 421-432.*

Bush R.R and Mosteller F.A (1951). A mathematical model for simple learning. *Psychological Review, 58*, 313-323.

Canestrari R.E (1963). Paced and Self-Paced Learning in Young and Elderly Adults. *Journal of Gerontology, 18*, 165-168.

Estes W.K (1950). Toward a Statistical Theory of Learning. *Psychological Review, 57*, 94-107.

Estes W.K (1960). Learning theory and the new mental chemistry. *Psychological Review, 67*, 207-223.

Fisher D.L (in press). State models of skill acquisition: Optimizing the training of older adults. In W.A Rogers, A.D Fisk and N.Walker (Eds.), *Aging and skilled performance: Advances in theory and applications.*

Kausler D.H (1992).Comments on aging memory and its everyday operations. In L.W Poon, D.C Rubin and B.W Wilson (Eds.), *Everyday cognition in* adulthood and late life. Cambridge University Press, pp. 483-495.

Kausler D.H and Puckett J.M (1980). Frequency judgements and correlated cognitive abilities in younger and elderly adults. *Journal of Gerontology, 35*, 376-382.

Korchin S.J and Basowitz H. (1957). Age differences in verbal learning. *Journal of Abnormal and Social Psychology, 54*, 64-69.

Monge R.H and Hultsch D.F (1971). Paired associate learning as a function of adult age and the length of the anticipation and inspection intervals. *Journal of Gerontology, 26*, 157-162.

Norman M.F (1963). Incremental learning on random trials. *Journal of Mathematical Psychology, 1*, 336-350.

Pullen M.T (1995). *An optimal training theory: A paired associate learning task.* Unpublished master's thesis. Amherst: University of Massachusetts.

Richardson J. and Erlebacher A. (1957). Associate connection between paired verbal items. *Journal of Experimental Psychology, 56*, 62-69.

Ruch F.L (1934). The differential effects of age upon human learning. *Journal of General Psychology, 11*, 261-286.

Salthouse T.A, Kausler D.H and Saults J.S (1988). Utilization of path analytic procedures to investigate the role of processing resources in cognitive aging. *Psychology and Aging, 3*, 158-166.

Young J.L (1971). Reinforcement test intervals in paired associate learning. *Journal of Mathematical Psychology, 8*, 58-81.

IDENTIFYING THE LEARNING CAPABILITIES OF OLDER ADULTS: ASSOCIATIVE AND PRIORITY LEARNING

Wendy A. Rogers and Arthur D. Fisk, Georgia Institute of Technology, Atlanta, GA
Leonard M. Giambra and Edwin H. Rosenberg, National Institute on Aging

ABSTRACT

Effective search performance is determined by two important factors: memory load and display load. Memory load factors can be reduced by "associative learning" where memory-set elements become unitized and the stimuli are compared as a "category" rather than serially. Display-load effects can be reduced by target-distractor differentiation, a process referred to as "priority learning". In this paper we describe a three-phased experiment conducted to examine how those factors affected search performance for young and old subjects (mean ages, 24 and 71). Subjects were first trained in two varied mapping (VM) conditions (Phase 1): (1) Associative - allowed unitization of the stimulus sets; (2) Nonassociative - inhibited unitization. In Phase 1 all subjects unitized the associative sets thus implying maintenance of associative learning. In Phase 2, the stimuli were consistently mapped (CM) thus allowing the opportunity for priority learning. Following CM training young adults' performance was qualitatively superior to old adults'. In Phase 3, the CM target and distractor roles were reversed to assess the strength of CM learning. The attention-capturing strength (a measure of priority learning) was age-dependent with young adults showing greater effects. The results demonstrate that age differences in perceptual learning are primarily a function of a disruption in priority learning.

INTRODUCTION

The current experiment was designed to assess age-related differences in perceptual learning. More specifically, we investigated the effectiveness of the learning mechanisms involved in memory scanning and visual search for young and old adults. A three-phase experiment was conducted to separately assess "associative learning" which is critical for memory search improvements and "priority learning" which allows for improvment in visual search.

Age differences in perceptual learning are frequently studied in memory scanning and visual search tasks for at least two reasons. First of all, there is a strong empirical and theoretical base of knowledge about the performance of young adults. The second, and perhaps more important reason, is that memory scanning and visual search are often important components in more complex real-world tasks. For example, driving an automobile contains such memory scanning aspects as recalling directions to your destination or trying to remember how best to brake on an icy surface. Visual search components of driving include looking for road signs or being aware of the locations and actions of other vehicles on the road. Previous research suggests that age differences in performance may be localized in the visual search components of a task (e.g., Fisk and Rogers, 1990; Plude and Doussard-Roosevelt, 1989; Plude and Hoyer, 1986). The present experiment was designed to specifically test the effectiveness of learning mechanisms involved in both memory search and visual search for young and old adults (see Schneider, 1985; Schneider and Detweiler, 1987, 1988). The design allowed a separate assessment of the two learning mechanisms and hence, more precise specification of the locus of the age differences previously reported.

To briefly review, previous work in cognitive aging reveals a different pattern of age effects in perceptual learning depending on whether the task involves memory scanning or visual search (Fisk and Rogers, 1990; Madden, 1982; Plude and Doussard-Roosevelt, 1989; Plude and Hoyer, 1986; Puglisi, 1986; Rogers and Fisk, in press). In consistently mapped search tasks,[1] where learning occurs, old adults show qualitatively equivalent memory search performance but qualitatively different visual search performance. These differential age effects may be explained by assuming that memory search and visual search improve as a function of different learning mechanisms. Memory search performance improves primarily due to the ability to unitize the memory set items into a single class of items and compare them in parallel rather than serially. This memory-set unitization is referred to as associative learning. Visual search performance, on the other hand, improves through learning to attend to the important relevant information in a display (i.e., target items) and ignore the irrelevant information (i.e., distractor items). With consistent practice, target items become "strengthened"; that is, their ability to attract attention is increased. Consistent distractors become weakened and do not attract attention. This target/distractor strength differentiation is referred to as priority learning (Schneider and Detweiler, 1987, 1988).

The fact that old adults generally reveal poorer performance compared to young adults in visual search tasks suggests that a deficiency in priority learning may accompany aging. In order to test this hypothesis it was necessary to utilize a paradigm in which associative learning and priority learning could be assessed separately. Based on previous findings, it was proposed that associative learning would be intact for old adults but there would be age-related differences in priority learning.

We used an adaptation of an experiment by Shiffrin and Schneider (1977, Exp. 3). During the first phase of the experiment subjects received training in two VM conditions: an associative condition designed to allow unitization of memory-set elements and a nonassociative condition in which the memory-set items could not be unitized. Due to the VM nature of the training, both conditions required attentive search and priority learning could not occur. During Phase 2 the task was changed to a CM task. This phase of training provided an opportunity for priority learning (i.e., target/distractor strength differentiation) to occur in both the associative and nonassociative conditions. Finally, the roles of the CM target and distractor items were reversed in Phase 3. The reversal of target and distractor items provides an estimate of the attention-capturing effects of previously trained CM target items (see Shiffrin and Schneider, 1977, Exp. 1).

Our previous research (e.g., Fisk and Rogers, 1990)

suggests that associative learning remains relatively intact but priority learning declines as a function of age. Based on these assumptions the following predictions were made for the present experiment. After VM training (Phase 1), both age groups should show an equal benefit from the opportunity to unitize the memory set in the associative condition and both age groups should show an increase in reaction time (RT) with an increase in memory-set size for the nonassociative condition. Consequently, the Age x Memory-set Size and the Age x Memory-set Size x VM Condition interactions should not be significant.

After CM training (Phase 2), the young adults should show the benefit of priority learning (i.e., target/distractor strength differentiation) in both the associative and non-associative conditions. If in fact there is a disruption in priority learning for old adults, they will be less able to benefit from the consistency of mapping; hence, there should be an Age x Memory-set Size interaction after CM training.

The reversal data obtained in Phase 3 will provide an assessment of the degree of target/distractor strength differentiation which occurred during the CM training as a function of priority learning. Greater disruption due to the reversal of target and distractor roles signifies greater target/distractor strength differentiation. Therefore, it is expected that the young adults will show greater disruptions in performance at reversal for both the associative and the nonassociative conditions.

METHOD

Subjects

Ten young adults (six females) aged 20 to 31 (mean age 24) and 10 old adults (five females) aged 70 to 74 (mean age 71.3) participated in the experiment. All subjects were compensated monetarily ($10.00 per hour) for their participation. All subjects were administered the vocabulary and forward digit-span subscales of the Wechsler Adult Intelligence Scale (except two young subjects who had previous experience with the test). The median raw scores were as follows: vocabulary - 60 (range 17 to 71) and 68 (range 54 to 77), respectively, for young and old; and forward digit span - 11.5 (range 8 to 13) and 8.5 (range 6 to 13), respectively, for young and old. The median education level was 16 years and 15.5 years, for young and old, respectively. All subjects were screened for psychotropic medicine use. Corrected or uncorrected visual acuity for distance was at least 20/40 for all subjects except one older adult with 20/50 and near vision was 20/40 except two older adults with 20/50.[2] Self-reports of health for both young and old subjects ranged from good to excellent.

Stimuli

The stimulus items consisted of four-letter words from the semantically unrelated categories (Collen, Wickens, and Daniele, 1975) of Fruits, Musical Instruments, Animals, Colors, Building Parts, Weapons, Earth Formations, Furniture, Body Parts, and Clothing. Four high-associate words were chosen from each category (Battig and Montague, 1969; Howard, 1980). The assignment of categories for each condition was counterbalanced by a Latin square and repeated across age groups.

Equipment

IBM ATs and STANDARD COMPAQs were programmed to present the appropriate stimuli, collect responses, and control timing of the display presentations. IBM 5154001 enhanced color monitors and COMPAQ 420 video graphics color monitors were used to present the stimuli. The standard IBM AT and enhanced COMPAQ keyboards were altered so that the '7', '4', and '1' numeric keypad keys were labeled 'T', 'M', and 'B's, respectively. Eighteen of the subjects were tested in the laboratory (one or two at a time), at individual, partitioned workstations which were monitored by a laboratory assistant. The remaining two subjects (old adults) were tested in their home under the supervision of an experimenter.

Procedure

During the first session subjects were given practice which consisted of 10 blocks of CM trials (32 per block). These orientation trials allowed the subjects to become familiar with the experimental protocol and also served to stabilize error rates. The words used for the orientation trials were five-letter words chosen from the categories listed above (these words were not used in the remainder of the experiment).

An individual trial consisted of the following sequence of events. The subject was presented with the memory set of two or four words which he/she was allowed to study for a maximum of 20 seconds. Subjects were instructed to press the space bar to initiate the trial. Three plus signs were then presented in a column for .5 seconds in the location of the display set (in the center of the screen) to allow the subject to localize his/her gaze. The plus signs were followed by the display set which consisted of three words presented in a column and the subject's task was to indicate the location of the target (i.e., top, middle, or bottom) by pressing the corresponding key (labeled 'T', 'M', or 'B'). If the subject did not respond within six seconds the trial was considered an error trial.

The subjects received the following performance feedback. After each correct trial the subject's RT was displayed in hundredths of a second. After incorrect trials an error tone was sounded and the correct response was displayed. Following each block of trials the subject was given his/her average RT and percent accuracy for that block; if a subject's accuracy fell below 90% in a block, a message was displayed instructing him/her to respond more carefully. (Subjects were encouraged to maintain an accuracy rate of 95% while responding as quickly as possible.) A target (i.e., a member of the memory set) was present on every trial.

Design

There were three phases of the experiment. In Phase 1 (VM), the subjects were trained in two VM conditions: associative and nonassociative. In the VM associative condition, there were two sets of words; each set contained four words from a single category (e.g., Fruits or Weapons). An item from the category of Fruits could appear as a target on one trial and as a distractor on another trial thus it was variably mapped. However, when a Fruit was a target item, words from the category of Weapons served as distractors. Conversely, when a Weapon was the target item, words from the category of Fruits served as distractors. On each trial one of the categories was chosen to serve as the target set and the other was designated the distractor set (with the restriction that each category serve as the target set an equal number of times).

In the VM nonassociative condition, stimulus items consisted of eight words chosen from distinct categories that were not used in the associative condition (e.g., Musical Instruments, Animals, Furniture, Body Parts, Earth Formations, Clothing, Colors, and Building Parts). Target and distractor items were chosen, with replacement, from this set of items.

The VM training phase lasted for eight one-hour sessions each of which consisted of 20 blocks of 32 trials. There were 10 blocks per condition during each session and the presentation of condition alternated by block. Each block contained 16 trials of Memory-set Size 2 and 16 trials of Memory-set Size 4 (presentation order was randomized). Each subject completed 5,120 VM trials (2,560 per condition).

Phase 2 of the experiment immediately followed Phase 1 and involved CM training. In the CM associative condition, one associative set was chosen to be the consistent target set and the other associative set became the distractor set. The items from the target set never appeared as distractors during this phase and the distractor items were never used as targets. In the CM nonassociative condition four of the eight words used in the VM nonassociative condition were randomly chosen to become the consistent target set and the remaining four words became the consistent distractor set. Hence, during the CM phase, the stimulus to response mapping was consistent in both the associative and the nonassociative conditions. The remainder of the procedure was the same as Phase 1. The CM training phase lasted for six one-hour sessions each of which consisted of 20 blocks (10 per condition). Each subject completed 3,840 CM trials (1,920 per condition).

The third phase of the experiment was the reversal phase. The reversal phase consisted of a single one-hour session of 20 blocks of trials. The first 4 blocks were exactly the same as the blocks in the CM phase (to serve as a baseline). However, during the remaining 16 blocks, the roles of the target set and distractor set were reversed within each condition. Thus the CM target items became distractor items and the CM distractor items became target items. Subjects completed 512 reversal trials (256 per condition).

The within-subject independent variables were: (1) VM conditions (associative and nonassociative); (2) CM conditions (associative and nonassociative); (3) Reversal conditions (associative and nonassociative); and (4) Memory-set size (2 and 4). VM, CM, and reversal conditions were manipulated between blocks and memory-set size was manipulated within a block; these variables were thus within-subjects variables. The quasi-experimental variable was age, young or old. The dependent variables were RT and accuracy.

RESULTS

Performance After VM Training - Phase 1

Reaction Time. Mean RT for correct trials were analyzed for the final session of VM training (i.e., the last 320 VM trials of each condition). These data are presented in Figure 1. An Age (young or old) x Memory-set Size (2 or 4) x VM Condition (associative or nonassociative) analysis of variance (ANOVA) was conducted. The main effects of Age, $F(1, 18) = 14.8$, Memory-set Size, $F(1, 18) = 266.98$, and VM Condition, $F(1, 18) = 24.58$, were significant, as was the interaction of Memory-set Size x VM Condition, $F(1, 18) = 35.12$.[3] The young subjects had faster verification times overall than did the old subjects, and performance was faster for Memory-set Size 2 and for the VM associative condition for all subjects. The source of the Memory-set size x VM Condition interaction is the fact that RT increased more from Memory-set Size 2 to 4 for the nonassociative condition relative to the associative condition.

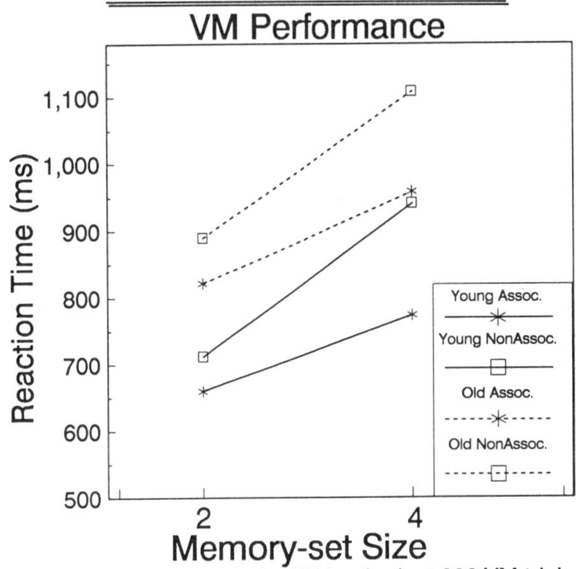

Figure 1. Correct trial RT for the last 320 VM trials.

Accuracy. An ANOVA conducted on performance accuracy showed that only the main effects of Age, $F(1, 18) = 4.9$, Memory-set Size, $F(1, 18) = 18.19$, and VM Condition, $F(1, 18) = 18.65$, were significant. Note that the accuracy differences between the age groups were quite small: the old subjects had an average accuracy rate of 98% whereas the average for the young subjects was 97%. Similarly the accuracy differences between Memory-set Sizes (98% and 97%, respectively, for Memory-set Sizes 2 and 4) and VM conditions (98% and 97%, respectively, for the associative and nonassociative conditions) were relatively small.

Summary of VM Data

As was predicted, neither the Age x Memory-set size condition ($F<1$) nor the Age x Memory-set Size x VM Condition ($F=1.03$) interactions were significant. The Memory-set Size x VM condition interaction suggests that performance for both young and old adults was superior in the associative condition and less affected by an increase in Memory-set Size (relative to the nonassociative condition). These results support our proposal that associative learning was facilitated in the associative condition for both young and old adults. Thus while the old subjects were slower than the young subjects, the pattern of performance after VM training was the same for both age groups.

Performance After CM Training - Phase 2

Reaction Time. Mean RT for correct trials were analyzed for the final session of CM training (i.e., the last 320 CM trials of each condition; see Figure 2). An Age (young or old) x Memory-set Size (2 or 4) x CM Condition (associative or nonassociative) ANOVA was conducted. The main effects of Age, $F(1, 18) = 17.15$, Memory-set Size, $F(1, 18) = 34.83$, and CM Condition, $F(1, 18) = 15.22$, were significant, as were the interactions of Age x Memory-set Size, $F(1, 18) = 4.27$, and Memory-set Size x CM Condition, $F(1, 18) = 4.47$. Again, the young subjects had faster verification times overall than did the old subjects, and performance was faster for Memory-set Size 2 and for the associative condition for all subjects. The source of the Memory-set size x CM Condition interaction is the fact that RT increased more from Memory-set Size 2 to 4 for the nonassociative condition relative to the associative condition. The Age x Memory-set Size interaction is due to the fact that an increase in Memory-set Size (i.e., from 2 to 4 words) resulted in a greater increase in RT for the old adults relative to the young adults.

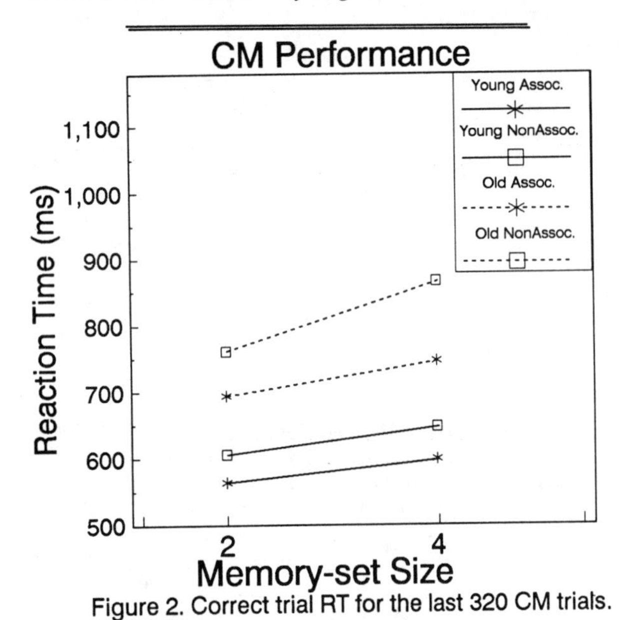

Figure 2. Correct trial RT for the last 320 CM trials.

Accuracy. An ANOVA conducted on performance accuracy revealed significant main effects of Age, $F(1, 18) = 8.66$, Memory-set Size, $F(1, 18) = 10.69$, and CM Condition, $F(1, 18) = 6.45$. Once more, the accuracy differences between the age groups (99% and 97%, respectively, for young and old), Memory-set Sizes (98% and 97%, respectively, for Memory-set Sizes 2 and 4) and CM conditions (98% and 97%, respectively, for the associative and nonassociative conditions) were quite small.

Summary of CM Data

These data support the predictions of an age-related disruption in the priority learning mechanism. The CM training enabled the young subjects to improve performance in both the associative and nonassociative conditions through priority learning (target/distractor strength differentiation). However, the Age x Memory-set size interaction suggests that the old adults were less able to differentially strengthen targets and distractors and hence were more slowed by an increase in Memory-set Size.

Performance at Reversal - Phase 3

Percentage Change. In order to assess the magnitude of disruption due to the reversal conditions, relative reversal scores were calculated for each individual subject ([Reversal RT - Training RT]/Training RT). The Training RT consisted of the average of the four blocks immediately prior to the reversal manipulation and the Reversal RT was the average of the 16 reversal blocks. This method of calculating change scores was chosen, a priori, because of the ubiquitous finding that old adults respond more slowly than young adults. Therefore, dividing by training RT adjusts for baseline differences (between groups and between subjects) and the resultant score provides a more meaningful estimate of change due to the experimental manipulation (cf. Plude and Hoyer, 1981; Roscoe and Williges, 1979; Salthouse, 1978).

The average change scores for each condition are presented in Figure 3. A positive change score indicates an increase in RT (i.e., slower performance) and, hence, disruption in performance. An Age (young or old) x Memory-set Size (2 or 4) x Reversal Condition (associative or nonassociative) ANOVA revealed significant main effects of Age, $F(1, 18) = 5.45$, Memory-set Size, $F(1, 18) = 15.56$, and Reversal Condition, $F(1, 18) = 7.20$, and a significant interaction of Age x Reversal Condition, $F(1, 18) = 4.55$. *The young subjects were more disrupted than the old subjects.* There was also a greater disruption for the Memory-set Size 4 condition. An analysis of simple effects revealed that the source of the Age x Reversal Condition was due to the fact the young adults were more disrupted in the nonassociative condition relative to the associative condition and the old adults' disruption scores were not different between the two conditions. Also, the age difference in disruption was evident only in the nonassociative condition.

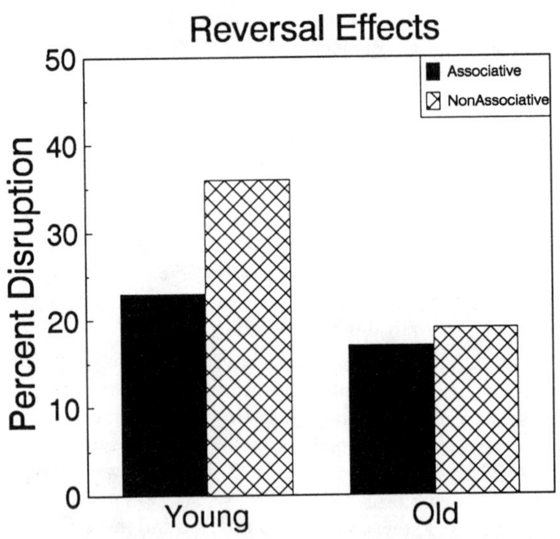

Figure 3. (Training RT - Reversal RT)/Training RT

Summary of Reversal Data

This analysis of the percentage change in performance and comparisons of the magnitude of these changes across age groups provides the most relevant analysis of age differences in priority learning. The greater disruption for young adults in the nonassociative condition suggests that the target/distractor strength differentiation learned during CM training disrupted performance at reversal when previously attended items had to be ignored and vice versa. The fact that the young subjects were disrupted less in the associative condition (and not more than old adults) was somewhat surprising. However, it is true that the associative reversal condition was not really novel because during VM training the items had appeared as both targets and distractors (although separated by category). Thus it is possible that due to the previous exposure to the particular target and distractor pairings, the young subjects were able to contextually prime targets in the associative reversal condition which reduced the interference. Similar context cuing effects have been reported in the literature (Fisk and Rogers, 1988; Rogers and Fisk, 1990).

DISCUSSION

The results of the present experiment suggest that both young and old adults benefit equally from a situation that allows the unitization of memory-set elements (presumably through associative learning). However, young and old subjects show differential effects under CM conditions which allow target/distractor strength differentiation through priority learning. Performance differences after CM training as well as differences in disruption for young and old adults at reversal support the proposal that old adults are deficient in priority learning.

These data have important implications for training and retraining issues, as well issues related to the safety and mobility of older adults. As mentioned above an important component of driving is the ability to scan the environment, detect relevant stimuli (brake lights, road signs, etc.), and to quickly act upon those stimuli. A disruption in visual search performance for old adults has important implications for individuals learning to drive at an advanced age. Furthermore, in the work environment, inspection activities (e.g., on an assembly line) may be affected for older workers especially if they are required to search for new items.

FOOTNOTES

[1]Consistent mapping (CM) refers to a situation in which a stimulus item is responded to invariably across situations (i.e., it is always attended to *or* always ignored). CM training yields significant performance improvements and in some cases automatic process development. Varied mapping (VM) training denotes a situation in which a stimulus is responded to differentially across situations (i.e., sometimes it is attended to and sometimes ignored). Consequently, VM training yields little performance improvement.

[2]The subjects with poorer vision did not show differential patterns of performance relative to the other subjects in their age group.

[3]Alpha level was set at $p < .05$ unless otherwise indicated.

REFERENCES

Battig, W. F., and Montague, W. E. (1969). Category norms for verbal items in 56 categories: A replication and extension of the Connecticut category norms. Journal of Experimental Psychology Monographs, 80(3, Pt. 2).

Collen, A., Wickens, D. D., and Daniele, L. (1975). The interrelationship of taxonomic categories. Journal of Experimental Psychology: Human Learning and Memory, 1, 629-633.

Fisk, A. D., and Rogers, W. A. (1988). The role of situational context in the development of high-performance skills. Human Factors, 30, 703-712.

Fisk, A. D., and Rogers, W. A. (1990). Toward an understanding of age-related memory and visual search effects. Manuscript submitted for publication.

Howard, D. V. (1980). Category norms: A comparison of the Battig and Montague (1969) norms with the responses of adults between the ages of 20 and 80. Journal of Gerontology, 35, 225-231.

Madden, D. J. (1982). Age differences and similarities in the improvement of controlled search. Experimental Aging Research, 8, 91-98.

Plude, D. J., and Doussard-Roosevelt, J. A. (1989). Aging, selective attention, and feature integration. Psychology and Aging, 4, 98-105.

Plude, D. J., and Hoyer, W. J. (1981). Adult age differences in visual search as a function of stimulus mapping and processing load. Journal of Gerontology, 36, 598-604.

Plude, D. J., and Hoyer, W. J. (1986). Age and the selectivity of visual information processing. Psychology and Aging, 1, 4-10.

Puglisi, J. T. (1986). Age-related slowing in memory search for three-dimensional objects. Journal of Gerontology, 41, 72-78.

Rogers, W. A., and Fisk, A. D. (1990). Priority learning in consistent mapping visual search: Localizing age differences in automatic process development. Manuscript submitted for publication.

Rogers, W. A., and Fisk, A. D. (in press). A reconsideration of age-related reaction time slowing from a learning perspective: Age-related slowing is not just complexity-based. Learning and Individual Differences.

Roscoe, S. N., and Williges, B. H. (1979). Measurement of transfer of training. In S. N. Roscoe (Ed.), Aviation Psychology. Ames, IA: Iowa State University.

Salthouse, T. A. (1978). The role of memory in the age decline in digit-symbol substitution performance. Journal of Gerontology, 33, 232-238.

Schneider, W. (1985). Toward a model of attention and the development of automatic processing. In M. I. Posner and O. S. Martin (Eds.), Attention and Performance XI (pp. 475-492), Hillside, NJ: Erlbaum.

Schneider, W., and Detweiler, M. (1987). A connectionist/control architecture for working memory. In G. H. Bower (Ed.), The psychology of learning and motivation (pp. 53-118), Volume 21. New York: Academic Press.

Schneider, W., and Detweiler, M. (1988). The role of practice in dual-task performance: Toward workload modeling in a connectionist/control architecture. Human Factors, 30, 539-566.

Shiffrin, R. M., and Schneider, W. (1977). Controlled and automatic human information processing: II. Perceptual learning, automatic attending, and a general theory. Psychological Review, 84, 127-190.

ABILITY-PERFORMANCE RELATIONSHIPS IN MEMORY SKILL TASKS FOR YOUNG AND OLD ADULTS

W. A. Rogers[1], D. K. Gilbert[1], and A. D. Fisk[2]

[1]Memphis State University [2]Georgia Institute of Technology

ABSTRACT

The present experiment investigated ability-performance relationships for two memory skills, each of which required associative learning. Evidence suggests that, after practice, young and old adults have equivalent associative learning abilities (Fisk and Rogers, 1991; Kausler, 1982). We provided 41 young and 52 old adults with extensive practice on consistently and varied versions of a memory search task and a noun pair look-up task (Ackerman and Woltz, 1993). Only consistent practice allows associative learning because the stimulus items are consistently paired; in varied practice, item pairings change across practice and associative learning is not possible. We also assessed a wide range of abilities for each subject and were thus able to investigate ability-performance relationships across practice conditions and across age groups. These relationships provide an indication of the underlying abilities related to task performance (Ackerman, 1988). The mean data suggested that both young and old adults demonstrated successful associative learning in the two CM tasks. The individual differences data suggest, however, that different abilities may be driving performance across the two age groups. These data have important implications for predicting whether or not older adults will successfully acquire a new skill. If the target skill requires associative learning, older adults, may perform as efficiently as young adults if they are provided with sufficient, consistent practice. The ability-performance data suggest that predictions about which individuals will be most successful at skills requiring associative learning, may be dependent on the age of the target population.

INTRODUCTION

The purpose of the present experiment was to investigate ability-performance relationships for two different memory skills, each of which required associative learning. Evidence suggests that, after practice, young and old adults have equivalent associative learning abilities (Fisk and Rogers, 1991; Kausler, 1982). Associative learning involves pairing new items together, either two items into one pair, or a group of items into an entire set. For example, learning item-price codes in a retail situation involves pairing two items whereas learning which items are taxable involves grouping a number of things into a single set. In the present study we provided subjects with extensive practice requiring both types of associative learning. Our goal was two-fold: (a) to investigate the patterns of performance improvements across practice for young and old adults; and (b) to assess the ability-performance relationships to determine whether similar abilities are related to performance not only across types of associative learning tasks, but also across age groups.

We provided extensive practice on consistently mapped (CM) and variably mapped (VM) versions of memory search task and noun pair look-up tasks (Ackerman and Woltz, 1993). CM practice allows associative learning because the stimulus items are consistently paired; in VM, item pairings change across practice and associative learning is not possible. Associative learning involved associating the memory set items (four total items) into a single set in the memory-search task, and learning word pairs in the noun pair look-up task. We assessed a wide range of abilities for each subject and were thus able to investigate ability-performance relationships across practice conditions and across age groups. These relationships provide an indication of the underlying abilities necessary for task performance (Ackerman, 1988).

METHOD

Subjects

Subjects were 41 young adults (aged 17 to 34) and 52 older adults (aged 62 to 78). Students received course credit for their participation and the remaining subjects received $5.00 per hour plus parking expenses.

Procedure

Subjects participated in five sessions. The first two consisted of paper-and-pencil ability tests. The latter three sessions consisted of computerized ability tests and training on the criterion memory skill tasks. The ability tests allowed estimation of the following constructs: fluid intelligence (g_f), verbal ability (*V*), perceptual speed (*PS*), psychomotor speed (*PM*), working memory (*WM*), associative memory (*AM*), short-term memory (*STM*), and semantic memory access (*SMA*).

Criterion Task - Noun Pair Look-up. The noun pair look-up task is analogous to a computerized version of digit-symbol substitution with word pairs. Nine word pairs were presented across the top of the computer screen and two words were presented in the center of the screen (Ackerman and Woltz, 1993). The subjects' task was to decide if the word pair presented in the center of the screen corresponded to one of the word pairs presented above. After each trial, subjects received feedback on the RT and accuracy for that trial. At the end of each block, the subjects received their mean RT and accuracy rate for that block. They were instructed to maintain an accuracy rate of 90%.

There were two versions of the noun-pair task: a CM version and a VM version. In the CM version, the noun pairs did not change throughout the task (although the location of the word pairs on the screen changed after each block of 18 trials). After every 5 blocks, subjects were given a test block on which the display of words did NOT appear at the top of the screen. Subjects simply saw a word pair and had to determine whether or not the word pair represented a match. Thus on test trials, they had to rely on memory for the word pairs. Subjects were instructed that their goal was to try to learn the word pairs so that they would be able to perform well on the test blocks. Subjects completed 810 trials of CM noun pair training and 10 test blocks.

In the VM version, the word pairs changed on every trial. There were not any word pairs to be learned, thus there were no recall tests for the VM task. Subject competed 540 VM trials.

Criterion Task - Semantic Category Memory Search. The second criterion task was a semantic category memory search task. During each trial subjects were presented with the memory set of one, two, or three category labels, which they were allowed to study for a maximum of 20 seconds. They were instructed to press the space bar to initiate the trial. A plus sign was then presented in the center of the screen to allow subjects to localize their gaze. After 500 ms the probe word was presented. The subjects' task was to indicate whether or not the probe word was a member of one of the memory-set categories. There was a target present on 50% of the trials. After each correct trial, the subjects' RT was displayed in milliseconds. After each incorrect trial, an error tone sounded and the correct word was displayed. Following each block of 60 trials, the subjects received their average RT and accuracy for that block. Subjects were instructed to maintain an accuracy rate of 95%.

Subjects received practice in two conditions: (a) CM - target and distractor items were drawn from distinct stimulus sets, and (b) VM - target and distractor items were drawn from the same stimulus set with replacement across trials. Each subject was assigned four categories as CM target categories and four different categories as CM distractor categories; the remaining eight categories served interchangeably as targets and distractors in the VM condition. Subjects received search practice first on the CM version of the task. Each subject completed 15 blocks of 60 trials for a total of 900 CM trials. Subsequent to the CM practice each subject completed 4 blocks of VM trials for a total of 240 VM trials.

RESULTS

Ability Tests. As is typically the case, the young adults performed significantly better on all of the ability tests with the exception of the tests that measured verbal ability (all t's $< .01$). The performance of the old adults was superior for two of the verbal tests ($p < .001$) and not different from the young adults on the third verbal test ($p = .86$). Thus the current sample of subjects conforms to the traditional pattern of age-related ability differences.

Noun Pair Look-up Task. At the end of CM practice, RTs for the young adults were faster than for the old adults, $t(91) = 7.31$, $p < .001$. However, a comparison of accuracy rate for the last test block did not reveal a significant age difference ($p = .14$). Thus, although the young adults responded faster, overall, than the older adults, both age groups learned the word pairs equally well. By the end of practice, the mean accuracy rates for the test blocks were 90% and 87%, respectively, for the young and old adults.

A comparison of VM RTs revealed an age difference at the end of practice, $t(91) = 7.33$, $p < .001$. In the VM task, it was not possible to learn the words pairs, thus the age differences remained significant even at the end of practice.

Memory Search Task. In terms of RT, there was a significant age difference at the end of practice, $t(91) = 7.27$, $p < .001$. Comparison slope estimates were calculated to represent the increase in RT corresponding to the increase in memory load. In the CM condition, both age groups showed a large reduction in slope across trials. The slopes for the young adults were reduced from 96 ms to 10 ms. For the older adults, CM slopes were reduced from 150 ms to 27 ms. Thus both age groups benefitted from the CM practice and were able to associatively connect the memory-set items.

In the VM condition, young adults were significantly faster than old adults, $t(91) = 7.27$, $p < .001$. However, both young and older adults had substantial slopes (222 ms and 256 ms, respectively for the young and older adults).

Summary of Normative Data. The ability data completely conformed to the predicted pattern. In addition, age-related and practice-related differences in the two criterion tasks was as expected. In both the noun pair look-up task and the memory search task, performance was superior under CM practice conditions. Moreover, both young and old adults showed performance improvements in the CM tasks and the characteristics of their improvements is suggestive of successful associative learning. In the following section, we will investigate the ability-performance relationships across practice conditions and age groups.

Ability-Performance Relationships. Due to space limits, we will highlight some of the interesting patterns of ability-performance relationships. Figure 1 presents the correlations between g_f, *WM*, *V*, and *PS* across practice for the CM and VM conditions of the memory search task. Looking first at g_f, it is clear that there was a higher correlation with CM performance for the old adults relative to the young adults. In addition, the correlation remained high across CM practice for the old adults, whereas it decreased with practice for the young adults. The g_f correlation with VM performance was relatively high for both age groups. This pattern suggests that the CM task was more demanding for the old adults, and the VM task is demanding for both age groups.

The patterns for the performance correlations with *WM* revealed that the CM correlations were high and stable for the old adults, and shallower and decreasing for the young adults. The VM correlation for *WM* was slightly higher for the older adults. These correlations indicate the importance of both *WM* for the memory search task and suggest that the working memory influence was greater for the older adults.

For the young adults the relationship of V to performance was near zero throughout CM practice and for the VM condition. However, for the old adults, initial CM performance showed initially significant correlations with V. Thus individual differences in verbal ability may be related to success in associative learning, at least for the older adults.

The *PS* correlations were high and stable for both age groups for the CM and VM conditions. There was little indication of a decreasing *PS* correlation across practice; thus, the influence of *PS* remained high throughout practice.

In Figure 2, the correlations between g_f, *V*, and *PS* are presented for both age groups for the CM and VM conditions of noun pair look-up. With respect to g_f, the patterns were the same for the VM task for the two age groups. However, in the CM task the correlations were shallower for the young relative to the old. This pattern suggests that the CM task may be relatively more demanding for the old adults.

The *V*-performance correlations revealed age-related patterns that were similar for the VM task, but for the CM task the correlations were higher for the older adults. These data suggest verbal ability predicted the ability to learn the word pairs for the old adults.

As in the memory search task, the PS correlations were high and quite stable for both age groups for CM and VM practice. There was some increase for the young adults across CM practice; there was no evidence of a decline in the PS correlation even late in practice.

Summary of Ability-Performance Relationships. There were several important age differences in the pattern of ability-performance relationships for the two criterion tasks. For example, in both CM tasks (memory search and noun pair), the influence of g_f was greater for the old adults relative to the young adults. In addition, *V* was a significant predictor for the old adults, whereas it was not for the young adults. Finally, in the CM memory search task, *WM* predicted performance throughout practice for the old adults whereas the correlations decreased with practice for the young adults.

The mean data suggested that young and old adults both demonstrated successful associative learning in the two CM tasks. The individual differences data suggest, however, that different abilities are predictive of performance differences across the age groups.

IMPLICATIONS

The present data provide support for the hypothesis that age differences are minimized when CM practice is provided for memory skill tasks which require associative learning. In both the noun pair look-up task and the memory search task, the performance of both age groups was indicative of successful associative learning. There are two caveats, however. The first is that older adults require more practice for associative learning. The second is that young adults continue to have faster overall response times even after extensive practice.

These data have important implications for predicting whether or not older adults will successfully acquire a new skill. If the target skill requires associative learning (e.g., learning new item-price relationships in a retail setting), older adults, as a whole, will perform as efficiently as young adults if they are provided with sufficient, consistent practice. The ability-performance data suggest that predictions about which individuals will be most successful at skills requiring associative learning may depend on the age of the target population.

ACKNOWLEDGMENTS

This research was supported by the National Institutes of Health Grant RO1AG07654 (NIA) and Faculty Research Grant #2-20595 from Memphis State University. Correspondence to W. A. Rogers, Dept. of Psychology, Memphis State University, Memphis, TN, 38152.

REFERENCES

Ackerman, P. L. (1988). Determinants of individual differences during skill acquisition: Cognitive abilities and information processing. Journal of Experimental Psychology: General, 117, 288-318.

Ackerman, P. L. and Woltz, D. J. (1993). Determinants of learning and performance in an associative memory/substitution task: Task constraints, individual differences, and volition. Unpublished manuscript.

Fisk, A. D., and Rogers, W. A. (1991). Toward an understanding of age-related memory and visual search effects. Journal of Experimental Psychology: General, 120, 131-149.

Kausler, D. H. (1982). Experimental psychology and human aging. New York: John Wiley.

Figure 1. Ability-Performance Relationships for Memory Search

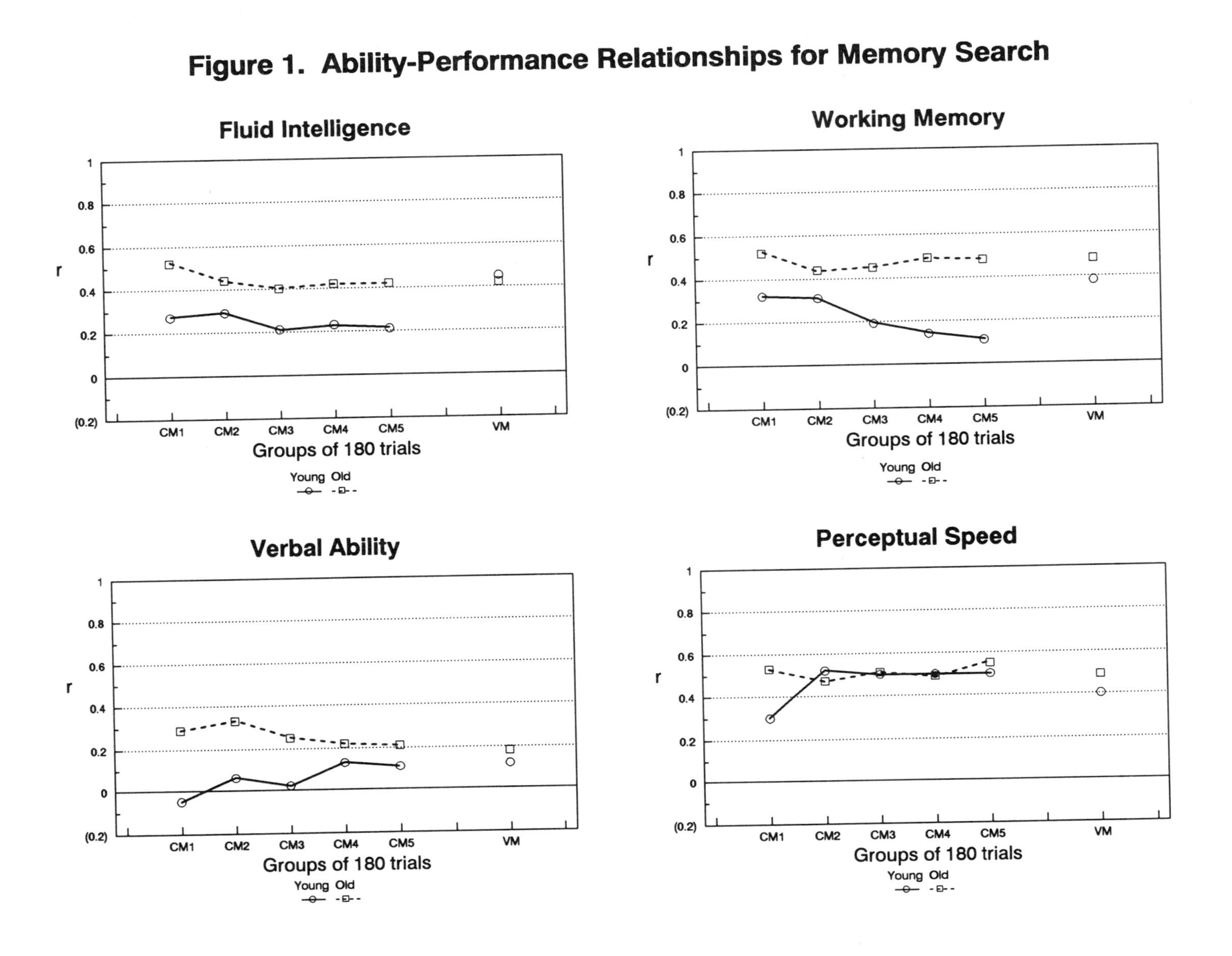

Figure 2. Ability-Performance Relationships for Noun-Pair Look-up.

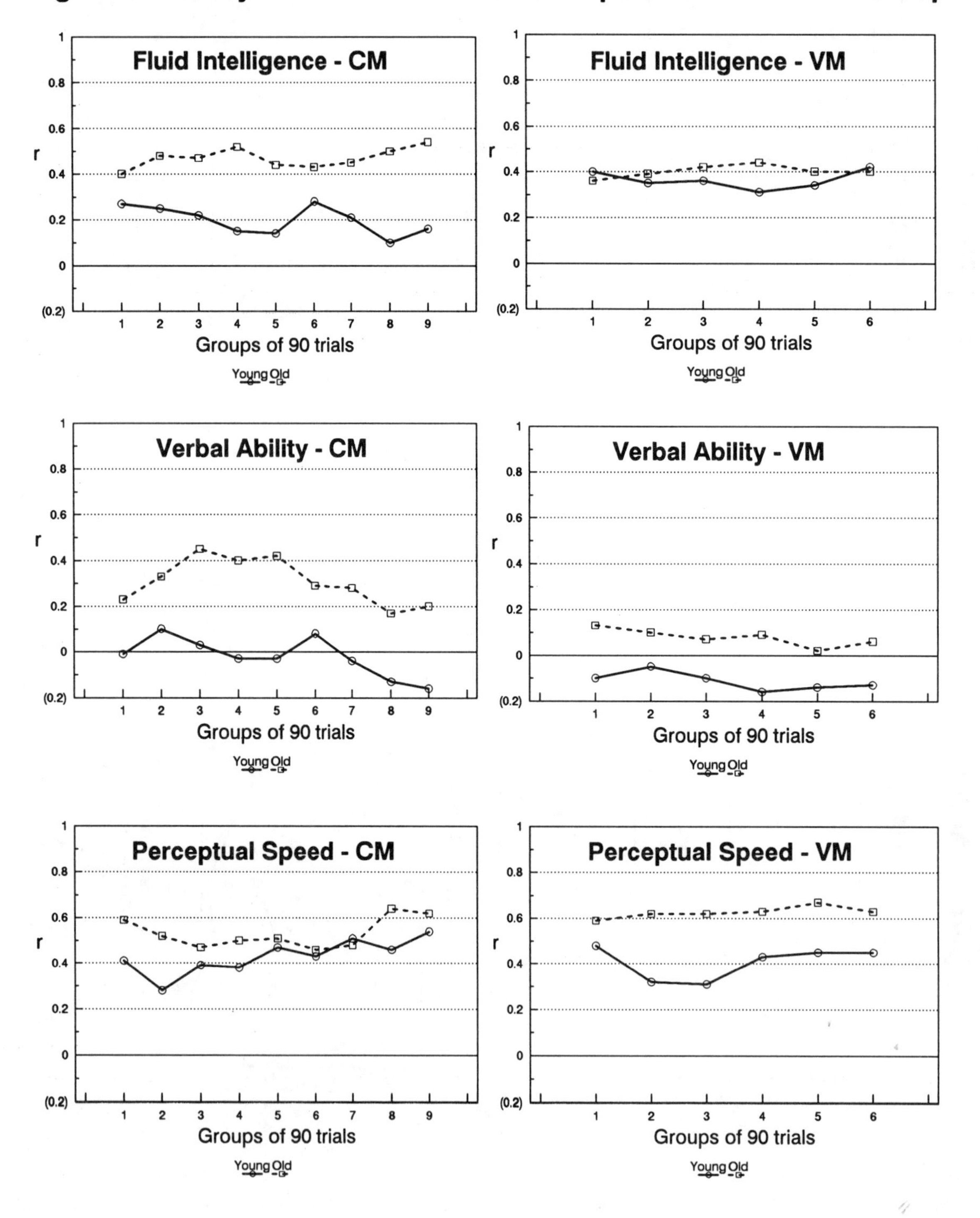

Age Similarities in Complex Memory Search: An Extension of Dual Process Theory

Brian P. Cooper and Arthur D. Fisk
Georgia Institute of Technology

Understanding age-related similarities and differences in development of cognitive skill is important as it can inform theories of cognitive aging as well as serve the pragmatic value of informing those individuals who are developing age-related interventions for numerous activities of daily living. We investigated both the performance and learning of skilled memory search, a task that has shown age-related similarity in performance if sufficient consistent practice is provided, to determine if training guidelines for this class of processing activities is applicable to both young and old adults. Old and young adults received memory search training, and then the participants were transferred to untrained exemplars of the trained memory set categories. The results suggest that both young and old adults are, at least to some extent, learning at the semantic-category level. This study provides additional evidence that training guidelines derived from an automatic and controlled processing framework can be applied to an older adult population in tasks which have memory search components.

Understanding the characteristic changes in information processing as one develops a cognitive skill has interested psychologists for quite some time. Also, understanding changes in information processing capability as one ages has been the subject of much scientific investigation. The intersection of these two important areas of investigation has received much less emphasis, although it has not gone without investigation (e.g., Fisk & Rogers, 1991; Rogers, Fisk, & Hertzog, 1994). Understanding age-related similarities and differences in development of cognitive skill is important as it can inform theories of cognitive aging as well as serve the pragmatic value of informing those individuals who are developing age-related interventions for numerous activities of daily living.

Studies of aging and skill development have begun to illuminate what aspects of the cognitive system change fundamentally with age as well as when cognitive efficiency may be enhanced with training regardless of normal aging. Indeed, given sufficient (and correct) practice, older adults have demonstrated that in many instances the aging cognitive system can appear to perform, qualitatively, like that of its younger counterpart. Careful, systematic study of components of age-related development of cognitive skill can add much to the theoretical and practical knowledge base.

An automatic and controlled processing framework (Shiffrin & Schneider, 1977) has often been applied to the design of complex training systems (Schneider, 1985; Eggemeier & Fisk, 1992). One aspect of dual process theory which makes it so applicable to training systems is that skill acquisition can occur with higher-order consistency (Fisk & Jones, 1992; Schneider & Fisk, 1984). However, these training guidelines have been applied almost exclusively to young adult populations. Whether such training guidelines are applicable to older adults depends on whether older adults benefit from higher-order consistency.

In the present study, we attempt to add to the understanding of age-related cognitive skill by investigating both the performance and learning of skilled memory search, a task that has shown age-related similarity in performance if sufficient consistent practice is provided. Unfortunately, we do not know if the similarity in performance between young and old adults is due to similar learning. Such an understanding is important because similar performance on different visual search tasks within age groups (Czerwinski, Lightfoot, & Shiffrin, 1992) and estimates of similar performance on the same task between age groups (Rogers et al., 1994) has been shown to be due to different learning mechanisms. Perhaps this is also the case for memory search.

The possible level of processing for memory search has not been determined as it has in visual search (Schneider & Fisk, 1984). For example, the extent to which the learning in semantic-category memory-search occurs at the category or word level for both young and old adults is unknown. In order to investigate these issues, subjects received consistently mapped memory search training in Experiment 1. Experiment 2 consisted of a transfer condition in which subjects were transferred to untrained exemplars of the trained memory set categories. The extent to which young and old adults are learning at the category level in memory search can be seen by the degree of transfer.

EXPERIMENT 1: TRAINING.

Method

Subjects. Eighty-seven young (ages 17-33) and 87 old (ages 63-82) adults were recruited from a large southeastern university and surrounding community. The subject's corrected or uncorrected visual acuity was at least 20/40 for both distance and near.

Stimuli. The stimuli for the memory search task were eight semantically unrelated categories (Collen, Wickens, & Daniele, 1975). Eight high associates from each category (Battig and Montague, 1969), four to seven letters long, were chosen as exemplars.

Design. The within-subject independent variables were: (a) Memory set size: one, two, or three categories; (b) Target type: target present or target absent; and (c) Blocks of Practice: 1-20. Age was a quasi-independent variable, young or old. The primary dependent variables were RT and accuracy.

Procedure. An experimental trial consisted of the following sequence of events. The subject was presented with a memory set consisting of either one, two, or three category labels which he/she was allowed to study for a maximum of 20 seconds. Subjects pressed the space bar to initiate the trial. A plus sign was then presented in the center of the screen to allow the subject to localize his/her gaze. After 500 ms, the display set was presented consisting of one word. The subject's task was to indicate whether or not the display word was a member of one of the categories in the memory set by pressing the appropriate key. Subjects were instructed to respond as quickly as possible while maintaining an accuracy level between 93 and 97%.

Subjects were trained on the semantic category memory search task for one session. Each subject was assigned four categories as the CM target set and another four categories as the CM distractor set. The training session consisted of 20 blocks of 60 trials. Memory set size and target type were manipulated within a block. Thus, there were 20 trials per memory set size within a block, and a target was present on half of the trials. At the conclusion of the session, each subject completed a total of 1200 CM trials.

Results and Discussion

A Memory Set Size x Trial Type x Blocks of Practice x Age ANOVA was performed on the data. All reported differences were significant at the $p<.05$ level. As expected, young adults were faster than old adults. Both groups responded faster with practice, but the older group improved more than the young. With practice, young adults improved from 702 ms to 505 ms, and the old adults improved from 948 ms to 642 ms. This is consistent with previous findings in which the old are initially slower than the young and thus show the largest improvement with practice. Response times for target absent trials were slower than target present trials, and this difference was greater for the older adults. The difference between target absent and target present trials decreased as a function of practice, and this decrease was larger for the older adults.

Reaction time increased as a function of memory set size, and this effect was larger for the old adults. The average RTs for the young adults were 510 ms, 545 ms, and 558 ms for memory set sizes 1, 2, and 3, respectively; the old adults' RTs were 635 ms, 698 ms, and 722 ms for memory set sizes 1, 2, and 3, respectively. However, the effect of memory set size attenuated with practice, and this improvement was greater for the older adults. For Block 1, the RTs for the young adults were 575 ms, 712 ms, and 821 ms, for memory set sizes 1, 2, and

3, respectively; for Block 20, the young adults' RTs were 496 ms, 509 ms, and 511 ms, for memory set sizes 1, 2, and 3, respectively. For Block 1, the RTs for the old adults were 781 ms, 966 ms, and 1097 ms for memory set sizes 1, 2, and 3, respectively; for Block 20, the old adults' RTs were 618 ms, 657 ms, and 651 ms for memory set sizes 1, 2, and 3, respectively.

This pattern of results is consistent with a transformation from a serial, terminating search to processing based on memory set unitization (Shiffrin, 1988; Fisk & Rogers, 1991) because the response times for higher memory set sizes are approaching the RT for memory set size 1. The greater reduction in comparison times for the old is consistent with the old performing qualitatively like the young. If the old are initially more effected by increasing memory set size, they must show more improvement in order to reach "near zero" comparison times. What needs to be evaluated in terms of the present study is whether or not the young and old are learning at the same level (i.e., category or word). This was evaluated in Experiment 2 where both young and old adults were transferred to untrained exemplars of the trained memory set categories.

EXPERIMENT 2: TRANSFER

Method

Stimuli. The stimuli for the memory search task were twelve semantically unrelated categories (Collen, Wickens, & Daniele, 1975). Eight high associates from each category (Battig & Montague, 1969), four to seven letters long, were chosen as exemplars.

Design. The within-subject independent variables were: (a) Memory set size: one, two, or three categories; and (b) Transfer condition: untrained categories (untrained/untrained), trained categories with untrained exemplars (trained/untrained), trained categories with trained exemplars (trained/trained), and baseline (from last block of training). The quasi-independent variable was age. The primary dependent variables were RT and accuracy.

Procedure. After training, subjects entered the memory search transfer phase. The same four categories which served as the target set during training were used as the trained categories. Four new categories were assigned to serve as the untrained target set, and four new categories served as the distractor set for all conditions. Subjects completed 3 blocks of 60 trials. The first two blocks consisted of the trained target categories. Half of the trials within these blocks used trained exemplars, and half of the trials used untrained exemplars. The third block consisted of the untrained target categories.

For the two blocks with trained target categories, memory set size and target type were manipulated within blocks. Half of the trials used trained exemplars, and half used untrained exemplars. Thus, for each block, there were 30 trials each for the trained exemplars and for the untrained exemplars (10 each memory set size). A target was present on half of the trials. For the block with untrained target categories, there were 60 trials (20 each memory set size), and a target was present on half of the trials. The procedure for individual trials was identical to Experiment 1.

Results and Discussion

Because all transfer conditions had identical distractors, only positive trials were analyzed to determine the amount of transfer. A Memory Set Size x Transfer Condition x Age ANOVA was performed on the data. All reported effects were significant at the $p<.05$ level. Young adults were faster than older adults. Response times varied as a function of transfer condition for both age groups. RTs were highest for the Untrained/Untrained condition and decreased for the Trained/Untrained, Trained/Trained, and Baseline conditions, respectively. For the young adults, the average RT was 500 ms, 539 ms, 633 ms, and 674 ms for the Baseline, Trained/Trained, Trained/Untrained, and Untrained/Untrained conditions, respectively. The old adults' average RT was 627 ms, 673 ms, 783 ms, and 903 ms for the Baseline, Trained/Trained, Trained/Untrained, and Untrained/Untrained conditions, respectively.

Considering that the Untrained/Untrained condition was novel, it is expected that response times were largest for this condition. Also, response times were fastest for the Baseline because it was from the last block of training. The fact that the Trained/Untrained condition is faster than the Untrained/Untrained condition yet slower than the Trained/Trained condition for both age groups suggests that there was some but not perfect transfer to untrained exemplars of the trained categories.

Response times increased as a function of memory set size, and this effect was larger for the older adults. The effect of memory set size was greatest for the Untrained/Untrained condition and decreased for the Trained/Untrained, the Trained/Trained, and the baseline conditions, respectively. This pattern of the memory set size effects suggests that transferring to untrained elements of trained categories resulted in some transfer, in terms of comparison times for both young and old adults. However, transfer was not perfect as the Trained/Trained condition was superior to the Trained/Untrained condition.

The response times for the baseline condition were faster than for the Trained/Trained condition, and this difference did not vary with age. Older adults are not differentially effected by moving from the Baseline to the Trained/Trained condition; therefore, the effect of changing the distractors does not vary with age group.

The RTs for the Trained/Trained condition were faster than for the Trained/Untrained condition, and this difference did not vary with age. Therefore, young and old adults are not differentially effected by changing to untrained exemplars of trained categories; if anything, old adults showed better transfer than the young.

The RTs for the Trained/Untrained condition were faster than for the Untrained/Untrained condition, and this effect was larger for the older adults. Thus, older adults were much more disrupted than the young adults when changing to a novel memory search condition.

CONCLUSION

The pattern of the data suggest that there is some transfer to untrained exemplars of trained categories for both young and old adults. Thus, both young and old adults are, at least to some extent, learning at the category level. However, the differences between the Trained/Trained and Trained/Untrained conditions show that learning is not totally at the category level for either the young or the old adults. The transfer in this current study was lower than Schneider & Fisk (1984) found for visual search using a young adult sample. It appears that both young and old adults are also learning at the word level. In fact, processing at the end of CM memory search training (at least after 1200 trials of training) appears to be a combination of word and category learning. The data from the present study are important because they show that young and old adults are learning at the same level in CM semantic category memory search. Old adults are only differentially effected when switching to a novel memory search situation.

These findings have several practical implications. First, many previous researchers have applied automatic and controlled processing theory (Shiffrin & Schneider, 1977) to the design of training systems. Schneider (1985) has outlined a number of guidelines for training high-performance skills. Myers & Fisk (1987) have determined that a dual process framework can be successfully applied to domains of complex, industrial tasks. By demonstrating the age equivalence in memory search learning, the present results show that training guidelines developed with an automatic/controlled processing framework can be successfully applied to an older adult population at least with tasks which have memory search components.

Secondly, Fisk and Lloyd (1988) found that, for young adults, the reduced set size effects with practice in a hybrid memory/visual search task using consistent stimulus-to-rule associations was similar to the results typically found using semantic categories. They concluded that training programs could capitalize not only on the consistency at the individual stimulus level but also on consistent patterns of information in complex tasks. Because the results of the present study reveal that the learning in pure memory search occurs at the semantic-category level for both young and old adults, it is reasonable to conclude that old adults can benefit from training interventions which capitalize on higher-level consistent relationships. However, even though both young and old adults can capitalize on higher-order consistency, designers of training systems need to make certain that the consistency at the desired level of learning (i.e. semantic-category, stimulus-to-rule, etc.) is salient. One way this can be accomplished is ensure that lower-level consistencies are made as nonsalient as possible during training on the higher-order skill (Fisk & Jones, 1992).

Finally, even though older adults have a working memory deficit (Salthouse & Somberg, 1982), they are able to improve in a qualitatively similar manner compared to young adults when the working memory deficit is eliminated through consistent training. Therefore, consistent training should be exploited in situations where working memory demand is high. The data also suggest that, for young and old adults, task consistency at a level higher than the specific stimuli can be utilized when training young and old adults. Hence, training can reduce working memory constraints for classes of situations (not just specific tasks) encountered during training.

ACKNOWLEDGMENTS

This research was supported by a grant from the National Institute on Aging (R01AG07654 and P50AG11715). Address correspondence to either author at the School of Psychology, Georgia Institute of Technology, Atlanta, GA 30332-0170.

REFERENCES

Battig, W. F., & Montague, W. E. (1969). Category norms for verbal items in 56 categories: A replication and extension of the Connecticut category norms. *Journal of Experimental Psychology Monograph, 80*, (Whole).

Collen, A., Wickens, D. D., & Daniele, L. (1975). The interrelationship of taxonomic categories. *Journal of Experimental Psychology: Human Learning, and Memory, 1*, 629-633.

Czerwinski, M., Lightfoot N., & Shiffrin, R. M. (1992). Automatization and training in visual search. *American Journal of Psychology, 105*, 271-315.

Eggemeier, F. T., & Fisk, A. D. (1992, March). *Automatic Information Processing and High Performance Skills.* (AL-TR-1992-0134). Dayton, OH: Human Resources Directorate, Logistics Research Division, WPAFB.

Fisk, A. D., & Jones, C. D. (1992). Global versus Local Consistency: Effects of Degree of Within-Category Consistency on Performance and Learning. *Human Factors, 34*, 693-705.

Fisk, A. D., & Lloyd, S. J. (1988). The role of stimulus-to-rule consistency in learning rapid application of spatial rules. *Human Factors, 30*, 35-49.

Fisk, A. D., & Rogers, W. A. (1991). Toward an understanding of age-related memory and visual search effects. *Journal of Experimental Psychology: General, 120*, 131-149.

Myers, G. L., & Fisk, A. D. (1987). Training consistent task components: application of automatic and controlled processing theory to industrial task training. *Human Factors, 29*, 255-268.

Rogers, W. A., & Fisk, A. D. (1991). Are age differences in consistent-mapping visual search due to feature learning or attention training? *Psychology and Aging, 6*, 542-550.

Rogers, W. A., Fisk, A. D., & Hertzog, C. (1994). Do ability-performance relationships differentiate age and practice effects in visual search? *Journal of Experimental Psychology: Learning, Memory, and Cognition, 20*, 710-738.

Salthouse, T. A., & Somberg, B. L. (1982). Skilled performance: Effects of adult age and experience on elementary processes. *Journal of Experimental Psychology: General, 111*, 176-207.

Schneider, W. (1985). Training high-performance skills: Fallacies and guidelines. *Human Factors, 27*, 285-300.

Schneider, W., & Fisk, A. D. (1984). Automatic category search and its transfer. *Journal of Experimental Psychology: Learning, Memory, and Cognition, 10*, 1-15.

Shiffrin, R. M. (1988). Attention. In R.C. Atkinson, R.J. Herrnstein, G. Lindzey, & R.D. Luce (Eds.), *Steven's handbook of experimental psychology* (2nd ed., pp. 739-811). New York: Wiley.

Shiffrin, R. M., & Schneider, W. (1977). Controlled and automatic human information processing: II. Perceptual learning, automatic attending, and a general theory. *Psychological Review, 84*, 127-190.

Age-Related Effects in Consistent Memory Search: Performance is the Same but What About Learning?

Brian P. Cooper, Mark D. Lee, Robert E. Goska, Marjo M. Anderson
Paul E. Gay, Jr., Lynne Ann Fickes, and Arthur D. Fisk

Georgia Institute of Technology

Two experiments were conducted to investigate the mechanisms which underlie the learning in consistently mapped (CM) memory search. In Experiment 1, old and young adults were trained in both CM and variably mapped (VM) category search. The training results replicate previous findings by Fisk and Rogers (1991). Even though older adults are initially at a disadvantage relative to young adults, the comparison times of young and old adults are near zero after CM training. For VM, older adults remain at a disadvantage relative to younger adults, even after extensive training. A full reversal manipulation was implemented in Experiment 2 to investigate the learning in memory search. Initially, the young subjects were less affected by the full reversal condition compared to the performance of the older adults. However, older subjects quickly recovered and both young and old were performing at trained CM levels within 60 trials of additional practice. These results suggest: (a) attention is not being trained in CM memory search; (b) automatic category activation does not contribute much, if at all, to the performance improvement in memory search; and (c) age-invariant learning mechanisms account for performance improvement in CM memory search.

INTRODUCTION

The qualitative changes associated with the acquisition of skill have been frequently investigated using search/detection tasks. The memory search paradigm is one of the search/detection tasks which has been frequently studied in the area of skill acquisition. Many researchers (e.g., Shiffrin & Schneider, 1977) have demonstrated that the degree of performance improvement is a function of the "consistency" of training. In a consistently mapped (CM) condition, target and distractor items are chosen from different sets of stimuli. This allows one to respond the same way (consistently) every time a particular stimulus is present in the display. In a varied mapping (VM) situation, the target and distractor items are selected from the same set of stimuli. When mapping is varied, one responds differently to a particular stimulus from trial to trial.

Fisk and Rogers (1991) have demonstrated that both young and old subjects demonstrate dramatic performance improvements with CM memory search training. Early in training, the comparison times of old adults are much slower than those of young adults. However, after extensive CM training, the comparison slopes of both old and young adults approximate zero. In VM memory search, old subjects remain at a disadvantage, relative to young subjects, in terms of VM memory comparison times. The comparison slopes of the older adults are much greater than those of younger subjects -- even after extensive training. Because both young and old adults exhibit similar performance improvements in CM memory search, the mechanisms underlying the learning are, presently, believed to be the same for both age groups.

According to Shiffrin (1988) the performance improvement observed in CM memory search may result from two mechanisms: (1) memory set unitization and (2) automatic category activation. In memory set unitization, all of the consistently trained targets are learned as a "super" category. Memory set unitization is assumed to be a form of associative learning. Associative learning is not a new concept and has been precisely specified by other investigators (see McClelland, Rumelhart, & Hinton, 1986; Schneider & Fisk, 1984; Shiffrin & Schneider, 1977). According to this view, memory is a large collection of interconnected nodes. Associative learning is reflected in the modification of the activation patterns between these nodes. Stimulus information which is concurrently activated in short-term storage will become associated when the co-activation consistently occurs across numerous training trials. Once a set of information nodes becomes associated, the set has been unitized, and a single representation can be extracted to represent the entire set. Automatic category activation is believed to be a form of attention training. After consistent training, a target appearing in the display will result in an automatic extraction of the "super" category. Automatic category activation is analogous to the "pop-out" effect in visual search and is believed to be a result of priority learning (Schneider, 1985). In addition to the two mechanisms previously described, it is assumed that some of the performance improvement in memory search is a result of learning task strategies. As subjects incorporate information from the instructions into the specific task, they learn about the characteristics of the display, response keys, and stimuli. This information about the task allows them to adopt encoding, search, and/or responding strategies which facilitate performance.

The present experiments were conducted to further explore the mechanisms underlying the performance improvements in CM memory search. In Experiment 1, young and old subjects received extensive training in a semantic category memory search task. It was predicted that the results will replicate Fisk and Rogers (1991). Older subjects are expected to exhibit higher VM comparison times than the younger adults, and this deficit should not diminish with training. At the beginning of training, the CM comparison slopes of the older adults should be higher than those of younger adults; however, after training, the comparison slopes of both young and old adults should approximate zero.

In Experiment 2, the extent to which automatic category activation is responsible for trained CM memory search performance was investigated by reversing the roles of targets and distractors. If automatic category activation is an important component to the performance improvements for both young and old subjects, reversing the roles of targets and distractors should lead to considerable disruption. If minimal disruption is observed, then automatic category activation will be ruled out as an important learning mechanism in memory search. Also, if the same mechanism is underlying the learning for both young and old subjects, similar patterns of disruption should be observed in both age groups.

EXPERIMENT 1

Method

Subjects and stimuli. Thirty-seven young subjects (ages 19 - 25) were recruited from a Southeastern university, and 17 old adults (ages 65 - 80) were recruited from the community. Subjects received course credit or $5 per hour for their participation.

The stimuli were words from sixteen semantically unrelated categories (Collen, Wickens, and Daniele, 1975). Eight high associates from each category (Battig and Montague, 1969), four to seven letters long, were chosen as exemplars.

Apparatus. Microcomputers were programmed to control the timing of the displays, present the stimuli, and collect the responses. The computer programs were developed using MEL software (Schneider, 1988). The data were collected using either EPSON Equity I+ microcomputers with EPSON MBM 2095 green monochrome monitors or EPSON Equity 286 Plus microcomputers with EPSON monochrome VGA monitors. Subjects used the same type of computer throughout the entire experiment. The '1' and '2' keys on the numeric keyboard were labeled 'P' and 'A', respectively. For all experimental sessions, pink noise was played at approximately 57 dB(A) sound pressure level to attenuate background noise.

Design. The within-subjects independent variables were: (a) Memory set size: two, three, or four categories; (b) Target type: target present or target absent; (c) Training condition: CM or VM. Age (young or old) was a quasi-independent variable. The dependent variables were RT and accuracy.

Procedure. An experimental trial consisted of the following sequence of events. A memory set of either two, three, or four category labels was displayed for a maximum of 20 seconds. Subjects pressed the space bar to initiate the trial. A plus sign was then presented in the center of the screen for 500 ms. The probe display, consisting of one word, then appeared. The subject's task was to indicate whether the display word was a member of one of the categories in the memory set. The subjects pressed the key labeled 'P' (for Present) for a positive response and the key labeled 'A' (for Absent) for a negative response. Subjects received feedback concerning speed and accuracy following each trial and each block of trials.

The subjects were trained on the semantic category memory search task for five 90 minute sessions. Each subject was assigned four categories as the CM target set and another four categories as the CM distractor set; the remaining eight categories served interchangeably as targets and distractors in the VM condition. Category assignment was partially counterbalanced using a Latin square. The first session of training consisted of 16 blocks of 60 trials per block. The second session consisted of 18 blocks of 60 trials, and the third, fourth, and fifth sessions each consisted of 20 blocks of 60 trials. Each subject completed a total of 2820 CM trials and 2820 VM trials. Memory set size and target type were manipulated within a block. Thus, there were 20 trials per memory set size within a block, and a target was present on half of the trials. Training condition (CM or VM) was manipulated between blocks. Each session began with a CM block, and CM and VM trial blocks alternated.

Results

For each subject, block median RT was calculated for each condition, and the slope of the RT by memory set size function (comparison slope) was calculated. As expected, CM performance showed rapid improvement for both age groups. Although the slopes of the older adults were initially larger than those of the younger adults, young and old adults were equivalent by session 2. For VM, neither the young nor the old adults improved much with practice. The VM comparison slopes of the older adults were consistently larger than those of the young subjects-- even after five sessions of training. Because the age differences in CM performance were eliminated in one session, the analyses will focus on the session 1 data.

Figure 1 contains the CM comparison slopes for each age group and trial type as a function of blocks of practice. Analyses reported below are significant at the .05 level unless otherwise specified. As can be seen in the figure, the comparison slopes of the older adults in the first block of practice are larger than those of young subjects, $F(1,52)=15.96$. Although the comparison slopes for the target present trials are smaller than those of target absent trials for both age groups, this result did not reach statistical significance. Figure 1 also reveals that much of the performance improvement occurs within the first block (60 trials) of training. By block 8, the comparison times of older adults are not statistically different from those of young adults,

$\underline{F}(1,52) < 1$.

Figure 2 contains the VM comparison slopes for each age group and trial type as a function of blocks of practice. As can be seen from the figure, the comparison times change very little as a function of practice. For session 1, the comparison slopes of the older subjects were greater than those of the younger subjects; however, this result did not achieve statistical significance (p=.122). However, for session 8, the comparison slopes of the older adults are significantly larger than those of the young adults, $\underline{F}(1,52)= 4.87$, $p=.032$. The nonsignificant result for the VM session 1 comparison slope is misleading. The comparison times of the older subjects is attenuated early in VM practice because the accuracy of VM memory set size 4 (66%) is much lower than that of memory set sizes two and three (87% and 82%, respectively). The reason for the age differences later in training may be attributed to a more accurate slope estimate resulting from elimination of speed/accuracy trade-off at the largest memory set size for the older subjects.

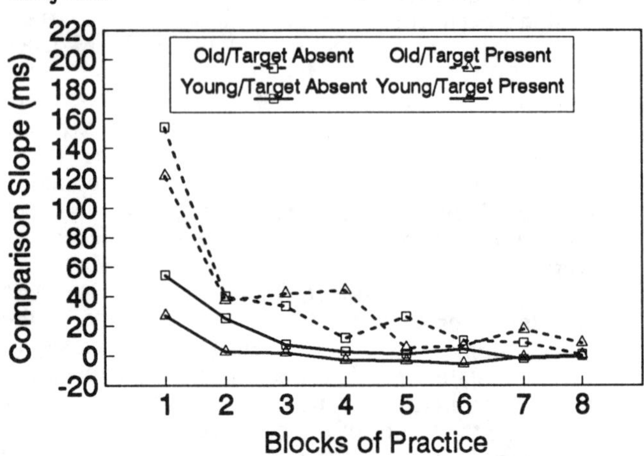

Figure 1: CM comparison slopes for each age group as a function of blocks of practice.

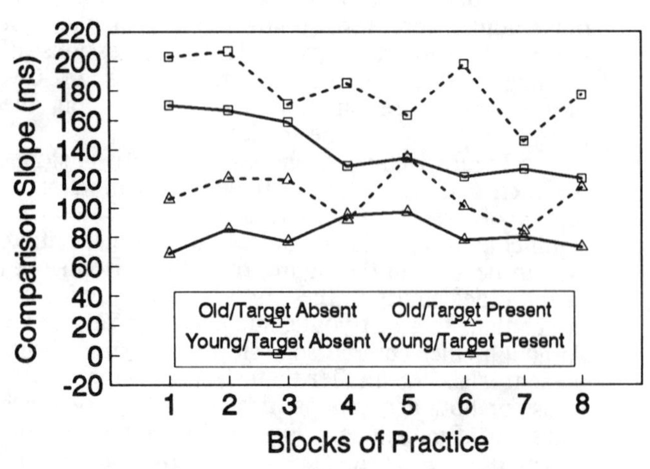

Figure 2: VM comparison slopes for each age group as a function of blocks of practice.

Discussion

The training results replicate previous research by Fisk and Rogers (1991). For CM memory search, both young and old subjects improved with practice. Although older subjects initially exhibited slower comparison times, after training, both young and old subjects displayed comparison times which were near zero. For VM, neither the young nor old subjects improved much with practice. The comparison slopes of the older subjects were consistently larger than those of the young subjects. The age-related differences are even more severe than indicated by the slope data. Older subject had much difficulty protecting accuracy for the VM memory set size 4 condition. Therefore, the comparison times reported are actually an underestimate of the actual comparison times which are exhibited by older adults.

It has been noted by several investigators (e.g., Salthouse and Somberg, 1982) that the performance of older subjects may suffer as a result of working memory deficits. As expected, when the working memory demands are decreased through consistent practice, older subjects achieve performance which is equivalent to that of the younger adults. However, the fast rate of learning brings into question the mechanisms underlying the performance improvement in pure memory search. In order to investigate the learning in CM memory search, a full reversal manipulation was implemented. Research examining the visual search (e.g., Fisk and Rogers, 1991) has demonstrated that older subjects do not modify the attention calling strength of the stimuli; thus, they are relatively unaffected by a reversal of the roles of targets and distractors. However, for young subjects, it appears that CM stimuli come to automatically attract attention because performance is severely disrupted when they are subjected to a full reversal in visual search. Hence, reversal of target/distractor roles has proven successful in demonstrating differences in learning in visual search. Observation of the patterns of disruption across age groups will allow us to determine if the same learning mechanisms underlie the performance for young and older adults in memory search.

EXPERIMENT 2

Method

The subjects, apparatus, and stimuli of this experiment were identical to Experiment 1. Subjects first received 5 blocks of CM practice (60 trials per block). These trials used the same CM target/distractor pairings that were used during Experiment 1. Following these baseline trials, the subjects were placed in the transfer phase. There were two conditions in this phase: (a) CM Reversal in which the roles of the CM targets and distractors were reversed such that the previous CM targets became distractors and the previous CM distractors became the targets; and (b) New CM condition in which four of the former VM categories composed the New CM target set and the remaining four of the former VM categories composed the New CM distractor set. Five 60-trial blocks of CM Reversal and five 60-trial blocks of New CM were presented alternately for a total of 300 CM Reversal trials (100 at each memory set size) and 300 New CM trials (100 at each memory set size). As in Experiment 1, a target was present on half of the trials within a block. The procedure for individual trials was the same as in Experiment 1.

Design. The within-subjects independent variables for the transfer phase were: (a) Memory set size: two, three, or four categories; (b) Target type: target present or target absent; (c) Training/Transfer Conditions: CM Baseline, CM Reversal, or New CM. Age was the quasi-independent variable. The dependent variables were RT and accuracy.

Results

For each subject, block median RT was calculated for each condition. Table 1 contains the comparison slopes and intercepts for each transfer condition for blocks 1 through 5. As can be seen from the table, the comparison slopes for the baseline condition are consistent with results from Experiment 1. Older adults are generally slower than younger adults as reflected by a significant difference in the intercepts at block 1, $F(1,52)=80.36$. However, the comparison times of both the young and old adults are near zero. Also, there was no difference between positive and negative trials with either the slopes or the intercepts ($p > .05$).

In order to test the amount of disruption relative to training, contrasts were performed on the slopes from block 1 for each age group. For the older subjects, the reversal comparison slopes were significantly greater than the baseline slopes, $F(1,16)=9.70$, and the comparison times of the New CM condition were significantly greater than those in the baseline, $F(1,16)=14.78$. The slopes from the New CM were about twice those of the reversal, although this effect did not reach statistical significance at the .05 level, $F(1,16)=4.01$, $p=.062$.

For the younger subjects, the reversal comparison times did not significantly differ from those of the baseline condition ($p>.05$). The comparison time for the New CM condition was about 12 times slower than trained performance; however, the effect did not reach statistical significance, $F(1,36)=3.72$, $p=.062$. Also, the difference between the slopes of the New CM and reversal conditions was not significant.

Observation of Table 1 reveals that, although the comparison times of older adults are more disrupted by the reversal and New CM conditions, the effects were temporary. By block 2, the reversal slopes of the older adults did not significantly differ from baseline performance. By block 5, the comparison slopes of both age groups were near zero for all conditions.

Contrasts were also performed on the intercept data for block 1. For the older subjects, reversal intercepts were greater than the baseline intercepts, $F(1,16)=15.98$. However, the difference between the New CM and baseline conditions and the difference between the reversal and New CM conditions were not significant. For the younger subjects, the reversal intercepts were significantly greater than the baseline intercepts, $F(1,16)=41.24$, and the New CM intercepts were greater than the baseline intercepts, $F(1,16)=67.38$. However, the difference between reversal and New CM was not significant.

BASELINE

	YOUNG		OLD	
BLK	SLOPE	INTERCEPT	SLOPE	INTERCEPT
1	-0.30	467.70	-8.74	616.03
2	-1.08	460.72	-4.22	596.01
3	0.07	456.78	-7.94	603.14
4	0.80	457.66	1.77	578.83
5	-0.36	454.24	-9.78	608.69

REVERSAL

	YOUNG		OLD	
BLK	SLOPE	INTERCEPT	SLOPE	INTERCEPT
1	5.93	626.71	32.60	774.51
2	-4.59	620.94	-3.40	766.79
3	-0.53	593.34	11.69	695.21
4	-4.18	592.02	-6.23	738.56
5	1.20	572.36	-3.73	725.83

NEW CM

	YOUNG		OLD	
BLK	SLOPE	INTERCEPT	SLOPE	INTERCEPT
1	12.49	640.09	65.17	679.83
2	5.96	594.62	34.85	667.11
3	0.22	597.91	24.03	649.46
4	3.66	564.50	11.18	659.31
5	-0.02	562.98	2.54	662.60

Table 1: Comparison slopes and intercepts for each transfer condition for blocks 1 through 5.

DISCUSSION

A full reversal condition was implemented in Experiment 2 in order to investigate the mechanisms underlying the fast performance improvement observed in Experiment 1. The most salient finding from Experiment 2 was the relatively minimal disruption occurring when the roles of the CM targets and distractors were reversed. Although, performance was somewhat disrupted for both young and old subjects in the reversal condition, the disruption was less than that found for the New CM stimuli. Disruption in the reversal condition, for the most part, was due to intercept effects not comparison slope effects. If automatic category activation (strengthening of the category to response links) were occurring in memory search, the present pattern of data would not be expected. If automatic category activation is a component of performance improvement in CM memory search, its influence is small.

The comparison slopes of the younger subjects were not affected much by the full reversal condition. However, the slopes of the older adults were initially larger in the reversal than in the CM baseline. These results contrast sharply with those found using a pure visual search paradigm (see Fisk and Rogers, 1991). In a visual search paradigm, young subjects are much more affected by a full reversal because their attention is being trained to strengthen targets and weaken distractors; in fact, reversal performance in visual search is much worse than in a New CM condition. Older subjects, who do not experience target/distractor strength differentiation, are relatively unaffected by a full reversal in visual search. Experiment 2 also revealed that the disruption experienced by the older adults was temporary. The reversal performance of the older subjects achieved baseline performance within 60 trials of additional practice. The patterns of disruption for both young and old subjects found in Experiment 2 combined with the fast rate of recovery suggests that attention is not being trained in a pure memory search and that automatic category activation is not an important mechanism underlying the performance improvement in CM memory search.

CONCLUSION

In conclusion, the results of Experiment 1 reveal that, when the working memory deficits of older subjects can be minimized by consistent training, the comparison times of young and older adults are equivalent. The patterns of disruption in the reversal condition of Experiment 2 combined with the fast rate of recovery suggests that automatic category activation is not an important mechanism underlying the learning in pure memory search. The initial difference between young and old subjects in the reversal condition was short lived relative to the young and old performance differences in the New CM condition. Hence, the initial age-related differences in the reversal condition were most likely due to strategy differences related to target responding (cf. Fisk, Rogers, & Giambra, 1990).

The fast rate of learning in Experiments 1 and 2 support the notion that, even though older subjects may have a working memory deficit, they are able to improve in a qualitatively similar manner compared to younger adults when the working memory deficit is eliminated through consistent training. Therefore, consistent training is an approach which should be exploited when training older adults in situations where working memory capacity is exceeded

ACKNOWLEDGMENT

This research was supported by the National Institutes of Health Grant R01AG07654 from the National Institute on Aging (NIA) to Arthur D. Fisk.

REFERENCES

Battig, W. F., & Montague, W. E. (1969). Category norms for verbal items in 56 categories: A replication and extension of the Connecticut category norms. Journal of Experimental Psychology Monograph, 80, (Whole).

Collen, A., Wickens, D. D., & Daniele, L. (1975). The interrelationship of taxonomic categories. Journal of Experimental Psychology: Human Learning, and Memory, 1, 629-633.

Fisk, A. D., & Rogers, W. A. (1991). Toward an understanding of age-related memory and visual search effects. Journal of Experimental Psychology: General, 120, 131-149.

Fisk, A. D., Rogers, W. A., & Giambra, L. M. (1990). Consistent and varied memory/visual search: Is there an interaction between age and response-set effects? Journal of Gerontology: Psychological Sciences, 45, P81-P87.

McClelland, J. L., Rumelhart, D. E., & Hinton, G. E. (1986). The appeal of parallel distributed processing. In D. E. Rumelhart & J. E. McClelland (Eds.), Parallel distributed processing: Explorations in the microstructure of cognition, Vol. 1 (pp 3 - 44). Cambridge, MA: MIT Press.

Salthouse, T. A., & Somberg, B. L. (1982). Skilled performance: Effects of adult age and experience on elementary processes. Journal of Experimental Psychology: General, 111, 176-207.

Schneider, W. (1985). Toward a model of attention and the development of automaticity. In M. I. Posner & O. S. Martin (Eds.), Attention and performance XI (pp. 475-492). Hillsdale, NJ: Erlbaum.

Schneider, W. (1988). Micro Experimental Laboratory: An integrated system for IBM PC compatibles. Behavior Research Methods, Instrumentation, & Computers, 20, 206-217.

Schneider, W. & Fisk, A. D. (1984). Automatic category search and its transfer. Journal of Experimental Psychology: Learning, Memory, and Cognition, 10, 1-15.

Shiffrin, R. M. (1988). Attention. In R.C. Atkinson, R.J. Herrnstein, G. Lindzey, & R.D. Luce (Eds.), Steven's handbook of experimental psychology (2nd ed., pp. 739-811). New York: Wiley.

Shiffrin, R. M., & Schneider, W. (1977). Controlled and automatic human information processing: II. Perceptual learning, automatic attending, and a general theory. Psychological Review, 84, 127-190.

COGNITIVE AGING: GENERAL VERSUS PROCESS-SPECIFIC SLOWING IN A VISUAL SEARCH TASK

Michael F. Gorman and Donald L. Fisher
University of Massachusetts Amherst

The fact that response times increase as one ages has long been established. Previously, a model of general slowing in the nonlexical domains has done a really good job of explaining the differences between older and younger adults. However, an alternative process-specific model has not been conclusively ruled out. This experiment tested general and process-specific models of slowing in the nonlexical domain using older and younger adults performing a visual search task. The task manipulated the presence of the target, the number of search items, and the structure of the display of the search items. It was found that a process-specific model explained significantly more of the variability than a general model of slowing. It was also discovered that the process most greatly affected was that of deciding to terminate a search when no target was present in the display.

It is well known that as humans age their performance decreases in a wide variety of tasks (Davies, Taylor & Dorn, 1992; Salthouse, 1991; Cerella, Poon & Williams, 1980). For example, performance decrements have been observed in tests of memory (Howard & Wiggs, 1993), intelligence (Hertzog, 1989; Schaie, 1989) and attention (Giambra, 1993; Madden & Plude, 1993).

In general there are two theories which attempt to explain the decline in performance of older adults. More specifically, we are talking about molar linear models of general slowing (and the related latent linear models of common slowing) and latent linear models of process-specific slowing (Fisher and Glaser, 1996). First, consider the molar linear model of general slowing. Defining O_i as the time on average for an older adult to respond, Y_i as the time on average for a younger adult to respond, and i as representing the task, the molar linear model of general slowing can be expressed by the equation:

$$O_i = \beta Y_i. \quad (1)$$

The model is a molar one because only the overall response times appear in the equation. The molar model is a general one because the same slowing factor, β, is used from one task to the next.

Now, consider the related latent linear model of common slowing. In order to define the latent model, we need to know the structure of the underlying cognitive processes in each task. Suppose that in one task the processes x_1, x_2 and x_3 with durations, respectively, of X_1, X_2 and X_3, were arranged in series. Then for younger adults, the latent model can be expressed as:

$$Y = X_1 + X_2 + X_3. \quad (2)$$

If the latent model is one of common slowing, then for older adults we can write:

$$O = \beta X_1 + \beta X_2 + \beta X_3. \quad (3)$$

The molar linear model of general slowing and latent linear model of common slowing are clearly conceptually identical to one another in this simple case. Specifically if all processes are arranged in series, then the slowing of each of the individual processes by a common factor β is identical to the slowing of the overall response time by a common factor β. However, the molar and latent models are no longer conceptually identical if each of the processes is slowed by a different factor. In this case (the latent linear model of process-specific slowing) we can write:

$$O = \beta_1 X_1 + \beta_2 X_2 + \beta_3 X_3. \quad (4)$$

The molar model of general slowing with its simplicity and elegance does a good job of predicting the differences between older and younger adults (Cerella et al. 1980). Support for this model has also been established by Salthouse and Somberg (1982) who showed a molar model of general slowing was able to explain 98.2% of the variability between older and younger adults. More recently, investigators have argued persuasively that slowing may not be general across all domains, but instead may be general only to broad domains such as the lexical (Lima, Hale and Myerson, 1991) and nonlexical (Lima, Hale and Myerson 1991; Hale, Myerson, Faust, and Fristoe, 1995) domains. We will focus here just on the nonlexical domain. Still, however well the molar models of general slowing explain the effect of age on performance in this domain, it is possible that a latent model of process-specific slowing may do a better job if more complete tests of the latter model are used (Fisk, Fisher & Rogers, 1992; Fisk & Fisher, 1994; Fisher, Fisk & Duffy, 1995). These tests we propose to undertake here. Specifically, our experiment attempts to determine whether a process-specific model will do a better job of predicting the effect of age on performance than a general slowing model by manipulating various factors of a visual search task. Simply, subjects were presented a target, and then asked to search a grid of letters and indicate whether or not the target was present.

METHODS

Subjects

Eleven male and 13 female, younger adults, between the ages of 19 and 30 with a mean age of 24.5, and 5 male and 19 female, older adults, between the ages of 60 and 90 with a mean age of 75.4 were each paid $30 for their participation in the study. The mean scores on the Wechsler (1955) Adult Intelligence Scale Vocabulary and Digit Span subtests were 34.46 and 12.63 for the younger adults, and 35.83 and 12.38 for the older adults. These were used to ensure that the groups had equivalent educational backgrounds.

Materials

A personal computer, running MS-DOS was used to run the experimental program (available from the author) written in C. All subjects were given the Digit Span and the first half of Vocabulary tests from the Wechsler Adult Intelligence Scale for adults.

Procedure

The experiment consisted of three sessions, which took place over three consecutive days. Before the first session, the subject read and signed a consent form describing the nature of the experiment. The experimental task was then described, to the subject, using a handout which showed what the various screens and parts of the tasks would look like. The subject was instructed to respond to each trial as quickly and accurately as possible, and allowed to sit at a distance from the screen which was comfortable to them. Day 1 was a practice session, consisting of fewer trials than the remaining two days, and contained a mix of both randomly and systematically presented search grids.

On the second and third days of the experiment, the experiment proceeded as follows. A target letter was presented in the center of the computer's screen for 1500ms. The target was then masked for 750ms, after which the screen was cleared for 500ms. Next, a search grid of 12, 16 or 20 letters was presented, consisting of 4 rows and 3 columns, 4 rows and 4 columns or 4 rows and 5 columns respectively. On one day the search grid was displayed systematically (i.e., the letters were arranged in straight rows and columns) and on the other day the search grid was displayed randomly (i.e., the letters were scattered over the screen). Over the course of the experiment, the target was present in the search grid on half of the trials. The subject then searched the grid for the target letter. If the subject saw the target letter they pressed the 'K' key for 'yes' , if they did not see the target they pressed the 'D' key for 'no'. The subject was then informed whether or not their response was correct and prompted to press any key for the next target. Response time was measured from when the search grid was displayed until the subject pressed a key in response. If an incorrect key was pressed as a response, the subject was informed by the computer and prompted to press one of the valid response keys. Every 10 trials the subject's accuracy was checked. If the accuracy dropped below 90%, a tone was generated and the subject was warned of the situation.

Design and Counterbalancing

The experimental session consisted of five repeated blocks. Each block consisted of all three grid sizes, 12, 16, and 20 search items. For each grid size the target appeared in each position, along with an equal number of trials where the target was absent. This resulted in 5 blocks X (12 present + 12 absent + 16 present + 16 absent + 20 present + 20 absent) = 480 trials. All the trials for each grid size within a block were presented together. There were six sequences in which the three grid sizes could be presented, and they were counterbalanced across subjects. The search grids were arranged in a display either systematically or randomly, the exact arrabgement depending on the day of the experiment. Half the subjects received the random grids first and the other half received the systematic grids first. Six, (absent and present) X (12, 16 or 20 items), lists of targets were randomly generated using four alphabets, such that the first 26 targets were from one alphabet, the next 26 from another alphabet and so on. Whether the target was present or absent on a given trial was determined randomly subject to the constraint that half of the trials in each grid size were absent trials. The position of the target when it was present was also selected randomly without replacement from the possible positions for the respective grid size. The search grids were centered on the screen. In the systematic grid, the search items were displayed in straight rows and columns 80 pixels apart on a 320x640 resolution screen. In the random search grids, each letter could occupy a position up to 30 pixels to the left or right, and 30 pixels up or down from where the letter would appear in a systematic grid. In a worst case situation, letters would still have a 20 pixel distance between them.

RESULTS

In order to obtain a more accurate representation of the data, first incorrect responses were removed, and then a grand mean and standard deviation. were computed to remove outliers. Response times more than 3 standard deviations from the grand mean and less than 100 ms were removed. Response times less than 100 ms were determined to be beyond human ability and attributed to equipment error. Equipment error was usually the result of a subject's hand inadvertently resting on a key and overflowing the keyboard buffer. This resulted in the computer getting a response immediately from the buffer, instead of from the subject.

Response times were analyzed using a mixed-factorial 2 (young or old) x 2 (random or systematic) x 3 (12, 16 or 20 items) x 2 (target absent or present) ANOVA. The analysis revealed main effects for age $F(1,46) = 36.37$, $p < .001$, grid size $F(2,92) = 131.09$, $p < .001$, and target presence $F(1,46) = 70.12$, $p < .001$. There was no main effect for the grid pattern (systematic or random) $F(1,46) = .27$, $p > .10$, nor any interaction between grid pattern and age $F(1,46) = 2.21$, $p > .10$. However, there were several significant interactions: grid size X age $F(2,92) = 28.00$, $p < .001$, target presence X age $F(1,46) = 34.95$, $p < .001$, grid size X target presence $F(2,92) = 82.59$, $p < .001$, and grid size X target presence X age $F(2,92) = 19.87$, $p < .001$. Greenhouse-Geisser corrections for sphericity did not change the significance of any of the above main effects or interactions

The condition means (Table 1) were also fit using both general and process-specific models of cognitive slowing (Table 2). For example, when the target was present and the search grid was systematic, the equation $Y = a + [(k+1)/2 * c]$ was derived as follows: a = the motor response time, $(k+1)/2$ = the number of items on average that will be examined in a serial self-terminating search, and c = the encoding and comparison time for each item. When the target was absent and the search grid random, the equation, $Y = a + (k * b * c) + n$, was derived similarly with k = number of items in the search grid, b = the proportion of items re-checked due to loosing one's place in the grid, and n = the time to terminate the search after all stimuli have been scanned.

To determine if the process-specific model explained significantly more variance than the general slowing model the following formula was used (Neter and Wasserman, 1974, pp 87-89):

$$\frac{\left(\dfrac{SSE(GS) - SSE(PSS)}{df(GS) - df(PSS)}\right)}{\left(\dfrac{SSE(PSS)}{df(PSS)}\right)} \qquad (5)$$

where $SSE(GS)$ is equal to the error sum of squares for the latent model of common slowing, $SSE(PSS)$ is equal to the error sum of squares for the latent model of process-specific slowing, $df(GS) = 24 - 5 = 19$ is equal to the degrees of freedom for the common model, and $df(PSS) = 24 - 8 = 17$ is equal to the degrees of freedom for the process-specific model. The result has an F distribution with $(df(GS) - df(PSS))$ and $df(PSS)$ degrees of freedom. The process-specific slowing model explained significantly more variability than the common slowing model $F(2,17) = 19.3164$, $p < .05$, with the common slowing model explaining 99.73% of the variance and the process-specific model explaining 99.88%. See Figures 1-2 for graphs of the observed data and the predictions made by the process-specific model.

Table 1
Mean Response Times in Seconds

Condition			Grid Size		
Age	Target Presence	Grid Display	12	16	20
Young	Absent	Systematic	1.397	1.775	2.097
Young	Absent	Random	1.432	1.803	2.170
Young	Present	Systematic	0.977	1.152	1.313
Young	Present	Random	1.004	1.166	1.275
Old	Absent	Systematic	2.421	2.991	3.556
Old	Absent	Random	2.514	3.077	3.638
Old	Present	Systematic	1.560	1.827	2.073
Old	Present	Random	1.583	1.843	2.053

Table 2
Equations for General Slowing and Process-Specific Models

General slowing

Age	Condition	Equation
Young	Present, Systematic	$Y = a + [(k+1)/2]\ c$
Young	Present, Random	$Y = a + \{[(k+1)/2]\ b\}\ c$
Young	Absent, Systematic	$Y = a + (k\ c) + n$
Young	Absent, Random	$Y = a + (k\ b\ c) + n$
Old		$O = \beta Y$

Coefficients
$a = 0.399$ $b = 1.029$ $c = 0.082$
$n = 0.076$ $\beta = 1.670$

Process-Specific

Age	Condition	Equation
Young	Pres., Sys.	$Y = a + [(k+1)/2]\ c$
Young	Pres., Rand.	$Y = a + \{[(k+1)/2]\ b\}\ c$
Young	Abs., Sys.	$Y = a + (k\ c) + n$
Young	Abs., Rand.	$Y = a + (k\ b\ c) + n$
Old	Pres., Sys.	$O = \beta_a\ (a) + [(k+1)/2]\ \beta_c\ (c)$
Old	Pres., Rand.	$O = \beta_a\ (a) + \{[(k+1)/2]\ b\}\ \beta_c\ (c)$
Old	Abs., Sys.	$O = \beta_a\ (a) + [k\ \beta_c\ (c)] + \beta_n\ (n)$
Old	Abs., Rand.	$O = \beta_a\ (a) + [k\ b\ \beta_c\ (c)] + \beta_n\ (n)$

Coefficients
$a = 0.413$ $b = 1.027$ $c = 0.084$
$n = 0.014$ $\beta_a = 1.583$ $\beta_c = 1.616$
$\beta_n = 12.695$

a = Motor response time
b = Proportion of stimuli that are rescanned
c = Time on average it takes to encode a stimulus and compare the stimulus to the target
k = Number of stimuli in the display
n = Time to decide that a stimulus is not present above the time it takes to say it is present

Figure 1. Predicted and observed response times of younger adults as a function of grid size. (Predictions come from Table 2, where YAS = observed response time when the target is absent and the grid is systematic, ypr = predicted response time when the target is present and the grid is random.)

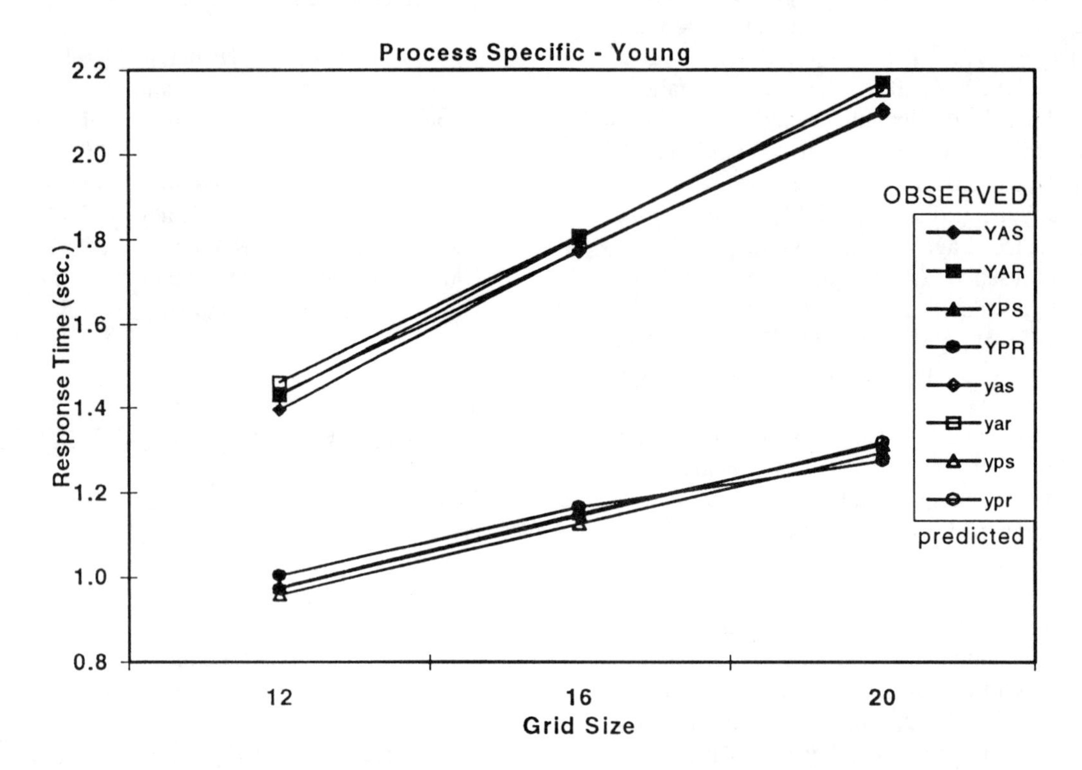

Figure 2. Predicted and observed response times of older adults as a function of grid size. (Predictions come from Table 2, where OAS = observed response time when the target is absent and the grid is systematic, opr = predicted response time when the target is present and the grid is random)

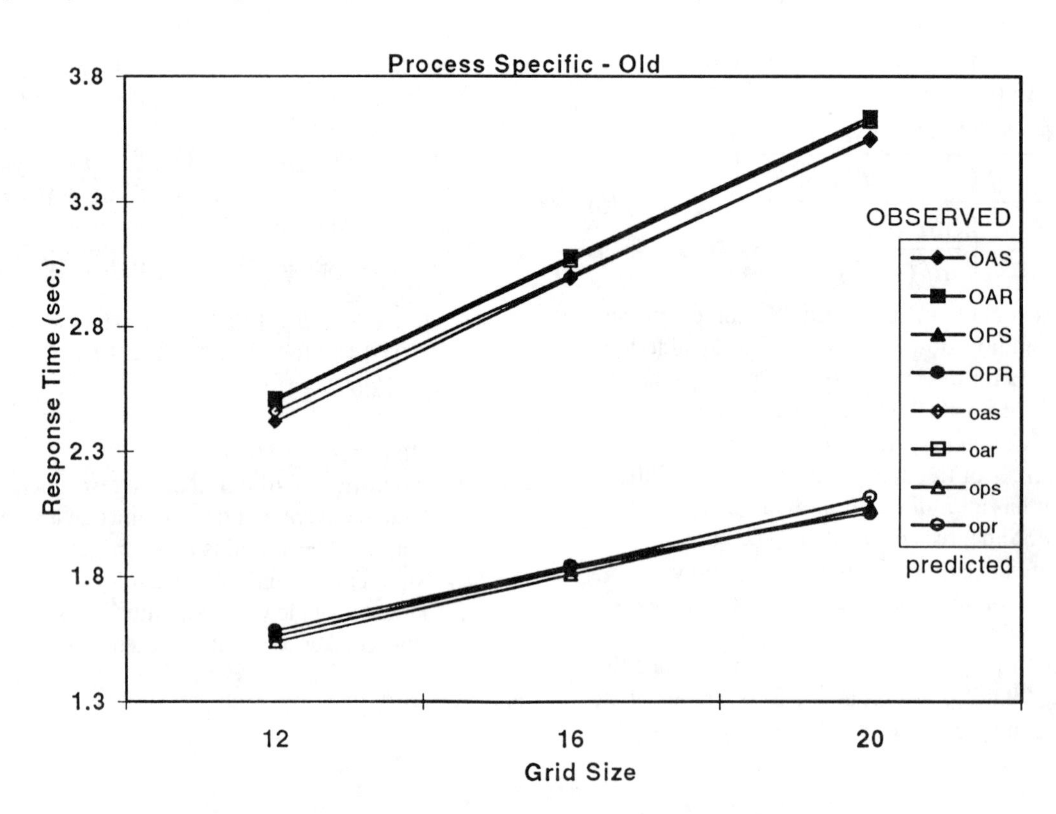

DISCUSSION

As the results indicate, the model of general (or common) slowing does a really good job of explaining the variability between the age groups, accounting for 99.73% of the variability. However, the process-specific model not only explains significantly more of the variability, it also provides us with additional insight as to what is occurring in this particular task. Looking at the coefficients of the models in Table 2, it is apparent that across models, the values are almost equal, except for β_n. That is, while β = 1.670, in the general model, and β_a = 1.583 and β_c = 1.616 in the process-specific model are almost equal, it is β_n = 12.695 that stands out and provides us with the important information. The fact that the slowing parameter for β_n, the time to terminate a search, is so much larger than the other slowing parameters indicates that this is the subprocess of the task that is giving older adults problems.

The above finding has a practical as well as theoretical import. For example, older adults are typically involved in many more accidents at signalized left turn intersections than younger adults (Staplin and Fisk, 1991). Perhaps it takes older adults so long to come to a decision that an oncoming car is not present after scanning the roadway, that other fast approaching cars that were not a threat become a threat during the long decision time. While, tests are now being proposed to screen older drivers, these tests are meaningful only if they require older adults to exercise the cognitive operations that are critical to safe driving. The critical cognitive operations can be known only if latent models like the ones we proposed are used to identify the problematic processes.

One finding we expected but did not observe was what we termed the "getting lost factor". We expected that in a random search grid, the older adults would have more difficulty remembering where they had been, and as a result would have to recheck some percentage of the search grid items. This however turned out not to be the case.

In conclusion, a model of general slowing often does a very good job of explaining the differences between older and younger adults. Although it is tempting to stop there and say good enough, in doing so we are missing out on the added insight and understanding that a process-specific model provides.

ACKNOWLEDGEMENTS

This research was supported in part by a grant from the National Institute of Aging (AG12461) to Donald L. Fisher.

REFERENCES

Cerella, J., Poon, L. W. & Williams, D. M. (1980). Age and the complexity hypothesis. In L. W. Poon (Ed.), *Aging in the 1980s: Psychological issues* (pp. 332-340). Washington DC: American Psychological Association.

Davies, D. R., Taylor, A. & Dorn, L. (1992). Aging and human performance. In A. P. Smith & D. M. Jones (Eds.), *Handbook of Human Performance: Volume 3 State and Trait* (pp. 25-61). San Diego, CA: Academic Press Limited.

Fisher, D. L., Fisk, A. D. & Duffy, S. A. (r1995). Why latent models are needed to test hypotheses about the slowing of word and language processes in older adults. In Ph. Allen & Th. R. Bashore (Eds.), *Age differences in word and language processing* (pp. 1-29). Elsevier Science B. V.

Fisher, D. L. & Glaser, R. (1996). Cognitive aging: Models of general, task-specific and process-specific slowing. *Psychonomic Bulletin and Review.* Accepted with revisions.

Fisk, A. D. & Fisher, D. L. (1994). Brinley plots and theories of aging: The explicit, muddled and implicit debates. *Journal of Gerontology: Psychological Sciences, 49*, P81-P89.

Fisk, A. D., Fisher, D. L. & Rogers, W. A. (1992). General slowing alone cannot explain age-related search effects: A reply to Cerella (1991). *Journal of Experimental Psychology: General, 121*, 73-78.

Giambra, L. M. (1993). Sustained attention in older adults: performance and processes (pp. 259-272). In J. Cerella, J. Rybash, W. Hoyer and M. L. Commons (Eds.), *Adult information processing: Limits on loss.* San Diego: Academic Press.

Hale, S., Myerson, J., Faust, M., and Fristoe, N. (1995). Converging evidence for domain-specific slowing form multiple nonlexical tasks and multiple analytic methods. *Journal of Gerontology: Psychological Sciences, 50B*, P202-P211.

Hertzog, C. (1989) Influences of cognitive slowing on age differences in intelligence. *Developmental Psychology, 5*, 636-651.

Howard, D. V. & Wiggs, C. L. (1993). Aging and learning: insights from implicit and explicit tests (pp. 512-528). In J. Cerella, J. Rybash, W. Hoyer and M. L. Commons (Eds.), *Adult information processing: Limits on loss.* San Diego: Academic Press.

Lima, S. D., Hale, S. & Myerson, J. (1991). How general is slowing? Evidence from the lexical domain. *Psychology and Aging, 6*, 416-425.

Madden, D. J. & Plude, D. J. (1993). Selective preservation of selective attention (pp. 273-302). In J. Cerella, J. Rybash, W. Hoyer and M. L. Commons (Eds.), *Adult information processing: Limits on loss.* San Diego: Academic Press.

Neter, J. and Wasserman, W. (1974). *Applied linear statistical models.* Homewood, Illinois: Richard D. Irwin.

Salthouse, T. A. (1991). *Theoretical perspectives on cognitive aging.* Hillsdale, NJ: Lawrence Erlbaum.

Salthouse, T. A. & Somberg, B. L. (1982). Isolating the age deficit in speeded performance. *Journal of Gerontology, 37*, 59-63.

Schaie, K. W. (1989). Perceptual speed in adulthood: Cross-sectional and longitudinal studies. *Psychology and Aging, 4*, 443-453.

Staplin, L. & Fisk, A. D. (1991). A cognitive engineering approach to improving signalized left turn intersections. *Human Factors, 33(5)*, 559-571.

Wechsler, D. (1955). *Manual for the Weschler adult intelligence scale.* New York: The Psychological Corporation.

DETERMINANTS OF VISUAL SEARCH PERFORMANCE: AGE AND PRACTICE EFFECTS.

W. A. Rogers[1], A. D. Fisk[2], and C. Hertzog[2]
[1]Memphis State University [2]Georgia Institute of Technology

ABSTRACT

In the present experiment, ability-performance relationships were used to assess changes in task requirements across practice. A variety of cognitive and speed ability measures were administered to each subject to measure the following factors: general, fluid, and crystallized intelligence; working memory; perceptual speed; semantic memory access speed; and psychomotor speed. Subsequently, ability-performance relationships were investigated across extensive practice on consistently mapped (CM) and variably mapped (VM) versions of a semantic category visual search task for young (17-30) and old (66-80) adults. The ability-performance relationships revealed similar patterns across CM and VM practice for both age groups. Namely, initial performance was predicted by general ability and semantic memory access, whereas later performance was predicted by perceptual speed. Thus although the mean data suggested that only the young adults had developed an automatic attention response in the CM condition, the locus of the differences between CM and VM or between age groups could not be localized through the ability-performance relationships. Only through a transfer manipulation designed to assess the automaticity of the response in the CM condition did we observe strikingly different ability-performance relationships for the young adults relative to the old adults.

INTRODUCTION

An understanding of the relationships between cognitive and speed abilities (e.g., general intelligence and reaction time) and performance on a task enables the prediction of which individuals will perform best on that task (Fleishman, 1972). For example, if we know that general intelligence is highly correlated with successful problem solving then we would predict that individuals with high general intelligence would perform well on a problem solving task. Several researchers have demonstrated that, while such ability-performance relationships do exist, they may change as a function of task practice (Ackerman, 1988; Fleishman, 1972; Kyllonen and Woltz, 1989). The purpose of the present experiment was to investigate ability-performance relationships across practice for young and old adults. The age comparison provides an opportunity to investigate an additional 'individual difference' that may contribute to performance differences either initially or after practice.

We investigated the relationships between a range of cognitive and speed abilities (described below) and performance improvements on a semantic category visual search task. In addition, we varied the type of practice provided on the task and we compared the ability-performance relationships for young adults and old adults. The choice of visual search was motivated by several factors. First, visual search is a perceptual skill which can become automatized after extensive consistent practice. Schneider and Shiffrin (1977; Shiffrin and Schneider, 1977) demonstrated that providing consistently mapped practice (CM) in which the items searched for (the targets) were completely distinct from the other items in the display (the distractors) resulted in the development of an automatic attention response (AAR) in a visual search tasks. CM practice may be contrasted with varied mapping (VM) practice in which targets and distractors are drawn from the same set and an AAR cannot develop because the item-to-response relationships changes across trials.

The second factor motivating the use of a visual search task in the present experiment is that such tasks have been used to investigate age differences in the ability to acquire new automatic processes. Fisk and Rogers (1991; Rogers, in press; Rogers and Fisk, 1991) have demonstrated that while the performance of old adults does improve with practice, the evidence suggests that they do not develop an automatic attention response. The empirical data suggest that even after extensive CM practice, older adults' performance remains attention-demanding and similar to VM performance (e.g., serial, more variable). The similarity between CM and VM performance, even after practice, indicates that old adults are perhaps learning search strategies but are unable to develop an AAR.

In an effort to understand the source of the age differences in performance improvements in visual search, we adopted an individual differences approach. Researchers have suggested that understanding the relationships between cognitive/speed abilities and performance can provide insight about the underlying processes involved (Ackerman, 1988; Kyllonen and Woltz, 1989). Moreover, a type of visual search has been used in the investigation of such ability-performance relationships across practice. Ackerman (1988) utilized a hybrid memory-visual search task to investigate the relationships of performance on the task to general ability (g) and perceptual speed (PS - the ability to make quick visual comparisons). He observed that g had a high correlation with initial performance, most likely due to the need for subjects to interpret the instructions and develop a search strategy. After practice, there was an increase in the relationships of PS and performance. Presumably, once subjects learned the mechanics of the task, the best predictor of performance was speed in making visual comparisons.

Changing ability-performance relationships across practice represent changes in the processes involved in task performance (Ackerman, 1988). That is, in a CM task in which an AAR can develop, performance is first determined by g due to the attention demands of the task. After some practice, PS determines performance as procedures are developed for performing the task (see Anderson, 1982). Final performance is determined by the quickness with which these procedures can be carried out. Ackerman's theory predicts that final-level, automatized performance would be determined by psychomotor speed (i.e., speed of reaction in the absence of cognitive demands) although this has not yet been empirically demonstrated with a search task.

Visual search provides a situation in which CM practice results in dramatic changes in performance, age differences have been documented, and ability-performance relationships have been demonstrated. In the present experiment, we extended the ability-performance analysis by including a wider range of abilities and providing a more extensive amount of practice. Ability-performance relationships have not previously been investigated in older adults and the inclusion of an age comparison allowed an assessment of whether previously observed age differences in visual search are accompanied by a different pattern of ability-performance relationships.

METHOD

Seventy young and 70 old subjects participated in the experiment. The experiment consisted of ten 90-minute sessions (Monday through Friday for two consecutive weeks). The subject characteristics and experimental procedure are presented in Table 1. A total of 20 computerized and paper-and-pencil ability tests were administered. Ability tests were chosen to enable the identification of the following constructs: general intelligence, fluid and crystallized intelligence, perceptual speed, psychomotor speed, working memory, and semantic memory access speed. The criterion task was semantic category visual search with display size equal to 2, 3, or 4 items. Subjects received training in two practice conditions: (a) CM - target items and distractor items were drawn from distinct stimulus sets, and (b) VM - target and distractor items were drawn from the same stimulus set with replacement across trials. Each subject completed a total of 3,000 CM trials (1,000 at each display size) and 3,000 VM trials (1,000 at each display size). The final session of the experiment consisted of a transfer condition designed to assess the degree of target strengthening and distractor weakening in the CM practice condition. There were two conditions in this phase: (a) CM Reversal in which the roles of the CM targets and distractors were reversed (i.e., the previous CM targets became the distractors and the previous CM distractors became the targets); and (b) a New CM condition which was created by pairing two of the former VM categories in a consistent mapping. Subjects completed a total of 420 CM Reversal trials (140 at each display size) and 420 New CM trials (140 at each display size). The procedure for individual trials was the same as during practice. The within-subject independent variables were: (a) Display size: two, three, or four words; (b) Training/Transfer Conditions: VM, CM, CM Reversal, or New CM. The primary dependent variable was RT; subjects were instructed to maintain accuracy at 95%.

Table 1. Subject Characteristics and Procedure

SUBJECTS	Young	Old
N	70	70
Male/Female	46/24	42/28
Age	20.83	70.68
Self-rated Health[a]	1.46	1.78
Formal Education[b]	13.48	15.04

[a] 1 - excellent, 5 - poor. [b] 12 - high school, 16 - college, etc.

PROCEDURE

Session 1
Extended Range Vocab.
Computation Span
Digit Symbol Substitution
Simple Reaction Time
Semantic Matching

Session 2
Eye Examination
Miller Analogies Test
Identical Pictures
Mathematics Aptitude
Making X's
Listening Span

Session 3
Raven's Matrices
Crossing Lines
Number Comparison
Information
Semantic Access

Session 4
Finding A's
Controlled Associations
Alphabet Span
Letter Sets
Synonym Matching

Sessions 5 through 9
1200 Trials Category Search
600 CM trials (3000 total)
600 VM trials (3000 total)

Session 10
300 Trials Trained CM
420 Trials CM Reversal
420 Trials New CM
Experimental Debriefing

RESULTS

Search Performance and Ability Test Performance. The mean age differences and CM/VM differences in performance on the criterion task are presented in detail elsewhere (Rogers, 1991; Rogers, Fisk and Hertzog, 1991). To summarize, differential CM/VM performance improvements were observed across the age groups suggesting that changes in performance as a function of practice may be driven by different mechanisms for young and old adults. Young adults conformed to the pattern of improvement predicted from previous category search results (e.g., Fisk and Schneider, 1983; Schneider and Fisk, 1984). Namely, they adults improved more under CM practice both in terms of RT scores as well as comparison slopes. Comparison slopes were reduced more in CM due to a greater reduction in RT for the larger display sizes with practice. For old adults, the performance differences between the CM and VM conditions were not as clear cut as those observed for young adults. There were overall differences in that CM performance was faster than VM performance and CM slopes were smaller than VM slopes. However, reductions in slope with practice did not differ between the CM and VM conditions.

The results of the transfer session also reflected differences in CM performance (and learning) for young and old adults. For both age groups, performance in the new CM condition was slowed relative to previous CM performance, although the slowing was greater for the young adults. The CM reversal condition resulted in a very large disruption for the young adults (nearly 60%) and amount of disruption was directly related to the number of "reversed" items in the display. Old adults were slowed at most by 20% and this was constant across display sizes. Disruption at reversal is a well-replicated finding for young adults (Fisk, Lee, and Rogers, 1991; Schneider and Shiffrin, 1977) and indicates the target/distractor differentiation that accrued during CM practice.

Means and standard deviations for the ability tests for each age group are also presented in Rogers (1991; Rogers et al., 1991). To summarize, the young adults performed better on all measures of psychomotor and perceptual speed, semantic memory access, working memory, and fluid intelligence whereas the old adults performed better on the vocabulary and information tests which are measures of crystallized intelligence.

Ability-Performance Relationships

Previous Analyses. The first set of analyses conducted on these data was designed to test general hypotheses about the ability-performance relationships. Rogers (in press; Rogers et al., 1991) conducted separate analyses in each of four contexts: young adults' CM performance, young adults' VM performance, old adults' CM performance, and old adults' CM performance. In a series of LISREL models it was determined that initial performance was related to g and SMA, and performance after practice was more related to PS. Importantly, the patterns of ability-performance relationships were quite similar across the CM and VM practice conditions and across the two age groups. It was only at transfer that the ability-performance relationships revealed the differential learning that had occurred for the young adults in the CM condition. For old adults, both CM Reversal and New CM performance was well-predicted by final-level CM performance. This was not the case for young adults; their transfer performance was predicted by SMA and PS, not by final-level CM performance. These results suggest that the transfer conditions were functionally different from final-level CM only for the young adults. The young adults had developed an AAR in the CM condition that could not be utilized at transfer.

One limitation of the analyses conducted by Rogers (1991, Rogers et al., 1991) was the fact that each of the situations (young CM, young VM, old CM, old VM) was analyzed in separate models. Thus direct statistical comparisons of the parameters across age groups or across training conditions was not possible. Moreover, the ability measurement models had not been constrained equal across the two age groups. The potential for differentially defined abilities could limit the interpretability of the ability-performance relationships. Subsequent analyses remedied some of these problems (Rogers, Hertzog, and Fisk, 1992). Rogers et al. conducted simultaneous analyses for the two age groups, albeit separately for the CM and VM conditions.

A simultaneous analysis of the measurement model of abilities revealed complete configural invariance across the age groups as well as an equivalence of the path coefficients of g_f, g_c, and WM to g. Due to the apparent differences in the interrelationships of speed to other variables, the factor loadings defining the SMA subfactor and the factor loading of SRT on g were freely estimated (i.e., not constrained equal). As a result, the definition of the SMA factor differs slightly between the young and old adults. Namely, for young adults SRT has a higher loading on SMA, whereas for older adults PS has a higher loading. The influence of g_c on SMA did not differ significantly between the age groups.

The Rogers et al. (1992) analyses succeeded in modeling comparable ability structures for the young and old adults as well as allowing for direct comparisons of the ability-performance relationships for young and old adults. Not surprisingly, the Rogers et al. (1992) results were comparable to the Rogers et al. (1991) results. Namely, the ability-performance relationships were quite similar across the two age groups and quite similar across the CM and VM training conditions. However, these analyses did not allow the direct statistical comparison of the ability-performance relationships across the CM and VM training conditions. That is the focus of the present set of analyses. Using a subset of the data (necessitated by the large number of parameters to be estimated) we estimate simultaneous models that include the CM and VM data for both young and old adults.

CM/VM Comparisons. Only the data from Sessions 1, 3, and 5 were included in the present set of analyses (trials 1-60, 1201-1260, 2401-2460, respectively). The ability structure was defined as in the previous simultaneous analysis (see Rogers et al., 1992, for details). In an effort to reduce the number of parameters to be estimated the search factors were based only on display sizes three and four (earlier analyses had been conducted on factors defined by all three display sizes).

An initial model was fit in which all of the regression coefficients for the ability-performance relationships were freely estimated, both across age groups and across CM/VM conditions. The statistics for this model (M1) are presented in Table 2. This model provides a substantially better fit to the data than a null factors model (N1) and a model in which all of the ability-performance relationships are constrained equal to zero (N2). The standardized regression coefficients for M1 are presented in Table 3 for young and old adults (note that the coefficients are standardized separately within each age group). The negative relationship of g to session 3 was included because a similar result was observed in the Rogers et al. (1992) analysis. The coefficient is significantly different from zero only for the old adults. Any explanation for this observation would clearly be post hoc (but see Cronbach and Snow, 1977, p. 124).

Table 2. Statistics from LISREL Models.

Model	df	X^2	GFI	CFI
N1 - Null Model	756	3337	.131	----
N2 - No ability-perf. relationships	681	1070	.701	.849
M1 - All coefficients freely estimated	667	871	.736	.921
M2 - CM = VM Young & Old free	673	880	.734	.920
M3 - CM = VM Young = Old	676	890	.731	.917
M4 - CM & VM free Young = Old	674	887	.730	.917

GFI - LISREL Goodness of Fit Index; CFI - Bentler (1990) Comparative Fit Index

Differences between the regression coefficients across the CM and VM conditions were investigated using a z-test (Usala and Hertzog, 1991). Within each age group, none of the regression coefficients were significantly different with the exception of the autoregressive coefficient (CM1 to CM3 vs. VM1 to VM3); for both age groups, the autoregressive coefficient was significantly higher for the VM condition. The only other coefficient that approached significance was the difference between the influence of SMA on CM and VM for the young adults. A z-test comparison yielded $z = 1.84$, where 1.96 is required for significance. Moreover, a comparison of a model in which SMA -> CM1 is constrained equal to SMA -> VM1 to Model 1 (in which they are freely estimated), yields a chi-square difference of 3.64 with one degree of freedom (a difference of 3.84 is required for significance). Given the modest sample size we may not have had sufficient power to detect the difference in the relationships between SMA and initial performance in the CM and VM conditions for young adults. However, the two significance tests (specific test of coefficients vs. model comparisons) did not reveal any other differences which were significant.

In Model 2 (M2, Table 2), the regression coefficients were constrained equal across the CM and VM conditions within each age group but freely estimated across age groups. A chi-square test of the difference between M2 and M1 revealed that constraining the coefficients equal across CM and VM did not yield a significant difference, change in $X^2(6, N=140)=8.4$, n.s. Thus the ability-performance relationships did not differ significantly between the CM and VM conditions within the age groups.

Age Comparisons. Model M3 was constructed to determine if the ability-performance relationships could be constrained equal across the two age groups without a significant loss of fit. The fact that model M3 provides a significantly worse fit then M2 suggests not (change in $X^2(3, N=140)=9.68$, $p<.05$). Thus the regression coefficients, as a whole, could not be constrained equal across young and old adults without resulting in a significantly worse fit of the model to the data.

In an effort to localize the source of the age differences, z-test comparisons were conducted on the regression coefficients and separate models were run, in turn constraining each ability-performance relationship equal across the age groups. Both sets of analyses converged on the same finding: the relationships of PS to CM3 and VM3 were significantly higher for the old adults and constraining only those paths equal across age yielded a significantly worse fit, change in $X^2(1, N=140)=7.02$, $p<.05$. Thus the only significant age difference in the ability-performance relationships across practice was the degree of influence of PS with the influence being greater for the old adults relative to younger adults.

Ability-transfer models. The regression coefficients presented in Table 4 are from Rogers (1991) for a model in which the coefficients were freely estimated across age groups. Only initial and final-level CM performance is included and the goal was to determine how best to predict performance in the CM Reversal and New CM conditions. The most striking difference is the degree to which transfer performance (either CM Reversal or New CM) is predicted by final-level CM performance. A z-test comparison of the coefficients for CM5 to CM Reversal and CM5 to New CM revealed that in both cases the degree of relationship was significantly higher for the old adults ($p<.05$). For the old adults, in both transfer conditions, performance was well-predicted from final-level CM performance. Moreover, the relationships of transfer performance to the other abilities in the model (PS or SMA) were not significantly different from zero. For the young adults, the pattern of ability-performance relationships was quite different. The regression coefficients from CM5 to CM Reversal or New CM were not significantly different from zero. However, there were significant relationships from PS and SMA to both CM Reversal and New CM.

The patterns of prediction are clearly different across the two age groups and the degree to which transfer performance could be predicted by final-level CM

Table 3. Standardized Regression Coefficients for Practice*

Coefficient	Young	Old
G --> CM1	.528	.382
G --> VM1	.317	.373
SMA --> CM1	.425	.570
SMA --> VM1	.699	.602
G --> VM3	-.270	-.346
PS --> CM3	.501	.585
PS --> VM3	.200	.678
CM1 --> CM3	.441	.344
VM1 --> VM3	.828	.598
CM3 --> CM5	.862	.965
VM3 --> VM5	.847	.979

*Within-group standardized solution.

Table 4. Standardized Regression Coefficients for Transfer*

Coefficient	Young	Old
G --> CM1	.422	.393
SMA --> CM1	.503	.535
PS --> CM5	.399	.604
CM1 --> CM5	.471	.356
SMA --> CM Rev	.438	.173
PS --> CM Rev	.337	-.107
CM5 --> CM Rev	.168	.930
SMA --> New CM	.467	.186
PS --> New CM	.322	-.149
CM5 --> New CM	.086	.954

*Within-group standardized solution.

performance did differ significantly across the age groups. The ability-transfer results suggest that, for young adults, the transfer tasks required different processes than did final level-CM performance; that is, they could not utilize the AAR learned in CM to successfully perform the transfer tasks. This was not the case for old adults. Their transfer performance was well-predicted by final-level CM performance presumably because the same processes were used to perform the tasks. These data are in accordance with the hypothesis that older adults do not develop an AAR, even after extensive consistent practice.

DISCUSSION

Initial performance for both young and old adults was influenced by g and SMA in both the CM and the VM practice conditions. The influence of g is comparable to previous results (e.g., Ackerman, 1988; Fleishman, 1972). The basic premise is that novice-level performance on a task is associated with general ability due to the requirements of the task: learning rules, remembering instructions, asking appropriate questions, making use of previous experiences with analogous task situations, and so on. For both young and old adults the influence of g was highest for the first block of practice and then decreased. This result implies that, at least initially, young and old adults were utilizing the same abilities to perform the task.

In addition to g, SMA also had a significant influence on initial performance. This ability-performance relationship has not previously been reported for search tasks because SMA ability has not been separately assessed (cf. Ackerman, 1988). The influence of SMA on initial performance may be an ability requirement which is specific to the semantic category search task; hence it remains to be seen if this ability will be important for other types of search tasks (e.g., memory search, letter search, search for spatial characters, etc.).

Performance later in practice was predicted by PS ability for both CM and VM conditions and for both young and old adults. Thus, the conclusion from these data is that PS predicts learning and is not necessarily an indicator of automatic process development. The individual differences in learning that occur in both CM and VM conditions, for both young and old adults are predicted by individual differences in PS.

Despite the fact that the ability-performance models are highly similar across CM and VM and across age groups, the transfer data suggest that there are differences in the underlying learning. The ability-transfer results (along with the mean data) suggest that only the young adults were successful at developing an AAR in the CM condition. This AAR could not be utilized in the transfer conditions, thus final level CM did not provide the best prediction of performance in the transfer conditions; instead SMA and PS were the best predictors of transfer performance.

The ability-transfer relationships for the old adults are straightforward. Final level CM performance predicts transfer performance in both the CM Reversal and the New CM conditions. This result is consistent with the idea that old adults have learned a general search strategy and have little difficulty transferring this strategy to new target distractor parings. The disruption that does occur in the transfer session is most likely related to a stimulus-specific search strategy which must be relearned for the new target/distractor pairings.

The results reported herein provide some important information about the ability-performance relationships across practice types and across age groups. Somewhat surprisingly, we found very similar ability-performance relationships for CM and VM practice and for young and old adults. This is surprising because there is a corpus of empirical data that suggests differential learning occurs in CM and VM practice (e.g., Schneider and Shiffrin, 1977) and for young and old adults (e.g., Fisk and Rogers, 1991). Furthermore, within the present study, the transfer data provide evidence that there was differential learning (i.e., of an AAR) only for CM practice and only for the young adults. Yet in all of the situations (young CM, young VM, old CM, and old VM) learning can and does occur. The fact that learning was related to PS in all situations is contrary to the idea that PS-performance relationships signify the compilation of production systems (e.g., Ackerman, 1988) or the development of automatic processing. Instead, a more general conclusion is warranted -- PS is related to learning in visual search tasks.

Acknowledgments

This research was supported, in part, by the National Institutes of Health Grant R01AG07654 from the National Institute of Aging. Correspondence concerning this article should be addressed to W. A. Rogers, Dept. of Psychology, Memphis State University, Memphis, TN, 38152 (e-mail to ROGERSWA @ memstvx1.memst.edu)

References

Ackerman, P. L. (1988). Determinants of individual differences during skill acquisition: Cognitive abilities and information processing. Journal of Experimental Psychology: General, 117, 288-318.

Anderson, J. R. (1982). Acquisition of cognitive skill. Psychological Review, 89, 369-406.

Cronbach, L. J., and Snow, R. E. (1977). Aptitudes and instructional methods. New York: Irvington.

Fisk, A. D., Lee, M. D., and Rogers, W. A. (1991). Recombination of automatic processing components: The effects of transfer, reversal, and conflict situations, Human Factors, 33, 267-280.

Fisk, A. D., and Rogers, W. A. (1991). Toward an understanding of age-related memory and visual search effects. Journal of Experimental Psychology: General, 120, 131-149.

Fisk, A. D., and Schneider, W. (1983). Category and word search: Generalizing search principles to complex processing. Journal of Experimental Psychology: Learning, Memory, and Cognition, 9, 177-195.

Fleishman, E. A. (1972). On the relation between abilities, learning, and human performance. American Psychologist, 27, 1017-1032.

Kyllonen, P. C., and Woltz, D. J. (1989). Role of cognitive factors in the acquisition of cognitive skill. In R. Kanfer, P. L. Ackerman, and R. Cudeck, (Eds.), Abilities, motivation, and methodology (239-280). Hillsdale, NJ: Erlbaum.

Rogers, W. A. (in press). Age differences in visual search: Target and distractor learning. Psychology and Aging.

Rogers, W. A. (1991). An analysis of ability/performance relationships as a function of practice and age. Unpublished doctoral dissertation, Georgia Institute of Technology, Atlanta, GA.

Rogers, W. A., and Fisk, A. D. (1991). Are age differences in consistent-mapping visual search due to feature learning or attention training? Psychology and Aging, 6, 542-550.

Rogers, W. A., Fisk, A. D., and Hertzog, C. (1991). Ability/performance relationships as a function of practice and age. Presented at the Annual Meeting of the Psychonomic Society, San Francisco.

Rogers, W. A., Hertzog, C., and Fisk, A. D. (1992). Models of age-related differences in ability/performance relationships for consistent and varied search tasks. Presented at the Fourth Cognitive Aging Conference, Atlanta.

Schneider, W., and Fisk, A. D. (1984). Automatic category search and its transfer. Journal of Experimental Psychology: Learning, Memory, and Cognition, 10, 1-15.

Schneider, W., and Shiffrin, R. M. (1977). Controlled and automatic human information processing: I. Detection, search, attention. Psychological Review, 84, 1-66.

Shiffrin, R. M., and Schneider, W. (1977). Controlled and automatic human information processing: II. Perceptual learning, automatic attending, and a general theory. Psychological Review, 84, 127-190.

Usala, P. D., and Hertzog, C. (1991). Evidence for differential stability of state and trait anxiety in adults. Journal of Personality and Social Psychology, 60, 471-479.

AGING AND DUAL-TASK TRAINING

John Larish, Arthur Kramer, Joseph DeAntona
University of Illinois at Urbana-Champaign

David Strayer
University of Utah

The efficacy of two methods of training dual-task skills was examined in this experiment. Thirty older subjects (Mean age = 67.8 years) were trained using either variable priority or fixed priority training. Subjects performed two tasks, a gauge monitoring task and a letter arithmetic task, both separately and together. Subjects in the variable priority group were trained to vary their processing priorities between the letter arithmetic and monitoring tasks. The fixed priority subjects were trained to devote equal priority to the two tasks. Subjects then transferred to a complex scheduling task which was paired with a paired-associates task. Variable priority subjects exhibited an initial performance cost relative to fixed priority subjects. By the end of training, however, variable priority subjects exhibited superior performance as compared to fixed priority subjects. The performance of variable priority subjects was also superior on transfer tasks with which the subjects had no prior experience, suggesting that variable priority training may involve a generalizable time-sharing skill.

With the aging of the population, there has been a rapid increase in the need to retrain older individuals in new skills. Frequently these skills involve divided attention or dual-task performance. Unfortunately, divided attention has generally demonstrated large age-related declines in performance (Crossley & Hiscock, 1992; Korteling, 1991; McDowd & Craik, 1988; Salthouse, Rogan, & Prill, 1984; but see Salthouse & Somberg, 1982; Wickens, Braune, & Stokes, 1987). A question that has not been clearly addressed, however, is whether these declines can be reduced or eliminated with training.

Several interventions have been proposed to alleviate the performance decrements associated with aging. Some researchers have found that aerobic exercise can reverse age-related performance decrements (Hawkins, Kramer, & Capaldi, 1992; Dustman, et al., 1984). For example, Hawkins, et al. (1992) found that a 10 week program of aerobic exercise specifically benefitted dual-task performance relative to non-exercise controls.

Extensive practice with computer video games has been found to lead to improved reaction time in response selection (Clark, Lamphear, & Riddick, 1987), and memory search (Dustman, et al., 1992) tasks for elderly subjects. Both of these studies, however, examined only single-task performance.

An intervention that has been found to enhance dual-task performance in young populations is variable priority training (Brickner & Gopher, 1981; Gopher, in press, Gopher, Weil, & Siegel, 1989). Gopher has argued that training subjects to vary their processing priorities among two or more tasks leads to enhanced multi-task performance on the trained tasks. Furthermore, Gopher and his associates have found that subjects trained with variable priority strategies exhibited superior dual-task performance when compared with subjects trained with fixed priority strategies even on dual-task pairs other than those on which they were initially trained. Thus, it would appear that variable priority training may lead to a generalizable timesharing skill.

Despite the impressive success that Gopher and colleagues have had with variable priority training there are, at present, still a number of gaps in our knowledge about this training strategy. First, the tasks that have been examined in the training paradigms have been largely psychomotor. Thus, one important question is whether the variable priority training strategy would be applicable to tasks that are more cognitive in nature. This issue is examined in our study by employing tasks which include substantial memory and decision making components. Second, the variable priority training strategy has only been employed with young adults. Thus, a second important question is whether this training strategy is appropriate for older adults. Finally, Gopher and colleagues have employed a restricted set of transfer tasks in their studies. For example, Brickner & Gopher (1981) transferred subjects to a dual-task pair with one old (pretrained) and one new task. In the present study, we employ two new tasks in our transfer conditions.

In the present experiment we test the effectiveness of variable priority training on an elderly population in comparison to a control group receiving fixed priority training. More specifically, we tested the hypothesis that older subjects trained with a variable priority strategy would attain high levels of mastery on the acquisition tasks and would also exhibit superior transfer to novel tasks than subjects trained with fixed priority strategies. Subjects were trained in using a complex gauge monitoring task (Humphrey & Kramer, in press) and a letter arithmetic task (Logan & Cowan, 1984). Subjects were then transferred to a dynamic scheduling and monitoring task (similar to Fuld, Liu, & Wickens, 1987) and a paired-associates running memory task.

METHOD

Subjects

Thirty subjects participated in this study. Subjects were randomly assigned to either variable priority (Mean age = 68.1 years) or fixed priority (Mean age = 67.5 years) training. Subjects were community dwelling and were paid $5.00 per hour. All subjects had corrected vision of 20-40 or better (Snellen chart) and excellent color vision (Ishihara color blindness test). Subjects in the two training conditions did not differ significantly in intelligence (Kaufman KBit), forward and backward digit span, or frontal lobe function (Wisconsin Card Sorting Task).

Experimental Tasks

Subjects were trained using a monitoring and a letter arithmetic task both separately and concurrently. Subjects then transferred to a scheduling and a running memory task which were also performed both separately and together.

Monitoring Task. The monitoring task required subjects to monitor a series of six gauges (see Figure 1) and reset each gauge when it reached the critical region (e.g. 9 and above) by pressing one of six keys on the computer keypad. Motion of the gauge cursors was driven by a pseudo-random forcing function. The motion of each column of gauge cursors was correlated, although initial phases were randomized for each trial. The position of only 1 cursor could be viewed at a time. By pressing one of six different keys on the keypad, the subject could view the position of a single gauge's cursor for 1.5 sec. If the subject failed to reset a gauge within 7.5 sec after the cursor entered the critical range, the computer reset the gauge and the event was scored as a miss.

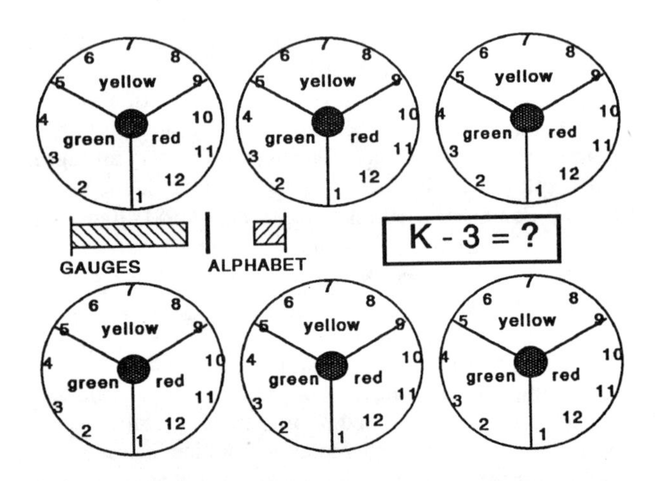

Figure 1. Dual task monitoring and letter arithmetic (training) tasks with feedback bargraphs.

Letter Arithmetic Task. The letter arithmetic task required subjects to solve problems of the form A + 2 = ?, where C is the correct answer, being 2 letters after A. A letter arithmetic problem is shown in the center-right of Figure 1. Half the subjects trained using the first half of the alphabet and half using the second half of the alphabet. After an initial trial in which the subject simply answered the problem on the screen, the subject was required to compare the answer to the current trial with the response made on the previous trial indicating the greater or lesser letter on the computer keyboard. Subjects were given feedback after each response after which the next trial was immediately presented.

Scheduling Task. This task required subjects to assign incoming boxes, illustrated in the top left corner of Figure 2, to one of four moving lines. The goal was to assign each new box to the shortest line (the line with the smallest area). In 25 percent of the trials, the subject manually assigned the box to a line. In the remaining 75 percent, the computer assigned the box to a line and the subject was to verify the correctness of the assignment. Subjects were required to assign/verify incoming boxes within 7 sec. New boxes appeared 2 sec after the last box was assigned. Difficulty was varied by using boxes of one height in the easy condition, three heights in the medium condition, and five heights in the difficult condition.

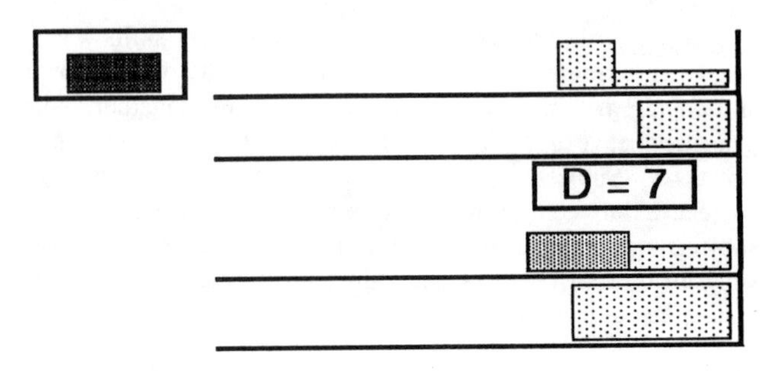

Figure 2. Dual task scheduling and running memory (transfer) tasks.

Running Memory Task. This task was a paired associates running memory task requiring the subject to associate the letters A-E with the numbers 1-8 (see the box in the center right of Figure 2). After between four and seven prime stimuli, a probe stimulus was presented and the subject was to indicate whether the probe was the most recent associate of the prime letter. Difficulty was varied by using the letters A-C, A-D, and A-E for the easy, medium, and difficult conditions respectively. This task was self-paced, with the subject pressing a key to view each new stimulus pair.

Procedure

The procedure is outlined in Table 1. In all sessions, blocks were five minutes long and feedback on average reaction time and accuracy was given following each block. Subjects performed the tasks on Dell 386 personal computers with VGA color monitors.

Introductory Testing. In sessions 1 and 2, subjects practiced the monitoring and letter arithmetic tasks alone. In session 3, subjects practiced both the single and dual tasks. These sessions were used to insure that variable priority and fixed priority subjects started with equivalent levels of single and dual-task performance.

Fixed/Variable Priority Training. In sessions 4-6, subjects were given continuous feedback in the form of a bargraph (center left in Figure 1) indicating their current performance (most recent five responses) on each task as a function of single task reaction time and accuracy from the previous session. Subjects were instructed to perform the tasks so as to keep each of the task feedback bars against the central vertical line. For subjects receiving fixed priority training (FP), equal priority was placed on both tasks at all times (e.g. the vertical line in the feedback display was located equidistant from the gauge and alphabet bars). For subjects assigned to variable priority training (VP), however, the priority placed on each task was varied between block requiring a 20-80, 35-65, 50-

50, 65-35, or 80-20 allocation of effort between the two tasks. Subjects also performed single-task versions of each task without continuous feedback.

Transfer. In sessions 7 and 8, subjects transferred to the scheduling and running memory tasks. These tasks were performed both alone and in dual-task conditions at three different levels of difficulty. Continuous feedback was not presented in these sessions, but subjects were asked to perform both tasks as well as possible.

Table 1. The conditions and their durations are listed for each session.

Preliminary Testing

Sessions 1-2.
- 1 - 90 second demonstration of monitoring and letter arithmetic tasks
- 4 - 5 minute single-task blocks of monitoring and letter arithmetic tasks

Session 3.
- 4 - 5 minute blocks for single-task monitoring and letter arithmetic tasks
- 6 - 5 minute blocks for dual-task monitoring and letter arithmetic tasks

Fixed/Variable Priority Training

Sessions 4-6.
- 10 - 5 minute blocks for dual-task monitoring and letter arithmetic tasks
- 2 - 5 minute blocks for single-task monitoring and letter arithmetic tasks

Transfer

Sessions 7-8.
- 1 - 5 minute block of each task (scheduling and running memory) in easy, medium, and difficult conditions
- 4 - 5 minute blocks of dual-task scheduling and running memory with easy-hard, hard-easy, and medium-medium difficulty pairings.

RESULTS

The results will be structured in the following way. First, we will present the response time and accuracy data collected in the last pre-training session (e.g. session 3). This data will reveal whether the subjects assigned to the two training groups exhibited similar performance prior to the training intervention. Second, we will present the data obtained during the three sessions which constituted the training intervention (e.g. sessions 4 through 6). This data will enable us to determine the relative benefits of fixed versus variable priority training on the acquisition tasks. Finally, the data collected in the transfer tasks (sessions 7 and 8) will be presented. This data will enable us to assess the extent to which any performance advantages associated with fixed or variable training strategies can be generalized to novel tasks.

Pre-training - Session 3. The response time and accuracy data is presented for the monitoring task in Table 2 and for the letter arithmetic task in Table 3. Each of the dependent variables was submitted to a two-way ANOVA with training group (FP and VP) as a between subjects factor and task condition (single and dual) as a within subjects factor. For the monitoring task there was a significant effect of task condition for the following variables: reaction time (RT), percent hits, and percent misses. Performance was poorer in the dual than in the single task version of the monitoring task. There were also marginal ($p < .08$) two-way interactions between training group and task conditions for percent hits and percent false alarms variables. Dual-task performance was poorer for the VP than for the FP subjects.

Table 2. Single and dual task performance in session 3 (pre-training) for FP and VP subjects in the monitoring task.

	Task Condition			
	Single		Dual	
Dependent Variables	FP	VP	FP	VP
Reaction Time (msec)	2680	2702	3258	3254
Percent Hits	59.9	60.5	44.9	36.6
Percent Misses	22.6	17.9	41.4	36.8
Percent False Alarms	17.9	22.3	13.7	26.6

Table 3. Single and dual task performances in session 3 (pre-training) for FP and VP subjects in the letter arithmetic task.

	Task Condition			
	Single		Dual	
Dependent Variables	FP	VP	FP	VP
Reaction Time (msec)	5670	6066	13436	12780
Percent Correct	85.7	88.8	86.2	84.5

For the letter arithmetic task there was a significant effect of task condition for RT ($p < .001$). RT was faster for the single than the dual task version of the task. In summary, pre-training performance was approximately equivalent for the two training groups. In fact, the nonsignificant trend was for better performance for the FP than for the VP subjects. Thus, any training advantages observed for the VP subjects would be underestimated.

FP and VP Training - Sessions 4 through 6. Tables 4 and 5 present the RT and accuracy data for the letter arithmetic and monitoring tasks, respectively. It is important to note that the comparison of training efficacy for the FP and VP strategies in the dual-task conditions is performed on the 50/50 dual-task emphasis blocks. This is necessary since emphasis/processing priority is equated across the two training conditions in these blocks. However, this puts the VP subjects at a distinct disadvantage since they receive 80% fewer practice trials in the 50/50 conditions than the FP subjects (e.g. the VP subjects perform the same number of practice trials as the FP subjects but across five different dual-task priority levels).

Given the enormity of the data set we will report only the significant effects of relevance to the training manipulation. For the letter arithmetic task there were two significant effects which involved the training manipulation (session x training, session x task condition x training). VP subject's RT's decreased across sessions, particularly in the dual-task conditions. The FP subjects failed to reduce their RT's across training sessions. For accuracy, there was a significant effect of training strategy. VP subjects performed more accurately than FP subjects in all of the conditions.

Table 4. Single and dual task reaction times and accuracy in sessions 4 through 6 for FP and VP subjects in the letter arithmetic task. Percent correct scores are in parentheses.

	Task Condition			
	Single		Dual	
Session	FP	VP	FP	VP
4	5378	5881	12305	14244
	(87.8)	(91.1)	(83.8)	(85.9)
5	4927	5598	12051	14158
	(88.2)	(92.2)	(85.3)	(89.6)
6	5287	5398	13121	11074
	(87.8)	(92.3)	(86.9)	(89.1)

Table 5 presents RT and accuracy data for the monitoring task. Although there was a non-significant trend (p < .10) for an interaction between training and session for RT (e.g. RT's decrease for VP and increase for FP with practice) none of the training main effects or interactions attained statistical significance. In summary, the differences that were observed as a function of training manipulation were in favor of the VP strategy. This is actually quite impressive given the fact that subjects in the VP conditions had 80% fewer practice trials in the dual-task blocks that were used in the analyses.

Table 5. Single and dual task reaction times and accuracy in sessions 4 through 6 for FP and VP subjects in the monitoring task. Percent correct scores are in parentheses.

	Task Condition			
	Single		Dual	
Session	FP	VP	FP	VP
4	2889	2863	3339	3491
	(69.3)	(62.0)	(42.5)	(35.9)
5	2984	2828	3300	3299
	(72.9)	(64.4)	(47.5)	(41.5)
6	3048	2764	3448	3263
	(71.3)	(67.1)	(46.6)	(40.8)

Transfer - Session 7 and 8. The RT and accuracy data for the scheduling and paired-associates task are presented in Tables 6 and 7, respectively. As with the analysis of the training data, we will limit the present discussion to the effects of training group. For the scheduling task, there was a significant three-way interaction among training, difficulty, and task condition (single and dual). RT's were faster for the VP than the FP subjects in all but the single-task easy condition of the scheduling task. For the paired-associates task, VP subjects performed more accurately than FP subjects in all of the experimental conditions.

Table 6. Single and dual task reaction times and accuracies in the scheduling (transfer) task. Percent correct scores are in parentheses.

	Task Difficulty					
	Easy		Intermediate		Difficult	
Group	Single	Dual	Single	Dual	Single	Dual
FP	1319	1716	1614	1749	1657	1647
	(78.9)	(73.9)	(67.9)	(67.3)	(70.7)	(67.3)
VP	1365	1565	1480	1690	1469	1443
	(86.7)	(79.3)	(66.6)	(75.0)	(69.7)	(73.9)

Table 7. Single and dual task reaction times and accuracies in the paired-associates (transfer) task. Percent correct scores are in parentheses.

	Task Difficulty					
	Easy		Intermediate		Difficult	
Group	Single	Dual	Single	Dual	Single	Dual
FP	2977	3291	3182	3202	3348	3440
	(81.2)	(76.1)	(76.9)	(72.2)	(72.4)	(66.8)
VP	2848	3116	2878	3374	3195	3290
	(83.3)	(84.7)	(80.9)	(81.0)	(75.8)	(76.2)

CONCLUSIONS

The present study was conducted to compare the efficacy of two different dual-task training techniques for older subjects. In previous studies (Brickner & Gopher, 1981) young subjects have been found to benefit from VP training over FP training in both the training task and to a more limited extent in transfer tasks. The present study suggests that the same pattern of effects can be obtained for older subjects.

There are a number of interesting aspects of the data. First, VP response times in the letter arithmetic task were longer at the beginning of training but significantly faster at the end of training (e.g. training sessions 4 through 6) than response times produced by the FP subjects. This finding suggests that the requirement to vary priorities between tasks can actually impede learning early in training. This could be due to the extra workload imposed on subjects to keep track of the priorities on a block by block basis. This extra planning and monitoring load is not imposed on subjects in the FP conditions since the priority is fixed throughout training. Interestingly, the fact that this initial cost turns into a benefit with additional practice is quite consistent with observations in motor and verbal learning that procedures which retard performance early in learning often result in the best long-term performance (Schmidt & Bjork, 1992).

Second, the task on which the VP subjects ultimately excelled during training, the letter arithmetic task, is strongly consistently mapped. Previous studies (Zbrodoff & Logan, 1986) have found that performance in this task becomes automatized with practice. Thus it would appear that VP training may be most useful for consistently mapped tasks within a dual-task context. In fact, we did obtain some data which suggests that automaticity was developing in the letter arithmetic task for the VP but not for the FP subjects. First, the VP subjects showed power function learning on the letter arithmetic task (R^2 = .91) while the FP subjects did not. Second, the difference in RT (e.g. slope) between the 2 and 3 in the addition and subtraction problems decreased significantly across training for the VP but not for the FP subjects. Power function learning and decreased slopes have both been taken as evidence for the development of automaticity (Logan, 1988; Schneider & Shiffrin, 1977).

Finally, the fact that VP subjects outperformed FP subjects on the transfer tasks suggests that the learning strategy is somewhat generalizable, at least with respect to the tasks employed in the present study. In summary, VP training appears to be a useful method by which to train older subjects in multi-task skills.

ACKNOWLEDGEMENTS

This research was supported by the National Institute of Aging (Grant No. AGO8435). We are grateful to David Burns, Heather Pringle and Ron Wolfman for their assistance in collecting the data.

REFERENCES

Brickner, M., & Gopher, D. (1981). Improving time-sharing performance by enhancing voluntary control on processing resources (Report No. HEIS-81-3). Haifa, Israel: Technion-III, Faculty of Industrial Engineering and Management, Research Center for Work Safety and Human Engineering.

Clark, J. E., Lamphear, A. K., & Riddick, C. C. (1987). The effects of videogame playing on the response selection processing of elderly adults. Journal of Gerontology, 42, 82-85.

Crossley, M., & Hiscock, M. (1992). Age-related differences in concurrent-task performance of normal adults: Evidence for a decline in processing resources. Psychology and Aging, 7, 499-506.

Dustman, R. E., Emmerson, R. Y., Steinhaus, L. A., Shearer, D. E., & Dustman, T. J. (1992). The effects of videogame playing on neuropsychological performance of elderly individuals. Journal of Gerontology, 47, 168-171.

Dustman, R. E., Ruhling, R. O., Russel, E. M., Shearer, D. E., Bonekat, H. W., Shigeoka, J. W., Wood, J. S., & Bradford, D. C. (1984). Neurobiology of Aging, 5, 35-42.

Fuld, R., Liu, Y., & Wickens, C. D. (1987). Computer monitoring vs. self monitoring: The impact of automation on error detection (Technical Report No. ARL-87-3/NASA-87-4). Savoy, IL: University of Illinois, Institute of Aviation, Aviation Research Laboratory.

Gopher, D. (in press). The skill of attention control: Acquisition and execution of attention strategies. To appear in Attention and Performance XIV. Hillsdale, NJ:Lawrence Erlbaum.

Gopher, D., Weil, M., & Siegel, D. (1989). Practice under changing priorities: An approach to training of complex skills. Acta Psychologica, 71, 147-179.

Hawkins, H. L., Kramer, A. F., & Capaldi, D. (1992). Aging, exercise, and attention. Psychology and Aging, 7, 643-653.

Humphrey, D. G. & Kramer, A. F. (in press). Towards a real-time measurement of mental workload. Human Factors.

Korteling, J. E. (1991). Effects of skill integration and perceptual competition on age-related differences in dual-task performance. Human Factors, 33, 35-44.

Logan, G. D. (1988). Toward an instance theory of automatization. Psychological Review, 95, 492-527.

Logan, G. D., & Cowan, W. B. (1984). On the ability to inhibit thought and action: A theory of an act of control. Psychological Review, 91, 295-327.

McDowd, J. M., & Craik, F. I. M. (1988). Effects of aging and task difficulty on divided attention performance. Journal of Experimental Psychology: Human Perception and Performance, 14, 267-280.

Salthouse, T. A., Rogan, J. D., & Prill, K. A. (1984). Division of attention: Age differences on a visually presented memory task. Memory & Cognition, 12, 613-620.

Schmidt, R., & Bjork, R. (1992). New conceptualizations of practice: Common principles in three paradigms suggest new concepts for training. Psychological Science, 3, 207-217.

Schneider, W., & Shiffrin, R. (1977). Controlled and automatic human information processing: I. Detection, search, and attention. Psychological Review, 84, 1-66.

Somberg, B. L. & Salthouse, T. A. (1982). Divided attention abilities in young and old adults. Journal of Experimental Psychology: Human Perception and Performance,8, 651-663.

Wickens, C. D., Braune, R., & Stokes, A. (1987). Age differences in the speed and capacity of information processing: 1. A dual-task approach. Psychology and Aging, 2, 70-78.

Zbrodoff, N. J., & Logan, G. D. (1986). On the autonomy of mental processes: A case study of arithmetic. Journal of Experimental Psychology: General, 115, 118-130.

RETENTION OF MULTIPLE-TASK PERFORMANCE: AGE-RELATED DIFFERENCES

Richard A. Sit and Arthur D. Fisk
Georgia Institute of Technology
Atlanta, GA

This study examined the relationship between retention of both multiple-task performance and the micro-components of a complex task. Young and older adults trained on a synthetic work task (Elsmore, 1994) with both groups acquiring skill in performing the complex task. After a five month retention period, older adults' initial performance on the multiple-task declined significantly more than younger adults. Both groups of adults regained their final trained level of performance after only four 5-minute trials. However, throughout the retention trials older adults only emphasized a single component of the complex task. Young adults successfully allocated attention to all task components. These and other aspects of the data suggest that a major locus of age-related decline in complex task performance is due to differential loss in strategic allocation of attention to component tasks. The data also show how measuring multiple-task performance may underestimate lack of component processing efficiency.

Over the last few years, the major objective of much research in cognition and aging has been to understand the moderating influences of skill acquisition (e.g., consistency, task structure, learning requirements, amount of practice) as they relate to models developed to describe and predict the influences of aging on performance improvement and learning underlying the acquisition of skill. This research has been successful in delineating age-related performance improvement and learning underlying acquired skills for several task domains (e.g., see Fisk & Rogers, 1991; Rogers, Fisk & Hertzog, 1994). However, a major gap in understanding concerns long-term retention of acquired knowledge and acquired skills. As a consequence, we know relatively little about how maintenance of knowledge and skills is affected by conditions of original learning, training strategies, intervening activities, the differential decay or survival of component processes of skill and so on.

Why is the Study of Retention Important?

Understanding retention of knowledge and skill is clearly important from both a pragmatic and a theoretical perspective. Pragmatically the importance is clear. In most activities of daily living, the opportunity and the need for performance after training are either delayed or infrequent. This is often the case when individuals learn how to use new technology. Another major area, emergency procedures have a low probability of occurrence; but, when they do occur, accurate and fluent performance is crucial. The goals of designing instructional and training procedures for skills and knowledge that must be maintained during periods of disuse might be quite different than goals for simply ensuring the attainment of mastery during training. From this perspective, it is crucial to have an accurate age-related account of how skills and knowledge decay during periods of disuse.

From a theoretical perspective, the importance of understanding age-related issues of retention are equally clear. Although the number of studies providing information on retention of skill after various intervals is small, the study of retention of learned materials has been important for theory development and refinement. For example, through the assessment of retention we have learned about the importance of massed and distributed practice and the differences in the learning that occurs in these situations.

What Do We Know About Age-Related Retention Capability?

Salthouse (1991) remarked that no firm conclusion could be reached concerning the presence or absence of age differences in mechanisms related to retention of information. It is not surprising that he would reach such a conclusion. Consider that of the 23 studies he reviewed, 12 showed similar rates of loss for young and old adults and 11 demonstrated greater loss for older adults. Currently, it is difficult to integrate age-related studies that measure the retention of knowledge and skills. The number of studies using long-term retention intervals is relatively small, in part due to the difficulty of performing such studies.

Yet, the picture is not as bleak as it might seem. Although the results of previous studies that have examined age-related maintenance of knowledge or skills are mixed, there seems to be the potential for an integrative framework. One potential explanation for the inconsistent findings may be due to the assessment of different original learning (either in degree or kind) between young and old adults on the target task (Fisk, Hertzog, Lee, Rogers, & Anderson-Garlach, 1994). Alternatively, old adults may truly be less able to retain trained performance for some types of learning but not others (Anderson-Garlach & Fisk, 1994; Rogers, Gilbert, & Fisk, 1994). Finally, activities subsequent to original learning of the target task may affect older adults more than younger adults (Fisk, Cooper, Hertzog, & Anderson-Garlach, 1995).

Our preliminary long-term retention studies examining maintenance of acquired search-detection skill have yielded interesting, yet theoretically challenging, results. Our studies suggest: (a) in some situations old and young adults retain an impressive amount of skill even after 16 months without exposure to the task; (b) retention performance declines within a three month period and that decline remains stable between three and six months for both age groups; (c) old and young adults equally retain general, task-relevant skills, at least when single task performance is evaluated; (d) old adults' performance declines more than young adults' for both extensively trained and moderately trained stimuli; (e) when an interfering processing activity is inserted prior to the retention interval, old adults' performance declines disproportionately more than young adults' performance especially when compared with a task not subjected to such interference; and (f) depending on the type of search task, for both age groups, the initial retention deficit is largely attenuated by the end of the retention retraining periods we have used (see Fisk, Hertzog, et al., 1994; Fisk, Copper, et al., 1994 for a review). This pattern of results suggests that a simple model of retention may not best describe age-related long-term retention. A difference in the qualitative nature of learning is sufficient to produce age-related differences in retention capability (Fisk, Hertzog et al., 1994). However, our data suggest that learning differences may not be a *necessary* condition for age-related retention effects.

The previous research involving age-related patterns of retention of skill has involved single task components of more complex tasks. The present research attempts to broaden our understanding of age-related retention capability by directly examining retention of performance in multiple-task situations. In today's society a growing number of occupational and daily living activities require the performance of multiple, simultaneous actions. In many work situations operators are required to monitor, control, and manipulate information via complex systems. The important questions for the present investigation concerned the relationship between retention of multiple-task performance and the micro-components of the task: Is it sufficient to know the retention of a single task component to predict multiple-task performance (is there an average, general decline)? Does multiple-task performance after some period of disuse give lawful insight into performance of the underlying micro-components? Finally, it is important to understand whether the same retention function, whatever it is, can be applied across age groups with equal accuracy.

Synthetic Work Environments

Synthetic work tasks have been found to be effective in simulating real-world multiple-task environments (e.g., Alluisi, 1967; Chiles, Alluisi, & Adams, 1968; Elsmore, 1994; Morgan & Alluisi, 1979). The goal of such simulations is to abstract the critical aspects of a wide range of activities (e.g., memory and classification of items, performance of a self-paced task, arithmetic problems; and monitoring and reaction to both visual and auditory information) while bringing the study under controlled laboratory conditions.

Purpose of the Study

The purpose of this study was to investigate the performance of young and older adults trained on a synthetic work task, after a 5-month retention interval. Specifically, this study examined the effects of Age, Task Type (memory, arithmetic, visual monitoring, and auditory monitoring), and Trial (performance before versus after a 5-month retention interval), on performance (multiple-task score and the four individual component task scores).

METHOD

Participants

Twelve young adults (*M*=32.0 years of age, *SD*=4.6) and 19 older adults (*M*=68.8, *SD*=5.8) participated in this retention study. This group was recruited from a sampling frame of 20 young and 26 older adults. These individuals were trained on the task in a previous study (Salthouse, Hambrick, Lukas, and Dell, in press). Subjects were paid $25.00 for their participation in retention testing, had at least the equivalent of 20/40 correct vision, and reported their health as good to excellent. Participants were administered several general ability tests which revealed age-related patterns consistent with the general population.

Apparatus

Synwork1. This study used Synwork1, a computer-based synthetic work task (Elsmore, 1994), and operated on 486 PC-compatible computers equipped with color monitors. All interaction with the testing program was done via a standard mouse. During a test trial, the computer screen is divided into four quadrants or "windows", each assigned to a different task. A small window in the center of the screen is used for displaying a composite "score" for performance on all of the tasks within the synthetic work environment. In the upper left of the screen is a memory task. The initial display in this task consists of a set of 5 letters. The set is then removed and followed by periodic displays of a probe letter which is to be classified as YES, a member of the set, or NO, not a member of the set. The upper right of the display contains an arithmetic task. In this task, two three-digit numbers are to be added by adjusting plus and minus buttons to produce the correct sum, in the row below the addends. This task is completely self-paced. The lower left quadrant of the display contains a visual monitoring task. In this task, the participant monitors the position of a pointer moving continuously along a horizontal scale, and attempts to reset it before it reaches the end of the scale. The lower right quadrant of the display contains an auditory monitoring task. High and low tones are presented periodically throughout the trial, and the task is to respond whenever a high tone occurs. On any given trial, all participants were exposed to the same sequence of events.

Procedure

Initial Training. The training portion of the study is described in Salthouse et al. (in press). To summarize, all participants came to the laboratory on three separate days within a 10-day period. Testing was conducted in groups of one to five individuals, with each participant seated at a microcomputer. On the first day participants performed a variety of abilities tests, performed eight trials maneuvering a pointer through a W-shaped maze, and read specific instructions on how to perform Synwork1. Each task was then presented in isolation for two 1-minute trials, followed by the four tasks presented together for one minute. The remainder of the first day was spent training on the four tasks together for five 5-minute trials. Over the next two days, participants trained on the multiple-task for 10 5-minute trials per day.

Retention Testing. After a five month retention interval, participants returned to the laboratory and the same procedure used for training was followed. Testing was conducted in groups of one to four individuals, with each participant seated at a computer workstation. Participants completed a series of abilities tests and were given verbal and written instructions on the study and Synwork1. Participants performed eight mouse maneuvering trials, the component tasks in isolation for two 1-minute trials each, and the four tasks presented together for one 1-minute trial. Next, all four tasks were performed together for four 5-minute trials. Participants performed four additional 5-minute trials following retention testing; however, those trials were part of a different investigation and will not be discussed.

The scoring parameters were as follows: in the memory task, probe stimuli occurred every 10 seconds, 10 points were awarded for correct responses, and 10 points were subtracted for incorrect responses and memory list retrievals after the initial display at the beginning of a trial. In the arithmetic task, 10 points were awarded for correct responses and five points were subtracted for incorrect responses. In the visual monitoring task, the participant was awarded one point for every 10 pixels the pointer (moved 6.7 pixels per second) was away from the middle of the scale at the time of reset and 10 points were deducted for every second that the pointer was at the end of the 100.5 pixel scale (15 seconds to reach the end of the scale). The auditory monitoring task required the participant to discriminate between a low (523 Hz) non-target tone and a high (2092 Hz) target (.2 probability) tone every five seconds. Ten points were awarded for a hit and 10 points were subtracted for a miss or false alarm.

RESULTS

Initial Training

Retention participants' performance during initial training was examined to determine if this subset of people had developed performance indicative of skill in the multiple-task environment. Presented in the first column of Table 1 are both adult group's percent improvement attained during initial training. These percentages represent the change in scores from initial training performance to final trained performance relative to initial training performance.

During initial training, at most, a participant could earn 1650 points on the multiple-task. Possible points for each of the component tasks were: memory (1200), arithmetic (130), visual monitoring (200), and auditory monitoring (120). During training, both groups of adults acquired skill on all four component tasks. After 20 training trials, young and old adults' mean multiple-task scores improved to 1365 and 951, and their mean component scores to: 1110 and 996 (memory), 38 and 9 (arithmetic), 119 and 7 (visual monitoring, and 98 and -62 (auditory monitoring), respectively.

An ANOVA test examining multiple-task performance across the initial training trials also indicated that both young and older adults significantly improved their performance on the complex task ($F(1, 29) = 106.19, p < .001$). This analysis also found a significant main effect of Age ($F(1, 29) = 15.13$, $p < .001$). Throughout training, young adults' multiple-task and component-task performance was higher than older adults' performance. These data are consistent with the entire sample tested by Salthouse et al. (in press).

Retention Performance

Initial- Multiple-Task Retention Performance. Multiple-task performance on the last trial prior to the retention interval and on the first trial subsequent to the five month retention interval (see Table 1) were examined. This analysis found main effects of Age ($F(1, 28) = 16.99, p < .001$), Task Type ($F(3, 84) = 43.66, p < .001$), and Trial ($F(1, 28) = 4.93, p < .05$). Also, there were significant Age x Trial ($F(1, 28) = 4.43$, $p < .05$) and Age x Task Type ($F(3, 84) = 3.18, p < .05$) interactions. Younger adults had better retention of previously acquired complex skills over a five month period than did older adults. Older adults' multiple-task performance declined 68.5%, whereas, younger adults' performance did not statistically change (actually improved 1.8%).

To better understand these significant main effects and interactions, it is necessary to examine the corresponding performance on individual task components as they were performed during multiple-task performance. An analysis of individual component scores (see Table 1) revealed strong performance trends. Older adults emphasized the memory search component. In fact, by the end of training they performed this task as well as younger adults (component score of 251 versus 238, respectively). After the retention interval, older adults obtained their observed multiple-task score by continuing to emphasize the memory search component. Although the older adults' memory search score declined 33% (from 251 to 168), this component score was 158% of the mean multiple-task score (the other component scores were negative). It is clear that older adults' decline on the other task components was quite dramatic. However, young adults successfully allocated attention to all four task components.

Table 1. Mean scores for multiple-task and individual component performance on the synthetic work task

	Percent Improvement During Initial Training	Final Trained Performance *(SD)*	5-Min. Trials After Retention Interval *(SD)* 1	2	3	4
		Young Adults (n = 12)				
Multiple-Task Score	453%	621 (182)	632 (162)	631 (149)	546 (220)	644 (167)
Component Score:						
Memory	111%	238 (55)	254 (70)	264 (48)	225 (77)	277 (19)
Math	538%	131 (91)	93 (84)	115 (83)	98 (83)	130 (86)
Visual	146%	149 (44)	176 (24)	160 (47)	127 (67)	135 (91)
Auditory	567%	101 (27)	91 (44)	92 (34)	98 (29)	100 (20)
		Older Adults (n = 19)				
Multiple-Task Score	240%	336 (279)	106 (394)	194 (313)	209 (200)	318 (239)
Component Score:						
Memory	254%	251 (85)	168 (121)	182 (131)	168 (85)	231 (83)
Math	150%	31 (55)	- 5 (46)	17 (38)	19 (38)	27 (48)
Visual	101%	73 (151)	- 9 (245)	59 (132)	51 (132)	113 (117)
Auditory	30%	- 19 (99)	- 52 (113)	- 61 (107)	- 30 (97)	- 53 (109)

Single Task Retention. Performance on isolated individual components before training and before retention testing were examined. When tasks was performed in isolation, there were only significant main effects of Age on the memory ($F(1, 29) = 7.86, p < .05$) and arithmetic tasks($F(1, 29) = 12.68, p < .001$). Younger adults consistently performed higher than older adults on all components performed in isolation. The most interesting finding from these analyses was that after the retention interval neither age group performed worse on the isolated components relative to the same performance assessed during initial training.

Relearning During Retention Testing. Throughout retention testing young adults maintained their multiple-task score and their component scores at or near the level of performance they attained at the end of training. Although the older adults demonstrated dramatic decline in performance when initially tested after the retention interval, by the fourth 5-minute trial their mean performance level was at or near their level of final trained multiple-task performance. However, older adults did exhibit a higher degree of inter-subject performance variability during retention testing than during the final 5-minute trial of training. The return to trained performance level for the older adult participants shows the relatively transient nature of the initial decline.

CONCLUSIONS

The main findings from this experiment examining age-related retention capability were: (a) On multiple-task performance the older adults' performance declined across the retention interval more than the younger adults. (b) The initial multiple-task performance attained by the older adults was possible only because the older adults seemed to focus mainly on one component. (c) Single task performance of each component did not suffer for either age group due to nonperformance during the retention interval. (d) Performance quickly returned to the final trained level for the older adults.

These data answer the initial questions that motivated the study and add to our growing understanding of age-related skills retention capability. First, the data suggest that, for complex task performance requiring multi-tasking, assessing retention of the combined task performance data (as well as retention of components performed in isolation) may underrepresent older adults performance on microcomponents of the multiple-task. Such data are important for suggesting the critical nature of refresher training for older adults learning a new task requiring coordination of attention allocation during task performance (cf. Korteling, 1994; Kramer & Larish, in press).

Second, the data point clearly to the locus of loss in these types of task (where each major task component is variably mapped and cannot benefit from automatic processing) as being related to the loss in strategies for allocation of attention and task coordination (Schneider & Detweiler, 1988). This conclusion is straightforward when performance of each single component performed in isolation is considered (no loss due to retention interval) and the quick reinstatement of performance to that seen during the final training trial. Given a loss of a strategy to perform the overall multiple-task, it is not surprising that older adults would maintain performance on the memory search task. During initial training (the first 20 5-minute trials) the memory search component was emphasized by giving that component more points for successful performance and increasing its demand frequency relative to the other components (see Salthouse et al., in press). Hence, what appears to have decayed across the retention interval was the older adults knowledge of how to maximize multiple-task score. Third, for this class of tasks, multi-component tasks requiring acquisition of a performance strategy that remains heavily dependent on control, attention demanding processes, it is clear that a generalized component decline function will not emerge.

ACKNOWLEDGMENTS

This research was supported in part by NIH Grant No. P50 AG11715 under the auspices of the Center for Applied Cognitive Aging Research on Aging (one of the Edward R. Roybal Centers for Research on Applied Gerontology) and in part by NIH Grant No. R01 AG 07654. We would like to thank T. Salthouse for making available the list of participants from the original study and for providing the original training data.

REFERENCES

Alluisi, E. A. (1967). Methodology in the use of synthetic tasks to assess complex performance. *Human Factors*, *9*, 375-384.

Anderson-Garlach, M. M., and Fisk, A. D. (1994, April). *Age-related retention of skilled performance: Within-subject examination of visual search, memory search, and lexical decision.* Presented at the Fifth Cognitive Aging Conference, Atlanta.

Chiles, W. D., Alluisi, E. A., and Adams, O. S. (1968). Work schedules and performance during confinement. *Human Factors*, *10*, 143-196.

Elsmore, T. F. (1994). SYNWORK1: A PC-based tool for assessment of performance in a simulated work environment. *Behavior Research Methods, Instrumentation, and Computers*, *26*, 421-426.

Fisk, A. D., and Rogers, W. A. (1991). Toward an understanding of age-related memory and visual search effects. *Journal of Experimental Psychology: General*, *120*, 131-149.

Fisk, A. D., Cooper, B. P., Hertzog, C., and Anderson-Garlach, M. M. (1995). Age-related retention of skilled memory search: Examination of associative learning, interference, and task-specific skills. *Journal of Gerontology: Psychological Sciences*, *50B*, P150-P161.

Fisk, A. D., Hertzog, C., Lee, M. D., Rogers, W. A., and Anderson-Garlach, M. M. (1994). Long-term retention of skilled visual search: Do young adults retain more than old adults? *Psychology and Aging*, *9*, 206-215.

Korteling, J. (1994). Effects of aging, skill modification, and demand alternation on multi-task performance. Human Factors, 36, 27-43.

Kramer, A. F., and Larish, J. L. (in press). Aging and dual-task performance. In W. R. Rogers, A. D. Fisk, and N. Walker (Eds.), *Aging and skilled performance: Advances in theory and application.* Hillsdale, NJ: Erlbaum.

Morgan, B. B., and Alluisi, E. A. (1979). Synthetic work: A methodology for assessment of human performance. *Perceptual and Motor Skills*, *35*, 835-845.

Rogers, W. A., Fisk, A. D., and Hertzog, C. (1994). Do ability-performance relationships differentiate age and practice effects in visual search? *Journal of Experimental Psychology: Learning Memory and Cognition*, *20*, 710-738.

Rogers, W. A., Gilbert, D. K., and Fisk, A. D. (1994, April). *Long-term retention of general skill and stimulus-specific abilities in associative learning: Age-related differences.* Presented at the Fifth Cognitive Aging Conference, Atlanta.

Salthouse, T. A. (1991). *Theoretical perspectives on cognitive aging.* Hillsdale, NJ: Erlbaum.

Salthouse, T. A., Hambrick, D. Z., Lukas, K. E., Dell, T. C. (in press). Determinants of adult age differences on synthetic work performance. *Journal of Experimental Psychology: Applied.*

Schneider, W, & Detweiler, M. (1988). The role of practice in dual-task performance: Toward workload modeling in a connectionist/control architecture. *Human Factors*, *30*(5), 539-566.

W. Poon, D. C. Rubin, and B. A. Wilson (Eds.), *Everyday cognition in adulthood and late life* (pp. 545-569). Cambridge, England: Cambridge University Press.

Interested readers should contact Richard Sit for information regarding minor changes in the data table and reported results.

AN EXAMINATION OF THE ADULT AGE DIFFERENCES ON THE RAVEN'S ADVANCED PROGRESSIVE MATRICES

Renee L. Babcock
University of Nebraska-Lincoln
Lincoln, Nebraska

The purpose of the current project was to examine the nature of the age-related differences on the Raven's Advanced Progressive Matrices (APM). Three components were hypothesized to be involved in correctly solving the APM problems. These included a rule-identification component, a rule-application component (involving a one-rule spatial transformation), and a rule-coordination component. The project was designed to examine the influence of each of the hypothesized components on the age-related variance on the APM. Two tests presumed to measure each hypothesized component were presented to 183 adults between the ages of 21 and 83. Hierarchical regression analyses indicated that although all of the hypothesized components accounted for a significant amount of the variance on the APM (approximately 50% each), only performance on the tasks measuring rule application accounted for a unique proportion of the age-related variance on the APM. Implications of the results in regards to following symbolic instructions in assembly of objects and in driving are discussed.

The purpose of the current project was to examine the nature of the age-related differences in performance on Raven's Advanced Progressive Matrices (APM). The APM is a type of series-completion test designed to measure an individual's capacity for reasoning and clear thought. There are two sets of the APM. Set I is designed to differentiate between dull, average, and bright individuals, whereas Set II is designed to provide more refined discriminations. An example of the type of problem presented in the APM is shown in Figure 1. The subject's task is to determine which of eight alternatives best fills in the missing cell of the matrix, such that both row and column rules are satisfied.

The APM is commonly referred to as a measure of intelligence because of its significant relationship with other tests of general intelligence (Burke, 1958; Raven, Court, & Raven, 1986). Nonetheless, attempts to determine more precisely what the Raven's actually measures have not been as conclusive. Evidence that young people perform better than older individuals has also been well documented (Denney & Heidrich, 1990; Edwards & Wine, 1963; Foulds, 1949; LaRue & D'Ella, 1985; Mergler & Hoyer, 1981; Panek & Stoner, 1980; Slater, 1947). However, although several suggestions have been made as to the nature of the age differences on the APM, none of these hypotheses have yet been supported with convincing evidence. Therefore, it is of interest in the current project to more precisely specify the nature of the adult age differences on the Raven's.

Three component processes were hypothesized to be involved in solving Raven's APM problems. Briefly, it is assumed that a subject

Figure 1: Example of Type of Problem Used in APM

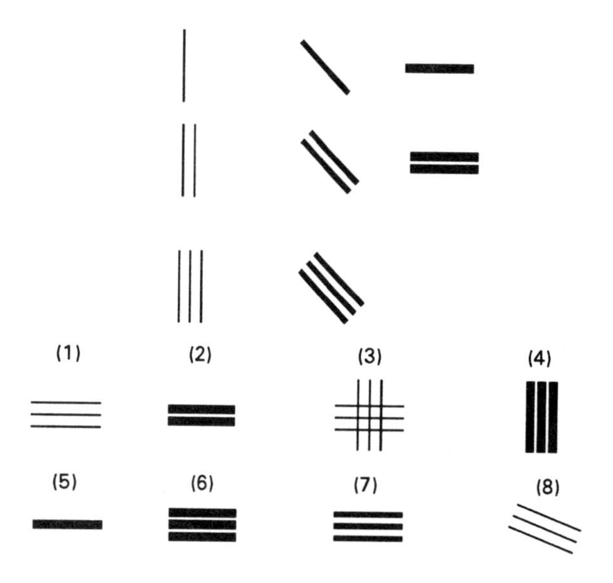

first attempts to determine the rules necessary to solve the problem. Once the rules have been identified, the subject must apply them to other rows or columns in order to verify that the correct rules have been determined. Because there is often more than one rule necessary to solve a Raven's problem, the subject must coordinate the rules while applying them (i.e., the problem in Figure 1 involves three rules that must be coordinated: change the thickness of lines, change the number of lines, change the orientation of lines).

The rule-identification process includes noting similarities and inferring relations between adjacent cells of the APM problems in order to generate rules for the rows or columns. Shipley's Abstraction Test and the Letter Sets test (Ekstrom, French, Harman, & Dermen, 1976) were chosen to measure the ability to identify rules. In the Letter Sets test, items consist of five sets of letters. Four of the five sets in each item are alike in that they follow a single rule. The subject is to decide which one of the five sets does not follow the rule. Shipley's Abstraction test is a series completion test in which subjects are asked to provide the digit(s) or word(s) that best complete the series.

After identifying the rules from one or more rows or columns, the subject must apply the rules to a new row or column, in a Rule-Application process. To do this, he or she must, among other things, mentally transform figures according to the previously determined rules. The ability to visually transform spatial information in this manner seems to be important to the application of rules, because successful performance on the Raven's depends partially on access to an accurately transformed image.

Two new tasks, the Geometric-Transformation Task and the Pattern-Transformation Task (see Figure 2), were developed to examine the ability to mentally transform geometric figures according to specified rules. The items in these tasks consist of a geometric figure in the Geometric-Transformation task or a line pattern in the Pattern-Transformation task which is the to-be-transformed figure. These items are followed by a representation of a transformation which should be mentally applied to the original figure. The possible transformations consist of the addition of a smaller figure or pattern to the original, subtraction of some portion of the original figure or pattern, and 90 or 180 degree rotations of the figure or pattern. The transformations are represented by a one word command (add, subtract, rotate) placed above a graphical representation of the to-be-added (or subtracted) figure, or the degree of rotation to be performed.

In the Coordination-of-Rules process, a subject must simultaneously remember and apply two or more sets of rules. This process occurs in the Raven's APM when the subject has to apply more than one rule to the final row or column in order to successfully solve the problem. The Calendar Test and Following Directions (Ekstrom et al., 1976) were chosen to measure the ability to coordinate rules. In the Calendar Test, "The subject is asked to select certain dates on a calendar by following a fairly complex set of directions (p. 88)." In the Following Directions test the subject is asked to select a letter from a pattern of letters according to a set of rules.

Figure 2: Example of Problems in Transformation Tasks

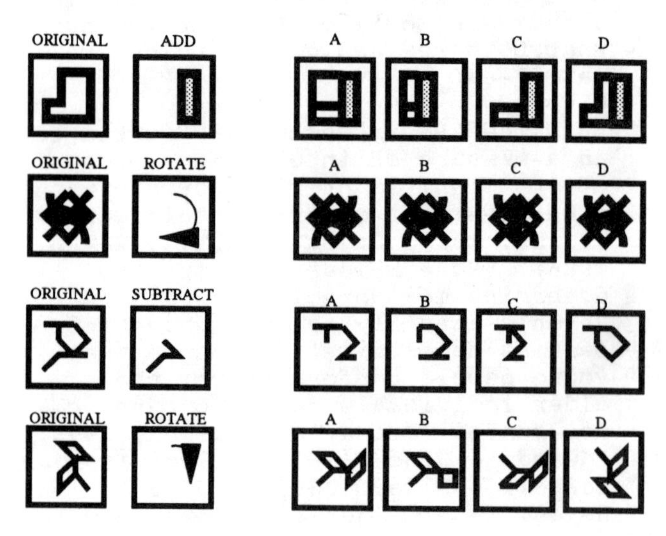

METHOD

Subjects

183 subjects, ranging from 21 to 83 years of age, participated in the current study. Subjects received $10.00 for participation in the project.

Procedure

Because the primary comparisons of interest were between-subjects correlations, all participants performed the tasks in the same order to avoid a confounding of subject and task order. The order of the tasks and their time limits were: Raven's APM - Set I = 5 min, Raven's APM - Set II = 20 min, Letter Sets = 5 min, Geometric-Transformation = 4 min, Calendar Test = 5 min, Following Directions = 5 min, Pattern-Transformation = 4 min, Shipley's Abstraction test = 5 min. Performance on the tests were recorded as number correct.

RESULTS

Aggregate scores for each hypothesized component were used in the subsequent analyses. The aggregate scores were computed for each component by taking the average of the z-scores of the two tests presumed to measure each component. For example, the average of the z-scores for Geometric Transformation and Pattern Transformation represented performance on Rule Application.

The correlations between age and the hypothesized components are presented in Table 1. As expected, age was negatively correlated with each of the hypothesized components. In addition, significant positive correlations were found between each of the hypothesized components.

The major goal in the current project was the determination of the relative importance of each of the hypothesized components to the age differences in performance on the Raven's APM. One method of examining this relationship involves using hierarchical regression to determine the amount of age-related variance in performance on the APM after first controlling for each of the hypothesized components. Results for this analysis are presented in Table 2. Age alone accounted for approximately 21% of the variance on the APM. The increment in R^2 indicates the amount of age-related variance on the APM after first controlling for one or more of the hypothesized components. Note that all three hypothesized components reduce the age-related variance to under 6%, and that Rule Application reduces the amount of age-related variance on the APM to under 3%.

The above analysis provides information about the age-related variance on the APM after first controlling for each of the hypothesized components. However, if there is considerable overlap among the hypothesized components with respect to the age-related variance accounted for on the APM, then according to the first method of analysis, the component which is entered into the regression equation first would explain the largest portion of the variance.

A more informative approach in determining the importance of each of the hypothesized components is to determine each component's unique contribution to the age-related variance on the APM. The hierarchical regression analyses were used to determine the amount of age-related variance after first controlling for two of the three components, and then after controlling for all three components (see Table 2). The decrement in the amount of variance associated with

Table 1: Correlations among age and the hypothesized components.

	Age	APM	ID	APP
APM	-.46			
ID	-.38	.73		
APP	-.48	.71	.64	
COORD	-.33	.70	.71	.71

Note:
APM = Raven's Advanced Progressive Matrices - II
ID = Rule Identification
APP = Rule Application
COORD = Rule Coordination

age in the latter equation will provide an indication of the amount of variance unique to age and the third component.

For example, to determine the amount of age-related variance on the APM that is unique to Rule Identification, it is first necessary to determine the amount of age-related variance that cannot be accounted for by the other two components (i.e., .015). When Rule Identification is added into the equation with the other two components, the amount of age-related variance left unexplained drops to .011. The subtraction method described suggests, therefore, that Rule Identification accounts uniquely for .4% of the age-related variance on the APM (i.e., .015-.011=.004).

Using this subtraction method, Rule Coordination also accounts for less than 1% of the age-related variance on the APM (.011-.011=.000). However, Rule Application still accounts uniquely for approximately 2% of the age-related variance on the APM (.027 - .011 = .016). Although 2% of the variance seems small in an absolute sense, when considered relative to the other hypothesized components, rule application remains a moderately strong predictor of the age-related variance on the APM.

DISCUSSION

The results from the current study suggest that although the three hypothesized components of the APM account for much of the variance in the APM (approximately 50% each), only Rule Application accounted for a unique proportion of the age-related variance. The Rule Application tasks involved manipulating a geometric figure according to a one-rule command. Because both the figure and the rule were present throughout the task, this transformation would seem to require minimal memory demands. The results suggest then that older adults have more difficulty visualizing the transformation of the figure and this inability leads to poorer performance on a much more complex cognitive task, such as the APM.

The implications from these results are many. For example, in the past several years, there has been a move from providing instructions to the general public in a verbal format to providing them in a symbolic format. For example, several companies now provide assembly instructions using only symbols, asking the consumer to make transformations of objects according to rules that are often quite complex. According to the results from the current project, symbolic instructions present a potential danger for older adults if certain objects are incorrectly assembled. Perhaps an even more dramatic example is the potential implication of the current results to the driving behavior of older adults. As with instructions from companies, many road signs today have been changed from verbal instructions to symbolic instructions (e.g., indicating curves

Table 2: Increment in R^2 for Age after statistical control of other variables (using alpha = .01 to determine significance).

After Control of:	R^2	Increment in R^2	F
None	.209	---	47.53
APP	.498	.025	9.38
COORD	.495	.056	22.19
ID	.539	.037	15.30
APP,ID	.634	.011	5.78*
APP,COORD	.624	.015	7.70
ID,COORD	.605	.028	13.60
APP,ID,COORD	.668	.010	5.86*

Note:
* = NOT SIGNIFICANT at $p < .01$
APM = Raven's Advanced Progressive Matrices - II
ID = Rule Identification
APP = Rule Application
COORD = Rule Coordination

or detours). If an older adult has difficulty applying the transformation on the sign to what happens on the road, there is a possibility that older people could be involved in more accidents simply because of their inability to apply one-rule transformations to objects.

ACKNOWLEDGEMENTS

This research was supported by National Institute on Aging Grant AG06826 to Timothy A. Salthouse. The author was also partially supported by a training grant from the National Institute on Aging.

REFERENCES

Burke, H.R. (1958). Raven's Progressive Matrices: A review and critical evaluation. The Journal of Genetics Psychology, 93, 199-228.

Denney, N.W. & Heidrich, S.M. (1990). Training effects on Raven's Progressive Matrices in young, middle-aged, and elderly adults. Psychology and Aging, 5, 144-145.

Edwards, A.E., & Wine, D.B. (1963). Personality changes with age: Their dependency on concomitant intellectual decline. Journal of Gerontology, 18, 182-184.

Ekstrom, R.B., French, J.W., Harman, H.H., & Dermen, D. (1976). Kit of Factor-Referenced Cognitive Tests. Princeton, NJ: Educational Testing Service.

Foulds, G.A. (1949). Variations in the intellectual activities of adults. American Journal of Psychology, 62, 238-246.

LaRue, A. & D'Ella, L.F. (1985). Anxiety and problem solving in middle-aged and elderly adults. Experimental Aging Research, 11, 215-220.

Mergler, N.L. & Hoyer, W.J. (1981). Effects of training on dimensional classification abilities: Adult age comparisons. Educational Gerontology, 6, 135-145.

Panek, P.E., & Stoner, S.B. (1980). Age differences on Raven's Coloured Progressive Matrices. Perceptual and Motor Skills, 50, 977-978.

Raven, J.C., Court, J.H., & Raven, J. (1986). Manual for Raven's Progressive Matrices and Vocabulary Scales: General Overview. London: H.K. Lewis & Co.

Slater, P. (1947). The association between age and score in the Progressive Matrices test. British Journal of Psychology: Statistical Section, 1, 64-69.

OLDER ADULTS SOMETIMES BENEFIT FROM ENVIRONMENTAL SUPPORT: EVIDENCE FROM READING DISTORTED TEXT

Raymond J. Shaw
Georgia Institute of Technology
Atlanta, Georgia

An implication of the generally assumed age-related deficit in processing resources is that older adults should be more reliant on support in the task environment than younger adults in any cognitive task. This notion, an extension of Craik's (1986) environmental support framework for memory, was investigated in four studies of reading. In Experiment 1, two distorted texts varied in the amount of irrelevant words created by rearrangement of the spaces between words. The irrelevant words were predicted to attract attention, and thus slow reading. Older adults were much more slowed than were younger adults by loss of support for reading. In Experiment 2, emphasized irrelevant words disrupted reading more than did de-emphasized ones, but not more so for older adults. These results were replicated in two further experiments using individual sentences rather than running texts. These results suggest that environmental support affects younger and older adults differently, but that this claim must be qualified by task-specific characteristics. Theoretical and practical implications of such qualifications are discussed.

For most people, reading is an important way to get a great deal of information, at work, at home, and in between. The ease with which reading can be performed is a primary factor in the level of success, pleasure, and personal safety that people enjoy. Studying the cognitive operations in reading, and how they are maintained across the lifespan, is of critical interest to human factors psychologists. The purpose of the present studies was to investigate the nature of age-related changes in the manner in which written information is apprehended.

A common assumption in the study of cognitive aging is that older adults have a reduction (relative to younger adults) in "processing resources" (e.g., Craik, 1986). A processing resource, in this context, can be thought of as "any internal input essential for processing ... that is available in quantities that are limited at any point in time" (Navon, 1984, p. 217). It is clear that reading, like any complex cognitive task, involves both "top-down" and "bottom-up" information processing (Carpenter and Just, 1987). That is, the efficiency of reading is determined by both internal and external input to the process. If older adults have a "weaker" internal input than younger adults do, then it seems likely that the nature of the external inputs will have a greater impact on the reading performance of older adults. Craik (e.g., 1986) used this general logic in his "environmental support" account of age-related changes in memory performance; it should apply as well to any cognitive task that requires internal processing resources. The basic idea is that the task environment "supports" performance in light of decreased internal influences.

An important external input in reading is the proper positioning of spaces between words (Duchnicky, 1986). If spacing is distorted by re-placing the spaces in a text so that word boundaries are ignored (asint his exa mple), irrelevant or "false" words may be formed from fragments of one or several of the words in the original (e.g., "his" in the example). In this event, the "environment" supports inappropriate parsing of the material, and readers find it especially difficult to read (Duchnicky, 1986). Given that words are encoded automatically (Just and Carpenter, 1987; MacLeod, in press), the irrelevant words in a distorted text attract attention away from the correct reading of the text. This slows down reading dramatically, and, as a negative environmental influence or "support" for reading, should be especially disruptive to older adults.

EXPERIMENT 1

In the first experiment, subjects read a normal text, and two types of distorted text. The texts were each approximately 300 words in length. The method of distortion was that outlined above; specifically, the spaces between words were removed, and then re-placed in the text without respect to the original word boundaries. In the Random distortion condition a few irrelevant (or False) words were created; in the False distortion condition, many such words were formed.

Relative to the normal, Control text, the distorted texts forced readers to control their own processing of information rather than be guided or supported by the information presented to them. Both distortions removed the support provided by spacing; in addition, the False condition supported the incompatible behavior of reading irrelevant words. Reading time was therefore expected to increase dramatically from Normal to Random to False, and more so for older than for younger adults.

Method

Subjects. Participants were 12 younger and 12 older adults; see Table 1 for subject characteristics.

Materials. Normal, Random, and False versions of three approximately-300-word passages were created, as described above. Each was typewritten in upper-case letters, double-spaced, on a single sheet of paper placed in a clear plastic cover. The Normal version of each text used normal spacing between words. For the two distorted versions, the spaces were shifted away from their original locations. The Random version was created by computer-generated random replacement of spaces. The False version was created by hand so that whenever possible, false (or irrelevant) words were physically separated by spaces. In addition to complete false words, fragments of such words and near misspellings of words were also created. Because the Random texts were completely random, a number of false words and false-word fragments also occurred in those texts.

Table 1. Subject characteristics by Experiment. Standard deviations are given in parentheses. Significant age differences were only in Vocabulary in Experiments 2 and 4.

Experiment	Age	Years of Education	Vocab[1]	#Females/ Total N
1 Old	69.3	14.0	16.1	7/12
	(5.0)	(2.7)	(5.2)	
Young	20.6	13.5	18.2	9/12
	(4.3)	(0.8)	(2.8)	
2 Old	72.1	12.3	.80	5/12
	(5.1)	(3.1)	(.08)	
Young	19.3	13.1	.69	8/12
	(1.4)	(0.3)	(.11)	
3 Old	74.1	13.3	55.6	7/12
	(4.0)	(3.2)	(8.4)	
Young	21.3	13.8	53.3	7/12
	(4.0)	(1.3)	(4.6)	
4 Old	74.4	15.0	59.8	9/12
	(4.9)	(3.5)	(7.2)	
Young	20.0	13.9	51.7	9/12
	(1.0)	(0.7)	(8.2)	

[1] Experiment 1: Subset of WAIS, maximum score was 24. Experiment 2: Subset or Full WAIS, proportion of maximum score. Experiments 3 & 4: Full WAIS, maximum score = 70.

Design and Procedure. Each subject read a Normal text first; half read a Random text followed by a False text, and the other half read a False text followed by a Random text. Twelve subjects were required to exhaust all possible orders of texts and conditions; each older adult read the same materials in the same order as one younger adult.

After practice with a sample of the False condition, subjects read three passages aloud, one of each type. Reading time was measured to the nearest 0.1 second with a stopwatch. In this and all of the following experiments, errors were corrected only after the reader was allowed to do so on his or her own. Reading time was not corrected for errors.

Results

Average reading times are given in Table 2. An Age x Text Type analysis of variance (ANOVA) confirmed that older adults read more slowly overall, $\underline{F}(1,22)=7.65$, $\underline{p}<.01$, and that reading time increased across text types, $\underline{F}(2,44)=106.25$, $\underline{p}<.001$. The interaction of Age and Text Type was also significant, $\underline{F}(2,44)=6.19$, $\underline{p}<.005$. Post-hoc tests revealed that the older adults showed increased reading times for each increase in negative support, whereas the younger adults only increased from Normal to distorted text, and their increase was smaller than that for the older adults.

Discussion

Consistent with the environmental support framework, older adults were more disturbed by the distortion of spacing than were younger adults. Further, the addition of false words in the least supportive texts led to an even greater age difference. Because the present sample size was rather small, the next experiment was conducted, in part, to replicate the present results.

EXPERIMENT 2

The primary goal of the present study was to magnify the effect of environmental support observed in Experiment 1. This was done by using two alternating colors of ink in the False-word texts from Experiment 1 to

Table 2. Average Reading Times (and SD's) for Experiment 1 (seconds/passage).

	Normal	Random	False
Old	108.6	677.0	945.5
	(24.3)	(263.7)	(423.2)
Young	88.9	461.7	594.9
	(12.9)	(112.2)	(156.7)

emphasize and support reading either the actual words (a Supportive version) or the false words (a Nonsupportive version). One obvious prediction was that reading time would be substantially lower in the Supportive compared to the Nonsupportive text, and that by the environmental support hypothesis, older adults would show a larger benefit from proper support than would younger adults.

An additional prediction is less obvious. In Experiment 1, the Random texts supported reading better than did the False texts primarily by eliminating most of the false words. In the present experiment, support was provided by merely emphasizing the actual words, but leaving the false words present, just de-emphasized. If older adults were more distracted because of the presence of false words, then even the Supportive texts in the present experiment should have a distracting effect, and the performance of older adults should be more disrupted than that of younger adults.

Method

Subjects. A new sample of 12 younger and 12 older adults participated; see Table 1.

Materials, Design, and Procedure. The design and procedure were identical to Experiment 1; the major change was in the materials. The Normal text was unchanged. The two distorted text types, however, were each printed in alternating red and black ink. Also, letters in red were always in lowercase type, and letters in black were in uppercase type. In the Supportive text, the actual words alternated color, and in the Nonsupportive text, color alternated to follow the distorted spacing. In the following, for example, let uppercase letters represent black ink, and lowercase letters represent red ink: A Nonsupportive text would read "**T** he **CATCH** as **EDIT** th **ROUG** hour **HO** use;" whereas a Supportive version of the same text would read "**T HE** cat**CH AS ED**it **TH ROUG H**our **HO USE**."

Results

Table 3 shows average reading time as a function of Age and Text Type. An ANOVA showed that only the main effect of Text Type was significant, $F(2,44)=94.8$, $p<.001$. Surprisingly, neither the effect of Age nor the interaction of Age and Text Type was significant.

However, based on the original hypotheses, two further, planned statistical analyses were conducted. One prediction was that the change from Normal to Supportive text would interact with Age, because older adults should be distracted by the presence of even de-emphasized false words. The other prediction was that the change from Supportive to Nonsupportive texts would also yield an interaction, because older adults should be more distracted by emphasized false words than are younger adults.

Table 3. Average Reading Times (and SD's) for Experiment 2 (seconds/passage).

	Normal	Support	Nonsupport
Old	96.0	361.7	928.7
	(13.1)	(150.5)	(490.8)
Young	77.7	209.3	733.8
	(9.9)	(47.9)	(199.0)

The first subset ANOVA, comparing Normal and Supportive text types, supported the relevant hypothesis. The main effect of Age was reliable, $F(1,22)=12.65$, $p<.005$, as was the main effect of Text Type, $F(1,22)=83.22$, $p<.001$. The interaction was also significant, $F(1,22)=9.48$, $p<.01$. Thus, older adults were more affected by the change from Normal to Supportive text than were younger adults.

The second subset ANOVA, however, did not confirm the relevant hypothesis, revealing only a main effect of Text Type, $F(1,22)=91.88$, $p<.001$. Neither the effect of Age nor the interaction was significant. Thus, older adults were not more affected by emphasized false words in the Nonsupportive texts.

Discussion

The results of the present experiment provided mixed support for the environmental support hypothesis. On the one hand, despite a color emphasis of real words in the Supportive texts, the residual presence of false words and the lack of proper spacing captured and slowed down the older adults more than the younger adults in comparison to Normal texts. This was the expected result.

However, the age difference in reading time for Supportive texts (which emphasized words to be read) was expected to be smaller than for Nonsupportive texts (which emphasized words to be ignored). The prediction was not confirmed: The age difference was nearly identical at the two levels of support.

EXPERIMENT 3

Given the surprising result from Experiment 2, two further experiments were conducted to directly replicate the first two with new materials. In these new experiments, individual sentences were used rather than running texts, with the advantage that more measures per condition could be made per subject. A second change was that, unlike Experiment 1, the Random distorted versions of each sentence contained no false words whatever, to allow a more accurate assessment of the effect of distorted spacing separately from the effect of false words in such texts. Each of these effects was expected to be greater in the older adults.

Method

Subjects. A new sample of 24 adults participated; see Table 1.

Materials. Thirty 13-word sentences were constructed, and two distorted forms of each sentence were created, Random and False, using the same methods as in Experiment 1. However, the Random version had no false words at all, and the False version always had 3 to 5 false words.

Procedure. Ten sentences were randomly assigned to each condition for each subject, and were presented on a computer monitor. Conditions were blocked; half of the subjects read Normal, Random, and then False condition sentences, the other half of subjects read Normal, False, then Random. Reading time for each sentence was recorded from the moment the program completed writing it to the screen. The experimenter indicated with a keypress when the reader was finished.

Results

Table 4 shows mean reading time as a function of Age and Sentence Type. An Age x Sentence Type ANOVA revealed significant main effects of Age, $F(1,22)=5.86$, $p<.05$, and of Sentence Type, $F(2,44)=53.21$, $p<.001$. The interaction of Age and Sentence Type was also significant, $F(2,44)=3.83$, $p<.05$. Post-hoc tests showed that an age difference occurred only on the False sentences; the younger and older adults did not differ on any other sentence type. Further, both groups showed a difference between Normal and Random sentence types, but only the older adults showed a difference between False and Random.

Discussion

The results of the present experiment suggested that removal of the support provided by spacing was not sufficient on its own to produce a greater difficulty for older adults; false words must be formed to draw attention away from the correct reading of the material.

Table 4. Average Reading Times (and SD's) for Experiment 3 (seconds/sentence).

	Normal	Random	False
Old	6.92	59.69	89.85
	(1.13)	(28.13)	(53.89)
Young	5.08	41.01	52.34
	(0.71)	(10.48)	(22.08)

EXPERIMENT 4

The present experiment was designed to replicate Experiment 2, using the materials and methods of Experiment 3. As in Experiment 2, a Nonsupportive version of each False sentence emphasized the false words by alternating color on the basis of spacing; a Supportive version of each False sentence emphasized and supported reading the correct words by alternating color.

The specific hypotheses were identical to those for Experiment 2. First, older adults were expected to be more disturbed than younger adults simply by the change from Normal to Supportive distorted sentences because of the presence of false words in these sentences (even though they were de-emphasized). Second, older adults were expected to be more affected than younger adults by the difference between Supportive and Nonsupportive sentences.

Method

Subjects. Characteristics of the final set of 24 adults are shown in Table 1.

Materials, Design, and Procedure. Correct spacing supported reading in the Normal version of each sentence; two distorted versions of each sentence used false-word spacing (from Experiment 3) to support distraction from reading. Color alternated (red and blue-white) between real words in the Normal and Supportive distorted sentences; in the Nonsupportive distorted version color alternation followed the false spacing, and thus supported distraction. The design and procedure were identical to Experiment 3.

Results

Table 5 shows average reading time as a function of Age and Sentence Type. An overall Age x Sentence Type ANOVA revealed significant main effects of Age, $F(1,22)=4.69$, $p<.05$, and of Sentence Type, $F(2,44)=97.30$, $p<.001$. The interaction was also significant, $F(2,44)=3.22$, $p<.05$.

Based on the original hypotheses, two additional ANOVAs were conducted, first on Normal and Supportive sentences, and then on Supportive and Nonsupportive sentences. The ANOVA on the Normal vs. Supportive revealed main effects of Age, $F(1,22)=8.93$, $p<.01$, and of Sentence Type, $F(1,22)=133.09$, $p<.001$, as well as a significant interaction, $F(1,22)=9.51$, $p<.01$. These results clearly indicate that the older adults were more slowed by the

Table 5. Average Reading Times (and SD's) for Experiment 4 (seconds/sentence).

	Normal	Support	Nonsupport
Old	5.84	18.53	71.68
	(1.06)	(6.28)	(30.35)
Young	5.21	12.55	50.32
	(0.50)	(2.59)	(23.70)

change from Normal to distorted spacing, regardless of support in the distorted version. The ANOVA on Supportive vs. Nonsupportive text, however, showed only main effects of Age, $F(1,22)=4.64$, $p<.05$, and of Sentence Type, $F(1,22)=88.34$, $p<.001$; the interaction was not significant, suggesting that the change in support for reading the two types of distorted text did not affect the older and younger adults differently.

Discussion

The results of the present experiment generally replicated Experiment 2. Reading times increased significantly with decreases in support for reading, and a comparison of Normal and Supportive (but distorted) sentences revealed a reliable interaction of Age and Sentence Type. Therefore, the results of the present experiment (and of Experiment 2, by extension) suggest that the presence of false words in these distorted texts presents a special difficulty for older adults. The effect of emphasizing false words over real words with color alternation does not differ between younger and older adults.

GENERAL DISCUSSION

The finding of age differences in the effect of distorted spacing demonstrates that environmental support is more critical for older adults than for younger adults in these reading tasks. However, the fact that the changes in salience of the false words did not affect older adults more than it did younger adults puts a serious qualification on the generality of the environmental support framework.

One interpretation of these results is that environmental support has differential effects only when the manipulation affects a central characteristic of the operations necessary to perform the task. In the present experiments, violations of spacing conventions and the creation of irrelevant words forced readers to alter the way in which they formed words from the visual array; this is inherent to any reading task.

Manipulation of the salience of the irrelevant words, as in Experiments 2 and 4, forced readers to ignore or suppress one aspect of the visual array, but in a manner that is extrinsic to reading per se. That is, that portion of the encoding process affected by irrelevant words was not altered by the salience of the irrelevant words. This type of manipulation of support may affect performance, as in the present experiments, but not differentially for younger and older adults.

An important theoretical implication of these findings, and the qualification on the environmental support framework, is that researchers on age differences in cognition must consider the nature of support offered by any presumed "supportive" difference between task conditions. The essential question is not whether a change in conditions supports performance, but the <u>way</u> in which the change supports performance, or the relevance of the supported operations to the task at hand.

The present series, and its general theoretical claims have important practical implications: Those attempting to provide living environments that will be especially beneficial for older adults must consider exactly what behaviors are to be supported. For example, simply making warning labels on medicines larger (to account for failing vision) does not necessarily support the critical behavior of adherence to the warnings. Simply making the environment conducive to living in general may not be especially beneficial to older adults, unless the target behaviors to be supported are specified and the environment is altered accordingly.

REFERENCES

Craik, F.I.M. (1986). A functional account of age differences in memory. In F. Klix and H. Hagendorf (Eds.), Human memory and cognitive capabilities: Mechanisms and performances (pp. 409-422). Amsterdam: Elsevier.

Duchnicky, R.L. (1986). Visual components of skilled reading: Evidence from studies of typographically transformed text. Unpublished Doctoral dissertation, University of Toronto.

Just, M.A., and Carpenter, P.A. (1987). The psychology of reading and comprehension. Boston: Allyn & Bacon.

MacLeod, C.M. (in press). Half a century of research on the Stroop effect: An integrative review. Psychological Bulletin.

Navon, D. (1984). Resources: A theoretical soup stone? Psychological Review, 91, 216-234.

Memory performance as a function of age, reattribution training and type of mnemonic strategy training.

Marilyn L. Turner
Wichita State University
Wichita, Kansas

This experiment investigated whether mnemonic strategy training, occurring over a two-month period, would result in improved memory performance when combined with reattribution training. It was also hypothesized that the old and young may differ in their ability to perform nonverbal and verbal mnemonics. Therefore, age-related differences in memory performance were investigated as a function of whether the mnemonic was verbal (Alphabet Search Method) or non-verbal (Method of Loci), and whether or not reattribution training was combined with mnemonic training. Subjects were 34 old (Mean age = 69.5) and 34 young (Mean age = 22.8) adults. Memory performance was measured on the California Verbal Learning Test, the Nelson-Denny Vocabulary Test, the Beck Depression Inventory and four memory span tasks, prior and following a two-month period of weekly mnemonic strategy training sessions. A third of the subjects were trained with the Method of Loci, a third with Alphabet Search, and the remaining third served as the waitlist control group. In addition, half the young and old subjects from each mnemonic group did, and half did not, participate in a reattribution training workshop. Results clearly showed that mnemonic strategy training was useful for the old and young. However, the combination of reattribution and mnemonic strategy training only enhanced old, not young, memory scores when the type of strategy required verbal skills (Alphabet Search). The implication was that mnemonic strategy training may be more effective for the old if combined with reattribution training, and, if the mnemonic requires verbal rather than non-verbal skills.

Attitude and depression were two of the critical psychological factors found to contribute to elderly cognitive impairment by the Office of Technology Assessment (1987). Recommended solutions included the development and implementation of accurate diagnoses of memory deficits, and appropriate and effective treatment intervention. One effective treatment intervention may be mnemonic strategy training. Learning a mnemonic technique (e.g., the Method of Loci; Bower, 1970) has substantially improved older adults' memory performance in serial word recall (e.g., Kliegl, Smith & Baltes, 1989; 1990). Thus, using a mnemonic could effectively improve memory performance required by the elderly for everyday functioning. However, research on memory performance following mnemonic training has shown that not all old people benefit from this treatment intervention. One reason that memory training may vary in its usefulness may be the inaccurate, but widely-held belief that aging inevitably results in memory loss. Hertzog, Dixon, Schulenberg & Hultsch (1987) suggested that negative beliefs about memory ability interferes with actual remembering behaviors. Further, according to Herrmann (1984), these beliefs bias the decision to participate in events requiring specific memories, e.g., names and faces of people at a party. Turner & Pinkston (In press) found their elderly subjects changed from an initially negative to a more positive belief regarding aging and memory following a two day memory workshop. Negative beliefs of the aging memory, therefore, may prevent the elderly from effectively using strategies that should improve memory performance. And, if so reattribution training may be a necessary prerequisite to mnemonic training (Cavanaugh, & Poon, 1989). Thus, the major question addressed in this experiment was whether or not reattribution training combined with mnemonic training resulted in improved old and young memory performance.

In addition, it was hypothesized that reattribution and mnemonic training would not completely attenuate differences found in recall of old and young. In fact, the additional information load, i.e., the mnemonic, may increase age-related differences. Robertson-Tchabo, Hausman & Arenberg (1976) suggested that using a mnemonic may initially overload memory. When using these strategies, additional information, such as adding a visual image or a verbal association, must be combined with the material to be remembered. The notion is that this additional information, non-verbal or verbal, may result in more of an overload for the old than young.

It is also possible that more of an overload is provided by non-verbal than verbal elaborations in the aging short-term, or working memory (Salthouse, 1991; Salthouse & Mitchell, 1990; Salthouse, Mitchell & Palmon, 1990). Aging has been found to negatively affect performance on non-verbal more than verbal tasks (Park, 1980; Tubi & Calev, 1989; Hultsch & Dixon, 1990; Hultsch, Hertzog & Dixon, 1990), and therefore, another reason memory training may vary in its usefulness could be that the old and young differ in their capacity to perform non-verbal and verbal mnemonics. That is, age-related memory deficits may be more pronounced when using non-verbal mnemonics, such as the Method of Loci, than verbal mnemonics, such as the Alphabet Search method. Therefore, this experiment investigated age-related differences in memory performance as a function of whether the mnemonic strategy was verbal or non-verbal, and was or was not combined with reattribution training.

METHOD

Subjects

Participating subjects were 34 old (Mean = 69.5) and 34 young (Mean = 22.8) individuals prescreened for clinical memory dysfunction and depression. Of the original 72 subjects, two elderly withdrew early for health reasons, and two young did not finish the training program.

Following a preliminary telephone prescreening ascertaining each subject was not taking a known memory-affecting medication, or electroconvulsive therapy, and did have an interest in memory training, the Mini-Mental State Examination (MMS; Folstein, Folstein, & McHugh, 1975) was administered. All participating subjects scored above 27 on the MMS (the criterion cutoff score for inclusion in this experiment), reported being in good health, were highly motivated, and perceived the training program as a personal challenge.

On the Nelson-Denny Vocabulary Test (ND; Nelson & Denny, 1973), the old adults performed at a level that was a half standard deviation above the average for their age group (*Mean* = 68.0 ; *SD* = 17.1), and the young adults performed at the average for their age group (*Mean* = 59.2; *SD* = 21.2).
On the Beck Depression Inventory (BDI; Beck, Ward, Mendelsohn & Erbaugh, 1961) the old scores (*Mean* = 9.5; *SD* = 16.2) were comparable with the young (*Mean* = 10.3; *SD* = 8.3). All subjects participated on a voluntary basis.

Design

Three independent variables were manipulated between subjects: (1) age (old/young), (2) type of mnemonic training (Method of Loci/Alphabet Search/Waiting list control), and (3) reattribution training (workshop participants/ non-participants), and one within subjects, i.e., time of testing (test/retest). Memory performance was measured for all 68 subjects at the beginning and end of a two-month period during which two-thirds (24 old and 24 young) participated in weekly mnemonic training sessions, with 10 old and 10 young adults serving as the waitlist control group. The prepost mnemonic measures were one short-term memory (STM) span task, three working memory (WM) span tasks (Turner & Engle, 1989) and the California Verbal Learning Test (CVLT; Delis, Kramer, Kaplan & Ober, 1987). In addition, the Metamemory in Adulthood Questionnaire (MIA; Dixon & Hultsch, 1983) was used to measure subjects' beliefs about the effects of aging on memory pre and post reattribution training in the workshop.

Training Materials

Mnemonic Training. The mnemonic that was taught was either the Method of Loci or the first-letter Alphabet Search Method. Following Bower (1970), subjects learning the **Method of Loci** were trained to image a mental map of rooms within a familiar place (home, office, etc), in the same sequential order. Each subject memorized these places, and the order in which they normally walked through them, until each was able to recite them forward and backward perfectly. Then they practiced using the mental map while vividly imaging and associating items to-be-remembered within each of the imaged places. Participants verbalized their images and the trainer occasionally offered suggestions. During recall subjects were then trained to image the same familiar places while recalling the items

Subjects trained to use the **Alphabet Search** mnemonic were instructed following the recommendations of Moffat (1984). Participants focused on the sound of the first letter of words to-be-remembered and repeatedly articulated the first letter's phonological sound aloud. During recall subjects were trained to use subvocal speech to verbally search through the alphabet for the first letter of the target words.

Reattribution training. The workshop consisted of interactive lectures on memory functions, normal and abnormal age-related changes, and non-age related factors that affect memory. The training was conducted by a cognitive psychologist, a graduate student in clinical psychology, assisted by two psychology undergraduates.

In three one-hour interactive lectures, accurate information was presented on the effects of age on memory, and importantly, on non-age-related factors, such as drugs, proper nutrition, and anxiety, that have been found to affect memory at any age. Diagrams, memory exercises and a movie illustrating *typical* forgetting enhanced the lectures and served to encourage active participation.

Prepost Materials

Prepost Reattribution materials. The MIA was administered at each subject's first and last session, prior to the workshop and/or mnemonic training, and at the end of the two month period, following all training sessions. The MIA consists of 120 items designed to rate beliefs and knowledge about memory and aging using a five point Likert scale, and differentiates eight, aging and memory factors; strategies, knowledge, capacity, change, activity, anxiety, motivation and control.

Prepost mnemonic materials. Four span tasks and the CVLT were given before the first and following the last mnemonic training sessions. The four *span tasks* were: (1) Word, wherein series of words were read and recalled, (2) Sentence-Word (SW), wherein series of sentences were read aloud, verified as to whether each sentence made sense, and sentence-last-words recalled, (3) Operation-Word (OW), wherein arithmetic operations were read aloud, answers verified, and the word following each operation recalled, and (4) Line-Shape (LS) non-verbal task wherein a pair of lines were identified as same/different and the shapes following the lines recalled.

Target words in the verbal Word, SW and OW tasks were selected from the most common, 4-6 letters, one-syllable concrete nouns published in Francis & Kucera (1982). Unrelated sentences for the SW task were generated to make sense using these words as the designated last-word. Half were "correct" (*The grades for our finals were posted outside the classroom door.*, and half "incorrect" (*The grades for our finals were classroom the outside posted door*). Arithmetic operations in the OW task consisted of two operations and a stated final answer of which half were "correct" [*(9 / 3) - 2 = 1*], and half "incorrect" [*(3 x 4) - 6 = 3*]. Shapes to be recalled in the LS task were triangles, squares, circles, arrows and hexagons presented individually in the center of 3x5 white index cards. Also, two parallel lines, 3/4 to 1 inch long, half the same and half differing by 1/4 inch, were presented in the upper left corner.

Procedure. The CVLT was administered prepost, consisting of free and cued recall and recognition trials. In all span tasks subjects individually viewed (LS) or read aloud (Word, SW, OW) series of stimuli presented one a time. The experimenter initiated the first trial, and as soon as subject finished viewing or reading aloud, the next item was presented. While reading or viewing items in the dual WM tasks (SW, OW, LS), subjects recorded whether that item was correct or incorrect (same/different in LS). After a series of items were read and verified, subjects were cued to serially recall the target items. There were three trials in Word, SW and OW at each set size, and four trials in LS at each set size. Set size increased from 2 to 5 in SW, OW and LS, and from 2 to 7 in Word. Separate spans were defined as the maximum number of items correctly recalled for each of the four tasks. Among the multiple recorded scores on the CVLT, subjects' total scores were reported herein.

RESULTS

Reattribution training. The workshop was more effective for the old than young. Planned prepost comparisons showed significant differences in the MIA subscale *locus,* $t(16) = 2.21$, $p < .041$ for elderly workshop participants, but not for non-participants. Young workshop participants, but not non-participants, improved from pre to post on anxiety subscale, $t(16) = 2.34$, $p < .05$. Also, over 65% of the elderly workshop participants improved post over pretest scores on the strategy, and change subscales, indicating reattribution training did affect elderly beliefs and knowledge of memory and aging.

Mnemonic strategy training. Although learning a mnemonic was effective for young and old, the verbal mnemonic was even more effective for the elderly when preceded by reattribution training. The elderly participating in mnemonic training had higher CVLT total post-training (M = 59.4) than pre-training (M = 53.1) scores, $F(2,31) = 31.64$, $p < .0001$. Young mnemonic participants also had higher CVLT post- (M = 64.6) than pre-training (M = 57.5) scores, $F(2,31) = 20.7$, $p < .0001$. However, neither the old nor young waitlist control groups improved CVLT scores from pre to post. Even more important, old (but not young) trained with alphabet search had higher CVLT posttest scores (M = 63.3) than pretest (M = 52.8) scores when they had also participated in reattribution training, $F(2,31) = 19.3$, $p < .001$. However, elderly using the Method of Loci, after participating in reattribution training did not show increased post over pre-training CVLT scores, $p > .3$.

The second question addressed by this experiment was whether the old and young differ in their capacity to perform non-verbal and/or verbal tasks. Tables 1 and 2 show post CVLT scores were similar for the old (M = 58) and young (M = 61), when trained with the verbal, Alphabet Search mnemonic. But, when trained with the non-verbal, Method of Loci the young post scores (M = 63.6) were higher than the old (M = 55.7), $F(2,63) = 30.4$, $p < .0001$. Further non-verbal and verbal age-related

TABLE 1: Memory performance as a function of workshop participation prior to mnemonic training with the *Method of Loci.*

	NON-WORKSHOP Young	Old	WORKSHOP Young	Old
*CVLT				
PRE	54.4	52.2	56.0	54.0
POST	64.8	54.2	62.5	57.3
**SW				
PRE	42.4	42.9	51.7	47.9
POST	47.1	42.4	63.6	41.2
**OW				
PRE	47.6	50.5	62.4	47.9
POST	55.5	51.9	70.2	38.8
**LS				
PRE	56.7	59.3	68.4	51.1
POST	78.4	57.0	70.0	50.7

*Scores are # correct. **Scores are % correct.

TABLE 2: Memory performance as a function of workshop participation prior to mnemonic training with the *Alphabet Search Method.*

	NON-WORKSHOP Young	Old	WORKSHOP Young	Old
*CVLT				
PRE	54.4	52.7	61.0	52.8
POST	54.1	54.5	68.0	63.3
**SW				
PRE	54.5	42.4	57.1	51.4
POST	54.8	43.3	55.7	49.0
**OW				
PRE	63.6	42.6	63.3	49.5
POST	59.8	44.5	70.4	56.7
**LS				
PRE	78.0	57.0	75.7	53.4
POST	65.6	53.1	83.3	61.3

*Scores are # correct. **Scores are % correct.

differences were found in WM span scores. Age differences were consistently found on the initial testing of the nonverbal LS spans (old=59%, young=72%, F(1,65)=4.01, p<.05), and on the retest (old=56%, young=76%, F(1,62)=10.12, p<.003). On the other hand, mean age differences were not found on the initial testing of the two verbal WM tasks (SW old=47%, young=50%, F(1,65)=0.94, p>.34; OW old=49%, young=55%, F(1,65)=2.63, p>.11), or in the retest (SW old=50%, young=56%, F(1,62)=2.22, p>14; OW old=53%, young=58%, F(1,62)=1.98, p>.32) spans. Finally, performance on the LS task correlated with stated use of Method of Loci (r=.44), and that on the SW with the Alphabet Search Method (r=.56).

DISCUSSION

Results clearly showed that mnemonic strategy training was useful, at least in the short run, for the old and young. These findings agreed with the reported gains of subjects trained with the Method of Loci (Kliegl, et al 1989; 1990). Reattribution training resulted in modest changes of elderly beliefs of age effects on memory. However, the combination of reattribution and mnemonic training only enhanced old, not young, memory scores, and only when the mnemonic required verbal cognitive abilities, i.e., the alphabet search method. The implication was that processing and/or prior knowledge limitations constrained non-verbal more than verbal performance.

Tubi & Calev (1989) have suggested that as people age there may be differences in the frequency with which verbal and nonverbal information is used; they may have more experience in processing verbal than nonverbal information. Age-related differences were consistently found on the prepost measures of nonverbal but not verbal WM span scores. These findings imply that a limited ability for the old to use imagery exists, perhaps due to a combination of little practical experience with non-verbal information,and the elderly having a more limited nonverbal WM capacity than the young. These limitations may be especially evident in the application of a mnemonic designed to improve memory that requires the use of imagery, such as the Method of Loci, and consequently provide an explanation of higher CVLT scores for the old trained with the Alphabet Search than Loci mnemonic.

The practical significance of these data cannot be underestimated. It has been shown that the majority of the elderly believe aging negatively affects memory ability, and that this negative belief can be altered by providing accurate knowledge of age-related and non-age-related factors affecting memory functions. Further, results showed that these negative beliefs can have an impact on one of the few methods that has been shown to improve memory in the elderly, i.e., mnemonic strategy training.

REFERENCES

Beck, A. T., Ward, C., Mendelsohn, M., Mock, J., & Erbaugh, J. (1961). An inventory for measuring depression. Archives of General Psychiatry, 4, 53-63.

Bower, G. H. (1970). Analysis of a mnemonic device. American Scientist, 58, 496-510.

Cavanaugh, J. C., & Poon, L. W. (1989). Metamemorial predictors of memory performance in young and older adults. Psychology of Aging, 4, 365-368.

Delis, D. C., Kramer, J. H., Kaplan, E., & Ober, B. A. (1987). The California Verbal Learning Test-Research Edition. New York: Psychological Corporation.

Dixon, R. A., & Hultsch, D. F. (1983). Structure and development of metamemory in adulthood. Journal of Gerontology, 38, 682-688.

Folstein, M. F., Folstein, S. E., & McHugh, P. R. (1975). Mini-Mental State: A practical method for grading the cognitive state of patients for the clinician. Journal of Psychiatric Research, 12, 189-198.

Francis, W. N., & Kucera, H. (1982). Frequency Analysis of English Language. Boston, MA: Houghton Mifflin.

Herrmann, D. J. (1984). Questionnaires about memory. In J. E. Harris & P. E. Morris (Eds.), Everyday memory actions and absent-mindedness (pp.133-151). London: Academic Press.

Hertzog, C., Dixon, R. A., Schulenberg, J. E., & Hultsch, D. F. (1987). On the differentiation of memory beliefs from memory knowledge: The factor structure of the metamemory in adulthood scale. Experimental Aging Research, 13, 101-107.

Hultsch, D. F., & Dixon, R. A. (1990). Learning and Memory in Aging. In J. E. Birren & K. W. Schaie (Eds.), Handbook of The Psychology of Aging, Third Edition. New York, NY, Academic Press, Inc.

Hultsch, D. F., Hertzog, C., & Dixon, R. A. (1990). Ability correlates of memory performance in adulthood and aging. Psychology and Aging, 5, 356-368.

Kliegl, R., Smith, J. & Baltes, P. B. (1989). Testing-the-limits and the study of adult age differences in cognitive plasticity of a mnemonic skill. Developmental Psychology, 25, 247-256.

Kliegl, R., Smith, J. & Baltes, P. B. (1990). On the locus and process of magnification of age differences during mnemonic training. Developmental Psychology, 26, 894-904.

Moffat, N. (1984). Strategies of memory therapy In B.A. Wilson & N. Moffat (Eds.), Clinical management of memory problems (pp.63-88). Rockville, Maryland: Aspen Systems Corp.

Nelson, M. S., & Denny, E. D. (1973). The Nelson-Denny Reading Test. Boston: Houghton Mifflin.

Office of Technology Assessment (1987). Losing a million minds. Washington, DC: U.S. Government Printing Office.

Park, D. C. (1980). Item and attribute storage of pictures and words in memory. American Journal of Psychology, 93, 603-615.

Robertson-Tchabo, E. A., Hausman, C. P., & Arenberg, D. (1976). A classical mnemonic for older learners: a trip that words! Educational Gerontology: An International Quarterly, 1, 215-226.

Salthouse, T. A. (1991). Theoretical Perspectives on Cognitive Aging. New Jersey. LEA.

Salthouse, T. A. & Mitchell, D. R. D. (1990). Effects of age and naturally occurring experience on spatial visualization performance. Developmental Psychology, 26, 845-854.

Salthouse, T. A., Mitchell, D. R. D., & Palmon,R. (1990). Memory and age differences in spatial manipulation ability. Psychology and Aging, 4, 480-486.

Tubi, N. & Calev, A. (1989). Verbal and visuospatial recall by younger and older subjects: use of matched tasks. Psychology and Aging, 4, 493-495.

Turner, M. L. & Engle, R. W. (1989). Is working memory capacity task dependent? Journal of Memory and Language, 28, 127-154.

Turner, M. L. & Pinkston, R. S.(1992). Effects of a memory and aging workshop on negative beliefs of memory loss in the elderly, Educational Gerontology, In Press.

AGING AND TECHNOLOGY: A DEVELOPMENTAL VIEW

J.L. Fozard

NIA Gerontology Research Center, Baltimore, MD 21224

Applications of technology to improve the living and working environment and medical care of aging and aged people define a newly developing discipline called gerontechnology. Both this field and the human factors applications to aging that are embedded in it require a developmental view of the relationship between a person and her/his environment. From a developmental viewpoint, technology can affect aging through prevention of chronic problems that limit mobility; enhancement of social activities, work, education and recreation, and compensation for impaired functioning. Integration of technology into the lives of aging persons reacquires a developmental approach to the design of products and environments, consumer involvement in design and significant changes in the infrastructure for technology development and dispersal.

AGING AND ERGONOMIC THEORY

Smith (1990) identified two broad models that have guided conceptual developments relative to aging in human factors and ergonomics theory. The first is a decrimental model (Faletti, 1984; Lawton and Nahemow, 1973) that emphasizes age associated decline in human sensory, cognitive and mobility function. The role of ergonomics is to improve the match between the declining human capacities and environmental demands; task analysis is the main tool used in determining how to achieve the match (Faletti, 1984; Clark, Czaja and Weber, 1990).

The second model emphasizes the total developmental process of aging which includes but is not limited to decline (Fozard, 1981; Fozard and Popkin, 1978). According to Smith (1990) the central ideas of Fozard's version of the developmental model would "...accommodate aspects of constancy and growth as well as decline; be sensitive to social and psychological needs as well as performance needs; and because both people and environments change, acknowledge the temporary nature of generalizations about aging." (p. 511). Smith points out that implementing this approach requires needs assessments, training and counseling as well as task analysis.

Both models are based on the systems theory that is central to human factors--the optimal utilization of human and machine capabilities to achieve best system performance. The implication of this view is that aging cannot be defined independently of the environment in which it occurs--age grading of human abilities only has meaning in reference to specified environmental challenges and supports. Accordingly the 1990 report of the National Research Council on Human Factors for an Aging population identified as the highest priority, the need for ergonomics data base on "...problems, tasks and abilities... so that task analyses can be performed where the benefit is likely to be the greatest..."

Adding Age to Ergonomic Analyses

Human factors analyses emphasize the reciprocal processes of a person receiving information from a machine and using that information to control the machine. The primary focus is on the interface between the person and the machine. Aging requires additional consideration of age differences in both the internal and external aspects of the environment that influence the interface. Important variations of the external environment include climate, lighting and acoustic factors-- all of which are age-sensitive. Important variations in the internal environment include age-related differences in organs and physiological systems that affect a person's performance, e.g., neural, cardiopulmonary, muscular-skeletal, hormonal, etc.

As indicated by Smith (1990), the developmental view adds a further dynamic to the person/environment analysis. Both technology and the environment of which it is a part are changing over time at the same time a person is aging; hence, personal aging and the epoch in time during which a person ages are interdependent. Thus, technology introduced at the present may be adapted to very differently by young and old people who have had different experiences, and, the technology in turn, may alter the course of aging itself for both the young and old.

Applications of the human factors analytic model just described sensitize the user to the heterogeneity of aging--a collection of universal processes that are certainly not uniform. Aging diminishes the similarity among coevals--people who are the same age by the calendar including identical twins. Differential exposure to disease, environmental pollutants, lifestyle choices, experiences with the built man-made environment--all combine with genetic differences to make each person's aging a very individual experience.

AGING AND GERONTECHNOLOGY

Gerontechnology is the multi disciplinary study of aging and technology for the benefit of a preferred living and working environment and adapted medical care of aging and aged people, and for care givers of elderly persons who cannot function without assistance. The term, gerontechnology, is a composite of gerontology, the scientific study of aging, and technology, the d evelopment and application of products, environments and services. The term was coined by Graafmans and Brouwers (1989), and has described in numerous publications including a book, e.g., Bouma and Graafmans (1992); Vercruyssen, et al (in press).

Many of the concepts and analytical tools of gerontechnology are the same as those used in human factors, particularly the developmental view of ergonomics described above and the application of the analysis of the person-environment interface. At the core of gerontechnology theory is the recognition of the heterogeneity of aging. As a discipline, gerontechnology is broader than ergonomics and aging inasmuch as it considers how technology oriented toward aging and aged people should be developed, dispersed and distributed (Bouma, 1992; Fozard, 1994). It also addresses how education and knowledge transfer in gerontechnology can best be accomplished (Vercruyssen et al, in press).

Research goals in gerontechnology

Research in gerontechnology is application driven and addresses the interaction of aging and aged persons with products and their technical or built environments. Consonant with the developmental view of aging described above, gerontechnology considers both the challenges and opportunities of normal and pathological human aging. "Challenges" refers to age-related declines in physical, physiological, perceptual, cognitive, and motor processes that may limit functioning during aging. "Opportunities" refer to such positive outcomes as increased time to pursue new activities in self-discovery, e.g., artistic activities, post-retirement or second career work activities; and relationships with grandchildren and others outside of the family.

A complementary goal is to understand how age affects the extent and pattern of use of technological devices, especially new ones. Older people may evaluate and adapt to new technology on the basis of previous experience with similar devices relatively more than younger people who have not had the same experiences.

The five goals of research in gerontechnology are to provide technology to: (1) improve the way in which aging is studied; (2) prevent the effects of declines in strength, flexibility endurance, perceptual and cognitively abilities that are associated commonly with aging; (3) enhance the performance of new roles (the opportunities) provided by aging; (4) compensate for declining capabilities (the challenges) of aging and (5) assist care givers.

1. *Improve aging research.* Examples of contributions of technology to improved research include computer-based imaging of organs and tissues, signal processing of neurological events, monitoring of blood flow, noninvasive acquisition of biochemical measures.

2. *Prevention.* Technology plays a role in the primary prevention of many 'problems of the elderly' that are modifiable through long range, nonmedical interventions involving nutrition, physical activity, strength training, behavior modification and life style remodeling which avoids exposure to chronically dangerous environmental conditions such as auditory noise, excessive alcohol and tobacco consumption, etc. The preventive role of technology includes the design of equipment to facilitate the interventions and the design of monitoring equipment that provides feedback about compliance with interventions and their effectiveness. Whether prevention is primary or secondary depends in part on the timing of the use of technology. For example, use of movement activated light switches near dangerous stairs or passages will prevent many falls from occurring. Where age-related declines have already occurred, monitoring devices may prevent additional problems.

3. *Enhancement.* Technology can enhance the performance of new roles (opportunities) provided by aging, including changes in work, leisure, living and social situations. Examples include adaptable housing that meets the differing needs of people during the life cycle of the family, user-friendly communication technology that remotely connect older persons with family and friends, technology that allows work at home, enhances the potential for artistic activities in music and art, etc.

4. *Compensation.* Examples of technology that compensate for declining abilities (challenges) of aging include products and techniques that offset the consequences of sensory losses, task redesign that speeds response time, and devices which can be operated with reduced strength and motor skill. This is the most developed area of gerontechnology.

5. *Care giver assistance.* While there a number of mechanical devices available for lifting and transporting persons who cannot move them selves, few are suitable for home use.Work related injuries particularly to the lower back occur very frequently in persons who lift and assist in the transport of less able elderly persons even when the helper has been trained in the proper techniques of lifting. The development of equipment suitable for home use deserves high priority.

Another significant development in home-based medical care is the use of complicated medical equipment for monitoring and administration of drugs by family and other nonprofessional care givers. Improvements in the ease and safety of use of existing equipment is a high priority for gerontechnological research and application.

Implementing gerontechnology

Turning gerontechnology's concepts into action require considerations of advocacy and the infrastructure for development, dispersal and distribution of technology.

Who are the advocates for gerontechnology--applications of technology directed toward aging and aged people? The heterogeneity of aging and the widespread ignoring or denying of the consequences of aging by most persons create a unique problem in advocacy.

Gerontechnology is differs from biomedical and environmental technology in the way people's interests are identified and used to shape the development of technology. In biomedical technology, physicians and scientists identify the needs of patient groups. People have a choice to accept or refuse treatments based on biomedical technology either through ethics committees or directly. In environmental technology, the needs of people are identified partly through public health , science, and public concerns for conservation of natural and human resources. Groups of people may benefit from noise reducing barriers between highways and residential areas, or reduction of air pollution. Working through elected public officials, people limit explication of natural resources and increase protection from environmental pollutants.

In gerontechnology, the needs of people are identified and articulated in two ways in the stream of events involved in technology development and dispersal. The first is feedback, the evaluations by people of available technology. Feedback requires an evaluation by users that goes beyond the success of sales, which is the most common criteria. To the extent that gerontechnology is successfully based on good scientific information about peoples needs, abilities and aspirations, feedback of such information to designers and manufacturers will also influence the development of technological application..

The second is feedforward, the identification of needs based on involvement by people in the developmental process itself--needs and preferences are identified by serving on focus groups or evaluators of prototypes. Examples of how these concepts are applied are given by Coleman and Pullinger (1993) and Pirkl (1994).

Changes in the infrastructure of the dispersal and distribution of technologically based products and services are required. In particular, client oriented approaches to the marketing of products are needed. Using a salon approach to marketing of lighting devices, vision enhancing devices and products such as television sets and computers are examples. The salon approach would be an alternative to the blend of warehouse and showroom that is emphasized in marketing today.

ACKNOWLEDGMENTS

Most of the concepts presented in this paper are shared intellectual property with many colleagues in Europe and the United States. I have tried to identify the major developers of theses ideas in the references, but the citation of references is a pale reflection of the discussions and arguments that are occurring in the development of the major concepts of gerontechnology.

REFERENCES

Bouma, H. (1992) Gerontechnology: Making technology relevant for the elderly. In H. Bouma, & J.A.M. Graafmans, (Eds) Gerontechnology. Amsterdam: IOS Press, pp. 2-5.

Bouma, H., & Graafmans, J.A.M. (Eds) (1992) Gerontechnology. Amsterdam: IOS Press.

Coleman, R., & Pullinger, D.J. (Special issue eds) (1993) Designing for our future selves. Applied Ergonomics, 24, 1-62.

Clark, M.C., Czaja, S.J., & Weber, R.A. (1990) Older adults and daily living task performance. Human Factors, 32, 537-549.

Faletti, M.V. (1984) Human factors research and functional environments for the aged. In A. Altman, M.P. Lawton, & J.F. Wohlwill (Eds), Elderly people and the environment. New York: Plenum, pp.191-237.

Fozard, J.L. (1981) Person-environment relationships in adulthood: Implications for human factors engineering. Human Factors, 23, 3-27.

Fozard, J.L. (1994). Future perspectives in gerontechnology. Report EUT/BMGT/94.689, Institute for Gerontechnology, Eindhoven University of Technology, PO Box 513, 5600MB Eindhoven, NL.

Fozard, J.L. & Popkin, S.J. (1978) Optimizing adult development: Ends and means of an applied psychology of aging. American Psychologist, 33, 975-989.

Graafmans, J.A.M., & Brouwers, A. (1989) Gerontechnology: The modeling of normal aging. Proceedings of the 33rd Annual Meeting of the Human Factors Society, Denver, CO.

Lawton, M.P. & Nahemow, L. (1973) Ecology and the aging process. In C. Eisdorfer & M.P. Lawton (Eds), The psychology of adult development and aging. Washington, D.: American Psychological Association, 1973.

National Research Council . (1990) Human Factors Research Needs for an Aging Population. Washington, D.C.: National Academy Press.

Pirkl, J.J. (1994) Transgenerational design: Products for an aging population. New York: Van Nostrand Reinhold.

Smith, D.B.D. (1990) Human factors and aging: An overview of research needs and application opportunities. Human Factors, 32, 509-526.

Vercruyssen, M., Graafmans, J.A.M., Fozard, J.L., Bouma, H., & Rietsema, J. (In press) Gerontechnology. In J.E. Birren (Ed), Encyclopedia of Gerontology. San Diego, CA: Academic Press.

MOUSE ACCELERATIONS AND PERFORMANCE OF OLDER COMPUTER USERS

Neff Walker, Jeff Millians & Aileen Worden
Georgia Institute of Technology
Atlanta, Georgia

ABSTRACT

In general, as people age, their movement control performance gets worse. Older adults take longer than younger adults to make similar movements. In this study we compared older and younger experienced computer users on their ability to use a mouse to position a cursor. The distance of the movements and the size of the targets were varied to represent a broad range of cursor control tasks that would be used on a computer. We also investigated the effects that dynamic gain adjustment had on performance for both age groups. Our results showed that older adults are both slower and less accurate when using the mouse. There was evidence that the age-related difference in performance was greater when the target size was smaller. Some of the difference in age-related performance could be ameliorated by using a specific dynamic gain function. The results are used to discuss possible age-related computer interface design guidelines

INTRODUCTION

There is large body of literature that shows that as people age, their movement control performance gets worse. Generally, older adults take longer than younger adults to make the same movements. While there are arguments about the specific factors and their relative importance as the source of this age-related decline in performance (see Jagacinski, Liao & Fayyad, 1995 for a discussion) almost all researchers believe that a major factor in the decline is due to an increase in motor noise with age (e.g., Welford, 1981).

Motor noise is defined as random, unintentional error that occurs during the transmission of the signal to the muscles that control the movement (e.g., Fitts, 1954). Motor noise is assumed to increase with increased force of movement (e.g., Schmidt, Zelasnik, Hawkins, Frank, & Quinn, 1978). The result is that degree of accuracy of the movement endpoint decreases as the force of the movement increases. This is the noise-to-force ratio. Older adults seem to have a larger noise-to-force ratio. In order to maintain the same level of accuracy in a movement task, they must slow down their movements.

It may be that older adults have more difficulty using a computer mouse due to their increased motor noise. In our previous work (Walker, Philbin & Fisk, unpublished) we found that older adults were much slower than younger adults when positioning a cursor with a mouse. There was also some evidence that there was a critical target size, below which the older adults had great difficulty in placing the cursor on the target. This may have been caused by the underlying motor noise. While the findings of our previous work were suggestive, the study did not use a wide range of target sizes nor did it have experienced mouse users. The primary purpose of this study was to determine if these findings would be obtained with experienced older mouse users.

A second purpose of the study was to investigate acceleration functions as a possible solution to this problem. One could implement software changes in the gain and acceleration profiles that translate mouse movement into cursor movement. In this study we evaluated the effects of different dynamic gain adjustments on older and younger computer users for a variety of cursor positioning tasks.

This study will seek to answer three questions. First, is there a critical minimum target size below which older adults cannot effectively use a mouse to position the cursor. Second, do different dynamic gain functions affect cursor positioning performance of older adults. Finally, do older adults have a larger noise-to-force ratio than younger adults and is the relationship between noise and force linear.

METHOD

Subjects

Twelve younger and twelve older adults participated in this experiment. The younger group consisted of undergraduates recruited from a Georgia Institute of Technology psychology course. Their ages ranged between 18 and 21 years (mean = 18.9). Five points of extra credit were given toward their course grade for participating in this experiment. The older adults were recruited from an organization that hosts enrichment classes for senior citizens. Their ages ranged between 63 and 79 (mean = 70.3). This group of subjects was paid $40 each for their participation. All subjects had a minimum of one year computer mouse experience. All subjects were informed that there was a $50 reward for the "best performer" in each group.

A drug questionnaire administered to both groups before the experiment revealed that none of the subjects were taking more than one drug rated to cause minimal effects on attention. All of the participants were required to have at least 20/40 natural or corrected vision.

Design

This study utilized a four factor, mixed design. The three within-subject variables were distance of movement to the target (50 pixels, 100 pixels, 200 pixels, 300 pixels, or 400 pixels), target width (3 pixels, 6 pixels 12 pixels or 24 pixels) and acceleration function (unaccelerated, low slope, medium slope and high slope). The corresponding slopes of these functions were 0.0, 0.1, 0.2, and 0.5 pixels to mouse movement ratio are shown in Figure 1. The between-subject variable was age (young and old). The two dependent measures were cursor positioning time and percentage of correct movement trials (hit rate).

Figure 1

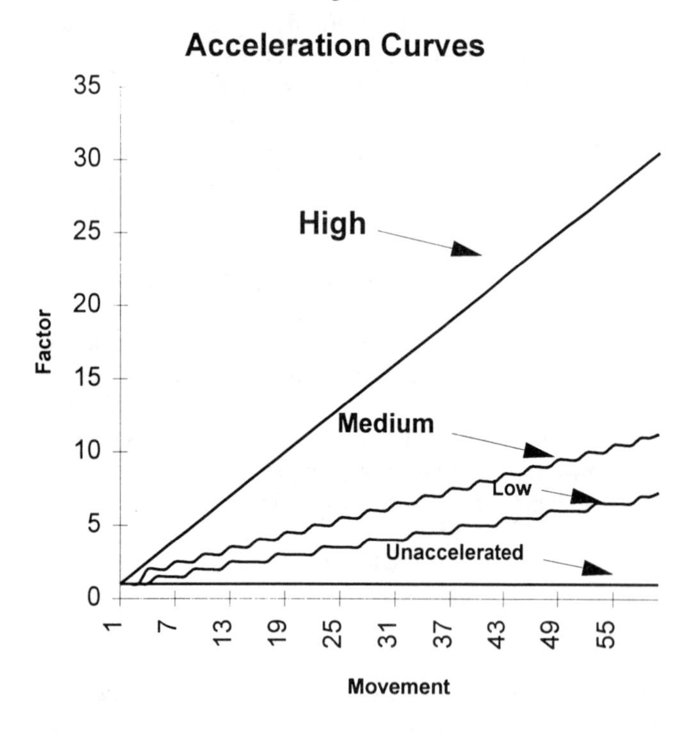

Apparatus

This study was run on PC's with 14 inch/color monitors. The input device was a Microsoft bus mouse that recorded mouse movements at 5 ms intervals. The program was run in graphics mode using 480 X 640 pixel resolution.

Procedure

The experiment was run over a maximum of 5 days. Each session lasted between 20 and 60 minutes. Some subjects participated in two session per day, but always had a 30 minute rest period between sessions. The first session was a training day. On this day, all of the subjects performed 3 correct trials with each of the distance x width x acceleration functions combinations for a total of 240 correct practice trials. For a correct trial, the subject clicked on the starting box which read "go" and moved the cursor across the screen to the target. When the target was reached a reverse video 10 X 10 pixel frame appeared and the participant released the mouse button. For an incorrect trial a beep signaled that the target was missed and the subjects repeated the trial.

Feedback for performance was given in terms of points after each trial and total points earned were given at the end of each session. Forty points were deducted for every incorrect trial. Points were also deducted or earned contingent upon the amount time it took to reach the target. The time-contingent points were based on established mean times for reaching targets. This established mean was based on previous studies that determined average times for reaching targets for young and old computer users. (Walker, et al., unpublished). Subjects that reached the target within 2.5% of the established mean did not receive or loose points. Deviations greater than 5% resulted in gained or lost points. One point was earned for every 5% deviation below the mean and one point was deducted for every 5% deviation above the mean. The point system was designed to reward speed and accuracy equally. The subjects were competing for a $50 reward that was given to the participant with the most points at the end of the experiment.

Each subject was tested on one acceleration function per session. They were presented with a different acceleration every session until all four accelerations were presented to each subject. Accelerations were assigned based on a Latin Square design to control for order effects.

Each session consisted of sixty blocks. There were five types of blocks corresponding to each target distance. There were four conditions within each block corresponding to each target width. Each type of block was repeated twelve times in random order resulting in 240 correct trials per day.

At the end of the last session participants performed 160 trials of a different task. They were instructed to aim for a target line and then to move the cursor toward the target as fast as possible and release the cursor as close as possible to the 1 pixel target without making any corrective movements. The emphasis was on the speed of the movement, with the only constraint being the aiming for the target. Subjects made 32 movements for each of the five movement distances. The purpose of this task was to generate a measure of the noise-to-force ratio. This was done by comparing mean maximum velocity of the movement (a measure of force) to the absolute value of the distance of the cursor from the target (a measure of error). The measures were based on the first submovement only.

RESULTS

The primary dependent measures that were analyzed were hit rate and mean movement time for correct trials. The mean movement time and hit rates were calculated for all trials that had the same acceleration function, movement distance, and target width for each subject. These resulting means were used in two age by distance by width by acceleration function analysis of variance.

The analysis of variance on mean movement time revealed eight significant effects. As expected, there were significant main effects of age ($F(1,22) = 36.88$, $p < .001$), movement distance ($F(4,88) = 210.34$, $p < .001$), and target width ($F(3,66) = 318.08$, $p < .001$). In addition, there was a significant main effect of acceleration function ($F(3,66) = 2.76$, $p < .05$. The interpretation of the main effects should be made in light of the significant interactions. There was a significant interaction of movement distance and age ($F(4,88) = 16.00$, $p < .001$). Follow-up analyses revealed that while younger adults always had shorter movement times than older adults, the size of this difference increased with movement distance.

There was a significant interaction of target width and age ($F(3,66) = 12.05$, $p < .001$). Follow-up analyses revealed that while younger adults always had shorter movement times than older adults, the difference due to age increased as target width decreased. There was also a significant interaction of movement distance and target width ($F(12,264) = 2.89$, $p < .001$).

Finally, there was a significant interaction of age by target width by acceleration function ($F(9,198) = 2.10$, $p < .05$). Follow-up analyses revealed that the locus of this interaction was in the effect that acceleration function had on movement time for the older adults when target width was smallest. For the older adults, movement time to the smallest targets was less when using acceleration function 2 (mean = 1773) than when using the other three acceleration functions (means for acceleration functions 1, 3, and 4 were 1882, 1870 and 1998, respectively). In general, there was a trend towards acceleration function 2 yielding shorter movement times for the older adults, especially when target width was less, while there was less of a difference in acceleration function for the younger adults.

The analysis of variance on hit rate revealed three significant main effects and a significant interaction. There was a significant main effect of acceleration function ($F(3,66) = 4.60$, $p < .01$). Pair-wise comparisons revealed that hit rate was significantly lower for acceleration function 3 than for acceleration function 4.

There was a significant main effect of age ($F(1,22) = 7.07$, $p < .05$). Younger adults had higher hit rates (mean = 97.2%) than did older adults (90.4%). In addition, there was a main effect of target width ($F(3,66) = 28.16$, $p < .001$). In general, hit rates went up as target width increased. However, the key to interpreting both of these main effects can be found in the interaction of age and target width ($F(3,66) = 9.67$, $p < .001$). Follow-up analyses revealed (and as is shown in Figure 2) hit rates for older adults was much lower when target width was smallest. Other than this condition, age and target width had little effect on hit rate.

Figure 2

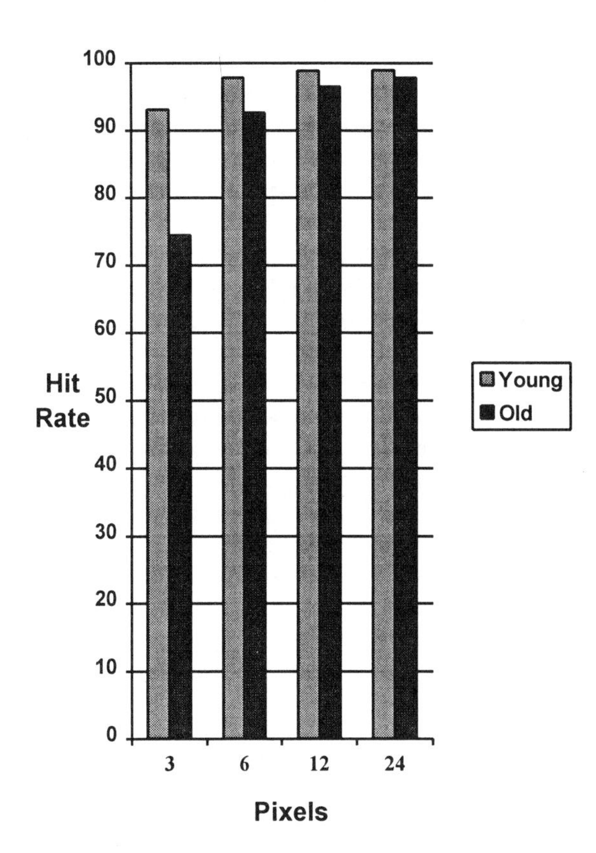

Finally, we ran regression analyses for both age groups using mean peak velocity as the predictor variable and mean standard deviation of the distance from the target as the outcome variable to test the underlying noise-to-force ratios of the two groups. Here only the data from the last 160 trials when there were no accuracy constraints were used. Regressions for both groups revealed significant equations (R^2 values of .66 and 0.62 for younger and older groups respectively). The key value was the slope functions as this gives the measure of how noise (standard deviation from the aimed for target) is affected by force (peak acceleration during movement). The slopes for the older adults were twice those of younger adults. This data supports the idea that age-related differences in the noise-to-force ratio does affect movement control performance.

DISCUSSION

A major finding of this study is that there is a critical minimum target size below which older adults cannot effectively use a mouse to position the cursor. In addition, the deleterious effects of small targets on accuracy increases significantly as distance to the target increases. Although hit rates for both age groups decreased as target size decreased a minimum critical target width and a distance-to-the-target / width-of-target interaction was not found for younger adults. These results suggest that even experienced older computer users have problems using a mouse to position a cursor.

The study also showed that different acceleration functions can be used to improve performance for older adults. When moving to smaller targets, older adults performed better with the second acceleration function. While the performance of the older adults was still worse than for the younger adults, the results suggest that adjustments of gain and acceleration may be a fruitful way to increase the usability of computer systems for older adults.

Finally, the study also showed that even experienced older computer users show a larger noise-to-force ratio when using a mouse to position a cursor. The overall results of this experiment can be used to evaluate possible solutions (e.g., use of area cursor, dynamic gain adjustment, sticky icons) to decrease the difficulty older computer users have in positioning a cursor and to begin to develop age-related interface design guidelines.

ACKNOWLEDGMENTS

This research was funded by a grant from the National Institute on Aging to the Southeastern Center for Applied Cognitive Aging Research. The center is one of the Edward R. Roybal Centers for Research on Applied Gerontology.

REFERENCES

Fitts, P. M. (1954). The information capacity of the human motor system in controlling the amplitude of movement. Journal of Experimental Psychology, 47, 381-391.

Jagacinski, R. J., Liao, M. J., & Fayyad, E. A. (1995). Generalized slowing in sinusoidal tracking by older adults. Psychology and Aging, 10, 8-19.

Schmidt, R. A., Zelasnik, H., Hawkins, B., Frank, J.S., & Quinn, J. T. (1979). Motor output variability: A theory for the accuracy of rapid motor acts. Psychological Review, 86, 415-451.

Walker, N., Philbin, D. A.., & Fisk, A. D. (unpublished manuscript). Age-related differences in movement control: Adjusting submovement structure to optimize performance.

Welford, A. T. (1981). Signal, noise, performance, and age. Human Factors, 23, 97-109.

The Processing of Synthetic Speech by Older and Younger Adults

Janan Al-Awar Smither
University of Central Florida
Orlando, Florida

This experiment investigated the demands synthetic speech places on short term memory by comparing performance of old and young adults on an ordinary short term memory task. Items presented were generated by a human speaker or by a text-to-speech computer synthesizer. Results were consistent with the idea that the comprehension of synthetic speech imposes increased resource demands on the short term memory system. Older subjects performed significantly more poorly than younger subjects, and both groups performed more poorly with synthetic than with human speech. Findings suggest that short term memory demands imposed by the processing of synthetic speech should be investigated further, particularly regarding the implementation of voice response systems in devices for the elderly.

INTRODUCTION

Advances in technology have led to the development of small, relatively inexpensive, programmable micro-chip devices that produce synthesized speech. Some current applications of synthesized speech include "extra features" on consumer products, sensory aids for the visually-impaired, feedback delivery systems for computer-aided instruction, and warning systems in appliances, cars, and aircraft cockpits. Despite its growing prevalence, however, little is known about how synthesized speech affects individuals of different ages. This paper considers the effects of synthesized speech on the performances of older and younger adults, two groups that are likely to differ in their ability to comprehend synthetic speech.

Some researchers (Luce, Feustel and Pisoni, 1983; Simpson and Navarro, 1987) have suggested that the perceptual encoding of synthetic speech requires more cognitive resources than natural speech. If so, then task performance under conditions of synthetic speech output would suffer when demands exceed cognitive resources available. Furthermore, because of age-related decrements in processing resources (Salthouse, 1988), performance decrements caused by synthetic speech are likely to be greater for older listeners than for younger.

The possibility that synthesized speech leads to a decline in performance is particularly important because synthetic speech may be used in technologies such as voice activated appliances designed to increase the functional independence of older adults. If, in fact, increased resource demands result in recall and performance deficits, elderly users may reject potentially beneficial voice response technologies. For this reason, it is important to understand the ability of older adults to recall information presented in synthetic speech.

Some experimenters (Luce et al., 1983, Waterworth and Holmes, 1986) have reported that recall of words or sentences is poorer when presented by a synthesized voice than when delivered by a human speaker. The synthetic speech used in these studies, however, was less intelligible than the natural speech, making it impossible to determine whether the recall deficits were due to memory problems or a failure to identify the stimuli correctly in the first place.

Luce and Pisoni (1983, experiment 2) examined memory for words presented by either a natural or synthetic voice under conditions where intelligibility was controlled. In this study, immediate serial recall of 10-word lists presented in either a synthetic or natural voice was compared. To equalize identification rates, they used only natural and synthetic words that could be accurately identified at a level of 98% or better. Under these more rigorous conditions, they replicated the finding of Luce et al. (1983, Experiment 3) that recall was significantly poorer when words were delivered by a synthetic, rather than human, voice.

Although this research suggests that performance with synthetic speech is poorer than with natural speech, and that perceptual encoding of synthetic speech requires greater resources, none of the studies included older adults as subjects. Research in memory and aging suggests that older adults seem to be relatively unimpaired on tasks tapping primary or short term memory unless these tasks require active manipulation of information or division of attention (Craik, 1977; Baddeley, 1989). Recent studies indicate that the age-related decrements in these tasks are due to the processing aspects of short term memory rather than the storage aspects (Hultsch and Dixon, 1990). It follows then that since the perceptual encoding of synthetic speech requires more cognitive resources than natural speech, and since the processing aspects of short term memory decline with age, the performance decrements caused by synthetic speech are likely to be greater for older listeners than for younger.

The purpose of this experiment was to investigate the demands placed on the short term memory systems of older subjects by the use of rule-based synthesized speech. The task chosen for this study was a typical short term memory serial recall paradigm. The paradigm used lists that approached, but did not greatly exceed, the span of immediate memory. Furthermore, the stimuli used were pretested to assure equivalent (and very high) intelligibility rates (correct identification rates of 100% for the natural speech, and 99.3% for the synthetic speech).

METHOD

Thirty-six volunteers participated in this study. The eighteen younger subjects were college-age students from introductory psychology classes at the University of Central Florida. The eighteen older subjects were seniors who were members of Premier Health, an organization for older adults at Florida Hospital in Orlando (Mean age=68.7 years). The subjects' task was to recall immediately, in the sequence in which they were presented, 8-digit strings of randomly ordered numbers. After each presentation, subjects were instructed to write down the number sequences in order and not to backtrack. A total of 78 strings were presented to each subject. The strings were divided into three blocks of 26 trials each. For half of the trials in each block, the digits were presented by a male human voice; the rest were presented by a text-to-speech synthetic voice generated by a VERT Plus add-on card (Telesensory Systems, Inc.) in a PC XT computer. Digits were presented at the rate of one per second, and the synthetic and human trials were interspersed randomly throughout the blocks.

RESULTS

The average number of correct responses for each subject at each serial position for the two groups was analyzed by a 2 (old, young) X 2 (human, synthetic) X 8 (serial position) analysis of variance. There was a significant main effect for age [$F(1,35)=89.25$; $p<.000$). Overall, younger subjects outperformed older subjects. There was also a significant main effect for voice type [$F(1,35)=8.66$; $p<.003$] and a significant main effect for serial position [$F(7,245)=26.82$; $p<.000$]. Overall, for both groups, memory was better for digits presented by the human voice than for those presented by the computer-generated voice, and performance was better on the first few serial positions and on the very last position. This performance reflects the typical finding that items in the early and final serial positions are recalled better than items in the middle, and also provides objective evidence of no hearing problems in the subjects tested.

The interaction for age by serial position was significant [$F(1,245)=3.63$; $p<.0008$] with younger subjects performing better on serial positions 4,5,6, and 7.

Subject performance was also evaluated in terms of the number of digit strings which were recalled in their entirety. The average number of correct strings for each subject over the three trials was analyzed by a 2 (young, old) X 3 (blocks) X 2 (voice type) analysis of variance. There was a significant main effect for age [$\underline{F}(1,35)=45.11$; $\underline{p}<.000$], a significant main effect for blocks [$\underline{F}(2,35)=4.01$; $\underline{p}<.019$], a significant main effect for voice type [$\underline{F}(1,35)=4.91$; $\underline{p}<.027$], but no interaction effects.

Overall, younger subjects performed better than older subjects; all subjects improved with practice over blocks; and all subjects performed better with human than with synthetic speech. The interaction effect of voice type and block did not approach significance ($\underline{F} < 1$), indicating that the performance difference between the natural and synthetic speech was not mitigated by general practice effects.

CONCLUSION

These results demonstrate that performance with highly intelligible synthetic speech is poorer for both younger and older adults. The results also indicate that although the processing of synthetic speech produces more demands on short term memory than natural speech, the performance decrement that occurs is not greater for older subjects. However, because older adults' performances are poorer to begin with, they are even poorer with synthetic speech, and, in practical terms, this may be very important.

Further research in the area of the processing of synthetic speech by older adults is needed, particularly regarding the implementation of voice response systems in devices for the elderly.

REFERENCES

Baddeley, A. (1989). Working Memory. Oxford University Press: New York.

Craik, F.I.M. (1977). Age differences in human memory. In J.E. Birren and W.K. Schaie (eds.), Handbook of the Psychology of Aging (pp. 384-420). Von Nostrand Reinhold: New York.

Hultsch, D.F., and Dixon, R.A. (1990). Learning and memory in aging. In J.E. Birren and W.K. Schaie (eds.), Handbook of the Psychology of Aging, 3rd edition (pp. 258-274). Academic Press, Inc.: San Diego.

Luce, P. A., and Pisoni, D. B. (1983). Capacity- demanding encoding of synthetic speech in serial-ordered recall. Research on Speech Perception, No. 9, Speech Research Laboratory, Indiana University, Bloomington, IN.

Luce, P. A., Feustel, T. C., and Pisoni, D. B. (1983). Capacity demands in short-term memory for synthetic and natural word lists. Human Factors, 25, 17-32.

Salthouse, T. A. (1988). The role of processing resources in cognitive aging. In M.L. Howe and C.J. Brainerd (eds), Cognitive development in adulthood: Progress in cognitive development research (pp. 185-239). New York: Springer-Verlag.

Simpson, C. A., and Navarro, T. N. (1984). Intelligibility of computer generated speech as a function of multiple factors. Proceedings of the National Aerospace and Electronics Conference (84Ch1984-7 NAECON). New York: IEEE.

Waterworth, J. A., and Holmes, W. J. (1986). Understanding machine speech. Current Psychological Research and Reviews, 5, 228-245.

UNDERSTANDING TIME-COMPRESSED SPEECH: THE EFFECTS OF AGE AND NATIVE LANGUAGE ON THE PERCEPTION OF AUDIOTEXT AND MENUS

Jenny DeGroot
Department of Psychology
University of Chicago
5848 S. University Avenue
Chicago, Illinois USA 60637

and

Eileen C. Schwab
Ameritech Services - 2C44
2000 Ameritech Center Drive
Hoffman Estates, Illinois USA 60196-1025
E-mail: schwab@hopper.center.il.ameritech.com

Time compression increases the rate of speech without altering its pitch. The present study investigated time compression as a means of improving the efficiency of audiotext applications for a variety of user populations. Subjects from three age groups (20-30, 40-50, and 60-70 years old) and two native language groups (native and nonnative English speakers) interacted with a prototype of an Interactive Voice Response system. Four prototypes were constructed, each containing speech compressed at a different rate: 30%, 20%, 10%, and uncompressed. Each subject telephoned one of the prototypes to learn how to use Call Forwarding and to order another telephone service feature. Compression rate did not significantly interact with age or native language. Across compression rates, 60-year-olds spent significantly more time on the phone than did 20- and 40-year-olds. Moreover, 60-year-olds were significantly less successful at forwarding phone calls, and reported more difficulty and confusion, than other subjects. Nonnative English speakers spent significantly more time on the phone than did native English speakers. Despite this difference, nonnative speakers were just as successful at forwarding phone calls, and rated the system and the announcer just as favorably as did native speakers of English. There was no main effect of compression rate on call duration; faster speech did not result in significantly shorter phone calls.

Time-compressed speech is faster than unaltered speech, but its pitch is the same (see Arons, 1992 for a review of compression techniques). Previous reports indicate that speech can be compressed by as much as 50% with little or no loss of comprehension (see Foulke, 1971 for a review). In fact, compressed radio commercials are often more engaging for consumers than their non-compressed versions (LaBarbera & MacLachlan, 1979). This study investigates the advantages and disadvantages of compressing speech in audiotext applications for different populations. Two potential advantages are (a) reducing the time required to provide information, thereby saving both customer and service-provider time and resources, and (b) enhancing customers' satisfaction with audiotext applications. A potential disadvantage is that too much compression may sound unpleasant or decrease comprehension. These effects may be more extreme for older customers, or for those who speak English as a second language. Moreover, compression might have different effects on the comprehension of long expository passages and of briefer items such as menu choices.

To investigate these issues, callers of different ages and language backgrounds interacted with time-compressed versions of Ameritech's Rapid Order® and QuickTeach® service. The service is an interactive voice response (IVR) system that provides recorded information about custom calling features and takes orders for features. The IVR contains several menus. For example, at the main menu, users select the ordering function or the information function, and at a sub-menu they select the custom calling feature that they want to order or learn about. During the ordering dialog, other menus include options such as confirming or canceling the order. The informational message about each feature is a one- to two-minute statement of its benefits and operation instructions. In this study, menu navigation, comprehension of instructions, and subjective evaluations were examined.

METHOD

Subjects

The subjects were 192 consumers, drawn equally from three age groups (20-30, 40-50, and 60-70 years old) and two language groups (native and nonnative speakers of English). A variety of native languages were represented among the nonnative English speakers, including Spanish, Tagalog, Italian, Gujarati (spoken in India), German, and Polish. Participants had no apparent difficulty communicating with the experimenter in English. They were paid $25 for a 20-60 minute session.

Materials

Four prototypes of the IVR were constructed, each containing speech compressed at a different rate: 0% (the original female voice-talent recordings, which varied from approximately 116 to 176 words per minute), 10%, 20%, and 30%. The four prototypes were otherwise identical.

Two questionnaires were constructed, with seven-point scale items and multiple-choice questions focusing on user opinions of the announcer, the recordings, and the IVR. One questionnaire included questions about the ordering function, and the other included questions about the information function.

Procedure

Subjects were told they would be testing a new toll-free service, which would allow callers to place orders for telephone service features or get information on features. They were initially given one of two assignments: (a) to place an order for Three-Way Calling (the ordering task), or (b) to learn how to use Call Forwarding (CF) and then forward the calls of a phone in the room (the information task). Subjects phoned the prototype and attempted to complete the assignment.

If a subject did not respond within eight seconds of the end of a prompt (i.e., a time-out), the prompt was repeated. As in the actual IVR, menus included a "repeat the menu" option. Some menu choices were disabled in the prototype; for example, features other than CF were not described in the information dialog. If subjects selected a disabled feature (i.e., a "wrong" keypress), they heard a message saying that the selection was not available and that they should select another, and the menu was repeated.

During calls, the start time and identity of each prompt and each keypress were logged automatically. During the information task, subjects' success at forwarding calls was recorded by the experimenter. After their first task, subjects completed the appropriate questionnaire. Each subject then completed the other task, followed by the other questionnaire. Task order was counterbalanced. Each subject heard one speech rate throughout both tasks.

RESULTS

Three classes of data were analyzed: (a) menu navigation data, (b) comprehension data, and (c) subjective evaluations. Menu navigation data included measures of subjects' efficiency in completing calls to the system. Comprehension was reflected in subjects' ability to forward calls after listening to the recorded instructions. Subjective evaluations were subjects' questionnaire responses. Within each of these three data sets, the effects of task order, compression rate, subject's age, and subject's native language were analyzed.

Menu Navigation

Three measures of navigability were calculated from the data logged during calls: call duration, number of prompts, and initial response time (RT). For each task, call duration was the time spent on the phone call to the prototype, ending when the subject hung up. If a subject made more than one call during a task, this measure included all of the time on the telephone and none of the time between calls. Number of prompts was a count of all prompts played during this duration. A prompt was played after each correct keypress, time-out, "repeat the menu" keypress, or wrong keypress. There was a fixed number of correct keypresses during each task, so variation in number of prompts was determined by occurrences of the other three responses. Initial RT was the time it took each subject to respond to the system's first prompt, which was, "If you are calling from a Touch-Tone phone, please press '1' now." Initial RT was measured from the end of the prompt, for each subject's first and second phone call.

Subjects who did not complete a task were omitted from the analysis of call duration and number of prompts for that task. Five subjects were excluded for this reason from the ordering task analysis only, three from the information task analysis only, and three from both. Nine of the omitted subjects were in the 60-70 year age group. Eight were nonnative English speakers. Five omitted subjects heard the 20% compression rate, and two heard each of the other rates. The initial RT analysis omitted six subjects who did not press "1" on the first try on both calls. These included four nonnative English speakers, four 60-year-olds, and one or two subjects at each compression rate.

Analysis of variance was used to examine the effects of task order, compression rate, age, and language on these measures. There were no significant two-, three-, or four-way interactions on any measure of navigability.

Task Order. There was a significant effect of task order on call duration for both the ordering task ($F(1, 134) = 20.36$, $p < .01$) and the information task ($F(1, 136) = 7.21$, $p < .01$). Each task took longer for those subjects who performed it as their first task, compared to those who performed it second. In addition, more prompts were played during the ordering task to the subjects who performed that task first, compared to those who performed it second, $F(1, 134) = 11.29$, $p < .01$. Moreover, initial RT was faster on subjects' second call, $F(1, 162) = 8.21$, $p < .01$. These findings suggest that experience with the speech and the menus facilitates system navigation.

Compression Rate. There was no main effect of compression rate on call duration for either task. The fact that compression did not shorten calls indicates that the time saved by compressing speech is offset by some combination of (a) increased RTs, (b) more time-outs, (c) more wrong keypresses, and (d) more "repeat the menu" keypresses.

In fact, the effect of compression rate on number of prompts did not reach significance, although the 30% and 20% compression rates were associated with the highest and second-highest numbers of prompts, respectively. The failure to reach significance may be due to within-group variation. Moreover, initial RT did not differ with compression rate. It is not straightforward to determine whether RTs to other prompts and menus were affected by compression rate, because of individual differences in navigating the system: Some subjects heard a given prompt only after many wrong keypresses and therefore many repeated menus. These subjects had more experience listening to compressed speech than did subjects who reached the same prompt more directly. Therefore, the effect of compression rate on RTs to later prompts cannot easily be gauged. In the end, although the present analyses do not provide evidence for a significant increase in RTs or number of prompts, it stands to reason that these factors or others are acting to offset the time saved by compressing announcements. If they were not, then call duration would be reduced by speech compression.

Age. The main effect of age on call duration was significant for both the ordering task ($F(2, 158) = 13.56$, $p < .01$) and the information task ($F(2, 160) = 17.08$, $p < .01$). On both tasks, 60-year-olds spent more time on the phone than 20- and 40-year olds. Correspondingly, there was a significant effect of age on number of prompts played in both tasks (ordering: $F(2, 158) = 6.86$, $p < .01$; information: $F(2, 160) = 23.95$, $p < .01$). Again, 60-year-olds heard more prompts than the younger subjects. Finally, initial RT was significantly affected by age ($F(2, 162) = 4.54$, $p = .012$), with the 20-year-olds responding more quickly than the 40- and 60-year-olds.

Language. Nonnative speakers of English spent significantly more time on calls than did native speakers (ordering task: $F(1, 158) = 23.51$, $p < .01$; information task: $F(1, 160)=5.93$, $p=.016$). They also heard more prompts than native speakers during the ordering task ($F(1, 158) = 7.99$, $p < .01$) and were slower to respond to the first prompt ($F(1, 162) = 6.38$, $p = .013$).

Comprehension

Comprehension was measured by observing whether subjects could forward phone calls after calling the prototype. Chi-square analysis showed a significant relationship between age and comprehension, $\chi^2_4 = 30.03$, $p < .001$. Sixty-year-olds were less successful than younger subjects at forwarding phone calls.

Chi-square analyses showed no significant relationship between compression rate and comprehension, between language and comprehension, or between task order and comprehension. Thus, while age and native language both affect system navigation, only age affects comprehension of CF instructions. Nonnative speakers of English comprehend the CF instructions as well as native speakers do.

The absence of a significant relationship between compression rate and either comprehension or navigation indicates that a decision to time-compress announcements can be guided largely by listener preferences. If listeners find a particular speech rate more interesting or less confusing than other rates, these results suggest that their preferences can be followed without affecting network load (represented here by call duration) or information transfer (as measured here by CF success).

Subjective Evaluations

Correlational analyses were conducted to examine relations among the seven-point scale items on the questionnaires. Several sets of highly intercorrelated items were identified. Responses in each set were averaged to yield composite measures, which were then examined with analysis of variance. Chi-square analysis was used to examine responses to multiple-choice items.

Task order. Subjects who placed an order after completing the information task found the ordering process easier than those who placed the order first, $F(1, 143) = 10.35, p < .01$. Practice and familiarity account for this finding. There was no significant main effect of task order on any other subjective measures.

Compression Rate. The more highly compressed announcements were more often rated as too fast or much too fast (ordering task: $\chi^2_6 = 58.18, p < .001$; information task: $\chi^2_6 = 39.87, p < .001$). There was also a significant effect of compression on ratings of the announcer, $F(3, 144) = 21.10, p < .01$. Subjects hearing 30% compressed speech gave less favorable ratings to the announcer than subjects hearing 0% or 10% compression.

Compression rate was not significantly related to any of the other subjective measures. Specifically, contrary to earlier findings on compressed speech in commercials, reported interest in the messages was not affected by speech rate. Subjects' reported impressions of CF were not affected by the compression rate of the CF feature description. Reported IVR clarity and ease of use were also not affected by rate.

Age. Several subjective measures were significantly affected by age. Sixty-year-olds reported more confusion and difficulty obtaining information than did younger subjects, $F(2, 142) = 9.92, p < .01$. Correspondingly, older subjects reported they would choose to get information from IVRs less often than did 20- or 40-year-olds, $\chi^2_6 = 15.49, p < .02$.

The effect of age on ratings of the announcer was significant as well, $F(2, 144) = 4.83, p < .01$. Pairwise comparisons showed that 60-year-olds gave higher ratings than 40-year-olds; 20- and 40-year-olds did not differ with each other. Age also significantly affected the reported interest level of the announcements, $F(2, 163) = 5.13, p < .01$. Sixty-year-olds rated them significantly more interesting than did 20-year-olds.

An item on which 60-year-olds were less favorable than other age groups was the rating of the length of the information message. Sixty-year-olds gave fewer "just right" responses than other subjects, $\chi^2_4 = 12.94, p < .02$. However, they were fairly evenly split between "(much) too short" and "(much) too long" responses. There was not a significant relationship between age and ratings of message speed.

Language. Native language was significantly related to subjects' preferred means of placing orders, $\chi^2_2 = 7.46, p < .05$. As compared to native English speakers, nonnative speakers more often reported a preference for using an IVR rather than speaking with a person over the phone. This may be due to prior difficulty with telephone conversations in their second language. There was no significant effect of native language on rated message speed, rated task difficulty, or any other subjective judgments. The fact that nonnative speakers took longer to place orders, but still rated the system as highly as the native speakers, may be due to their expecting some difficulty with any English interaction.

DISCUSSION

Age was the most consistent predictor of navigation ability, comprehension, and attitudes. Compared to younger subjects, 60-year-olds spent more time on the phone, were less successful at forwarding calls, and reported different attitudes toward the announcer, the system, and IVRs in general. These effects were consistent across compression rates. Even when speech was not compressed, the gap between younger and older subjects was wide. It appears that menus and longer text would have to be revised, and possibly the original speaking rate slowed, before 60-year-olds would be able to use this system to get feature information efficiently. Simplifying the text would probably make the system easier for younger callers to use as well. A slower speaking rate, however, might make the system needlessly slow for younger callers, who are the more likely users of such an IVR.

The effects of language background were less pronounced than those of age. Nonnative English speakers did not navigate the menus as efficiently as native speakers, but the two groups were equally successful at applying the CF instructions. Furthermore, the nonnative speakers' preference for IVRs was stronger than that of the native speakers. These effects did not vary with compression rate.

Compressing announcements does not reduce the time spent using the IVR, for any of the subject groups studied. The time saved by compression appears to be offset by slower responses and more keypresses to the compressed prompts, or by some other factor. On the other hand, time compression had few negative effects. Subjects' success at forwarding calls was the same, no matter what compression rate they heard. Moreover, compression rate did not significantly influence subjective judgments about Call Forwarding, the IVR, or the attractiveness of IVRs for placing orders or getting information. Even for young native English speakers, speech compression was not beneficial.

Although speech compression apparently does not save time in the one-time use of an IVR, compression would potentially save time in network announcements that do not require a user response. In those cases, call duration would not be affected by subject response times, so 10% compression would directly reduce call duration by 10%. At a 10% compression rate, listeners would still understand the announcements and would like the announcer just as much as in the non-compressed version.

The effect of task order suggests that familiarity with the system and the recordings facilitates navigation. It is thus possible that compression of a very familiar menu sequence would save time; highly practiced users might respond to compressed prompts as quickly and accurately as they respond to non-compressed prompts. As task order did not interact with age or native language, it appears that users of all types can benefit from practice.

REFERENCES

Arons, B. (1992). Techniques, perception, and applications of time-compressed speech. Proceedings of the 1992 Conference of the American Voice I/O Society, 169-177.

Foulke, E. (1971). The perception of time compressed speech. In D. L. Horton & J. J. Jenkins (Eds.), The perception of language (pp. 69-107). Columbus, OH: Charles E. Merrill.

LaBarbera, P., & MacLachlan, J. (1979). Time-compressed speech in radio advertising. Journal of Marketing, 43, 30-36.

ACCESSIBLE REMOTE CONTROLS FOR OLDER ADULTS WITH MILDLY IMPAIRED VISION

Juli J. Lin and Robert C. Williges
Virginia Polytechnic Institute and State University
Blacksburg, Virginia

Douglas B. Beaudet
Eastman Kodak Company
Rochester, New York

A three-phase methodology was used to design an accessible photo CD player for older adults with mildly impaired vision. During Phase I of the study, critical barriers to a photo CD player were identified that prevent older users with presbyopia from operating this product. These barriers included small Remote Control Unit (RCU) labeling, low label-background contrast, and inadequate feedback from the player system. During Phase II, cost-effective solutions were identified through research into existing literature and available technologies. Phase III evaluated the efficacy of these design modifications on both the accessibility and usability of the photo CD player. The results of the Phase III empirical study indicated that enlarging a RCU and using high contrast labeling significantly improved accessibility. Overall, these results support the use of such a three-phase methodology to design accessible consumer products for users with special needs.

INTRODUCTION

Visual impairment is one of the most prevalent conditions in adults age 65 and over. According to Berg and Cassells (1990), if one defines poor vision as 20/50 acuity or worse, 11% of the adults aged 65 to 73 who wear glasses are impaired and 26% of the adults aged 65 to 73 who do not wear glasses are also impaired. Interestingly, studies have reported that older adults as a group spend more time watching television than any other daily activity. Moss and Lawton (1982) examined the daily activity pattern of older adults across four lifestyles and found that television viewing was the most time-consuming activity for older adults next to sleeping. Czaja (1990) further indicated that many older adults spend as much as six hours a day watching television.

Despite the amount of time older adults spend watching television, studies have consistently cited the Remote Control Unit (RCU) as one of the products receiving very little attention from the human factors community (Parsons, Terner, and Kearsley, 1994). Users with visual impairments tend to complain about the small controls and the faint lettering of the RCU, while users with physical disabilities tend to praise the convenience of such devices and hope for more of such applications in consumer products (Phillips, 1990; Ward, 1990; 1991).

Currently only a few design guidelines exist in the literature to assist in designing consumer products for users with special needs (Pirkl and Babic, 1988; Vanderheiden and Vanderheiden, 1991). A better approach would be to provide product designers and human factors professionals with a systematic design process based on empirical testing (Williges, Williges, and Elkerton, 1987; Hix and Hartson, 1993). A three-phase methodology for designing accessible consumer products for individuals with special needs was created using basic human factors design approaches. This method allows designers to: (1) identify accessibility barriers on a product for users with special needs; (2) select and implement design solutions for eliminating or mitigating these barriers; and (3) evaluate empirically the accessibility and usability of design alternatives for the target user population before product production.

The study employed a three-phase methodology to design an accessible photo CD player for older adults with mildly impaired vision. The photo CD player is a relatively new consumer product capable of displaying digitized images on a television. Because of its capabilities, a photo CD player may provide individuals with visual impairments with an access to photos that were not viewable before. However, until such products are designed for accessibility, individuals with visual impairments will continue to be excluded from enjoying the potential benefits such technology can provide.

THREE-PHASE METHODOLOGY

Phase I

During Phase I of the study, a usability test was conducted to identify accessibility barriers to a photo CD player for older users with presbyopia (age-related loss of near point acuity).

Method. Seven males and five females between the age of 50 to 72 participated in the study. All participants had uncorrected presbyopia in the 20/40 to 20/67 range. An accessibility and usability evaluation of the photo CD player was conducted. Methods of data collection used include: verbal protocol, critical incidents and subjective Accessibility/Usability (AU) ratings.

Results. The primary accessibility barriers to a photo CD player for older adults with presbyopia were associated with the RCU that came with the product. The extent to which the RCU was a barrier was most apparent for older adults with presbyopia of 20/50 or worse. Problems with the readability of labels on the RCU were most commonly cited by the participants. Ten out of twelve participants reported difficulties using the RCU due to the size of the labeling and the low level of contrast between the labeling and the background. Participants with near vision worse than 20/50 assigned a significantly lower AU rating to the RCU labels (mean = 2.29; on a seven-point scale where 1 = low AU and 7 = high AU) than participants with better vision (mean = 5.31).

The inadequacy of the system feedback was another important accessibility barrier. Because critical feedback from the player system was primarily visual, eight participants failed repeatedly to perceive visual feedback on the photo CD player and became confused while operating the system. These participants also assigned a significantly lower AU rating to the system feedback (mean = 3.20) than participants with vision better than 20/50 (mean = 6.15). Although the color of the RCU was not an accessibility barrier, females in the study rated the pastel-colored RCU significantly more favorably (mean = 4.65) than their male counterparts (mean = 3.036).

Phase II

Based on Phase I results, three design solutions to improve accessibility of the photo CD player were proposed in Phase II: (1) improving the readability of labeling with enlarged label/RCU size; (2) enhancing label-background contrast with high contrast labels; and (3) augmenting the feedback system with voice confirmations.

Method. Potential solutions were gleamed from existing human factors design guidelines, recommendations from empirical studies, and available technologies. Selection of the best achievable design largely depended on the feasibility and the relative cost-effectiveness of each alternative. Solutions that were effective in removing barriers while minimizing costs (in time and actual dollars) were given greater consideration for implementation.

Results. The appropriate character stroke width and height were calculated for individuals with 20/67 near vision to read at a distance up to 30-inches. Results of the calculation indicated that a minimum of 15-point bold-face font is required. Since the current typeface on the RCU labeling is a sans serif font, as recommended in the literature, the proposed character design also employed this typeface. To ensure a uniform enlargement of labels while retaining as much of the original RCU design as possible, the entire RCU was enlarged proportionally in width and height to achieve the required 15-point bold-face.

To enhance the labeling contrast against the pastel background of the RCU, a black labeling was proposed to replace the current pale-gray labeling. A "talking remote control" design, similar to the concept of the "talking telephone" used by Nakatani and O'Conner (1980), was proposed for the augmented feedback system. The proposed RCU redesign speaks the names of the control buttons as they were pressed. This voice feedback system used quality digitized voice recordings to achieve the greatest legibility as recommended by Halstead-Nussloch (1989). In summary, the resulting RCU redesign consisted of a "talking" RCU with high contrast labeling that was two and one half times larger in width and height than the original design.

Phase III

An empirical study was conducted in Phase III to evaluate the effects of Phase II design modifications. These design modifications can be viewed as improving the RCU along three dimensions: A dimension of labeling/RCU size, a dimension of label-backgound contrast, and a dimension of feedback system. In addition to evaluating the effects of these design modifications, the following questions were also addressed:

1. Since women comprise 60% of population over 65, are there gender differences in using a consumer electronics product?
2. Since conflicting results on the effects of voice confirmation on user performance and preference have been found in the literature, does the inclusion of a voice feedback system in a consumer electronics product improve user performance and preference?
3. Does improving the accessibility of a product also lead to improving its usability?

Method. Thirty-two adults with presbyopic vision in the 20/29 - 20/100 range participated in the study. Participants were of age 50 to 75, and half were males and the other half were females. A 2^4 mixed-factor experimental design was employed to investigate the effects of gender, RCU size (large or small), label-background contrast (high or low), and voice feedback (present or absent). The RCU size was the within-subject factor, while gender, label-background contrast, and voice feedback were the between-subject variables. The order of presentation method (large or small RCU) was counterbalanced to control for the effect of learning.

The experimental treatment combinations of the RCU dimensions resulted in eight RCU design alternatives. These design alternatives were implemented on a Macintosh 840AV computer using a multimedia software package (Macromedia Director, Version 4.0). Instructions, training materials, and both the accessibility and usability evaluation tasks were also implemented using these tools. A *Wizard of Oz* simulation technique pioneered by Gould, Conti, and Hovanyecz (1983) was employed to simulate the operation of a photo CD player by eight RCU software prototypes. The *Wizard of Oz* setup involved having the experimenter operate the photo CD player remotely in response to participants' selections on the simulated RCU.

Each participant completed both the accessibility and usability evaluation tasks. The accessibility evaluation was conducted to evaluate the effects of the RCU design modifications on the selection of buttons on the RCU, while the purpose of the usability assessment was to evaluate the effects of the redesigns on the use of the RCU in operating the photo CD player.

The accessibility evaluation tasks involved having the participants "search and press" ten control buttons on the RCU according to digitized voice prompts. The task response time to ten search tasks and subjective accessibility ratings were collected to evaluate the accessibility of each RCU design alternative. The total task completion time for six photo manipulation tasks and subjective usability ratings were obtained to determine the usability of each RCU design alterative.

Results. An analysis of variance (ANOVA), using gender, the RCU size, the label-background contrast, and the feedback system as main effects, was used to evaluate both the performance and preference data. Consistent with findings reported by Mann, Ottenbacher, Tomita, and Packard (1994), this study found that enlarging a RCU and using a high contrast labeling significantly improved older users' performance and preference ratings. Furthermore, results of this study also support the hypothesis that improving accessibility of a product also leads to improving the usability of the product.

Both the RCU size and the label-backgound contrast had significant differential effects on the accessibility task response time ($F_{(1.24)} = 47.720$, $p<0.001$; and $F_{(1,24)} = 7.554$, $p<0.001$ respectively). As shown in Figure 1, participants took significantly less time under the large RCU condition (mean = 107.074 seconds) to perform the accessibility tasks than under the small RCU condition (mean = 219.783 seconds). Participants who received the high label-background contrast condition completed the accessibility tasks in a significantly shorter amount of time (mean = 118.896 seconds) than participants who received the low label-background contrast condition (mean = 207.961 seconds).

As depicted in Figure 2, under the large RCU condition, participants took significantly less time to complete the six usability tasks than the same tasks performed under the small RCU condition ($F_{(1,24)} = 7.051$, $p<0.015$). Participants who received the high label-background contrast condition also took significantly less time to complete the tasks than participants who received the low label-background contrast condition ($F_{(1,24)} = 6.003$, $p<0.025$).

Further support for an enlarged RCU with high contrasted labeling was provided by users' accessibility and usability ratings as shown in Figures 3 and 4, respectively. Overall, participants rated both the accessibility and usability of the player system significantly higher under the large RCU condition ($F_{(1,24)} = 162.692$, $p<0.001$). Participants who received the high label-background contrast condition also

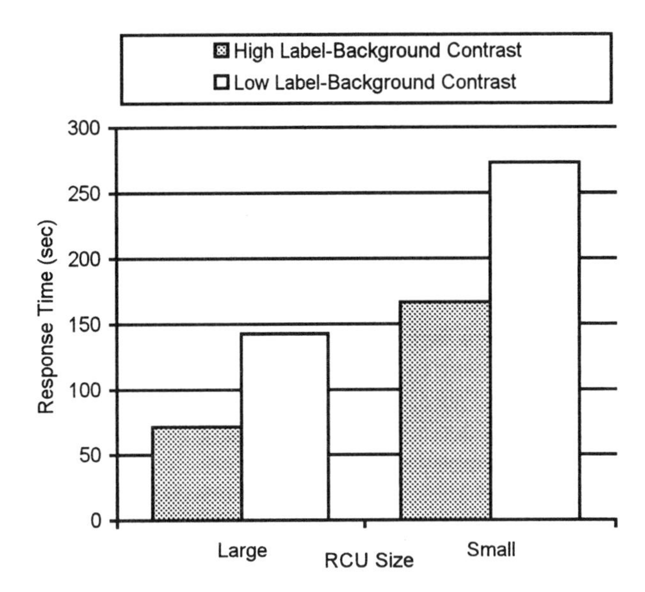

Figure 1. Mean accessibility task response time as a function of the RCU size and the label-background contrast.

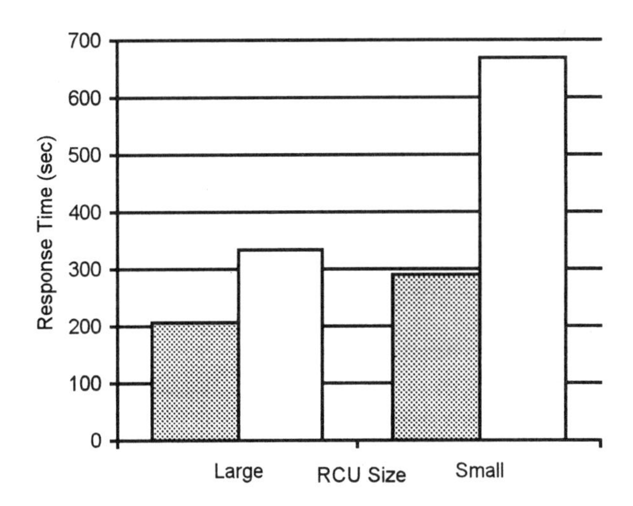

Figure 2. Mean usability task completion time as a function of the RCU size and the label-background contrast.

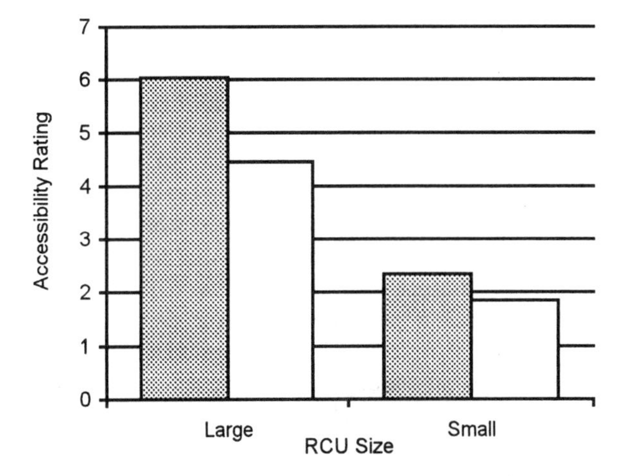

Figure 3. Average accessibility ratings as a function of the RCU size and the label-background contrast.

rated both the accessibility and usability of the player system significantly higher than participants who received the low label-background contrast condition ($F(1,24) = 6.743$, $p<0.02$).

Although the application of voice confirmations in a RCU did not significantly ($p>0.05$) improve users' performance in the present study, it was generally well received. No significant gender difference in the performance or preference ratings was observed. This indicates that both male and female users are equally proficient in operating consumer electronics products.

A post-hoc analysis of users' near visual acuity revealed that near vision had a significant effect on the participants' performance in the accessibility tasks ($F(2,20) = 10.0187$, $p<0.001$), and their accessibility ratings ($F(2,20) = 8.71$, $p<0.002$). Although participants across the vision groups did not differ significantly in their usability task performance, participants in the 20/67 - 20/100 vision group rated the usability of the photo CD player significantly less favorably than participants in the other two vision groups ($F(2,20) = 5.907$, $p<0.001$). As depicted in Figures 5 and 6, participants with better vision had superior performance and rated the accessibility of the Photo CD player more favorably than participants with lower vision.

DISCUSSION AND CONCLUSIONS

Several major findings from the data collected from Phase I and Phase III of this study have implications for the design of products used by older adults with mild visual impairments. The differential preferences for the colors on the RCU between gender obtained in Phase I of the study have important design implications, particularly if the products are to be used by older adults. Since females comprise 60% of older adults in the population, special care in the color selection must be taken whenever colors are used in a product. It is highly recommended that an iterative design process, such as the three-phase methodology presented in this study, be conducted before any decision for using a set of colors is made. Similarly, the strong user preference for the inclusion of voice confirmations on a RCU found in Phase III also have important design implications. However, given inconclusive results reported in the literature, the use of voice confirmations in consumer products warrants further empirical explorations.

Although the notion that products that are made more accessible to individuals with special needs usually also become easier for the able-bodied users is commonly cited in the literature, it is possible that products that are accessible for one group of users bar another from accessing the product. Therefore, such notions should not be accepted to apply in all cases but should be tested empirically. The effects of the RCU design modifications recommended for older adults in this study should be evaluated for the general user population before they are implemented in an actual product that is designed for the mass-market.

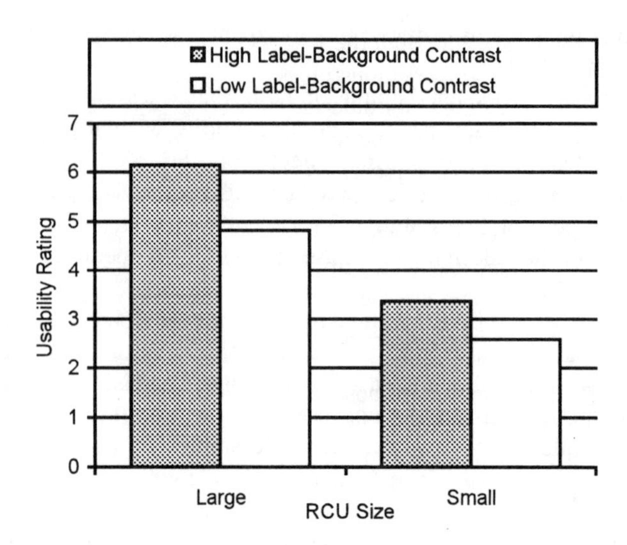

Figure 4. Average usability ratings as a function of the RCU size and the label-background contrast.

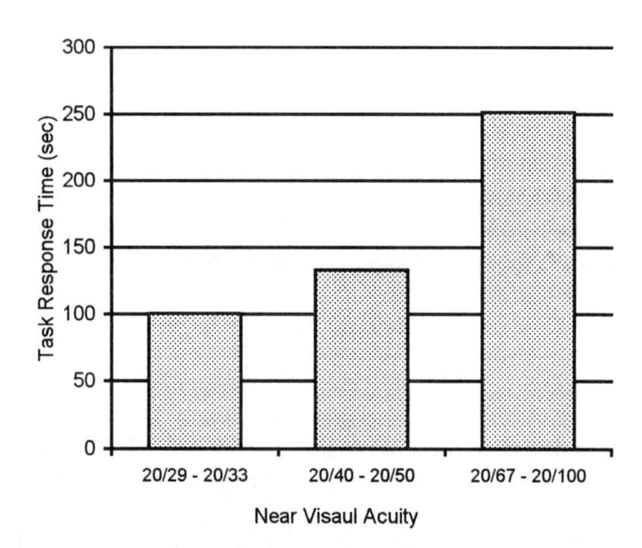

Figure 5. Mean accessibility task response time across vision groups.

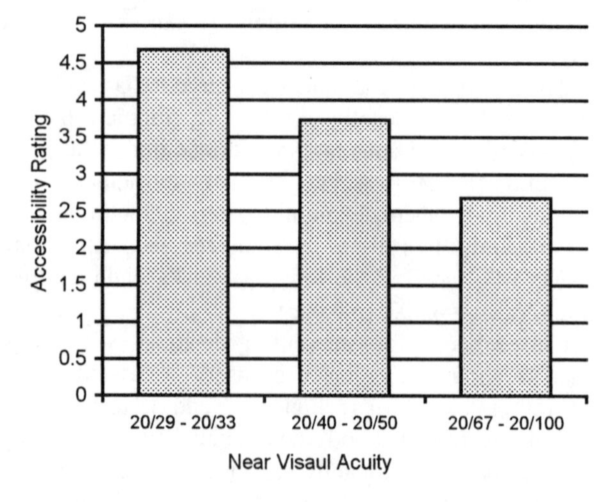

Figure 6. Average accessibility ratings across vision groups.

Furthermore, although the RCU design modification proposed in this study demonstrated improved accessibility and usability of a RCU for older adults, it is very unlikely that users would prefer such a large RCU in an actual operational setting. For example, a placard printed in this size may be useful as a reference or training aid. These results do not endorse nor recommend manufactures to produce an actual RCU with dimensions tested in this study. Rather, manufactures should use these design modifications as design guidelines to make trade-off decisions during the design process.

While the major purpose of this study focused on the photo CD player and older users with mildly impaired vision, a secondary purpose was to ascertain the effectiveness of the methodology for designing accessible products for users with special needs. These results strongly support the use of this three-phase methodology for designing accessible consumer products. The identification of accessibility barriers in Phase I led to the specification of critical barriers and the selection of cost-effective design solutions in Phase II. The empirical evaluation conducted in Phase III verified the effectiveness of the design modifications in improving the accessibility and usability of a photo CD player for older adults with mildly impaired near vision. The systematic application of relevant information on the target user population, an iterative design process, and empirical testing that center around appropriate users performing representative tasks are essential in designing accessible products for users with special needs.

ACKNOWLEDGMENT

This paper is based, in part, on a master's thesis prepared by the first author. The research was partially supported by funds provided under a graduate intern fellowship from Eastman Kodak Company. Research facilities at both Eastman Kodak Company and Virginia Tech were used in this project.

REFERENCES

Berge, R. L., and Cassells, J. S. (Eds.). (1990). *The second fifty years: Promoting health and preventing disability.* Washington, DC: National Academic Press.

Czaja, S. J. (Ed.). (1990). *Human factors research needs for an aging population.* Washington, DC: National Academic Press.

Gould, J. D., Conti, J., and Hovanyecz, C. T. (1983). Composing letters with a simulated listening typewriter. *Communications of the Association for Computing Machinery, 26* (4), 367-370.

Halstead-Nussloch, R. (1989). The design of phone-based interfaces for consumers. In *Proceedings of the CHI '89 Conference on Human Factors in Computing Systems* (pp. 347-352). New York, NY: Association of Computing Machinery.

Hix, D., and Hartson, H. R. (1993). Developing user interfaces: Ensuring usability through product and process. New York, NY: John Wiley and Sons, Inc.

Mann, W. C., Ottenbacher, K. J., Tomita, M. R., and Packard, S. (1994). Design of hand-held remotes for older persons with impairments. *Assisted Technology, 6* (2), 140-146.

Moss, M., and Lawton, M. P. (1982). The time budgets of older people: A window on four lifestyles. *Journal of Gerontology, 37*, 115-123.

Nakatani, L. H., and O'Conner, K. D. (1980). Speech feedback for touch-keying. *Ergonomics, 23* (7), 643-654.

Parsons, H. M., Terner, J., and Kearsley, G. (1994). Design of remote control units for seniors. *Experimental Aging Research, 20*, 211-218.

Pirkl, J. J., and Babic, A. L. (1988). *Guidelines and strategies for designing transgenerational products: An instructor's manual.* New York: Syracuse University Series in Gerontology Education.

Phillips, L. (1990). *Consumer needs assessment: A qualitative study of the needs of people with disabilities, results of the first year of a five year study* (Technical Report). Washington, DC: Electronic Industries Foundation.

Vanderheiden, G. C., and Vanderheiden, K. R. (1991). *Accessibility design of consumer products: Guidelines for the design of consumer products to increase their accessibility to people with disabilities or who are aging* (Technical Report). Madison, WI: University of Wisconsin-Madison.

Ward, C. (1990). *Design for all: Consumer needs assessment project year 2, results of the second year of a five year study* (Technical Report). Washington, DC: Electronic Industries Foundation.

Ward, C. (1991). *Increasing independence through technology, the views of older consumers with disabilities and their caregivers: Consumer needs assessment project year 3, results from the third year of a five year study* (Technical Report). Washington, DC: Electronic Industries Foundation.

Williges, R. C., Williges, B. H., and Elkerton, J. (1987). Software Interface Design. In G. Salvendy (Ed.), *Handbook of human factors* (pp. 495-504). New York, NY: John Wiley and Sons, Inc.

AN IN-DEPTH ANALYSIS OF AUTOMATIC TELLER MACHINE USAGE BY OLDER ADULTS

Wendy A. Rogers, The University of Georgia
D. Kristen Gilbert, The University of Memphis
Elizabeth Fraser Cabrera, Georgia Institute of Technology

The present study investigated the usage of Automatic Teller Machines (ATMs) by older adults. We conducted 100 telephone interviews of older adults wherein we queried subjects about their frequency of ATM usage. From this pool of individuals we chose eight frequent users and eight intermediate users to participate in an in-depth structured interview. The phone and structured interviews provided detailed information about usage patterns and general ATM knowledge of older adults. The interviewing technique provided insight into the concerns of older adults and the problems they encounter when using ATM technology. The results of this study provide information relevant to design and training for ATMs. Although the data are derived from a sample of older adults, any improvements of design, safety, or training will be beneficial to the population of users as a whole.

INTRODUCTION

There is some research to suggest that older adults are less likely to use new technology, relative to younger adults (e.g., Gilly and Zeithaml, 1985; Rogers, Cabrera, Walker, Gilbert, and Fisk, 1994; Zeithaml and Gilly, 1985;). In particular, these researchers have demonstrated that older adults are less likely to adopt Automatic Teller Machine technology. Questions remain, however, as to why older adults are averse to using new technologies and whether or not their attitudes and usage patterns can be changed. In the present study, we investigated these issues in terms of Automatic Teller Machines (ATMs). We conducted 100 telephone interviews of older adults in Memphis and Atlanta wherein we queried subjects about their frequency of ATM usage. From this pool of individuals we chose eight frequent users and eight intermediate users to participate in an in-depth structured interview. The phone and structured interviews provide detailed information about usage patterns and general ATM knowledge of older adults. In addition, we were able to determine the types of problems that older users experience and their suggestions for the improvement of ATMs.

METHOD AND RESULTS

The study was conducted in two phases. Phase one consisted of telephone interviews with 100 individuals. Phase two consisted of detailed structured interviews with 16 individuals who were either intermediate or frequent ATM users. The details of the two phases are provided below.

Phase 1 - Telephone Interviews

Subjects. One hundred older adults (50 residents of Memphis and 50 of Atlanta) participated in a telephone interview. There were 44 males and 56 females. The subjects ranged in age from 61 to 81 (mean = 71.80 years), their education level ranged from 7 years to 17 years (mean = 14.56 years), and their average self-rating of health was 2.95 (on a scale of 1 to 4, with 4 being excellent).
Procedure. The subjects were asked a series of demographic questions (e.g., age, sex, race, health,

education), as well as the following questions related to ATM usage:

Do you do the majority of banking for your family?
Are you familiar with ATMs?
Do you have an ATM card?
If not, why not?
If yes,
Do you use your ATM card?
If not, why not?
If yes, how often?
What percentage of your banking is done at an ATM?

Results. 91% of the subjects were familiar with ATMs and 78% had an ATM card (only 50% used them). Age was negatively correlated with usage (.25) and frequency of use (.42). Sex was correlated with knowledge of ATMs (.21), owning an ATM card (.21), using an ATM (.29), frequency of ATM use (.28), and percentage of banking done at an ATM (.29). Older males were more likely to be aware of and to use ATMs than older females. Neither health rating nor education was significantly correlated with any of the ATM variables.

Phase 2 - Structured Individual Interviews

The goal of this phase of the project was to probe older adult ATM users to determine some of the problems they have with ATMs and their concerns about using them. We were interested in those comments relevant to human factors concerns such as design and training issues. The focus was not on whether all of the subjects reported the same types of problems. Instead, this was an exploratory approach to determine if some users reported problems that might be useful to ATM designers. Moreover, we have repeatedly been informed by bank officers that training is not necessary for ATMs because they are inherently user friendly. However, the recent survey conducted by Rogers et al. (1994) suggest that this may not be the case. Therefore, we wanted to utilize the in-depth interview approach to learn more about the training needs of an older adult user population.

Subjects. From the pool of telephone interviews, 16 subjects (9 residents of Memphis and 7 residents of Atlanta) were chosen to participate in the structured interview. There were eight intermediate users, and eight frequent users. Usage determinations were made on the basis of the telephone interview; frequent users used the ATM more than once a week and intermediate users used an ATM less than once a week. The demographics of the subjects are presented in Table 1. There were no significant differences between the groups.

Table 1. Means and standard deviations of demographic variables for each user group.

Demographics	Frequent		Intermediate	
	M	SD	M	SD
Sex (M/F)	5/3		4/4	
Age	71.13	6.58	70.75	5.55
Education[1]	14.63	3.16	15.63	3.29
Health[2]	3.00	1.15	3.50	0.53

[1]Number of years of formal education (12=high school).
[2]Self-rated health (1 = poor to 4 =excellent).

Procedure. Subjects were asked a series of questions designed to elicit information about their familiarity with ATMs in general, their familiarity with specific procedures, the types of problems experienced in using ATMs, as well as their understanding of how ATMs functioned (some of the questions were adapted from Payne, 1991). Each interview was audiotaped and transcribed.

Results. Table 2 presents the types of transactions that the subjects conduct on the ATM. Clearly, the majority of the subjects use the ATM for making cash withdrawals. When asked if they ever used their cards for other transactions, two of the intermediate users claimed that they did not know how: "*No, I'm not sure I know how*" and "*No....I'm not real comfortable and I don't want to stand there in front of people and try to figure it out*". Two of the intermediate users said they conducted other transactions and the remaining 4 subjects said they never did. Of the frequent users,

3 conducted other transactions and the remaining 5 subjects said they never did. These data combine to suggest that training may in fact be useful for ATMs. First, many current ATMs are equipped to perform a variety of functions. The present subjects are obviously not taking advantage of the breadth of ATM services. Moreover, two of the users explicitly stated that they did not do so because of their lack of knowledge about how to use the system.

Table 2. What transactions do you use the ATM for? (listed by subject number)

Frequent Users

1	Withdrawals
2	Occasional Withdrawals, Deposit Checks (never cash)
3	Withdrawals
4	Withdrawals
5	Withdrawals
6	Withdrawals, Transfer Funds, Check Balance, Deposits
7	Withdrawals
8	Withdrawals, Check Balance

Intermediate Users

1	Withdrawals
2	Withdrawals
3	Withdrawals
4	Withdrawals
5	Make Deposits, Withdrawals
6	Deposits
7	Withdrawals
8	Withdrawals

The subjects' responses are in the order they listed them.

Problems with ATMs. We were interested in whether subjects recalled having difficulties using the ATM when they first began. It is important to qualify these results with the knowledge that all of the subjects had been using ATMs for at least 3 years, some as many as 15 years; consequently, memory may be an issue relevant to the following results. Nine of the frequent users and two of the intermediate users reported that they did not recall having any difficulties when they first started using the ATM. The difficulties reported by the other subjects are listed in Table 3. We also queried users as to how many times they had to use the ATM before they really felt comfortable and the mean response was 2.4 times (SD=1.4). This is comparable to the results of the Rogers et al. (1994) study where the majority of subjects responded 1 to 4 times (81.3%, 79.5%, 79.7%, and 70.1%, respectively, for the young, middle-aged, young-old, and old adults). Interestingly, 9% of the old adults in that study responded that they *never* felt comfortable using the ATM.

Table 3. Did you have any difficulties when you first started?

- *Yes, inserted card the wrong way. (frequent user)*
- *Yes...when I first started I couldn't make it work and I think that's the reason I don't go more often.. Yeah, [I had trouble with] the secret code. (intermediate user)*
- *Well, as you can imagine, I was unfamiliar with it. I just had to, I had to take the time, you know, you kind of feel intimidated when other people are just standing there waiting for the machine, because they've been doing it and they know how to do it. And to take the time to read all those instructions, you know, it took a time or two to get familiar with it, but after that it wasn't any problem. (intermediate user)*
- *Well, not really. Keeping up with the amount of money that was coming out of it. I wanted a fairly good size amount of money and I'd try to use that thing and it would come out thirty dollars at a time or I'd hit something wrong and then I'd have to stop and count. And that's something that made me uncomfortable, I didn't like doing that in a glassed place. (intermediate user)*
- *Just getting used to it, yes. (intermediate user)*
- *I felt very uncomfortable the first time I used it. I remember I thought, oh, I'm gonna mess up here and hit the wrong numbers and stuff...I forgot my code (intermediate user)*
- *Just getting the PIN [personal identification number] number that I wanted. I switched banks because my first bank wouldn't let me choose my own number. (intermediate user)*

Even something as seemingly trivial as understanding how to insert the card can be problematic. It would be easy for users to learn how to do this; however, it can be disconcerting when one is first trying to use the machine and people are waiting in line. Training would certainly alleviate this problem. Similarly, it would be helpful to encourage users to develop a mnemonic for remembering their code. Finally, some subjects do report having difficulties and having anxieties when first using an ATM. A training program offered by the bank would minimize such difficulties and might also increase the amount and variety of banking done at ATMs.

Subjects were also asked to describe any of the problems they have had with using an ATM. These data are presented in Table 4. It is important to keep in mind that even the intermediate users, although they don't use ATMs often, have been using them for a number of years.

Table 4. Have you had any problems using an ATM?

- *One time it...wouldn't give it [my card] back to me and they sent me another one. (frequent user)*
- *I put my card in and the money came out but the door wouldn't open....they[the bank] retrieved the money and they credited my account. (frequent user)*
- *Gave me $20 short once...of which we took care of, I always count my money right in front of the video camera, walked into the bank and said here's my problem. (frequent user)*
- *I think it's the [name of bank]...you do your thing then there's a "push if amount is correct"...the things are so situated that you can't tell which key to push. They don't line up properly and I have to get down and see which one they want...There's a yes and a no and three keys and I often push this key when I mean to push this one because I can't tell where it's focused and it doesn't do anything. It doesn't give you any feedback... (frequent user)*
- *Uh, I've [entered the wrong PIN], the [name of bank] ATM makes you go through the whole process, the withdrawing from checking and blah, blah, blah. You go through the whole process and then you get down and it sends the card back...I don't remember if it has a message...if it said...it's got some words...I don't know if it says "transaction rejected". I don't remember...Yeah, it makes you do everything...it says cancel or delete, I don't know if...it seems that after I've entered the money... I think maybe there's a chance to change your mind, but I'm not sure. (frequent user)*
- *One of the things I don't like about [it], I occasionally have problems with the ATM is the printing. Occasionally they'll run out of ink... (frequent user)*
- *Well, if you consider this a problem I guess...one time I forgot my secret code...then I've stopped at ATMs two or three times and it has been out of order. (frequent user)*
- *Once or twice it's been out of order. (intermediate user)*
- *...the only problem I've had is ...there'd be a little sign there and it should have been bigger I reckon, that said 'Temporarily Out of Order' and before I know it I've stuck my card in there and it's gone. But I just hit the clear and it'll come right back. (intermediate user)*
- *...sometimes I punch something wrong and...it'll come out when I'm not through because I have been slow in giving it the next command... (intermediate user)*
- *The only problem is when they are out of service, like last week I had to go to three, one in my neighborhood, stopped at another it was out of service too, finally ended up banking in the mountains. (intermediate user)*
- *One time $100 was lost, not machine's fault, not on bank statement, never said what had happened. (intermediate user)*
- *Not really, sometimes card gets bent, have to get another card. (intermediate user)*
- *I think the first time when I didn't get any money out of it, and I put my secret code and everything, I had to do it twice...I went inside and I said, look I had to go through this transaction twice...but I only got the money out one time, is it gonna charge me for double transaction...They said don't worry... (intermediate user).*

One subject mentioned a problem with seeing the keys and difficulties determining which keys line up with which button. The issue of being able to see the screen well was also reported as a problem by the larger sample in the Rogers et al. (1994) study. That particular subject, along with another subject, also mentioned the issue of the ATM not providing sufficient feedback. Both of these concerns could be addressed with design improvements.

Safety Issues. Not surprisingly, nearly all of the subjects expressed concerns about their safety when using ATMs. Their responses to our query about their concerns about using ATMs are presented in Table 5. In addition, we asked them if they knew about any security measures that were provided for ATMs. Ten of the 16 subjects responded that they were unaware of any security measures. ATM providers, designers and trainers thus have two concerns: (a) making the ATMs as safe as possible to use, and (b) educating the users about the security measures being used to make them more confident about their safety. Doing only the latter would not be sufficient because you would not want the users to become overly confident. It is evident, at least in this sample of older adults, that safety is an important issue for ATM users (see also Rogers et al. 1994).

Table 5. Do you have any concerns about using the ATM?

- *Safety...Personal safety, that kind of thing. The area I live in I don't really worry about it but I still have a concern and do not like to use it after...7 o'clock in the evening or after dark...I have a lot of qualms about using the [free standing ATMs]. (frequent user)*
- *In other places they're all outside, and I don't feel secure with things like they are now (frequent user)*
- *I make it my business to do it in the daytime if I can't at a time when there are other people around... (intermediate user)*
- *Primarily safety. And next, I have a little bit of concern with the accuracy...if it's really being handled accurately. However, that has never happened. Nothing has ever happened wrong. (intermediate user)*
- *Yes, I'm very concerned about safety because I read about it in the newspaper about these people that are robbed and in some cases killed... (intermediate user)*
- *Especially they caution the seniors about it [safety] at a lot of meetings that I go to...you know they're just easier prey than a young person...the locations of some [ATMs] are just terrible. (frequent user)*
- *My major concerns about ATMs and this, I believe would hold for the majority of them, they're in terrible locations, they're isolated, away from the main street. Safety is my biggest concern. Convenience and safety. Safety and convenience, I go for safe convenience. (intermediate user)*
- *[I] hear about people in dark coming to take your money, I've never worried about it. (frequent user)*
- *My only concern is safety and I generally do that in the daytime, in fact I always do it the daytime unless my husband is with me. (intermediate user)*
- *I try to be prudent about it, I would not use an ATM in a remote location that is dark at night. (frequent user)*

DISCUSSION

Although older adults may use newer technologies less than young adults, there are certainly a number of older adults who are willing to use ATMs, for example. However, it is also the case that older adults may have some difficulties using ATMs. They reported difficulty determining which keys to press, seeing the screen, understanding how to use the different types of ATMs, and being hesitant to learn under the stress of having other people watching and waiting. Improved design, and the development of a training program, could minimize the problems experienced by older adult users. Moreover, such improvements might increase their usage and enable them to take advantage of ATM features of which they are currently unaware.

The present study utilizes an interviewing technique to gain insight into the concerns of older adults and the problems they encounter when using ATM technology. In addition, we were able to gather information about why individuals might choose not to use ATMs. The results of this study provide information relevant to design and training for ATMs. Although the data are derived from a sample of older adults, any improvements of design, safety, or training will be beneficial to the population of users as a whole.

ACKNOWLEDGMENTS

This research was supported in part by a grant from National Institutes of Health (National Institute on Aging) Grant No. P50 AG11715 under the auspices of the Southeastern Center for Applied Cognitive Aging Research (one of the Edward R. Roybal Centers for Research on Applied Gerontology). We thank Jennifer Clark, Mary Cregger, Roderick Lilly, Jennifer Nestor, and Tom Rucker for assistance with various aspects of this project.

REFERENCES

Gilly, M. C., and Zeithaml, V. A. (1985). The elderly consumer and adoption of technologies. Journal of Consumer Research, 12, 353-357.

Payne, S. J. (1991). A descriptive study of mental models. Behaviour and information technology, 10, 3-21.

Rogers, W.A., Cabrera, E.F., Walker, N., Gilbert, K.G., and Fisk, A.D. (1994). Survey of Automatic Teller Machine usage across the adult lifespan. Manuscript submitted for publication.

Zeithaml, V. A., and Gilly, M. C. (1985). Characteristics affecting the acceptance of retail technologies: A comparison of elderly and nonelderly consumers. Journal of Retailing, 63, 49-68.

TRAINING NEW TECHNOLOGY: AUTOMATIC TELLER MACHINES AND OLDER ADULTS

Brian A. Jamieson, The University of Georgia
Elizabeth F. Cabrera, Georgia Institute of Technology
Sherry E. Mead, Georgia Institute of Technology
Gabriel K. Rousseau, The University of Georgia

The purpose of the present study was to assess the benefits of providing on-line training for an automatic teller machine (ATM). An ATM simulator was developed for the study, and older adults (65-80) served as the subjects. Subjects were assigned to one of two conditions. Half of the subjects were given a written description of how the ATM worked. The other half went through an on-line tutorial, which showed them how to perform transactions on the simulator. After performing 30 transactions on the simulator, subjects were transferred to a new ATM simulator that was topographically different. The subjects who received the on-line tutorial performed more transactions correctly during acquisition, and were better able to transfer their skills to a different ATM simulator and to novel transactions.

INTRODUCTION

Automatic Teller Machines (ATMs) are an easy and efficient way to perform many banking transactions. However, many older adults choose not to use this form of technology. A study by Rogers, Cabrera, Walker, Gilbert and Fisk (in press) found that only 33% of older adults use ATMs. One possible reason that elderly people avoid ATMs may be that they are hard to use. There is some evidence that suggests that older adults have difficulty in successfully operating an ATM (Hatta and Iiyama, 1991). However, almost 20% of the older adults in the Rogers et al. study said that they would be interested in learning how to use an ATM if training were offered. Most banks provide little or no instructions; when they do, the materials consist of only a brief outline of how to use an ATM. Another concern of older adults is safety. One possible way of dealing with these concerns is by training older adults how to use an ATM. Training would increase familiarity with the machine, and reduce the amount of time needed to complete a transaction.

The purpose of this research was to determine the effects of providing on-line training for novice older adults in using an ATM. Because banks don't provide any type of training at all, the basis of this study was to examine the benefits of training versus no training. A study by Adams and Thieben (1991) found that some types of training improved performance. However, all their training was done on various tasks designed to mimic operations performed on an ATM. They did not provide a training condition in which subjects were actually trained on the ATM simulator. Two conditions were used in the present study. The first was a description condition in which a written text described how the ATM worked. This condition was designed to mimic the typical information received from the bank. The second was an on-line tutorial condition that used the simulator to provide step-by-step training on how to perform transactions. Subjects in the tutorial condition received the description in addition to the on-line tutorial.

Subjects performed 30 different transactions on the ATM simulator. We then assessed their ability to transfer their knowledge to a different

ATM simulator (i.e., the topographical layout was different and novel transactions could be performed). We predicted that subjects who received training in the tutorial condition would perform the initial transactions faster and more accurately than those only reading the description. Having hands on experience with how the ATM worked should provide subjects with a much better idea of how to operate the ATM. In addition, we predicted that the tutorial group would show superior transfer to the new simulator.

METHOD

Subjects

Older adults who had never used an automatic teller machine participated in this study. The subjects ranged in age from 65 to 80 years (M = 69.3), and were comprised of 13 males and 15 females. Lack of prior ATM experience was determined through a telephone screening. Subjects were randomly assigned to one of two experimental groups, which differed in the type of instructional material they received for using an ATM. No significant differences were found in age, gender, education level, or health ratings between the groups. All subjects passed both near and far visual acuity tests. Subjects received $40 for their participation.

Materials

A computer simulated automatic teller machine (ATM1) was developed for the study. The ATM was designed to simulate the ATM of a particular bank, having all of the standard features and options of the majority of ATMs. These included five types of transactions: fast cash, withdrawal, deposit, transfer, and account information. A second ATM simulator (ATM2) was designed to test transfer of learning. ATM2 had the same features as ATM1, although the features appeared in different locations. In a near transfer condition, subjects perform the same transactions on ATM2 that they performed on ATM1. In addition to the features being rearranged, ATM2 offered options that are not normally offered at ATMs, but will likely be offered in the future. These include the options to buy tickets for a concert, check the gold exchange, and buy lottery tickets. These transactions were considered far transfer, because subjects had never seen them before.

Subjects had to use the computer mouse to perform transactions on the simulated ATM. The indicator on the computer screen appeared as a hand with the index finger extended. In order to use the simulated ATM, subjects had to move the mouse so that the finger was pointing to the command button or number on the keypad that he or she wanted to push and then had to click the mouse. A mouse trainer was designed to give subjects the opportunity to practice using a computer mouse before they used the mouse for the simulated ATM. The mouse training included practice performing two actions that the subject would perform with the ATM simulators. Subjects used the mouse to enter sequences of six numbers on a software keypad identical to the simulated ATM keypads. Subjects were required to re-enter each sequence until it was entered correctly before proceeding to the next sequence. All participants completed three blocks of ten trials. Blocks of keypad practice were alternated with blocks of practice clicking sets of ten software buttons that varied in size and appeared at random locations on the screen. Subjects had to successfully activate each button before another would appear. This was intended to provide practice moving the cursor to targets at various locations on the screen. Subjects received a total of 160 trials of practice moving and clicking the mouse.

A transaction began with the appearance of a text window containing a description of the transaction to be completed (e.g., Withdraw $100 using FastCash). The subject was required to click a button to dismiss the window before proceeding. In order to reduce the effects of memory for transaction descriptions on performance, a help function was included in both simulators. Pressing the 'H' key on the keyboard caused the transaction description window to reappear. All subjects were assigned a security code or Personal Identification Number (PIN) of 1234. All transactions proceeded as they do for the automatic teller machines after which the simulators were modeled. Experimenters reminded subjects to make use of the help function and assisted subjects with any difficulties specific to the simulator and not representative of interactions with real ATMs. All mouse clicks, simulator screens traversed, and response times were recorded.

Procedure

There were two different training groups: The description group and the tutorial group. The subjects in the description group read a brief document that presented a description of how an ATM worked. This was designed to mimic information that a new user is likely to receive from a bank. They received no specific instruction on how to use an ATM. Subjects in the tutorial group read the description and completed an interactive on-line tutorial that walked subjects through each of the five different transactions. Instructions appeared on the screen with an arrow pointing to the location where the subject needed to place the mouse and click. In this way, subjects were led to perform the transactions.

Subjects participated in this study during three consecutive days for approximately two hours each day. On the first day they filled out demographic information, a survey about ATM use, and learned how to use the mouse.

On the second day subjects were given the instructional materials for the group to which they had been randomly assigned. After the subjects had read all of their assigned instructional materials or completed the on-line tutorial, they began conducting a series of transactions on the first simulated ATM. Subjects completed four blocks of five transactions. Each block consisted of a fast cash, a withdrawal, a deposit, a transfer, and an account information query. Block one was identical to block three. Blocks two and four had the same order of transactions as blocks one and three, but differed in terms of dollar amounts and accounts accessed (see Table 1).

On the third day subjects completed two blocks of five transactions on ATM1. Block five was identical to block one completed on the previous day. Block six differed in terms of dollar amounts and accounts accessed. Subjects then switched to ATM2, where they completed four blocks of five transactions. Block seven was identical to block five, and block eight was identical to block six. Blocks nine and ten included novel (unpracticed) transactions such as ticket purchases and investment information requests. Blocks seven and eight assessed near transfer, while blocks nine and ten assessed far transfer. After subjects completed the ATM transactions, they were given a brochure on tips for using ATMs, including safety and record keeping, and were thoroughly debriefed.

Table 1.
Structure of Blocks.

Day	Two				Three					
Block	1*	2	3*	4	5*	6	7*	8	9	10
ATM	Version One						Version Two			
Transfer							Near		Far	

Note. *identical blocks

Dependent Variables

Two dependent variables are considered here: transaction correct and time per transaction. For a transaction to be considered correct, the subject had to perform the requested transaction using the specified account and amount of money. Subjects also had to take their card, receipt, and cash, if they were presented with any (e.g., a fast cash transaction presents all three, but a transfer only presents the card and receipt). If at any time a subject made a mistake during a transaction but corrected it, their performance on the transaction was considered correct. The time per transaction excluded system response time and time that the help screen was displayed.

RESULTS

The results for the proportion of transactions correct are presented in Table 2. The on-line group was superior to the description group throughout practice ($F(9,18)=4.2, p<.05$). Comparisons between the two groups were only significant for the second near transfer transaction on ATM2 (i.e., block eight). However, the on-line tutor group was significantly better on all far transfer transactions (i.e., blocks nine and ten) (p<.05).

Table 2.
Proportion of Transactions Correct by Practice Block.

	Tutorial		Description		
ATM1	M	SD	M	SD	t-value
Block 1	.41	.24	.23	.22	2.13*
Block 2	.51	.30	.27	.34	2.01
Block 3	.57	.32	.29	.32	2.35*
Block 4	.59	.33	.24	.33	2.78*
Block 5	.61	.30	.26	.30	3.17*
Block 6	.53	.32	.26	.33	2.22*
ATM2 Near Transfer					
Block 7	.37	.19	.26	.29	1.24
Block 8	.57	.28	.33	.38	1.92*
ATM2 Far Transfer					
Block 9	.36	.21	.16	.18	2.71*
Block 10	.27	.17	.11	.15	2.60*

Note. *p<.05

The average time per transaction within each block is presented in Table 3. The description and on-line tutorial groups were significantly different for three of the first four blocks of practice. Block three showed marginally significant differences. No significant differences were found for near or far transfer.

DISCUSSION

Our findings indicate that old subjects who received training in the form of an on-line tutorial performed better than those subjects who only received a description of how an ATM works. Subjects in the tutorial group were initially superior, and remained so for the duration of the experiment. One of the potential confounds in interpreting the results is that the tutorial group had an initial block of practice, which constituted the training, before the first block of the experiment. However, even after 30 transactions of practice the description group still had not reached the level of performance at which the tutorial group started. This implies that learning how to use an ATM requires more than just practice. Some type of training, such as an on-line tutorial, is necessary for older adults to learn how to use an ATM.

Table 3.
Time per Transaction (seconds) by Block.
(excludes system response time and time in help system)

	Tutorial		Description		
ATM1	M	SD	M	SD	t-value
Block 1	89.29	36.19	179.41	65.10	4.53*
Block 2	66.11	19.04	89.67	34.40	2.24*
Block 3	58.35	17.04	80.26	38.17	1.96
Block 4	54.36	13.14	78.80	37.51	2.30*
Block 5	70.68	24.81	89.06	55.46	1.13
Block 6	64.86	24.72	73.20	34.82	0.73
ATM2 Near Transfer					
Block 7	71.97	23.45	80.86	30.56	0.86
Block 8	57.57	16.35	64.72	20.01	1.04
ATM2 Far Transfer					
Block 9	95.78	32.27	112.18	29.08	1.41
Block 10	79.40	26.03	82.28	17.11	0.35

Note. *p<.05

Differences between the groups were also found on the transfer tasks. Being able to transfer to a different ATM is an important skill, because most people are not always able to use the ATM associated with their bank. They may have to use other ATMs that are slightly different. The tutorial group showed a decrease in correct transactions when they transferred to ATM2 on block seven, but their performance improved on block eight to equal that of their performance on ATM1. One possible explanation for the initial decrease is that subjects may have thought that the task was different in some way due to the new simulator. After subjects performed a few transactions they realized that the task was not different, and performance recovered. The description group was not adversely affected by the new ATM which seems to indicate that they had not learned ATM1 well enough to be disrupted when they switched to ATM2. That is, they were only 26% correct at the end of ATM1 practice and they were 26% correct at the beginning of ATM2.

Both groups declined when performing far transfer transactions. However, the tutorial group performed significantly better than the description group. The tutorial group appears to be more capable of applying what they learned during the previous trial blocks to novel transactions. Indeed, the description group was only performing roughly ten percent of the novel transactions correctly.

The time per transaction was only significantly better for the tutorial group on the initial training blocks. This advantage could be due to the on-line training the tutorial group received. Although time per transaction may not have been significantly different between the groups over near and far transfer, the tutorial group performed with more accuracy. Thus, even though training does not increase speed, it does increase accuracy, which means that ultimately less time would need to be spent at the ATM correcting mistakes.

Training older adults how to use an ATM appears to be beneficial. Even the minimal training of this study's on-line tutorial, which only provided practice on five transactions, resulted in a large increase in successful performance. The tutorial seems to equip elderly subjects with skills that are more efficient and more flexible than those learned by subjects who receive only a description of the task.

REFERENCES

Adams, A.S., and Thieben, K.A. (1991). Automatic teller machines and the older population. Applied Ergonomics, 22, 85-90.

Hatta, K., and Iiyama, Y. (1991). Ergonomic study of automatic teller machine operability. International Journal of Human-Computer Interaction, 3, 295-309.

Rogers, W.A., Cabrera, E.F., Walker, N., Gilbert, D.K., and Fisk, A.D. (in press). A survey of automatic teller machine usage across the adult lifespan. Human Factors.

ACKNOWLEDGMENTS

This research was supported in part by a grant from National Institutes of Health (National Institute on Aging) Grant No. P50 AG11715 under the auspices of the Southeastern Center for Applied Cognitive Aging Research (one of the Edward R. Roybal Centers for Research on Applied Gerontology). Project directors are Wendy A. Rogers and Arthur D. Fisk.

ONLINE LIBRARY CATALOGS: AGE-RELATED DIFFERENCES IN QUERY CONSTRUCTION AND ERROR RECOVERY

Sherry E. Mead[1], Brian A. Jamieson[2], Gabriel K. Rousseau[2], Richard A. Sit[1], and Wendy A. Rogers[2]
[1]Georgia Institute of Technology, Atlanta, GA
[2]University of Georgia, Athens, GA

Online library catalogs have become pervasive in today's library. Unfortunately, these systems have been developed by computer programmers or librarians with little analysis of user behavior on the system. The present study compared the search performance of younger and older adults with general computer experience who were novice online catalog users on a set of ten search tasks of varying difficulty. This study examined types of errors made by novice users in database query construction and subsequent error recovery. Younger adults achieved a higher overall success rate than did older adults and were more efficient in performing these searches. Older adults made more query construction errors and recovered from them less efficiently than did younger adults. These data have important implications for identifying the specific needs, limitations, and capabilities of online library catalog users and the design of online library catalog systems for adults of differing ages.

There is a lack of human factors studies that have focused on older adults and their use of technology, including online library catalogs. Online library catalogs provide fast access to library information from any location, using many access points, and powerful search commands. The few studies of older adults and their use of computer technology that do exist mostly have explored applications such as word processing or spreadsheets (Charness and Bosman, 1990; Zandri and Charness, 1994). A prominent finding among existing research studies is that older adults have problems with and do not use new forms of automated technology (e.g., Adams and Thieben, 1991; Dyck and Smither, 1992; Smither and Braun, 1994). Consequently, there is a need to explore and assemble basic data specifically on the needs, limitations, and capabilities of older adults with an increasingly common type of computer system in today's society, the online library catalog.

Seymour (1991) reviewed 16 studies of online library catalog users and concluded that such studies frequently employ inadequate methodology. She describes surveys used as poorly designed, samples as excessively small, and sampling methods as questionable. User observation studies employed few, if any, experimental controls. Lack of information about user goals make their results difficult to interpret.

The quantitative analysis of transaction logs and the introduction of rigorous experimental methods has been suggested by Lewis (1987) as a remedy for the limited usefulness and generalizability of earlier studies. The present study included measures of computer and database experience, and controlled for exposure to training materials, task difficulty, user goals, and system functionality. Additionally, transaction logs, which included all screens viewed and characters typed during searches, were recorded and analyzed. The selected online library catalog possesses many features common to most online library catalogs and computerized databases in general. For example, the present system allows character string matching (keyword search), specification of the field to be searched, and the use of Boolean AND and OR.

The issues that will be discussed here are database query construction errors and error recovery rates and strategies for older and younger adults. Other researchers have reported that character string matching is a difficult concept for new users to acquire (Elkerton and Williges, 1984); Boolean AND and OR can cause confusion since they are inconsistent with everyday usage (Avrahami and Kareev, 1993); and that searchers frequently employ sub-optimal cross-referencing and keyword selection strategies (Markey, 1984). Search algorithms (e.g., string matching vs. synonym matching) and the contents of specific database fields may contradict user expectations as evidenced by keywords selected and fields searched. Nearly all of these earlier studies have focused on younger adults. Sit (1994) studied 54 older adults and concluded that the categories of problems experienced by older adults were similar to those reported for younger adults. The present study will allow a direct comparison of the performance of older and younger adults.

Of interest also is whether the complex command syntax often required by mainframe computer systems poses problems for younger and older adult novice users. We will gain a better understanding of the needs, expectations, and capabilities of online library catalog users of differing ages and to identify aspects of this particular online library system in need of improved human factors design.

METHOD

Participants

Ten young adults aged 18 to 33 years ($M = 22$, $SD = 5.14$) and ten older adults aged 63 to 76 years ($M = 68.7$, $SD =$

5.08) participated in this study. Six of the younger adults and six of the older adults were women. Average educational attainment by older participants was college graduation (range: high school diploma to Master's degree). All younger adults reported having some college. Health ratings (range: 1 = poor to 4 = excellent) reported by older adults ($M = 3.4$, $SD = .52$) were similar to those reported by young adults ($M = 3.6$, $SD = .52$).

All participants passed a Snellen acuity test at a level of 20/40. Participants were able to read 10 point type on a reduced Snellen chart at a distance of approximately 20 cm.

Materials

Ability tests. The Digit-Symbol Substitution (Wechsler, 1981), the Extended Range Vocabulary Test (Ekstrom, French, Harman, and Derman, 1976) the Nelson-Denney Reading Comprehension Test (Brown, Bennet, and Hanna, 1981), and the Alphabet Span Test (Craik, 1986) were used to assess differences in perceptual speed, vocabulary, reading comprehension, and working memory capacity among the groups. Scores for younger and older adults are reported in Table 1.

Table 1. Mean ability test scores by age group.

	Older adults		Younger adults		p <
Vocabulary	30.8	(9.15)	18.2	(6.84)	.01
Digit-Symbol	48.5	(8.50)	72.5	(8.05)	.01
Reading Comprehension	24.4	(9.01)	32.6	(4.62)	.02
Reading Rate	271.3	(80.45)	252.4	(51.47)	.55

Online library system. The University of Georgia Academic Libraries Information Network (GALIN) has a command line interface and requires the user to specify a search method (e.g.,, keyword search, browse an alphabetized list, guided search) and the field to be searched. Boolean AND and OR are available. Online help consists of a menu-based help system and on-screen examples and suggestions.

A search command must take the form: Find (keyword search) or Browse (browse an alphabetized list) followed by a two-letter field code (e.g., AU = author) followed by a search term. Boolean AND or OR may be followed by another search term. A second field code must follow Boolean AND and may follow OR but is not required.

In general, search commands take the form:

Find *fieldCode searchTerm Boolean fieldCode searchTerm*

Search terms may include more than one keyword (e.g., "insomnia therapy" is a search term). Multi-keyword search terms will match a record if the keywords appear adjacent and in order in the specified field. That is, "insomnia therapy", "therapy insomnia", and "therapy for insomnia" will match different records. Spaces and punctuation (commas, hyphens) may be used interchangeably.

Computers. The system was accessed via an IBM-compatible PC with a 33 MHz 486DX CPU located in a cubicle in the laboratory and having a direct hardware connection to the IBM mainframe computer on which the system resides. Participants viewed 12 point green-on-black text on a 12 inch Super VGA monitor. Experimenters viewed the participant's display on a secondary monitor in an adjacent cubicle.

Training program. A training demonstration consistent with demonstrations given by University of Georgia reference librarians was developed. Each participant was given step-by-step instructions for performing each of four sample searches. Keyword searching of the Title, Author, Subject, and Notes fields; the use of Boolean operators; navigation commands; and accessing the online help system were demonstrated. Screen layout was explained and the location of useful on-screen information was pointed out. Database record organization and the type of information in the to-be-searched fields were described.

The paper documentation currently available in the UGA library was edited to remove references to functionality that participants were instructed not to use, specifically the browse an alphabetic list and guided search methods, changing databases, and quitting the system. The edited documentation was printed in black 12 point type on white paper, compiled into a booklet, and made available throughout the experiment. Experimenters recorded the number of references to paper documentation made by participants for each search task.

Search tasks. Ten search tasks were constructed. Each could be successfully completed via a single search command. There were three simple searches: Title search, Author search, and Subject search. A fourth search required participants to use the ALL field delimiter which searches the Notes field in addition to the Title, Author, and Subject fields. Three conjunctive searches required participants to locate records containing two target search terms in the same field (e.g., both "Frank Herbert" and "Bill Ransom" in the Author field). Three disjunctive searches required participants to locate records having one of two target search terms in the specified field (e.g., either "Susan Faludi" or "Naomi Wolf" in the Author field). The task descriptions given to participants and their optimal search commands are listed in Table 2.

Transaction log coding. In order to assess search efficiency and to categorize errors, participant command histories (transaction logs) were coded by two raters (not the experimenters). Transaction logs included each screen viewed by the participant, all command line entries, and all system messages.

Each component of a search command was coded separately. That is, if a participant entered "find au grisham, john", the search method, "find", the field code, "au", and the search term, "grisham, john", were coded individually.

Table 2. Search task descriptions and optimal search commands by search type.

Order	Task description	Optimal search command
Simple searches		
1	Look for 2 books by **John Grisham**.	f au grisham john
2	Look for a book called **Internet Navigator**.	f ti internet navigator
3	Look for 2 books about **insomnia therapy**.	f su insomnia therapy
ALL search		
4	Look for a book involving **Mount Olympus**.	f all mount olympus
Conjunctive searches		
5	Look for 3 books by **Frank Herbert** and **Bill Ransom**.	f au herbert frank and au ransom bill
7	Look for a book with the word **jelly** and the word **beer** in the title.	f ti jelly and ti beer
9	Look for a book about **abdomen muscles** and **back muscles**.	f su abdomen muscles and su back muscles
Disjunctive searches[1]		
6	Look for 3 books by **Susan Faludi** or **Naomi Wolf**.	f au faludi susan or au wolf naomi
8	Look for 4 books with the word **U-Haul** or the word **Bekins** in the title.	f ti u-haul or ti bekins
10	Look for 2 books about **Hershey chocolate** or **chocolate candy**.	f ti hershey chocolate or ti chocolate candy

[1]*Note.* The field code following OR is not required.

Each component was assigned a command type (e.g., field code, search term) and a descriptor (e.g., inefficient, target). Inter-rater agreement was 82% for descriptors and 93% for command types. All discrepancies were discussed until 100% agreement was reached.

Procedure

All participants were tested individually during a single three-hour session. The session began with vision and ability tests followed by demographic and technology use surveys.

Participants read all paper documentation. An experimenter conducted the training demonstration. Participants read over the task instructions and were informed that they had unlimited access to paper documentation and online help, but would receive no verbal assistance from the experimenter. Participants were told that all ten search tasks could be completed successfully and that they had an unlimited amount of time to do so. Breaks were encouraged.

RESULTS AND DISCUSSION

Self-reported relevant experience (from the technology use survey) is presented in Table 3. Although both age groups had computer experience, younger adults reported more relevant experience than did older adults. Therefore, performance differences between the age groups cannot be attributed exclusively to aging, per se, but may be due in part to differences in experience levels.

Scoring

Search tasks were scored as correct if the participant viewed the required number of target records (records that met the criteria given in the task descriptions) and recorded their call numbers on their answer sheets. Search commands were scored as optimal if they allowed the retrieval of target records via the minimum number of command line entries.

Table 3.
Computer, library, and other technology usage by age group.

	Older adults	Younger adults
Freq. of computer use	weekly	more than weekly
# applications used (e.g., word processor, spreadsheet)	1.8 (1.23)	5.5 (2.95)
# who use databases	n = 1	n = 4
Freq. of library use	monthly	more than monthly
# technologies used (e.g., photocopier, answering machine)	2.5 (3.89)	9.6 (4.79)

Search Success

Younger adults successfully completed more searches ($M = 9.2$, $SD = 1.55$) than did older adults ($M = 7.7$, $SD = 1.64$, $t(18) = 2.1$, $p < .05$). Specifically, younger adults completed more conjunctive searches and were more likely to complete the search requiring use of the ALL field delimiter. Older adults completed as many disjunctive and simple searches as did younger adults (See Table 4). Difficulty of conjunctive searches was increased by the required field delimiter following Boolean AND. The ALL search task was initially interpreted as a subject search by all participants in both age groups.

Table 4.
Percent correct search tasks by age group and search type.

Search type	Older adults	Younger adults
Simple	97%	100%
ALL	10%	70%
Conjunctive	70%	93%
Disjunctive	87%	90%

Search Efficiency

Optimal search commands retrieve a list containing all the target records for a given search task and no non-target records. Thus, each search task could be successfully completed by entering a single search command.

Older adults entered more search commands per task (M = 3.0, SD = .63) than did younger adults (M = 1.69, SD = .48, $t(18) = 5.21, p < .0001$). Younger adults entered at least one optimal search command on a higher proportion of search tasks (M = .85, SD = .21) than older did adults (M = .60, SD = .24; $t(18) = 2.47, p < .01$).

Rate of search success via sub-optimal commands was similar for younger and older adults. Since young adults successfully completed 92% of tasks and entered optimal search commands on 85%, they succeeded via sub-optimal commands on 13% of search tasks. Likewise, since older adults successfully completed 77% of tasks and entered optimal search commands on 60%, they succeeded via sub-optimal commands on 17% of search tasks. Thus, older adults were only slightly more likely to successfully employ sub-optimal search commands than were younger adults.

Boolean operators were required to optimally perform 6 of the 10 search tasks. Younger adults used Boolean operators to successfully complete more search tasks (M = 5.0, SD = 1.89) than did older adults (M = 2.3, SD = 2.26, $t(18) = 2.90$, $p < .01$). Older adults used Boolean operators unsuccessfully on more search tasks (M = 1.6, SD = 0.84) than did younger adults (M = 0.2, SD = 0.42, $t(18) = 4.70, p < .001$). This result is consistent with previous findings that Boolean AND and OR are problematic for database searchers. Further, they are more problematic for older searchers than for younger searchers.

Search Command Errors

Older adults entered more search commands that contained errors (M = 18.8, SD = 6.99) than did younger adults (M = 6.3, SD = 5.10, $t(18) = 4.57, p < .001$). Older adults were also more likely to search non-target fields (M = 9.0, SD = 5.33) than were younger adults (M = 3.3, SD = 3.23; $t(18) = 2.89, p < .01$). This result suggests that database structure, or the type of information contained in specific database fields, is inconsistent with older user's expectations.

Older adults were more likely to enter non-target search terms (M = 4.6, SD = 3.03) than were younger adults (M = .60, SD = 1.35; $t(18) = 3.82, p < .005$). Search commands entered by older adults were more likely to contain syntax or typographical errors (M = 5.4, SD = 4.70) than were commands entered by younger adults (M = 1.4, SD = 1.07; $t(18) = 2.63, p < .02$). Character string matching algorithms may be less intuitive for older users than for younger users. As mentioned earlier, differential experience with databases and with computers in general may exaggerate these effects.

Older adults were also more likely to leave out required command components, specifically, search methods (e.g., f or find) and field delimiters, (M = 14.0, SD = 4.44) than were younger adults (M = 1.0, SD = 2.31; $t(18) = 2.33, p < .04$). This result suggests that complex command structure is especially undesirable for older users.

Error Recovery

Older adults successfully completed 61% (SD = 0.21) of search tasks that included at least one search command error. Error recovery rate for younger adults was 86% (SD = 0.26, $t(18) = 2.32, p < .04$). Older adults entered more search commands when successfully recovering from an error (M = 2.69, SD = .95) than did young adults (M = 1.98, SD = .94), but this difference was not significant (p = .11).

Interestingly, if their first search command failed to retrieve target records, both younger and older adults were most likely to try searching a different field. Relative to younger adults, older adults were more likely to try different search terms and Boolean operators, and to reverse the order of two-word search terms. Table 5 lists specific changes made to search commands during error recovery attempts.

Although older adults tried a wider variety of error recovery strategies, younger adults were more likely to select the correct strategy. Of course, this result is due in part to the fact that older adults made more different types of search command errors than did younger adults.

Table 5. Mean number of error recovery attempts that included specific types of changes to search commands by age group.

	Older adults		Younger adults		$p <$
Changed field	4.6	(1.71)	3.0	(2.05)	.0
Changed search method	3.8	(4.49)	1.1	(2.13)	.1
Changed search term	4.0	(1.89)	1.2	(2.30)	.0
Changed Boolean	3.0	(1.41)	1.1	(1.29)	.0
Added missing command	4.1	(4.46)	0.8	(0.79)	.0
Reversed keyword order	1.1	(0.88)	0.3	(0.48)	.0
Corrected typo	0.6	(0.97)	1.0	(1.15)	.4

CONCLUSIONS

Searching the present online library system required the entry of complex, multi-term search commands which must be typed in a specific order and must exactly match database records in order to return any useful results. The results of this study indicate that this task is more difficult for older adults than for younger adults, even when both groups of participants had some previous computer experience. Errors made by novice users suggest several system characteristics that are especially problematic: 1) Requiring an exact match between character strings entered and character strings in records. More flexible string-matching algorithms and the inclusion of dictionary and thesaurus functions may be helpful as long as they do not create new tasks for the user; 2) Requiring entry of a list of disparate command elements in a specific order. Simplicity and flexibility, in combination with informative error messages are necessary; and, 3) Requiring the user to specify a field to search. System defaults may be especially helpful for older adults.

A tendency to search non-target fields and to search for non-target terms indicates that the structure of computerized databases and the nature of computerized search algorithms may not have been obvious to the older adults who participated in this study. A combination of free-text search capabilities and improved training and documentation may make keyword search of existing databases more productive and efficient, especially for older adults. In conclusion, this study provided a better understanding of the capabilities and limitations of online library catalog users of differing ages. Further, this study made several specific design recommendations based on user performance that promise to help improve both online library systems and other online information retrieval systems.

ACKNOWLEDGMENTS

This research was funded by NIH/NIA grant #P50AG11715 to the Center for Applied Cognitive Research on Aging, one of the Edward R. Roybal Centers for Applied Gerontology.

REFERENCES

Adams, A. S., & Thieben, K. A. (1991). Automatic teller machines and the older population. *Applied Ergonomics*, *22*, 85-90.

Avrahami, J. & Kareev, Y. (1993). What do you expect to get when you ask for a cup of coffee and a muffin or a croissant - on the interpretation of sentences containing multiple connectives. *International Journal of Man-Machine Studies*, *38*, 429-434.

Brown, J. I., Fishco, V. V., & Hanna, G. (1993). *Nelson-Denney Reading Test*. Chicago: Riverside.

Charness, N. & Bosman, E. A. (1990). Human factors design for older adults. In J. E. Birren & K. W. Schaie (Eds.), *Handbook of the Psychology of Aging* (3rd ed., pp. 446-463). New York: Van Nostrand Reinhold.

Dyck, J. L. & Smither, J. A. (1992). Computer anxiety and the older adult: Relationships with computer experience, gender, education, and age. *Proceedings of the Human Factors Society 36th Annual Meeting*, 185-189.

Ekstrom, R. B., French, J. W., Harman, H. H., & Derman, D. (1976). *Kit of factor-referenced cognitive tests*. Princeton, NJ: Educational Testing Service.

Elkerton, J. & Williges, R. C. (1984). Information retrieval strategies in a file search environment. *Human Factors*, *26*, 171-184.

Lewis, D. W. (1987). Research on the use of online catalogs and its implications for library practice. *The Journal of Academic Librarianship*, *13*, 152-157.

Markey, K. (1984). *Subject searching in library catalogs: Before and after the introduction of online catalogs*. Dublin, OH: OCLC Online Computer Library Center, Inc.

Seymour, S. (1991). Online public access user studies: A review of research methodologies, March 1986 - November 1989. *Library and Information Science Research*, *13*, 89-102.

Sit, R. A. (1994). *Relationships among Education, Experience/Expertise, User Performance, and User Satisfaction in Older Adult Online Public Access Catalog Users.* Unpublished master's thesis, University of Southern California, Los Angeles, California.

Smither, J. A. & Braun, C. C. (1994). Technology and older adults: Factors affecting the adoption of automatic teller machines. *Journal of General Psychology*, *121*, 381-389.

Wechsler, D. (1981). *Manual for the Wechsler Adult Intelligence Scale-Revised.* New York: Psychological Corporation.

Zandri, E., & Charness, N. (1989). Training older and younger adults to use software. Special Issue: Cognitive aging: Issues in research and application. *Educational Gerontology*, *15*, 615-631.

Editor's Note: The following reference was inadvertently omitted from the original proceedings publication:

Craik, F. I. M. (1986). A functional account of age differences in memory. In F. Klix & H. Hagendorf (Eds.), *Human memory and cognitive abilities* (pp. 409-422). Amsterdam: Elsevier.

A COMPARATIVE STUDY OF TEXT-EDITING PROGRAMS AMONG A SAMPLE OF OLDER ADULTS

Sara J. Czaja
Department of Industrial Engineering
State University of New York at Buffalo
J. Bonnie Joyce
Katka Hammond
Advanced Automation Concepts, Inc.

ABSTRACT

Research indicates that older adults have difficulty acquiring text-editing skills. The data suggest that the cognitive demands associated with text-editing programs create problems for older learners given the age-related changes in cognitive abilities. This study compared the learning efficiency of older adults for three text-editing programs which varied in format and command structure. A total of 45, computer naive, women ranging in age from 40 to 70 years participated. The results indicated significant differences in learning efficiency as a function of text-editing program. Participants using a full screen editor with pull down menus demonstrated significantly better performance than did those using other programs. Data was also collected regarding types of difficulties encountered by subjects during learning. This type of information can be used as input into the design of future software and training programs.

INTRODUCTION

Text-editing is one of the most widely used computer applications. In most offices, for example, the majority of routine typing and filing tasks are performed using a text-editing program. Typically, the text-editor is the first computer software to which people are exposed. Several investigators have found that older people encounter difficulties when attempting to learn text-editing skills. For example, Egan and Gomez (1985) found age to be a significant predictor of learning success for text-editing. Elias, Elias and Gage (1987) also found age differences in learning in a study which compared the effectiveness of alternative training methods to teach three age groups text-editing. Although the older people mastered the skill they took longer than the younger people and required the most instructor assistance.

A recent study, Czaja, Hammond and Joyce (1989) indicated that although training manipulations improve the learning performance of older adults there are still age differences in learning efficiency. These data suggest that current text-editors impose cognitive demands which are difficult for older learners to master given age related changes in cognitive abilities. For example, data indicates that spatial cognition is important to the mastery of text-editing and there is abundant evidence which shows that spatial skills decline with age. The objective of this study was to compare efficiency of older adults for three text-editing programs. The goal of the research was to gain specific information regarding formats and command structures which are problematic for older learners.

METHODS

A total of 45, computer naive, women ranging in age from 40 to 70 years participated in the study. All subjects had at least a high school diploma and were able to type 20 words per minute.

Subjects were randomly assigned to one of the three text-editing conditions. The text-editors included were: Wordstar Classic, Wordstar with pull-down menus and WordPerfect. Subjects participated for one and a half days. Initially, participants were given a basic introduction to computers and text-editing which included a lecture-demonstration and an interactive session. They were then given a goal-oriented training manual developed in previous studies and asked to use the manual to type a letter and make editorial changes to existing documents. Basic text-editing functions such as inserting and deleting text were presented on Day 1 and more advanced functions such as boldfacing and moving text were presented on Day 2. Subjects were instructed to complete the tasks at their own pace.

Subjects also completed a computer attitude questionnaire, the Building Memory Test (spatial memory) and the Diagramming Relations Test (logical reasoning).

RESULTS

Performance measures included number and types of questions asked each day and overall; time to complete each task; and number and types of errors per task. For the editing tasks, the error measure was computed as a proportion accounting for text processed or number of edits attempted. The data was analyzed using a one way analysis of variance with type of word processing program as the independent measure.

The results indicated that age, spatial memory and logical reasoning were important predictors of learning success. These variables were significantly correlated with task completion time, errors per task and number of questions asked (Table 1) and were included as covariates in the ANOVA for these measures. As shown in Table 1, individuals who demonstrated higher logical reasoning scores and higher spatial memory scores performed better on the text-editing tasks. Also time on task, errors per task and number of questions asked during the sessions increased as age increased.

TABLE 1

RELATIONSHIP AMONG PREDICTOR VARIABLES AND PERFORMANCE MEASURES

MEASURES	PREDICTORS			
	AGE	TYPING SKILL	LOGICAL REASONING	SPATIAL MEMORY
ERRORS/ TYPING	.48***	-.27	-.37**	-.48***
ERRORS EDITING	.28*	-.08	-.53***	-.62***
TIME/ TASK	.50***	-.22	-.41***	-.65***
EDITS	-.51***	.18	.38**	.70***
QUES.	.50**	-.16	-.51***	-.51***

* P< .05
** P< .01
*** P< .001

Overall, the results suggest that the structure of the text-editing program had an impact on the learning success of older adults. The data indicated a significant difference among the groups in average time per task (F (2,45)= 12.94, p<.001), errors on the editing tasks (F (2,45)= 5.73, P<.01), and total number of questions asked across Days 1 and 2 (F (2,45)=5.2, p<.01). As shown in Table 2, subjects using Wordstar Classic took, on average, significantly longer to complete each task than the other two groups (p<.01) and made more errors on the editing task. It appeared that people using Wordstar Classic spent more time reading the manual, whereas people using Wordstar Pull-Down spent the least amount of time reading the manual and instead used the on-screen information. Subjects using WordPerfect asked the greatest number of questions during the session and made more errors on the advanced editing tasks.

In order to gain insight into aspects of the text-editing programs that were particularly difficult for participants, types of errors committed and questions asked were examined for the three programs. Subjects using Wordstar Classic made more errors than did the other subjects inserting and deleting text (p<.01). They also asked questions regarding setting margins and tabs and underlining text. Individuals using WordPerfect seemed to have difficulty understanding the procedures involved in accessing or starting the program and had substantial difficulty understanding program codes. They also made the most errors associated with the layout of documents (p<.001). People using Wordstar Pull-Down had problems inserting text and asked questions regarding the use of the ENTER, space and tab keys. However, their questions were fewer in number than the other participants.

DISCUSSION

The results of the current study suggest that the design of the text-editing program is important to the learning efficiency of older adults. Subjects in this study were more successful using a program with pull-down menus rather than a program with on-screen menus or a full page editor which relied heavily on the use of function keys. It may be that the act of "pulling down" a menu created more active processing on the part of the user and thus focused attention on the on-screen information. The beneficial effect of active processing for older learners is consistent with other findings regarding older adults and cognitive interventions. Also, an effort was made in the pull-down program, in contrast to the on-screen program, to link the name of the menu to menu content. This strategy reduces the need for memorization which is beneficial for older people. Finally, the full screen editor (WordPerfect) relied on function keys, which could take on different meanings depending on the context in which they were used. This is problematic for all people as it creates a heavy memory load but especially for older adults who often have difficulty recalling new information which is confusing or complex.

TABLE 2

GLOBAL PERFORMANCE MEASURES IN RELATION TO WORD PROCESSING PROGRAM

	WORDSTAR CLASSIC	WORDSTAR PULLDOWN	WORDPERFECT
ERRORS TYPING TASK	X=7.08 SD=4.84	X=4.97 SD=3.35	X=9.25 SD=5.90
ERRORS TYPING TASK	X=.21 SD=.16	X=.11 SD=.11	X=.14 SD=.11
TIME TASK	X=31.46 SD= 7.81	X=29.02 SD=16.16	X=40.91 SD=29.70
TOTAL QUES.	X=50.67 SD=47.17	X=26.12 SD=16.16	X=37.13 SD=29.70

The findings from this study suggest an important area for future research. Efforts need to be directed towards systematically identifying software design elements that are difficult for older people to grasp and conversely those design features which aid performance. It is anticipated that this type of research will benefit all computer users not just older adults.

REFERENCES

Czaja, S.J., Hammond, K., and Joyce, B. (1989). Word processing training for older workers: Phase II final report. Prepared for the National Institute on Aging, National Institutes of Health, Washington, D.C.

Egan, D.E. and Gomez, L.M. (1985). Assaying, Isolating and Accommodating Individual Differences in Learning a Complex Skill. In R.F. Dillion (Ed.) Individual Differences in Cognition, Vol. 2, New York: Academic Press

Elias, P.K., Elias, M.F., Robbins, M.A. and Gage, P. (1987). Acquisition of Word-Processing Skills by Younger, Middle-Age and Older Adults, Psychology and Aging, 2, 340-348.

COMPUTER COMMUNICATION AMONG OLDER ADULTS

Sara J. Czaja
M. Cherie Clark
Ruth A. Weber
Stein Gerontological Institute
Miami, Florida

Daniel Nachbar
Bell Communications Research
Morristown, New Jersey

Currently an estimated 2.8 million people aged 65 years or older need some type of assistance in carrying out everyday activities. Therefore, there exists a need to identify strategies which enhance the functional independence of older adults. There are a number of computer and communication technologies which can be used to provide support. For the potential of these technologies to be realized, they must be easy to use, easily available and accepted by older adults. The goal of this research project was to evaluate the feasibility of having older people use computers to perform tasks in their own home environment and to identify design parameters which facilitate their interaction with these systems. The study involved installing a customized e-mail system in the homes of 38 elderly women. Additional features were added over the course of the project. Data collected included: frequency of use, number and type of messages sent, communication patterns, time distribution of messages and frequency of features used. Overall the results of the study indicate that older adults are willing and able to use computers in their own homes if the system is simple, features are added in an incremental fashion and they are provided with a supportive environment.

Introduction

In the next few decades the number of older people in the population is going to dramatically increase. Also, data regarding the living arrangements of older adults show that most older adults live at home either alone or with a spouse. These trends have significant implications for researchers, designers and policy makers as people in older age groups typically have greater problems maintaining functional independence and warrant more environmental and support service. For example, current estimates suggest that 2.8 million people aged 65 years or older need some type of assistance in carrying out everyday activities. Therefore there exists a need to identify strategies which enhance the functional independence of older people.

There are a number of computer and communication technologies which can be used to provide support for and enhance the quality of life of older adults. For example computers can be used for communication, health services, educational services, and entertainment services. There are currently over 200 older people using SeniorNet an "electronic city" developed specifically for seniors. SeniorNet includes applications in word processing, financial management, e-mail, computer conferencing and travel services. Current data indicates that the participants are highly enthusiastic and willing to use the system (Furlong, 1989). Despite the widespread use of computer network such as SeniorNet there are data which indicate that older people often have difficulty using current computer systems (eg. Czaja, Hammond, Blascovich and Swede, 1989).

In order for the potential of computer technologies to be realized they must be easy to use, available and accepted by older people. The goal of this project was to evaluate the feasibility of having older people use computers in their own homes, to perform routine tasks and to identify design parameters which facilitate their interactions with these systems. Another goal of the project was to identify computer applications that are of potential value to older people.

Method

The project involved installing a customized electronic mail system, developed by Bellcore in the homes of 38 women ranging in age from 50-94 years (mean age = 68). All of the participants who volunteered for the study had at least a high school education, and no prior experience with computer technology. The participants were recruited from the community of South Florida. Each subject had a terminal, modem and printer connected through telephone lines to a host system that uses a Compaq 386 computer within a XENIX environment. Initially the system was restricted to electronic mail with a simple editor, however several additional features such as news and weather were added over the course of the project. The system was designed so that it was easy to use and operating procedures were consistent across applications.

Participants were initially trained to use the system on a one-to-one basis in the lab at the Stein Institute. The system was then installed in their home and they were given a brief review. They were asked to use the message system to send messages to other participants and to project staff and to use the additional features as they were added. They were encouraged to use the system, but it was emphasized that they should use the system as frequently as they wished, there was no limit on the amount of messages that could be sent.

Participants were given several questionnaires throughout the course of the project to assess their perceptions of the system. In addition data was collected on frequency of use, number, type and length of messages, frequency of usage of features, communication patterns and time distribution of messages.

Results

Overall the data indicated that older people are willing and able to use computer communication systems if the system is simple to use, features are added in incremental fashions and they are provided with a supportive environment. Specifically, 87 percent of the sample found the system easy to use and only 3 percent reported having any difficulty remembering operating procedures. Also, errors committed while initiating a message, over all subjects, over a twelve month period, varied from a high of 29 percent to a low of 10 percent. The greatest number of errors occurred upon the introduction of a new feature. However it should be noted that number of errors committed was consistently low over time ($r = -.127$, $p > .05$) because of the design of the system. In fact most of the errors were typographical in nature.

The data also indicate that our sample like using computers and find them useful. Ninety-seven percent of our participants reported that they find it valuable to have a computer in their own home and 80 percent indicated that they would miss their computer if it were removed. Despite the participants' expressed interest in using computers, the results showed that actual usage significantly declined over time ($r = -.665$, $p = .036$). The average number of messages sent in the month of April was approximately 1,100 as compared to approximately 400 in the following January (Figure 1).

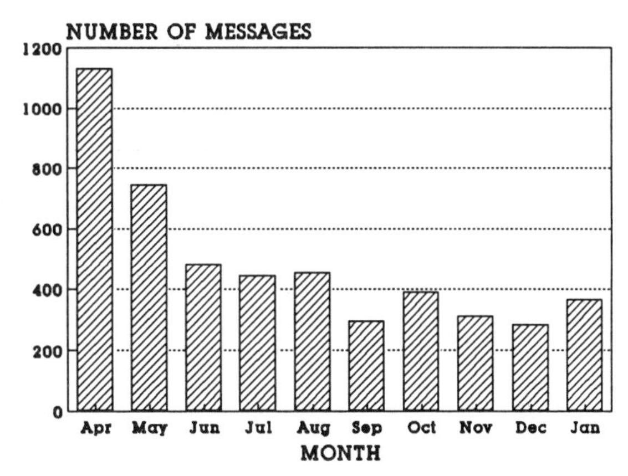

Figure 1. No. of Messages Sent over Time

The decrease in system usage may be due to a novelty effect and the simplicity of the sytem with respect to available features. In fact, the participants frequently asked to be provided with additional capabilities. Also 76 percent indicated that thye were becoming bored with sending messages to the same group of people. Specific requests included: word processing (79%), community service information (76%), health services (76%), and online shopping/banking (70%).

During the course of the project, several new features were added to the system. They included world news, weather information, entertainment news, and movie reviews. The usage rate increased each time a new feature was added. Of the features available, those that were used most frequently were e-mail, weather, and movie reviews.

The results also showed that system usage varied as a function of age and employment status. Persons aged 65-74 years consistently generated the largest number of messages ($F(2,30) = 5.13$, $p = .01$). Persons over 75 sent the fewest number of messages. These findings may be explained by employment status and health status. The data showed that people who were retired spent the most time using the system (Figure 2). Also, there was a significant relationship ($r = -.35$, $p = .06$) between visual and system usage. People with low acuity used the system less often. This may be the reason for the lower usage of the system by individuals over 75 years of age. As might be expected there was a significant relationship between age and visual acuity ($r = .66$, $p < .001$). This suggests that the design of the system might be improved by enlarging the type size for both the screen and hard copy print.

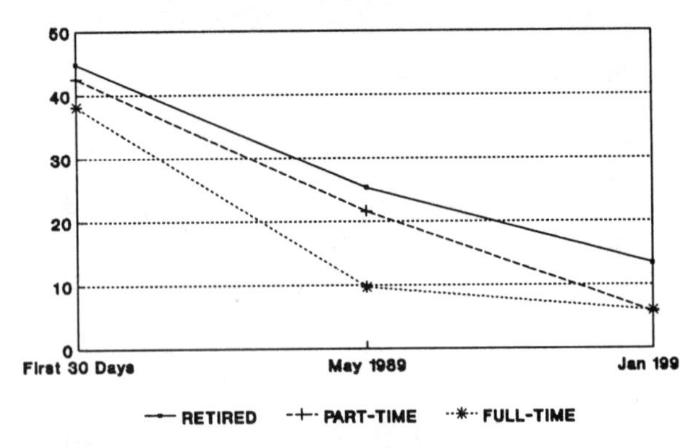

Figure 2. Number of Messages as a function of Employment Status

Other system reported by the participants included problems operating the printer and determing whether or not a message had in fact been sent. The participants also indicated that the printed message was sometimes difficult to read. However, the most frequent complaint was that the system was too simplistic and restricted in application.

Conclusions

The research reported demonstrates that computer technology can be very valuable in enhancing the quality of life for older adults. Our sample indicated that it was useful to have a computer in their home and also that computers provided them with a mental challenge and enhanced their social interaction.

However, our data suggests that for this population, computer systems must be easy to use; features should be added in an incremental fashion; and people should be provided with a supportive environment. An important aspect of our study was that participants were provided with one-on-one training and a help service was available at all times.

Another important finding of this study is that unless computer systems are equipped with features perceived as useful they will not be used. Features most desired by our population included: word processing, news, weather, community information, and health related services. In this regard, computers may be especially valuable for frail elderly who have restricted mobility. In fact, our data indicated that the system was used most frequently by people who spent considerable amounts of time at home. Also most people indicated that they made new friends using the system. Computer may also be used to help relieve health care and case management burdens associated with frail older people.

In addition to providing useful features, systems must be designed so that the needs of this population are accomodated. The data clearly showed , for example that print size is an important design consideration.

Clearly, more research is needed in this area. the data reported are based on a limited sample using a simplistic, closed-loop system with limited applications. However, current real world systems such as SeniorNet, which older people are likely to encounter are more complex, and are different in design. Most of these systems have been designed without consideration of the needs of older adults which may limit their success among this population. Thus, there is a need to evaluate the ability of older adults with varying capabilities to learn and use these types of systems.

References

Czaja, S. J., Hammond, K., Blascovich, J., and Swede, H. (1989). Age group differences in learning to use a text - editing system. Behavior and Information Technology, 8, 309-319

Furlong, M. S. (1989). Crafting an Electronic community: The SeniorNet story. International Journal of Technology and Aging, 2, 125-134

This research was done in collaboration with Bell Communications Research and supported by the John and Mary Markle Foundation.

SUBJECTS IN HUMAN FACTORS: WHO SHOULD THEY BE?

Tammy E. Fleming
Battelle Human Affairs Research Centers

Stephen J. Morrissey, Ph.D.
Industrial Ergonomics Consultant, Oregon OSHA

Rhonda A. Kinghorn
Battelle Human Affairs Research Centers

The already diverse workforce in America is expected to diversify at an even greater rate over the next decade. Projected workforce changes include those of age, gender, and race. The recently passed Americans with Disabilities Act also ensures that a growing number of persons with diverse physical needs will enter the workforce. Data from Moroney and Reising (1992) provide some clear indications of the types of subjects currently used in human factors experiments. Not surprisingly, these subjects represent a range of persons that is much less narrow than the range represented in the current and projected workforce. If not corrected, the differences between human factors subjects and those of the American workforce will increase at a magnified rate. To ensure that the results produced from human factors experiments are useful and valid, researchers should first analyze the diverse characteristics of their intended users and select subjects who possess these characteristics.

INTRODUCTION

Human factors research is generally aimed at improving the relationship between workers and their workplace. To be relevant, it is important that either (1) human factors subjects resemble the ultimate workers, or that (2) the differences between human factors subjects and the ultimate workers do not significantly affect the results of the research. This paper addresses the projected characteristics of workers in the next decade and their implications for the selection of human factors subjects.

It is first important to briefly discuss the current characteristics of human factors subjects, although this topic is discussed in greater detail in the paper presented in this session by Moroney and Reising. The two classic subject populations of the college sophomore (usually white and middle class) in an introductory psychology class and the military recruit (having been selected by less obvious physical, physiological, racial, sexual, and cultural criteria) are still the largest source of subjects for human factors studies. These populations are often used because of they are readily available and cost-effective.

It has long been recognized that, for a variety of reasons, the subjects in these two pools do not represent the general worker population, or those workers who will ultimately benefit from the research. As will be discussed below, this lack of representation is likely to worsen if sources of subject selection do not change due to the diversification of the American workforce.

CURRENT AND PROJECTED WORKER CHARACTERISTICS

The U.S. workforce is projected to change in a number of ways over the next decade. These projected changes include those of age, gender, race, and physical impairment. The following section discusses the major demographic projections and their likely impact on the American workforce.

One of the most marked changes concerns the characteristics of workers who will enter the workforce over the next decade. White males are projected to account for only 15 percent of net additions to the labor force between 1985 and 2000, while other worker groups will experience marked increases (Johnston & Packer, 1987). Women are expected to account for three-fifths of new employees in the next decade, and will constitute approximately 47 percent of the workforce by the year 2000. Non-white workers, who currently make up 13 percent of the labor force, are expected to account for nearly one-third of new employees between 1985 and 2000. Immigrants of all races, many of whom will have English language limitations, are expected to account for an additional percentage of jobs (Johnston & Packer, 1987).

A second important change that is projected to occur over the next decade concerns worker age. The workforce will be progressively "graying" as a result of two factors: the average age of workers is projected to rise as the baby boom generation ages, and workers are likely to delay retirement and work to a significantly older age over the next decade. The combination of these factors will produce a population with a median age of 36 by the year 2000, which is six years older than at any time in the history of the nation (Johnston & Packer). The median age of the workforce can be expected to rise in a similar fashion.

A third change concerns workers with physical impairments. Between 10 and 12 percent of the American workforce are considered to have physical or functional impairments that may affect their work (Vanderheiden, 1990). The recently passed Americans with Disabilities Act ensures that a growing number persons with physical and other impairments will be entering the workforce.

Each of the above aspects of workforce diversity directly affects the interaction between the workers and their workplace. The classical interpretation of these characteristics is of singular linear effects, such as an aging workforce requiring that equipment or tasks be changed to accommodate diminished hearing, vision, or reaction times. In reality, the various characteristics of the emerging workforce are highly interrelated and must be treated accordingly. For example, a disaggregation of the demographic data reveals concentrations of particular groups of workers in particular occupations (e.g., a high concentration of immigrant workers in agriculture). For this reason alone it can be argued that human factors subjects should mirror the characteristics of the workers for which the research is intended. As shown in the following section, however, there is further evidence that a failure to do so limits the extent to which the research results can be generalized.

CONSIDERATIONS FOR HUMAN FACTORS SUBJECT SELECTION

The extent to which physical differences vary by characteristics such as gender and race has been a controversial and hotly debated topic. While the exact effects of such characteristics vary according to a number of factors, there are some general considerations that should be taken into account when selecting human factors subjects. The following are some examples of some particular characteristics which, if left unaddressed, could limit the usefulness of human factors research for particular worker populations.

Age

While there are a number of studies on the characteristics of older workers, many of these studies sample workers who grew up in during the 30's, 40's and early 50's-- populations that do not necessarily have the same psychological and social goals and reward systems as do current workers, much less the physical and social-psychological characteristics of tomorrow's worker. Although Human Factors, Ergonomics, and other professional journals have had numerous papers (and special issues) devoted to human factors concerns of aging, these studies represent only a small amount of usable data for the emerging workforce.

Particular characteristics of older workers may be misrepresented by traditional subjects in human factors studies. For example, fine motor skills, endurance, mobility strength, and freedom of movement may decrease with age (Smith, 1990). Sensory capacities such as vision weaken, and auditory thresholds increase, as do touch, temperature, and odor thresholds (Kelly & Kroemer, 1990). In addition, as technology changes older workers may not have the same abilities to pick up new skills as do their younger counterparts, who constitute a large percentage of traditional subjects. These differences between young and old workers may lead to situations in the workplace that unnecessarily and unintentionally hamper the effectiveness of older workers.

Race

While a number of studies have documented significant and subtle differences in anthropometry and related physical features between human races, the number of usable cross-racial studies of perceptual-cognitive issues and anthropometry is small. The studies that do exist rarely deal with potential differences between native born and transplanted subjects or consider the effects of immigration on social, psychological and physical features. Much of the data that do exist is based on military populations of the native country that do not represent those workers coming to the United States.

The consideration of race is particularly important for two reasons. First, in addition to more obvious physical differences, some research suggests the possibility of more subtle differences between races that could have unintended workplace effects. DeAngelis (1991) reports on current research that has identified some possible differences in the rate and manner in which whites, blacks, and hispanics responded to particular medications. Second, with respect to immigrants, it is becoming increasingly important to consider the needs of workers who do not speak English. The use of subjects with limited English may be the best way to anticipate the types of problems these workers are likely to experience.

Functional Impairment

While a wide range of physical and mental capabilities exist among workers, this range of capabilities has not been addressed in studies using traditional subject populations. Although estimates vary, approximately 30 million persons in the United States have physical or functional impairments that result from injury, illness, or aging (Vanderheiden, 1990). Although the total number of affected persons is large, the number with any particular type or combinations of functional impairments is much smaller. The variation in impairments, which range from slight hearing loss to total paralysis, makes research on this group of workers particularly difficult.

The recently passed Americans with Disabilities Act includes a number of provisions for physically disabled workers, and is expected to result in the increased entry of physically and functionally impaired workers into the workforce. The inclusion of subjects with physical and functional impairments in human factors research will allow researchers to make better decisions on task and work station design, job aids, and training, and will help to ensure that these workers are maximally effective.

CONCLUSION

To be able to confidently generalize the results of human factors experiments to the emerging workforce, human factors subjects must represent the populations for which the research is ultimately intended. In addition to classical anthropometric, strength, and physiological data, the selection of human factors subjects should also consider perceptual-cognitive and sociological issues that could impact a worker's performance, perceptions or attitudes. These efforts should begin immediately so that the results can be used to design new equipment, materials, and policies that address the particular workforce characteristics of the next decade.

REFERENCES

DeAngelis, T. (1990). Ethnic groups respond differently to medication. APA Monitor, 27.

Johnston, W. & Packer, A. (1987). Workforce 2000: Work and Workers for the 21st Century. The Hudson Institute, Indianapolis, IN.

Kelly, P. L. and Kroemer, K. H. E. (1990). Anthropometry of the elderly: Status and recommendations. Human Factors, 32 (5), 571-595.

Smith, D. B. D. (1990). Human factors and aging: An overview of research needs and application opportunities. Human Factors, 32 (5), 509-526.

Vanderheiden, G. C. (1990). Thirty-something million: Should they be exceptions? Human Factors, 32 (4), 383-396.

INDUSTRIAL ACCIDENTS: DOES AGE MATTER?

Comila Shahani
Rice University
Houston, TX

ABSTRACT

This study examined the relationship between risk of accident involvement and the aging process. It was predicted that the relationships between age and accident frequency and severity would differ depending upon job context. The study also examined the extent to which progressive selection was a factor. 7,131 accidents that occurred over a five year span in a large Southwestern petrochemical facility were analyzed. In addition, information about age and employment history was obtained for the 3,015 employees at this plant. There were no differences in the proportions of employees in different age groups across job families indicating progressive selection was not a factor in this workforce. Younger workers had higher overall accident rates than older employees; but there were few differences between them in the proportion of severe accidents incurred. The relationship between age and accident frequency and severity did not differ across job families (except in the oldest age group, where the accident frequency rate declined for two of the five job families).

INTRODUCTION

Accidents can occur through the life span of an individual, and it is possible that an individual's risk of accident involvement increases with the aging process. The assumption that age related decrements in functional abilities leads to lowered performance is one of the strongest deterrents to hiring older workers in industry (Schmidt, 1985). It is possible that these declines in functional abilities significantly enhance the likelihood of a workers accident involvement. Given the recent and projected rise in the median age of the American workforce, it becomes especially important to understand this relationship.

Studies generally show, after controlling for occupation and industry, that as age increases the likelihood of accident involvement decreases. Thus, older workers are less likely to incur accidents than younger workers (Coates & Kirby, 1982). In his review of industrial accidents, Root (1981) concluded that though accident frequency tends to decrease with age, accident severity tends to increase with age. In other words, younger workers tend to be hurt more frequently, though not as seriously. In his study, Root (1981) also found that the occupational injury frequency declined steadily up to 64 years, and then dropped down even more sharply. Perhaps this reflected increased experience and skills possessed by older workers. Alternatively, this could reflect progressive selection, whereby people who were ill-suited for their jobs were separated from the organization due to retirement, voluntary selection, or termination (Welford, 1976).

Different studies have provided divergent findings. One reason is the inability to collect uniform data on accident exposure and incidence (Root, 1981). Another plausible reason for the discrepent findings is because the importance of job context has not been adequately considered. Griew (1958) found that accident rates in some jobs increased with age to a greater extent than others, and jobs with higher accident rates for older people were predominantly occupied by younger workers. Accidents that decreased with age were the ones that required experience and refined skills; whereas, some accidents that increased with age were due to a slowing of sensori-motor processes (King, 1958). It has often been found that middle aged employees tend to leave speeded or rigidly paced jobs for jobs that are self-paced (Belbin, 1953; Clark & Dunn, 1955; Forsman, 1976; Griew & Tucker, 1958; Powell, 1973).

Finally, Petree (1985) and Schmidt (1985) analyzed industrial accidents in a large petrochemical facility and confirmed the notion of situationally dependent functional declines in worker accidents. There was little evidence for the progressive selection process in their studies.

The present study attempted to extend the results of Petree (1985) and Schmidt (1985) working with an expanded database. The following hypotheses were examined:

Hypothesis 1: The proportion of employees in different age groups will differ across job categories providing evidence of progressive selection.

Hypothesis 2: Frequency of accidents will decrease with age; whereas, severity of accidents will increase in older age groups.

Hypothesis 3: Frequency and severity of accidents will differ for the various job categories.

Hypothesis 4: The relationship of accident frequency and severity with age will differ across job categories.

METHOD

The Data

Two databases (Laughery et al, 1983) were examined. The first contains reports of 7,131 accidents at Shell Oil Company's Deer Park Manufacturing Complex during 1981 to 1985. There were 1,839 different employees who had one or more accidents. Demographic, scenario, causal, and injury-severity information on each incident was contained in the data. The second database consists of information on employee job category, age, time with Shell, and time in current job for all 3,015 employees in the complex.

Age, accident severity and occupation were grouped in order to facilitate the analyses. The groupings took into account traditional and industrial gerontological findings, job requirements, and exposure to hazards.

Five age groups were constructed. The age group ranges in years were: 21-30, 31-40, 41-50, 51-60, 61+ years. An employee was counted as two separate observations if the employee moved from one age category to the next over the five year data collection period. This was done because the employee may have contributed accidents to the accident database in either, or both of the age categories. Thus, the number of observations increased to 3,810.

Five job families were constructed from 12 hourly pay positions, similar with respect to job requirements and work environment. The job family classifications were derived by Schmidt (1985):

OPERATIONS PERSONNEL: Operator and Lab tester (**OP**)
ELECTRICAL CRAFT: Electrical and instrument mechanic (**EC**)
PROCESS CRAFT: Pipefitter and machinist (**PC**)
MAINTENANCE CRAFT: Boilermaker and welder(**MC**)
MISCELLANEOUS CRAFT: Carpenter, insulator, painter, and garage mechanic (**MSC**)

Accident frequencies were computed for the different age and job classifications. These frequencies consisted of the number of accidents sustained by members of the five age groups in each job family. Frequency was divided by the number of employees in each category to obtain a rate measure: the number of accidents per employee.

Accidents were classified into three levels of severity depending upon the nature of injury received; no injury, minor injury, or major injury. No injury were accidents requiring no medical treatment and involving no injury. Minor injury were accidents requiring administration of first aid by a nurse or physician. Major injury were OSHA recordable accidents requiring treatment and follow up by a physician or resulting in restricted work capacity or lost time.

RESULTS

Chi-square analyses testing for significant differences in obtained and expected frequency distributions could not be performed because in general, the number of observations far exceeded the number of employees, thus violating a critical assumption of the chi-square statistic (the assumption of independence of observations).

Age Group and Job Family Classification

The distribution of employees in age groups for each of the five job families is presented in Table 1.

Table 1 Population Base Rates (Number of Employees)

	Age Group					
Job Family	21-30	31-40	41-50	51-60	61+	Total
OP	583	1136	348	269	74	2410
EC	35	253	119	52	13	472
PC	63	291	106	73	24	557
MC	33	118	29	23	10	213
MSC	28	77	19	25	9	158
Total	742	1875	621	442	130	3810

An examination of the table indicates that the number of employees in given age groups is disproportionate. Almost 50% of employees were between 31-40 years; whereas, only 3% of employees were over 61 years. In addition, the number of employees in the different job families was very disproportionate, with 63% of employees in operations personnel, and only 4% in miscellaneous crafts. Finally, the number of employees in each of the age groups was disproportionate across the five job families. It ranged from a high of 1,136 for the 31-40 year group in operations personnel to a low of 9 for the over 61 age group in miscellaneous craft.

There were not, however, substantial differences in proportional representation of different age categories across the job families nor in the proportion of workers in different jobs across age categories. In other words, the proportion of employees in the various age groups did not greatly differ across job families.

Age Group and Accident Frequencies.

The number of accidents per employee for the five age groups is presented in Table 2.

Table 2 Accident Frequency by Age Group

Age Group	Frequency	Accidents/Employee
21-30	1495	2.01
31-40	2988	1.60
41-50	693	1.12
51-60	723	1.63
61+	238	1.83

An examination of Table 2 indicates a range in accident rates for the different age groups. Employees between 21 to 30 years and over 61 years had the highest accident rates. In general, accidents tended to decrease across the first three age groups and then the trend took an upward turn. However, younger workers did have a higher frequency of accidents (2.01 per employee) than the oldest employees (1.83 per employee).

Age Group and Accident Severity

The proportion of employees in each age group who had major injuries as opposed to minor or no injuries was computed.

Table 3 Accident Severity by Age Group

Age Group	**Injury Severity** Major	Minor	No	Total
21-30	91	1373	99	1563
31-40	209	2967	74	3250
41-50	43	728	14	785
51-60	36	795	25	856
61+	16	268	4	288

The percentage of total accidents that were in the major category did not vary considerably between the age groups (range 4-6 %).

Job Family and Accident Frequency

As shown in Table 4, the frequency of accidents per employee for the five job families ranged from 1.04 for operations personnel to 3.7 for miscellaneous craft. Jobs with rates over 3.00 are traditionally considered to be high risk (Schmidt, 1985).

Table 4 Accident Frequency by Job Family

Job Family	Frequency	Accidents/Employee
OP	2498	1.04
EC	804	1.70
PC	1571	2.80
MC	687	3.30
MSC	577	3.70

Job Family and Accident Severity

Table 5 presents the number of accidents for each severity level for each job family. The percentage of employees in each job family that had major injuries as opposed to minor or no injuries did not vary across job families (range 5-7%).

Table 5 Number of Accidents by Severity and Job Family

Job Family	**Injury Severity** Major	Minor	No	Total
OP	147	2254	120	2521
EC	37	747	20	804
PC	105	1423	45	1573
MC	44	641	13	698
MSC	29	539	11	579

Accident Frequency by Age Group and Job Family

The number of accidents for the different age groups by each job family is presented in Table 6. In addition, the number of accidents per employee is shown.

Younger workers had the highest frequency of accidents across all job families. In general, accident frequency tended to decline with age until 51 years, at which point the trend took an upward turn. The accident rate for the job families showed there were few differences in the relationship between accident frequency and age across job families, except in the oldest age group. For two job families, electrical and maintenance craft the accident rate showed a decline for the oldest group.

Accident Severity by Age Group and Job Family

The proportion of total injuries that were major is shown in Table 7 for each age group within each job family.

Table 6 Accident Frequencies by Age Group and Job Family

Job Family	21-30	31-40	41-50	51-60	61+
OP	827 (1.42)	1102 (0.97)	238 (0.68)	247 (0.92)	84 (1.14)
EC	76 (2.17)	397 (1.57)	196 (1.65)	117 (2.25)	18 (1.38)
PC	301 (4.77)	850 (2.92)	165 (1.56)	190 (2.60)	65 (2.70)
MC	145 (4.40)	370 (3.14)	50 (1.72)	92 (4.00)	30 (3.00)
MSC	146 (5.20)	269 (3.50)	44 (2.30)	77 (3.10)	41 (4.60)

Note. Numbers in parentheses are rates.

Table 7
Proportion of Major Injuries by Age and Job Group

Job Family	21-30	31-40	41-50	51-60	61+
Operations Personnel	.06	.07	.05	.04	.05
Electrical Craft	.04	.05	.05	.03	.00
Process Craft	.06	.08	.07	.03	.09
Maintenance Craft	.06	.06	.10	.05	.07
Miscellaneous Craft	.08	.05	.00	.04	.00

Overall, the proportion of major injuries did not greatly vary within the five age groups across the different job families. The proportion of major accidents ranged from .00 to .10.

DISCUSSION

The present study was conducted to examine the relationship between age and industrial accidents. The first hypothesis examined whether there were differences in the proportion of people in different age groups across different jobs, i.e., was there evidence of progressive selection. If so, it would be expected that the proportion of older workers in high risk jobs would be less than in low risk jobs. Though, the number of employees in age groups varied, and the number of employees in different job families varied, it did not appear that the *proportion* of employees in different age groups differed across job families. Therefore, this study provides little evidence of progressive selection. In fact, only 3% of employes were over 61 years for the relatively low risk job family, operations personnel; whereas, 6% of employees in miscellaneous crafts were over 61.

It was predicted that accident frequency would decline with age, i.e. younger employees would have more accidents than older workers. In support of this, the relationship between accidents and age was negative. However, the declining trend for accident frequency took an upward turn in the the 51-60 group. In spite of this upward trend, younger workers did have a higher frequency of accidents (2.01 per employee) than the oldest workers (1.83 per employee). This trend was consistent with research conducted by Broberg (1984) in Sweden.

It was expected that the proportion of major accidents would be higher for older than younger workers. However, the proportion did not differ considerably between age groups. This finding is in contrast to previous research which has almost consistently shown that older employees have more severe accidents than younger employees. Thus Hypothesis 2 received only partial support.

The five job families in this study had markedly different accident frequency rates. In terms of accident severity, there was little difference in the proportion of major accidents across different job groups. High and low risk jobs did not greatly differ in proportion of severe accidents. Hypothesis 3 received partial support in this study.

Past research has generally ignored the importance of the situation or job context when examining the relationship of industrial accidents and age. This study predicted that the relationship of age-related functional ability declines and frequency and severity of industrial accidents would differ depending upon job family. However, regardless of the riskiness of occupational groups, the youngest age group had the highest frequency of accidents. The over 61 year group had the lowest rates for electrical, process, and maintenance crafts, whereas they had higher rates for miscellaneous craft- the highest risk occupational group. The accidents for all age groups increased for high risk jobs. In general, the accident rates for the job families showed there were few differences in the relationships between accident frequency and age across job families, except in the oldest age group. However, since there were not very many people in the oldest age groups it is difficult to make conclusive statements regarding these differences. There was no relationship between age and accident severity across different job families.

In general, the data in this experiment have not provided much evidence for Hypothesis 4. There were few differences in the relationship of accident frequency and severity with age across the different job families. The relationship of age and accidents did not differ for high and low risk occupational groups. However, an important factor to keep in mind is the relative homogenity of these job families. It will be important to replicate these results using a wider occupational sample.

No statistical tests were conducted to determine whether the differences were significant or merely due to chance. As pointed out earlier, chi square analyses were inappropriate for this data. However, in situations like this where there are a large number of cases in the database, statistical tests of significance may not be necessary to detect meaningful differences. Any interpretation of data finally involves some subjective judgment of the meaningfullness of such differences.

Since younger workers had higher accident rates than older workers, it was considered important to examine the common accident scenarios for the different age groups. Do younger and older workers differ in terms of the prior activities, accident events, and injury events that accompany accidents? Scenario analyses for different age groups were also examined within each job family. On the whole, there were few differences in the scenarios for different age groups, nor did age groups greatly differ in accident scenarios within different job families.

Given the increase in the median age of the American workforce, this is heartening news. It does not appear as though increased age is associated with an increase in the rate and severity of industrial accidents, at least for the current occupational sample.

REFERENCES

Belbin, R. (1953). Difficulties of older people in industry. Journal of Occupational Psychology, 27(2), 179-190.

Broberg, E. (1984). Use of census data combined with occuapation and accident data. Journal of Occupational Accidents, 6, 147-153.

Clark, F. & Dunn, A. (1955). Aging in Industry. London, Great Britain: The Nuffield Foundation.

Coates, G. & Kirby, R. (1982). Organism factors and individual differences in performance and productivity. In E. Allusi & E. Fleishman (Eds), Human performance and productivity, Vol 3: Stress and performance effectiveness. Hillsdale, NJ: Lawrence Erlbaum Associates.

Forsman, s. (1976). Occupational mobility and age. In H. Thomas (Ed.), Pattern of aging, (pp 174-197). New York, NY: S. Karger Publishing Co.

Griew, S. (1958). A study of accidents in relation to occupation and age. Ergonomics, 2(1), 17-25.

Griew, S, & Tucker, (1958). The identification of job activities associated with differences in the engineering industry. Journal of Applied Psychology, 42(4),278-282.

King, H. (1956). An age analysis of some agricultural accidents. Journal of Occupational Psychology, 29(2), 245-253.

Laughery, K., Petree, B., Schmidt, J., Schwartz, D., Imig, R., & Walsh, M. (1983). Scenario analysis of industrial accidents. Proceedings of the 6th International System Safety Conference, Houston, Tx., 82, 1-21.

Petree, B. (1985). Age and industrial accidents: A reexamination. Unpublished doctoral dissertation. University of Houston, TX.

Powell, M. (1973). Age and occupational change among coal miners. Journal of Occupational Psychology, 47(1), 37-49.

Root, R. (1981). Injuries at work are fewer among older employees. Monthly Labor Review, 104, 30-34.

Schmidt, J. (1985). The effects of age cohort and job family on industrial accident rates. Unpublished Masters thesis. University of Houston, TX

Welford, A. T. (1976). Thirty years of psychological research in age and work. Journal of Occupational Psychology, 49(1), 129-138.

PSYCHOLOGICAL DISTRESS IN RELATION TO EMPLOYEE AGE AND JOB TENURE

Lawrence R. Murphy
Applied Psychology & Ergonomics Branch
National Institute for Occupational Safety & Health (NIOSH)
Cincinnati, Ohio 45226

The present study examined psychological distress among workers at various stages of career development, with special reference to the first year of job tenure, the organizational socialization period. Measures of psychological and physical health, as well as demographic data, were obtained from 3,151 employed persons who participated in a national health interview survey conducted by the Census Bureau in 1978. Analysis of covariance was performed in which employee age and job tenure were predictor variables, and gender, marital status, educational level, and number of physical health conditions entered as covariates. The results indicated that distress was highest among workers with less than 6 months tenure, and distress levels decreased progressively with longer tenure. Employee age moderated these effects, however, in that older workers with less than 6 months tenure reported higher levels of distress than younger workers with similar tenure. Older workers also showed a delayed decrease in psychological distress with longer tenure than younger workers. The results identify organizational socialization as a critical period with respect to employee mental health.

Little empirical research has examined the relationship between job stress and stage of career development. The substantial amount research devoted to job stress and career development have developed largely as independent endeavors. For example, job stress studies typically treat age and tenure (i.e., career stage) as nuisance variables whose influence on health outcome variables is controlled statistically. Likewise, career development studies typically do not include a health outcome variable, preferring instead to focus on performance, withdrawal behaviors, job satisfaction. Consequently, there has been little cross-fertilization of ideas and theories in these areas.

The few studies which examined job stress across career stage have reported provocative findings. There is evidence that work conditions which are perceived as stressful, and the strategies used to cope with stress, differ among older and younger workers (Hurrell, McLaney, and Murphy, 1990). Also, levels of psychological distress appear to differ by age; older workers generally report lower distress than younger workers (Osipow and Doty, 1985). These results suggest an interaction between job stress and career stage which could have important implications for an aging U.S. workforce.

The purpose of the present study was to assess relationships among employee age, job tenure and psychological distress, using data obtained from a nationally representative health interview survey. Two hypotheses were tested: (1) Psychological distress will be highest among the lowest tenured workers, reflecting the stressful effects of the early organizational socialization period; and (2) the distress associated with the early socialization will be less pronounced for older newcomers because of their prior experience with socialization processes and their increased coping resources.

METHOD

The data were obtained from a health interview survey conducted in 1978 which was representative of noninstitutionalized persons aged 18-64 who resided in the continental United States. Response rate to the interview was 84.4%. A total of

3,151 employed persons comprised the sample for this study.

The survey contained 20 questions dealing with symptoms of psychological distress. The questions were similar to the "demoralization" scale developed by Dohrenwend, Shrout, Egri, and Mendelson (1980). The 20 items were subjected to principal components factor analysis. Twelve (12) of the 20 items factored together, forming an internally consistent measure of psychological distress (alpha= 0.78). Age, job tenure, education, and number of illnesses were continuous variables; gender, race, and marital status were dichotomous variables.

Analysis of covariance (ANCOVA) was used to identify variables related to distress, and to compare adjusted mean levels of distress for age and tenure groups. For these analyses, age was grouped into five strata (18-24 years, 25-34 years, 35-44 years, 45-54 years, and 55-64 years) and job tenure was grouped into seven strata (0-5 months, 6-11 months, 1-2 years, 3-5 years, 6-10 years, 11-19 years, and 20+ years). Also, marital status was dummy coded as Marital1 (formerly married vs. currently married) and Marital2 (never married vs. currently married).

Means, standard deviations, and correlations among the study variables are shown in Table 1. The mean age of respondents was 37.93 years. Fifty-six percent (56%) were male and 89% were white. Average educational level was 12.42 years, and the average job tenure was 7.69 years.

The ANCOVA model contained main effects for age and job tenure group, the age by tenure interaction, and the covariates (gender, race, marital status, educational level, and number of illnesses). The full ANCOVA model was significant [$F(38, 2994)=35.30$, $p<=0.001$] and explained 31% of the variance in psychological distress scores. Employee age was not significant in the model ($p>0.05$) but tenure and the interaction of age with tenure were significant ($p<=0.05$). All of the covariates except Marital2 (never married) were significant in the model ($p<=0.05$).

Workers with the 0-5 months tenure had the highest distress scores for all age groups except the middle age group (35-44 years), whose distress levels did not vary with tenure and were similar to the sample grand mean. Workers in the three longest tenure groups reported average or below average levels of distress, compared to the sample grand mean. Older workers with low tenure (i.e., experienced newcomers) had the highest levels of distress, especially newcomers in the 55-64 group.

Whereas distress decreased with increasing tenure for the 18-24 and 25-34 age groups, distress remained high with increasing tenure for the oldest two age groups (45-54 and 55-64). This effect was especially prominent in the oldest group, where distress did not decrease to the sample grand mean until these workers accrued more than 5 years tenure.

The pattern for the oldest age group was noteworthy. Levels of distress were very high in the lowest tenure group, but distress decreased after 6-11 months tenure, although the mean was still substantially above the overall sample average. However, the number of workers in the lower two tenure groups (0-5 months and 6-11 months) was very low ($n=5$ and 8, respectively). If the two tenure groups were combined for the oldest age group, the adjusted mean distress score becomes 2.93 (s.d.=0.64), which is significantly above the sample grand mean ($p<= 0.05$). Unlike all other age groups, distress among the oldest age group increased after 1-2 years of job, and remained high in the 3-5 year tenure group. Only after five years of tenure did the level of distress in oldest age group decrease to the sample grand mean, resembling that of the younger age groups. The 45-54 age group also showed a delayed decline in distress with tenure, but the decline occurred after 2 years of tenure, instead of after 11 months as was found for the younger two age groups.

DISCUSSION

This study assessed the relationships among employee age, job tenure and psychological distress. Analyses of national health interview survey data indicated the importance of the first few months of job tenure as a critical period with respect to worker emotional health. Which specific aspects of the early months of job tenure were responsible for the high levels of distress could not be ascertained in the present study because no measures of task, role, and interpersonal demands were contained in the questionnaire survey. However, Katz (1985) has proposed a number of testable hypotheses regarding the mechanisms which may underlie the stressful effects of organizational socialization processes (e.g., relationships with supervisor, clarity of demands and worker discretion to meet the demands, and the congruence of job demands with organizational rewards.

Beyond the early socialization period, psychological distress decreased monotonically with tenure, and the effect varied with age group in an orderly fashion. The decline in distress occurred earliest in the 18-24 age group (after 5 months tenure), after 11 months tenure in the 25-34 group, after 2 years tenure in 45-54 group, and after 5 years tenure in the 55-64 group. The only exception was the middle age group (35-44), whose distress levels were invariant with tenure and were not different from the sample grand mean.

The elevated levels of psychological distress during early job tenure, and the delayed decrease in distress among older newcomers to an organization, have practical implications for organizational socialization programs. Such programs might profitably target older newcomers for greater attention to reduce the negative psychological effects associated with entry into organizations. Reduction of distress among these employees should result in improved work performance, better employee health, and lower health care costs.

REFERENCES

Dohrenwend, BG Shrout, P, Egri, G, and Mendelson, JS: (1980). Nonspecific psychological distress and other dimensions of psychopathology: Measures for use in the general population. Archives of General Psychology, 37, 1229-1236.

Hurrell, J.J., Jr., McLaney, A., and Murphy, L.R. (1990). The middle years: Career stage differences. Prevention in Human Services, 8, 179-203.

Katz, D. (1985). Organizational stress and early socialization experiences. In Beehr, T. and Bhagat, R. (Eds) Human Stress and Cognition in Organizations. New York: John Wiley and Sons, pp. 117-139.

Osipow, S. and Doty, L. (1985). Job stress across career stage. Journal of Vocational Behavior, 16, 26-32.

EFFECTS OF TASK DEMANDS AND AGE ON VIGILANCE AND SUBJECTIVE WORKLOAD

John E. Deaton
United States Navy

Raja Parasuraman
Catholic University of America

ABSTRACT

Sensory and cognitive vigilance were directly compared in two experiments. The question of whether sensory and cognitive vigilance task demands can be differentiated on the basis of perceived workload was also addressed. A third focus of the study was to investigate changes in sensory and cognitive vigilance across the adult life span. In Experiment 1 60 subjects from three age categories--young, middle, and elderly were studied. Experiment 2 consisted of 20 subjects from only the young and old age categories. Subjects performed a visual sensory and a cognitive vigilance task at low and high event rates. Each task used identical stimulus sets (pairs of digits) and differed only in the definition of a critical target. Task demands were a major determinant of vigilance performance. Cognitive vigilance was more resistant to decrement over time than sensory vigilance. On the other hand, the cognitive task was more adversely affected by high event rate than the sensory task. Older subjects had lower hit rates than young and middle-aged subjects on the cognitive task, particularly at the high event rate. Subjective workload results suggested that the increased mental demands required of the cognitive task at the high event rate were associated with performance differences between sensory and cognitive tasks. However, the results also revealed an apparent dissociation between performance and subjective workload measures. Implications of the results for display design and assessment of individual differences in monitoring capability are discussed.

INTRODUCTION

Automation in military and industrial environments has led to increasing demands for prolonged monitoring of system status on the part of human operators (Wiener, 1984). There has been considerable research on a broad spectrum of issues concerned with the ability of human monitors to maintain attention over prolonged periods in order to detect infrequent targets (vigilance). However, most of this research has investigated vigilance for sensory characteristics of displayed stimuli (Davies & Parasuraman, 1982). The ability to distinguish cognitive features and to perform higher-level cognitive operations is becoming increasingly crucial with the advent of more complex displays requiring monitors to make decisions based on a multitude of variables often symbolic in nature (Parasuraman, 1987). Despite the fact that human operators are increasingly employed in occupations requiring the processing of complex, symbolic information rather than sensory stimuli, little research has been conducted in the area of cognitive vigilance. Preliminary evidence suggests there may be differences in performance between sensory and cognitive tasks, and that cognitive tasks may be more resistant to vigilance decrement than sensory tasks (Loeb, Noonan, Ash, & Holding, 1987; Warm, Dember, Lanzetta, Bowers, & Lysaght, 1985). However, direct comparison of the two kinds of tasks has not been carried out previously, and was the focus of the present study. The question of whether sensory and cognitive vigilance task demands can be differentiated on the basis of perceived (subjective) workload was also addressed. Finally, the current and projected "greying" of the population (working and retired) suggests the need for examining vigilance capabilities in different age categories. Hence, a third focus of the present study was to investigate changes in sensory and cognitive vigilance across the adult life span.

METHOD

Subjects

Sixty volunteers, 27 men and 33 women, participated as subjects in Experiment 1 and were obtained from the University subject pool as well as the surrounding metropolitan area. Subjects were equally divided into three age categories: (1) Young--18 to 29 years; (2) Middle-aged--40 to 55 years; and (3) Old--65 to 85 years. Experiment 2 consisted of twenty volunteers, 8 men and 12 women, who were paid $15 for participating in the study. Subjects were equally divided into two age categories: (1) Young--18 to 29 years; and (2) Old--67 to 76 years. Subjects in both experiments were reported to be in good health and had corrected vision of at least 20/30. No subjects were included that reported the use of any medications resulting in drowsiness or lowered alertness.

Procedure

All subjects completed both a visual sensory and cognitive vigilance task. The sensory task required the subject to discriminate differences in the size of two digits displayed on a microcomputer video monitor. Subjects were required to detect digit pairs which included one digit physically smaller in dimension relative to the other digit. The cognitive task was identical to the sensory task except that signals were defined by the relative numeric values of the digits rather than by their physical sizes--i.e. a pair containing one even and one odd digit. Thus, identical sets of stimuli were used in both tasks, with only the definition of a critical signal being manipulated. Stimuli were presented for 300 msec at one of two event rates, high (40 events/minute) or low (15 events/minute). Subjects were given a short practice session prior to each task, and then were administered the main task continuously for 32.4 minutes. Experiment 2 was identical to Experiment 1 except for the additional requirement that each subject evaluate task workload at two different time periods for each task--once after the practice session and once immediately after the main task was completed. In order to obtain the workload measures, subjects completed the NASA Task Load Index (TLX), a multi-dimensional rating procedure that provided an overall workload score based on a weighted average of ratings on six subscales (Hart & Staveland, in press): (1) Mental Demands; (2) Physical Demands; (3) Temporal Demands; (4) Own Performance; (5) Effort; and (6) Frustration. Performance data consisted of correct and incorrect detection responses, obtained in three time period measurements of 10.8 minutes each for a total of 32.4 minutes. These data were used to calculate hit and false alarm rates for each subject by time period.

RESULTS

Experiment 1

Hit rate data were submitted to a 3x2x2x3 (age group x event rate x task x time period) analysis of variance with repeated measures on the last two factors. The analysis revealed a highly significant main effect for event rate ($F(1,54)=16.65$, $p<.001$). That is, the detection rate was higher for the low event rate task than for the high event rate task. The age group main effect and the age group x event rate interaction were not significant. Tests of the within subjects effects revealed a highly significant event rate x task interaction ($F(1,54)=11.28$, $p<.001$). Moreover, a highly significant effect was found for the time period main effect ($F(2,108)=9.40$, $p<.001$), thus revealing a vigilance decrement in all conditions over time. The event rate x time period ($F(2,108)=4.52$, $p<.05$), the task x time period ($F(2,108)=3.55$, $p<.05$), and the age group x task x time period interactions ($F(4,108)=3.80$, $p<.01$), were also significant.

The event rate x task interaction shows that the effects of event rate were greater for the cognitive task than for the sensory task. Mean hit rates for the sensory task were 89.1 and 83.6 for low and high event rate conditions, respectively; the corresponding values for the cognitive task at low and high event rates were 96.1 and 79.6. Tests of the simple effects of event rate for each task revealed that performance on the cognitive task was significantly poorer at the high event rate ($F(1,116)=27.93$, $p<.001$), whereas sensory task performance was not significantly affected by event rate. Detection performance was higher in the cognitive task than in the sensory task at the low event rate ($F(1,58)=9.68$, $p<.01$); whereas there was a nonsignificant trend for sensory task performance to be higher than the cognitive task at the high event rate.

The task x time period interaction (see Figure 1) shows that sensory task performance declined over time, whereas cognitive task performance remained relatively stable, and in fact, increased slightly in the final time period. Tests of the simple effects of task for each time period revealed a significant effect for task at the third time period ($F(1,177)=4.00$, $p<.05$). That is, cognitive task performance was significantly higher than sensory task performance at the final time period measurement. Previous to that, there was no difference in performance between the tasks. Tests of the simple effects of time period for each task revealed a significant effect only for the sensory task ($F(2,177)=4.25$, $p<.05$). Thus, decrements in performance over time were found only for the sensory task.

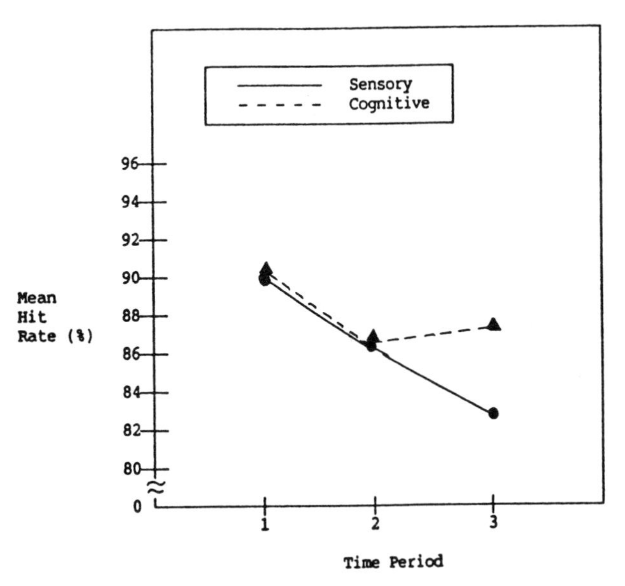

Figure 1. Mean hit rate as a function of task and time period.

Overall main effects for age in hit rate were not found. However, there was a clear tendency for older subjects to show poorer performance in the high event rate condition, particularly for the cognitive task. Older subjects also had higher false alarm rates than younger and middle-aged subjects ($F(2,54)=20.11$, $p<.001$), particularly for the sensory task.

While overall age differences in hit rate were not found in Experiment 1, results suggested that age differences were greatest for the high event rate condition. Hence, Experiment 2 was conducted using only the high event rate schedule.

Experiment 2

Hit rate data were submitted to a 2x2x3 (age group x task x time period) analysis of variance with repeated measures on the last two factors. A highly significant time period main effect was found ($F(2,36)=8.98$, $p<.001$). In addition, there was a highly significant main effect for age ($F(1,18)=25.49$, $p<.001$). Hit rate performance for the young group was significantly higher than the elderly group. False alarm results indicated that the main effect for group was significant ($F(1,18)=8.59$, $p<.01$), older subjects having a higher false detection rate than young subjects. The main effect for task was significant ($F(1,18)=5.77$, $p<.05$). Higher false alarm rates were found for the sensory task.

Table 1 shows mean subjective workload scores for each subscale of the TLX. The main effect of task was significant for the "Mental Demand" subscale ($F(1,18)=4.54$, $p<.05$). The main effect for test time was also significant ($F(1,18)=14.13$, $p<.001$), as was the group x test time interaction ($F(1,18)=5.13$, $p<.05$). Mental demands thus increased over the course of the vigil (especially for the elderly group), and the cognitive task was associated with a higher level of mental demand than the sensory task. The only subscale to show a significant age effect was "Own Performance" ($F(1,18)=4.85$, $p<.05$). That is, older subjects experienced greater demands contingent upon how well they were performing on the task. Finally, product-moment correlations between subjective measures and sensory and cognitive task performance showed that subjective workload was more closely associated with cognitive task performance than with sensory task performance.

Table 1

Mean Task Load Score by Group, Task, and Test Time

	Sensory		Cognitive	
	Pre	Post	Pre	Post
Mental Demand				
Young	56.5	58.0	69.4	74.5
Old	57.9	73.0	69.9	81.4
Physical Demand				
Young	37.9	48.5	29.0	40.5
Old	29.5	45.0	36.9	44.0
Temporal Demand				
Young	62.5	58.0	61.9	60.5
Old	66.5	74.0	68.0	69.9
Own Performance				
Young	30.0	45.5	36.0	38.5
Old	57.9	63.4	49.0	60.0
Effort				
Young	65.5	63.5	73.4	74.9
Old	65.0	71.5	73.0	82.8
Frustration				
Young	43.5	49.0	47.5	49.5
Old	49.9	63.5	54.3	61.5
Weighted Score				
Young	59.5	62.6	62.8	65.4
Old	62.7	74.0	65.0	74.4

DISCUSSION

The present study is the first to compare sensory and cognitive vigilance directly in the same group of subjects. The difficulty in the past has been in designing appropriate sensory and cognitive tasks that allowed for systematic comparisons between tasks which vary on only a limited, specified number of dimensions. In previous studies sensory and cognitive tasks have been examined in different experiments and have varied in a number of stimulus, task, and subject factors. The present study overcame this problem by manipulating the basis for defining critical signals for an identical set of stimuli. The results showed that sensory task performance declined over time, whereas cognitive task performance remained relatively stable, and in fact, increased slightly in the final time period of the vigil. The slight increase in performance over time for the cognitive task may be an instance of what Warm et al (1985) and others have called a vigilance increment. The results provide direct support for the view that cognitive vigilance is more resistant to decrement over time than sensory vigilance.

While cognitive task performance was less likely to demonstrate vigilance decrement, it however, was more affected by high event rate than the sensory task. Previous studies have shown that a high event rate necessitates effortful processing in vigilance (Parasuraman, 1985). The greater effort associated with the high event rate condition, coupled with the increased mental demands of the cognitive task, may have lead to an enhanced degradation of performance on the cognitive task.

The failure to find an event rate effect for the sensory task was surprising. This finding was unexpected given previous research indicating that detection rate generally decreases as event rate increases (Parasuraman, 1985). However, unlike previous studies, the present study used symbolic stimuli, although a physical discrimination was still required as in other sensory tasks. It is possible, therefore, that failure to find an event rate effect for the sensory task is associated with specific features of the stimulus set. Another possibility for this finding lies in the distinction between data-limited and resource-limited processing (Norman & Bobrow, 1975). Given that the sensory task involves mainly data-limited processing, and the cognitive task requires primarily resource-limited processing, the increased resource demands of the high event rate condition would be more likely to affect the cognitive task.

The subjective workload measures provided an interesting pattern of association and dissociation with the objective task measures. In some respects, the two were found to be associated. That is, subjective workload was greater for old subjects, who also performed poorer than young subjects. However, subjective workload was reported greater for the cognitive task, but performance differences for cognitive and sensory tasks were not found in hit rate. Yeh & Wickens (1988) proposed a theory which suggests that motivational incentives may produce dissociation between performance and subjective measures of workload. Their theory predicts that dissociation on the cognitive task was due to subjects finding this kind of task more interesting. Thus, subjects were motivated to expend greater resources and performance improved, even though the subjective measures suggested that the task was quite demanding. Dissociation was also produced on the sensory task because the subject's motivational level was low due to a discrepancy between what the subject would like to achieve and what was perceived as possible given the constraints (data-limitations) of the sensory task. Thus, performance dropped as did subjective measures of workload. These results suggest that when comparing the two tasks in terms of workload levels, subjective measures of workload would erroneously recommend the selection of the sensory task to the workload practitioner, while performance outcomes favor the cognitive task. Finally, the finding that subjective workload was more closely associated with cognitive task performance may be a function of the degree to which a task includes controlled processing requirements (Gopher & Donchin, 1986). Tasks which are data-driven and automatic (such as the sensory task) place few limits on the limited capacity system and will not closely parallel subjective workload evaluations. Research on verbal protocals (Ericsson & Simon, 1980) supports the view that subjective workload evaluations are accurate only in tasks involving information manipulated in short-term memory (e.g. the cognitive task).

Experiment 1 did not show significant differences in hit rate between age groups. Experiment 2, however, found that age effects were maximized when a high event rate condition was employed. The effect of high event rate on hit rate performance was most evident on the cognitive task. It was clear that the elderly performed more poorly on the cognitive task than the younger subjects. Since older adults are generally at a greater disadvantage relative to young subjects as processing load increases (Plude & Hoyer, 1981), the magnitude of age differences between young and old may increase directly with the number or complexity of operations involved. Both experiments showed age effects for false detection rate. Older adults also demonstrated that the ability to make fine-grained discriminations may deteriorate with age. Support for this idea comes from Rabbitt (1965) and by Hoyer, Rebok, & Sved (1979) who showed that the elderly have more difficulty in ignoring irrelevant information and are thus at a disadvantage in learning to discriminate between patterns.

The results of the present study suggest some implications for display design and individual differences in monitoring capability. First, it is apparent that tasks requiring the monitoring of symbolic stimuli can maximize performance when stimulus events are presented at a slow rate; whereas tasks requiring discriminations of physical characteristics remain somewhat insensitive to differences in event rates. Second, if design considerations permit a choice between tasks that require sensory discriminations or cognitive decisions, select the latter. Third, the elderly will be at a disadvantage when monitoring displays requiring fine discriminations. Young individuals are better candidates to serve as monitors in tasks requiring cognitive abilities, particularly when event rates are high. At low event rates, differences between young and elderly are minimal. And finally, results of subjective measures of workload obtained from elderly subjects may erroneously favor the very tasks for which they are most ill-suited: sensory tasks. If subjective measures are to be used, the age of the evaluators should be considered when making system design decisions.

REFERENCES

Davies, J., & Parasuraman, R. (1982). The psychology of vigilance. London: Wiley.

Ericsson, K.A., & Simon, H.A. (1980). Verbal reports as data. Psychological Review, 87(3), 215-251.

Gopher, D., & Donchin, E. (1986). Workload--an examination of the concept. In L. Kaufman & J. Thomas (Eds.), Handbook of perception and human performance VII. New York: Wiley.

Hart, S.G., & Staveland, L.E. (in press). Development of a multidimensional workload rating scale: Results of empirical and theoretical research. In P.A. Hancock & N. Meshkati (Eds.), Human mental workload. Amsterdam: Elsevier.

Hoyer, W.J., Rebok, G.W., & Sved, S.M. (1979). Effects of varying irrelevant information on adult age differences in problem solving. Journal of Gerontology, 34, 553-560.

Loeb, M., Noonan, T.K., Ash, D.W., & Holding, D.H. (1987). Limitations of the cognitive vigilance increment. Human Factors, 29(6), 661-674.

Norman, D., & Bobrow, D. (1975). On data-limited and resource-limited processing. Cognitive Psychology, 1, 44-60.

Parasuraman, R. (1985). Sustained attention: A multifactorial approach. In M. Posner (Ed.), Attention and performance: XI. New Jersey: Erlbaum.

Parasuraman, R. (1987). Human-computer monitoring. Human Factors, 29(6), 695-706.

Plude, D.J., & Hoyer, W.J. (1981). Adult age differences in visual search as a function of stimulus mapping and processing load. Journal of Gerontology, 36, 598-604.

Rabbitt, P.M.A. (1965). An age-decrement in the ability to ignore irrelevant information. Journal of Gerontology, 20, 233-238.

Warm, J.S., Dember, W.N., Lanzetta, T.M., Bowers, J.C., & Lysaght, R.J. (1985). Information processing in vigilance performance: Complexity revisited. In Proceedings of the Human Factors Society 30th Annual Meeting. Santa Monica, CA: Human Factors Society.

Wiener, E.L. (1984). Vigilance and inspection. In J.S. Warm (Ed.), Sustained attention in human performance. London: Wiley.

Yeh, Y., & Wickens, C.D. (1988). Dissociation of performance and subjective measures of workload. Human Factors, 30(1), 111-120.

ACKNOWLEDGEMENTS

This investigation was supported in part by contract N62269-88-M-076 from the Naval Air Development Center, Warminster, PA., to Raja Parasuraman. We thank Len Giambra, Steve Kerst, and Joel Warm for comments and suggestions concerning this research.

EFFECTS OF AGING ON SUBJECTIVE WORKLOAD AND PERFORMANCE

Douglas L. Boyer
Jay G. Pollack
F. Thomas Eggemeier
University of Dayton
Dayton, Ohio

Demographics indicate that the population in the United States and other industrialized nations is growing older, and that the number of older workers and systems users can be expected to increase substantially over the next several decades. In order to assess possible differences between age groups the mental workload experienced by older adults as compared to that experienced by younger adults was investigated. Two tasks were utilized to assess short term memory (continuous recognition) and psychomotor (first-order unstable tracking) performance. The workload of each task was assessed with the Subjective Workload Assessment Technique (SWAT). Memory task performance measures and subjective workload ratings indicated a decrement in performance and an increase in workload for the older group relative to the younger group. Psychomotor task performance measures and subjective workload ratings indicated no difference between the age groups. It is hypothesized that the memory task makes greater demands on central processing resources than the psychomotor task used in this study. In support of this hypothesis, an analysis of the changes in ratings on the individual SWAT dimensions of time, mental effort and psychological stress revealed that an increase occurred only on the mental effort dimension for the memory task. This study implies that designers should 1) reduce or provide design features that lessen memory laden task performance for older workers, and 2) give more weight to the reduction of central processing resource requirements in trade-off studies.

INTRODUCTION

Significant growth is expected in the older population as the baby-boomer cohort grows older (Russell, 1982). In many industrialized nations, cohorts following the baby boom are not as large and have led to predictions of future labor shortages in nearly every profession and vocation (Jones, 1977; Wattenburg, 1987). Conceivably, as the number of young persons entering job markets decreases, the call for older workers will become greater. Therefore, guidelines for systems designed to accommodate this population represents a central human factors issue in the coming decades. This study provides data concerning perceived mental workload and actual task performance for younger and older subjects.

Mental workload can be defined as expenditure of processing resources in task accomplishment (Eggemeier, 1988). This construct of mental workload is built upon a limited resources model of the human information processing system. Three categories of factors contributing to workload are: input load (environmental, situational, and procedural elements); operator effort (processing and memory resources); and work result (feedback and error perception) (Jahns, 1973). System interfaces and input devices (i.e. mouse, joystick or keyboard) can contribute to an operator's workload perception.

Differences in younger and older populations may be such that the amount of workload experienced by each population differs. As a result, systems may need to be designed accordingly if they are to accommodate both young and old operators. A model of aging (e.g., Graafmans and Brouwers, 1989), which utilizes sensorial decline associated with age might be useful in the design process. Although this model takes into account elements of perception and motor response included in the workload construct, it does not indicate what happens to cognitive functioning or workload. It must be inferred from such a model that as perceptive or response ability declines workload might increase.

Age-related behavioral slowing is a very robust finding in aging research and does not appear to be primarily due to peripheral processes, but rather to processes in the central nervous system (Botwinick, Brinley, and Robbin, 1959; Simon, 1968; Welford, 1980). Such slowing is important when considering workload in older individuals. Although literature concerning aging and abilities of older individuals abounds with examples of performance measures (see, for example, Braune, Wickens, Strayer, and Stokes, 1985; Ponds, Brouwer, and van Wolffelaar, 1988; Salthouse, Mitchell, Skovronek, and Babcock, 1989), measurement of workload in older people has been somewhat limited (but see, Diaz, 1986; Mertens and Collins, 1986).

This experiment examined subjective workload and performance measures in older and younger groups with a first-order unstable tracking (UT) task and a continuous recognition (CR) task. The CR task was principally intended to tap short term memory function, while the UT task was intended to tap motor output resources. The

Subjective Workload Assessment Technique (SWAT) (Reid and Nygren, 1988) was used to assess workload. SWAT requires ratings on three dimensions which contribute to the subjective workload associated with task performance: time load, mental effort load, and psychological stress load. Ratings on the three dimensions provided by subjects at the completion of task performance are subsequently combined to yield an overall workload score for each rated task condition. Since manual and memory tasks are reportedly affected by slowing in the human information processing system, age-related effects were expected to be similar for performance and workload for each task.

METHOD

Subjects

Two groups of 12 subjects participated: a volunteer group of younger subjects (mean age = 22.6 years), six of whom received course credit at the University of Dayton and six from the community at large; and a volunteer group of older subjects (mean age = 64.9 yrs) recruited from the community at large. Subjects were screened for near vision using digits and letters presented on the monitor used in the experiment. Subjects were screened for medications known to affect cognitive and psychomotor skills. The groups did not differ on years of education, verbal ability, or on departure from their respective norms for psychomotor speed.

Apparatus and Materials

A Commodore 64 computer with a 13 inch Commodore 1702 color monitor was used for task presentation. Viewing distance was approximately 40 cm. The UT task and CR task are part of the Criterion Task Set (CTS) (Amell, Eggemeier, and Acton, 1987; Shingledecker, 1984). Performance data, consisting of root-mean square error (RMSE) and number of control losses (CL) for the UT task, and response time (RT) and percent correct (PC) for the CR task, were collected by computer. SWAT ratings were collected via a paper and pencil form for each set of trials.

Procedure

Subjects participated in two 120-minute sessions with no less than one day or more than seven days separating sessions. The first session consisted of background data collection and SWAT scale development.

SWAT scale development (Reid and Nygren, 1988) consists of a card sort procedure, whereby subjects rank order the subjective workload associated with each combination of three level sof time, mental effort, and psychological stress load. These data are necessary for development of the overall workload scale which results from the SWAT procedure. During the second session, subjects performed both UT task trials at each of three difficulty levels and CR task trials at each of three difficulty levels. Each level of difficulty was presented three times for a total of nine trials per task per subject. Half the subjects performed the UT task first, while the other half performed the CR task first. Order of presentation of difficulty levels was counter balanced using a Latin square. One subject from each age group was assigned to each order of presentation. Each subject was allowed one practice trial at each difficulty level before beginning data collection trials.

The objective of the UT task is to center a cursor on the screen, reducing horizontal movement by means of a rotary control. Demand levels are created by varying lambda values in the equation which controls task instability. The task version employed (Amell, Eggemeier, and acton, 1987) utilized lambda values of 1.0, 2.0, and 3.0, creating three demand levels. In the CR task, the objective is to test short term memory of numbers by having the subject update a set of digits and compare a probe digit to a previously presented digit at a designated number of positions back in the series. The task is to decide whether a probe digit matches a digit in memory. Responses were made via two keys labeled "same" or "different". Difficulty was manipulated by varying comparison digit position in the series. Difficulty levels employed were 1, 2 and 3 positions back which required keeping in memory a set of 2, 3 or 4 digits, respectively.

RESULTS

A 2 X 3 (age group by difficulty level) multivariate analysis of variance (MANOVA) was conducted on the performance measures for each task. All MANOVA results reported are for Pillias' V test of significance. A 2 X 3 (age group by difficulty level) analysis of variance (ANOVA) was conducted on the SWAT scores and on the raw SWAT ratings for each dimension for each task. A significant main effect of difficulty level was observed on all measures in all analyses. (All effects reported are significant at the .05 level or greater.)

Table 1 contains the means and standard deviations for CR task performance measures and subjective workload ratings. CR performance MANOVA results showed the expected age differences [$F(2,21) = 2.253$)]. Univariate ANOVA found that the older group had slower RTs [$F(1,22) = 11.718$] and a greater number of errors [$F(1,22) = 23.441$] than the younger group. A significant multivariate interaction between age group and difficulty level [$F(4,88) = 2.922$] was observed. Univariate ANOVA results revealed that a significant interaction [$F(1,22) = 3.478$] occurred for RTs but not for errors. Analyses of simple effects on the RT data indicated significant differences between age groups for low [$F(1,22) = 51.877$] and moderate [$F(1,22) = 5.958$] levels of difficulty, but not for the high difficulty level.

CR task SWAT scores analysis revealed a difference between age groups [$\underline{F}(1,22) = 4.357$] indicating that the older group rated the task as imposing more workload than did the younger group. The interaction between age group and difficulty level was not significant. Analyses of SWAT dimensions of time, mental effort and psychological stress revealed a difference between groups on the mental effort dimension [$\underline{F}(1,22) = 11.099$]. Time and psychological stress dimension analyses found no significant differences between groups. No significant interaction of age group and difficulty level was found for any SWAT dimension.

Table 1.
Continuous Recognition Task Results

Difficulty		Low	Moderate	High
Response Time (msec)				
Young	M	1002.285	1902.960	2139.954
	SD	(276.901)	(811.509)	(969.753)
Old	M	2190.656	2597.579	2726.640
	SD	(499.999)	(559.639)	(743.801)
Percent Correct				
Young	M	96.109	86.571	82.662
	SD	(3.605)	(9.809)	(10.712)
Old	M	79.633	62.364	53.826
	SD	(22.912)	(17.623)	(12.425)
SWAT Overall Workload Scores				
Young	M	22.250	51.758	68.108
	SD	(20.700)	(18.766)	(26.843)
Old	M	33.967	65.425	87.708
	SD	(29.164)	(25.557)	(15.354)
SWAT Mental Effort Dimension Ratings				
Young	M	1.500	2.333	2.333
	SD	(0.522)	(0.492)	(0.651)
Old	M	1.917	2.500	3.000
	SD	(0.515)	(0.522)	(0.000)

Means and standard deviations for UT performance measures and subjective workload ratings are reported in Table 2. UT performance MANOVA results did not show the expected age-related effects [$\underline{F}(2,21) = 3.302$, $p > .05$].

A significant multivariate interaction between age group and difficulty level was observed [$\underline{F}(4,88) = 3.690$]. Univariate ANOVA revealed that significant interactions occurred for CL [$\underline{F}(1,22) = 4.114$] and RMSE [$\underline{F}(1,22) = 4.479$]. Analyses of simple effects on CL data indicated that age groups did not differ at low or moderate difficulty levels but did differ significantly at the high difficulty level [$\underline{F}(1,22) = 4.528$]. Analyses of simple effects on RMSE data revealed no significant differences at any difficulty level. Inspection of effect sizes (Eta^2) revealed an increase in the size of the age group effect across difficulty levels.

Analyses of SWAT scores and raw ratings for each SWAT dimension showed no significant differences between groups for the UT task. No age group by difficulty level interactions were found to be significant.

Table 2.
Unstable Tracking Task Results

Difficulty		Low	Moderate	High
Root Mean Square Error				
Young	M	10.631	22.000	36.075
	S	(8.677)	(9.957)	(7.922)
Old	M	9.922	30.461	41.947
	SD	(6.457)	(14.543)	(7.731)
Control Losses				
Young	M	1.472	13.417	70.694
	SD	(2.894)	(26.764)	(62.872)
Old	M	1.250	46.333	133.583
	SD	(2.753)	(48.035)	(80.803)
SWAT Overall Workload Scores				
Young	M	17.733	36.783	64.933
	SD	(26.387)	(19.343)	(25.397)
Old	M	15.992	52.725	72.733
	SD	(25.350)	(41.280)	(34.953)
SWAT Mental Effort Dimension Ratings				
Young	M	1.417	1.667	2.250
	SD	(0.669)	(0.651)	(0.754)
Old	M	1.250	2.167	2.500
	SD	(0.452)	(0.835)	(0.674)

DISCUSSION

Effects of difficulty level were observed for all measures and indicate increasing degradation in performance and increasing subjective workload as difficulty increased. This result parallels the findings of Amell et al. (1987) who tested the same versions of the present tasks with young subjects.

CR task performance analysis found the expected differences between age groups and indicated a clear advantage in favor of the younger group. UT task performance analysis found no significant main effect of age group, but did show differences in CL which favored the younger group at the high level of tracking difficulty. The analysis also indicated that younger subjects showed less increase in RMSE than did older subjects as tracking difficulty increased.

Although UT performance results did not precisely match the findings of other age-related research which used second-order tracking tasks (Braune et al., 1985; Ponds et al., 1988), the findings were nevertheless indicative of some younger group performance advantages. No age-related research which utilized the CTS version of the UT task has been reported in the literature.

Differences between our results and those of earlier studies might reflect the type of first-order tracking task used here. For exampke, performance of first-order tracking tasks has been shown to be superior to performance of higher-order tracking tasks (e.g., Wickens, 1986, 1992).

CR task SWAT scores supported the hypothesis that the older group would experience a higher level of workload than the younger group. Analyses of individual SWAT dimensions revealed a difference between groups only for the mental effort dimension. The noted differences in workload are therefore principally attributed to perceived differences in mental effort associated with CR task performance.

Predictions that subjective workload scores would reflect an age-related difference in UT task performance were not substantiated. Although the performance measures revealed some advantage in favor of the younger group, SWAT scores for the UT task did not indicate a difference between groups. Analysis of each SWAT dimension likewise revealed no effect of age group.

A possible explanation of these seemingly disparate results lies in the type of processing resources being utilized by the UT task versus the CR task. The CR task primarily loads memory resources while the UT task primarily loads psychomotor resources. As was seen in the CR task workload analyses, the difference in workload was principally attributed to differences in mental effort, the dimension which can be assumed to most directly assess working memory load. No such differences were observed for mental effort in UT task workload analyses, perhaps because the UT task does not tap memory resources as heavily as the CR task. Further support for this hypothesis comes from the fact that in terms of predicting cursor movement, first-order tracking tasks impose relatively little cognitive load compared to the higher-order tracking tasks used in previous age-related studies.

Implications for system design for older workers are that memory-laden tasks should be supported for such workers while first-order tracking psychomotor tasks require less additional support. Tasks requiring heavy short term memory use might be enhanced with memory aiding. For example, users of systems involving computer displays might be aided through a system which retains a chain of previous commands on the display. Extrapolations of possible actions based on previous commands could also be displayed to reduce working memory load.

Further research should examine several types of tracking task demand manipulations to determine if age differences exist in subjective workload for various motor control manipulations. Because manipulation of control order could be expected to more directly involve central processing resources than the manipulation of tracking element stability that was used here, the former type of manipulation might show a greater age effect than the latter.

REFERENCES

Amell, J. R., Eggemeier, F. T., & Acton, W. H. (1987). The Criterion Task Set: An Updated Battery. Proceedings of the Human Factors Society 31st Annual Meeting, 405-409.

Botwinick, J., Brinley, J. F., and Robbin, J. S. (1959). Modulation of Speed of Response With Age. Journal of Genetic Psychology, 95, 137-144.

Braune, R., Wickens, C. D., Strayer, D., and Stokes, A. F. (1985). Age-Dependent Changes in Information Processing Abilities Between 20 and 60 Years. Proceedings of the Human Factors Society 29th Annual Meeting, 226-230.

Diaz, M. F. (1986). Age and the Measurement of Workload by a Secondary Task. Proceedings of the Human Factors Society 30th Annual Meeting, 1154-1158.

Eggemeier, F. T. (1988). Properties of Workload Assessment Techniques. In P. A. Hancock & N. Meshkati (Eds.), Human Mental Workload (pp. 41-62). North Holland: Elsevier Science Publications.

Graafmans, J. A. M., and Brouwers, T. (1989). Gerontechnology, The Modelling of Normal Aging. Proceedings of the Human Factors Society 33rd Annual Meeting, 187-190.

Jahns, D. W. (1973). Operator Workload: What Is It and How Should It Be Measured? In K. D. Cross and J. J. McGrath (Eds.) Crew System Design (pp. 281-287). Santa Barbara, CA: Anacapa Sciences.

Jones, R. (1977). The Other Generation: The New Power of Older People. Englewood Cliffs, NJ: Prentice Hall.

Mertens, H. W., and Collins, W. E. (1986). The Effects of Age, Sleep Deprivation, and Altitude on Complex Performance. Human Factors, 28(5), 541-551.

Ponds, R. W. H. M., Brouwer, W. H., and van Wolffelaar, P. C. (1988). Age Differences in Divided Attention in a Simulated Driving Task. Journal of Gerontology, 43, 151-156.

Reid, G. B., and Nygren, T. E. (1988). The Subjective Workload Assessment Technique: A Scaling Procedure For Measuring Mental Workload. In P. A. Hancock and N. Meshkati (Eds.), Human Mental Workload (pp. 185-218). North Holland: Elsevier Science Publishers.

Russell, L. B. (1982). The Baby Boom Generation and the Economy. Washington, DC: The Brookings Institution.

Salthouse, T. A., Mitchell, D. R. D., Skovronek, E., and Babcock, R. L. (1989). Effects of Adult Age and Working Memory on Reasoning and Spatial Abilities. Journal of Experimental Psychology: Learning, Memory, and Cognition, 15(3), 507-516.

Shingledecker, C. A. (1984). A Task Battery for Applied Human Performance Assessment Research (Technical Report AFAMRL-TR-84-071). Wright-Patterson AFB, OH: Air Force Aerospace Medical Research Laboratory.

Simon, J. R. (1968). Signal Processing Time as a Function of Aging. Journal of Experimental Psychology, 78(1), 76-80.

Wattenburg, B. J. (1987). The Birth Dearth. New York, NY: Pharos.

Welford, A. T. (1980). Relationships Between Reaction Times and Fatigue, Stress, and Sex. In A. T. Welford (Ed.) Reaction Times (pp. 321-354). London: Academic Press.

Wickens, C. D. (1986). The Effects of Control Dynamics on Performance. In K. L. Boff, L. Kaufman, and J. P. Thomas (Eds.), Handbook of Perception and Human Performance, (pp.(39)1-(39)60). New York, NY: Wiley.

Wickens, C. D. (1992). Engineering Psychology and Human Performance, 2nd Ed. New York, NY: Harper Collins.

EXPERTISE AND AGE EFFECTS ON PILOT MENTAL WORKLOAD IN A SIMULATED AVIATION TASK

Donald L. Lassiter
Methodist College
Fayetteville, NC

Daniel G. Morrow
University of New Hampshire
Durham, NH

Gary E. Hinson
Methodist College
Fayetteville, NC

Michael Miller
Catholic University of America
Washington, DC

David Z. Hambrick
Georgia Institute of Technology
Atlanta, GA

ABSTRACT

This study investigated the effects of expertise and age on cognitive resources relevant to mental workload of pilots engaged in simulated aviation tasks. A secondary task workload assessment methodology was used, with a PC-based flying task as the primary task, and a Sternberg choice reaction time task as the secondary task. A mixed design using repeated measures was employed, with age and expertise as between-subjects factors and workload as the within-subjects factor. Pilots ranging in age from 21 to 79 years and 28 to 11,817 hours of flight time served as subjects. Of interest was whether expertise would mitigate the adverse effect of aging on pilots' mental workload handling ability as defined by two measures of secondary task performance: choice reaction time and accuracy. Results indicated that expertise did mitigate the effects of age regarding secondary task accuracy. Implications of results are discussed, and directions for future research are presented.

INTRODUCTION

The task of flying involves the time-sharing of several tasks, i.e., it is a multiple-task situation, placing a great deal of mental workload demand on the pilot. The concept of mental workload implies that limitations exist in the human information processing framework (Gopher & Donchin, 1986). It has often been reported in the aging literature that physical abilities, perceptual processes, and memory processes decrease with age (e.g., see Wickens, Braune, & Stokes, 1987; Tsang, 1992). To some degree, laboratory research with non-pilots has indicated that aging detrimentally effects the ability of the human information processing system to handle appreciable amounts of mental workload.

A frequent finding in the expertise literature is that expertise improves performance of domain-relevant tasks by reducing workload demands on short-term memory capacity (e.g., see Yekovich, Walker, Ogle, & Thompson, 1990). Domain-relevant tasks are tasks that tap the specific encapsulated knowledge within an expert's particular area of expertise, or domain (Rybash, Hoyer, & Roodin, 1986), whereas domain-general tasks do not tap a specific area of knowledge, but rather draw on generalized background knowledge and ability. To access the knowledge to perform a domain-relevant task might require less effort for an expert than a novice, who may not possess the required knowledge in as useful a form to perform the same task. This is probably because the knowledge possessed by the novice is far less structured and automatized than that of the expert (Rybash et al., 1986).

Flying can be considered a well-defined task domain, so it may be that expertise in flying could counteract the detrimental effects of aging resulting from increases in mental workload of domain-relevant aviation tasks. The number of studies looking at the relationships among expertise, aging, and workload involving pilots as subjects has been small, but is now increasing (see Morrow, Leirer, & Altieri, 1992; Morrow, Leirer, Altieri, & Fitzsimmons, 1994; Tsang & Shaner, 1994; Tsang, 1995). However, there is still a need for systematic research programs involving pilots to investigate the relationships among these variables (Tsang, 1992).

Two specific aims of the current research effort were to determine: 1) if aging adversely affects the ability to handle increases in mental workload in a simulated aviation task; and 2) if expertise in piloting (as defined by the number of hours of flight time) can mitigate the adverse effect of aging on the ability to handle increases in mental workload. Such mitigation would be demonstrated by a significant interaction between expertise and age regarding the ability to handle mental workload.

METHOD

This study utilized the secondary task workload methodology (i.e., the subsidiary task technique) using a Sternberg (1969) choice reaction time task as the secondary task, and flying courses of two different levels of difficulty on Microsoft's Flight Simulator 4.0 (Microsoft, 1990) as the primary task. The Sternberg task has been used extensively as a secondary task in mental workload research and has shown sufficient sensitivity and diagnosticity (e.g., Wickens et al., 1987). The rationale for picking this task was that it may tap cognitive resources related to working memory and monitoring and compete with the primary task for these resources. The major independent variables in this study were age (in years), expertise (in hours of total flight time), and workload. Workload consisted of six levels based on combinations of levels of difficulty of the primary task (3 levels) and secondary task (2 levels). The three levels of primary task were zero (i.e., it was not performed with the secondary task; the secondary task was performed alone); easy course, and difficult course. The two levels of secondary task were 2 letters (low load) and 4 letters (high load) in the memory set. Therefore, the six levels of workload were: 1) secondary task alone (low load); 2) secondary task alone (high load); 3) primary task (easy course) with secondary task (low load); 4) primary task (easy course) with secondary task (high load); 5) primary task (difficult course) with secondary task (low load); and 6) primary task (difficult course) with secondary task (high load). The two major dependent variables in the study reflected secondary task performance: choice reaction time (in msecs) and accuracy (percent correct responses).

Subjects

A total of 42 paid volunteers served as subjects in this study. All had general aviation experience. The age range was 21 to 79 years, while the range of expertise was 28 to 11,817 hours of flight time. Subjects were recruited from the local general aviation community, primarily from a large local flying club. Other subjects came from the sizable active/retired military aviation community in the area. All subjects had general aviation experience, and all but three subjects had such experience within the twelve months prior to participation in this study. None of the subjects reported having prior laboratory testing, although those with military experience reported having simulator time. None of the subjects reported having any experience with the PC-based flying simulation used in this study.

Apparatus

Equipment for this study consisted of two computers for presenting the primary and secondary tasks, as well as collecting and analyzing data. The lab was partitioned into subject and experimenter stations. The subject station (see Figure 1) had two monitors, one each for the primary and secondary tasks, as well as a flight yoke and pedals for performing the primary task. Two large telegraph keys were mounted next to the yoke for responding to the secondary task. The experimenter station had two monitors which allowed the experimenter to see what the subject viewed, as well as mice/keyboards for controlling the simulation.

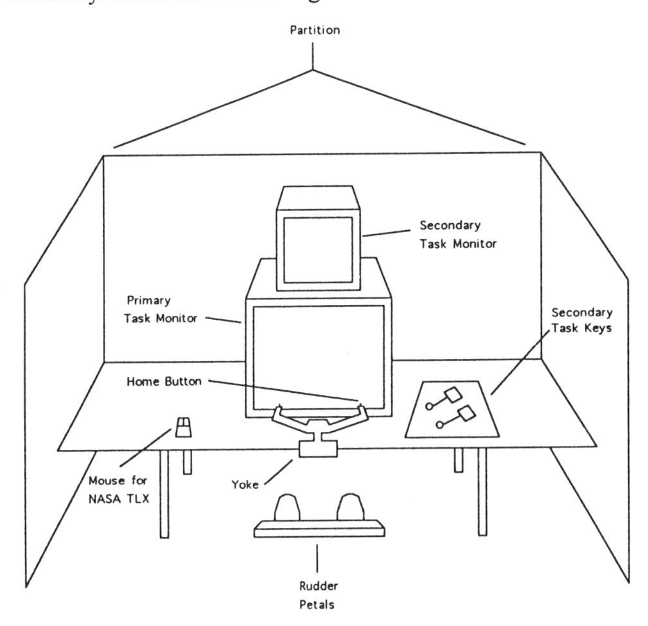

Figure 1. The subject station. The mouse was used to collect data not reported here in another phase of the study.

Procedure

Subjects participated in a total of seven two-hour experimental sessions: five practice sessions and two data collection sessions. Subjects were run one at a time, one session per subject per day. The first practice session consisted of a standardized briefing (as the subject read along), updating the subject's biographical information form, and the subject signing a consent form. Following instructions read aloud by the experimenter, the subject became familiar with the simulator apparatus and primary task software. Then the subject practiced the primary task, flying eight ten-minute courses once (four easy and four difficult courses). Before practicing each course, the subject was given a map of the course and the experimenter read aloud detailed instructions describing that course. During this first session, the software program monitoring the subject's performance did not interrupt the session if performance was not up to criterion. Feedback concerning performance was given verbally to the subject. After this first session, the subject was given maps of the eight courses and instructed to memorize them before the second session. An example of a course map is shown in Figure 2. The purpose of memorizing these courses was to have subjects use their working memory to call up and maintain a representation of the course as they flew it, therefore competing with working memory demand of the secondary task (described below). The subject's memory was tested before flying each course in subsequent sessions by having the

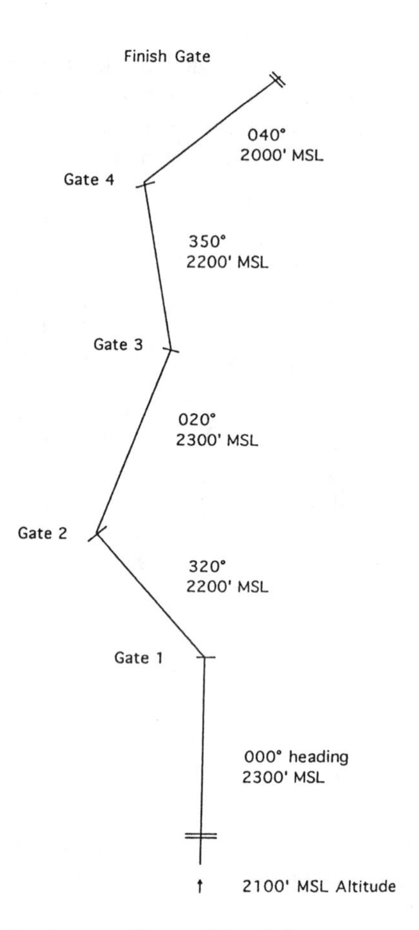

Figure 2. A map of one of the eight courses used in the study.

subject sketch the course map from memory without error. As sessions proceeded, subjects had no difficulty reproducing the maps. An easy course consisted of five two-minute splits, where at each of the four middle splits, the subject had to execute a change in heading of 60 degrees. Gates located at the positions of the course changes provided guidance, but were not visible to the subject until they were only a short distance away to insure that the subject relied on instruments and memory to fly the course. For these easy courses, all gates were at the same altitude. A difficult course was the same length as an easy course, but the subject had to make different changes in heading at each gate, as well as change altitude. Four variants of the courses within each level of difficulty were constructed by using mirror and inverted rotations. The second and third practice sessions each gave the subject ten courses to fly. From the ninth course flown in the second session onward, the software program monitoring the subject's performance interrupted and terminated the flight if the subject did not fly to criterion, and the subject was required to start that course over. The criterion was defined as flying each 20-mile course with no more than a 2000 feet deviation from ideal course, a 20 degree deviation from course (to allow for turns), a 200 feet deviation from assigned altitude, and a 10kts/hr deviation from assigned airspeed. The fourth practice session contained four additional ten-minute runs of primary task practice. Thus, over the first three and a half sessions of the experiment, 32 total practice runs of the primary task were performed. The purpose of the extensive practice was to insure that subjects flew the primary task at criterion and could maintain that performance during the dual-task sessions of primary and secondary task performance that came later. (The secondary task must not intrude on primary task performance; otherwise, any differences in secondary task performance between single- and dual-task conditions would not be interpretable). Then, after a standardized set of instructions read aloud, the subject was given four 10-minute blocks of practice on the secondary task by itself (two blocks each of memory sets of two and four letters). The subject was given the letters to memorize right before each run. On a trial, the subject was presented a brief flash of a single letter on the secondary task monitor. While keeping the right thumb on the red "home" button located on the right handle of the flight yoke, the subject decided if the letter presented was a member of the memory set. The subject then released the home button and pressed either the "yes" or "no" telegraph key located next to the flight yoke. Subjects were instructed to make their decisions and key presses as quickly as possible. Choice reaction time (CRT) was operationally defined as the time between onset of the letter flash and release of the home button. Movement time was defined as the time between release of the home button and pressing the "yes" or "no" telegraph key. Of interest in this study was CRT. Also, accuracy of the subject's responses was recorded. An accurate response was defined as either responding "yes" when a memory set letter was presented (a "hit") or "no" when it was not (a "correct rejection"). All other response types (including infrequent non-responses) were considered incorrect responses. The percentage of correct responses for each run of secondary task performance in the study was calculated. The fifth practice session consisted of practicing the secondary task alone for two runs, then practicing the primary and secondary tasks together for four runs in a dual-task situation. These dual-task runs consisted of combining both levels of primary task (easy and difficult course) and both levels of secondary task (memory set of two and four letters). The sixth and seventh sessions were each comprised of six data collection runs like those in session five. All seven sessions utilized counterbalancing of primary task courses and secondary task stimuli across the runs to minimize possible order effects. After the last session, the subject was debriefed and compensated.

RESULTS

The data were analyzed by a mixed design ANOVA with two between subjects variables (Age and Expertise) and one repeated measures (within subjects) variable (Workload) with six levels. Age and Expertise were allowed to freely vary in the subject sample because a regression analysis had been originally planned. To perform the ANOVA, however, Age and Expertise were partitioned into two levels each using the median split technique to form four groups of subjects (young, less expertise; young, more expertise; old, less expertise; and old, more expertise). Initial analysis of the data from the entire sample revealed a virtual absence of significant findings involving expertise, perhaps because a median split was used to partition expertise into two levels - a weak manipulation. Therefore, another analysis was performed on a subset of the data consisting of those subjects with extreme amounts of expertise (i.e., least and most hours of flight time) within each age group. Within each age group, ten subjects were selected: the five with the most and the five with the least flight time. For this subset of subjects, the age range was 21 to 75 years, while the expertise range was 28 to 11,817 hours of flight time. For the groups, the mean ages and hours of flight time were: 1) young, less expertise: 27.2 years and 60.02 hours; 2) young, more expertise: 36.6 years and 5277.88 hours; 3) old, less expertise: 58 years and 217.6 hours; and 4) old, more expertise: 52.6 years and 7323.4 hours.

First, analyses were conducted to test if the secondary task intruded on primary task performance in the dual-task conditions (i.e., workload levels 3 - 6). These analyses were done because a requirement of the secondary task methodology is that the secondary task does not intrude on primary task performance. Separate ANOVAs were conducted comparing primary task performance measures in the dual-task conditions with those same measures collected during the last practice runs of performing the primary task alone. Primary task performance measures analyzed were deviations from criterion values (i.e., root-mean-square-errors, or RMSEs) for heading, altitude, airspeed, and distance off course. None of these analyses found a significant main effect for task condition (i.e., primary task alone vs. dual-task) on any of these measures, thus indicating that the addition of the secondary task did not intrude on these measures of primary task performance. Analyses were then performed to see if primary task performance varied across the different levels of workload in the dual-task conditions. Results indicated that as workload increased, distance off course (a general summary measure of primary task performance) and heading were unaffected, but altitude ($F[3, 48] = 3.31$, $p < .03$) and airspeed ($F[3, 48) = 25.62$, $p < .001$) deviations increased with workload.

Next, mixed design ANOVAs were conducted separately on CRT and percent correct responses on the secondary task for those conditions where subjects performed the secondary task alone during the last two sessions. Here, since the primary task was not performed in these conditions, Workload was equivalent to memory load of the secondary task (2 or 4 letters), and thus had two levels. For CRT, the results indicated significant effects for Age ($F[1, 16] = 8.30$, $p < .011$) and Workload ($F[1, 16] = 21.12$, $p < .001$). No other effects were observed. The presence of an age effect was expected, as the literature has repeatedly shown that age slows CRT. For percent correct responses, the results indicated only a significant effect for Workload ($F[1, 16] = 10.35$, $p < .005$). The absence of an expertise effect and interactions involving age and expertise in these analyses demonstrated that the Sternberg task was domain-general, an important finding indicating that the expert pilots in this study were not simply more capable people in terms of secondary task performance.

Next, mixed design ANOVAs with Workload as a six-level within-subjects factor were conducted separately on CRT and percent correct responses on the secondary task. For CRT, the following results were observed (see Figure 3a). Age ($F[1, 16] = 11.53$, $p < .004$), Workload ($F[5, 80] = 99.14$, $p < .001$), and Expertise ($F[1, 16] = 6.84$, $p < .019$) all had significant main effects. Also, there was a significant Expertise x Workload interaction ($F[5, 80] = 4.90$, $p < .001$), indicating that the CRT of pilots with more expertise was less affected by increases in workload. Regarding CRT, however, expertise did not significantly mitigate the effects of age on the ability to handle increases in workload, as indicated by the absence of a significant three-way interaction of Age, Expertise, and Workload (although it approached significance). For percent correct (see Figure 3b), Age ($F[1, 16] = 13.32$, $p < .002$), Expertise ($F[1, 16] = 11.06$, $p < .004$), and Workload ($F[5, 80] = 16.39$, $p < .001$) were all significant. Also, these interactions

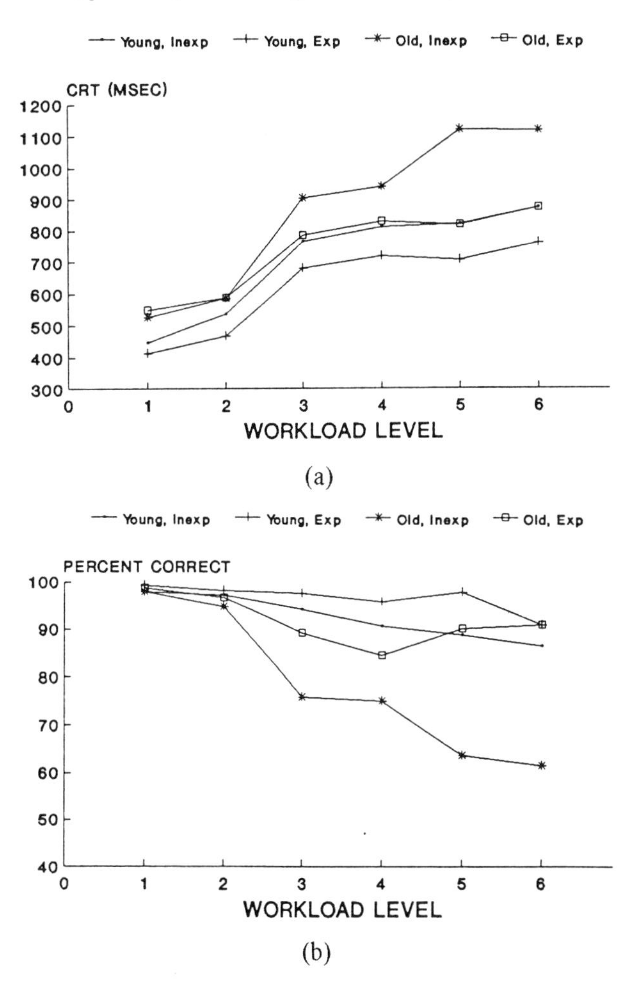

Figure 3. Secondary task (a) choice reaction time performance and (b) percent correct performance of subjects as a function of workload level. See text for subject group and workload level definitions.

were significant: Age x Workload ($F[5, 80] = 4.58$, $p < .001$); Expertise by Workload ($F[5, 80] = 5.22$, $p < .001$); and a three-way interaction of Age x Expertise x Workload (F[5, 80]

= 2.52, p < .036). This three-way interaction is discussed below. Also, repeated measures, mixed design ANOVAs with Workload as a four-level within-subjects factor (i.e., restricted to dual-task conditions) were conducted. For CRT, Age (F[1, 16] = 10.30, p < .005), Expertise (F[1, 16] = 9.09, p < .008), and Workload (F[3, 48] = 8.65, p < .001) were all significant. For percent correct, Age (F[1, 16] = 14.24, p < .002) and Expertise (F[1, 16] = 11.62, p < .004) were significant.

DISCUSSION

The goal of this study was to find evidence that expertise mitigates the affects of aging on pilot performance; specifically, a three-way interaction among age, expertise, and workload. Evidence of this interaction was provided by the analysis of subjects with extreme levels of flight expertise, with six levels of workload, for percent correct (accuracy) but not choice reaction time (although it approached significance; see Figures 3b and 3a, respectively). Figure 3b shows that: 1) the accuracy of younger subjects with more expertise was least affected by increases in workload; 2) the accuracy of older subjects with less expertise was most affected as workload increased; and 3) younger subjects with less expertise and older subjects with more expertise were moderately affected by the increase in workload. So it would seem that this subset of the data indicates that in going from a single-task situation performing a choice reaction time task to a dual-task situation performing a simulated aviation task and choice reaction time task, expertise (i.e., hours of flight time) mitigated the detrimental effects of age to some degree regarding cognitive resources involved in working memory and monitoring. This important finding coincides with those reported by Tsang and Shaner (1994) and Morrow et al. (1994). Although this three-way interaction was not found in the strictly dual-task workload manipulations, strengthening these manipulations in future studies may uncover it.

To improve the research, the following refinements will be incorporated: 1) an extreme groups design to better manipulate age and expertise; 2) stronger primary task workload manipulations; 3) more primary task practice for subjects to stabilize specific measures of primary task performance across levels of workload; 4) embedding the secondary task display into the primary task display; and 5) higher fidelity simulation to increase domain relevancy and support for domain-specific strategies used by expert pilots.

The results of this preliminary study provide a partial indication that expertise mitigates the adverse effects of aging regarding the ability to handle mental workload (regarding working memory and monitoring) in a simulation task. These findings will help "fine tune" the methodology and direct further research investigating additional cognitive resources involved in flying that may be affected by age, expertise, and workload. Ultimately this research may provide support for development of: 1) performance criteria for aging pilots to augment the age criteria currently in place in several branches of aviation; and 2) training regimens for older pilots to help them maintain their skills and certification.

ACKNOWLEDGEMENTS

This research was supported by National Institute of Aging Grant AG12388. We would like to thank John Demos for his help in data collection.

REFERENCES

Gopher, D., & Donchin, E. (1986). Workload: An examination of the concept. In K. R. Boff, L. Kaufmen, & J. P. Thomas (Eds.), Handbook of Perception and Human Performance (pp. 41-1, 41-44). New York: John Wiley & Sons.

Microsoft (1990). Flight Simulator 4.0. Redmond, WA: Microsoft Corporation.

Morrow, D. G., Leirer, V. O., & Altieri, P. A. (1992). Aging, expertise, and narrative processing. Psychology and Aging, 7, 376-388.

Morrow, D. G., Leirer, V. O., Altieri, P. A., & Fitzsimmons, C. (1994). When expertise reduces age differences in performance. Psychology and Aging, 9, 134-148.

Rybash, J. M., Hoyer, W. J., & Roodin, P. A. (1986). Adult cognition and aging. New York: Pergammon Press.

Sternberg, S. (1969). On the discovery of processing stages: Some extensions of Donders' method. Acta Psychologica, 30, 276-315.

Tsang, P. S. (1992). A reappraisal of aging and pilot performance. The International Journal of Aviation Psychology, 2(3), 193-212.

Tsang, P. S. (1995, April). Boundaries of cognitive performance as a function of age and piloting expertise. Paper given at the Eighth International Symposium on Aviation Psychology. Columbus, OH.

Tsang, P. S., & Shaner, T. L. (1994, March). Age and expertise in time-sharing performance. Paper given at the Biennial Cognitive Aging Conference. Atlanta, GA.

Wickens, C. D., Braune, R., & Stokes, A. (1987). Age differences in the speed and capacity of information processing: 1. A dual-task approach. Psychology and Aging, 2, 70-78.

Yekovich, F., Walker, C., Ogle, L. & Thompson, M. (1990). The influence of domain knowledge on inferencing in low aptitude individuals. In A. Graesser & G. Bower (Eds.), The psychology of learning and motivation: Advances in research and theory (Vol. 24, pp. 259-278). San Diego, CA: Academic Press.

AGE DIFFERENCES IN PERCEPTION OF WORKLOAD FOR A COMPUTER TASK

Sara J. Czaja
Department of Industrial Engineering
Miami Center on Human Factors and Aging Research
One of the Edward R. Roybal Centers for Applied Gerontological Research
University of Miami
Joseph Sharit
Department of Industrial Engineering
State University of New York at Buffalo
Sankaran N. Nair
Miami Center on Human Factors and Aging Research
University of Miami

Research concerned with age and work activities is an important area of investigation since the workforce is aging and there are concerns regarding economic dependency as well as labor shortages for certain occupations. Previous work by the research team indicated age differences in the performance and perceptions of task difficulty and fatigue for three simulated real-world computer tasks. This study is an extension of that research and is investigating the extent to which age differences in performance and perceptions of workload are moderated by experience and task practice. One hundred and twenty subjects aged 25 yrs. to 75 yrs. performed a real-world data entry task. Data will be presented regarding age differences in the perception of workload, stress, discomfort, and attitudes towards computers. The implications of these results for design interventions will be discussed.

INTRODUCTION

In recent years there has been renewed interest in the older worker due to the aging of the workforce and concerns regarding economic dependency and the potential shortage of younger people for some occupations. The current data regarding aging and work is limited, especially for present day jobs, and inconclusive with respect to findings regarding aging and job performance. For example, most workers need to interact with some form of computer technology to perform their jobs, yet there is little data available on the potential impact of this technology for older workers. This is an important issue as these types of jobs are characterized by their cognitive demands and there are age-related changes in most component cognitive abilities (Park, 1992). In order to design jobs and interfaces which are suitable for older workers we need to understand if there are age differences in performance for these types of tasks and the underlying source of these differences. Further, given that the current cohort of older adults is less likely to have had computer experience, we need to understand the degree to which age differences in abilities are moderated by experience or offset by compensatory strategies. As noted by Salthouse (1990) more research is needed to understand the degree to which experience offsets age declines in abilities and the nature of the experience effect. Finally, examination of age and computer task performance also requires consideration of the "cost of work." It could be that while the elderly are capable of performing these types of tasks, they are also experiencing greater stress as the result of the need to compensate for declines in basic abilities.

Previous research (Czaja and Sharit, 1993) indicated age differences in the performance of computer-based tasks as evidenced by longer response times and more errors on the part of the older participants. The older people also reported greater fatigue and found the tasks to be more difficult. The current research is part of a larger study which is an extension of the previous research, and concerns examining the relationship between cognitive abilities and performance as well as how age performance differences are moderated by experience and task practice. These issues are being addressed using real-world computer tasks that vary in cognitive demands and that are representative of tasks performed across a variety of occupations.

This paper reports on the perception of workload, discomfort, and stress as a function of age and task experience for one of these tasks, a data entry task. As noted, our previous research indicated age differences in these measures for computer-based work. However, in that study task performance was limited to a single task session. In this study we are examining how these age effects are moderated by experience on the task.

When considering aging and work, examination of the "cost of work" is a critical issue given the increased vulnerability of older adults to health related problems. Poor health is one of the most frequent factors contributing to reduced labor force participation among older people (Sterns, Laier, and Dorsett, 1994). If in fact older adults experience significantly greater stress and fatigue while performing computer tasks, the issue of job design needs to be carefully evaluated.

In order to enhance employment opportunities for older adults, jobs must be designed so that older workers are able to meet job requirements in a safe and comfortable manner.

METHOD

One hundred and twenty six subjects participated in the study. There were 54 males and 72 females, and their ages ranged from 20 - 75 yrs. Subjects were divided into three age groups: 20 - 39 yrs. (N= 44), 40 - 59 yrs. (N=40), and 60 - 75 yrs. (N= 42). All subjects were recruited from the Miami community, had at least a high school education, were literate, and had no prior experience performing the job being investigated. Amount of prior experience with computers was assessed via a scale which assesses duration, frequency of use, and breadth of experience with computers.

All subjects participated for five days for four to five hours per day. On day 1 they completed an extensive cognitive battery, a computer attitude questionnaire, and the computer experience scale. On day 2 they were trained on the task to criterion level, and on days 3 to 5 they performed the task for a period of 3 hours/day. In addition, they completed the Modified Stress Arousal Checklist (Mackay, Cox, Burrows, and Lazzerini, 1978) and Body Part Discomfort Scale (Corlett and Bishop, 1978) pre- and post-task, and the NASA Task Load Index (TLX scale) (Hart and Staveland, 1988) following task performance. The TLX scale provides a measure of overall workload as well as ratings of the contribution of six factors (mental demands, physical demands, temporal demands, performance, effort, frustration) to workload. Subjects also completed a task knowledge questionnaire at the end of training and at the end of each task session.

The task consisted of a data entry task which involved entering trip record information from truck driver records into pre-formatted computer screens. Subjects were told that the task performance criteria included both speed and accuracy. The task is an actual job performed at a large transportation corporation. However, it is also representative of a common computer task (data entry) performed in a vast array of work environments. Performance data included measures of work output, accuracy, and job knowledge.

RESULTS

The data indicated age differences (χ^2 (6,110) = 16.65, $p < .05$) in prior experience with computers, with the younger participants having more computer experience than the older participants. Forty-nine percent of the older people had no prior experience with computers as compared to 23% and 28% of the younger and middle-aged participants respectively. Prior experience with computers was found to be related to pre-task attitudes towards computers ($r = .22$, $p < .05$) such that people with more computer experience had more positive attitudes towards computers prior to participation in the study. Computer experience was not found to be related to post-task computer attitudes, perceptions of stress, arousal, or workload.

Age differences in attitudes towards computers (pre- and post-task), perceptions of workload, stress/arousal, and discomfort were assessed using a Repeated Measures ANOVA (SPSS) with age at 3 levels as the between factor and days on task (experience) at three levels as the within factor. It should be noted that some subjects (N=16) had to be dropped from the analyses as some of the middle-aged and older participants had difficultly completing the workload scale.

There were no age differences in perception of overall workload. Moreover, perceptions of workload did not vary as a function of experience (days) and there was no age by experience interaction. However, there were some age effects with respect to the ratings of the factors contributing to workload (Table 1). There was a significant age by experience interaction ($F(4,214) = 2.45$, $p < .05$) and a significant experience effect ($F(2,214) = 6.19$, $p < .01$) for ratings of physical demand. Ratings of physical demand increased for all participants across the three days, however, the increase was greater for the middle-aged and older subjects.

		Age Group						Overall	
		20 - 39 Years		40 - 59 Years		60 - 75 Years			
		X	SD	X	SD	X	SD	X	SD
Physical Demand	Day 3	3.92	6.38	2.61	3.91	2.56	4.82	3.12	5.24
	Day 4	4.77	7.31	3.97	5.10	2.02	2.87	3.72	5.69
	Day 5	4.23	5.94	5.28	6.31	4.46	7.21	4.63	6.40
	Overall	4.31	6.06	3.95	4.43	3.02	4.13	3.82	5.04
Mental Demand	Day 3	12.29	8.63	14.30	7.81	14.45	8.65	13.54	8.36
	Day 4	10.03	7.93	13.32	8.45	15.08	8.44	12.50	8.44
	Day 5	10.46	8.67	13.60	9.20	15.45	7.91	12.86	8.82
	Overall	10.93	7.41	13.74	7.27	15.00	7.51	12.97	7.53
Performance	Day 3	16.59	7.48	15.35	8.14	13.99	8.61	15.44	8.03
	Day 4	15.57	8.26	13.72	8.56	14.04	8.63	14.54	8.42
	Day 5	14.89	8.83	13.31	8.47	16.75	9.12	14.93	8.82
	Overall	15.68	6.82	14.13	7.58	14.93	7.54	14.97	7.24
Frustration Level	Day 3	5.03	7.53	6.37	8.38	5.83	8.91	5.69	8.17
	Day 4	4.61	8.54	5.41	8.13	4.97	8.19	4.97	8.24
	Day 5	3.64	6.80	3.92	6.25	3.68	6.85	3.70	6.58
	Overall	4.43	5.67	5.23	6.52	4.83	6.05	4.80	6.01

Table 1. Workload Component Ratings as a Function of Age and Task Experience

There was also a significant age by experience interaction ($F(4,214) = 2.36$, $p < .05$) for ratings of performance. The younger and middle-aged subjects rated their performance lower over time whereas the older subjects perceived that their performance improved with time. The data also indicated a significant difference in ratings of frustration as a function of experience with the task ($F(2,214) = 3.13$, $p < .05$). As people became more experienced with the task they found it to be less frustrating. There was a strong trend in the data indicating age differences in ratings of mental demands ($F(2,105)=2.95$, $p = .056$), with the older adults finding the task to be more mentally demanding than the other subjects. There were no differences for ratings of temporal demands or effort.

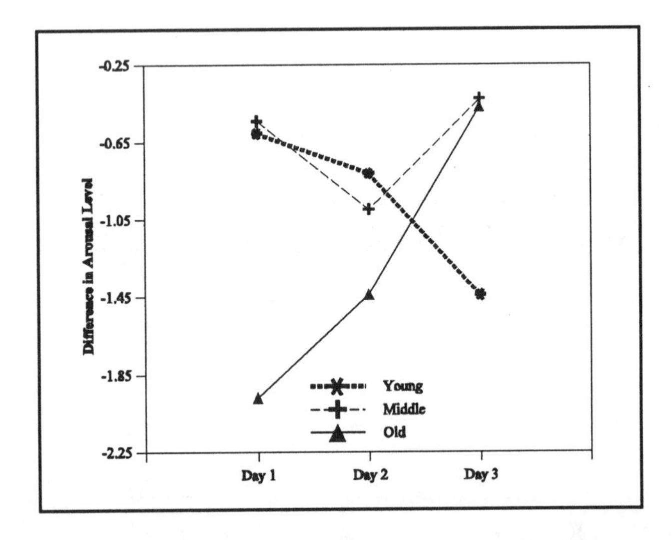

Figure 1. Difference in Arousal Levels Observed

The data also indicated a significant age by experience interaction for changes in level of arousal ($F(4,214) = 2.66$, $p < .05$). As shown in Figure 1 the younger subjects experienced a decrease in level of arousal across the three days whereas the older subjects experienced an increase in level of arousal. The were no differences in ratings of stress. With respect to discomfort there were no differences in overall ratings of discomfort as a function of age or experience on the task. However, of those participants who reported discomfort, severity of discomfort increased as a function of task experience ($F(2,210) = 4.82$, $p<.01$). In order to evaluate the source of body discomfort the ratings for upper body discomfort and trunk were analyzed given that the task involved sitting and entering data via a keyboard. Findings indicated that ratings of upper body discomfort increased for all participants over time ($F(2,210) = 3.12$, $p <.05$). Also of those subjects who experienced upper body discomfort, the severity of discomfort increased with time ($F(2,210) = 4.39$, $p <.01$). Finally there were no age differences in attitudes towards computers pre- or post-task even when controlling for differences in computer experience.

DISCUSSION

Overall, the data indicate that task experience has an influence on perceptions of workload, arousal, and discomfort, and that this effect varies as a function of age. While all participants indicated that the task became more physically demanding as they became more experienced, this perception was greater for the middle-aged and older participants. This result is not surprising given that the emphasis in a data entry task is on speed and accuracy of performance. It will be interesting to compare the workload ratings across the three tasks in view of the fact that the tasks comprising the study vary in information processing requirements.

The older participants also found the task to be more mentally demanding than the other subjects. This may be due to age differences in speed of processing. Age-related slowing has been shown to affect the performance of other cognitive operations (Salthouse, 1993). Interestingly, the older people also perceived that their performance improved with time whereas the younger and middle-aged people felt that their performance got worse. These findings are consistent with ratings in level of arousal (Figure 1). One possible explanation is that subjective appraisal of performance over time is a function of perceived level of arousal, which in turn depends on the degree to which the task is perceived to be mentally demanding. The overall implication is that there may be age differences in "cost of work" for this type of task. Note however, that despite these age differences all subjects found the task to be less frustrating with experience, implying that perceptions of performance on this task were not likely to be related to feelings of frustration. Finally there were no age differences in attitudes towards computers pre- or post-task.

The findings from this study have implications for job and workplace design. For example, work-rest schedules are particularly important for older workers. In addition, the design and layout of the workplace is important for these types of tasks. The participants indicated that frequency and severity of upper body discomfort increased with task experience. However, before design guidelines can be specified these results need to be interpreted within the context of performance data. This data is currently be analyzed and will be reported in the near future.

REFERENCES

Czaja, S.J. and Sharit, J. (1993). Age differences in the performance of computer-based work. Psychology and Aging, 8, 59-67.

Hart, S.G. and Staveland, L.E. (1988). Development of NASA-TLX (Task Load Index): Results of empirical and theoretical research. In P.A. and N. Meshkati (Eds.), Human Mental Workload (pp. 139-183). Amsterdam: North-Holland.

Mackay, C., Cox, T., Burrows, G., and Lazzerini, T. (1978). An inventory for the measurement of self-reported stress and arousal. British Journal of Social and Clinical Psychology, 17, 283-284.

Park, D. (1992). Applied cognitive aging research. In F.I.M. Craik and T.A. Salthouse (Eds.) The Handbook of Cognitive Aging. New York: John Wiley, 449-494.

Salthouse, T.A. (1990). Influence of experience on age differences in cognitive functioning. Human Factors, 32, 551-570.

Salthouse, T.A. (1993). Speed and knowledge as determinants of adult age differences in verbal tasks. Journal of Gerontology, 48, 29-36.

Sterns, H.L., Laier, M.P., and Dorsett, J.G. (1994). Work and retirement. In B.R. Bonder and M.B. Wagner (Eds.), Functional Performance in Older Adults (pp. 148-164). Philadelphia: F.A. Davis Company.

PREDICTORS OF ALTERNATIVE SCHEDULING: AGE, LEVEL, & TENURE

Thomas M. Franz, M. A.
Oak Ridge Associated Universities
Oakridge, Tennessee

Employees (N = 3,129) were surveyed to determine their likelihood to use three types of alternative work schedules: fewer hours (e. g., part-time), at home hours (e. g., occasional or part-time work at home) and more flexible hours (e. g., a flexitime schedule) if offered by the employer, in the next five years. It was found that half of the employees indicated that the schedules might apply to them (n = 1,290). Regression analyses tested the relationship of need for childcare, age, organizational level, and tenure with the alternative schedules. It was found that respondents who needed childcare were more likely to use all three types of alternative work schedules than those who did not need childcare. Younger respondents were more likely to use all three types of alternative work schedules than were older respondents. Higher level employees were found to be as likely to use alternative schedules as lower level employees. While employees of differing tenure were equally likely to work alternative schedules, many who indicated that the schedules did not apply to them were of higher tenure, thus biasing the final sample on which the tenure conclusion was based.

INTRODUCTION

A major demographic trend in the workforce today is the changing average age of the employee. The baby boom generation dominates the workforce today, lowering the average age from 40 in 1970 to 34 in 1980 (Bureau of Labor Statistics, 1980). However, population growth has slowed, and life expectancy is increasing, presenting us with what has come to be called the "graying of the workforce." Cahill and Salmone (1987) projected that continued full-time employment for older employees will probably not be norm and that future options will include self-employment, part-time work, job sharing and other alternative work schedules.

Need for Childcare

While older employees might be more likely to use part-time and more flexible scheduling than younger employees, it should be noted that need for childcare, regardless of an employee's age, may also be a predictor of AWS use. Although past literature usually indicated that shiftwork schedules were disruptive of family life, past studies may be out-of-date as in most two parent families both parents typically work today.

In the present study respondents were asked to indicate if they had sought assistance in finding childcare within the past four years Respondents who had sought assistance were expected to indicate that they would be more likely to use AWS's than those employees that had not sought assistance in finding childcare, because childcare was an obvious issue for them and AWSs might help them with this issue.

Age

It was predicted that age would account for a significant increment of likelihood to work fewer hours and likelihood to use more flexible scheduling. Based on previous research, it was predicted that older employees would be more likely to use AWSs than younger employees. Although the preceding paragraphs indicate that older employees might be more likely to use part-time and more flexible scheduling than younger employees, it should be noted that need for childcare, regardless of an employee's age, was also a predictor of AWS use.

It would be difficult to unambiguously evaluate the effect of age on likelihood to use AWS's, if one were not controlling for "need for childcare" (which would typically be found among younger employees). For this reason, the predictor, "sought assistance in finding childcare" was entered prior to the variable "age" in the regression equations. **Hypothesis 1: Age was expected to account for significant increments of variance in likelihood to work fewer hours and to use more flexible scheduling, over that which was accounted for by need for childcare.**

Level and Tenure

Dunham and Hawk (1977) described employees that most preferred alternative schedules as having low level jobs and little tenure, typically seeking a partial escape from a negative work environment. Therefore, in the present study a significant amount of variance in likelihood to use alternative schedules was expected to be accounted for by age and other demographics, as well as by job level and tenure. Specifically, employees with low level jobs and little tenure were expected to seek the potential escape offered (a three or four day weekend) by this schedule.

It was determined that the variables level and tenure be entered into the regression equation after the demographic variables (sought assistance for childcare, age, and gender) for the following reasons: (1) If organizational demographics (such as level and tenure) were entered prior to age, gender, etc., it could be argued that employees at different levels and with more or less tenure may have been different demographically; and (2) Demographic variables have been documented predictors of AWS use, therefore it seemed reasonable to predict that such differences would better predict likelihood to use AWS's than differences such as tenure and level. Tenure was entered prior to level because employees with more tenure are commonly selected for higher level positions.

Hypothesis 2: Employee level and tenure were expected to account for a significant increment in variance of employee self-reported likelihood to use more flexible scheduling over that accounted for by employee demographics.

METHOD

Survey data were obtained via mail from 3,129 randomly chosen employees (a 72% response rate) of an Southern high-tech manufacturing firm (Company ABC). Of these respondents, nearly half either indicated that some of the alternative work schedules did not apply to them (coded 9) or left the item responses blank (coded missing). For example, one AWS item asked how likely it was that the employee would take more than one year leave of absence to care for children. Of the 3,129 respondents only 1,490 (52%) were coded as having useable responses on this item. On another item respondents indicated how likely they would be to work at home. A full 20% of the respondents indicated that such a schedule would not apply to them. Thus, all of the hypotheses were tested only on employees who were open to the potential of caring for children, and working at home.

The employees with useable data in the dependent measures (n = 1,490) were compared to the population being sampled. Respondents with useable data were younger (55% of the sample were 34 years of age or younger, compared to 36% of the population), had less tenure, (59% of the sample had less than ten years compared to 45% of the population) and that the sample included more women than the population (an 8% difference). However, the respondents were representative of the population organizational levels (i.e., exempt, non-exempt, first and second level management).

Principal component analyses with varimax rotation were used to reduce the eleven AWS items into components of likelihood to work alternative schedules. A three three factor solution of the Alternative Work schedule Index was achieved with this PCA. Items in the first component of AWS were all judged to be examples of likelihood to work fewer hours. The second component was labeled at home hours, and the third component was judged to represent a desire for more flexible scheduling. Regression analyses were used to test the hypotheses of the study.

RESULTS / DISCUSSION

Childcare Needs

It was found that need for childcare accounted for a significant amount of variance in all three dependent measures. That is, employees with a need for childcare services were more likely to use AWS's than those not needing such services.

Age

The first hypothesis stated that employee age would be predictive of reported likelihood to use two types of AWS's. Based upon previous research (Bogart, 1983; Cahill & Salmone, 1987; Bosworth & Holden, 1983; Gustman & Steinmeier, 1984) it was expected that older employees would be more likely to use new schedules or more flexible schedules than younger employees.

For those with useable responses, age did prove to be a significant predictor of likelihood to use AWS's. However, it was found that *younger employees* were more likely to use all three types of AWS's than were older employees. At first blush it would appear that these findings are in conflict with previous reports that older individuals (i.e., retirees) were more likely to work AWSs. However, the present sample excluded many older employees who had reported that the AWS's "would not apply to them" (e. g., AWS's linked with childcare) and did not include any retired individuals.

In contrast, previous researchers examined individual interest in AWS's when near retirement, or those interested in partial retirement arrangements. Therefore, when future researchers examine the relationship of age against likelihood to work AWS's, it will be important that all age groups of employees are represented in the sample. It should be noted that controlling for the effects of need for child care indicated that there are other reasons (such as education or different life styles) that younger employees may choose AWSs over traditional hours.

These findings may indicate that as employees age, they become more accustomed to a traditional 9 to 5 - Monday through Friday schedule. The data also suggested that older employees tend to have fewer conflicts with their on-duty hours. Based on these results, practitioners could expect that, in general, all types of AWS's would appeal to younger employees, and especially to younger employees with childcare needs.

Tenure

For tenure, no relationship was found between tenure and likelihood to work alternative schedules. Unlike previous reports that less tenured employees preferred AWS's, this data indicated no such relationship existed. The discrepancy can be accounted for by noting that Dunham and Hawk did not control for age when reporting on the relationship of tenure with working AWS's. In the present work, age (which was correlated at .76 with tenure) was controlled when examining the tenure/AWS relationship, and a clearer picture emerged of the relationship. That is, for respondents with available data, the younger were more likely to try all types of alternative schedules than the older respondents.

It should be noted that many of the employees with more years of service with ABC were excluded from the sample, and this may have contributed to the failure to account for variance in the AWS indices. However, this indirectly offers support for previous research, as those with less tenure were more willing to "consider" AWS then their more tenured counterparts. As previously stated, future researchers should take care to design AWS items that will not exclude those who feel the schedule will "not apply" in their case.

Level

Level was not linked with likelihood to work alternative schedules. This finding failed to support Dunham and Hawk's report that lower level employees were more likely to try alternative scheduling than higher level employees. Dunham and Hawk (1977) had described employees who most preferred flexible schedules as young, with low job levels and little tenure, and as seeking a partial escape from a negative work environment. Such a claim seemed to imply that employees of all ages, levels, and amounts of tenure were surveyed in their research. Unfortunately, this was not the case, as the conclusions reached by Dunham and Hawk were based upon a survey of mid-level exempt personnel.

In the present study, all levels of employees were represented, although employees with fewer years of tenure were over represented. Therefore we can conclude for the respondents with useable data, all levels were equally likely to use the alternative schedules. Although the present sample was limited, the findings demonstrate the importance of controlling for need for childcare and age when testing the relationship of likelihood to use AWSs and level.

REFERENCES

Bogart, A. (1983) Part-Time Employment Makes Retirees A Valuable Resource. Personnel Administrator,28,June,35.

Bosworth, T. W., and Holden, K. C. (1983) The Role Part-Time Job Options Play in the Retirement Timing of Older Wisconsin State Employees.Aging andWork,6,31-36.

Cahill, M., and Salomone, P. R. (1987) Career Counseling for Work Life Extension: Integrating the Older Worker into the Work Force. The Career Development Quarterly, 35, 188-196.

Dubinsky, A. J., and Skinner, S. J. (1984) Job Status and Employees' Responses: Effects of Demographic Characteristics. Psychological Reports, 55, 323-328.

Dunham, R. B., Hawk, D. L. (1977) The Four-day/Forty-hour Workweek: WhoWants it? Academy of Management Journal, 20 . 644-655.

Gustman, A. L., and Steinmeier, T. C. (1984) Partial Retirement and the Analysis of Retirement Behavior. Industrial and Labor Relations Review, 37, (3) April, 403-415.

AGE DIFFERENCES IN THE ADJUSTMENT TO SHIFTWORK

by

Christopher M. Keran and James C. Duchon

U.S. Bureau of Mines
Minneapolis, Minnesota

Due to the aging U.S. workforce and the increase in the percentage of shiftworkers in the workforce, an understanding of the effects of shiftwork on older employees is considered important to the U.S. Bureau of Mines. As part of a larger study on shiftwork by the Bureau of Mines, survey data were obtained from 295 rotating shiftworkers. The workers were categorized into three age groups with means of: 27.4, n=76; 36.4, n=177; and 49.3, n=42. ANOVAs revealed that older workers reported greater frequency of 6 out of 23 physical symptoms than younger workers. Older workers also reported themselves to be more "morning" type than younger workers, which may help explain some of the Age by Shift differences. MANOVAs were used to determine Age by Shift interactions of sleeping habits, stress, and physical symptoms. Though most of the workers have problems with the night shift, the older worker seemed to have more trouble adjusting to it, and also the afternoon shift, than the younger worker. For the most part, however, the older worker seemed to adjust to the day shift as well if not better than the younger worker.

INTRODUCTION

The U.S. work force will continue to grow older in the 21st Century (Smith, 1988), and if the popularity of continuous operations persists, many more of these workers will work "irregular" hours, or shiftwork. Studies have found that the older worker has greater difficulty working rotating shifts than the younger worker (Koller, 1983). Few studies, however, have looked at age differences in the adjustment to the particular shift worked. One such study (Akerstedt & Torsvall, 1981), found older workers slept less and had poorer quality sleep than younger workers on the night and afternoon shift, but just the opposite was true on the day shift. However, the sample in this study included Swedish steel workers on a rapidly rotating schedule (shift changes every 3 days) and early shift start times (445, 1300, 2115); schedules that are uncommon to the U.S. workforce.

This study will analyze data obtained from the Bureau of Mines Shiftwork Survey to determine adjustment to particular shifts by older shiftworkers compared to younger shiftworkers. Adjustment is evaluated in terms of sleeping habits and physical and mental stress on the day, afternoon, and night shift. Physical symptoms and diurnal type (without regard to shift) will also be looked at.

METHOD

Apparatus

The Bureau of Mines Shiftwork Survey includes demographic factors, sleep quantity and quality, satisfaction with the work schedule and eating habits, stress, health complaints, meal frequency and timing, alertness and performance, accidents, and diurnal type (morningness/eveningness). The validity of the survey, and the sleep and morningness/eveningness questions in particular, have been discussed previously (Duchon, 1988; Duchon & Keran, 1990).

Subjects

As part of a larger study on shiftwork by the U.S. Bureau of Mines, data were obtained from a sample of 295 (a response rate of 92 pct) rotating shiftworkers (mean age = 35.9, ranging from 22 to 60; 95.2 pct male; and 80.0 pct married). Subjects were classified into one of three age groups:

	age range	frequency	mean	sd
Young	<= 30	n=76	27.4	2.1
Middle	30 < x < 45	n=177	36.4	3.6
Old	>=45	n=42	49.3	4.2

The data are aggregated from subjects who work at three different industrial operations. Two of the settings (power generating companies) are continuous operations rotating every week with shift start times of 630, 1430, and 2230. One operation rotates in a forward direction (day to afternoon to night), the other in a backward direction (day to night to afternoon). The third operation (a surface mine) rotates every two weeks in a backward direction (day to night to afternoon) with shift starting times of 800, 1600, and 2400 with all weekends off.

RESULTS

The data were analyzed using either the ANOVA (shift-independent questions) or the MANOVA procedure (repeated measures for shift-dependent questions) in SPSS/PC+. Due to the significant correlation ($r=.39$, $p < .000$) between age and shiftwork experience (total experience on the night shift), experience was entered as a covariate.

General Health Items

The following questions were asked in order to compare the general health of the different age groups without regard to any particular shift.

An Analysis of Variance procedure, revealed significant differences between the three age groups. Older workers reported a higher frequency, on average, than younger workers for 6 out of 23 physical symptoms:

shortness of breath/
trouble breathing...... F=10.20, df=2, p <.000
periods of severe
fatigue/exhaustion...... F=3.53, df=2, p <.05
feeling sweaty/trembly.. F=4.00, df=2, p <.05
leg cramps.............. F=3.35, df=2, p <.05
tight stomach........... F=4.41, df=2, p <.01
trouble digesting food.. F=5.19, df=2, p <.01
colds/sore throat......................... ns
swollen or painful muscles/joints.......... ns
back pain................................. ns
headaches................................. ns
nervous/shaking inside.................... ns
racing/pounding heart..................... ns
acid indigestion/heartburn/acid stomach.... ns
diarrhea.................................. ns
gas/gas pains............................. ns
nausea/vomiting........................... ns
constipation.............................. ns
bloated/full feeling...................... ns
feeling of pressure in the neck........... ns
dryness in the mouth...................... ns
stomach pains............................. ns
belching.................................. ns
difficulty with feet/legs
when standing for long periods............ ns

Significant differences were also found in the reported diurnal type of the worker (F=7.60, df=2, p <.001). Older workers tended to describe themselves to be more "morning" type than younger workers.

Sleep Habits

Questions were asked about routine sleeping habits for each of the three shifts. Consistent with other studies (Webb, 1982; Monk & Folkard, 1985; Tune, 1969; Torsvall *et al.*, 1981; Foret *et al.*, 1981) a significant negative relationship (F=12.3, df=2, p <.000) was found between sleep length and age (see Figure 1). However, the significant interaction of Age x Shift (F=5.23, df=6, p <.000) indicates that older workers, on the afternoon and night shifts, and on Free days (when workers are totally free to plan their day), slept less than younger workers, although older workers on the day shift slept as long as younger workers. One-Way ANOVAs revealed that older workers on the afternoon shift woke up earlier in the morning than younger workers (F=3.63, df=2, p <.05), and older workers on the night shift started sleeping later in the morning than younger workers (F=11.4, df=2, p <.000).

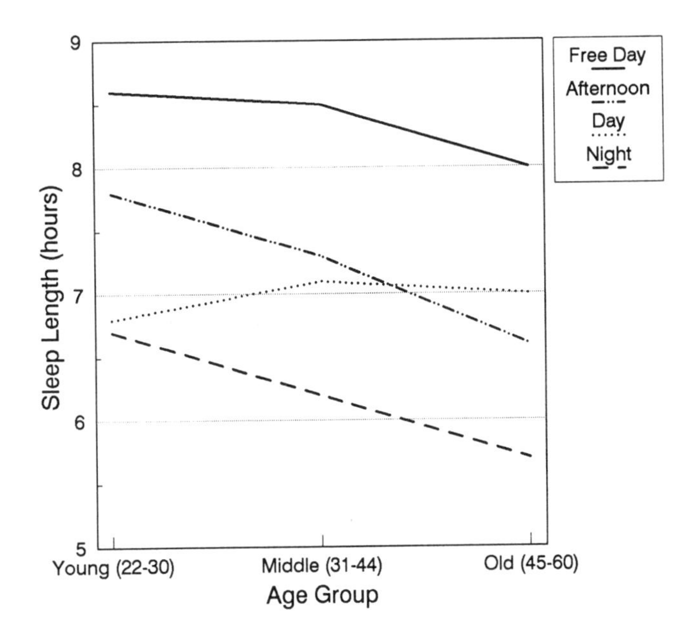

Figure 1. Sleep lengths by age group by shift.

Although there was not a significant main effect for age for satisfaction "...with the amount of sleep you get," the Age x Shift interaction was significant (F=2.80, df=4, p <.05). The trend for the afternoon and night shifts was that the older workers were less satisfied with their sleep quantity than the younger workers. On the day shift, however, older workers were slightly more satisfied than were younger workers.

Physical and Mental Stress at Work

Questions were asked about fatigue and physical and mental stress experienced at work and after work, and also whether the stress level became better or worse as the workweek went on (from Monday to Friday) for each shift.

All workers reported greater sleepiness on the night shift than either the day or afternoon shift (F=216.8, df=2, p <.000). However, the significant Age x Shift interaction (F=2.6, df=4, p <.05) identified that on the afternoon shift, older workers tended to feel somewhat more sleepy than younger workers, and on the day shift, older workers felt slightly more awake than younger workers.

All workers reported being more physically tired after work on the night shift than either the day or afternoon shift (F=181.3, df=2, p <.000). There were no significant age differences nor interactions found for how physically tired after work nor physical exhaustion as the workweek went on. However, the Age x Shift interaction as the workweek went on was significant ($F=2.94$, $df=4$, $p<.05$). The physical exhaustion of older workers, compared to younger workers, tended to become worse on the afternoon and night shifts and better on the day shift (see Figure 2).

Compared to younger workers, older workers reported experiencing more aches and pains after work on all shifts ($F=6.22$, $df=2$, $p<.01$). Although there was no significant main effect for age for aches and pains as the workweek went on, the Age x Shift interaction was significant ($F=3.12$, $df=4$, $p<.05$). The amount of minor aches and pains for older workers, compared to younger workers, tended to get better on the day shift and worse on the afternoon shift.

The work performance of older workers as the workweek went on tended to get worse on the afternoon and night shifts compared to that of the younger worker ($F=3.84$, $df=2$, $p<.05$). Though not significant, there was a tendency for the work performance and mental exhaustion of older workers to improve throughout the workweek of day shifts, and to get worse on the afternoon and night shifts.

There were no significant main age effects nor shift interaction effects for either difficulty falling asleep nor for difficulty staying asleep. However, all workers had more difficulty falling and staying asleep on the night shift than the day and afternoon shifts.

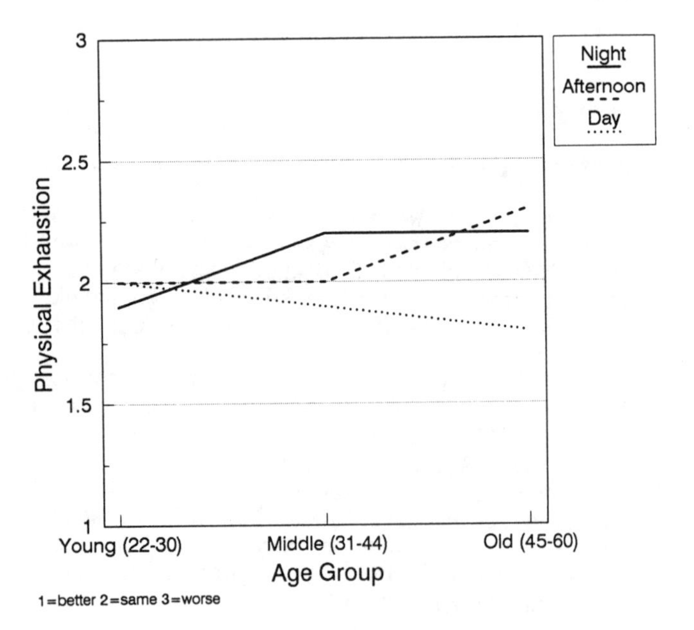

Figure 2. Physical exhaustion as the workweek goes on by age group by shift.

DISCUSSION

The purpose of this study was to see if older shiftworkers reported differential adjustment to the day, afternoon, and night shifts, compared to younger shiftworkers. An analysis of the sleep and physical and mental adjustment of shiftworkers indicates that the ability to tolerate different shifts is to some degree dependent upon both the age of the shiftworker and the particular shift worked. For instance, the common finding that sleep lengths are longest during the afternoon shift and shortest during the night shift, with the day shift falling in between, is misleading when age is not considered, at least for the workers in this study. In fact, this study found that on the afternoon shift the average sleep length for the older group was only 6.6 hrs, compared to the day shift, when their sleep length was 7 hrs.

Several findings indicate that older workers, on the afternoon shift and to some extent the night shift, may have a problem sleeping that does not exist on the day shift. While older workers, when on the afternoon and night shifts, slept less than younger workers, when on the day shift, older workers slept just as long as younger workers, and they slept an average of 8 hours on their Free days. Older workers also report less satisfaction with their amount of sleep than younger workers on the afternoon and night shifts, but just the opposite is true on the day shift. Furthermore, older workers were less sleepy after work on the day shift and more sleepy after work on the afternoon shift than were younger workers.

The tendencies of older groups reporting more problems on the afternoon and night shifts and less on the day shift are further supported by reports of physical and mental stress. On the day shift, older workers tended to get better as the workweek went on, compared to younger workers, in terms of physical exhaustion and minor aches and pains. However, they tended to get worse, compared to younger workers, as the workweek went on for the afternoon and night shifts for these same complaints. Older workers, compared to younger workers, described their work performance as getting worse as the workweek went on for both the afternoon and night shift but not the day shift. There was a similar tendency for mental exhaustion, though not significant. Although workers in the older age group reported experiencing more aches and pains after work on all shifts, they did not describe themselves to be more physically tired than did younger workers for any shift.

Why are older shiftworkers having more difficulty on the afternoon and night shifts but apparently less difficulty on the day shift? One hypothesis is that because the older workers were more morning type (morning active) than younger workers, they may have more difficulty sleeping during the morning hours than do younger workers. In fact, the data showed that on the afternoon shift, older workers tended to wake up earlier in the morning than younger workers, and on the night shift they tended to go to sleep later in the morning than younger workers. Another possibility, offered by Akerstedt & Torsvall (1981), is that the "...proneness to internal desynchronization of the circadian system increases with age, making it more susceptible to disturbances of the sleep/wake pattern."

CONCLUSIONS

Further research is needed to substantiate the age by shift differences in sleep lengths. However, it is evident that reporting sleep lengths for different shifts, without regard to the mediating effects of age, may be incomplete.

Though most all of the workers have problems adjusting to the night shift, the older workers seemed to have more trouble adjusting to the night shift and the afternoon shift than the younger workers. Inspite of the fact that older workers were less healthy than younger workers, (they reported more occurrences of 6 out of 23 symptoms), the older workers seemed to adjust to the day shift as well if not better than the younger workers.

Because the significant age differences were inconsistent across the three shifts, a generalized increase in stress (i.e., reported physical and mental exhaustion) with age is unsupported. Due to the cross-sectional nature of this study, however, self-selection and cohort biases are not ruled out.

REFERENCES

Akerstedt, T., & Torsvall, L. (1981). Shift-dependent well-being and individual differences. Ergonomics, **24**, No. 4, 265-273.

Duchon, J.C. (1988). Evaluation of two work schedules in a mining operation. Paper in: Trends in Ergonomics/Human Factors V, ed. by F. Aghazadeh, Elsevier Science Pub., (151-160).

Duchon, J.C., & Keran, C.M. (1990). Relationships among shiftworker eating habits, eating satisfaction, and self-reported health in a population of miners. Work and Stress (in press).

Foret, J., Bensimon, G., Benoit, O., Vieux, N. (1981). Quality of sleep as a function of age and shift work. Proceedings of the Fifth International Symposium on Night and Shift Work (149-154).

Koller, M. (1983). Health risks related to shift work: An example of time-contingent effects of long-term stress. International Archives of Occupational and Environmental Health, **53**, 59-75.

Monk, T.H., & Folkard, S. (1985). Individual differences in shiftwork adjustment. In S. Folkard and T.H. Monk (eds.) Hours of Work. New York: John Wiley & Sons, Ltd., (227-252).

Smith, B.D. (1988). The older and disabled population: Forensic issues in accidents and age discrimination. Proceedings of the Human Factors Society: 32nd Annual Meeting (213-214).

Torsvall, L., Akerstedt, T., & Gillberg, M. (1981). Age, sleep and irregular workhours. Scandinavian Journal of Work and Environmental Health, **7**, 196-203.

Tune, G.S. (1969). The influence of age and temperament on the adult human sleep-wakefulness pattern. British Journal of Psychology, **60**, No. 4, 431-441.

Webb, W.B. (1982). Sleep and biological rhythms. In W.B. Webb (ed.) Biological Rhythms, Sleep, and Performance. New York: John Wiley & Sons, Ltd., (87-110).

A STUDY OF LIFE EXPECTANCY FOR A SAMPLE OF RETIRED AIRLINE PILOTS

Robert O. Besco
Professional Performance
Improvement, Inc.
Dallas, TX

Satya P. Sangal
Wright State University
School of Aerospace Medicine
Dayton, OH

Thomas E. Nesthus and
Stephen J. H. Veronneau
Federal Aviation Administration
Oklahoma City, OK

There is a popular belief in the aviation industry that retired pilots die at a younger age than their counterparts in the general population. If this is true, research into factors associated with this career would be of interest to the FAA as indicators of possible health factors to be monitored in the pilot population. A sample of 1494 pilots who retired at age 60 from a major U.S. airline between the study dates of April 1968 to July 1993 were surveyed. The Life Table Method was chosen as the most suitable approach to analyze the pattern of mortality for this data set. Comparisons were made with the U.S. general population of 60 year-old white males in 1980. A difference in life expectancy of more than 5 years longer was found for our sample of retired airline pilots. Half of the pilots in this sample retiring at age 60 were expected to live past 83.8 years of age, compared to 77.4 years for the general population of 60 year-old white males in 1980. The authors concluded that the question of lowered life expectancy for airline cockpit crews was not supported by the results of these data.

INTRODUCTION

A question that has been discussed, pondered, and argued for years in the cockpits, briefing rooms, negotiating tables, and watering holes of commercial aviation is "What is the life expectancy of the typical retired airline pilot?" The conventional, operational wisdom has pointed to those instances where a retired airline pilot in excellent health and physical condition, died within the first few years, and perhaps within a few months after retirement. Every airline pilot can recount several personal anecdotes of colleagues who died very soon after the mandatory retirement age of 60. But do these examples reflect a real trend or just an externalized concern for their own mortality? Some people believe that factors associated with this career may have precipitating effects on mortality.

Professional airline pilots should be in far better health at all ages compared to the general population. Every six months, airline pilots must pass a Class One Flight Physical, which is defined and required by the Federal Aviation Administration (FAA). Airline pilots reach their mandatory-retirement date at age 60 following an entire career of active health monitoring and maintenance. Many people might conclude that the life expectancy should be considerably longer than their 60 year-old counterpart in the general population due to the continual medical scrutiny, early detection and treatment of disorders.

However, the "flight line talk" of the aviation industry contends that pilots die at a younger age than the general population. The underlying premise of this lowered life expectancy is thought to be concerned with the unknown but potentially negative effects of environmental and job related factors that airline pilots are exposed to during their careers (Besco and Smith, 1990; Mohler, 1993; Morgan, 1992; Reinhart, 1988). Every time an airline pilot dies in the first few years after retirement, the question of premature death among this group is rekindled and reinforced in the minds of the observers.

Survival patterns of retired airline pilots have a far-reaching impact on pilots' careers, life insurance, and retirement benefits (Mayhew, 1992; Muhanna and Shakallis, 1992; and U.S. Congress, 1990). However, our literature review found little defensible evidence for or against the hypothesis of premature death among retired airline pilots. In one relevant preliminary study suggesting a trend of premature deaths (Muhanna and Shakallis, 1992), we found that the authors assessed retired pilot mortality in a manner that does not account for the changing nature of the sample, i.e., taking into consideration all of the individuals who enter (retire) and leave (die) during each year of the study period (i.e., a cohort study). Also, their rates were expressed as percentages which presented the mortality data in an interesting manner, however, we felt it was an inappropriate method for analysis of realistic trends. In fact, all of their data were presented as percentages. No actual sample sizes were reported nor were any survial statistics.

Several other studies adressed questions of pilot mortality but not longevity and survival after retirement. Kaji, Asukata, Tajima, Yamamoto and Hokari (1993) found that the mortality rate from natural causes was lower in active Japanese airline pilots than in the general Japanese population. However, they did not have a large enough sample of retired pilots to determine reliable estimates of post retirement life expectancy. Irvine and Davies (1992) conducted a proportional mortality study of active and retired British Airways pilots but did not report any survival data for retired pilots. Salisbury, Band, Threlfall, and Gallagher (1991) conducted an epidemiological study of deceased British Columbia pilots. They reported an elevated Proportional Cancer Mortality Ratio (PCMR) for airline pilots. Band, Spinelli, Ng, Math, Moody, and Gallagher (1990) conducted a study on causes of death in 891 Canadian Pacific Airlines pilots. They reported elevated PCMRs for the pilots in their sample. Hoiberg and Blood (1983) in a study of age-specific morbidity concluded that Navy pilots are in much better health than the normal population. The oldest pilots in their study were under 54 years of age. None of these authors studied survival patterns of retired pilots. So, this preliminary survey was proposed to address the hypothesis of premature death or extended life span among retired airline pilots. The purpose of the study therefore, was to investigate the survival pattern of a large enough sample of retired pilots to provide reliable and valid estimates of post retirement survival. No attempt was made to determine cause of death due to the need to preserve anonymity and secure the cooperation of the airline providing the data.

METHODS

We received cooperation from American Airlines (AAL), the Allied Pilots Association, and the Grey Eagles (Retired AAL Pilots). The records for our sample were on a computerized data base. The data received were anonymous. No data were made available to conduct any epidemiological investigations. This study's initial sample was based on 2209 of the pilots and flight engineers who retired from active service with American Airlines in the 25 year-period between April 1968, and July 1993. Out of the 2209 retirees, 360 had retired before age 60, probably for medical reasons, and 355 retired after their 60th birthday, probably as flight engineers. The early and late retirees were excluded from the sample, leaving 1494 pilots retiring at age 60. Of this group, 1298 (87%) survived throughout the 25 year study period.

One popular and practical technique for describing survival experience over time is the actuarial or life table method. Described by some authors in detail (Griswold, Wilder, Cutler, and Pollack, 1955; Pearl, 1923), Cutler and Ederer (1958) described one very important advantage of the life table method, which is why we felt the technique was the most suitable approach to analyze the pattern of mortality for this data set. Specifically, this technique utilizes all survival information and data on partial exposure to the risk of dying to provide the best estimates of survival. It allows subjects to enter (retire) or leave (die) the study at different points of time during the 25 years of the study. Also, the life table approach is non-parametric and requires no assumptions about the distribution of the survival function.

The life table analysis provides estimates of probabilities of surviving a given number of years after retirement and emphasizes the advantage gained by including survival information on individuals entering the series too late to have had the opportunity to survive the extent of the 25 year study. In addition, it estimates median residual lifetime or median remaining life expectancy for the end of each year survived after retirement. This method indicates that among those pilots who survive a given number of years past retirement, half of the survivors will die before the median life expectancy is reached and half will live beyond this median residual lifetime age.

RESULTS

Table 1 contains the results of the life table analysis. An explanation of some entries in Table 1 might clarify and help in understanding the table:

Years After Retirement Interval Column (1) (x to x+1) i.e., 0--1 refers to the first year after retirement at age 60; 1--2, the second year, etc. Each interval is a cohort for the survival analysis function.

Alive at Beginning of Interval Column (2) (l_x) is the number of cases that have survived to the beginning of the current interval. All 1494 were alive during their first year after retirement. Successive entries in this column are obtained by this formula: $(l_x+1) = l_x - (d_x + w_x)$.

Number Died Column (3) (d_x) gives the number of pilots who died during the interval. For example ten pilots died during their first year after retirement. Fourteen during the second year, etc.

Withdrawn Alive During Interval Column (4) (w_x) refers to the pilots who were known to be alive at the close of the study, but were in the study for the maximum duration denoted by the upper limit of the interval. At the conclusion of the 25-year study period in July 1993, 172 surviving pilots had been retired for less than 1 year during the first interval. Thus, each cohort (in each year of the 25-year study) contributes some information to our knowledge of survival during the interval.

Effective Sample Size Column (5) It is assummed that pilots lost or withdrawn from observation during an interval were exposed to the risk of dying, on the average, for one-half the interval. Thus, $l_x' = l_x - (w_x \div 2)$.

Proportion Dying During Interval Column (6) An estimate of the probability of dying during the interval. It is obtained by dividing the number of deaths by the effective number exposed to risk: $q_x = d_x \div l_x'$.

Propotion Surviving the Interval Column (7) is referred to, alternately as the probability of surviving the interval, or the survival rate. It is obtained by subtracting the proportion dying during the interval from unity: $p_x = 1 - q_x$

Cumulative Proportion Surviving the Interval Column (8) is generally referred to as the cumulative survival rate, and gives the probability of a pilot surviving to the end of the specified yearly interval after retirement. Calculated by cumulatively multiplying the proportion surviving each interval: $P_x = p_1 \times p_2 \times p_3 \times ...p_x$. Table 1 indicates that the probability of a pilot surviving for 2 years after retirement is 0.98. For a pilot surviving 20 and 25 years after retirement is 0.60 and 0.49, respectively. Figure 1 depicts this column with the 95% confidence interval derived from the standard error.

Survival Standard Error Column (10) provides a measure of confidence with which one may interpret the statistical results. Refer to Culter and Ederer, 1958 for computional formula.

Median Residual Lifetime (MRL) Column (11) An estimate of the time point at which the value of the the cumulative survival function is 0.50. That is, it is the time point by which half of the pilots retiring at age 60 are expected to have died. Table 1 shows that this point for our sample occurs during the 23--24 year interval. Linear interpolation is used to calculate the precise year value. So, half of our sample is expected to live 23.81 years (after interpolation) beyond retirement age, i.e. to 83.81 years of age (60 + 23.81). For those surviving the first year after retirement a median life expectancy of 22.98 more years was shown, to an age of 83.98 years old. For those surviving two years after retirement, their median life expectancy was increased to 21.99 more years or to an age of 83.99 years old. MRLs for surviving beyond this point in our sample cannot be accurately estimated because an insufficient percentage of retirees had died to compute reliable estimates.

Since virtually all of the retired pilots in our sample were white males (exceptions were unknown due to our anonymous sample), comparisons were made with American white males in the general population. Mortality statistics were surveyed from the years 1980, 1985, and 1989 (U.S. Dept. of Health and Human Services, 1984, 1988, and 1991). The 1980 statistics were selected for comparisons because they represented data from approximately the middle of our 25-year study. It can be seen in Figure 1, that the survival probability curve for the 1980 U.S. white male population, is entirely below the lower 95% confidence interval for pilot survival. The pilots in this sample, therefore, live significantly longer than other white American males who were age 60 in the year 1980. The median survival age for pilots is over 83 years. For the typical 60 year-old white American male, the median survival ages for the years 1980, 1985, and 1989 were 77.4, 78.1, and 79.0, respectively (U.S. Dept. of Health and

Table 1. Combined Life Table and Survival Estimates of Pilots Retiring at Age 60 From April 1968 to July 1993

Years After Retirement Interval	Alive at Beginning of Interval	Died During Interval	Withdrawn Alive During Interval*	Effective Number Exposed to Risk of Dying	Proportion Dying (Col. 3÷ Col. 5)	Proportion Surviving (1-Col. 6)	Cumulative Proportion Surviving Interval	Survival Standard Error	Median Residual Lifetime (MRL)	MRL Standard Error
(1)	(2)	(3)	(4)	(5)	(6)	(7)	(9)	(10)	(11)	(12)
x to x+1	l_x	d_x	w_x	l'_x	q_x	p_x	P_x			
0 -- 1	1494	10	172	1408.0	0.0071	0.9929	0.9929	0.0022	23.812	0.25
1 -- 2	1312	14	147	1238.5	0.0113	0.9887	0.9817	0.0037	22.979	0.28
2 -- 3	1151	7	114	1094.0	0.0064	0.9936	0.9754	0.0043	21.986	0.28
3 -- 4	1030	7	69	995.5	0.0070	0.9930	0.9685	0.0050	NA‡	NA
4 -- 5	954	13	42	933.0	0.0139	0.9861	0.9550	0.0062	.	.
5 -- 6	899	11	31	883.5	0.0125	0.9875	0.9431	0.0071	.	.
6 -- 7	857	17	44	835.0	0.0204	0.9796	0.9239	0.0084	.	.
7 -- 8	796	8	31	780.5	0.0102	0.9898	0.9145	0.0089	.	.
8 -- 9	757	9	60	727.0	0.0124	0.9876	0.9032	0.0095	.	.
9 -- 10	688	12	68	654.0	0.0183	0.9817	0.8866	0.0105	.	.
10 -- 11	608	7	61	577.5	0.0121	0.9879	0.8758	0.0111	.	.
11 -- 12	540	8	79	500.5	0.0160	0.9840	0.8618	0.0120	.	.
12 -- 13	453	9	77	414.5	0.0217	0.9783	0.8431	0.0133	.	.
13 -- 14	367	7	57	338.5	0.0207	0.9793	0.8257	0.0145	.	.
14 -- 15	303	14	53	276.5	0.0506	0.9494	0.7839	0.0176	.	.
15 -- 16	236	9	41	215.5	0.0418	0.9582	0.7511	0.0200	.	.
16 -- 17	186	5	23	174.5	0.0287	0.9713	0.7296	0.0216	.	.
17 -- 18	158	9	19	148.5	0.0606	0.9394	0.6854	0.0248	.	.
18 -- 19	130	6	21	119.5	0.0502	0.9498	0.6510	0.0272	.	.
19 -- 20	103	7	21	92.5	0.0757	0.9243	0.6017	0.0309	.	.
20 -- 21	75	3	23	63.5	0.0472	0.9528	0.5733	0.0335	.	.
21 -- 22	49	1	10	44.0	0.0227	0.9773	0.5603	0.0352	.	.
22 -- 23	38	1	11	32.5	0.0308	0.9692	0.5430	0.0381	.	.
23 -- 24	26	2	11	20.5	0.0976	0.9024	0.4901	0.0495	.	.
24 -- 25	13	0	11	7.5	0.0000	1.0000	0.4901	0.0495	.	.
25 -- 26	2	0	2	1.0	0.0000	1.0000	0.4901	0.0495	.	.

Notes: * Those individuals who were alive at the conclusion of the 25-year study, in July 1993.

‡ NA denotes data is not available.

Human Services, 1984, 1988, and 1991). In our sample, the retired pilots have more than a 5 year advantage of median life expectancy compared to the 60 year-old U.S. white male population.

DISCUSSION

The hypothesis of premature mortality among retired airline pilots compared to their counterparts in the general population was not supported by the data in this study. Retired pilots in this sample appeared to enjoy a life expectancy of more than 5 years longer than the 1980 U.S. general population of 60 year-old white males. However, before it can be concluded that this is true for all retired airline pilots, the adequacy of this sample to represent the population of retired airline pilots should be determined. This sample represents only one industry airline.

It could be argued that, compared with other airline pilots, this sample may have been subjected to higher medical standards at the time of initial hiring and monitored more closely throughout their careers. It could be hypothesized that an even greater increase in life expectancy should have been realized, but was not because of the effects of the environmental and personal stress factors associated with this occupation.

A question concerning the adequacy of our general population sample could also be raised. Because our pilot sample was anonymous, we could not match personal characteristics (e.g., socio-economic status, education, health consciousness, family history or specific health attributes) that could be influential in promoting longevity to the general population we sampled.

Questions such as these are recommended for investigation and that a follow-up or updated survey of AAL pilots be conducted on a regular basis to track changes in the median residual lifetime estimates of this surviving pilot population. As the age of this sample matures, more accurate life expectancies for each year following retirement will become available. If the survival distribution changes to indicate an earlier mortality, a need for epidemiological study and the investigation into factors associated with this career would be warranted and of interest to the FAA.

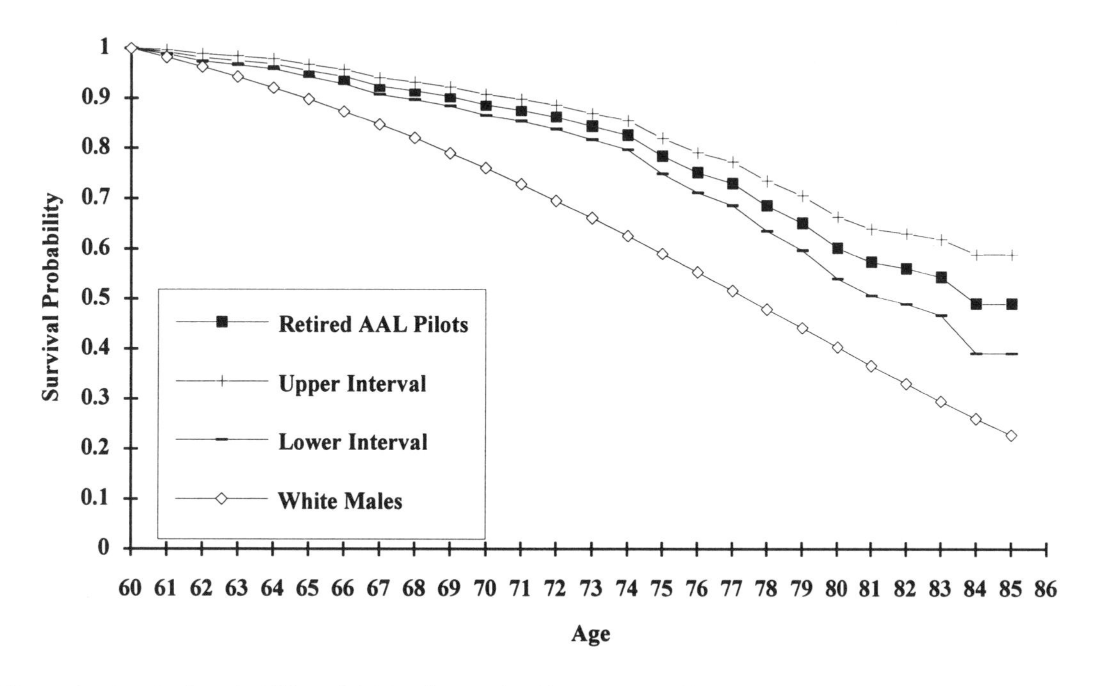

Figure 1. Survival probability of AAL pilots retired between April 1968 to July 1993 with the ± 95% confidence interval, and the survival probability, after age 60, for U.S. white males in 1980.

CONCLUSIONS

The hypothesis that retired airline pilots die at younger ages than their general population counterparts was not supported by this study. On the contrary, this study revealed a significantly longer life expectancy for the sample of retired American Airlines pilots compared to their 60 year-old counterparts in the 1980 U.S. general population. The airline pilots' longevity for this sample was greater than 5 years longer than the U.S. white male population.

ACKNOWLEDGMENTS

The authors wish to acknowledge the cooperation of American Airlines: the late Captain Bill James, Vice President of Flight; Dr. Jeff Davis, Corporate Medical Director; Gayla Coffelt, Manager, Life, Health and Disability Plan Administration; and Bobbie Rystad, Life and Disability Administrator. We are also grateful for the contributions from the Allied Pilots Association: Captain Brian Mayhew, Special Projects Director; and Mike Knoerr, Benefits Manager. We also give special thanks to Captain Jack Webb (AAL, Ret.), Secretary of the Grey Eagles and the FAA's Civil Aeromedical Institute, particularly, Dr. Bob Blanchard, Manager, Human Factors Research Laboratory

REFERENCES

Band, P. R., Spinelli, J. J., Ng, V. T. Y., Moody, J., and Gallagher, R. P. (1990). Mortality and cancer incidence in a cohort of commercial airline pilots. Aviation, Space, and Environmental Medicine, 61(4), 299-302.

Besco, R.O. and Smith, C.B. (1990) Minimizing Diurnal Desyncrhronization. Flight Safety Foundation, Human Factors & Aviation Medicine, 37(6), November/December.

Cutler, S. J., and Ederer, F. (1978). Maximum utilization of life table methods on analyzing survival. Journal of Chronic Diseases, 8, 699-712.

Griswold, M. H., Wilder, C. S., Cutler, S. J., and Pollack, E. S. (1955). Cancer in Connecticut, 1935-1951. Hartford Connecticut State Department of Health, 112-113.

Hoiberg, A., and Blood, C. (1983). Age-specific morbidity among Navy pilots. Aviation,Space, and Environmental Medicine, 54(10), 912-918.

Irvine, D., and Davies, D. M. (1992). The mortality of British Airways pilots, 1966-1989: A proportional mortality study. Aviation, Space, and Environmental Medicine, 63, 276-279.

Kaji, M., Tango, T., Asukata, I., Tajima, N., Yamamoto, K., Yamamoto, Y., and Hokari, M. (1993). Mortality experience of cockpit crew members from Japan Airlines. Aviation, Space and Environmental Medicine, 64, 748-750.

Mayhew, B. A. (1992, November). Special report: The impact of an increase in the age 60 rule on the membership of the Allied Pilots Association. Allied Pilots Association, Arlington, TX.

Mohler, S.R. (1993). Airline Crew Members Suffer High Rate of Occupational Injuries. Flight Safety Foundation, Human Factors & Aviation Medicine, 40, 3. May/June.

Morgan, L. (1992). Deadly stress. Flying, December, 106-107.

Muhanna, I. E. and Shakallis, A. (1992). Preliminary study confirms that pilots die at younger age than general population. Flight Safety Foundation - Flight Safety Digest, 11(6), 1-6.

Pearl, R. (1923). Introduction to Medical Biometry and Statistics. Philadelphia and London, W.B. Saunders Company.

Reinhart, R.O. (1988). Pilots Must Be As Airworthy As Their Aircraft. Flight Safety Foundation, Human Factors & Aviation Medicine, 35(1), January/February.

Salisbury, D. A., Band, P. R., Threlfall, W. J., and Gallagher, R. P. (1991). Mortality among British Columbia pilots. Aviation, Space, and Environmental Medicine, 62, 352-352.

U.S. Congress, Office of Technology Assessment (1990). Medical risk assessment and the age 60 rule for airline pilots. Washington DC: Subcommittee on Investigations and Oversight, Committee on Public Works and Transportation. US Government Printing Office.

U.S. Department of Health and Human Services. (1984, 1988, 1991). Vital Statistics of the United States [1980, 1985, 1989] Volume II, Mortality Part S, Section 6, Life Tables. Washington DC: Author.

DISTRIBUTION OF DISCRETIONARY TIME BY RETIRED PEOPLE

John A. Modrick, Susan Meyers, and Robert Papke
Honeywell Retiree Volunteer Project
Minneapolis, Minnesota

This study is a survey of the activities of retirees of a Fortune 500 manufacturing company. The objective was to provide a baseline description of activities as people adjust to changes associated with retirement and decremental changes of aging. A sample of 40 retirees was interviewed to obtain information on 1) demographics, living conditions, and health; 2) distribution of time over activities of work, recreation, family, and volunteering; 3) changes associated with aging and socio-economic factors. It is intended to provide an activity picture that will be a context for identifying behavioral problems and devices, organizations, and supporting aids to enhance quality of life.

Respondents represent the gamut of occupations and skill levels, socio-economic conditions, and ethnic diversity characteristics of a large, midwestern manufacturing corporation. The distribution of ages in the retiree population and sample were equivalent. The activities surveyed include reading, watching television, household chores and maintenance, skill or craft hobbies, intentional exercise, care-giving, volunteer work, and work for pay. Analyses include differences and changes in activities as a function of gender, age, length of retirement, health, and living situation. Volunteer work is analyzed in detail concerning type, location of the work, and time spent on it.

INTRODUCTION

This study was undertaken to determine what older people in this society actually do, to develop an idea of their lifestyle, and to learn how the lifestyle is modified to adapt to psychological and physical changes that occur with retirement and aging. This activity picture will ultimately provide a context for identifying behavior problems and devices, organizations, and support activities to enhance the quality of life.

We assume that aging is characterized by a progression of deteriorative changes in capacities, abilities, and health that result in the progressive degradation of performance, restriction in activity, and compromise of one's independence. These changes are a series of transitional challenges to which the individual must respond. The major challenge in these aging transitions is to focus one's activities selectively and adapt one's behavior to optimize performance and satisfaction. The initial challenge of retirement is the choice of activities to replace the activities and socialization formerly available through the workplace.

This study is a survey of the activities of retired employees of a Fortune 500 manufacturing company in the metropolitan area of Minneapolis-St. Paul. It is an exploration of how one group of retirees adjusts to the challenge of distributing an increased amount of discretionary time. The study is predominantly empirical to gain information on what people of the ages from 50–60 years and beyond actually do, their conditions of living, and the problems they face. We were especially interested in the extent of participation in volunteer activities and the role of volunteering as a source of personal satisfaction. The objective was to determine the distribution of discretionary time over activities of work, recreation, family, socializing, and volunteering and changes in type and distribution of activities with changes or differences in age, education, and health.

Research on older adults tends to emphasize basic sensory/motor capacities, memory, and major problems of health. Some recent studies of activities have dealt with daily maintenance routines such as grocery shopping, meal preparation, and personal grooming (Weber et al., 1989). Each activity is described in terms of task analytic actions such as lifting, holding, and rotating. The most common activities are grocery shopping and laundry (45%) while lift/lower and push/pull constitute 58% of the actions.

Type, frequency, and functional significance of more molar activities have received little attention. These molar activities include household upkeep, recreation, socializing, volunteer work, and work in which one engages for income, achievement, expression of interest, or sense of contribution to society. Sears (1977) used a categorization of six areas of life experiences as sources of satisfactions: occupation, family life, friendship, richness of cultural life, service to society, and joy in living. These areas provide a basis for classification of activities. Volunteering by retirees is widespread and organized but factors influencing personal choices are largely undocumented.

Retirement precipitates a transitional adjustment to a new role, status, and economic condition. Prior to retirement, work is a source of both activity and social interaction. The need for activity is satisfied mainly through work and to a lesser degree through leisure pursuits such as sports and crafts in the physical domain and reading and entertainment that exercise thinking, imagination and problem solving in the intellectual domain. Socialization is satisfied mainly through interaction with family members, friends, and acquaintances.

The activity and socialization derived from work are markedly reduced with retirement. The individual must reorganize his/her activities to fill this void. Available alternatives include spending more time on existing leisure pursuits or adding new pursuits, increasing time spent in family social activities such as more time with grandchildren, increasing social activities outside the family (clubs, lunches, senior center), increasing involvement in volunteer activities and continuing to work for pay on a part- or full time schedule. One may also utilize the increased discretionary time by reading, watching television and movies, or participating in artistic and cultural events.

The need for activity may decrease with advanced age. As one's energy decreases, one becomes less agile and health problems compromise one's autonomy. Health factors, both personal and of a significant other, will impinge on activity, reducing the energy and time available for activity. Although the same level of activity may not be required, some activity is necessary. Recent medical evidence is suggestive that continued physical and mental activity is conducive to good health and perhaps longevity (Anonymous, 1990; Portman, 1962).

The theoretical/philosophical context in which this research was formulated was the model of successful aging as selective optimization with compensation (Baltes and Baltes, 1990). In addition, we wanted to generate information on the challenges, transitions, adjustments, adaptive processes, and outcomes that characterize the age span from retirement to old age. We intend to use this information to begin an extension of the adult development model by Levinson et al. (1978) to provide a framework for developmental processes of retirement and post-retirement. The Levinson et al. model deals with development from post-adolescence to middle adulthood as a series of stages through which one must pass, each stage presenting challenges to which one must adapt. Late adulthood from approximately 50 yeas to old age is sketchily treated.

We approached the study with the following hypotheses or expectations:

1. Activity will decline with age, decreasing in amount and shifting from active to passive activity with increasing age. The choice of activities will shift to more sedentary ones with age to accommodate reduced physical capacity and energy. Thus, golf, tennis, gardening, and household repairs, for example, will be replaced by increased reading and watching television.

2. Some individuals will work for pay, typically on a part-time basis. These individuals will continue to work on a limited schedule in volunteer work for non-profit organizations.

3. The amount of volunteer activity will decrease with age.

4. Activity and volunteering will decrease with illness and disability. The effect will be greatest for conditions that limit mobility and independence.

5. Amount of activity and volunteering will decrease if one is caring for another person who has an illness or disability.

METHOD

Four hundred retirees were selected randomly from a retiree population of 4825 to be interviewed by telephone using a structured interview guide. Retirees in the sample were divided among 17 interviewers. Interviewers were volunteers from the staff or graduate students from the All University Council on Aging at the University of Minnesota or the Retiree Volunteer Program of Honeywell Inc. All interviewers had some experience in telephone interviewing. Training sessions were conducted on use of the interview guide, objectives of specific questions, guidelines to standardize interaction with respondents, and data recording. Ten attempts were made to contact an individual. If unsuccessful, that person was dropped from the study. Three hundred eight good protocols were obtained.

The coverage of the interview included five areas:

1. Demographic information, such as address, gender, birth date, marital status, and educational level. Topics in the interview include data not available from employee records such as living situation, state of health of self and cohabiting significant others.

2. Retirement data, such as date of retirement, age at retirement, and length of time retired.

3. Living conditions, such as location of residence, length of residence, type of housing, living with others or alone.

4. Health status of retiree and significant other, if applicable.

5. Type of activity and time spent in each activity. The activities explicitly surveyed are reading, watching television, household chores and tasks, home maintenance and decorating, sleeping, walking/jogging or intentional exercise, recreational travel, church attendance, paid work, helping or caregiving to another especially a dependent, engaging in skills or crafts as a hobby, and volunteer projects. Other activities were also solicited in open-ended questions. Respondents identified the most meaningful of the reported activities.

RESULTS

Space limitations restrict reporting of the analysis to selected segments of the data. A report containing more extensive analysis and interpretation is in preparation.

Demographic Information

Respondents represent the gamut of occupations, skill levels, and socio-economic strata, and ethnic diversity characteristics of a large, Midwestern manufacturing corporation. The sample consisted of 152 women and 154 men.

The age range in the population of 4825 was 36.23–101.32 years with a mean age of 70.76 years. Young retirees (the 36 year old person) were retired early for medical reasons and not included in the sample. The ages in the sample ranged from 56.75–91.42 years with a mean average of 790.34 years. Means and standard deviations of age across gender were equivalent.

The marital status of the respondents is 63% married (43% men and 20% women), 21% widowed (3% men and 17% women), 7% divorced and single (0.6% man and 6%

women), and 7% single, never married (2% men and 6% women).

The educational level of the respondents ranges from eighth grade or less to doctoral degrees. The median educational level falls between High School Graduate and Technical/Junior College. Twenty percent of the men are college graduates vs. 3% of the women. Eighty-one percent of the men are in the educational categories between High School Graduate and College Graduate. The largest educational category for women is High School Graduate (37%) and 75% of the women have educations of High School Graduate to Eighth Grade or Less.

Retirement Data

The mean age at retirement was 62.04 years (61.87 years for women and 62.22 years for men). The range of age at retirement was 50.67–70.08 years. The average length of retirement was 8.28 years ranging from 1.25–26.17 years.

Living Conditions

The predominant type of housing is a single or duplex dwelling (71%) followed by the apartment (13%) and condominium (6%). The mean length of residence in their dwellings is 13.63 years with a standard deviation of 14.98; the maximum length of residence is 77 years. Twenty percent of the retirees made a recent change in residence.

The geographical distribution of residences by ZIP codes was explored. The idea behind this analysis was to explore the possibility of a blue collar/white collar correlation or a distribution by inner city, near suburbs, far suburbs, and rural. However, there was no discernible pattern in the location of residences. Recent real estate practices such as the gentrification of inner city and transitional neighborhoods may have minimized socio-economic differences between ZIP code areas.

Fifty-two percent (n = 158) of our respondents live with their spouses. However, widowed, divorced/single, and single/never married predominately live alone; 76% of the widowed, all of the divorced/single, and 62% of the single/never married. The 18 cases of living with someone other than one's spouse are distributed over the groups of children, siblings, other relatives, and friends. Forty-four percent were living with children.

Health Status

Eighty percent of the respondents report excellent or good health; five percent report poor or very poor health. Forty percent report that they have a chronic health condition and 18% report that a significant other with whom they are living has a chronic condition.

Sixteen percent (n = 48, 29 men and 19 women) provide care for another individual. The relationship of the care recipient to care giver include spouse (11; M = 9, W = 2), parent (10; M = 8, W = 2), sibling (5; M = 3, W = 2), offspring (6; M = 5, W = 1), and friend (1).

Activities

Activities are divided into two broad categories. Reading, watching television, and church attendance are classed as passive activities; the remainder are classed as active activities.

Summary information on activities called out specifically by the interviewer is presented in the following.

Reading: Ninety percent of the respondents reported engaging in reading; the mean amount of time spent on this activity is 12.46 hours/week.

Television Viewing: Eighty-two percent of the respondents reported engaging in watching television; the mean amount of time spent on this activity is 19.43 hours/week.

Church Attendance: Seventy-two percent of the respondents reported that they attended church or synagogue; the mean amount of time spent attending church is 2.8 hours/week.

Household Tasks: Eighty-eight percent of the respondents reported engaging in household tasks; the mean amount of time spent on this activity is 45.85 hours/week.

Maintenance/Decorating: Forty-seven percent of our respondents reported engaging in maintenance and decorating of their home and property; the mean amount of time spent on this activity is 18.92 hours/week.

Exercise: Sixty-five percent of the respondents reported engaging in a regular, explicit program of exercise; the mean amount of time spent on exercise is 7.79 hours/week.

Skills/Crafts: Thirty-six percent of the respondents reported engaging in skills and crafts activities; the mean amount of time spent on this activity is 7.56 hours/week.

Dependent Care: Sixteen percent of the respondents reported that they were providing care for another individual; the mean amount of time spent on this activity is 10.34 hours/week.

Recreational Travel: Sixty-seven percent of the respondents reported that they engaged in recreational travel; the mean amount of time spent traveling is 45.19 days/year. However, travel turned out to be a complex variable. Time spent on travel ranged from nine months to less than a week. The modal amount of travel is 21 days (11% of the travelers) and there is a clustering of times around the weekly intervals of one, two, and three weeks. Seven percent of the respondents report 30 days of travel per year. Another 8% report 13 weeks, the typical time of the winter flight to southern climates.

Several people reported their seasonal trips to their "lake cabins" or resort as recreational travel. Typically, only the travel time to and from the cabin is reported as travel.

We have found that the responses to questions about the amount of recreational travel include four types of travel:

1. Conventional vacation trips of one to four weeks duration.
2. Extended seasonal sojourns of two to four months.
3. Dual residency of dividing the year between two locations.
4. Periodic travel during the summer from one's home to a nearby lake or resort area, usually within two or three hours driving time, when one maintains permanent accommodations.

Each of these kinds of recreational travel would be associated with socio-economic differences among the retirees.

Paid Work: Sixteen percent of our respondents work for pay. A third of this group are women. Sixty percent of the jobs are seasonal. The type of work ranges over managerial, technical, financial, support services, and consulting. There is a company president, director, general manager, and supervisor. Representative technical jobs include glass specialist, watch repair, senior laboratory technician, and test instrument technician. Support services include instructor; shipping clerk, and child care.

Volunteer Projects: Forty-six percent of respondents (47% of men and 44% of women) were engaged in volunteer work at retirement. Two-thirds of the sample were in volunteer work at the time of the interview. Some volunteer work is informal consisting of helping neighbors, friends, and relatives by providing transportation, assistance in shopping and house work, or filling out tax forms. Some retirees participate in programs for elderly, handicapped, and poor such as Meals on Wheels, Loaves and Fishes, and Food Share. The church or synagogue is the organization through which the largest number of people volunteer (35%). Community senior centers and organizations in health care and social welfare each draw 6–7% of the volunteers

Time spent in work on formal volunteer projects was asked for both the time spent last week and in an average month. The mean amount of time spent last week is 5.49 hours with a maximum of 40 hours; the mean amount of time spent in an average month is 17.66 hours with a maximum of 180 hours.

Volunteering by educational level was examined. The trend is non-linear. The largest category is high school graduate accounting for 28%. Less than high school graduate represents 21%, some college 18%, technical/junior college 16%, and college graduate 11%. The number of advanced degree people is too small to provide an estimate.

Volunteering by age is also a non-linear relationship. In the ages 62–70 it runs at levels of 44%–58%. In the ages 71–75 the level ranges over 26%–40%. It rises again in the late 70s and declines to the mid-20% level in the 80s.

Most Meaningful Activities

Respondents were asked to indicate which of their activities is the most meaningful to them. These activities were clustered using a key word sort.

The most frequently cited activities are:

Reading: n = 31 (10%)
Travel: n = 21 (7%)
Volunteering: n = 17 (6%)
Church: n = 16 (5%)
Garden/Yard Work: n = 14 (5%)
Golf: n = 8 (3%)
Fishing: n = 7 (2%)

When diverse activities are collated into broad categories such as sports, arts and craft including hobbies, and social such as card playing and dancing, the resulting frequencies are:

Arts/Crafts: n = 28 (9%)
Sports: n = 24 (8%)
Social: n = 9 (3%)

REFERENCES

Anonymous (1962). Don't Take it Easy—Exercise! *Exercise*, National Institute on Aging, Bethesda, MD.

Baltes, P.B. and Baltes, M.M. (1990). Psychological Perspectives on Successful Aging: The Model of Selective Optimization with Compensation. In Baltes, P.B. and Baltes, M.M. (Eds.) *Successful Aging*, Cambridge University Press, New York, NY.

Levinson, D.J., Darrow, C.N., Klein, E.B., Levinson, M.H., and McKee, B. (1978). *The Seasons of a Man's Life*. Ballantine Books, New York, NY.

Portman, J. (1990). "Pumping Iron" Benefits Frail 90-year Olds. *Healthline*, National Institutes of Health, Bethesda, MD.

Sears, R.S. (1977). Sources of life satisfaction of the Terman gifted men. *American Psychologist*, 32, 119-128.

Weber, R., Czaja, S., and Bishu, R. (1989). Activities of daily living of the elders: A task analytic approach. *Proceedings, Human Factors Society 33rd Annual Meeting*, Human Factors Society, Santa Monica, CA.

DESIGNING MEDICATION INSTRUCTIONS FOR OLDER ADULTS

Daniel Morrow, Von Leirer, & Jill Andrassy
Decision Systems
Los Altos, California

We examined if medication instructions were better remembered when organized in terms of older adults' pre-existing schemes for taking medication. A preliminary study suggested that older adults share a general scheme with medication information grouped into 3 categories: (a) General Information (e.g., medication purpose), (b) How to take (dose), and (c) Possible Outcomes (side-effects). In the present study, we investigated age differences in this scheme and in instruction recall. We also examined if individual differences in organization related to cognitive abilities, health care beliefs, and medication taking experience. For the most part, the results provided further evidence that older adults share a scheme for taking medication and revealed few age differences in this organization. Verbal ability was more important than health attitudes for predicting individual differences in instruction organization. Most important, older and younger subjects preferred and better remembered instructions that were organized in terms of their medication taking scheme.

INTRODUCTION

Medication nonadherence is a complex problem related to many cultural, patient, and medication factors. Several factors relate to poor communication between health professionals and patients. For example, people may not take their medication as prescribed because incomplete and poorly organized medication instructions do not clearly describe the task (Morrell, Park, & Poon, 1989; Morrow, Leirer, & Sheikh, 1988). We are investigating expanded medication instructions, which give people more complete information than typically appears on a container label. If designed with older adults' limitations and strengths in mind, expanded medication instructions should effectively guide adherence behavior. Instructions may minimize demands on cognitive resources by building on any relevant prior knowledge. This prediction is consistent with the observation that while aging is associated with a gradual decline in cognitive resources such as processing rate or working memory capacity, general knowledge such as vocabulary and domain-specific knowledge remains stable or even increases (Salthouse, 1990).

This principle may be particularly relevant for instructions or other materials about health-related tasks. Taking medication is all too often a part of older adults' daily routine. Therefore, elders may be very familiar with this general kind of task, and this knowledge may be represented as a scheme in long term memory. If a specific set of instructions matches a pre-existing medication-taking scheme, people will easily interpret the new information in terms of their scheme in order to create a mental model of how to take the particular medication. This "compatibility" principle for procedural instructions is supported by research in human-computer interaction. For example, people perform computer tasks such as searching data bases more quickly and accurately when computer displays match user knowledge of the domain (e.g., Holland & Merikle, 1987).

Our previous research identified 10 items that a complete medication instruction set should contain (Morrow, et al., 1988). Older adults in a preliminary study indicated which order the items should appear in a set of instructions. Hierarchical cluster analysis of the preferred orders suggested that subjects shared a general scheme for taking medication, with the items grouped into 3 categories: (a) General Information (e.g., Medication Name and Purpose), (b) How to take (Dose and Schedule), and (c) Possible Outcomes (e.g., Side effects). A second group of older subjects more accurately recalled instructions matching this preferred order, suggesting that instructions are easier to understand and remember when designed to take advantage of elders' knowledge about the medication task (Morrow, Leirer, Altieri, & Tanke, 1991).

In the present study, we wanted to provide further evidence for this medication taking scheme and examine if there are age differences in these knowledge structures. Previous research suggests that older and younger adults have similar knowledge structures, although older adults may benefit to a less extent from these structures on some tasks (Hess, 1990). We also examined if individual differences in instruction organization relate to cognitive abilities, health care beliefs, and medication taking experience.

EXPERIMENT 1: THE MEDICATION SCHEME METHOD

A sample of 42 older adults (mean=69.5, 60-87 years old) and 42 younger adults (mean=24.6, 20-30 years old) participated in the study. The two groups did not differ in terms of education (Older=15.0 Younger=15.7, $t(82)=1.5$, $p > .10$), Quick vocabulary scores (Older=82.4 Younger=77.1, $t(82)=1.7$, $p=.09$), and self-reported health (7-point scale, 7=very healthy, Older=5.3 Younger=5.5, $t(82) < 1.0$).

As in the preliminary study, subjects first performed a sort task. They were given 10 items, each on a separate card, and sorted the items that they thought belonged together into separate groups. They did this for three medications. Next, they completed the order task. For each medication, they ordered the items according to how they should appear in an instruction set. Sort data (the frequency with which each pair of items were grouped together) and order data (the distance between each pair of items in the preferred orders) were examined by hierarchical cluster analyses with Ward's and Complete linkage methods. Individual differences in organization were also examined. For each subject and medication, each item's position was subtracted from the group (or standard) position. The mean deviation score across the 10 items of the subject's instruction set summarized the difference between the subject's order and the standard order.

Health beliefs were measured with Robinson-Whelen and Storandt (1992)'s 22 item questionnaire, which combines Wallston's Multidimensional Health Locus of Control Scale and Krantz's Health Opinion Survey. In this study, factor analyses of scores from samples of older and younger adults produced 5 sub-scales, 3 from the Wallston questionnaire (Internality, Chance, and Powerful Others) and 2 from the Krantz questionnaire (Behavioral Involvement in health care, and Need for Health Information). We also asked about any medication taking strategies such as using pill organizers or schedules and the frequency of consulting references about medication.

Results

We found more evidence for the medication scheme and showed that younger as well as older adults organize medication information in terms of this scheme.

(a) Medication Scheme. Cluster analyses of the order data showed that older subjects in the present study organized most medication items the same as in the previous study. Table 1 shows the positions where items were most frequently placed on the third trial of the order task. Items are also grouped according to the clusters produced by analysis of the order preferences (Ward's linkage method).

Table 1 shows that the only age differences in preferences related to doctor name. Older subjects grouped this item with possible outcomes (ninth position) while younger subjects grouped it with general information (third position), as in the first study. This disagreement may reflect the fact that the doctor name can be viewed as general information or it can be related more specifically to the severe side-effects, which mentions calling the doctor if these side-effects appear. Both age groups appeared to split between these interpretations. Preferences did not change across the three trials of the sort and order tasks, suggesting that the procedure tapped a pre-existing scheme rather than teaching subjects a new scheme.

(b) Health beliefs and medication taking experience. Consistent with previous research, older subjects were more likely than younger subjects to believe that powerful others control their health ($t(82)=6.0$, $p < .001$), and to report less desire for involvement in their health care ($t(82)=3.8$, $p < .001$). There were no age differences in desire for information about health care or belief in chance or internality. Health beliefs among older subjects predicted some self-reported medication taking behaviors. For

Table 1
Percent Subjects Choosing Item Positions
(Clusters based on third trial order preferences, Ward's linkage method)

Older Subjects

Item	Position	Percent Subjects Choosing Position
Medication Name	**1**	**64**
	2	21
Purpose	**2**	**41**
	3	29
Dose	**3**	**21**
	4	26
	5	19
Schedule	3	21
	4	**29**
	5	21
Duration	**5**	**33**
	6	29
*Warnings	**6**	**19**
	7	29
Mild Side-Effects	**7**	**33**
	8	29
Severe	**8**	**33**
*Doctor Name	1	19
	2	19
	9	**26**
Emergency	**10**	**64**

* = Item in different cluster than Morrow et al., (1991)

Younger Subjects		
Medication Name	**1**	**81**
Purpose	**2**	**57**
	3	26
Doctor Name	1	17
	2	19
	3	**14**
	9	38
Dose	3	24
	4	**38**
	5	21
Schedule	4	41
	5	**31**
Duration	5	31
	6	**41**
Warnings	6	31
	7	**26**
Mild Side-Effects	7	48
	8	**31**
Severe Side-Effects	8	50
	9	**36**
Emergency	**10**	**86**

example, older subjects who reported a greater desire for information about their health care also reportedhigher frequency of actually seeking information about their medications (i.e., asking for handouts about their medication, reading magazine articles; $r=.57$, $p<.001$).

Older subjects reported taking more medications than younger subjects at the time of the study (2.1 v 0.4 medications). Older subjects who took more medications rated themselves as in poorer health ($r=-.51$, $p < .001$) and as less likely to take their medications as prescribed ($r=-.40$, $p < .01$).

(c) Individual differences in medication instruction organization. Regressing demographic and health belief scores on the individual subject organization deviation scores showed that only verbal ability accounted for a significant amount of variability ($R^2=.12$, $p < .01$). Subjects with higher vocabulary scores had smaller deviation scores ($r=-.43$, $p < .001$), suggesting that they created more standard medication instructions.

Experiment 1 shows that older and younger subjects have similar schemes for taking medication. We next examined if instructions are better remembered when they are made compatible with this scheme by presenting medication information in the preferred grouping and order.

EXPERIMENT 2: RECALL OF MEDICATION INSTRUCTIONS

Method

A sample of 42 older adults (mean=70.4, 60-87 years old) and 42 younger adults (mean=23.6, 20-30 years old) participated in the study. While older and younger subjects did not differ in education and self-reported health, older subjects had higher vocabulary scores (82.5 v. 74.9, $t(82)=4.6$, $p < .001$). We compared standard instructions (items presented in preferred order and grouping), category instructions (items were grouped but the categories presented in nonstandard order), and scrambled instructions (all items in nonpreferred positions). Subjects performed the following tasks for each of three instructions (one from each instruction type): (a) Read each instruction for 90 sec, (b) Complete intervening task for approximately 2 minutes, (c) Recall instruction, (d) Rate level of free recall (7=Excellent, 1=Very Poor), (e) Answer specific memory questions (e.g., "What was the dose?"), (f) Rank order the three instructions in terms of preferences.

Results

The proportion of items recalled from each instruction was analyzed by a mixed design ANOVA with Age as a between-groups variable and Instruction as a repeated measure. Figure 1 shows that younger subjects recalled more information from the instructions (Younger=.64 Older= .47 items, $F(1,82)=25.0$, $p < .001$). As in Morrow, et al., (1991), more compatible instructions were recalled more accurately (Standard=.59, Category=.57, Scrambled= .52; linear trend $F(1,82)=11.2$, $p < .01$; Standard v. Scrambled, $t(83)=3.4$, $p < .001$). Age did not interact with Instruction, suggesting that younger and older subjects benefitted from well-organized instructions to the same extent. The same pattern was found for accuracy of answering the memory questions.

Age and Instruction also had similar effects on recall self-assessment. Younger subjects rated their recall more highly than older subjects (4.9 v. 4.3, $F(1,82)=5.7$, $p < .05$) and all subjects' ratings were higher for more compatible instructions (Standard=4.8, Category=4.7, Scrambled=4.4; linear $F(1,82)=8.5$, $p < .01$). Age did not interact with Instruction, suggesting both older and younger adults were attuned to the benefit of schematic organization on recall. Subjects also preferred the more compatible instructions.

FIGURE 1
MEDICATION INSTRUCTION RECALL
FOR OLDER AND YOUNGER SUBJECTS

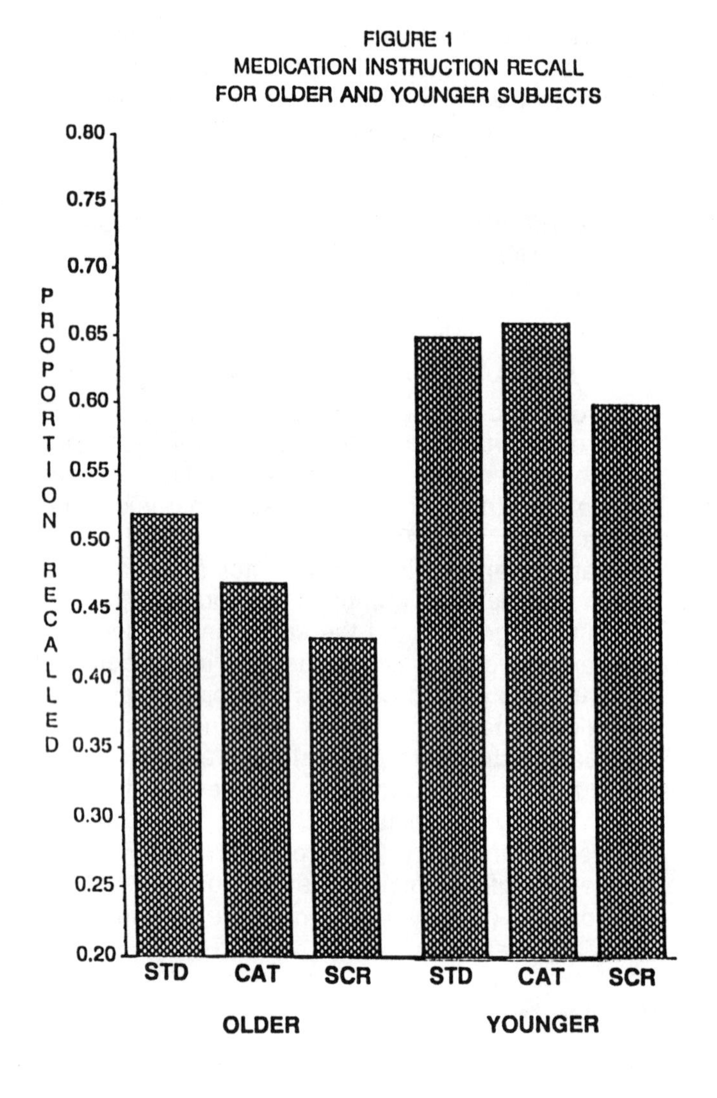

STD = Standard CAT = Category SCR = Scrambled

Compared to scrambled instructions, subjects thought the standard version was easier to understand (mean rank scores with 1 most preferred and 3 least preferred: 1.8 v. 2.3, t=3.9 $p < .01$); did the best job of describing how to take the pills (1.9 v. 2.3, t=2.6, $p < .05$), and would be the best instruction to receive from their pharmacist (1.7 v. 2.4, t=5.1, $p < .001$).

DISCUSSION

The present study suggests that older and younger adults share a scheme for taking medication. We find few age differences in this knowledge organization, which is consistent with other research finding that knowledge organization is largely age-invariant (Hess, 1990). Instructions that were compatible with the medication taking scheme were recalled more accurately than instructions that presented information in a nonstandard order. The benefit of the organization was similar for older and younger subjects even though overall recall was higher for younger subjects. Both groups were also aware that well-organized instructions were better remembered, and they preferred these instructions. These findings suggest that older as well as younger adults are likely to use expanded medication instructions as a guide for taking medication, thus reducing nonadherence.

We also found typical age-related differences in health beliefs-- older adults tended to be more externally directed and have less desire to be involved in their health care. Nonetheless, elders with a high desire for information about their health care reported that they frequently gathered information about their medications. This suggests that certain older adults may benefit more than others from the additional information provided by expanded instruction sets. It may also be possible to include motivating information in expanded instructions that overcome age declines in self-efficacy beliefs, or beliefs in the value of health care involvement. Finally, verbal ability appeared to be more important than health beliefs for predicting individual differences in instruction organization. This may reflect the ability to spontaneously or actively attend to and organize information about health activities into knowledge structures.

Taken together, these findings help to integrate theories of instruction design, knowledge organization, and adherence to health-related tasks.

ACKNOWLEDGMENTS

Preparation of this paper was supported by NIA grant R01 AGO9254.

REFERENCES

Hess T. (1990) Aging and Semantic Influences on Memory. In T.M. Hess (Ed) **Aging and Cognition: Knowledge Organization and Utilization.** Amsterdam: North-Holland.

Holland, J, & Merikle P. (1987) Menu Organization and User Expertise in Information Search Tasks. **Human Factors, 29,** 577-586.

Morrell R, Park D, & Poon L. (1989) Quality of Instructions on Prescription Drug Labels: Effects on Memory and Comprehension in Young and Old Adults. **The Gerontologist, 29,** 345-354.

Morrow D, Leirer V, Altieri, P & Tanke E. (1991). Elders' Schema for Taking Medication: Implications for Instruction Design. **Journal of Gerontology: Psychological Sciences. 48,** P378.

Morrow D, Leirer V & Sheikh J. (1988). Adherence and medication instructions: Review and recommendations. **Journal of the American Geriatric Society, 36,** 1147-1160.

Robinson-Whelen, S., & Storandt, M. (1992). Factorial structure of two health belief measures among older adults. **Psychology and Aging**, **7**, 209-213.

Salthouse, T.A. (1990). Influence of experience on age differences in cognitive functioning. **Human Factors**, **32**, 551-569.

Facilitating Information Acquisition for Over-the-Counter Drugs using Supplemental Labels

Michael S. Wogalter, Amy Barlow Magurno,
Kevin L. Scott, and David A. Dietrich

Department of Psychology
North Carolina State University
Raleigh, NC 27695-7801

ABSTRACT

This study examined the effect of the presence and color of a supplemental cap label on medication information acquisition and container preference. Participants were 75 elders from a retirement community who were asked to examine one of five manipulated labels on a fictitious but realistic-appearing over-the-counter (OTC) pharmaceutical product container and then to respond to questions concerning their knowledge about the medication. Later they were shown all five manipulated bottle labels and asked which they preferred in effectively communicating medication information. The five bottles differed in the use of labeled surface area and color. Two bottles, displaying labels only on the body of the bottle, served as controls. One control had only a front label, and the other control was conventionally labeled with printed information on the front, back and sides of the bottle. The three other bottles were identical to the conventionally labeled control bottle except they included a supplemental cap label that reprinted the most critical product-use information in large type on three different colored backgrounds. The results showed greater medication-related knowledge for the bottles with supplemental cap labels compared to bottles without the supplemental cap label, with no significant difference among the different colored caps. Participants indicated a strong preference for the bottles with supplemental cap labels over the two control bottles. A distinctive cap color (different from the main label color) was most preferred. Making use of extended surface areas on medication containers to print important information in a more noticeable, legible form benefits elders' knowledge about proper use and hazards.

INTRODUCTION

Pharmaceutical products sold over-the-counter (OTC) in the U.S. generally have labeling displaying directions for use, contraindications, warnings, and other information. The information may be on the product container itself, on inserts, or on exterior packaging. The information included with the product is often the only way for many consumers to learn about the characteristics of OTC medications. In order to provide a complete set of relevant information, most OTC labels contain so much information that the text size must be substantially reduced to fit on the available surface of the container or packaging. Individuals with vision problems have difficulty reading the reduced print (Vanderplas and Vanderplas, 1980; Zuccollo and Liddell, 1985). The elderly, who tend to have age-related visual difficulties (e.g., presbyopia, cataracts), are more likely than other age groups to take more medicines. As a consequence, seniors with these age-related visual conditions have difficulty reading important information about the drugs that they take. Although additional space may be available on insert sheets and on exterior packages to make the print larger, this is seldom done. Moreover, these separate and unattached items are frequently discarded after initial use of the product. As a result, these materials are of little help when the product is used at a later time (Wogalter, Forbes, and Barlow, 1993).

One possible solution to this labeling-communication problem is to enlarge the surface space of the container to attach a more legible label. The added surface area could also be used for more elaborative instructions and warnings. In one study (Wogalter and Young, 1994), the surface area of a small glue container was expanded by using an extended tag label. The tag allowed the use of larger fonts than the original label. Results showed that compliance behavior (wearing protective gloves) increased with the tag label compared to a control label without the tag.

Barlow and Wogalter (1991) and Wogalter et al. (1993) found that the elderly preferred glue containers having labels with increased surface area. One of the most preferred bottle designs by this population was a wings (or fin) design that not only provided more surface area for print information but also made it easier to hold and turn the cap. Recently, several drug manufacturers have begun to package OTC pain medications in easy-open containers with caps having extended fins. This new design makes it easier for someone with arthritis or with a hand/arm disability to open the container. Current versions of the easy-open feature lack child resistance, however.

This new easy-open container design also increases the usable surface area of the container (relative to other containers) that could allow the printing of larger instructions and warnings. Wogalter and Dietrich (1995) examined the effect of making use of this added surface area by reprinting and extending some of the most important warnings and instructions for use onto the container cap section. The product used was Motrin IB® (Upjohn Co, Kalamazoo, MI), a national brand of ibuprofen in an easy-open bottle. They found that the elderly participants preferred the bottles with the cap label over control bottles without the cap labels. Of the four cap labels that differed in background color (orange, white, two-toned orange and white, and fluorescent green), they most preferred the green cap label. These subjective evaluations indicate a preference for more readable and noticeable labels afforded by the enlarged print and use of color. However, these preference judgments may not reflect any actual performance advantages for these labels (i.e., increased transmission of information). Moreover, the reason for the color effect is unclear. The main body label of the Motrin IB® was orange and white and the lesser preferred caps used those same colors. Thus, the green preference might be due to distinctiveness with respect to the main label, simple color preference, or because it was a fluorescent hue.

The current research attempts to clarify the Wogalter and Dietrich (1995) study by examining the effect of the supplemental cap label and its color using performance and preference measures. Addressed are whether the supplemental label facilitates performance in a knowledge acquisition task and whether color adds to this effect. Additionally, the study attempts to replicate the preference findings of the earlier study examining whether color distinctiveness is responsible for the elevated evaluations.

Preliminary testing with the Motrin IB® container labels used by Wogalter and Dietrich (1995) failed to show differences in knowledge after brief exposure. High levels of product knowledge was found for all conditions, suggesting the presence of a ceiling effect that was probably due to the elderly population's high level of familiarity with this particular medication. The present study uses an unfamiliar product to reduce the possibility of a ceiling effect by reducing the influence of pre-existing knowledge that might mask differences between different versions of bottle labels.

It was expected that the bottles with supplemental cap labels would produce greater knowledge about the medication and be preferred compared to bottles without supplemental labels. Two distinctive cap colors (fluorescent yellow and fluorescent green) were expected to be preferred and produce greater knowledge than a fluorescent orange cap that matched the primary color of the main label.

METHOD

Participants

Seventy-five volunteers from a retirement community in Chapel Hill, North Carolina participated. Mean age of participants was 79 years (*SD* = 5.8, ranging from 69 to 90), 77% were females, and all were Caucasian. A monetary contribution was made to the community fund in appreciation for residents' participation.

Materials

In a preliminary trial, a purchased bottle of the pain reliever Datril® (Bristol-Myers Co, New York, NY), a brand-name acetominophen product, was shown to participants. The trial run was used to get participants acquainted with the main experimental procedure.

For the main experimental trials, "Marvine," a fictitious OTC motion sickness preparation, was created. Most people are relatively unfamiliar with motion sickness preparations, and thus pre-existing knowledge that could enable participants to score highly on a subsequent knowledge test without having examined the experimental labels was expected to be limited. Although the product was fictitious with respect to active ingredient and manufacturer, the text on the labels was constructed to be plausible and realistic containing information from currently available motion sickness preparations and information from the Physicians' Desk Reference (1993).

The Marvine container shown in Figure 1 had a physical height of 4.2 cm and a width of 4.0 cm. The cap height was 2.4 cm with a circumference of 13.5 cm. The main label (front, back and sides), as shown in Figure 2, was attached around the the body of the container, as is typical. The text of the main label was formatted to be similar to other OTC labels currently on the market. The front panel contained the product name, chemical name, indications for use, and other typical principal display panel information, printed in sizes

Figure 1. Representation of the Marvine Container and Label Placement.

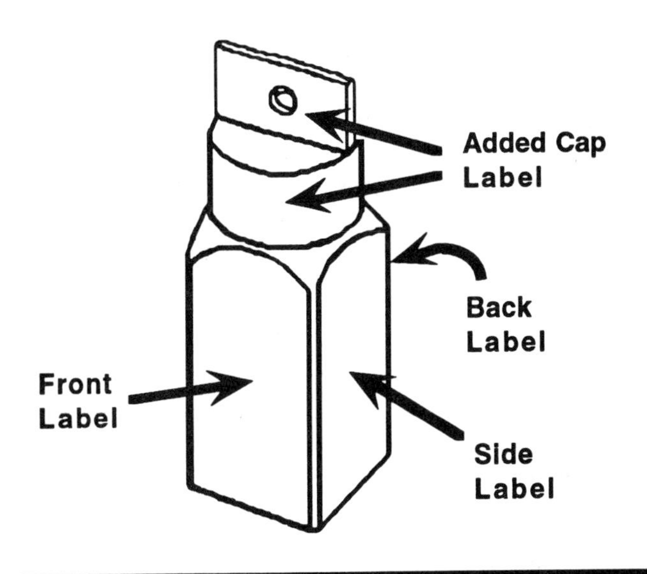

ranging from 7 to 14 point black type. On the back and side panels, type size (4 point) was the same as is found on other commercially-available OTC products (e.g., Motrin IB®). The main label background color was fluorescent orange.

All bottles except for one of the controls had the complete main label on all four sides of the bottle. The control condition without the complete main label had only the front label attached to the bottle (i.e., it lacked the back and side panels). The other (conventional) control had the complete main label on all four sides of the bottle but lacked the supplemental cap label, as is typical for products using this container design.

The three other bottles were the same as the conventional control, with labels on all four sides of the bottle, but also displayed a supplemental cap label. The textual content and layout of the three cap labels was identical, varying only in the background color of the label: orange, yellow, and green. All were fluorescent hues. The orange cap label was identical to the background color of the main label, while the other two cap labels were different. The cap label text was black print, in New Helvetica Narrow font, having type sizes ranging from 7 to 17 point.

The information on the cap label was extracted from the main label text and was chosen, based on consultation with a pharmacist, to reflect the most important cautions and directions for proper, safe use of the product. The supplemental cap labels were composed of three parts. One was positioned on the front extended tab on the cap top. This section contained the WARNING signal word and the signal alert icon (an exclamation point surrounded by a triangle). The other two sections of the cap label completely wrapped around the base of the cap so that one part faced the front and the other part faced the back of the bottle. Important cautions were printed on the front and dosage information on the back. The text of these labels is shown in Figure 3.

All labels were produced on a 600 dpi laser printer. The labels were attached bottles with glue and overlayed with clear plastic lamination.

A medication knowledge test was developed concerning information on the Marzine label. The 12-item test (with a total of 42 subparts) consisted of open-ended and probe-type questions concerning what the drug treats, when and how much of the drug to take, when not to take the drug, signs/indications of overdose, side effects, whether the drug can be given to children of various ages, and the bottle's adequacy of child proofing. The test was administered in an interview format, with the experimenter recording everything the participant said in response to each item.

Procedure

The study was conducted in a conference room at the retirement community. Participants arrived at prearranged times, and were tested individually. The experimenter explained to participants that the purpose of the study was to investigate their impressions of labels on medicine bottles. Participants were told they would be shown drug containers, and then would be asked questions about the medications. Participants signed a consent form before beginning the study.

The first phase of the study was a trial run intended to acquaint participants with the type of task they would be performing in the main experiment. The experimenter read aloud a scenario in which participants were to assume they were shopping for a pain reliever to be used by family members of various ages and with various medical histories. They were asked to read the label for information on how to use the product, and who should or should not take it. The participant was handed a Datril® bottle, and after 60 s the bottle was removed and the participant was asked three questions about the product's use by children, by someone allergic to aspirin, and by someone with a peptic ulcer. Responses from this phase were recorded but not analyzed.

Figure 2. Text of the Main Label (Front, Back and Side) in Actual Size.

MARVINE

(Promethazine Dimenhydrinate)

For nausea, dizziness, & motion sickness

No Prescription Needed

PACKAGE NOT CHILD RESISTANT

100 TABLETS 50 mg each

DO NOT USE IF PRINTED PLASTIC OVERWRAP OR FOIL INNER SEAL IS BROKEN

WARNINGS: DO NOT TAKE THIS PRODUCT, UNLESS DIRECTED BY A DOCTOR, If you have a breathing problem such as emphysema, asthma, shortness of breath, or chronic bronchitis, If you have glaucoma, If you have sleep apnea (periods when breathing stops), or if you have difficulty in urination due to enlargement of the prostate gland. May cause marked drowsiness, dryness of the mouth, or temporary blurring of the vision. Other possible effects include disorientation, memory disturbances, dizziness, restlessness, hallucinations, confusion, headaches, ringing in ears, difficulty urinating, skin rashes or redness. Do not drive, operate dangerous equipment, or participate in activities that require full mental alertness while taking this product. Do not take this medication if you: are pregnant or nursing a baby, have (or have had) any metabolic, liver, or kidney disease, have any obstruction of the stomach or intestine, have any skin allergy or have had a skin reaction to any drug or chemical or food substance—unless directed by your doctor. Keep this and all drugs out of the reach of children. Children and the elderly may be particularly sensitive to the effects of this medication. Marvine should not be used in children under 8 years of age. The safety and effectiveness of this product for use in children has not been determined. Seniors (over 60) may be more likely to experience dizziness, drowsiness, confusion, overstimulation, or lowered blood pressure. Not for frequent or prolonged use except on the advice of a doctor. Do not exceed the recommended

DRUG INTERACTION PRECAUTION: Do not take this product if you are taking sedatives, sleeping medications, tranquilizers, or MAO inhibitors without first consulting your doctor. Also avoid alcoholic beverages while taking this product, as the effects of alcohol may be intensified and the action of the Marvine may be changed.

INDICATIONS: For the prevention and treatment of nausea, dizziness and vomiting associated with motion sickness due to car, boat, or airplane travel, or to amusement park rides. Marvine is a H1 histamine receptor blocker. It exhibits its action by an effect on the Central Nervous System, possibly by its ability to block muscarinic receptors in the brain. In addition to its antihistaminic action, it provides clinically useful sedative and antiemetic effects.

DIRECTIONS: To prevent motion sickness, the first dose should be taken one half to one hour before starting activity. ADULTS: 1 to 2 tablets every 4 to 6 hours, not to exceed 8 tablets in 24 hours, or as directed by a doctor. CHILDREN 8 TO UNDER 12: 1/2 to 1 tablet every 6 to 8 hours, not to exceed 3 tablets in 24 hours, or as directed by a doctor.

ACTIVE INGREDIENT: Each tablet contains promethazine dimenhydrinate 50 mg.

INACTIVE INGREDIENT: Corn and potato starch, dextrin, lactose, and magnesium stearate

STORAGE: Keep below 86 F (30 C) in a dry place and protect from light.

dosage; overdose may cause serious consequences such as death. Symptoms of overdose include breathing difficulties, flushing, loss of consciousness, very low blood pressure, stomach and intestinal problems. In the event of overdose, seek professional medical assistance or contact a poison control center immediately.

If you have Questions or Comments: Call us toll free at 1-800-535-1726 between 9 AM and 5 PM Eastern Time.

Lincoln Pharmaceuticals Inc.
Marcon, MA
Made in USA

Lot: 0559962

Expiration date: 5/95

Patent 7,813,121

Figure 3. Text of the Supplemental Cap Label in Actual Size.

⚠ WARNING
Container is NOT Child Resistant.
Keep Out of Reach of Children at ALL Times.
CAN CAUSE DROWSINESS.
Do not take with alcohol or other drugs as the effect may be intensified.

Labels Surrounding the Circumference of the Cap (Front and Back)

Directions: Take 30 to 60 minutes before starting activity. *Adults:* Take 1 to 2 tablets every 4 to 6 hours, but no more than 8 in 1 day. *Children (8 to 12 years):* 1/2 to 1 tablet every 6 to 8 hours, but no more than 3 in 1 day. Do not give to children under 8 years. Doctor may advise other dosages.

CAUTION. **Before using, see a physician if you are:** (1) pregnant or nursing a baby; (2) have any of the medical conditions listed on the main label (such as breathing difficulties, glaucoma, sleep apnea, enlarged prostate, liver/kidney disease, skin allergy); or (3) are taking sleep aids, sedatives, tranquilizers or MAO inhibitors.

In the second (main experimental) phase, another scenario was presented to participants. They were asked to assume they were buying a motion sickness medication in preparation for a one-day bus trip on winding mountain roads. Further, they were to assume that fellow travelers would have a variety of medical conditions and would be of different ages. The purpose of the scenario was to provide realism as well as relevance to encourage careful examination of the label for a broad range of purposes. Each participant was then presented with one of the five Marvine bottles(depending on the condition for which they were randomly assigned) and asked to examine the label so that they could answer questions on the medication. After three minutes elapsed, the Marvine bottle was removed and the knowledge test was given. Participants were encouraged to give answers to all the questions, based on the information viewed on the label and any background knowledge they had regarding the motion sickness medication.

In the final phase, participants were given all five bottles of Marvine and asked to rank the bottles from most preferred to least preferred. Participants were instructed to consider overall effectiveness in communicating important medication information, ease in reading, likelihood of reading, and likelihood of purchase. Later, participants were debriefed and thanked for their assistance.

RESULTS

Knowledge Test

Each subpart item of the knowledge test was scored as correct = 1 or incorrect = 0. Data analysis used the mean proportion correct score for each participant (based on a total of 42 subpart items). Prior to analysis, a second judge re-graded the open-ended responses. Inter-rater reliability (calculated as number of agreements/total X 100) was nearly perfect (99.75%).

The mean proportion correct knowledge scores are shown at the top of Table 1. A one-way between-subjects analysis of variance showed a significant effect of bottle label conditions, $F(4, 70) = 17.82$, $p < .0001$). Comparisons using Newman-Keuls Multiple-Range test showed that the three cap label conditions produced significantly higher scores than the two control label conditions ($p < .05$). The three cap label conditions did not differ among themselves ($p > .05$). The 4-panel (front, back, and sides) control label condition produced significantly higher knowledge scores than the front-panel-only control condition.

Preference Ranks

Mean preference ranks for the five label conditions are shown at the bottom of Table 1. Lower rank means indicate greater preference. The data were analyzed using the non-parametric repeated-measures Friedman test. This test showed a significant effect of label condition, $\chi^2(4, N = 75) = 223.05$, $p < .0001$). Paired comparisons were performed using the Wilcoxon Matched-Pair Signed-Rank test. The table shows that the yellow cap received the lowest mean rank, and was significantly preferred over the orange cap ($p < .05$). The green cap was intermediate between these two conditions but did not significantly differ from the two other cap colors ($p > .05$). All three cap conditions were judged significantly better than the two no-cap label controls ($ps < .05$). The 4-panel control label was significantly preferred over the front-panel-only control ($p < .05$).

DISCUSSION

This study showed improved knowledge acquisition with the addition of the cap labels. These labels reprinted the most critical information on the product's use and warnings. Apparently this method was more effective in providing this information than the conventional label method.

The different colors of the additional label did not make a

Table 1. Mean Knowledge Score and Preference Rank of Container Label Configurations..

	Control Containers		Experimental Containers with Front Back & Side Labels and Supplemental Information on Color Cap		
	Front Panel Only No Back or Side Labels	With Front, Back & Side Labels	Orange	Green	Yellow
Knowledge					
Mean	.15	.38	.57	.51	.55
SD	.11	.23	.15	.16	.12
Preference Rank					
Mean	4.97	3.85	2.33	2.05	1.79
SD	.16	.69	.88	.77	.86

Note. ***Higher comprehension and lower ranks indicate better scores.***

difference in knowledge acquisition scores, but did have an effect on preference. Wogalter and Dietrich (1995) found that participants preferred a fluorescent green cap that was distinctly different from the color of the main bottle label (orange). In the present study, yellow was preferred over orange, with green intermediate between these two. This finding partially clarifies an unresolved issue of color preference in the earlier study. Although the green cap in the present study was not significantly preferred over the orange cap, the yellow cap was. This suggests that the effect found earlier was not due to the particular color green, but appears to be its distinctiveness from the main label.

The fact that the two control conditions (front panel-only vs. front, back, and sides) differed for both knowledge acquisition and preference indicates that at least some of the elders used some of the information the additional panels provide. Nevertheless, several participants spontaneously commented that the back and side panels were very difficult for them to read. Some participants reported that they were not able to read the main label at all (meaning only the information on the cap labels was available to them). Some participants stated that they would be less likely to purchase a product if the print was too small for them to read. Some noted that they routinely carry with them a magnifier.

The easy-open OTC container provides additional surface area to present important information. Direct attachment of the label to the medication container avoids the pitfalls of methods like inserts and external packaging which might be discarded or lost after initial use of the product. It also allows the use of enhancements such as larger print size and pictorials (Kalsher, Wogalter, and Racicot, 1996). Informal discussions with participants in the post-experiment debriefing phase suggested that being able to read labels is important to elderly consumers, and that this need is not being met by current labels.

More research is needed on ways to increase the legibility and understandability of pharmaceutical labels. Necessarily, considerable information needs to be communicated, and increasing the surface area of medication containers is one way to enable use of larger, more legible print. There are other methods of extending the surface area of labels. One, a foldout method, is used on some labels of Aleve® (Procter & Gamble, Cincinnati, OH) pain reliever. Additional research will help to determine the factors that enhance people's knowledge of pharmaceuticals.

REFERENCES

Barlow, T., and Wogalter, M. S. (1991). Increasing the surface area on small product containers to facilitate communication of label information and warnings. *Proceedings of Interface 91* (pp. 88-93). Santa Monica, CA: Human Factors Society.

Kalsher, M. J., Wogalter, M. S., and Racicot, B. M. (1996). Pharmaceutical container labels and warnings: Preference and perceived readability of alternative designs and pictorials. *International Journal of Industrial Ergonomics,* in press.

Physicians' Desk Reference (1993). *PDR* (47th edition). Oradell, NJ: Medical Economics.

Vanderplas, J. M. and Vanderplas, J. H. (1980). Some factors affecting legibility of printed materials for older adults. *Perceptual and Motor Skills, 50,* 923-932.

Wogalter, M. S., Forbes, R. M., and Barlow, T. (1993). Alternative product label designs: Increasing the surface area and print size. *Proceedings of Interface 93* (pp. 181-186). Santa Monica, CA: Human Factors Society.

Wogalter, M. S., and Young, S. L. (1994). Enhancing warning compliance through alternative product label designs. *Applied Ergonomics, 25,* 53-57.

Wogalter, M. S., and Dietrich, D. A. (1995). Enhancing label readability for over-the-counter pharmaceuticals by elderly consumers. *Proceedings of the Human Factors and Ergonomics Society 39th Annual Meeting* (pp.143-147). Santa Monica, CA: Human Factors and Ergonomics Society.

Zuccollo, G., and Liddell, H. (1985). The elderly and the medication label: Doing it better. *Age and Ageing, 14,* 371-376.

Measurement Techniques and Level of Analysis of Medication Adherence Behaviors across the Life Span

Denise C. Park, Roger W. Morrell, David Frieske, Christine L. Gaines, & Gary Lautenschlager
University of Georgia

We have collected data on medication adherence in several studies from samples of younger as well as elderly adults. Samples have included hypertensive adults, adults taking medications for a range of illnesses, and adults with osteoarthritis. The time range for collecting adherence data has varied from two weeks to two months, and the level of analysis has varied from examination of individuals medications across 60 days to monthly estimates of overall adherence rates. Finally, our research group has extensive experience with two microelectronic techniques for measuring adherence: the Videx time wand system which relies on bar code scanners to measure adherence, and the Medication Event Monitoring System (MEMS) which involves pressure sensitive lids that record the date and time a lid is removed from a prescription medication. Issues involved in measurement of adherence are presented and various techniques for presenting and analyzing data are discussed.

There are a number of recently developed techniques for medication adherence which rely on microelectronic monitoring devices which permit more precise measurement of medication adherence behaviors than has occurred in the past. Deyo, Inui, and Sullivan (1981) have noted that the absence of a "gold standard" for measuring adherence behaviors has contributed to the confusion that surrounds adherence issues in the literature. The most commonly-relied upon measure for studying adherence is verbal reports, which likely results in the underreporting of nonadherence behaviors. Other measures which are somewhat more reliable include pill counting and blood assays. Pill-counting techniques do not permit an examination of daily patterns of adherence. An individual who takes too much of a medication on one day and too little the next, would have exactly the right amount of medication remaining, and be classified as highly adherent with pill count as a measure. Blood assays only permit a few medications to be detected, are expensive and can typically only establish the presence or absence of a medication rather than amount of medication, and provide only crude measures of adherence. Newer techniques utilized in our laboratory permit extraordinarily accurate assessment of adherence behaviors and are an improvement over these earlier methods.

The first technique, the Videx time wand, relies on small, credit-card sized bar-coding scanners. Subjects are given the tiny scanner along with a wallet containing clearly labeled bar codes for each medication they are taking. Whenever the subject takes a particular drug, he or she glides the scanner across that drug's bar code, recording the date and time, which is later up-loaded into our computer system. Another technique is the Medication Event Monitoring System (MEMS) which involves the use of medication bottles that have a microchip in the cap which records the date and time when the cap is removed from the bottle. The MEMS system is less obtrusive than the Videx system, but is extremely expensive to use.

VALIDITY OF MEASUREMENT

It needs to be accepted that every system for measuring adherence has a certain amount of error associated with it. Although both the Videx time wands and the MEMS bottle cap system are almost certainly improvements over other techniques, they are not error-free. With the Videx system, as in the collection of diary data, the subject is highly sensitized to the fact that adherence behaviors are being monitored. For this reason, we believe that the Videx system may generally provide an underestimate of adherence behavior, particularly in the early weeks of monitoring. Using the Videx system, Park, Morrell, Frieske, and Kincaid (1992) reported a nonadherence rate of 7.6% in the first week and 10.0% the second week, rates that are substantially lower than the 20% or greater than has been reported in the literature by others (Deyo, 1982; Epstein & Cluss, 1982, Cramer, Scheyer, & Mattson, 1990). The important point about the Videx system, however, is that, although it may provide an underestimate of errors, it does appear to be sensitive to intervention. Park et al. (1992) reported a significant improvement in adherence rates when oldest-old subjects were given cognitive aids to support adherence, using the Videx system as a monitoring technique.

In contrast, the MEMS system probably provides an overestimate of nonadherence. The strength of the system is its reliability as well as its low salience as a monitoring technique. The system is unobtrusive as the bottle caps are only slightly larger than standard prescription caps and the microchip is invisible. We have also found that the MEMS system is highly reliable and have had no chip failures after testing 40 subjects for two months. Nevertheless, the system does have problems. First, a subject may not replace the cap tightly, and the system may not be activated, although the software will record this. More problematic is the fact that many people do not keep individual medications in the bottles in which they are initially packaged. Many individuals combine multiple medications into a single prescription bottle, transfer a few doses of medications into pockets and purses, or routinely load their medications into medication organizers. Thus, an individual might accurately take medications that have been previously removed from the container, and these episodes would be recorded as omissions, providing an overestimate of nonadherent behavior.

LENGTH OF MONITORING PERIOD

We have monitored subjects for period ranging from two weeks to two months. Because we found that adherence was higher with the Videx system in Week 1 compared to Week 2 (Park et al., 1992), we recently completed a study with 48 hypertensive patients where we monitored them for two months. We examined changes in adherence behavior across the eight week period and found no evidence that adherence changed across this interval. For this reason, we recommend that two weeks appears to be long enough to establish a stable baseline period to test the effect of various interventions on adherence behaviors.

TYPES OF ADHERENCE MEASURES

Adherence data can be analyzed in many ways. We have typically categorized the data according to types of errors: (a) omission errors--a dosage is missed or not taken; (b) commission errors--an extra dosage is taken; (c) quantity errors--the wrong number of pills is taken; (d) time errors--the medication is taken at the wrong time, (e) total errors--the sum of the proceeding proportions. We have found it essential to always utilize conditional probabilities in calculating adherence rates. Overall, we have found that omission errors are nearly isomorphic with total errors. By far, the most common error is omission. Commission occurs infrequently and quantity errors are rare, as most medications are taken as single pills. Finally, we have found that it is not productive to calculate time errors, as it is difficult to determine objectively when a medication is taken too close or to far apart relative to another dose of the same medication. For example, if a medication is to be taken once a day and it is taken at 8 a.m. one day and 12 p.m. the next, is this a time error? It might be for a medication that regulates heart beat but be of no consequence for a prescription vitamin.

LEVEL OF ANALYSIS

A more substantive issue relates to the level of analysis of the data. Although we typically represent group data, we have also found it most informative to represent the data of individual subjects. Figure 1 illustrates the performance of 15 subjects who had rheumatoid arthritis across three six-day intervals and were monitored with the Videx system. A medication organizer was introduced on Day 13, and the figures present the variable response of subjects to the organizer which was hypothesized to act as a cognitive prosthesis. Figure 2 illustrates the performance of the same 15 subjects as a group. Figure 2 illustrates that the intervention introduced on Day 13 did have an effect relative to performance from Days 7-12. Analysis of variance (with means for Days 6-12 compared to Days 13-18 as the unit of analysis) indicated that this effect was statistically significant when one highly variable subjects was excluded ($F(1, 13) = 9.20$, $p = .009$). Figure 2 also illustrates the higher levels of adherence that occurred in the first few days of monitoring. At the same time, the complex patterns of behavior observed in the individual graphs are completely lost when examining the group data.

Figure 3 is taken from the same data set of 15 rheumatoid arthritis patients and illustrates the usefulness of breaking adherence data down medication by medication across subjects. There was a significant main effect of medication type for commission errors (F (4, 41) = 3.997, p = .008). Subjects were more likely to take extra doses of methotrexate, a powerful immunosuppressant, than any other medication. This is likely because methotrexate has an odd once or twice a week dosing schedule and that subjects either cannot remember if they took the medication or believe that by taking more, they will improve the drug's effect.

Finally, Figures 4 and 5 suggest how important it can be to look at the data of individual subject's drug by drug. These figures present the data of two subjects with hypertension across 48 days of monitoring with the Videx time wand. Both figures demonstrate a phenomenon we have called "selective nonadherence." Both subjects omit their vitamin supplements regularly that are prescribed by their physician but show reliable adherence to their hypertension medication.

SUMMARY

To summarize, the data presented suggest that microelectronic techniques are a reliable mechanisms for measuring medication adherence. The microelectronic data appears to be sensitive to interventions, and two weeks of monitoring appears to be long enough to establish baseline. There are many ways to examine and analyze the data. Techniques that we have found useful include examining individual data day by day to look at medication effects as well as intervention effects. We have also found it useful to examine the data as a function of medication category for homogeneous groups of subjects. Finally, we have successfully used analysis of variance techniques with data collapsed over weeks to characterize group behavior and the effects of interventions.

REFERENCES

Coe, R., Prendergast, C., & Psalhas, G. (1984). Strategies for obtaining compliance with medication regiments. *Journal of the American Geriatrics Society, 32*, 589-594.

Cohen, G. (1981). Inferential reasoning in old age. *Cognition, 9*, 59-72.

Cramer, J.A., Mattson, R.H., Prevey, M.L., Scheyer, R.D., & Ouellette, V.L. (1989). How often is medication taken as prescribed? A novel assessment technique. *Journal of the American Medical Association, 261*, 3273-3277.

Cramer, J.A., Scheyer, R.D., & Mattson, R.H. (1990). Compliance declines between clinic visits. *Archives of Internal Medicine, 150*, 1509-1510.

Deyo, R.A. (1982). Compliance with therapeutic regimens in arthritis: Issues, current status, and a future agenda. *Seminars in Arthritis and Rheumatism, 12*, 233-244

Deyo, R.A., Inui, T.S., & Sullivan, B. (1981). Compliance with arthritis drugs: Magnitude, correlates, and clinical implications. *Journal of Rheumatology, 8*, 931-936.

Einstein, G.O., & McDaniel, M.A. (1990). Normal aging and prospective memory. *Journal of Experimental Psychology: Learning, Memory, and Cognition, 16*, 717-726.

Epstein, L.H., & Cluss, P.A. A (1982). A behavioral medicine perspective on adherence to long-term medical regimens. *Journal of Consulting and Clinical Psychology, 50*, 950-971.\

Gabriel, M., Gagnon, J., & Bryan, C. (1977). Improved patient compliance through use of a daily drug reminder chart. *American Journal of Public Health, 67*, 968-969.

Gardner, E., & Monge, R. (1977). Adult age differences in cognitive abilities and educational background. *Experimental Aging Research, 3*, 337-383.

Kass, M.A., Gordon, M., & Meltzer, D. (1986). Can ophthalmologists correctly identify patients defaulting from pilocarpine therapy? *American Journal of Ophthalmology, 101*, 524-530.

Kendrick, R., & Bayne, J.R. (1982). Compliance with prescribed medication by elderly patients. *Canadian Medical Association Journal, 127*, 961-962.

Leirer, V.O., Morrow, D.G., Pariante, G.M., & Sheikh, J.I. (1988). Elders' nonadherence, its assessment, and computer assisted instruction for medication recall training. *Journal of the American Geriatric Society, 36*, 877-884.

Morrell, R.W., Park, D.C., & Poon, L.W. (1989). Quality of instructions on prescription drug labels: Effects on memory and comprehension in young and old adults. *The Gerontologist, 29*, 345-354.

Morrell, R.W., Park, D.C., & Poon, L.W. (1990). Effects of labeling techniques on memory and comprehension of prescription information in young and older adults. *Journal of Gerontology, 45*, 166-172.

Park, D.C. (1992). Applied cognitive aging research. In F.I.M. Craik & T.A. Salthouse (Eds.), *Handbook of Aging and Cognition.* Hillsdale, NJ: Erlbaum.

Park, D.C., Morrell, R.W., Frieske, D.A., Blackburn, A.B., & Birchmore, D. (1991). Cognitive factors and the use of over-the-counter medication organizers by arthritis patients. *Human Factors, 33*, 57-67.

Park, D.C., Smith, A.D., Morrell, R.W., Puglisi, J.T., & Dudley, W.N. (1990). Effects of contextual integration on recall of pictures in older adults. *Journal of Gerontology: Psychological Sciences, 45*, 557.

Rehder, T., McCoy, L., Blackwell, W., Whitehead, W., & Robinson, A. (1980). Improving medication compliance by counseling and special prescription calendar. *American Journal of Hospital Pharmacy, 37*, 379-385.

Salthouse, T.A., & Mitchell, D.R.D. (1989). Structural and operational capacities in integrative spatial ability. *Psychology and Aging, 4*, 18-25.

World Health Organization. (1981). Health care in the elderly: Report of the Technical Group on use of medications by the elderly. *Drugs, 22*, 279-294.

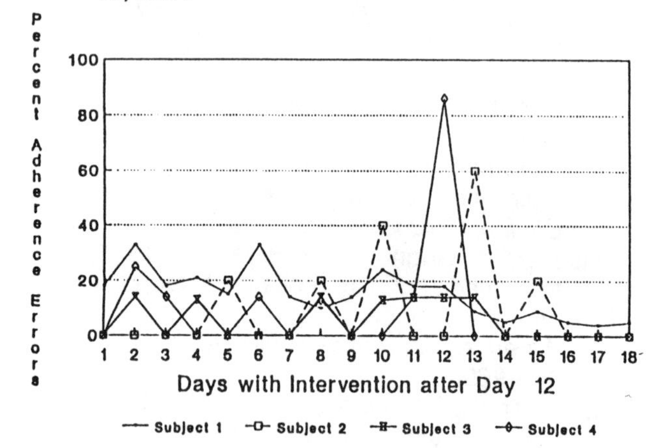

Marked Improvement When Organizer Was Introduced (Four Subjects)

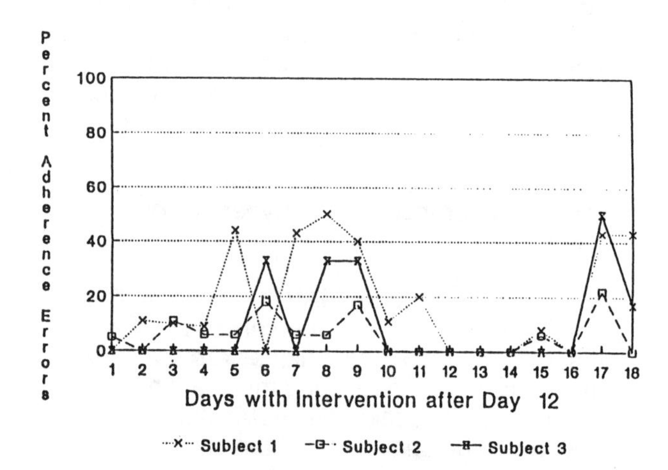

Short-term Improvement When Organizer Was Introduced (Three Subjects)

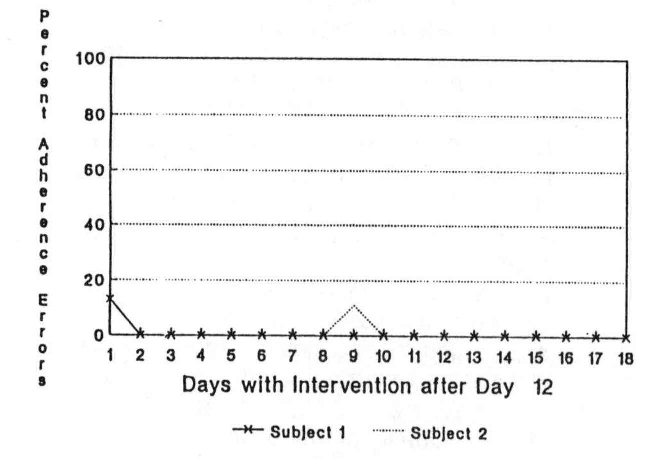

High Adherence That Cannot be Improved (Two Subjects)

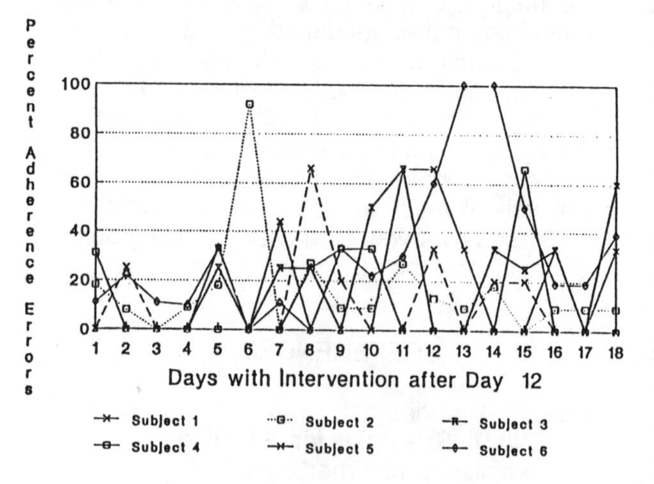

No Obvious Effect of Organizer Despite Substantial Nonadherence (Six Subjects)

Figure 1

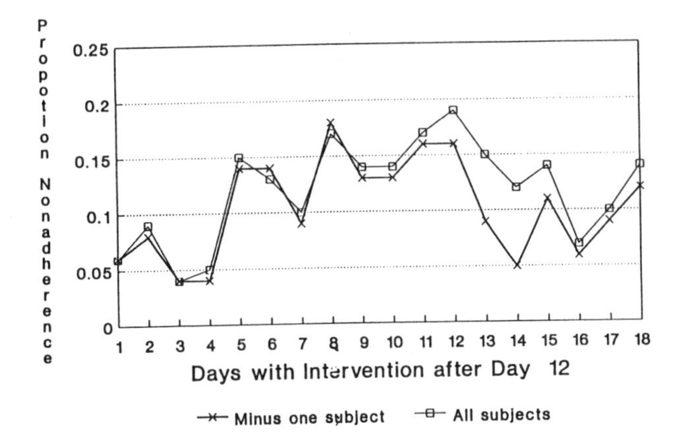

Figure 2

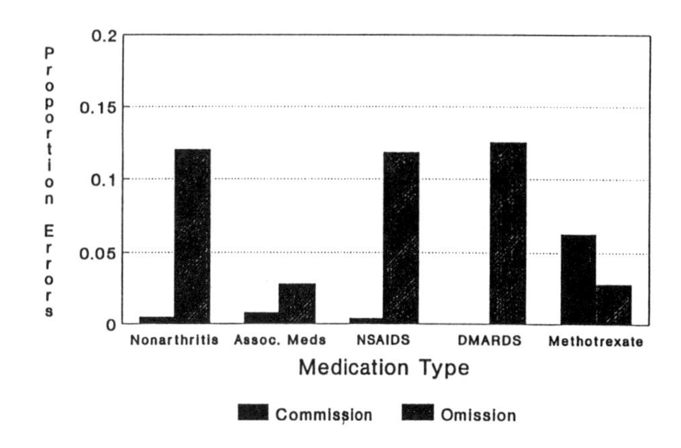

Figure 3

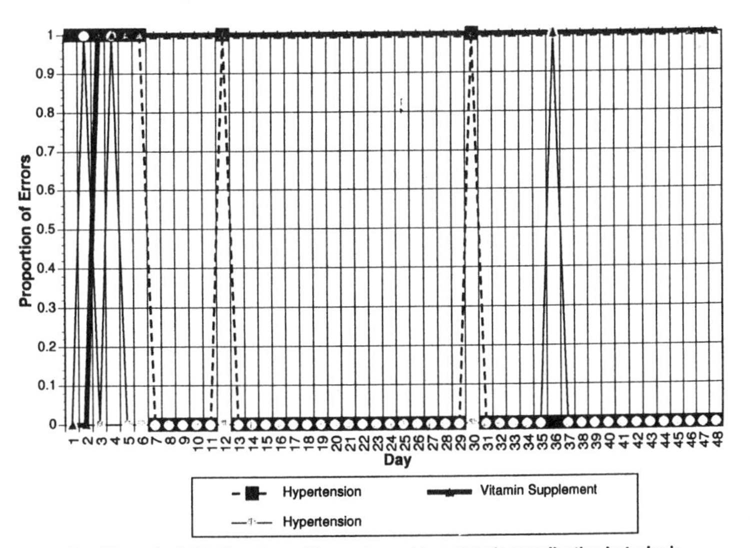

A subject who is highly adherent to one type of hypertension medication but who is selectively non-adherent to another hypertension medication and highly non-adherent to his/her vitamin supplement

Figure 4

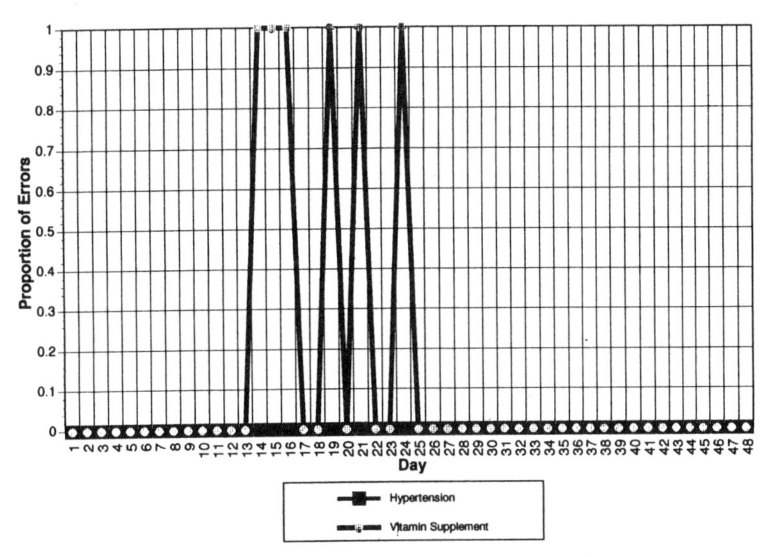

A subject who was highly adherent to his/her hypertension medication but was selectively non-adherent to his/her vitamin supplement.

Figure 5

DEVELOPMENT OF A MEMORY AID DESIGN CONCEPT FOR OLDER USERS

Mark Kirkpatrick, Ph.D.
Randy M. Perse
Lisa A. Dutra
CARLOW INTERNATIONAL INCORPORATED
Falls Church, Virginia

Michael A. Creedon, D.S.W
THE CREEDON GROUP
Vienna, Virginia

Jiska Cohen-Mansfield, Ph.D.
GEORGETOWN UNIVERSITY CENTER ON AGING
Washington, DC

This study was conducted to develop a design concept for an electronic memory device to enhance medication compliance in older users. The effort was supported by a Phase I Small Business Innovation Research (SBIR) grant from the National Institute on Aging (NIA).

A user-oriented approach was used to develop a design concept for a memory device for older users. One hundred seniors were interviewed to identify their physical, physiological and cognitive capabilities and limitations, as well as their preferences for memory aid functions. Specific design requirements were gathered from user testing of six currently available memory aids with 30 of the original 100 elderly subjects. The interview and user testing results were consolidated to provide the basis for tradeoff criteria for memory aid interface concepts, and for the development specifications for an optimal interface design for a memory aid designed specifically for the elderly user.

A design concept was developed for a medication device that would be easy to use, would reduce the likelihood of scheduling errors, and would be non-threatening to older users who might otherwise be intimidated by an electronic device. The Phase I effort focused on enhancing medication compliance, which is a priority issue with the senior population.

BACKGROUND

The population of persons over the age of 65 was 31 million in 1989 (Hollman, 1990). More than 4.4 million of these elders living in the community were experiencing difficulties in one or more Instrumental Activities of Daily Living (IADL) (Aging America: Trends and Projections, 1991).

The average community-based elder takes four to five prescription medications regularly, while one in nine takes 10 or more medications (Ansello, Lamy and Gondek, 1990). Stuck and Tamai (1991) report that the main responsibility of executing the prescribed therapy for community-based geriatric patients rests on the patient or caregiver, with very limited direct supervision by health care professionals.

In order to enable community-based seniors to maintain their independence, it is necessary to offer interventions to ameliorate these self-care problems that are commonly associated with old age. High technology products, which are being integrated into nearly every aspect of daily living, are therefore natural candidates for research tools for addressing these issues.

Although there has been very little research performed on technological interventions for medication compliance or memory enhancement for the elderly, the few studies that have been performed report promising findings. Voice-mail was used successfully to electronically schedule calls to elders to remind them that they should take their medications (Leirer, Morrow, Tanke, and Pariante, 1991). Research on the use of computers by older adults (Czaja, 1990, and Middleton, 1991), and findings from focus groups (Ward, 1991) suggest that high tech memory aids, when designed properly, could be a useful addition to the array of available interventions that can enhance medication schedule compliance and independence.

OBJECTIVES

The objectives of this effort were to identify functional and interface design requirements for an electronic memory aid for medication compliance that was based entirely on the preferences, attitudes, requirements, and physical and cognitive characteristics of older users, and to develop a design concept that met the identified requirements.

Three potential populations were identified that would likely benefit from a memory aid. They are the community-based "healthy elderly" who believe they need a memory aid; the elderly in the early stages of dementia who are capable of functioning on their own, but require a memory aid for daily activities; and the frail elderly who are cared for by a caregiver in the home. Due to the time constraint of the Phase I grant, this effort concentrated on the "healthy elderly" population, with the intention of expanding analyses to the remaining populations in Phase II.

METHODS

Interviews

A user profile was established through interviews with 100 seniors in the Washington, DC metropolitan area. The questionnaire instrument was developed to obtain information on the physical and cognitive capabilities, limitations and characteristics of memory aid users, and to identify users' needs, expectations and performance requirements associated with the use of memory aids. Supplementary information on user characteristics and activities of daily living were gathered using portions of the Preliminary Questionnaire of the Older Americans Resources and Services (OARS) Community Survey Questionnaire (Ernst, 1984). Subjective cognitive and physical status ratings were assigned by the experimenter, using standard five-point subjective rating scales, with a score of five representing the highest level of function.

User Testing

User testing was performed using 30 randomly selected subjects from the initial 100 interviewed. The only criteria for selection were that the subjects had indicated a willingness to participate, had a cognitive rating of four or above, and that there were equal numbers of males and females. Devices for user testing were selected through assessment of the state-of-the-art in memory devices. Specifications for several products were obtained, and a taxonomy of the features of these devices, which included electronic "diaries" and medication devices, was developed and implemented for device selection. Six devices were selected that together encompassed all the design and functional features necessary for testing. Two multifunctional electronic diaries, three medication devices, and one talking calculator were obtained.

A task analysis was performed on the operating tasks for each of the selected devices to identify functional and design issues that should be addressed in the design concept. These task performance issues provided the basis for performance measures that would be used to evaluate the user testing of the devices.

The user testing experimental design is illustrated in Table 1. The order of presentation of the devices was randomly determined. Tasks were written out in a standardized format for all devices for all participants. All subjects listened to alarms, opened all devices and read all displays. They also reviewed a list of commonly used icons, and selected those which best represented alarms, prescriptions, and the telephone. Tasks which were common to groups of devices were analyzed for errors and time to complete tasks. Subjective ratings were gathered at the end of each session. In addition to task time data, observational notes were taken by the experimenter throughout each session. User testing data and interview data were integrated to generate criteria for a design concept. At the end of the session, each subject was asked to answer seven questions regarding the design and functionality of the devices they tested.

Table 1. Experimental Design

Subject Group	Experimental Design		Measures		
			Response Time	Errors	Subjective Rating
1	Read Displays	Medication Devices/ Talking Calculator			
2	Hear Alarms	Electronic "Diary" #1			
3	Select Icons	Electronic "Diary" #2			

Storyboards

Design concepts were developed in a storyboard format. Tradeoffs were systematically conducted as the design concept evolved from initial sketches to a final storyboard concept. A final design concept was selected, and detailed storyboards were developed.

DISCUSSION

Interviews

User profile. Participants were 100 elderly volunteers recruited mostly from senior centers and congregate housing. The mean age was 78 with a range of 65 to 91. Seventy six were females, and 24 were males. The majority (87) were white, 10 were black, and 3 were from other minorities. In spite of attempts to recruit subjects from varied ethnic and socio-economic backgrounds, ethnic minorities were underrepresented in the subject sample. The mean number of prescriptions taken daily was three. Their educational levels ranged from incomplete elementary school to postgraduate studies. More than half the participants were widowed (54), with nearly a quarter (21) divorced, and 14 married, 10 single, and 1 separated. Most (76) lived alone; all the married lived with their spouses, and the others (as well as one couple) lived with relatives such as children, grandchildren and siblings.

Interest in using electronic memory aids. Fifty-eight percent of the subjects stated that they would use an electronic memory aid. These subjects tended to own more electronic devices such as VCRs and microwaves, they tended to forget more of their daily activities such as appointments and addresses, and they tended to use more types of reminders such as calendars, phone/address books, notes, etc, than their cohorts who were not interested in using the devices.

Some 75 percent stated a willingness to learn how to use such a device. They were more likely to find the different functions of an electronic memory aid (i.e., monitor medication, appointments, etc.) more important, and they were less likely to have reservations about using the device, than those who did not want to learn how to use it. They were also more likely to have someone available to help them use a new electronic device (78% of those who wanted to learn vs. 55% of those who did not).

User requirements. In addressing perceived needs for memory enhancement, fewer than half of the participants reported any problems with remembering daily tasks. Most frequent was forgetting birthdays and least frequent was forgetting to pay bills.

Forgetting in regard to medication was specifically queried. Most frequent was taking medication at the wrong time or missing a dose (reported by 38 of the participants).

Currently used memory aids. Memory aids, such as calendars, phone books, paper notes, other people, and alarm clocks, were used by ninety-two percent of the participants.

The most frequently mentioned problems reported by 15 participants were forgetting to look, update or use current reminders. Several had specific problems, ranging from physiological problems to device design issues.

Medication Reminding. Thirty-three percent of subjects experienced no problems with medication scheduling. Nearly twenty percent used compartmentalized pill boxes or pill dispensers, or boxes with compliance caps which they filled regularly. A large proportion, twenty-two percent, used environmental reminders, such as putting the bottle on the counter, on a white surface where it will be visible, or near the food of the meal with which the medication should be taken.

Desired Device Characteristics. The most important anticipated function of the memory aid was that of remembering appointments, after which came monitoring medications, and then remembering addresses and phone numbers.

A small device was preferred by 73 participants. A large device was preferred by 17, and 10 did not have a preference. The predominant reason for preferring a small device was portability, so the device would fit in a purse or a pocket. This may partially reflect the preference to be able to conceal a device, which was expressly stated by seven participants. Additionally, four participants wanted a small device because of lack of space to accommodate a large one. The reasons for choosing a large device included the need for a large display in order to see because of vision problems (n=22), the need for large buttons to manipulate (n=7), and the need for a large device to prevent losing it (n=1).

Caregivers. The importance of significant others was probed and 36 percent of the sample stated that they were reminded by others, most frequently by a spouse or adult child. Interestingly, 10 of the 18 who described themselves as caregivers stated willingness to use electronic memory aids in their caregiving role. Since adult family members could set such aids and monitor their use, there may be an even wider use rate than our findings might suggest, that is, family members may purchase such aids either as gifts for their elderly relative or as an element of their own role as caregivers to them.

Task Analysis

The task analysis identified critical tasks, which if performed incorrectly, could have life-threatening effects. These were related primarily with medication schedule calculations, which become increasingly complex as the number of daily prescriptions increases. The required information is not always readily available or well-understood by the senior. The limiting factor in determining a medication schedule is not necessarily the number of prescriptions that are taken, but rather the number of different daily doses that are prescribed. For example, a person taking two medications, one three times daily, and the other four times daily, must calculate a daily regimen for five separate dosage times. Design issues regarding labels, buttons, displays and alarms were also identified, and are addressed in the Storyboard Design Guidelines section.

User Testing

Response time. Response times were measured for all tasks with the devices. They appeared to be a function of the number of key presses or steps to perform the task (i.e., the complexity of the task), the degree of familiarity with the keyboard arrangements, labels, the size and types of buttons, and the complexity of displays. The response time data for equivalent tasks across devices was analyzed to determine if significant differences were present, and whether a performance difference was achieved between the devices. Performance of the "Set Alarms" task was significantly better (i.e., shorter times to finish the task) using the devices which required fewer steps for task completion. Response times for "verifying an alarm" were largely dependent upon the presence of a sliding switch on one device that most seniors had difficulty activating.

The two multi-functional "electronic diaries" required more complex operation than the medication devices and talking calculator. One device had a layered menu approach which took considerably longer to navigate through than the simple button presses of its counterpart. The presence of buttons with dual labels also proved confusing on the electronic diaries and the talking calculator. Complex and cluttered displays contributed to operator confusion, and increased response times. When asked to open each device, subjects had considerable difficulty with those with latches that were stiff, small, poorly marked, or that opened in an unexpected manner.

Errors. Each individual task was rated as a success or a failure in the eyes of the experimenter. Tasks were not rated as failures if the subject realized that an error was being committed and was able to overcome the problem. Of the 16 instances in which the subjects were unable to complete a task, 40% were due to the time-delay feature on the two medication devices that caused the display to revert to the clock mode after five seconds of inactivity; subjects would continue performing the task for setting the device, unaware that they were no longer in the set mode.

Subjects read the displays on six devices. There was no performance difference between the two text sizes, 1/8" and 3/8". Smaller text sizes, however, as well as ambiguous and inappropriate placement of labels, caused errors in reading displays. Figure 1 is an example of one such device.

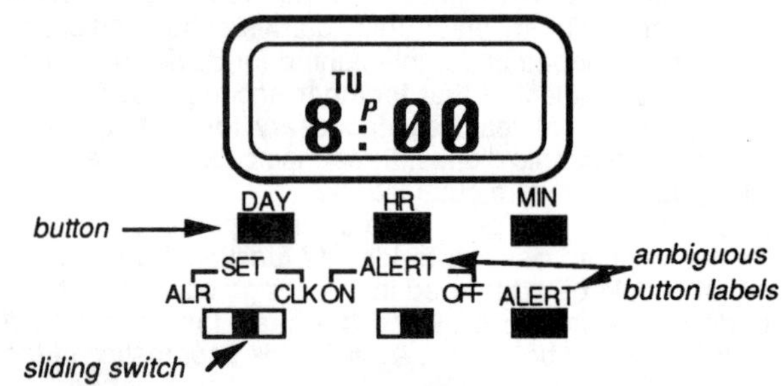

Figure 1. Sample-state-of-the-art layout

Subjects were asked to respond to voice, music, and multiple beeping alarms which are used in currently available devices. Eight subjects were unable to hear the beeping alarm of one device and two were unable to hear the beep of another. Three were unable to decipher the voice alarm of the talking calculator and all subjects were able to hear its music alarm. Subjects were asked to rate the alarm types of beep, voice, and music. Those who preferred the beep rated the music as their least favorable alarm and vice versa. Music was rated as the second favorable alarm by 17 of the 30 subjects.

The rest of the failed tasks were due to the inability to read information on the display, the difficulty of sliding switches, and the problems of accidentally pressing the wrong button or of pressing buttons twice. A few of the failures were due to the subjects' inability to understand the task they were asked to perform.

Subjective Ratings. Subjective ratings of the usability of the device were frequently inconsistent with observed performance. Subjects tended to be lenient in their ratings, even though they struggled with the tasks. Although the ratings tended to be conservative, they do point out problems that the subjects had using the devices. The subjective data revealed that of those subjects who used the devices with complex keyboards, 47.4 percent disagreed with the statement "The buttons are easy to find." On the other hand, of those subjects who were tested using the devices with fewer buttons and keys, only 10.0 percent disagreed with the same statement. It is clear that complex keyboards result in increased response times and errors and in increased user confusion, frustration or self doubt. Subjects were more critical of the physical characteristics of the devices than of the functional characteristics of the devices, such as "Setting the alarm is easy."

Storyboard Design Guidelines

Design Criteria. Designing a device that meets specific design criteria and that does not intimidate the older user is a challenge. Criteria were established based on the literature, interviews and user testing data.

Displays. Displays must present information, and should meet the following criteria: text should be large enough to read, layout should appear uncluttered, contrast should have a uniform acceptable level, dates and times should be presented in a standard form and consistently formatted, icons should be unambiguous, and visual cues should be used to assist in the interpretation of the display.

Buttons and Switches. Tactile and auditory feedback should be distinct and provided with each button press.

Activation should occur with a discrete single button press, spacing should be generous for increased accuracy, buttons should be guarded from accidental activation and the number of buttons should be kept to a minimum. If switches are used in place of buttons, they should also be large and easy to slide.

Pill compartment. The pill compartments should be easy to open and should provide easy access without risking undue contamination or spillage of pills. They should be large enough to hold many large pills and there should be enough compartments to hold at least one full day's supply of medication. The proper compartment should be identified when the alarm sounds or when the device is being refilled, and lastly, a means should prevent unintentional opening of the incorrect compartment.

Labeling. In general, labels should be large and easy to read, and their meaning should be clear and unambiguous. Buttons should not have more than one label, and labels on buttons should be large and easy to understand.

Alarms. Alarms should be designed to be heard by persons with typical age-related hearing loss and by wearers of hearing aids. All important auditory information should be provided in visual form as well. If possible, combining an auditory alarm with a tactile alarm is an optimal combination (Vanderheiden and Vanderheiden, 1991).

Design Concept

The design concept meets user requirements for portability and size of displays. The device, when open, measures 6.5" by 6.5" and 1/2" in depth. This allows for large buttons and a large display for readability and overall usability, yet when closed it becomes a compact 6.5" by 3.25" by 1.0" in depth, which affords the user a compact, portable device. Larger buttons assist users with visual deficiencies and also address the ambiguity problem that users had when the button labels were written on the panel between two buttons. The design concept has large text and icon labels directly on the buttons which assist the subjects in determining the correct association between the buttons and labels (Figure 2).

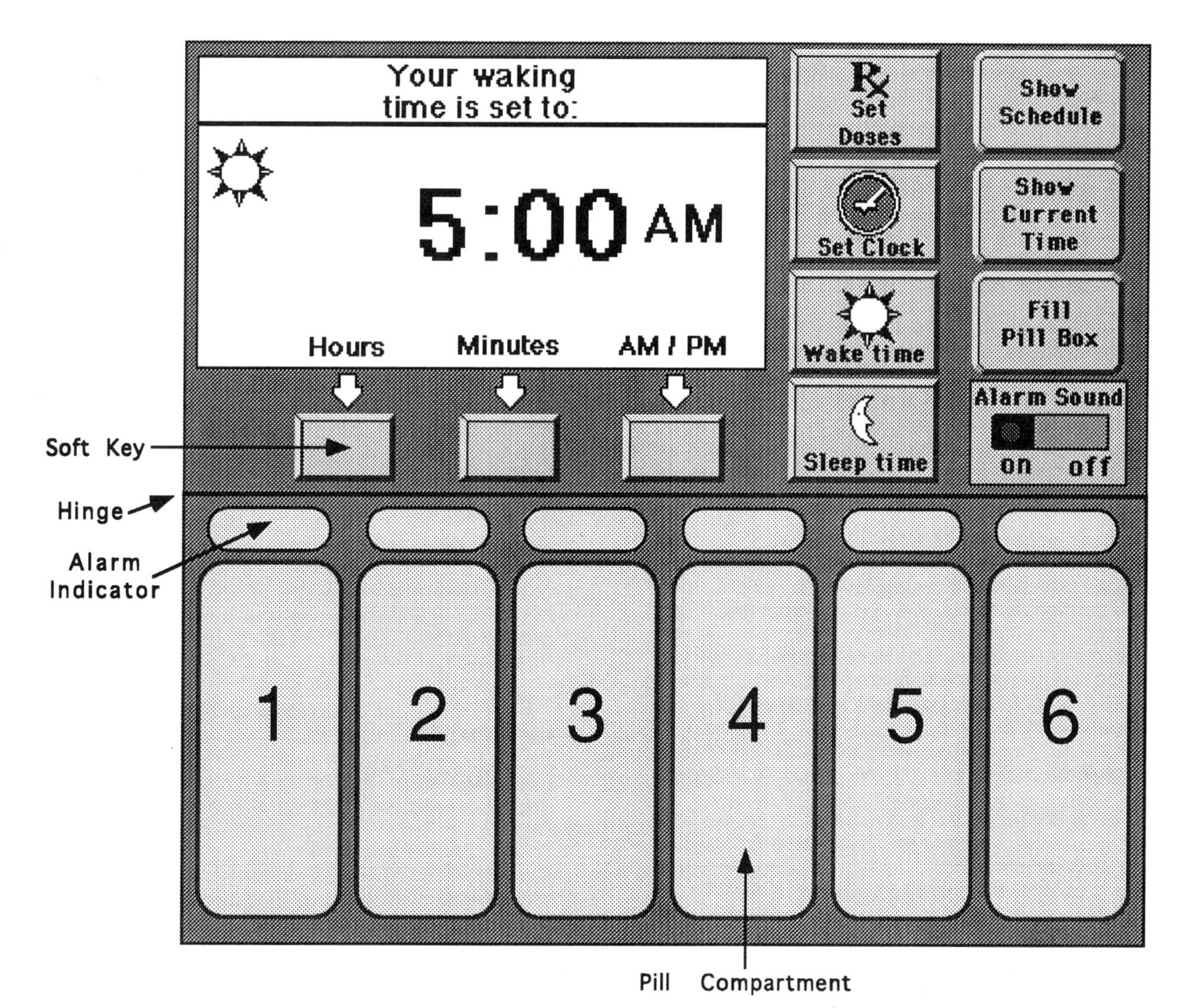

Figure 2. Medication Device Design Concept

As mentioned previously, the response times seemed to be a function of the complexity of the task and of the complexity of the keyboard. Earlier versions of the storyboard concept had 12 fixed function buttons. To use 12 buttons, yet still maintain a size large enough for visibility and easy activation, was a challenge. At this point the concept evolved into one which uses soft keys. As the mode or task at hand changes, soft key labels, which are located on the display, change to reflect the associated change in function. Actuating buttons are located beneath the soft key labels. This not only allows the device to be smaller since the function of the keys vary according to the task the user is performing, but also reduces the time it takes the user to scan the device in search of the appropriate button. The use of shape to code buttons, and icons and text to label buttons, assists the user in forming associations between a button and the function that it performs. When the user presses a button to perform a task, the icon associated with the button appears on the screen to facilitate the mental model the user has of the device.

The research and task analysis identified the calculation of the medication schedule as a significant source of human error. For example, a patient taking one medication three times a day and another medication four times a day would need to develop a regimen for taking medication at five intervals during a 24 hour period. The design concept seeks to avoid this potentially dangerous situation by allocating the creation of the medication schedule to the device rather than to the user. The device calculates medication schedules according to the user's waking hours, and assigns the medications to the correct compartments. There are enough compartments to accommodate medications taken up to six times daily. At the dosage time, an alarm sounds and a light above the appropriate compartment flashes. All other compartments remained locked.

The supporting research for deciding how large text should be for a product, states that the amount of enlargement that can be used for the lettering of any given product is a decision that must be made as part of the design process for that product (Vanderheiden and Vanderheiden, 1991). This is due to the fact that the device can be modified to be more accessible but there will always be people who cannot use it. The visual acuity of the population for which we are designing must be specified so that these issues can be resolved. The text size for the design concept is 1/8" or larger on buttons and on the display. This allows the displays and buttons to be easily distinguished and at the same time allows the device to remain small enough to be portable.

CONCLUSIONS

It became obvious through the survey of the state-of-the-art electronic memory devices, that manufacturers are not designing their products with the needs, capabilities and limitations of the older population in mind. In future studies, we hope to validate our design concept, and to expand the research to include designs for a broader senior population. Color will be an interesting area to investigate as a technique for coding the association of the buttons, their functionality, and the presentation of information. Research will be conducted on medication scheduling so that guidelines may be established for tailoring a medication schedule, and for compensating for late or missed doses.

ACKNOWLEDGEMENTS

This publication was made possible under grant number 1R43AG09358-01A1 from the National Institute on Aging.

REFERENCES

Aging America: Trends and projections. US Department of Health and Human Services Publ. No.(FCOA) 91-28001. 1991.

Ansello, E.F., Lamy, P.P, Gondek, K. (1990). Generational caregivers, pharmacists and medications: Their contributions to social and personality variables. Psychology and Aging, 30.

Czaja, S.J., Clark, M. C., Weber, R.A. (1990). Computer communication among older adults. Proceedings of the 1990 Human Factors Society Annual Meeting. Santa Monica, CA: Human Factors Society.

Ernst, M. and Ernst, N.S. (1984). Functional capacity. In. D.J Mangen and W.A. Peterson (Eds.), Health, program evaluation, and demography Minneapolis: University of Minnesota Press.

Hollman, Frederick W. (1990, March). United States population estimates by age, sex, race and Hispanic origin. Current Population Reports. Series P-25, No. 1057.

Leirer,V.O., Morrow, D.G., Tanke, E.D., and Pariante, G.M. (1991). Elders non-adherence: Its assessment and medication reminding by voice mail. The Gerontologist. 31 (4), 514-520.

Middleton, Francesca (1991). Computers for seniors. Resourceful Aging: Today and Tomorrow: Vol V. Lifelong Education (pp. 75-78). AARP. Washington, D.C.

Stuck, Andreas E., and Tamai, Irene Y. (1991). Medication management in the home. Clinics in Geriatric Home Care, 7 (4), 733-48.

Vanderheiden, G.C. , Vanderheiden, K.R. (1991). Accessible design of consumer products. Guidelines for the design of consumer products to increase their accessibility to people with disabilities or who are aging. Preliminary/working draft. Trace R&D Center (p.16). University of Wisconsin-Madison.

Ward, C. (1991). Increasing independence through technology: The views of older consumers with disabilities and their caregivers. Consumer Needs Assessment Project Year 3. Results of the Third Year of a Five Year Study. Electronic Industries Foundation, Rehabilitation Engineering Center, Washington, D.C.

USE OF AUTOMATED TELEPHONE REMINDERS TO INCREASE ELDERLY PATIENTS' ADHERENCE TO TUBERCULOSIS MEDICATION APPOINTMENTS

Elizabeth Decker Tanke and Von O. Leirer
Decision Systems
Los Altos, California

Elderly patients (N=617) with scheduled appointments in a public health clinic tuberculosis clinic either received or did not receive an automated telephone reminder the evening before their appointment. Patients in this population were primarily non-English speaking immigrants who received reminders in their own language. Automated reminders decreased nonadherence 21% (from 29% to 23%), and this impact did not differ across ethnic groups.

With the development of more effective and short-term regimens for treatment and prevention of tuberculosis, the major determinant of prevention and treatment success is now patient adherence (American Thoracic Society, 1986). Unfortunately, low levels of adherence are common in all aspects of tuberculosis control, including screening (Glassroth, Bailey, Hopewell, Schecter & Harden, 1990; Haefner, Kegeles, Kirscht & Rosenstock, 1967), preventative therapy (Comstock;, 1983; Snider & Hutton, 1989) and treatment (Addington, 1979; Farer, 1986). For example, reports submitted to the Centers for Disease Control indicate that 22% of patients with tuberculosis fail to complete treatment in a one-year period, and among those prescribed preventative treatment, 33% fail to pick up all needed medications (Snider & Hutton, 1989). This high rate of nonadherence to medical regimens creates multiple problems; it affects the patient's health, wastes valuable resources (medication, provider time, etc.), and contributes significantly to overall treatment costs (Barron, 1980; Ulmer, 1987). Failure to complete medication regimens frequently results in increased resistance to the medication and necessitates use of significantly more expensive and more toxic forms of therapy (Harding & Bailey, 1988; Glassroth, Robins & Snider, 1980). Finally, because of the infectious nature of tuberculosis, nonadherence thwarts attempts to control its spread (Conference Report, 1986; Snider, Kelly, Cauthen, Thompson, & Kilburn, 1985). These obstacles to tuberculosis control created by nonadherence have a disproportional impact on older individuals. Tuberculosis case rates are highest among the elderly in all racial and ethnic groups, and minority, immigrant, and poorer elders are particularly affected (CDC, 1989).

One potential cause of nonadherence in the area of tuberculosis is failure of prospective memory; i.e., patients may simply forget to attend a medical appointment or take a prescribed medication. "Forgetting" has been cited in some studies as a primary reason given both for failure to attend appointments (Deyo & Inui, 1980; Oppenheim, Bergman, & English, 1979) and failure to take medication (Wardman, Knox, Muers, & Page, 1988). Studies reporting greater nonattendance with a longer period between scheduling and appointment dates (e.g., Benjamin-Bauman, Reiss & Bailey, 1984; Levy & Claraval, 1977; Nazarian, Mechaber, Charney & Coulter, 1974) also supports this explanation of nonadherence. In addition, there now exists a fairly extensive literature on the effectiveness of reminders such as phone calls and mailed reminders for increasing appointment attendance (Frankel and Hovell; 1978) and adherence with medication regimens (Kirscht, Kirscht & Rosenstock, 1981; Leirer, Morrow, Tanke, & Pariante, 1991; Smith, Weinberger, Katz & Moore, 1988).

This research suggests that reminders may be effective in reducing nonadherence in tuberculosis, and, in fact, recommendations written for providers suggest the use of such reminders (Snider & Hutton, 1989). However, at least two factors make it practically impossible for tuberculosis clinics to implement such interventions. First, they require a significant expenditure of money and staff time, both of which are frequently unavailable in tuberculosis programs dealing with an increased demand for services and insufficient resources for existing programs (Leff & Leff, 1989; Reichman, 1993). Second, they frequently require the availability of staff who can speak the language of the patient; 24% of TB cases reported in 1989 were among foreign-born individuals (CDC, 1989).

The present study addresses these problems by using a specialized form of telecommunications technology, variously referred to as "automated notification", "outbound voice mail", and "voice messaging". This technology allows human voice messages to be digitally recorded and saved in a computer's hard disk storage system. The computer then phones patients at designated times and gives them the "spoken" message. Because the system is automated, it does not require extensive staff time to operate. In addition, because reminders can be recorded in different languages, it has the capacity for communicating important information to patients in a form that they can understand.

The technology we developed and evaluated in the present study may be applied to the problem of medication nonadherence in two ways. First, automated reminders may be used to remind patients of medication appointments. Second, automated reminders may be used to call and remind patients to take these medications. The present research was designed to evaluate the effectiveness of an automated patient reminder system called TeleMinder® on elderly tuberculosis patients' adherence to medication appointments.

METHODS

Subjects

Subjects were 617 patients aged 60 and over with scheduled appointments in the Tuberculosis Control Program of Santa Clara County Health Department. This patient population consisted primarily of non-English speaking immigrants. With respect to ethnicity, 30% were Vietnamese, 26% were Filipino, 20% were Chinese, and 24% represented other groups. Only 8% indicated English as their primary language. The average age of this population was 67.60 years, with a standard deviation of 5.75 years and a range of 60-89. Fifty-one percent of the patients were male. Although information about income of these patients was not directly available, a survey of 464 clinic patients conducted during the period of the study found the population served by the clinic to be largely indigent, with 59% reporting household incomes of under $10,000. Patients in this sample were scheduled for appointments to follow up on diagnostic tests for tuberculosis or to receive medication for either the active treatment or prevention of tuberculosis.

Research Design

This study was conducted during two time periods. During the first time period patients received reminder calls on alternate days to control for effects related to the day of the week or the time of the year the appointment was scheduled. During the second time period, minor variations in the exact wording of the message were used and message variations (including a no message control) were again balanced across days of the week. In all cases the reminder message (a) identified itself as a pre-recorded message from the county health department; (b) indicated that the patient had an appointment the following day; and (c) gave the address and phone number of the clinic twice. The message concluded with the statement "We are looking forward to seeing you tomorrow" followed by an instruction to listeners that they could hear the message repeated by remaining on the line. (The message did not refer to tuberculosis because of the possible stigma attached to the disease.) In the first time period, reminders were translated and recorded in English, Vietnamese, Spanish, Mandarin, Cantonese, Lao, Cambodian, and Tagalog. In second time period only the four most frequent languages of the total subject population (Spanish, Vietnamese, Tagalog, and English) were included.

Apparatus

Messages were recorded and delivered with the TeleMinder® system. TeleMinder is an interactive voice messaging system based on 80386/80486 CPU architecture, running under the MS-DOS operating system. The system includes a database and software enabling the user to maintain a patient database and to create and deliver telephone messages[1].

Procedure

The appropriate recorded message was sent to patients between 6:00 and 9:00 P.M. the evening before the scheduled appointment. Parameters on TeleMinder were set to allow a message to be left on an answering machine if one were detected and to call back up to 5 times at half hour intervals if a patient's phone were busy or if there was no answer after 8 rings. The message was sent in the patient's primary language when possible; if the primary language of the household was unavailable or not specified, the message was sent in English.

Attendance or nonattendance at the scheduled appointment was obtained from clinic records. We also used these records to collect information about patient age, sex, and ethnicity.

Statistical Analysis

Because preliminary analyses indicated no differences in adherence patterns with message variation used, the different message variations were combined and compared to the no reminder control group over both time periods. A logistic regression analysis was used to assess the effects on adherence of the automated reminders, time period, and patient age, sex and ethnicity. The model-building strategy we employed first assessed the main effect of all predictor variables taken together, and then assessed the effect of each individual predictor variable by means of the likelihood ratio test, comparing the resulting G statistic to the appropriate Chi square distribution. The effects of interactions between message variations and the other predictor variables were then assessed with the same procedure.

RESULTS

This analysis revealed a significant effect of automated reminders. Nonattendance decreased by 21% (from 29% to 23%) when automated reminders were used ($\chi^2(1)=9.21$, $p < .01$). There was also a significant difference in overall attendance between time periods, with attendance higher during the first time period. Attendance was not significantly related to patient age, sex or ethnicity. Finally, there were no significant interactions between the use of reminders and either time period or patient demographics; i.e., the effectiveness of the reminders was not significantly different for different time periods or different groups of patients.

DISCUSSION

The results of this study suggest that automated appointment reminders may play an important part in reducing elder's nonadherence to medication appointments in the area of tuberculosis. The reduction in appointment nonadherence of 21% was accomplished with minimal staff time; manual entry of patient information into the TeleMinder® system in our study averaged less than half a minute per patient, and such entry was necessary only once for patients with multiple appointments. In addition, the ability to deliver reminders in the patient's own language resulted in reminders being effective among members of different ethnic groups. This 21% decrease in nonadherence was obtained using a simple reminder that provided only basic information about the day and location of the appointment. Nonadherence might be decreased still further with additional enhancements to messages, or with the use of multiple reminders.

Decreases in appointment nonadherence with automated reminders have both direct and indirect effects on elder's health care. By increasing patient's ability to come to appointments, reminders increase the likelihood that patients will receive medication as soon as it is needed, and with the appropriate medical care. In addition, decreases in patient nonattendance result in more efficient use of clinic resources, and make more resources available for other aspects of patient care. For example, medical personnel who do not have to call non-attenders can spend more time in other tasks related to adherence, such as communicating information about the proper procedures for taking medication.

ACKNOWLEDGEMENTS

This research was supported by SBIR grants #2 R44 AI31750-02 from the National Institute of Allergy and Infectious Diseases and #2 R43 AG10659-01 from the National Institute on Aging. We thank the Santa Clara County Health Department for their cooperation and assistance.

FOOTNOTE

1. Further information about the specifications and capabilities of the TeleMinder® voice messaging system can be obtained by contacting the authors at Decision Systems, 318 State St., Los Altos, CA 94022.

REFERENCES

Addington, W. W. (1979). Patient compliance: The most serious remaining problem in the control of tuberculosis in the United States. Chest, 76, 741-743.

American Thoracic Society. (1974). Preventive therapy of tuberculosis infection. American Review of Respiratory Disease, 110, 371-374.

Barron, W. M. (1980). Failed appointments: Who misses them, why they are missed, and what can be done. Primary Care, 7, 563-574.

Benjamin-Bauman, J., Reiss, M. L., & Bailey, J. S. (1984). Increasing appointment keeping by reducing the call-appointment interval. Journal of Applied Behavior Analysis, 17, 295-301.

Centers for Disease Control. (1989). A strategic plan for the elimination of tuberculosis in the United States. Mortality and Morbidity Weekly Report, 38, 1-25.

Comstock, G. W. (1983). New data on preventive treatment with isoniazid. Annals of Internal Medicine, 98, 663-665.

Conference Report. (1986). Future research in tuberculosis: Prospects and priorities for elimination. American Review of Respiratory Disease, 134 (Supplement), 401-423.

Deyo, R. A., & Inui, T. S. (1980). Dropouts and broken appointments: A literature review and agenda for future research. Medical Care, 18, 1146-1157.

Farer, L. S. (1986). Tuberculosis: What the physician should know. New York: American Lung Association.

Frankel, B. S., & Hovell, M. F. (1978). Health service appointment keeping: A behavioral view and critical review. Behavior Modification, 2, 435-464.

Glassroth, J., Bailey, W. C., Hopewell, P. C., Schecter, G., & Harden, J. W. (1990). Why tuberculosis is not prevented. American Review of Respiratory Disease, 141, 1236-1240.

Glassroth, J., Robins, A. G., & Snider, D. E. (1980). Tuberculosis in the 1980's. New England Journal of Medicine, 302, 1441-1450.

Haefner, D. P., Kegeles, S. S., Kirscht, J., & Rosenstock, I. M. (1967). Preventive actions in dental disease, tuberculosis, and cancer. Public Health Reports, 82, 451-459.

Harding, S. M., & Bailey, W. C. (1988). Chemotherapy of tuberculosis. In D. Schlossberg (Ed.), Tuberculosis). New York: Springer Verlag.

Kirscht, J. P., Kirscht, J. L., & Rosenstock, I. M. (1981). A test of interventions to increase adherence to hypertensive medical regimens. Health Education Quarterly, 8, 261-272.

Leff, D., & Leff, A. R. (1989). Tuberculosis control policies in major metropolitan health departments in the United States. American Review of Respiratory Disease, 139, 1350-1355.

Leirer, V. O., Morrow, D. G., Tanke, E. D., & Pariante, G. M. (1991). Elders' nonadherence, its assessment, and medication reminding by voice mail. The Gerontologist, 31, 514-520.

Levy, R., & Claravall, V. (1977). Differential effects of a phone reminder on appointment keeping for patients with long and short between-visit intervals. Medical Care, 15, 435-438.

Nazarian, L. F., Mechaber, J., Charney, E., & Coulter, M. P. (1974). Effect of a mailed appointment reminder on appointment keeping. Pediatrics, 53, 349-352.

Oppenheim, G. L., Bergman, J. J., & English, E. C. (1979). Failed appointments: A review. Journal of Family Practice, 8, 789-796.

Reichman, L. B. (1993). Fear, embarrassment, and relief: The tuberculosis epidemic and public health. American Journal of Public Health, 83, 639-641.

Smith, D. M., Weinberger, M., Katz, B. P., & Moore, P. S. (1988). Postdischarge care and readmissions. Medical Care, 26, 699-708.

Snider, D. E., Jr., & Hutton, M. D. (1989). Improving patient compliance in tuberculosis treatment programs. Atlanta: Centers for Disease Control.

Ulmer, R. A. (1987). Editorial: Patient noncompliance and health care costs. Journal of Compliance in Health Care, 2, 3-4.

Wardman, A. G., Knox, A. J., Muers, M. F., Page, R. L. (1988). Profiles of non-compliance with antituberculosis therapy. British Journal of Diseases of the Chest, 82, 285-289.

Recognizability and Effectiveness of Warning Symbols and Pictorials

David L. Mayer and Lila F. Laux
Department of Psychology
Rice University
Houston, Texas 77251

Abstract

In this study we sought to determine the relative effectiveness of pictograms for a group of 139 subjects ranging in age from 17 to 83. We gave a pictogram identification task for 16 pictograms from the Westinghouse Product Safety Label Handbook (1981) to subjects. Pictogram identification ranged from 100% to completely unrecognizable. Generally, pictorials which depicted simple, clearly identifiable hazards or protective equipment were more identifiable than symbols. Pictograms which showed the injury occurring to a hand rather than the entire human figure were also more recognizable. Finally, to explore more than simple pictograms identification, we presented subjects with three pictograms: We asked half of the subjects to list all of the ways they could be hurt, injured or killed as well as any precautions they would take while using a product displaying one of the pictograms. The other half of the subjects endorsed precautions that they would observe on a checklist of possible precautions. In general, subjects were able to name at least one of the hazards associated with each graphic, but they generally did not name all of the hazards for a given pictogram. Sex and age effects are commented on in the paper.

Introduction

Interest in the use of pictorials and symbols to communicate hazard information has been increasing due to a growing recognition that a large minority of Americans are not functionally literate in English. Some symbols, such as the circle with the slash across it representing "do not," are understood in many countries. Other symbols which have been around for an equally long time (the symbol for radiation, for instance) seem to have gained international use, but are not universally recognized. Many guide-books and standards on the design and development of warning labels recommend the use of pictorial information whenever possible to reduce ambiguity, thus encouraging the proliferation of pictorials and symbols. But is ambiguity always reduced?

The difference between a symbol and a pictorial is not always clear in practice, but is made here for the purpose of study. A pictorial is a picture rather than an abstraction—the meaning is represented in the picture. A graphic displaying eye protection is one example. A pictorial might depict the pinching of a hand in moving equipment. A symbol is an abstract graphic whose meaning must be learned. Examples are the skull-and-crossbones and radiation hazard symbol. Collectively, symbols and pictorials are referred to here as pictograms.

ANSI standard Z535.3 (ANSI, 1987) recommends an 85% recognition rate for a pictogram to be considered acceptable for general use. But, it is important to determine the extent to which symbols and pictorials are recognized *and* the degree to which those who recognize them also understand the associated hazards. Thus, it may be that many adults would recognize the skull-and-crossbones symbol as identifying a poison, but would they also understand that the hazard may be from inhalation or dermal absorption of the poisonous substance as well as simple ingestion? Furthermore, to what extent does recognition of the hazard correlate with the understanding of the appropriate risk-reducing safety behaviors.

This study had two parts: In part one, we showed subjects 16 symbols and pictorials from the *Westinghouse Product Safety Label Handbook* (1981). For each, we asked subjects to describe the hazard that such a symbol or pictogram would be used to warn about. Recognition ranged from 100% to completely unrecognizable. In general, pictorials which depicted simple, clearly identifiable hazards or protective devices were more recognizable than symbols.

In part two, we presented subjects with three of the Westinghouse symbols: *Electric Shock* (wire shocking hand), *Poison* (skull-and-crossbones), and

Flammability (flames). We asked half of the subjects to list all of the ways that they could be hurt, injured or killed while using a product displaying one of these symbols. They were also asked to list the precautions they would take if using a product with this pictogram on it. The remaining subjects simply endorsed precautions that they would observe on a checklist. In general subjects were able to name at least one of the hazards associated with each symbol, but they generally did not not name all of the hazards for a given pictogram.

Subjects for these experiments were recruited from undergraduate psychology courses at the University of Houston as well as from a sample of older drivers assembled for an unrelated research effort. Research with the university students has appeared previously—see Note 1. The effects of gender as well as age were studied. For statistical purposes, the sample was split into two groups to study age effects: a group of subjects under age 40 and a group of subjects age 40 or more.

Part One

Subjects

Subjects were 139 male and female participants. Subjects ranged in age from 17 to 83 with an mean age of 34.4. There were 88 females 51 males. Ninety-three subjects were under age 40, and 46 subjects were age 40 or over.

Method

We showed each subject 16 symbols and pictorials from the *Westinghouse Product Safety Label Handbook* (1981) as noted below:

Biohazard	Flammability
Corrosive liquid	Fumes / Vapors
Ear protection	Hot
Electricity (3)	Laser
Explosion	Moving parts (2)
Eye protection	Radiation
Falling object	

Two different pictograms were used for moving parts and three for electrical hazards as the *Handbook* recommended more than one. The moving parts pictograms differed in the type of machinery (*i.e., Moving Gears* as opposed to *Rollers*). The electrical hazard pictograms differed in level of abstraction. One showed a simple lightening bolt; another depicted a lightening bolt shocking a human figure; the last showed a hand being shocked by a wire. We wanted to know if these graphics were differentially effective or meaningful.

Results for Part One

Subjects responses were scored with very lenient criteria. Subjects who successfully described the hazard depicted by a given pictogram were considered to have correctly recognized that pictogram. The most recognizable pictograms by far were those for *Flammability* and *Corrosive*. The *Flammability* pictogram was recognized by 96.4% of the subjects and the *Corrosive* pictogram was recognized by 92.2% of the subjects.

Between 78% and 85% of subjects correctly identified the *Fumes, Hearing Protection, Eye Protection, Hand Electric Hazard* and *Moving Gears* pictograms. There were no notable effects of sex or age on the identification of these pictograms. The *Hot* pictogram was also very identifiable (70.5%), but younger subjects were more likely to correctly identify it. The *Explosive* pictogram was correctly identified by about 70% of subjects, but fared significantly better with males and the under 40 group.

The following table lists each pictogram and the overall percentage of subjects who correctly identified each one in parentheses. Further

	% correct			
	by sex		by age	
PICTOGRAM (%)	F	M	<40	≥40
Flammable (96.4)	95.5	98.0	98.9	91.3
Chemical (92.2)	89.2	97.6	92.5	91.0
Fumes (84.7)	79.2	95.1*	83.9	88.0
Ear Protection (81.4)	80.5	83.0	79.6	88.0
Eye Protection (80.0)	77.0	85.4	78.5	86.4
Hand Electric (78.4)	76.1	82.4	74.2	87.0
Moving Gears (78.2)	79.7	75.6	79.6	72.3
Hot (70.5)	68.2	75.0	79.6	52.2**
Explosive (69.7)	61.4	84.3**	76.3	56.5*
Body Electric (66.2)	64.8	68.6	60.2	78.3*
Moving Rollers (64.7)	59.1	74.5	73.1	48.0**
Radiation (56.1)	43.2	78.4***	62.4	43.5*
Lightening Elec (37.4)	38.6	35.3	43.0	26.1
Falling Object (32.4)	21.6	51.0***	33.3	30.4
Laser (3.4)	1.3	7.3	1.1	12.0**
Biohazard (0.0)	0.0	0.0	0.0	0.0

breakdowns by sex and age group are also given. Note 2 at the end of the paper explains the statistic used as well as the convention adopted for denoting significance levels. It should be noted that only the top 3 pictograms in the table were found to be recognizable by 85% or more subjects. Under the guidelines of ANSI Z535.3, only these 3 could be considered acceptable for general use.

The *Rollers* pictogram had an overall identification rate of 64.7%, but fared poorly among the over 40 age group. The *Falling Objects* pictogram was best identified by the younger males, but had an unacceptable overall identification rate of 32.4%.

With regard to the three electrical hazard pictograms, the *Hand Electric Hazard* pictogram—which depicts a human hand being shocked by a live wire—was the most effective across groups, 78.4% of subjects correctly identified it. The pictogram which depicts a lightening bolt shocking a human body was correctly identified by 66.2% and the lightening bolt of electricity was identified as meaning an electrical hazard by only 37.4% of subjects.

The *Radiation* pictogram was correctly identified by 56.1% of subjects, but was more likely to be correctly identified by male subjects and subjects under age 40. The *Laser* and *Biohazard* pictograms were almost uniformly mysterious to subjects. Many of those who misidentified these abstract pictorials inferred physical action such as from blades or rotating machinery. This suggests that the ANSI standard which requires that these pictograms be accompanied by additional verbal information is a good practice, but it acknowledges the ineffectiveness of these symbols with regard to illiterate or non-English speaking users.

Part Two

Subjects

Subjects were 122 male and female participants. There were 68 females and 54 males. Subjects ranged in age from 17 to 83. Seventy-five of the subjects were under age 40, and 47 were age 40 or over. Subjects were randomly assigned to one of two groups: hazard recall or hazard recognition. There were 62 subjects in the recognition group with a mean age of 37.6 years. There were 60 subjects in the recall group. Their mean age was 35.2 years. Fifty percent of the recognition group was female, while 60% of the recall group was female.

Method

All subjects were presented with the same three pictograms from the *Westinghouse Product Safety Label Handbook* (1981). Pictograms used were *Electric Shock* (wire shocking hand), *Poison* (skull-and-crossbones), and *Flammability* (flames). Subjects were asked to assume that they were preparing to use a product displaying a given symbol. Those in the recognition group were presented with a checklist of 19 precautionary behaviors for each hazard. They were asked to endorse those precautions that they would observe if they were preparing to use such a product. Some behaviors were generally good practice for any product *(e.g., I would look for instructions on the product)* and others were more specific *(e.g., I would not smoke while using it)*. As opposed to checklists, subjects in the recall group were asked to list four possible hazards and four precautions that they would take if they were planning to use such a product.

Results for Part Two

Results for the hazard recognition (checklist) portion of the experiment are summarized for each pictogram by sex. Interesting age effects are commented on as appropriate.

ELECTRICAL HAZARD	% endorsing F	M
I would not use it if the cord was frayed	90.3	83.9
I would not get it wet	83.9	90.3
I would avoid exposed wires	90.3	87.1
I would not repair it myself if it broke	64.5	29.0*

This *Hand Electric Hazard* pictogram seems to suggest several appropriate precautionary behaviors which should be associated with electrical hazards. Significantly more females responded that they would not attempt to repair it themselves if it broke compared to the males. This difference was not observed between the age groups.

ELECTRICAL HAZARD	% endorsing <40	≥40
I would not repair it myself if it broke	48.7	43.5
I would look for warnings on the package	84.6	47.8**
I would look for warnings on the product	94.9	60.9**

An interesting age difference was noted: Those subjects under 40 were significantly more likely to indicate that they would look for warnings on the package or product than were the subjects over 40 years old.

POISON	% endorsing F	M
I would wear rubber gloves	32.3	35.5
I would not smoke while using it	51.6	38.7
I would not use it in a poorly ventilated room	51.6	67.7
I would look for instructions on the package	90.3	96.8
I would look for warnings on the package	58.1	61.3

It is interesting to note that such advisable behaviors as wearing rubber gloves, not smoking and using a poison only in a well ventilated room are not universally suggested by the skull-and-crossbones. Further, while most people indicated that they would look for instructions, fewer subjects indicated that they would look for warnings on the product. This may suggest that warnings and instructions should appear in close proximity on products of this sort. As is noted below, subjects over 40 were much less likely to indicate that they would look for on-package warnings.

POISON	% endorsing <40	≥40
I would not drink or eat it	100.0	43.5***
I would wash my hands after using it	100.0	65.2***
I would look for instructions on the package	97.4	87.0
I would look for warnings on the package	92.3	4.4***

It is also interesting to consider significant differences on two fundamental statements about handling poisons: While subjects under 40 uniformly indicated that they would not purposefully ingest the poison, less than half of the older subjects said this. A similar age difference is noted with regard to washing one's hands after use.

FLAMMABILITY	% endorsing F	M
I would not smoke while using it	64.5	67.7
I would keep it away from sparks or flames	71.0	71.0
I would store it at a cool temperature	54.8	45.2

While the *Flammability* pictogram did a reasonable job suggesting to subjects that smoking while using a flammable product or using it near any other source of ignition or storing it at a hot temperature were unsafe behaviors, the numbers are not impressively high. Further, some striking effects of age are noted:

FLAMMABILITY	% endorsing <40	≥40
I would not smoke while using it	94.9	17.4***
I would keep it away from sparks or flames	89.7	39.1***
I would store it at a cool temperature	69.2	17.4***

Subjects in the over 40 group seem to be much less knowledgeable about the hazards associated with flammable products, or less able to recall these hazards when presented with the pictogram. Either explanation is cause for concern.

For the hazard recall phase of the experiment, subjects' responses were evaluated on the basis of three key precautionary behaviors. Subjects' responses, scored on this basis, are presented below by sex and age group:

For the *Electric Shock* pictogram, most subjects' responses indicated that they realized that water and frayed wiring were hazards associated with electricity. A small number of subjects indicated that they would not attempt to repair an electrical device themselves. Significantly more males and younger subjects gave this response.

With regard to the *Poison* pictogram, most subjects knew that ingestion of a poison is hazardous, but significantly fewer subjects in the over 40 group gave this response. More males than females indicated that poisoning could occur by inhalation, while fewer subjects in the over 40 group indicated that dermal absorption of a poison could be hazardous. This pictogram did not seem to communicate the same information to all subjects.

PICTORIAL	Gender		Age Group	
•hazards	F	M	<40	≥40
ELECTRIC SHOCK				
•wet	64.9	82.6	80.6	58.3
•frayed	54.1	56.5	61.1	45.8
•repairs	10.8	30.4*	30.6	0.0**
POISON				
•ingestion	83.8	95.7	97.2	75.0**
•inhalation	46.0	73.9*	66.7	41.7
•absorption	59.5	82.6	80.6	50.0*
FLAMMABILITY				
•burn	73.0	78.3	83.3	62.5
•explosive vapor	16.2	21.8	22.2	12.5
•location/storage	27.0	56.5	44.4	29.2

On the other hand, the flammability pictogram did a better job of uniform communication. Though few subjects' responses indicated that they knew about fire hazards associated with explosive vapor.

General Discussion

One cannot expect all users to derive the same meaning from a given pictorial or symbol. These graphics convey a varying amount of information depending on a variety of factors. We assume that much of the differential effectiveness for males and females as well as that observed between age groups for the symbols and pictorials tested here was due mainly to differences in subjects' experience. Manufacturers and designers of warnings must remember that people of differing levels of experience and training will be using their products and interacting with their symbols and pictorials, and design pictograms and warning systems accordingly.

There is no universally effective graphic, but steps to measure and increase meaningfulness can be taken. Intuitively similar pictograms may have disparate rates of effectiveness. Every change to a graphic must be studied to ensure that it does not detract from meaningfulness. And it must also be borne in mind that simple recognition does not ensure *meaningfulness*.

Complex or overly detailed human figures may detract from the effectiveness of the pictogram by directing attention away from the hazard. But parts of human figures (such as hands or feet) are important to pictograms as they can provide a frame of reference. We found good recognizability for pictograms depicting protective equipment as well as others with simple messages. Those pictograms requiring the interpretation of motion (such as *Falling Object* or *Rollers)* or those with more complex abstractions such as *Electrical Hazard* were more difficult for subjects to interpret.

As the population ages, more and more older Americans will be called upon to interact with pictogram-based warning systems. Results from the present work suggest that they may be less likely to expect these warnings. This finding may have implications for the design of warnings and graphics for this population.

Most importantly, these responses indicate that there are two problems which must be addressed in the development of symbols and pictorials: Will the user recognize the hazard? and Will the recognition of the hazard suggest proper precautionary behavior? If symbols and pictorials are to be used successfully in informing and warning illiterates and non-English speakers, they must fulfill both of these responsibilities. It cannot be assumed that a user's recognition of a pictogram's meaning will also mean that the user have the necessary knowledge to avoid the hazard.

Notes

1. Some of this work has appeared previously in Laux, L.F., Mayer, D.L. and Thompson, N.B. (1989). Usefulness of symbols and pictorials to communicate hazard information. In *Interface 89: The Sixth Symposium on Human Factors and Industrial Design in Consumer Products*, (pp. 79-83). Santa Monica, CA: The Human Factors Society.

2. Statistics used to show differences between groups are Yeats corrected Chi-Squares with one degree of freedom. Significance levels for differences are noted in tables by *, ** or *** meaning $p<.05$, $p<.01$ and $p<.001$ respectively.

References

Westinghouse Electric Corp. (1981). *Westinghouse Product Safety Label Handbook.* Author.

ANSI Z535. (1987). *Criteria for safety symbols.* American National Standards Institute.

PERCEPTION OF SAFETY HAZARDS ACROSS THE ADULT LIFE-SPAN

David B. D. Smith, Ph. D. and James R. Watzke, Ph.D.
University of Southern California and
Research and Training Center on Aging,
Rancho Los Amigos Medical Center

We conducted a survey to study the risk perception and assessment of safety hazards as a function of age. The survey asked subjects (N=200) to answer a 25 item survey about their feelings, experience and knowledge concerning specific (falls, burns, and medication) and general safety hazards. Results of an ANOVA, contrast and principle component analysis each supported the conclusion that the perception and evaluation of risk plays an increasingly important role in safety behavior and hazard avoidance as people age. These results are consistent with previous work on cautiousness as a personality trait in older people, and indicate this cautiousness begins in the middle years of life. The principle components analysis indicated a core of 12 items from the original 25 that have a distinct age-risk-safety structure. This may have potential as a profile useful in research and clinical contexts for predicting and intervening with persons at high risk for accidents.

It is generally acknowledged that accident causation is multi-factorial. One approach, especially appropriate to a human factors and aging concern, conceptualizes this interaction of multiple factors as a sequence model of information processing and psychomotor events. Much of the research about the accidents of older people has been concerned with how age changes in this sequence contributes to accident occurrence. In large part this research has focused either on sensory deficits causing failure to detect a hazard, the initial stage of information processing, or the inability to avoid a hazard, the latest or psychomotor stage. To date, central stages of this sequence, such as risk perception and assessment, have received little study in age-accident research. Nevertheless, there is good reason to believe they should. For example, accident statistics suggest older people intentionally avoid environments that may exceed their capabilities, such as driving and climbing stairs, indicating an important role for the perception of risk in their safety behavior. Also, there is growing appreciation of the significance of central factors in accident causation in general (Barrett, Mikal, Panek, Sterns, and Alexander, 1977; Watzke, 1990) and in particular some evidence (Wooley, 1989) for the involvement of central age-related decrements.

The present study focused on these factors of perception, assessment and avoidance of environmental hazards over the greater part of the adult life-span. To do this we capitalized, by use of a survey, on the reported feelings, experience and knowledge of our subjects concerning specific and general safety hazards.

Methods

Subjects

Since our primary interest was age-related, subjects (total N=200) were sought in young (under 30, N=44), middle-aged (30-59, N=42), young-old (60-75, N=59), and old (75+, N=55) categories. The young and middle-aged were enrolled in upper division undergraduate psychology courses at two Southern California universities. The older subject groups were volunteers from a community service program which provides free door/window locks and grab bars for home-based elderly. These older people were ambulatory and living independently. The sample was predominately female, 145 to 55 males.

Safety Survey

The survey (Smith, 1988) consisted of questions developed from a review of the age-related accident literature (Smith, 1987; 1990). It contained 25 items in three safety areas of growing concern as people age, namely falls (10 items), medication problems (3 items) and burns (4 items). In addition, there were 8 items on general safety knowledge and concerns. Each item was in statement form accompanied by a 4 category Likert scale with no neutral point. The 4 categories were, agree strongly, agree, disagree and disagree strongly. In addition to the survey, subjects were asked two questions about their accident history. First, if they had received medical treatment from a fall in the past year and second, if in the past year they had any other fall that did not require medical treatment, but did cause them concern. The same two questions were repeated for medication problems and for burns.

Methods

The two younger age groups, composed of students, filled in the survey in class. The two older age groups were given the survey as part of the community service program and returned it by mail. A standard instruction sheet accompanied the survey.

TABLE 1
Abbreviated Survey Items and Summary of ANOVA and Factor Analysis

Item/Variable	F[1] Test	Principle Component Factors 1	2	3	4	5	6	7
Age-group factor loadings		**-.75**	**-.06**	**.16**	**-.04**	**.19**	**.00**	**-.15**
1. I sometimes feel I may lose balance and fall.	S	.79						
2. Medicine bottles are difficult to read.	S	.62						
3. One should only climb status where...handrails.	S	.72						
4. Know what your doing, O.K. to smoke in bed.	NS							.86
5. Wrong medicine is something I worry about.	S		.53					
6. I worry about riding in a car at night.	S		.54					
7. When retiring close/lock the doors/windows	NS				.71			
8. Overall, I'm more cautious about accidents than others.	S				.60			
9. My eyes are especially sensitive to glare.	NS		.69					
10. Less dangerous to go upstairs than down.	NS		.54		.42			
11. Cooking... dangerous to wear long sleeve nightwear.	S	.52						
12. In bath...I am worried I may fall.	S	.81						
13. I worry using a step stool to reach things.	S	.83						
14. Safer to take elevator than stairs, even 1 flight.	S	.59						
15. Prefer bright lights in my home.	S	.41						
16. Worry about tripping over carpet/cords.	S	.80						
17. Feel helpless/weak to prevent/avoid accidents.	S	.52		.51				
18. Not much can prevent accidents...fate.	S			.79				
19. Difficult to tell if hot water will burn/scald.	S		.55	.50				
20. Seldom think about poisoning by medication.	S			-.49			.59	
21. Careful person can prevent almost all accidents.	NS					.74		
22. Exercise, nutritious foods could help avoid falls.	NS					.67		
23. Old people grab bars advisable, not required.	NS						.81	
24. Afraid of falling when house is messy.	S	.58						
25. Worried with more than one appliance in socket.	S	.44						

1. S = Significant; NS = Non Significant; $p < .02$

Analysis

Numerical values from 1 (strongly agree) to 4 (strongly disagree) were assigned to the qualitative choices on the Likert scale. These scores were analyzed by ANOVA for age effects over the 4 age groups. Factor analysis was done to look at the conceptual structure of the survey. Because precise age data was missing on some of our subjects we used the age-grouping variable, again assigning a numerical score, to enter age into the factor analytic procedure.

Results

Analysis by ANOVA

The 25 survey items are listed in somewhat abbreviated form in Table 1. One way ANOVA on responses to the Likert scale revealed a significant age effect ($p < .02$) for 18 of the 25 items in the survey (See Table 1). In each case these differences were in the direction of a greater concern for safety with advancing age. These significant age effects were present for each kind of accident (falls, medication problems and burns).

Predominate among the non-significant items were those related to knowledge about safety. For example items such as, "It is less dangerous to go up stairs than to go downstairs", or "For older people grab bars in the bathtub are advisable but not required". Items expressing personal concern or worry about safety, especially falls, produced the most pronounced age differences. Examples would be, "Taking the wrong medicine is something I worry about.", or "I worry about using a step stool to reach things in high cupboards."

There were no significant differences between males and females for any of the 25 items.

ANOVA Contrasts

To look at specific age-group differences, separate variance estimates (ANOVA contrasts) were carried out for the 18 items found significant in the one-way analysis of variance. Table 2 gives a summary of the comparisons which reached the .02 level of significance. For comparisons between either of the two younger and the two older groups most of the 18 items were significant. When comparisons were between the under 30 and the middle-aged (30-59) seven of the items were statistically significant. Finally, only 1 statistical difference resulted when the two oldest groups, the young-old (60-75) and the old (75+), were compared.

Factor Analysis

Next we made use of principle component analysis to study the conceptual structure of the survey and how this structure relates to age. The analysis for the 25 items and for age with varimax rotation, extracted 7 orthogonal factors selected for eigen values greater than 1 and accounting for 61.7% of the variance. Table 1 gives the loadings on each of the 7 factors across all survey items and for the age-grouping. Only those loadings .40 or greater are reported, accept for the age-group variable which is given for all 7 factors.

Factor 1 accounted for 32% of the variance and was the only factor with an appreciable age loading. This factor seems to reflect *safety concerns related to aging* with the emphasis on falls. Supporting the strong association with age is the fact that the items in this factor were all

TABLE 2
ANOVA Contrasts for the 18 Survey Items Significant in the One-Way ANOVA

Contrast Comparison[1]	Items Significant for this Contrast
Young vs. old	All 18 significant at $p < .02$
Young vs. Young-old	All but item 20 significant at $p < .02$
Middle-aged vs. old	All 18 significant at $p < .02$
Middle-aged vs Young-old	All but items 15 and 6 significant at $p < .02$
Young vs. Middle-aged	Items 1, 2, 3, 14, 16, 24, 25, significant at $p < .02$
Young-old vs. Old	Item 23 significant at $p < .02$

1. Young = under 30; Middle-aged = 30-59; Young-old = 60-75; Old = 75+

significant for age in the overall ANOVA analysis. Also the 7 items that discriminated the young vs. the middle-aged in Table 2 were all loaded on this factor.

Factor 2, accounting for 6.6% of the variance, seemed to us to reflect a more general *concern with issues of safety and security not necessarily related to aging.* In this case two of the items loading on this factor (9&10), were not significant for age in the ANOVA analysis (Table 1).

Factor 3 (variance=6.1%) was slightly related to age, and the 4 items in this factor were all significant for age by the ANOVA analysis (Table 1). We interpreted this factor to reflect the *chance nature of accidents and the difficulty of avoiding them.* In this regard older people were more fatalistic about having an accident than were younger ages.

Factor 4-7 attracted only a few of the items and primarily the ones non-significant by ANOVA for age. With the exception of factor 5 which loaded on the two questions dealing with *prevention of accidents,* the interpretation of these several factors was problematic. Individually they were each associated with between 4-5% of the variance.

Accidents

The number of reported accidents occurring within the last year, either requiring or not requiring medical attention, was small for this group of subjects. Thus, it was not surprising that the number of accidents did not differ statistically across the age-groups.

Discussion

The overall results of this study are consistent with the view that the perception and evaluation of risk plays an increasingly important role in safety behavior and hazard avoidance as people age. Of course this is highly intuitive, is in agreement with accident statistics (Smith, 1987), and parallels the wide spread belief by society that conservatism and cautiousness are, in many areas of life, correlates of increasing age.

It is tempting to presume that this "concern for personal safety" reflects a recognition by the older person of declining psychomotor capabilities and a loss of confidence in the capacity to avoid accidents. However, two aspects of the present results question a necessary and simple link between these age-related phenomena. First, we might expect if concern for safety is an outcome of declining abilities then the oldest, therefore the most frail, should be most concerned. On the contrary, for comparisons between the young-old (60-75) and the old (75+) we found only one of the 25 items in the survey to indicate a significantly greater concern by the old group (75+). Second, the results here indicate that a growing concern for safety appears in the middle-aged years, not a time in which age changes in psychomotor abilities are thought to be very important - at least with regard to accidents. These kind of results suggest that the antecedents of age change in accident risk perception are multi-factorial, including perhaps cultural, attitudinal and motivational variables as well as physical capability.

The results here on risk perception are likely related to work in gerontology on risk and cautious behavior in older people (Botwinick, 1966, 1984: Okun, Stock, Ceurvorst, 1980). Carefulness and caution have been used as explanatory concepts for the observed difference in performance between young and old in studies involving perception, learning, intelligence and psychomotor tasks (Basowitz and Korchin, 1957; Korchin and Basowitz, 1957; Eisdorfer, 1965). Also, direct study on reported risk-taking behavior (Wallach and Kogan, 1961: Botwinick, 1966) found older people do respond in a generally more cautious way but not always. In summarizing these studies Botwinick (1984) concluded that the elderly avoid risk where possible but when it is not possible their risk-taking does not differ from young ages. Our research indicates this cautiousness with age extends to issues of personal safety and begins in the middle years of life. Whether age differences in safety concerns are dependent on the situation as Botwinick (1984) argues, will need future research. If so, this may be of major significance for the study of accidents in older persons, especially the frail elderly.

It remains to be seen whether our self-reported concern for personal safety is associated in any way with actual accident history. We tried to test this but our sample reported few accidents, serious or otherwise. Perhaps one year is insufficient time to sample accident history. However, we question whether the self-report of accidents is an adequate source of information about their true incidence (Watzke, 1990). We plan to pursue other more direct ways of obtaining accident history to test behavioral correlates of the safety survey.

One useful approach might be to use the results of the factor analysis that produced from the original 25 items a core of 12 with a distinct age-risk-safety structure. Individual scores on these items could be taken as a profile and tested as a predictor of accidents in older persons. If successful, such a profile could be clinically useful, e.g., an older person who scores less cautious than peers might, if showing evidence of frailty, be a candidate for safety training or environmental intervention.

References

Barrett, G.V., Mikal W. L., Panek, P. E., Sterns, H. E. and Alexander, R. A. (1977). Information processing skills predictive of accident involvement for younger and older commercial drivers. Industrial Gerontology, Summer, 171-182.

Basowitz, H., and Korchin, S. J. (1957). Age differences in the perception of closure. Journal or Abnormal and Social Psychology, 54, 93-97.

Botwinick, J. (1966). Cautiousness in advanced age. Journal of Gerontology, 21, 347-353.

Botwinick, J. (1984). Aging and behavior. Springer: New York, New York.

Eisdorfer, C. (1965). Verbal learning and response time in the aged. Journal of Genetic Psychology, 107, 15-22.

Korchin, S. J. and Basowitz, H. (1956). The judgement of ambiguous stimuli as an index of cognitive functioning in aging. Journal of Personality, 54, 64-69.

Okun, M. A., Stock, W. A. and Ceurvorst, R. W. (1980). Risk taking through the adult life span. Experimental Aging Research, 6, 463-473.

Smith, D.B.D. (1987). Safety and security: Human factors issues for older people. Paper for the National Research Council/National Academy of Science Panel on Human Factors for an Aged Population, Washington, D.C.

Smith, D.B.D. (1988). Safety awareness scale for the elderly. Unpublished work. Human Factors Department, University of Southern California.

Smith, D.B.D. (in press). Human factors and aging. An overview of research needs and application opportunities. Human Factors.

Wallach, M. A. and Kogan, N. (1961). Aspects of judgement and decision making: interrelationships and changes with age. Behavioral Sciences, 6, 23-26.

Watzke, J. R. (1990). Falls in home based elderly: A multi-factorial problem. Unpublished work. Research and Training Center on Aging, Rancho Los Amigos Medical Center.

Wooley, S. (1989). An assessment of falls in the elderly. Unpublished doctoral dissertation. State University of New York, Buffalo.

RESIDENTIAL FIRE SAFETY NEEDS OF OLDER ADULTS

Neil D. Lerner & Richard W. Huey
COMSIS Corporation, Human Factors Division
Silver Spring, Maryland

This paper concerns the fire safety needs of older people living in private residences. It includes consideration of the limitations of current consumer fire safety products (primarily smoke detectors) and design improvements that better meet the human factors requirements of this population. Older people suffer especially high residential fire death rates, and most of these deaths occur in private homes, where the resident/victim had primary responsibility for fire safety. The literature on human behavior during fires indicates certain differences in the behavior of older and younger adults, but this is based on limited data from often inappropriate populations (institutional settings). A review of currently available products, and their use, found that these products did not adequately meet the needs, nor match the capabilities, of older people. In fact, the need to climb to ceiling level to install, test, and maintain single station smoke alarms introduces a significant safety hazard of its own. Focus discussion groups of older homeowners identified various attitudes, problems, and behaviors, important for improved product design. A review of existing products and technologies identified various features which could benefit elderly users, and practical technologies that could be adapted from other applications (e.g., security systems). Based on the literature review, product/technology review, and focus groups, a set of fire-safety product needs and desirable features was developed, including a set of seventeen specific design features/functions. Substantial improvement in fire safety product usefulness for older users can be achieved in a cost-effective manner.

INTRODUCTION

Home fire safety is a serious safety concern for older Americans. The United States and Canada have fire death rates substantially higher than those of other developed nations (Cote, 1986), and the elderly suffer a far greater fire fatality rate than other groups within this population. Data reviewed by Lerner, Huey, Morrison, and Mowrer (1990) show the "young-old" (middle-sixties to early-seventies) to have a fire death rate about double that of other adults, while the "old-old" may have five to ten times the death rate of younger adults.

Most older Americans live with relative independence in private homes, and in fact are more likely than younger groups to own that home. The vast majority (e.g., 80-90%; Cote, 1986; Gulaid, Sattin, and Waxweiller, 1988) of fire deaths occur in private dwellings. Therefore most older people are largely personally responsible for their own fire safety. There are a variety of important products related to home fire safety, most notably the smoke detector. Ironically the elderly, the group most in need of the protection such devices provide, are poorly matched to the design characteristics of smoke detectors. They risk falls if attempting to reach the product during testing or maintenance; have difficulty physically manipulating covers, batteries, and controls; and the acoustic signal characteristics are not optimal. Furthermore, there are fire alerting and egress needs that typical home smoke detector simply do not address.

This paper will examine the residential fire safety problems of older people, with special emphasis on the adequacy of smoke detectors in meeting these problems. The report will describe the literature on fire-related behavior, particularly for older people; describe the range of current home smoke detector products and features; present information obtained from focus group panels of older homeowners; and present recommendations for improved smoke detector design to better meet the needs of elderly residents.

LITERATURE REVIEW: FIRE BEHAVIOR, ALERTING, AND AGING

The literature on human behavior in fires was reviewed, with particular emphasis on findings related to (a) older adults, or (b) aspects that might be important for the design of fire safety products, or (c) effective alerting. There are some unusual aspects to the literature on people's fire-related behavior. First, there is an emphasis on public or group settings and on major disasters, even though most fatalities come one or a few at a time in individual residences. Second, while there was a spate of research activity in the 1970's, relatively little has appeared in the past decade. Third, despite the number of articles published, a detailed knowledge of fire behaviors has been limited by the inherent research difficulties. Many of the findings come from delayed interviews with fire survivors, and controlled experimental research is constrained by obvious ethical considerations. Fourth, very little of the fire safety literature has explicitly concerned older adults. To the extent "the elderly" were addressed, they have typically been residents of institutional care facilities. Reports on the behavior of these populations (often anecdotal in nature) have too often been treated as being typical of older people in general. Summarizing these limitations of the literature, there is a lack of detailed knowledge about behavior during fires, and this is especially true for older people and residential fires. If typical behaviors have changed at all with the advent of widespread smoke detector use or due to recent public fire safety programs, the literature may also be somewhat dated now.

Given the breadth of the literature review, only a few key findings can be cited here. A full presentation can be found in Lerner et al. (1990). Specific reference citations are not included in these summary statements, but can be found in Lerner et al.

- The elderly are over-represented in fires of every type of origin except those started by children or those of suspicious origin.

- The old are more likely to be "intimately involved" in starting the fire and are more likely to be in close proximity to the fire.

- While somewhat over half of all home fires fatalities occur between midnight and 8 A.M., only 38% of fatalities aged 65 and older occur during this period.

- Alcohol is a significant factor in fire deaths, but much less so for the elderly.

- The old suffer greater vulnerability to toxic combustive products, but this does not appear to be a primary explanation for their greatly increased incidence of fire deaths.

- The literature is ambiguous with respect to the effects of age on the ease of arousal from sleep by alarms.

- Once initial cues to the existence of a fire are detected, the situation is usually characterized by ambiguity. The resident's initial response is normally to investigate and confirm the fire, which may bring him or her into closer proximity to the fire.

- Once the situation has been defined as an actual fire, various actions may take place (e.g., fight the fire, alert others, call fire department, leave the building, get dressed, etc.). No single behavior dominates as the "typical" first action.

- "Panic" is not typical of fire behavior, despite popular images.

- Males and older age groups are somewhat less likely to attempt to leave the building. There have been reports that older people may "freeze" and refuse to exit, but this is mainly based on anecdotal observations of institutionalized populations and does not appear supported by major studies of home fires.

- Egress takes longer for the elderly, and particularly in a fire situation, the old may suffer difficulties due to age-related declines in vision, audition, olfaction, decreased strength and mobility, slower walking speeds, greater vulnerability to fire products, less efficient information processing, poorer balance, and disorientation.

- There is a surprisingly high rate of re-entry into a building after successful egress, even though the fire is still in progress.

- Little information was found relating age to pre-fire incident factors, such as fire safety attitudes and practices.

RESIDENTIAL SMOKE DETECTORS

Home smoke detector use has increased dramatically since the early 1970's; recent studies have shown that four of five homes in the U.S. have at least one smoke detector (Hall, 1988). Although an estimated one third of these detectors are not operational, the risk of death if a fire occurs in homes with smoke detectors is approximately half that in homes without detectors.

Single-station, battery-operated detectors are perhaps the most common for home use, but today's technology provides many options, ranging from basic units to complex systems. There are two basic sensing methods used for detecting smoke; photoelectric and ionization. Ionization detectors are generally more sensitive to the invisible products of flaming fires, while photoelectric detectors are generally more sensitive to smoldering fires and the visible products of combustion. Normally these sensing methods are used independently. However, combining both methods provides a higher degree of accuracy in identifying a fire. Some detectors have special circuitry or features which can help to lessen the effects of various false alarm sources such as power transients and insects.

Fire protection codes now normally require that new homes have AC-powered detector systems. This frees the owner from having to test and maintain the batteries; one factor in the high proportion of non-operational units is that people remove batteries to replace them or because of the annoyance of "chirping" low-battery signals, or due to nuisance alarms. However, power failures could render the AC models ineffective so some units include a battery backup power source. Newer DC models provide innovative means of testing and easing maintenance problems of consumers. Most units provide a simple test button (in various sizes and shapes), while some allow the use of a magnet, flashlight, hand-clapping or artificial aerosol smoke to test the detectors' operation. Some manufacturers have resorted to hinged or twist-on covers or removable battery drawers to make the job of battery replacement less problematic.

The minimum intensity of smoke detector signals is defined by fire codes to be 85 dBA at 10 feet, and most manufacturers claim to adhere to this level. Some manufacturers, however, offer systems with higher level output to aid in user detection. The sound frequency of the signals used by most manufacturers is bimodal in the 2000 and 4000 Hz areas of the spectrum, meaning that they are fairly high and may be difficult for some elderly people to hear. Detection of low batteries can be a problem as well, especially for older consumers, since the "chirping" indicator may be difficult to hear, and in some cases, not understood when it is detected.

Although auditory signals are the most common, secondary signalling may be employed to aid individuals with sensory deficiencies. Such signalling is often used for the visually impaired. Visual strobes have demonstrated alerting capabilities for deaf people comparable to that provided hearing people by audible smoke alarms. Sensory stimulation through the use of vibrators is also possible, though often more troublesome to install and maintain effectiveness. Stimulation using fans may also be possible, but the potential to accelerate a fire's progress with this method makes it less feasible.

Networking and remote signalling of smoke alarm signals is also common. The ability for one detector to trigger the actuation of other detectors or a remote signalling device may add significantly to the time available for egress during a fire situation. Early detection is the key to safe escape.

Escape lights represent another feature available with some systems. Since sometime during a fire situation escape may be necessary, such a light can facilitate the maneuver by illuminating hazards along the escape path. Depending on the mounting location, some of these systems may be hampered by the location of smoke and fire products, which often accumulate near the ceiling and floor. Some products attempt to overcome this problem with lights mounted in lower positions (e.g., on a door) or luminescent markings along the egress route.

Among the most elaborate systems available to consumers are those which provide constant monitoring by an outside party. These systems are generally part of a more extensive personal emergency response system which may include security or medical emergency components. These systems normally require subscription to a monitoring service which is responsible for contacting the relevant authorities in case of an emergency; direct connection to the fire department is not permitted. The subscription rates vary among vendors. Although a 24-hour monitoring service provides greater safety coverage, the costs (for equipment, installation, and monitoring service) far exceed those associated with single-station smoke alarms.

FOCUS GROUP DISCUSSIONS

In order to get feedback and opinion on fire safety behaviors and products from older residents themselves, a series of focus group discussions were held. The focus group technique provides a method for obtaining qualitative information from people in an informal and non-threatening environment, where perceptions and opinions can be shared in a group setting. The question path for the discussions began with people's general perceptions of fire threat and the actions they have taken, then focused more specifically on views concerning smoke detectors, and finally dealt with specific product features. Participants also filled out a brief questionnaire form at the end of the session.

Three discussion groups were held, totalling 23 paid participants. All were homeowners in the greater Washington, DC area, recruited through senior activity centers, churches, and senior newsletters. The mean age was 74 (range of 64-87). Six reported having been involved in a fire at some time in their lives, but none within the past 16 years.

The homeowner focus groups proved to be a rich source of information regarding fire safety attitudes and practices, knowledge of smoke detector products and features, and consumer response to new product features and concepts. These older people showed a very clear concern for fire safety, which expressed itself in a variety of actions. However, most of the concerns and behaviors related to ignition sources, rather than fire awareness and escape. Examples included unplugging televisions and appliances when away, prompt removal of newspapers, non-use or very cautious use of heating pads and electric blankets, delay in emptying ashtrays (no embers), and so forth. Some expressed the opinion that an increased concern with fire safety came with advanced age.

The participants were aware of the importance of smoke detectors, and all but one of the homes had them (mean of 1.9 detectors per house). These were normally purchased by the homeowners, who had resided in their current homes for an average of 34 years, and none reported that their house was equipped with smoke detectors when purchased. When asked (in the questionnaire) to rank the usefulness of various fire safety items (fire extinguisher, sprinkler system, escape plan, etc.), smoke detectors clearly emerged as the most important. Despite the group's appreciation and use of smoke detectors, there was not good recognition of the need for proper location of the device. They were not aware of normal recommendations to locate detectors near sleeping areas and on all floors of the home, at or near ceiling level. They favored locations near likely sources of ignition (e.g., kitchen, furnace). False/nuisance alarms were a consensus problem, and this may be at least partly due to poor detector location. Several people had difficulties with units near the kitchen, and they interpreted this as a problem of the particular alarm being "too sensitive," rather than a problem of location.

People coped with false/nuisance alarms in various ways, including disabling the units through battery removal.

Some current smoke detectors have a false alarm cancel feature, which allows a transient change in sensitivity to override an unwanted alarm. If any participant's had this feature, they were unaware of it. The groups were generally surprised to learn of this feature and felt it was a good idea.

The discussions also identified problems in smoke detector testing and maintenance. These involved both reaching the product at ceiling level and manipulating the covers and batteries. However, there was also a general distrust of AC-powered smoke detectors, because of doubts about this power source in case of a power outage during a fire; battery operated units were clearly preferred. A wide range of additional issues were raised during the focus group sessions, and more detail on these can be found in Lerner et al. (1990). Many of the findings are reflected in the recommended smoke detector design features and functions, presented below.

PRODUCT-RELATED PROBLEMS AND DESIGN RECOMMENDATIONS

Many of the problems related to smoke detectors can be thought of as simple product design deficiencies based on a failure to include the extremes of the user population. The very location of the systems, e.g., on ceilings or in stair wells, provides a hazard potential for older consumers. Installation, let alone maintenance and testing, can put seniors at risk for dangerous falls. Characteristics and credibility of the alarm signals can also be a problem, since false alarm frequency and insufficient intensity may cause valid alarms to be missed or intentionally ignored. These types of problems are especially critical for the elderly since decreased mobility will likely put them at a disadvantage for timely egress. Also, though more difficult to document, the elderly victims of fire may have a greater instance of premature re-entry than the population at large. This, too, is an area which might be improved by design modifications to smoke alarms.

Based on all of the preceding findings and analysis, a smoke detector designed specifically to meet the needs of older adults should provide the features and functions listed below. Note that none of these are necessarily expensive features, and many of them already are available on some current products.

1. All controls and batteries should be easily accessible without climbing and operable even by those with limited hand strength or dexterity.

2. The product should be easily and inexpensively installed and not intimidating to older buyers.

3. The trouble indicator should be readily detectable and understandable. There should be some indicator of system failure even after the batteries have expired.

4. There should be a positive indicator of system status when the system is properly working.

5. The system should be self-testing.

6. The product should include an alarm-cancel feature, with the presence and function of control clearly apparent to the user.

7. The acoustic signal should be detectable by the normal range of older adults, including those suffering presbycusis, under all reasonably expected home listening conditions.

8. The alarm should be remotely distributed throughout the home, and not restricted to only the site of the activated detector (unless the home is small enough to be properly served by a single smoke detector).

9. Presuming activation of all alarms in a multi-story home, there should ideally be annunciation of the location of the active detector.

10. False alarm rates should be minimized, through product features and choice of location, to enhance alarm credibility.

11. Clear and simple guidance should be provided in selecting the appropriate number and location of smoke detector installation sites.

12. The system should be modular and flexible, so that the number of communicating smoke detectors, and other peripheral units such as remote alarms, can be tailored to the needs of the particular home.

13. The product should include the potential to be integrated with other safety and security products, in order to take advantage of the more costly features and technologies provided (such as automatic alerting of a 24 hour monitoring service).

14. The system should include aids to egress and rescue (e.g., detachable flashlight with an audible horn).

15. The product should address the problem of re-entry after escape, by printed or voice messages if there are no other means.

16. The product should be battery powered (or include obvious battery backup), because older consumer acceptance of AC-powered units appears questionable.

17. Aesthetic concerns may be very important to many older homeowners, and all aspects of the system must be designed for appeal, or at least acceptance, by this age group.

CONCLUSION

Fire casualty statistics make it clear that improved fire protection is necessary for older people living in their own private residences. The smoke detector is the single most important and effective home fire safety product. Significant improvements could be made to smoke detectors so that they are more compatible with the capabilities and needs of the elderly. These design changes are not necessarily costly and some are available on certain products now. This paper has described some of the issues in fire safety for older people, and has provided recommendations for a number of specific design improvements.

ACKNOWLEDGEMENTS

This research was funded by the National Institute on Aging, under grant 1 R43 AG08151-01.

REFERENCES

Cote, A.E. (1986). Assessing the magnitude of the fire problem. In A.E. Cote and J.L. Linville (Eds.), *Fire Protection Handbook*, Quincy, MA: National Fire Protection Association.

Gulaid, J.A., Sattin, R.W., and Waxweiller, R.J. (1988). Deaths from residential fires, 1978-1984. *Morbidity and Mortality Weekly Report, 37(SS-1)*, 39-45.

Lerner, N., Huey, R., Morrison, M., and Mowrer, F. (1990). *Residential Smoke Alarms and Fire Safety for Older Adults*. Report under Grant No. 1 R43 AG08151-01. Bethesda, MD: The National Institute on Aging, National Institutes of Health.

OLDER DRIVERS' VISIBILITY AND COMFORT IN NIGHT DRIVING: VEHICLE DESIGN FACTORS

Rudolf G. Mortimer
University of Illinois at Urbana-Champaign

ABSTRACT

Older persons are a growing proportion in the population, among drivers and those involved in traffic accidents. Changes in visual abilities of older persons are pertinent to night driving in which they need greater brightness contrast to see and minimum glare. Vehicle headlighting and related factors are reviewed which affect visibility and comfort in night driving. Older drivers, in particular, would be aided at night by: increasing the reflectivity of objects, limiting the mounting height of headlamps, appropriate reflectivities of mirrors for control of glare, automatic headlamp alignment, automatic headlamp cleaning and beam patterns that emphasize glare control.

INTRODUCTION

Relatively little is known about the effects of aging on driving performance. But a substantial amount of information is available concerning the effects of aging upon the human visual system. For example, the maximum area of the iris of the eye of persons aged 60 years is about half that of those aged 20, which results in less light reaching the retinae of older persons (Weale, 1963). It is for this reason, among others, that older persons require much more light to see than younger ones. Wolf (1960) showed that the contrast threshold of persons aged 20 is about one tenth that of persons aged 60 and about one thirtieth that of 80 year old persons.

Older persons also have much less tolerance for glare (Pulling et al. 1980) and are more susceptible to disability glare (e.g. Christie & Fisher, 1966) which reduces their visibility at night.

Notwithstanding these limitations in their visibility, which would be particularly notable in night driving conditions, a recent analysis (Mortimer & Fell, 1988) found that the fatal nighttime crash involvement of drivers aged 65 or more, as a group, was lower than that of drivers aged 24 or less.

THE HEADLIGHTING SYSTEM

The headlighting beam pattern that is used in the United States and some other countries is similar in many respects to that used in European countries. The main difference is that the U.S. meeting beam permits greater glaring intensities to oncoming drivers than the European beam. In other respects the two beams are quite similar and function about equally well. The main problem, in headlighting development, is of course to be able to provide illumination on the roadway and its surroundings while minimizing glare.

All motor vehicle headlamps in use in the United States must meet certain photometric criteria. The standards for motorcycles and mopeds are not the same as those for automobiles and trucks.

Computer simulations were made of hypothetical meetings on a straight, two-lane road between a truck, on which the low-beam headlamps were mounted at a height of 48 inches above the road, and a passenger car on which the same type of headlamps were mounted at a height of 24 inches above the road, with each one meeting another car. The truck driver had an initial visibility, before glare had an effect, of approximately 330 feet (101 m) for a 12% reflectance object on the right side of the road while the car driver had about 280 feet (85 m). For the object located on the centerline the truck driver had an initial visibility of about 250 feet (76 m) and the car driver 200 feet (61 m). The minimum visibilities during the meetings, due to glare, were about 300 feet (91 m) and 260 feet (79 m) for the target on the right of the lane, or a reduction of about 10%, and 185 feet (56 m) and 125 feet (38 m) for the object on the centerline, a reduction of 25% and 38%, for the truck and car drivers, respectively. The greater eye height of the truck driver reduces the glare to which the truck driver is exposed. The reason why the initial visibility is better for the truck driver than the car driver is due almost entirely to the higher mounting height of the headlamps rather than to the difference in eye heights of the drivers in the two vehicles. The reason that the visibility for both drivers is less when the target is on the centerline (i.e. on the left of the lane) is that the conventional low beam headlamps provide far less illumination to the left of the vehicle than to the right in order to minimize glare to oncoming drivers. The glare is substantially greater when the drivers are looking along the centerline than when they are looking for an object along the right edge. This is to be expected because when they are looking along the centerline they are looking closer to the headlamps of the oncoming vehicle and, hence, the glare effect is greater.

Mounting Height of Headlamps

As shown above, higher mounting heights of headlamps produce greater visibility distances and, therefore, some minimum mounting height must be specified. At the present time, the headlighting regulations in the U.S. indicate that headlamps may be mounted with their centers no less than 22 inches (0.56 m) nor more than 54 inches (1.4 m) above the pavement. The current minimum should not be reduced in the interest of preserving visibility.

Higher headlamp mounting heights provide greater visibility distances for the driver behind the headlamps, but they have a detrimental effect upon the drivers of preceding vehicles. As the headlamp mounting height is raised, the amount of illumination falling on the interior and exterior mirrors of vehicles ahead of the headlamps is also increased, thereby increasing both disability and discomfort glare in night driving. Table 1 (Mortimer, 1974) shows the illumination in an automobile driver's eyes, due to the headlamps of a vehicle which is following in the same lane at a distance of 100 feet. The vehicle has one interior mirror and one exterior mirror on the left side of the driver.

A low-beam headlamp on the following vehicle mounted at a height of 24 inches (0.6 m), properly aimed, will provide 0.13 foot-candles at the eyes of the preceding driver. For correctly aimed headlamps mounted at 48 inches (1.22 m), such as on a large truck, the illumination in the driver's eyes of the preceding vehicle rises to 0.45 foot-candles. Thus, increasing the mounting height of the headlamps considerably increases the glare illumination. Research has shown that the glare caused by light reflected in the interior and exterior mirrors can frequently exceed that caused by the headlamps of oncoming vehicles (Miller et al., 1974). Since glare is a particular problem for older drivers, methods should be incorporated in the design of vehicles to minimize this effect. One approach is to limit the maximum mounting height of headlamps on large trucks as well as other vehicles to about 30 inches (0.76 m). Table 1 shows that

Table 1

Illumination in Driver's Eyes by Low Beams Reflected in Interior and Exterior Mirrors from a Vehicle at 100 Feet to the Rear, by Mounting Height.

Headlamp Mounting Height (in)	Foot-Candles at Eyes
24	0.13
30	0.16
48	0.45

this produces about 0.16 f-c from following low beams at 100 feet, or only about one-third that from lamps mounted at 48 inches, as is common on many large trucks in the U.S.

A recent survey of the mounting height of headlamps on European trucks and buses indicated that mounting heights varied between 23-40 inches (0.58-1.02 m). Therefore, there appears to be no technical problem in reducing the maximum mounting height of headlamps to similar heights as on automobiles.

Glare from Rearview Mirrors

As just pointed out the glare from the headlamps of following vehicles reflected in rear view mirrors is often substantial and is present for long periods of time in dense traffic and frequently exceeds the levels occurring from oncoming traffic. It has already been suggested that the mounting height of headlamps on following vehicles should be reduced in order to reduce the glare illumination from mirrors. However, the mirror system itself can also be changed. Already many vehicles are equipped with interior mirrors which have a day and night position. In the night position the reflectivity is only about 4% and provides substantial relief from headlamp glare of following vehicles. However, many drivers feel that the 4% reflectivity is too low and does not provide them with enough other information as to what is happening to the rear of their vehicle. Hence, they tend not to use the night position. Research (Mansour, 1971; Olson et al, 1974) has suggested that the lower reflectivity of the interior mirror should be approximately 10-20%.

The exterior mirror on motor vehicles has a reflectivity of about 55% and does not yet have a night dimming feature. Headlamps of vehicles in the passing lane and those directly behind the exterior mirror are a substantial contributor to disability and discomfort glare in night driving. Steps need to be taken to provide driver control or automatic control of the reflectivity of exterior mirrors in order to control headlight glare. It appears that the best compromise is a reflectivity of 30-40% for a mirror without dimming capability (Mansour, 1971; Olson et al., 1974).

Cleanliness of Headlamps

Headlamps are normally mounted at the front of vehicles and are subject to the spray from passing vehicles so that they collect road dirt, snow and slush on their lenses. Dirt can substantially reduce the illumination provided by headlamps and change the distribution of the light to actually increase glare. Drivers tend not to become aware of this degradation in the light output from headlamps until 60% or more of the light output is lost (Rumar, 1970). The greatest collection of dirt will occur under inclement weather conditions when the roadway is wet due to the spray from following vehicles so that under the very conditions when visibility is poorest headlamp output is most degraded.

For these reasons it is important that headlamp cleaning systems are incorporated into motor vehicles. This could be done in various ways but one suggestion would be that, whenever the headlamps are lighted and the driver actuates the windshield washing system, the headlamp washing system is also activated.

Headlamp Aim

The biggest single factor constraining the improvement of visibility provided by current headlamp meeting beams is the aim of the headlamps in actual practice. Variations in headlamp aim of as little as 0.25 degrees can cause relatively large changes in beam effectiveness in terms of illumination as well as glare. There are many factors that lead to improper aim including initial improper aim when the lamp is mounted on the vehicle, although this a relatively small factor because of automated methods used at the factory; misaim induced when headlamps are replaced in service facilities because of improper use or failure to use aiming equipment; and static and dynamic factors associated with vehicle suspension systems.

Manual alignment of the headlamp in the vertical direction has been proposed to allow the driver to account for variations in vehicle loading. In addition, automatic headlamp alignment systems have been developed and tested to account for dynamic changes in vehicle pitch, but none have been put into production. There is little opportunity for further improving the headlamp beam until better control of the aim of the headlamp can be achieved under both static and dynamic variations in vehicle pitching motions.

Even if such automatic alignment systems are used and correct initial alignment is assured, glare will still accrue to drivers because of changes in roadway vertical and horizontal alignment, but, at least on straight sections of roadway, glare levels would be within design tolerances.

Beam Shaping

Until better control of headlamp aim can be achieved, no practical benefit can be obtained by changes to the low beam patterns that are currently in use. However, contemplated changes in the motor vehicle lighting regulations (USDOT, 1989) will enable manufacturers to use any combination of lamps of different sizes, possibly each producing a different beam pattern, to provide an overall beam pattern similar to the one currently in use. The advantage of such a regulation is to free manufacturers from the present constraints of using lamps having particular shapes and dimensions so that the front area of motor vehicles can be reduced for better aerodynamics and styling. On the other hand, the danger is that as more individual lamps are used to create a beam pattern they will each offer the opportunity of being out of tolerance and, therefore, the result could be a greater degree of misaim in the motor vehicle population than is true at this time. There would also be inherently greater cost of the headlamps of multi-lamp systems and probably greater difficulty in replacing and properly reaiming such lamps when they need to be replaced because of failure or damage, and there is greater likelihood that one lamp that fails out of a number of headlamps on a vehicle may not be noticed or replaced by owners as quickly.

Computerization in motor vehicles may allow an advantage of multi-lamp configurations to be realized. For example, substantial research has been done on a so-called "mid-beam" which would be particularly suitable for use on straight sections of divided highway (Mortimer, 1974). A disadvantage of that concept was that drivers would have to learn to use three beams instead of two, and there were some substantial issues raised as to the mode of switching between the beams and whether drivers would correctly use the beams under the appropriate conditions. If beam selection could be accomplished automatically, a three-beam or a multiple-beam headlighting system would become much more feasible. This would be a way in which greater roadway illumination could be provided while at the same time automatically controlling glare levels to other road users to produce an overall benefit in visibility in night driving.

Non-Vehicle Factors

Because of the inherent limitations in current headlighting systems, it is likely that the single most effective approach to improve night driving visibility lies in increasing the reflectivity of important objects in the night driving environment. Doubling the reflectance of an object is relatively simple to do and produces a doubling of its brightness. To achieve a doubling in brightness by headlighting requires doubling the beam intensity, which is very difficult to do with a low-beam headlamp. Thus, improved techniques of roadway delineation by the use of more reflective materials is a step in the direction that should be taken. Similarly, using reflectorized materials on objects on or about the road including lane markers, or to outline vehicles that are otherwise difficult to see, such as large trucks or bicycles, and on pedestrians, will greatly improve visibility for older drivers as well as all others. Such techniques are within the current state of the art and only require wider and uniform application to produce large increases in visibility.

CONCLUSIONS

There are a number of steps that can be taken by which improvements can be obtained in night driving visibility and comfort particularly for older drivers. The maximum mounting location for headlamps should be limited to about 30 inches while not permitted to be reduced below current levels; rearview mirrors should be provided with lower reflectivities for use at night than used in daytime to counteract the headlamp glare of following vehicles; headlamp cleaning systems should be provided to automatically operate with windshield washing systems to maximize the light output and retain the integrity of the beam pattern; improvements are also feasible in initial headlamp aim and in aim retention by automatic aim and alignment systems which in turn would permit some small changes in the meeting beam patterns; ultimately continuously variable beam patterns, computer controlled, could be incorporated to maximize the illumination at all times while controlling glare to other road users; innovative techniques of headlighting can still be developed which eliminate or reduce the glare element; and finally, non-vehicle factors probably offer the greatest potential for increasing the visibility of the roadway and objects around it without any increase in current glare levels, which may be the most expedient way to improve the night driving environment short of providing fixed illumination on the nation's roads and highways.

As the number of older drivers in the population grows, their mobility will be restricted, as it currently appears to be on a voluntary basis, because of the undesirable aspects associated with night driving. Efforts need to be made to implement measures such as are suggested and others yet to be developed in order to improve the night driving environment for better visibility, glare reduction and increases in overall night driving comfort and safety.

REFERENCES

Christie, M.A. & Fisher, A.J. The effect of glare from streetlighting lanterns on vision of drivers of different ages. Transactions of The Illuminating Engineering Society (London), 31, No. 4, 93-107, 1966.

Mansour, T.M. Driving evaluation study of rear view mirror reflectance levels. Society of Automotive Engineers Report 710542, 1971.

Miller, N.D., Baumgardner, D. & Mortimer, R.G. An evaluation of glare in nighttime driving caused by headlights reflected from rearview mirrors. Society of Automotive Engineers, Report 74096, 1974.

Mortimer, R.G. Some effects of road, truck and headlamp characteristics on visibility and glare in night driving. Society of Automotive Engineers, Report 740615, 1974.

Mortimer, R.G. & Fell, J.C. Older drivers: their night fatal crash involvement and risk. Accid. Anal. & Prev., 21, No. 3, 273-282, 1989.

Olson, P.L., Jorgeson, C.M. & Mortimer, R.G. Effects of rearview mirror reflectivity on drivers' comfort and performance. Report UM-HSRI-HF-74-22, University of Michigan, 1974.

Pulling, N.H., Wolf, E., Sturgis, S., Vaillancourt, D.R. and Dolliver, J.J. Headlight glare resistance and driver age. Human Factors, 22, No.1, 10-102, 1980.

Rumar, K. Dirty headlights - Frequency and visibility distances. Uppsala University, Sweden, 1970.

USDOT. Notice of proposed rulemaking, U.S. Department of Transportation, Federal Register, 54, No. 88, May 1989.

Weale, R.A. The Aging Eye. New York: Harper & Row, 1963.

Wolf, E. Glare and age. Archives of Ophthalmology, 64, 502-514, 1960.

AGE AND THE PERCEPTION OF A MODULATING TRAFFIC SIGNAL LIGHT IN A FIELD LOCATION

Lawrence T. Guzy and Nancy Pena-Reynolds
State University of New York at Oneonta
Richard D. Brugger
Industrial Control Associates, Erie, PA.
Herschel W. Leibowitz
The Pennsylvania State University

ABSTRACT

Young, middle, and older drivers were motored towards a traffic signal face and were required to report whether the illuminated signal lens was modulating. No significant differences were found among the three age groups in the distance at which modulation was first reported nor in the change of the modulation pattern as they approached the signal face. The modulating of the illuminated green lens was perceived significantly further away than was the red lens. These distances increased during trials associated with inclement weather. Seventy-four percent of the observers indicated that the modulation attracted their attention. Of these, 64% reported a preference for the use of the modulating light in the traffic signals as compared with the standard light source. Implications for improving safety at intersections for the older driver are discussed.

INTRODUCTION

At intersections controlled by a traffic signal light the ability to detect the presence of a signal face and respond appropriately is a critical activity in a vehicular oriented environment. However, on occasion, drivers and pedestrians may respond inappropriately. These situations may be the result of risk taking, inattention, distraction, or looking but failing to see (Hills, 1980), and these are compounded by competing environmental arrays that reduce the conspicuity of the illuminated signal lens.

The use of flashing lights has been successfully used to improve the conspicuity of two traffic control devices. Strybel & Nassi (1986) improved the recognition of a lane reversing signal under conditions of sun glare by placing a rotating beacon above the signal; and a strobe light strip has been attached to the face of the red lens and pulses when the red signal is illuminated (MUTCD, 1988). Although effective, a potential problem exists. These alerting devices are tangential to the signals and their failure could result in misidentification of the signal lights from observers who are familiar with their operation.

Recently, Brugger (1989) designed a dual-filament traffic signal light bulb where the warning was inherent in the signal's normal lighting pattern. By alternately energizing each filament, a modulation pattern was generated which observers describe as a "flickering or shimmering" light. Preliminary research with observers who were not informed of the shimmering light responded appropriately to the red and green signal lights. A rate of four hertz was identified as this rate produced the maximum distance for viewing modulation and was not confused with a "blinking" traffic signal light. The observers independently reported that the modulation pattern was qualitatively different from what they have previously experienced. Guzy, Suib, Leibowitz, and Brugger (1990) examined the effects of this light source in a traffic signal face under conditions of simulated sun backlighting. No errors involving recognition and driving were found with the modulating light source in the red or green lenses. Several errors were found with the standard traffic signal bulb. Here, the drivers incorrectly identified the illuminated lens and either stopped for a green light or passed a red light.

As compared with younger drivers, older drivers may benefit more from the modulating light source placed in the signal head. Whereas accidents involving younger drivers are attributed to such factors as speeding and driving while intoxicated, the older driver fails to heed signs, yield the right of way, or turn properly (Planek, 1973; Huston & Janke, 1986). The modulating light bulb may improve signal recognition and reduce such driving errors. The present research was designed to examine whether age related visual changes affect the recognition, distance, and quality of the modulation effect. As older drivers prefer daylight hours for driving, the research was conducted under photopic conditions.

Method

Subjects. Twenty-eight licensed men and women drivers were paid $5.00 for their participation. Subjects formed three age groups: a) young-ages 20 to 32, n = 9, mean age of 22 years, b) mid-ages 38 to 54, n = 9, mean age of 46 years, and c) older-ages 59 to 80, n = 10, mean age of 66 years. Nine men were found to be red-green color deficient. Each age group had at least two red-green color deficient observers.

Visual Screening. Visual acuity was determined with a Bausch and Lomb Orthorater using only the far visual acuity plates. Color deficiencies were identified with the Dvorine Pseudo-Isochromatic Plates color vision test.

Procedure. Observers sat on the passenger side of a full-sized vehicle driven at a speed of approximately 26 km. toward a three light traffic signal face. The bottom of the signal face was 4.92 m. from the road surface. Green and red lenses, 30.5 cm. dia., of the traffic light were illuminated by either a 150 watt Hytron bulb or a dual filament bulb modulating at a rate of 4 hz. Bulbs were matched for brightness.

Subjects began viewing the traffic light from a distance of 262 m. and were required to identify the color and amount of flicker, if any, using a five point scale, and any other unusual aspects of the traffic light. Distances from the light were recorded with a NuMetrics Roadstar Model 20 computer connected to the vehicle's transmission.

Eight different lighting conditions were administered twice for a total of sixteen trials. Each colored lens was placed in the top position of the traffic signal face for eight trials and in the bottom for the remaining 8 trials. The modulating and standard bulbs appeared equally often in each of these positions. Trials were block randomized. A post-test interview focused on observers impressions of the modulating light bulb.

RESULTS

The modulating effect of the dual-filament bulb produced reports of both quantitative and qualitative changes as subjects approached the traffic signal face. No significant difference were found among the three age groups. Within each age group large intersubject variability was found.

With the Orthorater, visual performance on the far distance plates decrease with increasing age. However, the perception of the modulation pattern showed no relationship with Orthorater performance.

The modulating pattern in the green and red lenses produced differential distance effects. The green lens was perceived as showing "some" modulation significantly further away (171 m., sd = 38.8) than was the red lens (130.8 m., sd = 53.5), $F(1,54)=16.2$, $p<.001$). A change in the reported modulation pattern to "more" followed a similar pattern, $\underline{F}(1,51)=8.6$, $p<.05$.

During inclement weather, i.e., rain, fog, and overcast skies, the modulation pattern was seen further away than under sunny conditions. A significant inverse relationship was found between the report of "some" modulation and weather for both the green and red lenses, $r=-.47$, $df=26$, $p<.01$ and $r=-.35$, $df=26$, $p<.05$, respectively.

A post-test interview showed that 64% of the subjects preferred the modulating light over the standard light. Two observers reported that the modulating light "attracted their attention" and found it "annoying".

Of the nine red/green color deficient participants, none incorrectly identified the color of the illuminated lens regardless of the lens' position in the traffic signal face. All of these observers reported a preference for the modulating red lens while the normal color visioned subjects were mixed with respect to their preference.

DISCUSSION

The modulation effect showed no differences among the three age groups in either the range of modulation or in the qualitative changes associated with the modulation pattern.

The large variabilities found within each age group may be attributed to the difficulty in conveying precise terminology to characterize the modulating light pattern and any subsequent changes as viewing distance decreased. When asked to supply their own descriptors to define the effect, observers showed little consistency. The most commonly elicited terms included "shimmering", "flickering", and "modulating".

Changes in the weather from photopic to mesopic conditions increased the range of influence of the modulating light. Regardless of the weather conditions, the overall range of the modulating light was limited. However, perception of the modulation would allow amply time for drivers to initiate an approriate response under both normal and degraded viewing conditions.

In the present study, with uncompromised viewing conditions, 64% of the subjects indicated a preference for the modulating light in the traffic signal face. In a previous study which used a degraded viewing condition involving simulated sun backlighting of the traffic signal lenses, twelve of fourteen subjects who reported seeing the modulation were unanimous in their preference for the modulation. Of the two subjects who could not recall seeing the modulation, one committed several recognition and driving errors with the standard light source, but not with the modulating light (Guzy, et al., 1990).

The difference in preferences may be attributable to a difference in procedures. In the earlier study, subjects were required to drive a three-wheeled vehicle along a prescribed route, identify signs placed along a simulated roadway, report the color of the illuminated lens, and at a simulated intersection respond accordingly, i.e., drive through or stop. The procedure in this study required the subject to identify the presence of modulation and to report any changes. Several of the passengers indicated that they had fixated on the illuminated lens for the entire duration of the trial (approx. 35 seconds) to the exclusion of all other visual information. Without these distractor tasks, the singular task of viewing the illuminated signal may account for the reduced preference and in two cases, annoyance.

Given the difficulties associated with the older drivers, eg., heeding signs and responding to information in the periphery (Kosnik, Sekuler, & Kline, 1990), the use of the modulating light may possibly improve intersection safety by enhancing the detection of the traffic signal face and recognition of the illuminated lens.

ACKNOWLEDGMENTS

The authors wish to thank Joyce A. Blake-Guzy for her invaluable assistance in data collection and to the Pennsylvania Transportation Institute for the use of their automotive test track for preliminary testing.

The research was supported by a grant from the University Transportation Research Center, Federal Region 2.

For specific information on modulating bulb characteristics contact Richard Brugger, P.E., Industrial Control Associates, P.O. Box 82,Erie, Pa. 16512.

REFERENCES

Brugger, R. D. (1989). Traffic Signal. *United States Patent Office*. Pat. No. 4,799,060. Patent Date 1/17/89.

Guzy, L.T., Suib, S., Leibowitz, H.W., and Brugger, R. D. (1990). Improving the conspicuity of traffic signal lights. "Work in Progress". *Human Factors Society Conference*. Orlando, FL.

Hills, B. (1980). Vision, visibility, and perception in driving. *Perception*. *9*. 183-216.

Kosnik, W. D., Sekuler, R., and Kline, D. W. (1990). Self-reported visual problems of older drivers. *Human Factors*, *32*, 597-608.

Huston, R. and Janke, M. (1986). Senior driver facts. *Tech. Report CAL-DMV-RSS-86-82*. Sacramento, CA. Department of Motor Vehicle.

Manual on Uniform Traffic Control Devices (1988). Government Printing Office, Superintendent of Documents, Washington, D.C. 20402.

Planek, T. W. (1973). The aging driver in today's traffic: A critical review. In *Aging and highway safety: The elderly in a mobile society. North Carolina Symposium on Highway Safety. V.7.* Chapel Hill: University of North Carolina, Safety Research Center.

Strybel, T. Z. and Nassi, R. (1986). Daylight conspicuity of reversible-lane signal systems. Human Factors. *28*, 83-89.

Age Differences in the Legibility of Symbol Highway Signs as a Function of Luminance and Glare Level: A Preliminary Report

by

Frank Schieber
Department of Psychology
University of South Dakota
Vermillion, SD 57069
U.S.A.

Donald W. Kline
Department of Psychology
University of Calgary
Calgary, Alberta T2N 1N4
CANADA

Three experiments were conducted to investigate the effects of adult aging upon the legibility of simulated symbol highway signs. Each experiment employed a different set of lighting conditions: (1) daytime luminance, (2) nighttime luminance, and (3) nighttime luminance with glare. Young (ages 18-25) and middle-aged (ages 40-55) observers demonstrated small reductions in legibility when luminance was reduced from daytime to nighttime levels. However, older (ages 65-79) observers demonstrated marked losses in legibility distance with reductions in sign luminance. The introduction of a glare source (equivalent to approaching automobile headlights at 30 m) reduced sign legibility distance for the older observers but had no deleterious effects upon their young and middle-aged counterparts. The relative magnitude of the observed age, luminance and glare effects appeared to be equivalent across all signs examined.

INTRODUCTION

In 1988, the Transportation Research Board (TRB) published the two-volume Special Report 218 which was titled *Transportation in an Aging Society: Improving the Mobility and Safety of Older Persons* (TRB, 1988). The reviews and recommendations contained in Special Report 218 have served as a *blue print* for guiding research and development on older driver issues at both the Federal Highway Administration (FHWA) and the National Highway and Traffic Safety Administration (NHTSA). One of those recommendations called for the study of how symbol (i.e., "pictorial") highway signs might be improved to accommodate the visual deficits which accompany advanced adult age. Pursuant to this recommendation, the FHWA sponsored a comprehensive research initiative known as the *Symbol Signing for Older Drivers* project. This large-scale, multi-site project examined the nature of age-differences in symbol sign visibility and comprehension, explored the factors mediating these age-differences and developed computer-based techniques for optimizing the visibility of symbol highway signs. Data collection and analyses have just recently been completed on the project (see Dewar, Kline, Mark and Schieber, 1994). Currently, numerous reports are under preparation for sharing the results of this work with the research and engineering community. This report presents preliminary findings from one segment of the project; namely, laboratory research examining age-differences in the maximum legibility distance of symbol highway signs as a function of daytime vs. nighttime viewing conditions.

The visual functioning of older adults - as quantified via the contrast sensitivity function - is highly dependent upon prevailing lighting conditions. For example, age-related losses in visual sensitivity are relatively minor given luminance levels which approximate daytime viewing conditions. However, the magnitude of these age-related losses is greatly exacerbated at low luminance levels and in the presence of a glare source - conditions which often prevail while attempting to read highway signs at night. For these reasons, age-differences in the ability to read symbol highway signs were examined under three widely varying lighting conditions: (1) daytime - i.e., high photopic luminance, (2) nighttime - i.e., low photopic luminance, and (3) nighttime luminance with the addition of a narrow angle glare source.

METHOD

Subjects. Each experiment (one for each of the 3 lighting conditions) employed three age groups consisting of: 12 young (ages 18-25), 12 middle-aged (ages 40-55) and 18 older (ages 65-79) adult volunteers. All subjects were community-residents, held a valid driver's license and were screened for good general and ocular health.

Apparatus and stimuli. Test stimuli consisted of 18 modified or redesigned symbol highway signs selected to represent the range of legibility distances found in a previous study of the 85 symbol signs depicted in the *Manual of Uniform Traffic Controls and Devices*. These signs were digitized, scaled and stored electronically using an Apple Macintosh IIci computer system. The computer-graphic representations of the signs were presented - at programmablly variable sizes - on an Apple 13-inch color monitor viewed at a distance of 5.5 m. The variable size signs were presented in the center of the monitor against a uniform white background maintained at a high photopic luminance of 77 cd/m2 (i.e., daytime level). For both the nighttime and nighttime with glare conditions, the luminance of the screen was attenuated to 5 cd/m2 with a large optical filter. The glare source was mounted 5 deg to the left of sign stimuli and consisted of a single 40 watt high-diffusion incandescent lamp which subtended a visual angle of 0.6 degrees. The lamp was mounted in a housing which prevented the extraocular mixing of the glare light with that emitted from the stimulus display monitor. At the entrance pupil to the eye, the illuminance of the glare source was 8 lux - approximating the intensity of a pair of automobile headlamps viewed at a distance of 30 m.

Procedure. After preliminary assessments of visual acuity and contrast sensitivity, the subjects were introduced to the experimental task using a practice sign. To simulate increases in a sign's angular size as it is approached by a driver, the size of the sign - initially too small to be described - was incremented in 7% steps. After each step, observer's were requested to identify all of the sign's features that they could discern. The smallest size at which the participant could describe the structure of all of the critical features of a sign (according to predetermined scoring criteria) was recorded as its threshold. These threshold angular sizes were then geometrically transformed to yield the equivalent maximum legibility distance for each of the 18 sign stimuli. The signs were presented to each subject using one of six predetermined random orders.

RESULTS AND DISCUSSION

Preliminary analyses revealed that reducing luminance from daytime to nighttime levels had a small but significant effect upon the legibility distances achieved by young and middle-aged adults (see Table 1 below). The addition of the glare source, however, failed to yield reliable reductions in performance for either of these age groups. Large losses in legibility distance were observed for the older subjects when luminance was reduced from daytime to nighttime levels. The magnitude of this age-related performance decrement was deepened given the introduction of the glare source. Yet, the size of the glare effect was relatively small. The nature of these age-related deficits is depicted in Figure 1 which presents the "gold standard" performance of the young observers under daytime luminance conditions along with the performance curves for the old groups under daytime, nighttime, and nighttime with glare lighting conditions. Although the figure shows wide sign-by-sign differences in legibility distance, the size of the age, luminance and glare effects appear to be of the same order of

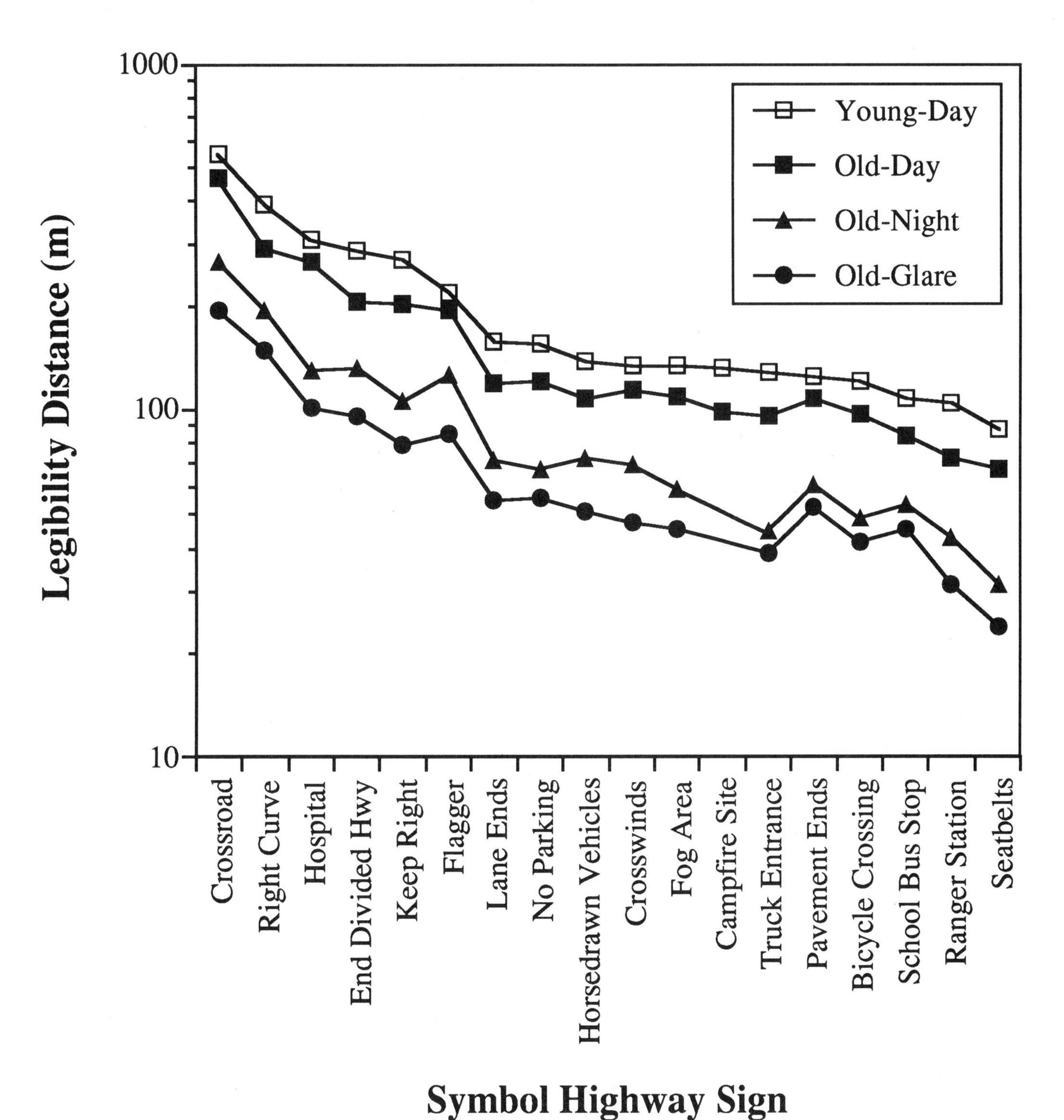

Figure 1. Legibility distances of older observers for a set of 18 symbol highway signs as a function of lighting condition. The ordinal position of the highway signs was determined by their rank-ordered legibility distances obtained from the sample of young observers under daytime viewing conditions.

Age Group	*Lighting Condition*		
	Daytime	**Nighttime**	**Nighttime + Glare**
Young	**1.00**	**0.70**	**0.69**
Middle-aged	**0.88**	**0.60**	**0.60**
Old	**0.80**	**0.46**	**0.34**

Table 1. Relative reduction in average sign legibility distance as a function of observer age and lighting condition. (Normalized to average performance of young observers under daytime viewing conditions).

magnitude regardless of which sign is observed (Note the roughly parallel nature of the performance functions in log legibility distance space).

In order to convey the general magnitude of the effects of age and lighting conditions upon symbol sign legibility, a relative legibility factor - normalized to the average legibility of all signs viewed under daylight conditions by the young group - was calculated for each experimental condition. That is, the average legibility distance across all 18 signs was calculated for each of the 9 performance functions (i.e., (3) age x (3) lighting conditions). Each of these averages was then expressed as a ratio to the average performance of the young observers under daytime viewing conditions. These relative legibility factors are presented in Table 1. Hence, Table 1 depicts the relative reduction in general sign legibility distance observed as age was increased or lighting quality was decreased.

CONCLUDING REMARKS

The *Symbol Signing for Older Drivers* project has generated much new information about the effectiveness of symbol highway signs for drivers of all ages. The project has generated data supporting the application of new techniques for predicting and optimizing the visibility of this class of signs and related materials. The project team, headed by Bob Dewar, Don Kline and Frank Schieber, is currently engaged in preparing the results of our research for widespread public dissemination.

REFERENCES

Dewar, R.E., Kline, D.W., Mark, I. and Schieber, F. (1994) Symbol signing design for older drivers. Final Report. [DTFH-61-91-C-0018]. McLean, VA: Federal Highway Administration.

Transportation Research Board (1988). Transportation in an aging society: Improving mobility and safety for older persons. Volumes 1 and 2. Washington, D.C.: National Research Council.

EMAIL CORRESPONDENCE
schieber@charlie.usd.edu

DETERMINING LEGIBILITY DISTANCE FOR HIGHWAY SIGNS: IS THE WITHIN SUBJECT VARIABILITY BEING OVERLOOKED?

Frances A. Greene, Rodger J. Koppa, Ronald D. Zellner, Jerome J. Congleton
Texas A&M University
College Station, Texas

Laboratory studies of warning symbol signs have been shown to underestimate legibility distances by up to a factor of two when compared with field studies. However, this research suggests it is more than simply experimental setting contributing to disparity in research findings. Using a group of old and young drivers, six symbol signs were investigated in both settings. With six trials per sign, legibility distances, defined as the distance at which the sign is correctly identified from a menu, were collected.

Large within subject variability was discovered in both age groups. This variability led to alternative ways of defining the dependent variable equivalent to designs of past studies examining legibility distances of the same signs. Different results arose out of the subsets created. The consideration is not just should a field-based versus laboratory-based methodology be used. An argument is posed that recommended distances at which signs are placed must be determined from a "worst-case" scenario. This premise requires a reexamination of our research methodologies for determining placement of highway signs.

INTRODUCTION

Research on traffic sign legibility spans over five decades. A dichotomy exists in paradigms used in this research; on one hand, laboratory-based studies and on the other, field studies. For many years researchers have discussed and debated the merits and problems associated with methods chosen for field and laboratory studies to determine legibility distances for highway signs. Dewar (1973) reviews the categories of methods used for traffic sign research. Dewar (1973) performs an excellent critique of both laboratory and field studies' methodologies; many of the criticisms still valid two decades later. Recently, Zwahlen, Hu, Sunkara and Duffus (1991), report discrepancies found between legibility distances from laboratory versus field studies for the same signs. Zwahlen et al. (1991) compare the legibility distances for warning symbol signs found by different researchers. The authors find that for studies conducted in a laboratory, the same signs have shorter legibility distances than those reported from field studies. Zwahlen et al. (1991) postulate these results could be a consequence of laboratory methodologies' lack of real-world fidelity. They note that many laboratory researchers do not completely describe such factors as display resolution, luminances, colors and contrast values used.

Thus, Zwahlen et al.'s (1991) research poses an interesting question. Which studies are reporting the "best" legibility distances for warning symbol signs? Or stated another way, what is the "correct" methodology to choose when collecting data to be used for traffic signs' placement guidance? A review of the literature does not find one study which evaluates the relative effectiveness of the two methodologies, field and laboratory-based, in order to make a clear-cut choice of which to choose. A case can be made that field-generated data is preferable, as it has maximum face validity for application of results. Is the decision to use a field-based versus laboratory-based methodology the only question a researcher must consider? Taking this assertion one step further, the present research asks the question: disregarding whether a field or laboratory study is performed, are experimental design and analyses of data without regard to within subject variability inherent in a driver having a dramatic impact on the reported results? It will be argued that recommended distances at which signs are placed to be recognized should be determined from a "worst-case" scenario. This supposition requires a reexamination of our research methodologies for determining guidelines for placement of symbol signs.

This study applies different analyses of the legibility distances collected from young and old drivers in both daytime field and laboratory studies. The research is aimed at exploring not only the experimental setting, but also implications for the methodology used to collect data.

FIELD STUDY

Method

Subjects. The same twenty-four subjects served as participants in both studies. Two age groups were used: 12 young drivers (7 males and 5 females), mean age of 23.7 years, and 12 older drivers (6 males and 6 females), mean age of 73.2 years. Visual acuity (corrected) for the young drivers ranged from 20/13 to 20/25 and from 20/20 to 20/50 for the older group.

Procedure. Participants received a training session where a menu of 27 possible symbol signs were reviewed to a 100% identification criterion. The subjects had the menu to refer to at all times while driving. Subjects were instructed to respond by sign number when they were certain as to the sign identification. Their certainty was stressed to be equated to willingness to make a speed reduction or vehicle maneuver in response to the sign (*Manual on Uniform Traffic Control Devices*, 1988). After at least two practice trials, subjects drove a vehicle instrumented with distance measuring equipment toward a warning symbol sign at a speed of 40 km/h on a 610 m straight taxiway on Riverside Campus, Texas A&M University. The distance readout, as well as the subjects' verbal sign identification, were recorded on video tape.

Stimuli. Nineteen new, standard-size warning symbol signs served as stimuli. Only six of the 19 signs were of interest for the purposes of collecting and analyzing legibility distances. The six target signs of interest, along with their *Manual on Uniform Traffic Control Devices,* 1988, (MUTCD) designation are shown in Table 1.

TABLE 1

Signs used in field and laboratory studies

Warning Sign	MUTCD Designation
Bicycle Crossing	W11-1
Deer Crossing	W11-3
Slippery When Wet	W8-5
Two-Way Traffic	W6-3
T-Junction	W2-4
Crossroad	W2-1

The six signs chosen were selected because they had been researched previously in the literature and a direct comparison could be made with other findings. Thirteen additional signs served as distractors. They were: *Divided Road Ends*, *Hill*, *Right Turn*, *Left Turn*, *Reverse Curve Right*, *Reverse Curve Left*, *Winding Road Right* , *Winding Road Left*, *Side Road Right 90 Degrees*, *Side Road Right 45 Degrees*, *Y Intersection* , *Right Land Ends Merge Left* and *Narrow Bridge*. All warning signs were mounted on a quick-mount pole such that the placement was in accordance with MUTCD requirements.

Six trials for each of the six target signs were collected for each subject. With a random order of target and distractor signs, a total of 60 trials were run. Signs were viewed during daytime conditions. Each sign was northerly facing to preclude glare from the sign surface. All subjects performed the field trials first, with the laboratory study occurring nine weeks later.

Dependent variables. Gender, age group and subjects nested within gender and age, formed the between subject variables. The within subject variables were individual sign and condition (field or laboratory). A repeated measures design was used.

Independent variable. Legibility distance, as defined as the distance from the warning sign where the subject was certain as to its identification, was the independent variable collected. Criterion for this response is discussed above under Procedure.

LABORATORY STUDY

The subjects, target and distractor signs, number of replications and trials in the laboratory study were exactly the same as described above. Dependent and independent variables were also identical.

Experimental apparatus. A Macintosh IIci computer and Viewsonic 43.18 cm high-resolution color monitor were used to present a "simulation" of the field study. Graphics and animation software programs were used to develop and present the simulation. Hypercard™ was used to control the experimental session, as well as collect and store the data.

Procedure. A similar treed background scene from the Riverside campus was videotaped, frame captured and digitized at 24 bit resolution. The background was nearly identical to the one against which field study signs were viewed. Using a graphics program, images of the signs, drawn in high (32 bit) resolution color and scaled to the correct percentage for size, were created to replicate retinal image size in 3.05 m increments from 610 m to 24.4 m. The result was 184 graphic sign elements, each embedded one at a time in the digitized scene background. An animation program placed these elements, representing the correct size sign, in the background. The resultant animation of these graphic sign elements created a simulation of approaching a sign at a speed of 40 km/h. Subjects were seated 6.1 m from the monitor displaying the simulation.

Using the same training standard and procedure as the field study, the equivalent criterion for "certain" legibility distance response was invoked. The experimenter input the subjects' responses via a keyboard template. The computer verified the correctness of the sign response and recorded legibility distance for each of the 60 trials.

RESULTS

Legibility Distance

For both the field ($F(1,20) = 10.38$, $p<0.004$) and laboratory ($F(1,20) = 10.51$, $p<0.004$) conditions, the main effect of age was significant. In the field, older drivers' mean legibility distance was 68% that of young drivers, with young drivers' legibility distance at 342.11 m, while older drivers' overall mean legibility distance was 232.64 m. The same means for the laboratory study were 372.2 m for young drivers and 263.6 m for older drivers, or 71% of the legibility distance of young drivers.

The main effect of individual sign was also significant for both field and laboratory studies. The rank order of signs by legibility distance was the same for field and laboratory studies. Figure 1 plots mean legibility distance by sign for both conditions. As Figure 1 displays, legibility distance for all signs was longer in the laboratory than the field, in contradiction to Zwahlen et al.'s (1991) findings.

Within Subject Variability

Individual plots of legibility distance for the six replications for each sign for each subject revealed large within subject variability. Due to the large amount of within subject variability among the six responses for both the field and laboratory conditions, a coefficient of variation (CV, computed as mean/standard deviation * 100) was computed for each condition, age group and sign.

The CVs ranged from values 2.04 to 49.17 for the field condition and 2.8 to 41.4 for the laboratory condition. Older subjects were more variable in their responses than the younger group ($M = 17.25$ for older subjects; $M = 12.53$ for the young group). The *Crossroad* sign had the least amount of variability ($M = 9.85$); while *Bicycle Crossing* had the most ($M = 20.39$).

This large within subject variability suggests that using the mean of the six replications of legibility distance for each sign could lead to misleading results and drawing erroneous conclusions.

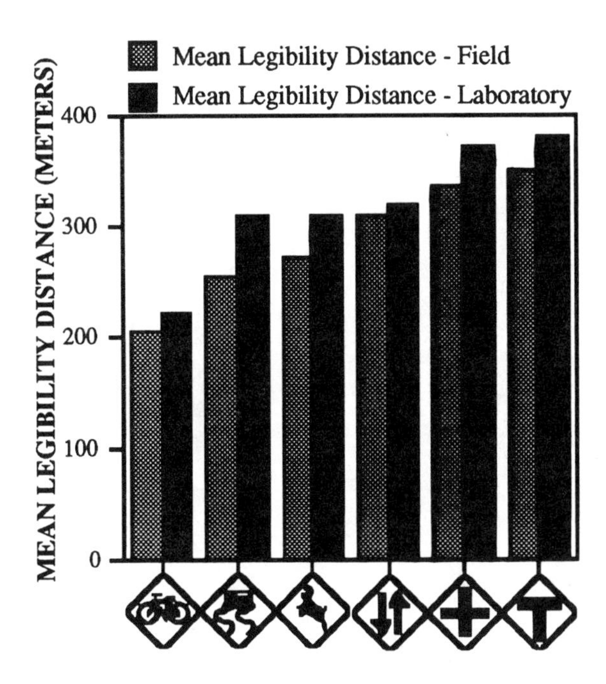

Figure 1. *Legibility distance by sign and condition*

Therefore, additional analyses of the six replications were examined. Analyses included the following subsets of the original legibility distances:

a. First two responses
b. Minimum of the first two responses
c. Overall minimum response

Of course, all six replications were also included in the analyses.

Figures 2 and 3 show the legibility distance obtained for each of dependent variables described above for two signs: *Bicycle Crossing* and *Crossroad*. Figure 2 summarizes the data for older subjects only, and Figure 3 for young subjects only. The two signs chosen for analyses represented extremes in variability, with *Bicycle Crossing* having the most in both field and laboratory studies.

Correlation of Field and Laboratory Legibility Distances

Examining the different subsets of response data, Pearson product moment correlation coefficients were calculated between the field and laboratory legibility distances. Using all six responses, the correlation calculated between the field and laboratory legibility distances ranged between 0.64 and 0.85; first response only between 0.61 and 0.80; mean of first two responses between 0.56 and 0.82; minimum of all six responses between 0.80 and 0.86. In all cases, *Bicycle Crossing* had the lowest correlation coefficient, while the largest was for *T-Junction*.

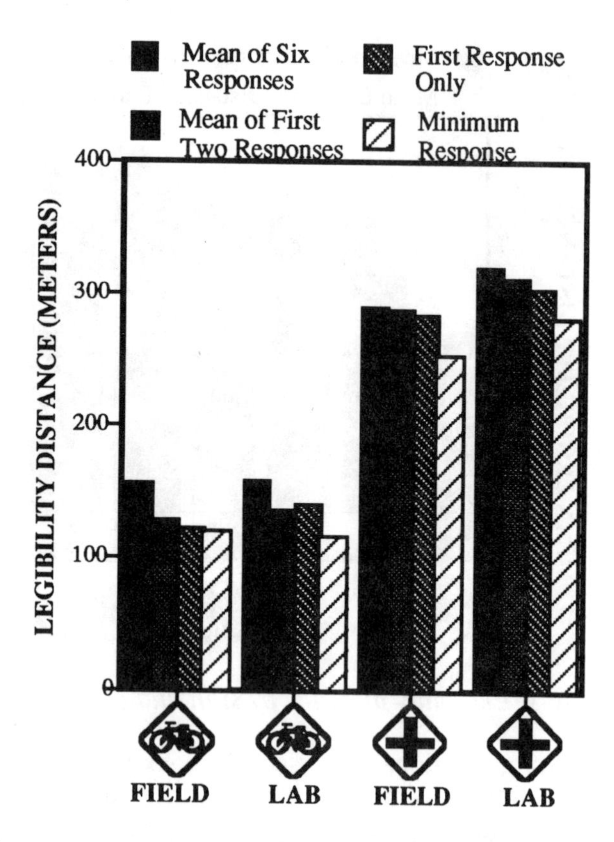

Figure 2. *Older drivers only, subsets of data responses, both conditions*

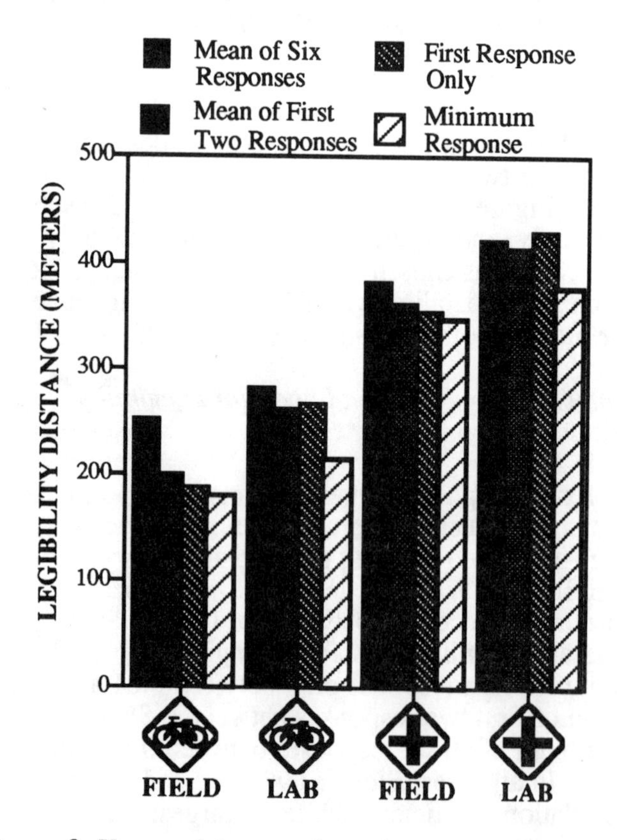

Figure 3. *Young drivers only, subsets of data responses, both conditions*

DISCUSSION/CONCLUSIONS

The data depicted in Figure 2 (older drivers only) dramatize the differences in conclusions which would be drawn for legibility distances of the six warning symbol signs. The data in the four columns for each warning sign represent determinations one would make for recommended legibility distances for the six signs in the following manner:

Column 1 - Mean of all six responses - the conclusions drawn from this study

Column 2 - First response only - equivalent to Paniati (1988) and any study which included no replications

Column 3 - Mean of first two responses - Zwahlen et al. (1991), since he collected two data points on each of his young subjects.

Column 4 - Worst-case scenario. This minimum of six responses from this study is the shortest legibility distance and represents an extremely conservative recommendation. These signs would be legible to a representative sample of older drivers (over the age of 65).

Figures 2 and 3 depict graphically that depending on the subset of dependent measures used in the analyses, the legibility distances reported are quite different. This difference is attributable to the large within subject variability.

Most studies either do not report the number of replications or do not include any. In Paniati (1988), only one data point is collected and the average legibility distance is reported for each sign as a function of subject age. Kline, Ghali, Kline and Brown (1990) also only collected one data point per person. Their design is comparable to reporting the "First response only" column of data from this study. Zwahlen et al. (1991) include two replications in their study. They also plot the mean, standard deviation, minimum and maximum for legibility distance for 12 signs investigated. However, no tabular data is reported and it is difficult to interpolate the data from their figure plot. Zwahlen et al.'s (1991) study results would equate to the reporting of the second column, "Mean of the first two responses " data from this study.

By not examining the within subject variability and reporting the mean of the six responses for older drivers' field condition overestimates the legibility distance for *Bicycle Crossing* by almost 28% over the first response only and 22% over the mean of the first two responses. *Bicycle Crossing* was the sign with the most within subject variability.

RECOMMENDATIONS

It was the close examination of data leading to the discovery of tremendous within subject variability for all signs, age groups and both conditions, when questions arose about what is the appropriate subset of dependent variable to use. This question raises the issue of experimental design and offers a suggestion of designing for a worst-case scenario.

It has been shown repeatedly that older subjects have shorter legibility distances for symbol and text signs than their young driver counterparts (Allen, Parseghian and Van Valkenburgh, 1984; Evans and Ginsburg, 1985; Kline, Ghali, Kline and Brown, 1990; Lum, Roberts, DiMarco and Allen, 1983; Paniati, 1988; Sivak, Olson and Pastalan, 1981). This study shows the same result. Therefore, it follows that legibility distances should be determined using a representative sample of older drivers, not young drivers with often better than 20/20 visual acuity.

It is recommended that the area of within subject variability in legibility distance responses be given serious examination and recognition for its existence. Also, the ramifications this variance has on predictions must be reconciled.

Researchers must come to grips with three issues of foremost importance: first, a representative sample of older drivers must be included in every study. The literature supports time and time again problems the aging process has on the visual system and reaction time. To continue the practice of using only young subjects to generate recommendations for sign placement is not recommended and ignores an important and growing segment of our population, the older driver.

Second, replications are vital and must be included in every study, whether it is field- or laboratory-based. Without replications, we, as practicing transportation researchers, are fooling ourselves if we think we can collect just one or maybe two observations and make recommendations. The conclusions in the literature which are based on only one observation or perhaps two are very likely misleading and possibly in error. Without replications, the author is overlooking that large and consequential within subject variation contribution.

Finally, authors must completely define their response variable. The literature is full of different independent variables like recognition, legibility, visibility and identification distance. Most are not adequately defined, or defined at all.

Replications appear to be very relevant to the results which are reported. The research community must first recognize the existence of and then focus on the source of this within subject variation phenomenon. Without a study of this factor, perhaps that which is being reported in the literature is more a function of the experimental design than the multitude of independent variables being painstakingly manipulated.

REFERENCES

Allen, R.W., Parseghian, Z., and Van Valkenburgh, P.G. (1980). *Age effects on symbol sign recognition* (Tech. Report FHWA/RD-80/126). Hawthorne, CA: Systems Technology, Inc.

Dewar, R.E. (1973). *Psychological factors in the perception of traffic signs.* Road and Motor Vehicle Traffic Safety Branch, Department of Transport. Calgary, Canada: University of Calgary, Psychology Department

Evans, D.A., and Ginsburg, A.P. (1985). Contrast sensitivity predicts age-related differences in highway-sign discriminability. *Human Factors, 27*(6), 637-642.

Federal Highway Administration (1988). *Manual on Uniform Traffic Control Devices for streets and highways.* Washington, DC: U.S. Department of Transportation.

Kline, T.J.B., Ghali, L.M., Kline, D., and Brown, S. (1990). Visibility distance of highway signs among young, middle-aged, and older observers: icons are better than text. *Human Factors, 32*(5), 609-619.

Lum, H.S., Roberts, K.M., DiMarco, R.J., and Allen, R.W. (1983). A highway simulator analysis of background colors for advance warning signs. *Public Roads, 47*(3), 89-96.

Paniati, J.F. (1988). Recognition and comprehension of traffic sign symbols. In *Proceedings of the Human Factors Society 32nd Annual Meeting* (pp. 568-572). Santa Monica, CA: Human Factors Society.

Sivak, M., Olson, P.L., and Pastalan, L.A. (1985). *Optimal and minimal luminance characteristics for retroreflective highway signs* (Transportation Research Record 1027), pp. 53-57. Washington DC: National Research Council, Transportation Research Board.

Zwahlen, H.T., Hu, X., Sunkara, M. and Duffus, L.M. (1991). Recognition of traffic sign symbols in the field during daytime and nighttime. In *Proceedings of the Human Factors Society 35th Annual Meeting* (pp. 1058-1062). Santa Monica, CA: Human Factors Society.

AGE DIFFERENCES IN VISUAL ABILITIES IN NIGHTTIME DRIVING FIELD CONDITIONS

Susan T. Chrysler, Suzanne M. Danielson, & Virginia M. Kirby
3M Company
St. Paul, Minnesota

Abstract

This study was conducted to provide field data on age differences in sign legibility and object detection. Two age groups of healthy drivers with normal vision were tested for nighttime visual ability. The older group had an average age of 65.6 years and the younger group averaged 22.5 years. The field study was conducted on a private road with the subjects seated in the front passenger seat. Subjects performed a Landolt ring legibility task for four types of signs; positive and negative contrast, new and worn material. Subjects also performed object detection tasks using a small object and a pedestrian target appearing in average and low reflectance. In addition, sign legibility and object detection were completed for some trials using a simulated inclement weather visor to create a worst-case scenario. The object detection task was also completed in the presence of glare from oncoming headlamps. Results showed that older driver's legibility distances were 65% those of the younger drivers. Age differences in the object detection task ranged from a 20% to a 45% reduction for older drivers across visibility conditions.

We're all getting older and as we age our visual abilities naturally degrade. While there are large individual differences in vision throughout the life span, certain aspects of vision clearly suffer with age. Several of these aspects, in particular contrast sensitivity and recovery from glare, play an important role in the driving task (Schieber, 1992). Research on driving is often done with a part-task approach. Clinical measurements of visual function are made and inferences are drawn about the implications of age differences and driving safety (e.g. Ball & Owsley, 1991; Shinar & Schieber, 1991). This focus on laboratory experiments is no doubt largely due to the expense and logistics of conducting field studies. As a result, there are few studies in the literature which contain actual driving data or field visibility measurements. This research project was intended to provide basic field data on the nighttime visibility of traffic signs and objects in the roadway.

Previous field studies of sign legibility have found that older drivers' legibility distances were 65 - 77% those of younger drivers (Sivak, Olson, & Pastalan, 1981). This field study, as well as previous laboratory studies examining age differences (Olson & Bernstein, 1977), used Landolt rings to assess legibility. This approach was taken in the current study as well. Field studies of hazard detection have primarily focused on the effects of roadway lighting (e.g. Janoff & Staplin, 1987). The current study is focused on establishing baseline age differences in object detection distances without influences of roadway lighting. For this reason, headlamp illumination alone was used.

This study was conducted to assess the magnitude of age differences across a wide variety of stimuli found in the driving scene and across a range of nighttime visibility conditions. Legibility of new and worn signs with white and green backgrounds was measured. Object detection distances for small road hazards and pedestrians in average and low reflectance conditions were also measured. In addition, the visibility conditions were varied by presenting oncoming headlamp glare and in separate trials by simulating reduced visibility due to inclement weather through the use of a visor.

METHOD

Subjects

Two groups of ten subjects were tested. The Older Driver group ranged in age from 53 to 75, with an average of 65.6 years. The Younger Driver group's ages ranged from 19 to 25, with an average of 22.5 years. All subjects had bilateral corrected acuity of 20/25 or better, with the majority scoring 20/15 or better. Pelli-Robson Contrast Sensitivity scores were statistically equivalent for the two age groups. Subjects were recruited through a local University and through a local retirees association. Subjects were paid $50.00 for their participation in the vision and field testing.

Field Site and Conditions

A private test track was used for this study. The roadway and markings met standards for public roads. The course consisted of three legs. Road A was used for sign legibility assessment and had a length of 2000 feet. There were two sign test positions spaced 1000 feet apart. Small road hazard detection was measured on Road B which had a length of 500 feet. Road C was used for the pedestrian detection task and measured 1000 feet. The testing began 1 hour after sunset and was conducted only in clear weather. Photometric measurements of ambient light were made each night of testing.

Apparatus

A 1993 Ford Taurus was used as the observation vehicle for all subjects. The low beam headlights were used. Each night prior to testing the vehicle's headlights were aimed and the windshield was cleaned. A Numetrics Nitestar distance measuring instrument was used to record visibility distances.

Three visibility conditions were tested: standard clear weather, simulated inclement weather, and glare. The simulated inclement weather visibility condition was achieved by having the subject wear a plastic visor which blurred and reduced the contrast of the scene. This was intended to simulate poor visibility found in heavy rain or fog. The glare visibility condition consisted of a set of standard headlamps mounted oncoming in the opposite travel lane. The headlamps were illuminated on low-beam for those laps where glare was to be present for the object detection tasks.

The legibility task consisted of identifying the gap position on an 8" Landolt ring placed on a 24" sign background. Four signs were used for this task. Positive contrast signs of a white ring on a green background and negative contrast signs of a black ring on a white background. The retroreflectivity, and therefore contrast ratio, of the sign materials was varied for both positive and negative contrast signs. Values were chosen which represent new and 7-year old signs on a typical enclosed lens sign sheeting material.

Table 1 shows the coefficients of retroreflectivity and resulting contrast ratios for the four signs. Note that for the positive contrast signs, the Worn sign resulted in a higher contrast ratio due to the different relative degradation of white and green material.

The object detection task consisted of two objects. The small road hazard object was an upside-down bowl measuring 7" high and 13" wide. The pedestrian object was a child-sized mannequin, 3'6" tall. Both objects appeared in average and low reflectance conditions. The average reflectance condition was achieved by dressing the objects in gray clothing (cap Y = 35), while the low reflectance clothing was dark red (cap Y = 9).

Experimental Design

The experiment consisted of two types of tasks: sign legibility and object detection. On each lap through the test course, subjects completed four tasks: 2 sign legibility and 2 object detection. Each subject completed 2 practice laps and 16 experimental laps. The order of treatments and correct responses was counterbalanced for each task separately. The dependent variable for all tasks was correct legibility / detection distance. Table 2 shows the number of observations for each subject for each stimulus.

The visibility conditions were intended to test a worst-case scenario for drivers. The simulated inclement weather condition was presented only with the worst-case version of each stimuli, i.e. the worn signs and the low contrast objects. The glare condition was used only for the object detection task and only with the low contrast objects.

Table 1. Retroreflectivity measurements for signs. The coefficient of retroreflectivity (R_A) is given in candelas/lux/square meter (cd/lx/m^2).

Sign Type	Background	Landolt Ring & Border	Contrast Ratio based on R_A
New Positive Contrast	15 (Green)	104 (White)	7 : 1
Worn Positive Contrast	3.7 (Green)	43 (White)	11.6 : 1
New Negative Contrast	104 (White)	0 (Black)	104 : 1 (approx.)
Worn Negative Contrast	35 (White)	0 (Black)	35 : 1 (approx.)

Table 2. Experimental design and number of observations per subject for each stimulus type in each of three visibility conditions (Clear, Simulated Inclement Weather, Glare).

	Legibility Task				*Object Detection*					
	Positive Contrast		Negative Contrast		Small Hazard			Pedestrian		
Visibility	(White on Green)		(Black on White)		Average	Low	Null	Average	Low	Null
Condition	New	Worn	New	Worn	(Gray)	(Red)		(Gray)	(Red)	
Clear	6	6	6	6	2	2	4	2	2	4
S. I.W.	---	4	---	4	---	2	2	---	2	2
Glare	---	---	---	---	---	2	2	---	2	2

The Landolt C gap position was counter-balanced within each sign type. Half of the subjects in each age group saw white signs at Sign Position 1 for the first eight trials, and half saw green.

For the detection task, the subject was to indicate whether the object was on the right or left side of the road. The objects could appear in one of four positions relative to the position of the glare headlights. Half of trials were nulls, where no object was present to provide a check for guessing. The other four positions were equally represented for each object type within a subject. The presentation order of color and position of objects, as well as the presentation of null trials was randomized for each subject.

Procedure

Subjects were seated in the front passenger seat of the test vehicle. The car was driven at a steady rate of 20 mph by an experimenter. Another experimenter was seated in the back seat with the Numetrics device. Subjects were instructed to call out the position of the gap or the position of the object as soon as they could. Subjects were encouraged to guess. As soon as the subject made a correct response the experimenter would press the Numetrics and record the distance. The distances recorded were slightly shorter than when the subject actually responded due to the time lag of the subject speaking and the experimenter pressing the button. To correct for this, prior to road testing a paired reaction time test was conducted to establish the response time for the subject - experimenter pair. This reaction time test consisted of a random presentation of arrows visible only to the subject. The subject called out the direction of the arrows and the experimenter responded to the verbal cue by pressing a mouse button, the same response regardless of the direction of the arrow. This arrangement provided a baseline response time which was used to correct the distances recorded in the field based on a constant travel speed of 20 mph. All results reported are these corrected distances.

RESULTS

Sign Legibility

Figure 1 shows sign legibility distances for all sign types as a function of age group. Across all four sign types in clear weather, young subjects correctly identified the Landolt gap 147 feet further away than the older group. The younger group average was 467 feet and the older group averaged 320 feet, yielding the result that older subjects could read the signs at 65% of the distance of the younger subjects. This effect was significant in a repeated-measures ANOVA at the 0.05 level. The Worn-White sign had significantly shorter legibility distances than the other three signs across age groups.

The simulated inclement weather visor served to reduce the visibility of the signs for both age groups. The age difference found in the Clear condition persisted in this reduced visibility condition. The young people's average was 192 feet while the older group was 126 feet. The older group's distance as a percent of the younger group's performance was again 65%. There were no significant main effects or interactions of sign type (i.e. white or green) for legibility distance under simulated inclement weather. The legibility distance with the visor was, on average across sign type and age group, 41% that of Clear conditions.

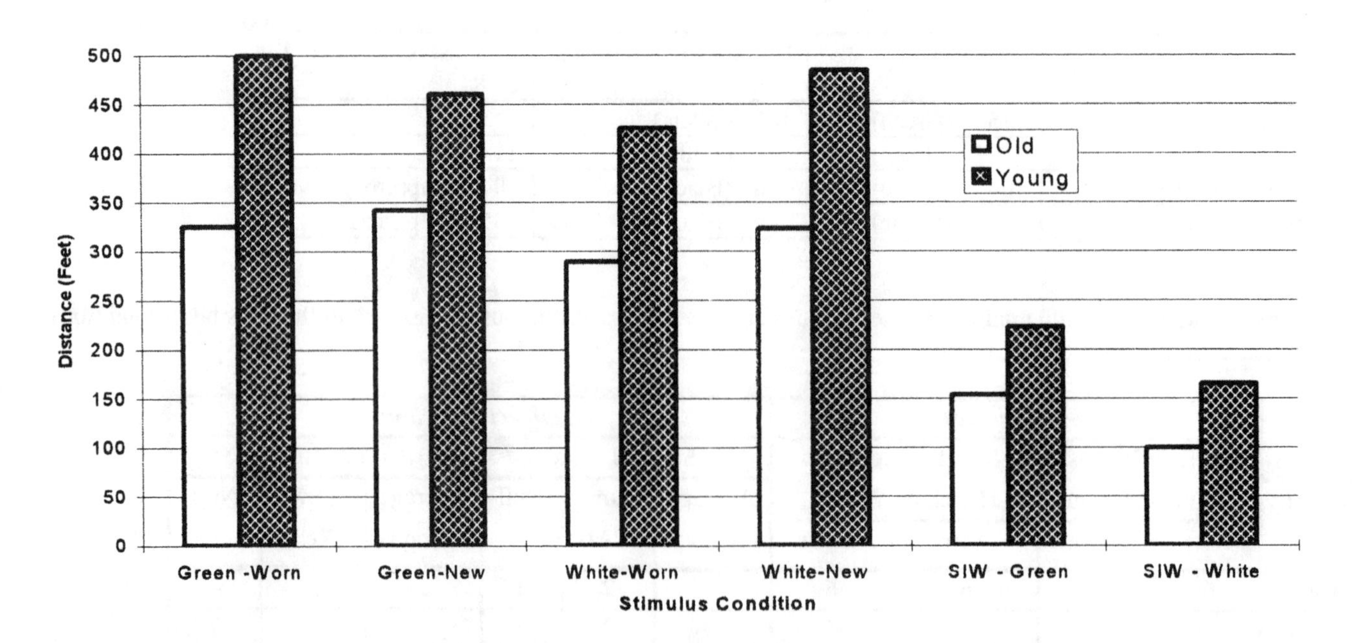

Figure 1. Legibility distances for an 8" Landolt Ring for two age groups. Stimuli were new and worn enclosed lens sheeting, and worn sheeting viewed through the simulated inclement weather visor (SIW).

Small Road Hazard Detection

Object detection distances by age for all stimulus conditions are presented in Figure 2. In the clear visibility condition, there was a trend (p=0.066) toward longer detection distances for the gray object across age groups. The age difference seen in the sign legibility task was again present for small object detection, though a smaller degradation was present. This age effect was significant, with the younger group detecting the objects at an average of 295 feet and the older group at an average of 230 feet, or 78% of the younger group's distance.

The reduced visibility conditions worsened the performance for both age groups by 51%. The age difference in the simulated inclement weather condition, a reduction of 20% for the older group, failed to reach significance. The presence of glare headlights significantly hurt the detection distances of both ages, an average 37% reduction from the clear condition. The age difference was quite pronounced in the glare condition; 136 feet for the older group and 198 feet for the younger subjects. The size of this detection distance difference with the elders reaching just 68% of the younger groups distances was of the same magnitude of the difference seen in sign legibility.

Pedestrian Detection

The results for the pedestrian detection are presented in Figure 3 in the same manner as those for the small road hazard object. For the pedestrian targets, the gray clothing resulted in significantly longer detection distances for both age groups (gray = 351 feet vs. red = 274 feet). Age was a significant factor as well. The older group's detection distances were 75% those of the younger subjects, with the older group mean being 265 feet and the younger group mean being 360 feet.

The simulated inclement weather visor reduced the detection of pedestrian targets for both groups by 37%, slightly less reduction than seen with the small road hazards. Again, the age difference within this visibility condition failed to reach significance. The glare condition showed a 45% reduction in detection distance for the pedestrian targets. The age difference of 149 feet for older and 191 feet for younger subjects, a 22% reduction, was significant.

DISCUSSION

The results of this field study clearly show that older drivers with normal clinical vision have reduced legibility and object detection distances when compared to a younger group of people. These 20 - 35% reductions held across tasks and visibility conditions, with the exception of the simulated inclement weather visor. This device may have reduced vision so severely for both groups, that the age difference was mitigated. The glare conditions resulted in particularly severe pedestrian detection losses for the older subjects. With higher profile targets, subjects were evidently conducting their visual search for the target in a range which forced them to look more directly into the glare headlights. This sensitivity to glare is consistent with previous findings concerning older drivers (Shinar & Schieber, 1991).

One concern in the design of this study was that older subjects would be more conservative in their guessing, thus producing shorter distances. An examination of error rates on both the legibility and detection tasks, did not support this

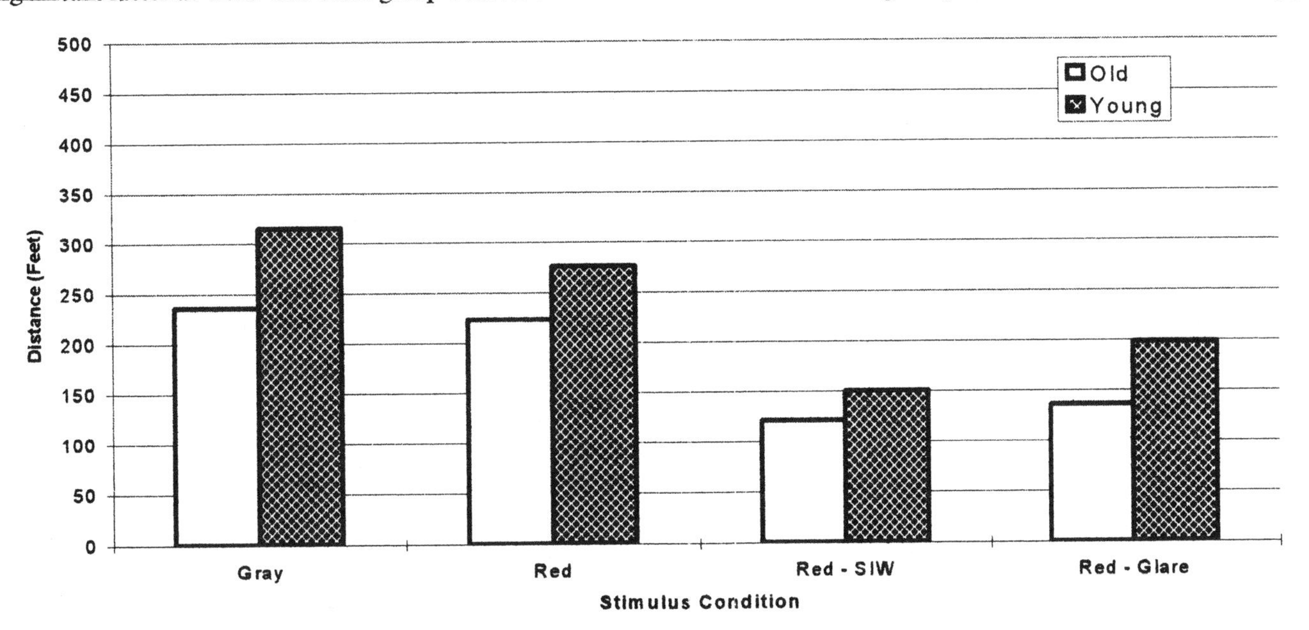

Figure 2. Small road hazard detection distances for two age groups. Stimuli were a gray and a red object, the red object viewed through the simulated inclement weather visor, and the red object viewed with oncoming headlamp glare.

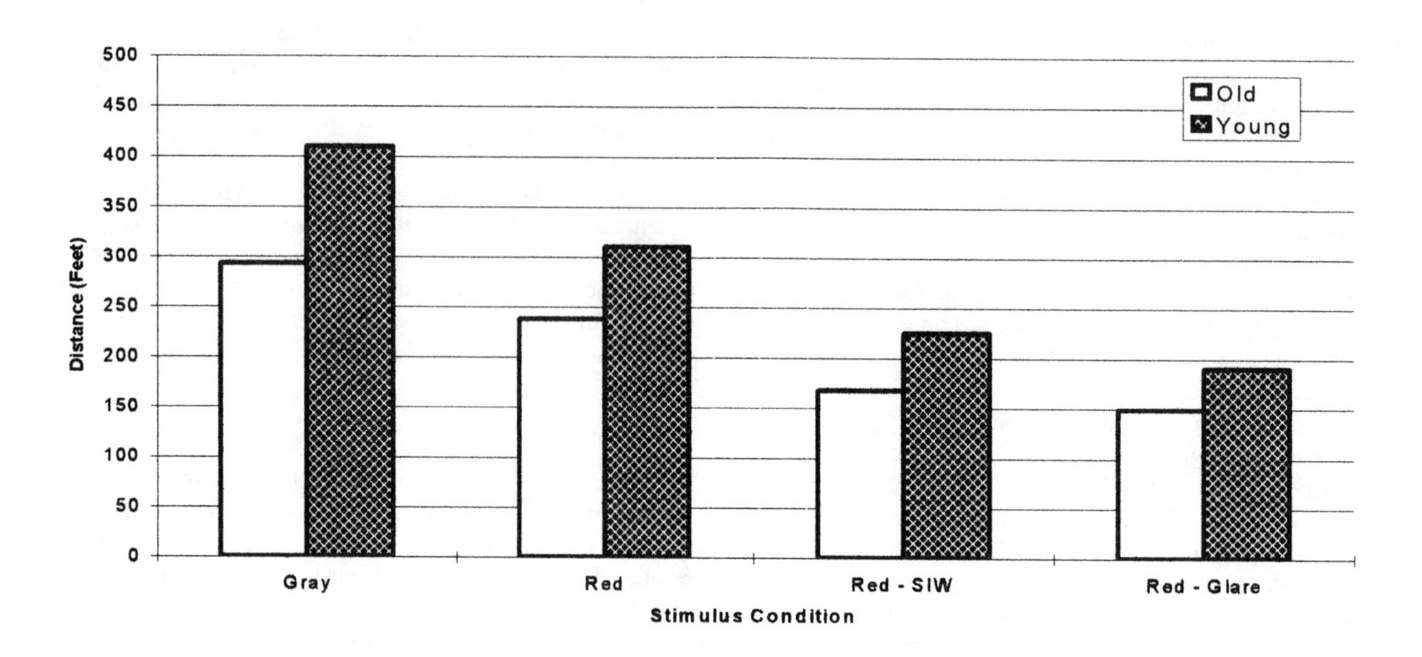

Figure 3. Pedestrian distances for two age groups. Stimuli were a gray and a red child- sized mannequin, the red mannequin viewed through the simulated inclement weather visor, and the red mannequin viewed with oncoming headlamp glare.

hypothesis. Both younger and older groups had very low (<0.01) error rates.

While demonstrating that older drivers have reduced visual abilities, this study does not speak to the safety implications of these findings. While vision is undoubtedly an important factor in driving, other factors including cognitive and motor skill deficits may play an even more important role in determining crash involvement (e.g. Ball & Owsley, 1991). As the population as a whole grows older, research in this area will only gain in importance and relevance to society. Highway and vehicle designers may have to change their system designs to accommodate the large group of older users who will be reluctant to cease driving.

References

Ball, K. & Owsley, C. (1991). Identifying correlates of accident involvement for the older driver. *Human Factors*, 33(5), 583-595.

Janoff, M.S. & Staplin, L.K. (1987). Effect of alternative reduced lighting techniques on hazard detection. *Transportation Research Record 1149*, Transportation Research Board.

Olson, P.L. & Bersntein, A. (1977). Determine the luminous requirements of retroreflective highway signing. UMTRI Report 77-6, University of Michigan.

Schieber, F. (1992). Aging and the senses. *Handbook of Mental Health and Aging, 2nd Ed.*, Academic Press.

Shinar, D. & Schieber, F. (1991). Visual requirements for safety and mobility of older drivers. . *Human Factors*, 33(5), 507-519.

Sivak, M., Olson, P.L, & Pastalan, L.A. (1981). Effect of driver's age on nighttime legibility of highway signs. *Human Factors*, 23(1), 59-64.

AGE DIFFERENCES IN JUDGEMENTS OF VEHICLE VELOCITY AND DISTANCE

Charles T. Scialfa
The Pennsylvania State University

Donald W. Kline
The University of Calgary

Brian J. Lyman
The University of Notre Dame

William Kosnik
Texas A & M University

ABSTRACT

The purpose of this study was to determine if older adults have more difficulty than younger adults in judging either the distance or speed of approaching vehicles. Eighteen elderly and 27 younger adults made judgements of the speed and distance of a video-taped automobile. Velocity judgements were made of 5 s segments of the car moving at 20, 30, 40, 50, and 60 mph. Distance judgements were based on 5 static sequences of the same test vehicle at 190, 235, 300, 360, and 480 ft. It was found that older women gave significantly higher estimates of car velocity than did younger ones. Older males also gave disproportionately high estimates of the car's distance. To the extent that these simulation data can be generalized to real-life settings, they suggest that older drivers and pedestrians (particularly older males) would view it as relatively safer than younger drivers to enter or cross the lane of an approaching car. Future research might be directed to a determination of age differences in distance perception under three-dimensional viewing conditions.

The number of elderly drivers on U.S. roads and highways is high. In California, for example, 77% of residents aged 65 to 69 hold a valid driver's licence (State of California Department of Motor Vehicles, 1982). As more people move into old age, particularly younger cohorts of women who are more likely to drive, the proportion of older drivers will continue to increase. Although elderly people as a group (i.e., per 1,000 drivers) do not contribute excessively to overall number of accidents, when rate (i.e., incidents per 100,000 miles) is considered, the older driver (past about 55 for men and 60 for women) participates disproportionately in both accidents and cited traffic violations (Harrington & McBride, 1985; Planek, 1973; State of California Department of Motor Vehicles, 1982). Further, their profile of involvement in such incidents is unique. Unlike younger drivers, they are seldom involved in accidents attributable to speed, equipment or major violations (e.g., reckless driving, DUI). On the other hand, they are significantly more likely to be involved in accidents and traffic citations involving failure to heed signs, yield right-of-way, or turn safely (Planek, 1973, State of California Department of Motor Vehicles,]982). The degree, however, to which accidents of these types can be attributed to the visual declines that occur with aging (cf., Kline & Schieber, 1985) has yet to be determined.

Although it has been commonly estimated that over 90% of the information used in driving is visual (Hills, 1980), research attempting to relate visual functions to driving effectiveness has achieved only modest success (Burg, 1967; Hills, 1980). The correlations between visual measures and safe driving have proved to be somewhat higher among old drivers than young ones, due in part, perhaps, to the greater range in quality of vision among the elderly. For example, modest but systematic correlations between accident rate and static acuity have been seen for individuals past 54 years of age (Hills & Burg, 1977). Likewise, dynamic visual acuity (DVA), which declines progressively with age (Scialfa, et al., 1987), appears to be particularly related to automotive driving record among elderly drivers (Burg, 1967).

The specific role(s) of such visual changes in accidents of the type experienced by older drivers in not clear. Can they be ascribed to impaired discriminability of critical stimuli? Is the judgement of velocity and/or distance impaired? Such deficits appear consistent with the "right-of-way" type of problems that older drivers have. For example, angular velocity of high-speed vehicles, which is likely to be underestimated by older drivers (Hills & Johnson, 1980, cited in Hills 1981) has also been related to accident involvement (Henderson & Burg, 1974). Depth judgement, at least that based on stereopsis, also appears to decline with age (Bell, Wolf & Bernholz, 1972). It is likely that drivers who perceive approaching traffic as moving more slowly and/or being more distant will also be more likely to judge it as safer to cross or join a traffic flow. The present investigation was carried out to determine possible perceptual contributors to the relatively high rate of right-of-way accidents involving elderly drivers. Specifically, this study examined age differences in estimated "approach" speed of, and distance to an automobile as viewed from a video-presented intersection scene.

METHOD

Subjects

Participants were recruited from visitors

at a community-based health fair. All subjects were independently-living and in self-reported good health. All older participants, and all but one of the younger adults were licensed drivers, who reported that they currently drove.

Young subjects (13 males, 14 females) ranged in age from 16 to 45 yrs. (*M*=32.6 yrs.). Older adults (9 males, 9 females) ranged in age from 54 to 79 yrs. (*M*=63.8 yrs.). Average (Binocular far visual) acuity (Bausch/Lomb Orthorator) or younger subjects was 20/20 (range = 20/17 to 20/33), and for older subjects was 20/22 (range=20/18 to 20/40). There was a small, but statistically significant age difference (p=.03) in acuity. There were no differences in acuity for men and women, nor did gender interact with age on acuity measures.

Apparatus and Materials

Display sequences were taped using a Panasonic video camera and recorder. Display sequences were viewed on a 25 in Sony color television monitor. Camera position was fixed and chosen to present a line of view similar to that of a driver stopped at an intersection, and looking left toward oncoming traffic. The camera was positioned at a height of 42 in above the road surface, 13 ft from the road edge, and 24.5 ft from the road's center line. The sequences were filmed at mid-day under sunny conditions.

The stimuli for the velocity judgements were 5 s video segments of an automobile (1984 Mazda 4-door sedan) moving at speeds of 20, 30, 40, 50, and 60 mph. Distance judgements were made viewing the same vehicle for a period of 5 sec at distances of 190, 235, 300, 360, and 480 ft.

Speed and distance sequences were filmed on a straight, two-lane, rural road. There was little visual clutter near the roadway (e.g., buildings, roadsigns, or other structures). In all sequences the test vehicle was the only car in the scene.

Procedure

Participants were tested in groups, of 2-6 persons, in a single experimental session that lasted approximately 15 min. Participants were tested for acuity prior to the distance-, and velocity-judgement tasks. For both distance and velocity judgements, subjects were informed that the film sequences would be brief and that they should make their judgements as quickly as possible during a short interdisplay interval. During this pause a filmed message instructed subjects to make judgements in miles per hour (speed estimates) or in feet (distance estimates).

All subjects made a total of 20 judgements based on 2 presentations of each of the 5 speeds and 5 distances. Order of stimulus presentation was random within each judgement task, with the restriction that each of the 5 speeds or distances were presented once before any were repeated. Task order was counter-balanced, with one-half of the subjects making speed judgements first, and the other subjects making distance judgements first. Subjects recorded their estimates on individual answer sheets; their 2 estimates were averaged for each speed and distance.

RESULTS

Prior to conducting any significance tests, we determined whether observers could reliably scale speed and distance. For both speed and distance scales, alpha coefficients were calculated for each age group, collapsing across ratings. The reliability coefficients were very high, both for younger (speed=.91, distance = .88), and older adults (speed=.82, distance =.93).

To protect against inflated Type I error rates for subsequent significance tests, these data were subjected to a doubly-multivariate analysis of variance (MANOVA) which considered speed and distance judgements simultaneously as separate dependent measures, with age (2) and sex (2) as between-subjects factors, and the 5 levels of speed and distance as a repeated factor. For this and subsequent analyses, only significant effects will be reported.

In the omnibus analysis, the main effect of age was significant, multF(2, 40)=10.43, p=.003. Because the sex effect was significant, the subsequent analyses were conducted separately for each sex.

Mean speed judgements for each age and sex group are shown in Figure 1. It can be seen that while old and young men did not differ in velocity judgements, older women displayed a tendency to overestimate velocity relative to their younger counterparts. As would be expected, the velocity estimates of both age groups increased with actual velocity. These observations were confirmed in an age(2) X stimulus level(5) MANOVA conducted separately for males and females. For men, there was an effect for velocity level, multF(4,17)=85.29, $p < .001$. For women, the main effect of age, F(1,21)=8.91, $p < .007$, and velocity level effect, multF(4,18)=50.70, $p < .001$ were significant.

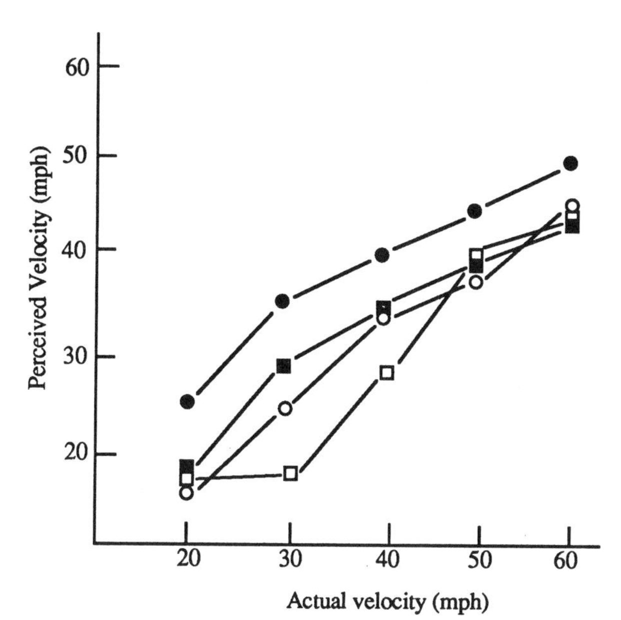

Figure 1. Perceived velocity as a function of actual velocity, age, and gender. □—□, young males; ■—■, older males; ○—○, young females; ●—●, older females.

A very different pattern of results was obtained for distance judgements shown in Figure 2. As would be expected, estimated distance is seen to increase with actual distance. However, whereas older and younger women tend to estimate distance similarly, older men give greater distance estimates than younger men. Further, among men, size of the age difference increases with stimulus distance.

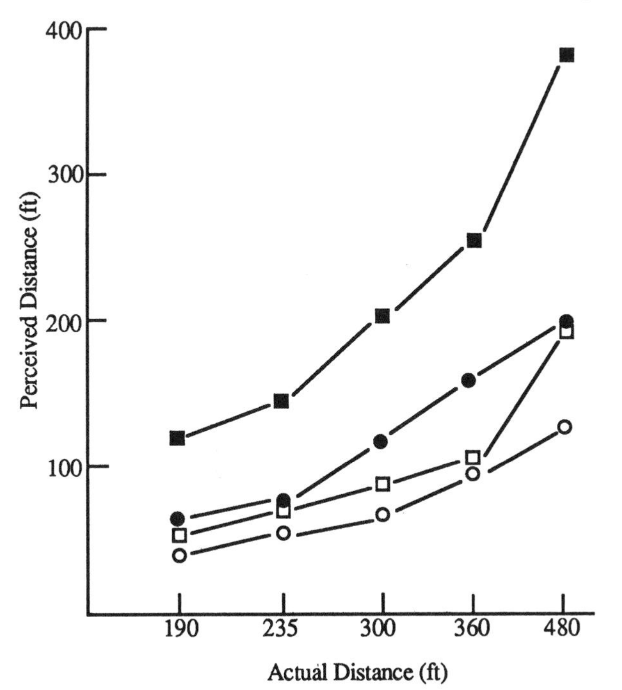

Figure 2. Perceived distance as a function of actual distance, age, and gender. □—□, young males; ■—■, older males; ○—○, young females; ●—●, older females.

Again, a repeated-measures MANOVA substantiates these observations. For women, the only significant effect was that due to stimulus distance, multF(4,18)=5.91, p .003. For men, the distance level effect was also significant, multF(4,17)=18.83, p .001. In addition, for men, both the age effect, F(1,20)=12.71, p .002, and the age X distance level interaction, multF(4,17)=9.08, p .001 were significant.

Although older subjects tend to overestimate both speed and distance, the most salient feature of these data was the striking age difference in distance judgements (Figure 2) versus those which occurred for speed (Figure 1), especially among men. In order to examine this apparent disproportionality for age X gender, differences in distance-speed ratios (defined as mean distance estimate/mean speed estimate) were tested. Averages were used because they provided the best estimates of speed and distance judgements, given the lack of a direct correspondence between particular level of distance and speed. Mean distance-speed ratios for each age and sex group were as follows: young males, 3.26; older males, 6.57; young females, 2.26; older females, 3.29. ANOVAs revealed a significant age effect for men, F(1,20)=9.00, p .007. There was no age effect for women (p=.259).

Due to the small, statistically significant age difference in acuity an analysis of covarience (ANCOVA) was then conducted on distance-speed ratios, using acuity as a covariate. The ANCOVAs yielded the same results as those analyses which did not adjust ratios on the basis of acuity. For women the age effect (p=.069); approached but did not reach significance; for men, after controlling for acuity, the age effect was still significant, F(1,19)=10.25, p=.005.

DISCUSSION

Results obtained suggest that there are reliable individual differences in the perceived velocity of a moving vehicle, and the distance of a static vehicle, when they are viewed in video-taped sequences. These differences are systematically related to both the age and gender of the observer. Specifically, the data suggest that older women judge cars to be travelling more quickly than men or younger women. Further, older men judge non-moving vehicles to be more distant than young men, or adult women of any age. Finally, a speed-distance ratio analysis suggests that, as observers, older men judge cars to be at disproportionately greater distances, even when differences in speed estimates are taken into account. This tendency to overestimate distance may lead older men to make decisions that put them at greater risk of accident involvement. For example, as pedestrians, overestimates of approaching vehicle distance might lead to roadway crossing when automobiles are, in fact, too close. Similarly, a "too distant" judgement of approach vehicles when attempting to merge or cross an intersection may result in a higher probability that older drivers would be involved in these types of accidents. In fact, older males involved in accidents do tend to make more perceptual errors than older women or young adults (Storie, 1977). These findings taken together, suggest that age-related patterns of accident type (California State Dept of Motor Vehicles, 1982) may be due to perceptual differences among older and younger adults.

Assuming that the differences obtained are replicable, and that they extend to three-dimensional scenes (an issue that will be addressed below), one might ask if they could plausibly be related to some age-sensitive visual mechanism. Such an explanation seems unlikely. Older adults were in self-reported good visual health, and possessed very good acuity relative to task demands. Secondly, it is unlikely that either the perception of speed of distance is strongly related to acuity per se. Finally, the ANCOVA suggested that obtained results could not be accounted for by individual differences in acuity.

Rather, it seems that age and gender differences obtained are perceptual in nature, and may relate to experience in the driving task. Recall that older females, who at present have less driving experience than males, judged vehicle velocity to be faster than any other group. Storie (1977) has shown how experience in driving can determine the accuracy of speed estimates. Older men, on the other hand, judged cars to be more distant that any other group. A perceptual learning hypothesis could also account for this observation by arguing that older men have more driving experience with large cars, and when faced with static scenes of a small car, assume prototype size constancy, and so judge the vehicles to be more distant. The size constancy hypothesis is currently being investigated.

Finally, it is important to acknowledge that results obtained with video presentations, which lack binocular depth cues, may not generalize to three-dimensional scenes. It is certainly the case that binocular cues influence depth perception, and presumably velocity judgements as well. By depriving observers of these cues, the present design may have produced individual differences which do not exist in many 3-D scenes. On the other hand, to the extent that 2-D simulations are being used with increased frequency, such a lack of external validity is important in its own right, and must be examined in future research.

REFERENCES

Bell, F., Wolf, E., & Bernholz, C. D. (1972). Depth perception as a function of age. Aging and Human Development, 3, 77-81.

Burg, A. (1971). Vision and driving: A report on research. Human Factors, 13, 79-87.

California Department of Motor Vehicles (1982). Senior driver facts. Report CAL-DMV-RS5-82-B2-HS-033 073 (January).

Harrington, D. M., & McBride, R. S. (1970). Traffic violations by type, age, sex, and marital status. Accident Analysis and Prevention, 11, 67-79.

Henderson, B. L., & Burg, A. (1974). Vision and audition in driving. System Development Corporation Final Report No. DOT H5801265, Santa Monica, CA.

Hills, B. L. (1980). Vision, visibility, and driving. Perception, 9, 183-213.

Hills, B. L. (1981). Some studies on movement perception, age, and accidents. Transportation and Road Research Laboratory, SR-137.

Hills, B. L., & Burg, A. (1977). A reanalysis of the California driver vision data: General findings. Transportation and Road Research Laboratory, LR-768.

Kline, D. W., & Schieber, F. J. (1985). Vision and aging. In, J. E. Birren & K. W. Schaie (Eds.). Handbook of the psychology of aging (2nd). VanNostrand, New York.

Scialfa, C. T., Garvey, P., Gish, K., Goebel, C., Deering, L., & Leibowitz, H. W. (1977). Some relationships among measures of static and dynamic sensitivity. Under review.

Storie, V. J. (1977). Male and female car drivers: Differences observed in accidents. Transportation and Road Research Laboratory, LR-761.

THE EFFECTS OF AGE AND TARGET LOCATION UNCERTAINTY ON DECISION MAKING IN A SIMULATED DRIVING TASK

Thomas A. Ranney
Liberty Mutual Research Center
Hopkinton, Massachusetts

Lucinda A. S. Simmons
Northbridge, Massachusetts

Spatial localization has been identified as an age-sensitive process in selective attention. Because visual search in driving involves uncertainty concerning the location of information necessary for maneuvering decisions, an experiment was conducted to examine the effects of age and target location uncertainty on a simulated driving task. Seventeen younger subjects (aged 30 to 45 years) and 13 older subjects (aged 65 to 75 years) completed three tasks including two reaction-time tasks and a simulated driving task. The reaction-time tasks included three conditions (simple left, simple right, and two-choice) in a laboratory and in a stationary vehicle. The simulated driving task was conducted on a closed driving course while subjects sat in a stationary vehicle. Subjects were required to select one of two lanes using information presented either on a changeable-message sign or on traffic signals. In the high-certainty condition, subjects were told where to look for relevant information; in the low-certainty condition, they were told that information could appear in either place. Response times were measured from sign or traffic signal onset to the subject's activation of the vehicle turn signal. The results indicated small, non-significant differences between age groups for the reaction-time tasks. Significant age-related differences were found in the simulated lane-selection task. Older subjects were 15% slower overall than the younger subjects. Uncertainty concerning the location of relevant information slowed decision-making speed for all subjects, but proportionately more for the older subjects (16% versus 11% for the younger age group). Uncertainty slowed responses to the changeable message sign more than to traffic signals for subjects in both age groups. The results are consistent with the spatial localization hypothesis, and suggest that older drivers may have more difficulty than younger drivers locating targets in visual search while driving. The results also suggest that effective use of changeable-message signs requires placement in locations with high expectancy, and allowing drivers sufficient time to locate the sign before reading the scrolling message.

INTRODUCTION

Age deficits in selective attention have been widely reported. Using a laboratory visual search paradigm, Plude and Hoyer (1985) had young and old subjects identify target letters in conditions which differed according to the number of distractors and whether or not subjects knew the location of the target. They found that the magnitude of the selective attention deficit differed according to whether subjects knew the location of the target. The selective attention deficit was defined as the difference in response times attributable to the presence of distractors. Specifically, there was no selective attention deficit in their nonsearch condition, in which subjects knew the location of the target, but a significant age-related deficit in their search condition, in which the target location was unknown. To explain these findings, they proposed the spatial localization hypothesis, according to which uncertainty about the location of task-relevant information has proportionately larger effects with increasing age (Plude and Hoyer, 1985).

Visual search is an essential component of driving. Unlike most laboratory paradigms where targets are identified in advance, a significant percentage of visual search in driving involves uncertainty concerning the target identity and location (Theeuwes, 1989). Therefore, if spatial localization effects generalize to driving, age-related deficits in driving may reflect difficulties in visual scanning due to increased uncertainty about the location of relevant target information. If this is true, the often-cited overrepresentation of older drivers in intersection accidents involving left turns (Transportation Research Board, 1988) may result in part from difficulties in locating conflicting vehicles in sufficient time to decide correctly whether or not to proceed into an intersection.

The objective of this study was to examine the effects of target location uncertainty and age on decision-making speed in a simulated driving task. This allows evaluation of the generality of the spatial localization hypothesis, which predicts that target location uncertainty will have a proportionately larger effect on older subjects than on younger subjects. A relatively simple two-choice lane-selection task was simulated using a stationary vehicle on a closed driving course. The vehicle was instrumented to record response times based on turn-signal activation. The task required subjects to select the right or left lane using information presented either on traffic signals or a changeable-message sign.

METHOD

Subjects

Subject age group categories were based on differences in mileage-based accident rates. Accident rates are lowest for drivers between the ages of 30 and 60, and begin to increase for drivers over age 65 (Williams and Carsten, 1989). Seventeen subjects aged 30 to 45 (M=38.7, SD=4.5) comprised the younger group; thirteen subjects aged 65 to 75 (M=68.4, SD=2.9) comprised the older group. All subjects were active drivers with at least ten years of driving experience. Vision was screened with a

Titmus Vision Tester. Four of the older subjects and two of the younger subjects failed to meet the standard far visual acuity requirement (20/40) for licensing in Massachusetts. However, all subjects satisfied the minimum visual acuity required to read stimulus material in the present experiment (20/80).[1] No subject reported difficulty completing any of the tasks. Subjects were recruited with newspaper advertisements and paid $10.00 per hour for participation.

Apparatus

The lane-selection task was implemented on a closed driving course. At one end of the course is a simulated intersection, equipped with two standard three-head traffic signals and a changeable-message sign (ADDCO 16 x 128 Dot-sign). The display area of the sign is approximately 30.5 by 162.5 cm (12 x 64 in). Letter height of words presented on the sign was 17.8 cm (7 in) to ensure that the words could easily be identified by all subjects independent of individual differences in visual acuity. The letter height was more than twice the minimum necessary to ensure legibility for drivers with 20/40 visual acuity.[1] The visual angle subtended by the traffic signals and changeable message sign was 9.6 degrees.

A 1988 Ford Crown Victoria LTD was equipped with sensors to record turn-signal activity. Two photo-transistors were connected to an FM transmitter located in the vehicle trunk. The transistors were taped over the rear signal lights and upon turn-signal activation a signal was transmitted to the main computer controlling the task. A schematic of the stimuli and stationary vehicle is presented in Figure 1.

Figure 1. Stimuli layout and stationary-vehicle position for simulated lane-selection task

One experimenter observed subjects' responses from inside the stationary vehicle. A second experimenter controlled the data collection from an instrumentation van located beside the traffic signals. The experiment was controlled by a PC-AT computer inside the van. Tasks were programmed using the Micro-experimental laboratory (MEL) software developed by Schneider (1988).

The laboratory task was implemented on a PC-AT computer connected to a Gerbrands reaction time response box which consists of two telegraph keys, a warning light, and two stimulus lights. A similar stimulus box was placed on the vehicle dashboard for the stationary-vehicle reaction time task.

Procedure

All subjects completed three tasks in one 3-hour session. The three tasks included: (1) reaction time (RT) in the laboratory, (2) RT in the stationary instrumented vehicle, and (3) the two-choice lane selection task in the stationary vehicle.

In the laboratory RT task, subjects activated telegraph keys in response to green or red lights displayed in the center of the stimulus box. There were three conditions: simple left, simple right and two-choice. The stationary-vehicle RT task included the same three conditions. Subjects activated the turn signal lever upward for right or downward for left, as in normal driving. There were 5 practice and 40 trials in each condition.

For the lane-selection task, subjects were seated in the stationary vehicle located between the two approach lanes at a distance of 48.8 m (160 ft) from the intersection. This distance corresponds to the point at which the driver would first see the sign if approaching the intersection in a moving vehicle. Subjects were required to select either the right or left lane by activating the turn signal in the appropriate direction. There were two types of trials: traffic signal and sign, corresponding to the location of the information necessary to make the right-left decision.

At the beginning of each trial, both traffic signals were green and subjects were instructed to fix their gaze on an orange traffic cone, located beyond the intersection at a point midway between the traffic signals and the changeable-message sign, from the perspective of the subject in the stationary vehicle. After a variable delay the stimulus was activated. On traffic-signal trials one traffic signal turned to red, while the other remained green. Subjects were required to select the lane associated with the signal that remained green. The red light indicated that the lane was closed. On sign trials, the changeable-message sign scrolled either the word "LEFT" or "RIGHT".

Subjects completed eight blocks of 32 trials, under two conditions of certainty. In the high-certainty condition, which consisted of two blocks of sign-only trials and two blocks of traffic-signal-only trials, subjects were told where to look for the information. In the low-certainty condition, which consisted of four blocks of mixed trials, they were not told where to look. This resulted in an equal number of trials with each stimulus in each condition. The order of the blocks was counterbalanced to distribute learning effects over all conditions. Subjects were instructed to respond as quickly as possible, without making errors.

RESULTS

Laboratory RT task. Error rates for the laboratory task were less than 1%. There were no differences between age groups. Because there were no differences in response times, the simple left and simple right conditions were combined. Error trials were removed and cell means were computed for each combination of independent variables, which included agegroup (younger, older), condition (simple, choice), and blocks (4 blocks of 10 trials). An

Analysis of Variance (ANOVA) was computed using PROC GLM in SAS/STAT (1989). Overall, the older subjects were approximately 10% slower than the younger subjects, however this difference was not statistically significant ($F(1,28)=3.56$, p=.07). The main effect of condition was significant ($F(1,28)=435.02$, p=.0001), reflecting faster RTs in the simple condition. Means and standard deviations for each combination of agegroup and task condition are presented in Table 1.

	Younger		Older	
Cond.	M	SD	M	SD
Simple	293	47	327	60
Choice	451	69	492	72

Table 1. Means and standard deviations for the laboratory RT task (ms)

The main effect of block was statistically significant ($F(3,84)=5.35$, p=.002), as was the Agegroup x Block interaction ($F(3,84)=6.15$, p=.0008). The interaction effect is shown in Figure 2. Post hoc analyses revealed that differences between the age groups were significant in the first two blocks, but not in the third and fourth blocks.

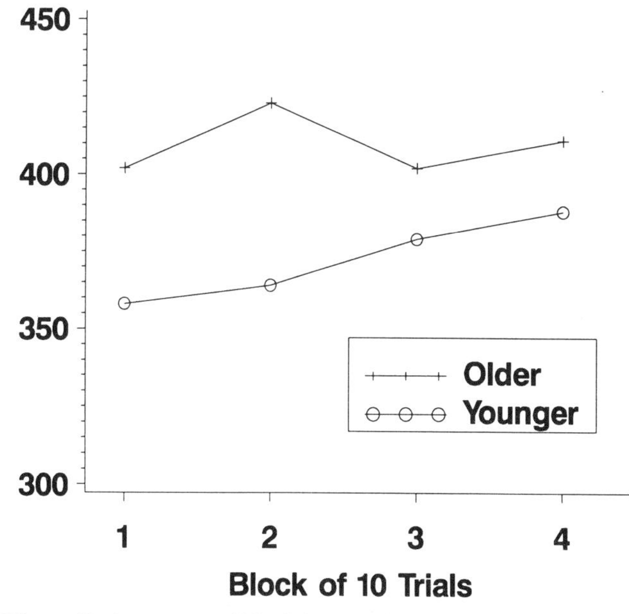

Figure 2. Agegroup x Block interaction for laboratory RT task (ms)

Stationary-vehicle RT task. As with the laboratory task, error rates were negligible for this task and exhibited no differences between the two age groups. Error trials were removed and an ANOVA was computed on the cell means. Independent variables were the same as above, with the exception that condition had three levels (simple, left, right). Overall, the older subjects were approximately 5% slower than the younger subjects. This difference was not statistically significant ($F(1,28)=1.54$, p=.22). The main effect of condition was significant ($F(2,28)=151.71$, p=.0001), reflecting significant differences between all three conditions. Means and standard deviations for each combination of agegroup and task condition are presented in Table 2.

	Younger		Older	
Cond.	M	SD	M	SD
Left	466	62	516	87
Right	565	117	568	131
Choice	729	76	779	104

Table 2. Means and standard deviations for stationary-vehicle RT task (ms)

The main effect of block was significant ($F(3,28)=10.40$, p=.0001), indicating a reduction in response times between the first and subsequent blocks. The Agegroup x Block interaction was not significant ($F(3,28)=1.43$, p=.24)

Lane-selection task. The overall error rate was 1.1%, with no differences between age groups. Error trials were removed and cell means were computed for each combination of independent variables, which included certainty (high, low), format (sign, traffic signals), block (1,2) and agegroup (younger, older). An ANOVA was computed on the cell means. The results indicated a significant main effect of agegroup ($F(1,28)=12.36$, p=.0015). Overall, the older subjects were approximately 15% slower (M=1000 ms, SD=160 ms) than the younger subjects (M=869 ms, SD=132 ms). The significant main effect of certainty ($F(1,29)=72.99$, p=.0001), reflected slower responses in the low certainty condition (M=983 ms, SD=154 ms) than in the high certainty condition (M=868 ms, SD=141 ms). The significant main effect of format ($F(1,29)=41.08$, p=.0001) reflected the fact that responses were faster to the changeable message sign than to the traffic signals (M=869 ms, SD=150 vs. M=982 ms, SD=147). Responses were slightly faster in the second block (M=916 ms vs. M=935 ms in the first block), as indicated by the block main effect ($F(1,28)$= 28.76, p=.0001).

The Agegroup x Certainty interaction, shown in Figure 3, was statistically significant ($F(1,29)=4.87$, p=.04). The Certainty x Format interaction was also significant ($F(1,28)$= 9.05, p=.006). The Agegroup x Format ($F(1,29)=0.00$, p=.97) and Agegroup x Block ($F(1,28)=0.79$, p=.38) interactions were not significant.

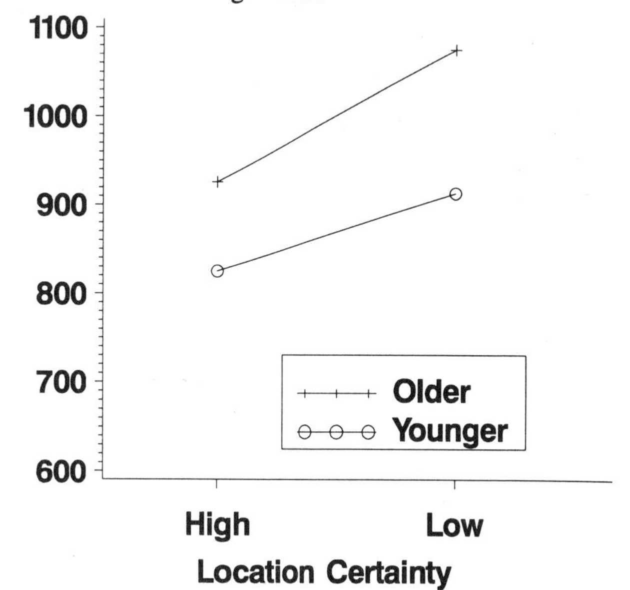

Figure 3. Agegroup x Certainty interaction (ms)

DISCUSSION

The major objective of the experiment was to evaluate the hypotheses that (1) uncertainty about the location of relevant information would impair decision-making speed in a simulated lane-selection task, and (2) that this effect would be proportionately greater for older subjects. The results supported both hypotheses, indicating that uncertainty slowed decision-making speed by approximately 13%, on average. This effect was greater for older subjects than for younger subjects, representing an increase of approximately 16% for the older subjects and 11% for the younger subjects. This finding is consistent with the spatial localization hypothesis of Plude and Hoyer (1985).

The lane-selection task used in the current study differed from that used by Plude and Hoyer (1985) in several respects. First, while Plude and Hoyer used a simple letter-detection task, the lane-selection task required subjects to interpret patterns of traffic signals and to read scrolling sign messages. The layout of stimuli in the signalized intersection required scanning of a larger visual array than in the laboratory study, which is also more consistent with everyday driving. Secondly, the performance measures used in the two studies differed. Plude and Hoyer evaluated the effects of location uncertainty on the selective attention deficit, defined as the difference in response times between conditions involving the presence and absence of distractors. The present study evaluated the effect of target location uncertainty more directly, using decision-making speed as the performance measure.

The absence of age differences in the laboratory and stationary-vehicle RT tasks is not consistent with the well-established slowing of behavior with increasing age (e.g., Salthouse, 1985). Possible explanations for this discrepancy include the relatively small sample sizes used in the present study, the lack of representativeness of the current samples, or instability of the RT scores due to an insufficient amount of practice. The sensitivity of the tasks is not suspected since they have previously elicited age differences (Ranney and Pulling, 1990). Differences obtained in this earlier study for the identical simple lab RT condition were larger than in the current study (16% vs. 12%). However, the previous results were obtained with older subjects (74+) and larger samples than used in the current study.

Our experience with volunteer subjects supports the possibility that the current samples were not representative of their respective populations. Unlike the majority of the population, our test subjects are all actively reading newspaper ads and seeking employment. Whether this could bias the samples in terms of the skills necessary for the present tasks is unknown, however motivational differences between the age groups are often apparent. The older subjects we recruit are typically healthy, confident and seeking a new experience. Our conclusion that they represent an above-average segment of their population is reinforced by our difficulties in recruiting unwilling spouses or friends. In contrast, many of the younger subjects are unemployed and motivated by their need to earn money. These subjects are more likely to be distracted and impatient, especially with the more tedious parts of our protocol.

Some support for these arguments is provided by the changes in lab RT scores observed over blocks of trials (see Figure 2). The gradual increase in RTs over blocks observed for the younger group is consistent with increasing distraction and impatience. In contrast, the more variable pattern exhibited by the older subjects suggests slower adaptation to the task requirements, consistent with Salthouse's (1985) recommendation that a minimum of 20 to 30 trials is required to obtain stable RTs in simple tasks. If this interpretation is correct, it may be extremely difficult to select an optimal number of trials for use with both age groups, and clearly the number of trials used can influence the magnitude of differences observed.

The pattern of results from the stationary-vehicle RT task differs from that observed in the lab task, in that the absence of differences between the two age groups was consistent over the blocks of trials. Both groups exhibited a gradual reduction in RTs, consistent with a learning effect. One reason for including this task was to provide practice in the rapid activation of the turn signal lever, a response which differs from the everyday use of vehicle turn signals. The considerably slower RTs in the right (vs. left) condition (see Table 2), observed for all subjects, reflect the more complicated motor response involved in activating the lever upward to indicate right. The absence of age-related differences in this condition is consistent with our hypothesis concerning the generally good health (e.g. absence of impaired motor performance due to arthritis) of our older test subjects. The 11% difference between age groups in the left condition is comparable to the difference observed in the simple condition in the lab RT task.

Significant differences between age groups were observed in the lane-selection task. The overall difference was 15%, as compared to the 5%-10% differences observed in the two RT tasks. If it can be assumed that the three tasks involving choice (lab choice RT, stationary-vehicle choice RT, lane-selection) consist of the same three successive independent processes, the logic of the Subtraction Method (e.g., Pachella, 1974) can be used to estimate the duration of the components of the lane-selection task and the magnitude of the age effect on each component. Specifically, we assume that all three tasks include stages for: encoding; decision and response selection; and response execution. Secondly, we assume that the lab simple RT condition differs from the lab choice RT condition only by the absence of the decision component. Thirdly, we assume that the stationary-vehicle RT task differs from the lab task only by the added requirement of turn-signal activation. Finally, we assume that the lane-selection task differs from the stationary-vehicle RT task only in the perceptual encoding requirement of reading sign messages and/or interpreting patterns of traffic signals. Based on these assumptions, we define the components of the lane-selection task as shown in Table 3. It is important to note that these components do not correspond directly to stages of information-processing. For example, the sign/signal reading component does not include all encoding, but only the incremental encoding beyond that required in the choice RT tasks. Similarly, the turn-signal activation component does not include all response-execution time, but only that unique to turn-signal activation. The information provided by the current experiment is not sufficient for further decomposition using deterministic methods, however we are currently verifying and extending this approach using estimation techniques.

Component of Lane-Selection Task	Def.	Mean RTs Older	Young	Contribution to Age Effect	
		ms	ms	diff	%
Simple reaction time	A	327	293	34	26
Two-choice decision	B-A	165	158	7	5
Read sign/traffic signals	D-C	221	140	81	62
Activate turn signals	C-B	287	278	9	7
Total	C	1000	869	131	100

Tasks: A: Lab simple RT, B: Lab choice RT,
C: Stationary-vehicle choice RT, D: Lane-selection

Table 3. Component contribution to age effect on lane-selection task

Using these definitions, we see that the 62% of the age difference observed in the lane-selection task was associated with reading the changeable-message sign and/or interpreting the traffic signals. The second largest contribution (26%) was associated with simple reaction time. Smaller contributions were associated with the left/right decision (5%) and turn-signal activation (7%).

Results of the lane-selection task allow examination of differences between subjects' responses to changeable-message signs and traffic signals. Overall, responses were faster to the sign messages than to the traffic signals. This difference reflects, in part, the fact that response times were computed from the time at which the sign completed top-to-bottom scrolling of the message. During this interval of approximately 0.75 s, subjects could use partial information to start their decision. The absence of the Agegroup x Format interaction indicates no differences between the age groups in ability to take advantage of partial information. However, as reflected by the significant Certainty x Format interaction, uncertainty about which stimulus would activate was more detrimental on the sign trials than on traffic-signal trials. This suggests that the subjects were less able to take advantage of partial sign information during scrolling, when they were uncertain where the information would appear.

Several conclusions can be drawn from the present results. First, uncertainty about the location of relevant information clearly slowed decision-making speed. The fact that this effect was proportionately greater for older subjects suggests that difficulties in locating targets, such as other vehicles or pedestrians, may impair older drivers' abilities to make appropriate decisions while driving. Furthermore, if the differences observed in the lab and stationary-vehicle RT tasks understate actual differences between the respective populations of middle-aged and older adults, then the observed differences between age groups in the lane-selection task may also underestimate the actual effects that exist in the respective populations. Replication of the current experiment with non-volunteer subjects would be necessary to evaluate this hypothesis.

With regard to selection and placement of traffic control devices, the results indicate that older subjects were slowed equally by the requirements of interpreting traffic signals and the changeable-message sign. However, the disproportionate effect of uncertainty on responses to the changeable-message sign, observed for subjects in both age groups, suggests that efficient interpretation of changeable-sign messages requires signs to be placed in predictable locations. In addition to sufficient time for reading of scrolling messages, traffic engineers must allow additional time for drivers to locate and direct their attention to changeable-message signs. This is especially important when signs are used in unpredictable locations, such as in workzones.

REFERENCES

Pachella, R.G. (1974). The interpretation of reaction time in information-processing research. In B.H. Kantowitz (Ed.), *Human information processing*. Hillsdale, NJ: Erlbaum Associates.

Plude, D.J., and Hoyer, W.J. (1985). Attention and performance: Identifying and localizing age deficits. In N. Charness (Ed.) *Aging and Human Performance*. New York: J. Wiley and Sons.

Ranney, T.A. and Pulling, N.H. (1990). Performance differences on driving and laboratory tasks between drivers of different ages. *Transportation Research Record No. 1281.*

Salthouse, T.A. (1985). Speed of behavior and its implications for cognition. In J.E. Birren and K.W. Schaie, *Handbook of the Psychology of Aging, Second Edition.* New York: Van Nostrand Reinhold.

SAS Institute (1989). SAS/STAT User's Guide, Version 6, Fourth Edition, Volume 1. Cary, ND: Sas Institute Inc.

Schneider, W. (1988). Micro experimental laboratory: An integrated system for IBM PC compatibles. *Behavior Research Methods, Instruments, & Computers, Vol.* 20, No. 2. 206-217.

Shinar, D. (1978). *Psychology on the Road: The Human Factor in Traffic Safety*. New York: J. Wiley and Sons.

Theeuwes, J. (1989). Visual selection: Exogenous and endogenous control; A review of the literature. Soesterberg, the Netherlands: TNO Institute for Perception: Report IZF-1989 C-3.

Transportation Research Board (1988). *Transportation in an aging society*. (Vol. 1; Special Report 218). Washington D.C.: National Research Council.

Williams, A.F. and Carsten, O. (1989). Driver age and crash involvement. *American Journal of Public Health, Vol.* 79, No. 3, 326-327.

Footnote

[1]Traffic engineering guidelines recommend one inch of letter height for every 50 feet of viewing distance (Shinar, 1978). The 7-inch letters used in this experiment would accommodate drivers with 20/40 vision at a distance of up to 350 feet. Drivers with 20/80 vision could be expected to read the sign at a distance of 175 feet.

BRAKE PERCEPTION-REACTION TIMES OF OLDER AND YOUNGER DRIVERS

Neil D. Lerner
COMSIS Corporation
Silver Spring, MD

The time drivers require to react in braking situations underlies many practices in highway design and operations. There is concern whether the perception-reaction time (PRT) values used in current practice adequately meet the requirements of many older drivers. This study compared on-the-road brake PRTs for unsuspecting drivers in three age groups: 20-40, 65-69, and 70-plus years old. The method included features to enhance the ecological validity of the observed reactions: subjects drove their own vehicles in their normal manner; driving was on actual roadways; extended preliminary driving put the driver at ease and without expectation of unusual events at the time of the braking incident; the incident occurred at a location lacking features that might enhance alertness (e.g., curves, crests, driveways). Subjects drove an extended route, under the guise that they were making periodic judgments about "road quality." At one point, a large crash barrel was remotely released from behind brush on a berm and rolled toward the driver's path. Although most of the fastest observed PRTs were from the young group, there were no differences in central tendency (mean = 1.5 s) or upper percentile values (85th percentile = 1.9 s) among the age groups. Furthermore, the current highway design value of 2.5 seconds for brake PRT appears adequate to cover the full range of drivers.

INTRODUCTION

Driver perception-reaction time (PRT) is a fundamental concept in highway design and safety. PRT refers to the time required to perceive, interpret, decide, and initiate a response to some stimulus. Different driving situations are characterized by different assumed values of PRT. For example, the PRT associated with reacting to an obstacle in the vehicle's path (brake reaction time) is different from the PRT associated with determining that it is safe to begin moving through an intersection, or with the time required to judge whether it is safe to initiate a passing maneuver. One of the most basic concepts for highway design and operations is that of stopping sight distance (SSD). SSD is the distance traveled before coming to a stop when a driver, travelling at roadway design speed, must brake in reaction to an unexpected obstacle in his or her path. Driver PRT is a key determinant of SSD, so that the specific PRT value assumed for design purposes influences many aspects of the roadway. Currently, the assumed value for design equations is 2.5 s (AASHTO, 1990). This parameter ultimately influences such roadway features as horizontal curve radius, vertical curves (crests), approaches to intersections, sign placement, traffic signal visibility and phasing, and other common roadway aspects.

Recently, there has been a good deal of concern about whether the assumed PRT values in design equations are adequate to meet the requirements of older drivers (Lerner, 1991). Although assumed values are only loosely linked to realistic driving data, to the extent they are empirically based, the underlying research has not appropriately considered elderly drivers. The general slowing of responses with age is a broad and well established laboratory finding (e.g., Salthouse, 1985), including studies showing slower foot pedal reaction times for older subjects (e.g., West Virginia University, 1988). To the extent that older drivers require longer PRTs than those established for highway design practice, they may be at a serious disadvantage for numerous driving situations. For safety reasons, it is important that design and operational practices meet the needs of all segments of the driving population. However, increasing the assumed PRT value for highway design would have extensive ramifications, and substantial costs.

Although PRT generally slows with age, there are some reasons for presuming that current sight distance criteria may be adequate for older drivers and do not require change. Lerner (1991) has discussed a number of these. Among the more interesting is the finding that the few studies that have measured on-the-road PRTs for unsuspecting drivers of different ages have <u>not</u> obtained meaningful differences between age groups (Olson, Cleveland, Fancher, Kostyniuk, and Schneider, 1984; Korteling, 1990; Hostetter, McGee, Crowley, Sequin, and Dauber, 1986). Furthermore, even if older drivers require somewhat longer response times, the "cushion" built into current design parameters may be adequate to cover their behavior.

The purpose of the present study, funded by the Federal Highway Administration, was to measure realistic, on-the-road braking PRTs for unsuspecting older and younger drivers, and to determine whether the currently assumed design value of 2.5 s is adequate for drivers of all ages. The study was designed to provide ecologically meaningful data, avoiding some of the limitations of past research. Additional detail on methods and analysis can be found in a project final report to the Federal Highway Administration (Lerner, Huey, McGee, and Sullivan, in preparation).

METHOD

The purpose of the study was to elicit and measure realistic brake reaction times from drivers. Given this, several methodological features were deemed important. The subjects should be driving on actual roadways. They should be driving normally and at ease. They should have no expectation of an emergency braking event and should not be leery of the motives of the experimenter. They should be driving their own vehicles, rather than trying to adapt to an experimenter-provided vehicle. This was of particular concern for older drivers, who may have selected or adapted their vehicles to meet their needs, and who may not adapt to a test vehicle as quickly.

In order to meet these methodological requirements, subjects drove their own vehicles over an extended route, under the guise that they were participating in a study that was recording their judgments of "road quality." Periodically (at stop signs), they made judgments about the quality of the road sections they had just travelled. Two slightly different procedures were involved. For some subjects, the initial part of the drive was incorporated with another experiment, and the subject had been driving about an hour prior to encountering the braking event. For the other subjects, the prior part of the drive was briefer, covering about three miles. (Brake reaction times for the two groups were not significantly different, t=0.08, and the datasets were combined for analysis). It was only at the completion of the route that the subjects encountered the site of the braking event; from their perspectives, they were simply continuing the ride, and the procedure, that had been in effect all along. They were driving in a relatively relaxed, normal manner, with no expectation of any unusual event (post-session debriefings confirmed this). The subjects turned onto a four lane divided highway. This highway provides access to an interstate highway, and then continues on for 0.7 miles beyond the freeway entrance. This extended stub of roadway is a functional, fully delineated roadway, but is closed to normal traffic by the use of barricades. Subjects were instructed that we had permission to continue on this road and drove around the barricades, and were told the appropriate speed was 40 mph. When the vehicle reached a location near the midpoint of the roadway section, a large yellow crash barrel, hidden on a berm behind some brush, was remotely released and suddenly became visible rolling toward the roadway. Although it appeared to be rolling directly into the road, a set of chains held the barrel to the shoulder area. The barrel emerged into view approximately 200 feet in front of the vehicle; this provided a time-to-collision of about 3.4 s at the target speed of 40 mph.

The data were recorded using a video-based data collection system. For subjects with the first procedure, the emergence of the barrel was recorded by an in-vehicle camera, and the occurrence of the brake response by activation of a pressure-sensitive tape switch attached to the brake pedal. In the other procedure, a hidden roadside microcamera recorded both the emergence of the barrel and the activation of the vehicle's brake lamps. The PRT was the interval between the emergence of the barrel and the initiation of braking.

All sessions were run in daytime and clear weather. Subjects were recruited in three age groups: 20-40 years old, 65-69 years old, and 70 or older. To minimize the selection bias toward more capable elderly, older subjects were recruited in the greater Washington, DC area, working through senior centers, churches, retirement communities, and so forth. Rather than placing initiative on the subject to volunteer, as much as possible we worked with directors of the institutions to help identify and approach individuals with a wide range of capabilities, and to provide social support and incentive for taking part. Although there can be no claim that the sample was representative, and while it is likely that those at the extreme lowest limits of ability and confidence tended to exclude themselves, the older group did appear to provide a broadly suitable range, and certainly included many individuals who would have been unlikely to participate without more active recruiting strategies.

Although over 200 subjects participated in the study, there was a very high rate of data loss due to a combination of factors, including equipment failures, video problems, weather, experimental error, inappropriate subject behavior, unauthorized traffic at the site, and so forth. Valid trials and records were obtained for 116 subjects; this included 30 in the 20-40 year old group, and 43 in each of the two older groups.

RESULTS

Nearly all (87%) of the 116 drivers made some overt vehicle maneuver in response to the emergence of the barrel. Of these, about 43% both steered and braked, 36% steered only, and 8% braked only. Thus, just over half of the drivers (51%) reacted by braking. This is consistent with various other on-the-road studies, which have found steering to be a more reliable reaction, and with the percentage of braking drivers varying from about 30% to about 80% (e.g., Triggs and Harris, 1982). Measurable brake reaction times were obtained for 56 subjects; this included 14 of the 20-40 year olds, 18 of the 65-69 year olds, and 24 of those 70 or older.

The mean brake PRT for all subjects was 1.5 s, with a standard deviation of 0.4 s. An analysis of variance revealed no significant main effect of age or gender, with the interaction of age with gender of borderline significance ($p=0.055$). The interaction reflects the particularly short mean reaction time of the young female group (1.22 s), while other age-by-gender group means ranged from 1.40 to 1.65 s. The age groups not only showed little difference in central tendency, but also in terms of upper percentile values. The 85th percentile PRT (a level often used in developing highway design values) was about 1.9 s for all age groups. Virtually all responses were captured by the 2.5 s design value; the longest observed time was 2.54 s and the next longest was 2.39 s.

However, the absence of differences in central tendency or 85th percentile values does not necessarily imply that older and younger groups were responding in the same manner. Most of the fast reaction times (e.g., <1.25 s) were provided by the young drivers. However, the distribution of reaction times for young drivers was bi-modal. Most were less than 1.45 s, but a few were quite long (over 1.9 s). In contrast, for the older groups, about half of the cases fell in the roughly half-second interval between these values. Given the relatively small number of observations for the young age group, the differences between these distributions must be viewed with caution; a chi square test approached, but did not reach, conventional statistical significance levels (Chi square = 11.2, 6 df; $0.05<p<0.10$).

Because subjects drove their own vehicles, it is possible that differences (or the absence of differences) between age groups could be attributable to differences in the vehicles they drive. The major difference noted was that almost all of the older driver's vehicles had automatic transmissions, while only about two-thirds of the younger group did. However, post hoc comparisons indicated that transmission type had no discernible effect on either the nature of the reaction to the barrel (brake, steer) or the speed of braking.

DISCUSSION

This study provided unique data on the brake PRTs of drivers of different ages. It used real drivers, of known ages, in their own vehicles, driving on actual roadways, under conditions where they were not expecting any unusual (emergency braking) event. None of the previous research has met all of these criteria, which are important for deriving "absolute" measures of brake reaction time that are ecologically valid.

Consistent with some earlier on-the-road studies, there was no indication of meaningfully slower braking by older groups, as measured by central tendency (mean, median) or upper percentile values (e.g., 85th percentile). Young drivers responded quickly more often, but also showed a higher proportion of slow reaction times. There may be a variety of explanations for this. Furthermore, this study only measured the time to initiate braking, and did not measure the braking profile (deceleration rate) or degree of driver control. The age groups could differ in these regards. However, based strictly on PRT, older groups were not slower to react. The original research plan anticipated sufficient data so that reliable frequency distributions could be obtained for each age group. However, due to data loss and the fact that only about half the drivers in such experiments respond by braking, actual brake reaction times were measured for only 56 of the original participants. Therefore, a great deal of precision should not be attached to the observed values. Nonetheless, this sample is sufficient to indicate that there is no important difference in the response times of the various age groups.

The 2.5 s value used for PRT in highway design applications appears to provide adequate coverage for the full range of driver age. The longest response time recorded was at this value. The mean was a full second faster, with 2.5 s being more than two standard deviations slower, and the estimate of the 85th percentile being more than a half-second faster than this value. The findings of this research are consistent with a number of other on-road studies that have observed brake or steering response upper percentile (e.g., 85th percentile) times of less than 2 s (e.g., Olson et al., 1984; Triggs and Harris, 1982; Sivak, Olson, and Farmer, 1982; Sivak, Post, Olson, and Donohue, 1981; Allen Corporation, 1978; Summala, 1981). There have been some studies that have observed longer brake response times, but these have not been for situations appropriate to SSD. For example, longer times may be observed in response to stimuli such as highway signs or changes in signal phase, or where there are more difficult maneuver requirements related to complex roadway geometries (treated as "decision sight distance" situations in traffic engineering terms). Some reviewers have failed to discriminate these different situations. There have also been calls to increase the PRT design value beyond 2.5 s in order to compensate for the effects of factors such as age, fatigue, or impairment. Based on typical research findings to date, the 2.5 s value already provides some design cushion (of over 0.5 s); it is not clear what the empirical basis for extending the design value further is. While any lengthening of the design value might provide some additional marginal increment in protection, it would necessitate ubiquitous changes, with substantial costs involved, throughout the highway system. Thus the safety benefits of increasing the assumed PRT for SSD beyond 2.5 s should be demonstrably significant.

There are some reasons why the present procedure may arguably have led to relatively conservative estimates of PRT. An analysis of the brake reaction times of drivers who also showed substantial steering in reaction to the barrel, compared to those that braked only, showed that those who steered had somewhat slower brake PRTs (about a quarter-second slower). Thus by providing an opportunity for avoidance steering, as well as braking, the estimate of brake reaction time may be somewhat longer than for a high-emergency situation in which braking is the only clear alternative. Furthermore, the "hazard" emerged from a hidden location off the side of the road, rather than being in more central view. Previous studies may also have used situations in which there was greater expectancy of the potential need for braking, either because of a geometric feature (coming over the crest of a hill, where sight distance was obscured) or because of the need to monitor leading traffic (where the stimulus to brake was the illumination of the brake lamps of a vehicle immediately ahead). In the present experiment, subjects were simply driving on a straight section of roadway, so that the occurrence of any conflict was quite unexpected. Whatever the reason, this study observed mean times that were somewhat slower (by about 0.15 to 0.35 s) than those reported in other on-road studies, such as those cited above. Despite this, the 2.5 s PRT design value covered the range of observed brake times.

The absence of substantially slower brake PRTs among older groups provides an illustration of how the factors of expertise and compensation in complex skills can maintain performance even in the face of reduced capabilities. Virtually all of the component psychomotor processes that underlie PRT -- information processing rate, visual search time, response initiation, movement time, etc. -- have been shown to slow with age, in laboratory studies. Age-related compensation is poorly understood for driving, as it is for various other skilled tasks. Furthermore, the mechanisms involved in compensation for one aspect of performance might be related to degradation in other aspects of performance (e.g., braking may be more "all-or-none," providing a greater risk of rear end collisions or loss of vehicle control). One possible explanation for the absence of a difference between age groups is that older drivers might be responding in a more reflexive, stereotyped manner. Younger drivers may be prone to do more evaluation before responding, or respond in a more gradual or controlled manner, using their faster information processing capabilities to refine the response, rather than quicken it. It was the subjective opinion of the primary research assistant who accompanied the drivers that the older drivers tended to make more evident and dramatic foot movements while braking, although he did not notice a subjective sense of more severe deceleration. This observation is also consistent with the data of Olson et al. (1984), whose instrumentation allowed the total PRT to be segmented into a "perception time" (from first target visibility to release of the accelerator) and a "response time" (from release of the accelerator to stepping on the brake pedal). Although there was little difference between age groups in the total PRT, the older group actually had faster "response times" (estimating from figures, about 0.1 s faster at the 50th percentile and about 0.2 s faster at the 90th percentile). Thus based on our observations and the Olson et al. findings, it may be that in a situation where there is a surprise need for possible braking, older drivers are more consistent in making a rapid move to the brake pedal, once the hazard has been recognized. The response may be more stereotyped, and subject to less evaluation and modulation. Whatever the reason for the absence of observed differences in overall brake PRT between age groups for SSD situations, it is none the less clear that most older drivers can continue to react with appropriate swiftness, even to an unanticipated braking event. It should be noted, however, that where the stimulus events and required driving maneuvers are more complex and ambiguous than emergency braking, there might be more deleterious effects of age.

REFERENCES

AASHTO (1990). *A Policy on Geometric Design of Highways and Streets*. Washington, DC: American Association of State Highway and Transportation Officials.

Allen Corporation (1978). *Field Validation of Tail Lights -- Report on Phase I*. Report under contract DOT-HS-7-01756. Washington, DC: National Highway Traffic Safety Administration.

Hostetter, R., McGee, H., Crowley, K., Sequin, E., and Dauber, G. (1986). *Improved Perception-Reaction Time Information for Intersection Sight Distance*. FHWA/RD-87/015. Washington, DC: Federal Highway Administration.

Korteling, J. (1990). Perception-response speed and driving capabilities of brain-damaged and older drivers. *Human Factors*, *32*, 95-108.

Lerner, N. (1991). Older driver perception-reaction time and sight distance design criteria. *ITE 1991 Compendium of Technical Papers*. Washington, DC: Institute of Transportation Engineers.

Lerner, N., Huey, R., McGee, H., and Sullivan, A. (in preparation). *Older Driver Perception-Reaction Time For Intersection Sight Distance and Object Detection*. Final report on Contract DTFH61-90-C-00038. Washington, DC: Federal Highway Administration.

Olson, P., Cleveland, D., Fancher, P., Kostyniuk, L., and Schneider, L. (1984). *Parameters Affecting Stopping Sight Distance*. NCHRP Report 270. Washington, DC: Transportation Research Board.

Salthouse, T. (1985). Speed of behavior and its implication for cognition. In J. Birren and K. Schaie (Ed.s), *Handbook of the Psychology of Aging*. New York: Van Nostrand Reinhold Co.

Sivak, M., Olson, P., and Farmer, K. (1982). Radar-measured reaction times of unalerted drivers to brake signals. *Perceptual and Motor Skills*, *55*, 594.

Sivak, M., Post, D., Olson, P., and Donohue, J. (1981). Driver responses to high-mounted brake lights in actual traffic. *Human Factors*, *23*, 231-236.

Summala, H. (1981). Driver/vehicle steering response latencies. *Human Factors*, *23*, 683-692.

Triggs, T. and Harris, W. (1982). *Reaction Times of Drivers to Road Stimuli*. Melbourne, Australia: Monash University, Department of Psychology.

West Virginia University (1988). *Physical Fitness and the Aging Driver, Phase I*. Washington, DC: AAA Foundation for Traffic Safety.

BRAKING RESPONSE TIMES FOR 100 DRIVERS IN THE AVOIDANCE OF AN UNEXPECTED OBSTACLE AS MEASURED IN A DRIVING SIMULATOR

Nancy L. Broen
Dean P. Chiang
Dynamic Research, Inc
Torrance, California

This study examined the effect of brake and accelerator pedal configuration on braking response time to an unexpected obstacle. One hundred subjects drove in the Dynamic Research, Inc, (DRI) Interactive Driving Simulator through a simulated neighborhood 21 times, each time with a different pedal configuration. Each subject was presented with an unexpected obstacle only one time, for one of three previously selected pedal configurations, to which he or she was instructed to brake as quickly as possible. Foot movements were recorded with a video camera mounted above the pedals. Data were analyzed manually, using time and course location information superimposed on the video data. Response times were analyzed using ANOVA to determine effects of pedal configuration and various driver factors. Response times ranged from 0.81 sec to 2.44 sec with a mean of 1.33 sec and a standard deviation of 0.27 sec. There was no significant effect of pedal configuration on response time. Driver age was significant, with increased age corresponding to increased response time. Car normally driven, gender, driver height, and shoe size had no significant effect.

INTRODUCTION

This paper summarizes the braking response time to an unexpected obstacle in the roadway, measured for 100 driver subjects in an interactive driving simulator. The braking response data were extracted from a large-scale 7 month study in which 100 subjects were used to evaluate the effects of 21 pedal configurations and other driver workspace variables on driver performance, comfort, and preference. The unexpected obstacle braking task was one of a number of tasks embedded in the overall driving scenario.

For each pedal configuration, the driver subjects were asked to drive through a simulated suburban neighborhood, operating at the posted speed limits and stopping for signals. While driving, the subject's foot locations and movements were videotaped with the run number, vehicle speed, x-y position on the roadway, heading angle, and the time superimposed at the top of the video data screen. Event lights were located above the accelerator pedal and the brake pedal to precisely indicate pedal actuation in the video data.

Each driver saw the unexpected obstacle only once. This occurred for one of three specified pedal configurations, unknown to the driver. As the driver was maintaining the posted speed limit of 25 mph (foot on the accelerator pedal), two pedestrians moved into the street, unexpectedly, and the driver responded by braking as quickly as possible.

The resulting unexpected obstacle braking response time data were analyzed manually, using the event lights and other digital information from the videotape.

EXPERIMENTAL METHOD

This study was conducted using the DRI Interactive Driving Simulator. The elements of the simulator included: the cab, a 1987 mid-sized sedan with automatic transmission adapted for use in the simulator; a graphics computer which hosted the equations of motion, the roadway display generator driving the projection displays on screens, one in front of and one to the rear of the cab; a one degree of freedom (longitudinal) motion system to simulate acceleration cue onset; an electromechanical steering feel loader; an audio system; and supporting hardware and software.

The brake and accelerator pedal locations were repositioned between runs, using digital servos, so that each subject drove each of the 21 pedal configurations for one warmup run and then one test run. The 21 pedal configurations were presented to the driver subjects in randomized order. The pedal positions were defined by the lateral location of the right edge of the brake pedal relative to the hip point center (HPC); the gap between the brake and accelerator pedals; and the height of the brake pedal face above the accelerator pedal face. Three of the 21 configurations were selected for the presentation of the unexpected obstacle task, one of the three used with each subject. The three pedal configurations were: lateral locations of 10, 30, and 60 mm; gap of 65 mm; and height of 40 mm. They are shown in Figure 1. The steering column lateral offset relative to the HPC was 5 mm to the right, for all pedal configurations.

Prior to beginning a run, the subjects were instructed to obey all traffic signs and signals, and to stay in the center of the lane. As an incentive, they were rewarded monetarily for good path and speed maintenance performance. They were also told that occasionally, an unexpected obstacle may appear in the vehicle's path, and in that event they should step on the brake and stop as quickly as possible.

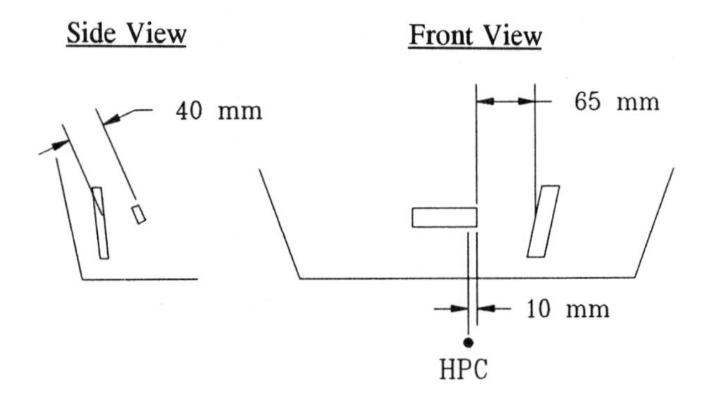

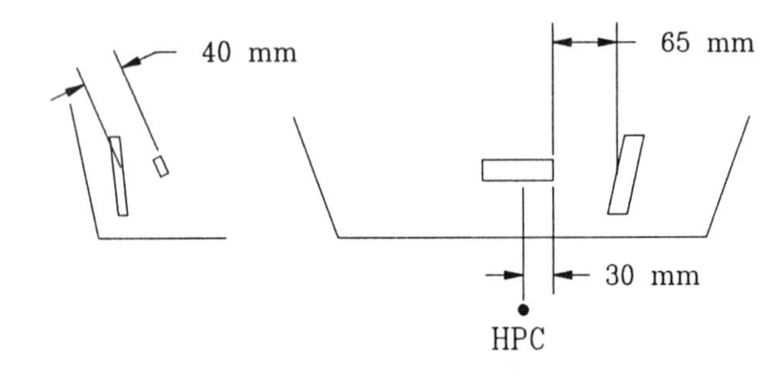

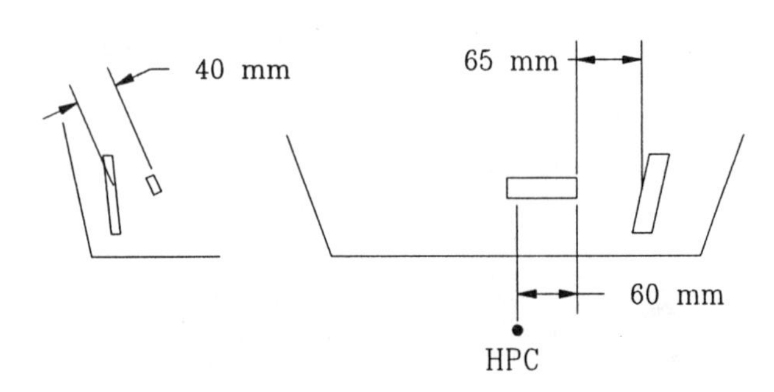

Figure 1. Three Pedal Configurations Selected for Unexpected Obstacle Avoidance

Each subject's first run was a familiarization run in which the experimenter rode in the passenger seat to explain the task, the course, and the controls. For the remainder of the experiment, the subject drove alone in the cab through a simulated neighborhood, sketched in Figure 2. The subjects were in continuous verbal communication with the experimenter. Each subject drove the warmup and test run for each of the 21 pedal configurations, with the unexpected obstacle run presented randomly, only one time.

The unexpected obstacle was located at a specified position along the course, defined by the x coordinate, which was superimposed on the video data. As instructed, when the obstacle appeared the drivers were going 25 mph with their right foot on the accelerator pedal and the accelerator event light illuminated.

The 100 driver subjects who participated in this experiment were recruited from the general population through a marketing research firm. The subjects were chosen based on age, gender, and car normally driven. The participants had the following characteristics:

- Age (years):
  - 18-30 (17 drivers)
  - 31-50 (57 drivers)
  - 51 and over (26 drivers)
- Gender: 50 F and 50 M
- Car normally driven:
  - Brand A: 15 F and 15 M
  - Brand B: 10 F and 10 M
  - Brand C: 25 F and 25 M

All of the drivers were predominantly right foot brakers.

The driver subject foot movement data were collected on videotape using a NEC Model TI-23A video camera mounted under the instrument panel and focussed on the pedal area. Superimposed on the videotape was the digital information described above.

The independent variables for this portion of the study included the pedal configuration (3) and the driver subject characteristics listed above. The dependent variable considered in this paper was the braking response time to the unexpected obstacle. Other extraneous variables considered in the response time data analysis were driver height and shoe size.

RESULTS

For this study, the braking response time was defined as the sum of the reaction time and the movement time. Reaction time was defined as beginning when the pedestrians stepped into the lane of traffic, and ending when the driver initiated a foot movement. For data reduction purposes, these times corresponded to the vehicle x coordinate superimposed on the videotape and the extinction of the accelerator pedal event light, respectively. The movement time was defined as the time from initiation of foot movement until illumination of the brake pedal event light.

The video data were reduced manually on a professional quality VCR, using the event lights, the x-y coordinates, and the time.

The braking response and component times for the three pedal configurations over the 100 driver subjects are summarized in Table 1. The cumulative frequency distributions of the braking response times, for the 3 pedal configurations are plotted in Figure 3, grouped in 0.1 sec bins.

The effects of the independent variables on the braking response times were analyzed using analysis of variance (ANOVA). The results showed no significant effect of pedal configuration at the .05 significance level, $F(2,97) = 2.44$, although a trend is apparent in Figure 3

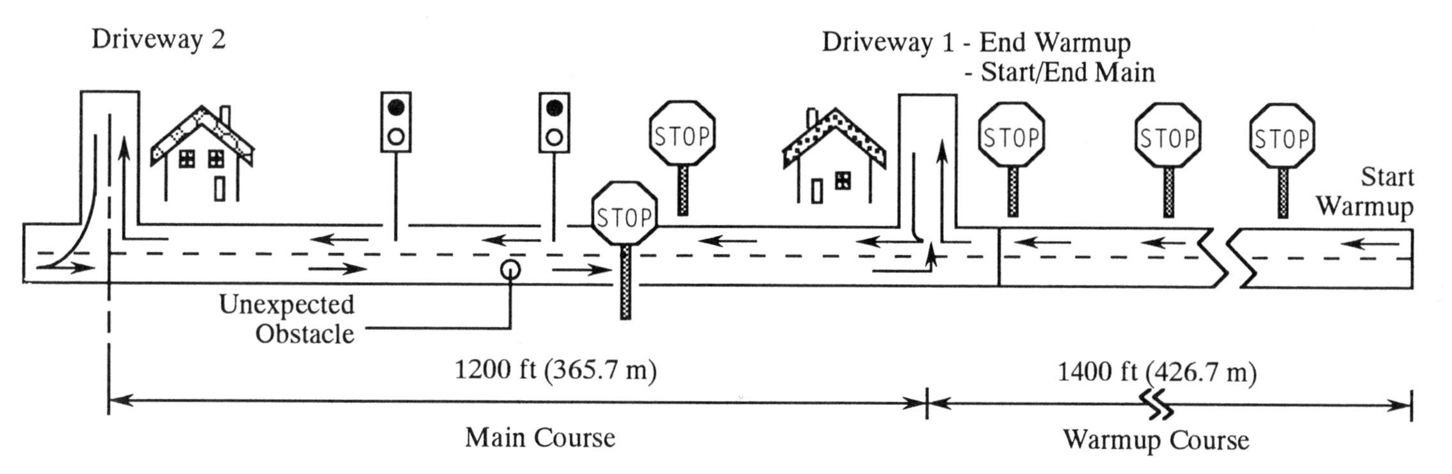

Figure 2. Sketch of Roadway Course Used for Braking Response Time Study

Table 1. Braking Response and Component Times

Reaction Time (sec)

Lateral separation (mm)	n	Range	Mean	Std. Dev.
10	32	0.87-1.76	1.20	0.22
30	29	0.86-1.47	1.19	0.17
60	39	0.57-1.76	1.12	0.27
Overall	100	0.57-1.76	1.16	0.22

Movement Time (sec)

Lateral separation (mm)	n	Range	Mean	Std. Dev.
10	32	0.10-1.14	0.20	0.18
30	29	0.10-0.47	0.16	0.07
60	39	0.07-0.41	0.17	0.07
Overall	100	0.07-1.41	0.17	0.11

Braking Response Time (sec)

Lateral separation (mm)	n	Range	Mean	Std. Dev.
10	32	1.00-2.44	1.40	0.30
30	29	1.03-1.83	1.34	0.20
60	39	0.81-2.10	1.28	0.29
Overall	100	0.81-2.44	1.33	0.27

and Table 1. The braking response time for the 60 mm lateral location appears to be generally between 0.1 sec and 0.2 sec faster than for the other 2 configurations. In fact, the pedal configuration does show a significant effect at the .1 significance level, $F(2,97) = 2.44$, $p=.09$.

Based on the data listed in Table 1, an ANOVA was performed to determine the effect of the pedal configuration on movement time. No significant effect was found.

Neither gender nor car normally driven showed a significant effect on braking response time.

The only significant effect on braking response time was age, $F(2,97) = 4.20$, $p < .05$. These results are shown in Figure 4, and summarized in Table 2. Generally, the drivers in the older age group (51 and above) had slower response times than did the other two age groups.

The possible effects of the extraneous variables of driver height and shoe size were also analyzed by ANOVA. No effect was found for either at the .05 significance level.

DISCUSSION

This study showed that after identifying an unexpected obstacle in their path, the drivers in this study, traveling at 25 mph in a driving simulator, were able to step on the brake pedal in an average of 1.33 sec.

Although the pedal configuration was not found to have a effect at the .05 significance level, there was a distinct trend toward a faster braking response time as the

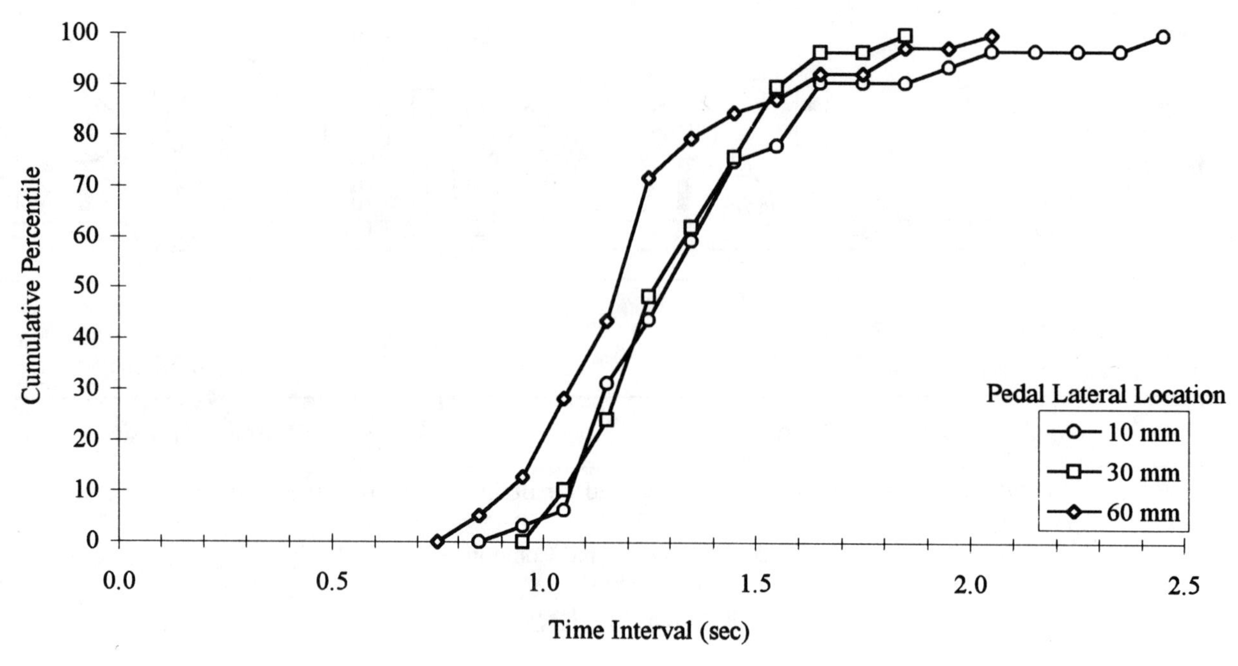

Figure 3. Brake Pedal Response Time to the Unexpected Obstacle

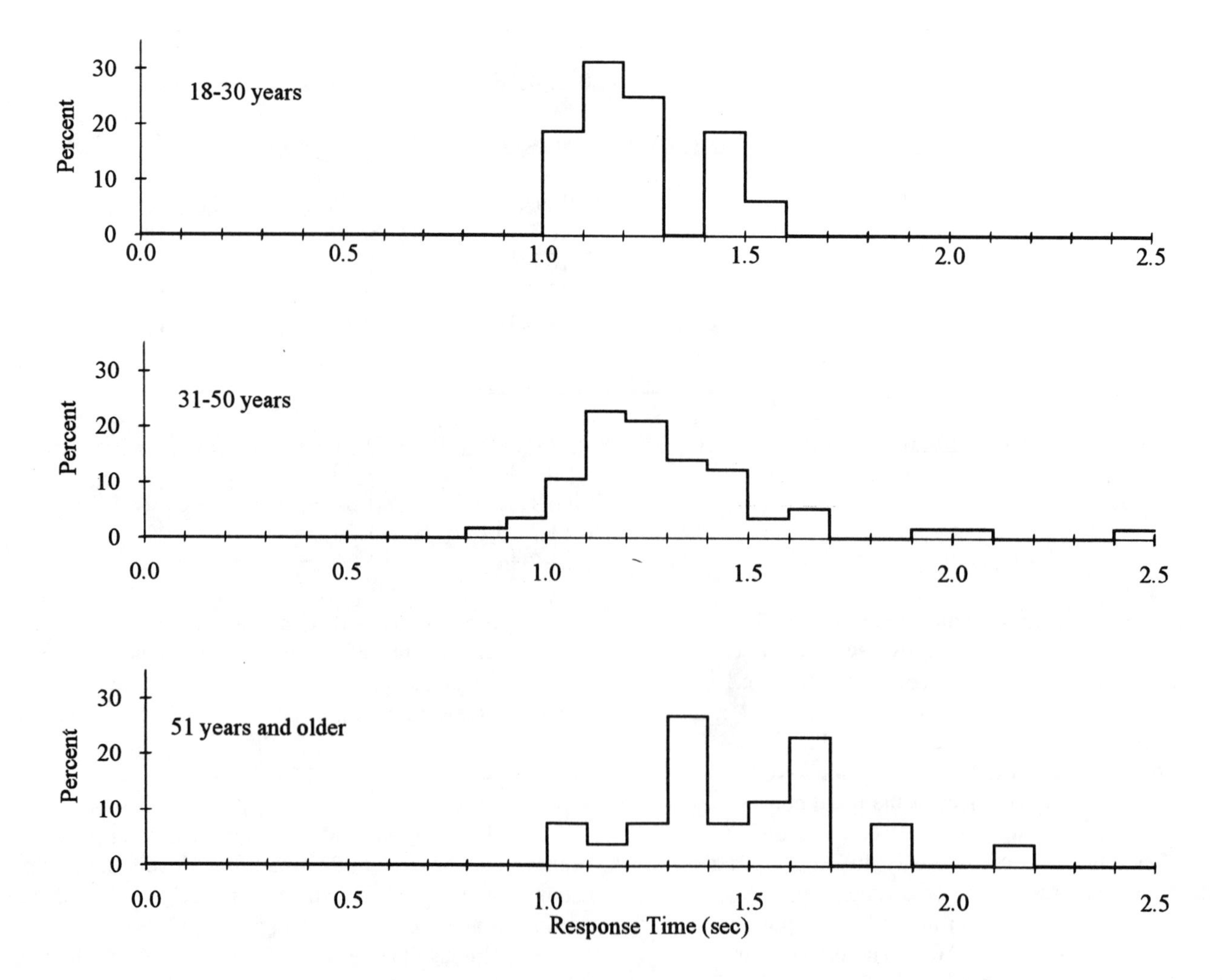

Figure 4. Effects of Age Group on Braking Response Time

Table 2. Braking Response Times by Age Groups

		Braking Response Time (sec)		
Age group (years)	n	Range	Mean	Std. Dev.
18 to 30	17	1.00-1.83	1.27	0.21
31 to 50	57	0.81-2.44	1.30	0.27
51 and older	26	1.00-2.10	1.46	0.25
Overall	100	0.81-2.44	1.33	0.27

lateral location of the pedal cluster was moved farther to the right of the HPC. Figure 3 shows that the response time for the 60 mm lateral location may have been faster than for the other two locations. However, the statistical variability in the measured times was relatively large, which made it difficult to establish fine grain correlations betweed pedal layout details and differences in movement times. In order to determine any statistical differences, a more focused study might be required, in which the test procedure is defined in more detail in order to better control extraneous variables.

A comparison of these data with those from other, previous studies shows a general agreement. Olson and Sivak (1986) performed an experiment in which driver braking responses to an unexpected obstacle were measured electronically in full scale tests for 64 driver subjects. They divided their subjects into (49) younger and (15) older drivers, 18 - 40 years and 50 - 84 years, respectively. They reported a response time ranging from 0.82 sec to 1.78 sec, with no significant difference between younger and older drivers.

Johansson and Rumac (1971) measured braking reaction times, only, for 321 drivers, to an auditory stimulus in over the road tests. Their reaction times, that is, the time from stimulus until foot movement, ranged from 0.3 sec to 2.0 sec, with a mean of 0.65 sec.. This compares with a range of 0.57 sec to 1.76 sec and mean of 1.16 measured in this study.

Other studies which were reviewed measured and reported only driver foot movement times, and most studied the effects of relative pedal location or seat position. Morrison, et al., (1986) reported movement times between 0.15 sec and 0.26 sec. Snyder (1976) reported movement times between 0.15 sec and 0.20 sec. Davies and Watts performed two similar studies; one with 10 males and one with 10 females (1969 and 1970). They found that movement times ranged from 0.13 sec to 0.30 sec. Generally, the ranges of these results were somewhat smaller than those recorded in this study. However, the means and standard deviations presented here are comparable.

Overall, the results of this driving simulator study using the DRI Interactive Driving Simulator agree with the results of previous studies. Generally, the driver braking response times for all the studies reviewed ranged from about 0.8 sec to 2.5 sec. Such results have implications for highway and vehicle design. The data reported here, and those reviewed are in agreement with the U.S. standard for perceptual reaction time of 2.5 sec, as used by traffic engineers in highway design (e.g. Oglesby, 1975).

REFERENCES

Olson, P. L. and M. Sivak, (1986). "Perception Response Time to Unexpected Roadway Hazards", Human Factors, 28 (1), pp. 91-96.

Johansson, G. and K. Rumar, (1971). "Drivers' Brake Reaction Times", Human Factors, 13 (1), pp.23-28.

Morrison, R. W., J. G. Swope and C. G. Halcomb, (1986). "Movement Times and Brake Pedal Placement", Human Factors, 28 (2), pp. 241-246.

Snyder, H. L., (1976). "Braking Movement Times and Accelerator - Brake Separation", Human Factors, 18 (2), pp. 201-204.

Davies, B. T. and J. M. Watts, (1969). "Preliminary Investigation of the Movement Time Between Brake and Accelerator Pedals in Automobiles", Human Factors, 11 (4), pp. 407-410.

Davies, B. T. and J. M. Watts, (1970). "Further Investigation of the Movement Time Between Brake and Accelerator Pedals in Automobiles", Human Factors, 12 (6), pp. 559-562.

Oglesby, C., (1975). Highway Engineering, Wiley, New York.

AGE AND DRIVER TIME REQUIREMENTS AT INTERSECTIONS

Neil Lerner
COMSIS Corporation
Silver Spring, Maryland

Current highway design models for required sight distance at stop-sign controlled intersections assume that the perception-reaction time (PRT) required is 2.0 seconds. That is, a 2.0 second interval to perceive, evaluate, decide, and initiate a response, is adequate to cover the range of time it takes real drivers to do this. This experiment evaluated the adequacy of the 2.0 second PRT assumption, including specific consideration of older drivers, who are known to experience relatively greater difficulty at intersections. Subjects in three age groups (20-40; 65-69; and 70+ years old) drove their own vehicles (fitted with a computer-controlled video-based data collection system) over a route that included 14 stop-controlled intersections. At each stop sign, they were required to make ratings of "road quality;" this broke visual search, and provided an opportunity for the experimenter to precisely define the initiation of search and the initiation of forward movement (thus defining PRT). The 2.0 second PRT assumption was found to work reasonably well for all age groups, and corresponded to roughly the 85th percentile PRT for all subjects. PRTs for older subjects were slightly (but significantly) briefer than for younger drivers. Reasons for not observing a slowing of intersection PRT with advancing age are discussed. The findings are also compared to gap acceptance data from another experiment. Even though the present experiment did not find objective evidence of older drivers requiring longer decision times, older subjects nonetheless demanded longer gaps in traffic in order to judge it safe to enter traffic.

INTRODUCTION

Driver decision making requires time. For the highway environment, these time requirements are translated into sight distance requirements; that is, how far along a road a driver must be able to see in order to have enough time to make a desired maneuver. One of the most important sight distance requirements is that for intersection sight distance. For drivers at a stop-controlled intersection, wishing to cross or turn onto a major road, this refers to the distance along the major road that must be visible to the driver. Sight distance equations are based on two components: (1) a perception-reaction time (PRT; the time required to perceive, evaluate, decide, and initiate a response); and (2) a time or distance required to make a given maneuver. For intersection sight distance, current design equations use a PRT of 2.0 seconds. This assumes a period of 2.0 seconds will be adequate to cover the range of times it actually takes drivers to search, perceive, and decide that it is safe to proceed into the intersection.

The experiment described here directly measured the PRT taken by older and younger drivers at stop-controlled intersections. It was part of larger project, described in Lerner, Huey, McGee, and Sullivan (1993), that examined PRT assumptions in design equations for various highway sight distance situations; different PRT values are assumed for different driving situations (e.g., emergency braking, complex lane change maneuvers). The purpose of the intersection sight distance experiment was to determine whether the assumed 2.0 second PRT covers drivers adequately, and in particular whether older drivers require longer times. There are several reasons why the adequacy of the PRT design value should be examined specifically for older drivers:

1. Based on traffic accident data, experimental studies, self-report, and other sources, it is well-recognized that older drivers suffer comparatively greater difficulty at intersections; in fact, it is probably the most critical age-related deficit in highway driving (e.g., TRB, 1988). While this does not necessarily implicate PRT as the causal basis,

intersection design criteria should be closely considered for these drivers.

2. Reviewers have noted that highway design criteria are often not based on actual driver performance data, and to the extent they are based on research, the studies have usually excluded or seriously under-represented older drivers (e.g., McGee, Hooper, Hughes, and Benson, 1983; Shapiro, Upchurch, Loewen, and Siaurusaitis, 1986). For this reason, current practice may not reflect the performance capabilities of older motorists.

3. Many behaviors are known to slow with age, a research finding that is well-established across many tasks and environments (e.g., Birren, Woods, and Williams, 1980; Salthouse, 1985). Search and attention allocation tasks are also known to be affected by age (e.g., Parasuraman and Nestor, 1991), and presumably are related to intersection negotiation in a vehicle. Laboratory simulations of driver reaction time tasks have observed slower responses for older subjects (Lerner, Ratte', Huey, McGee, and Hussain, 1990). Therefore, slower search and decision times may be expected on the road as well (however, Lerner, 1992, has noted that the few on-road studies that have examined age differences in PRT have not observed such age effects).

For these reasons, the study described here directly examined age differences in PRT at intersections and compared them to design criteria. The findings of the experiment were also compared with other data regarding the judged adequacy of various size gaps occurring in the traffic stream, for providing a safe margin for making a maneuver. Gap acceptance models may provide an alternative to PRT models for determining intersection sight distance requirements.

METHOD

Subjects were from three age groups: 20-40 (n=25), 65-69 (n=27), and 70+ (n=29) years old. They were recruited primarily in suburban Washington, DC. Recruiting procedures for older subjects were designed to minimize self-selection biases. Experienced recruiters worked through senior centers, retirement communities, churches, and so forth. Rather than placing initiative on the subject to volunteer, the recruiter worked as much as possible with the director of the institution to help identify and approach individuals with a wide range of capabilities, and to provide social support and incentive for taking part. While the sample cannot be considered "random" and no doubt under-represents the lower extreme of ability and confidence, it did appear to provide a broadly suitable range.

Subjects drove their own vehicles along a 56 mile route that included 14 stop-controlled intersections at which data were recorded. The intersections included a variety of geometric and traffic characteristics, and left turn, right turn, and crossing maneuvers. A computer-controlled video system recorded the view forward and to each side of the vehicle, and also recorded the driver's head and torso. The PRT design model assumes that it takes 2.0 seconds from the point when the stopped driver begins to search, until the time a decision is made and forward movement is initiated. This experiment directly measured this PRT interval. However, normal driver behavior frequently does not match this model; drivers begin scanning on approach to the intersection, may not stop completely, may show multiple starts and stops, and may continue scanning well into the maneuver. The experimental procedure brought driver behavior into better compliance with this model, by requiring the driver to make judgments of "road quality" each time they came to a stop (subjects thought this was the primary purpose of the experiment). They had to look down at a keyboard to make the rating, breaking off any search. They could not look up until they received the "ready" signal from the experimenter. They acknowledged this signal by pressing a button, at which point they could look up and proceed. Thus the moment that search began could be precisely defined, as could the moment forward movement began (determined from the video record). The logic of this experiment was therefore, given that a driver is behaving in a manner consistent with the PRT model, what is the range of PRTs observed?

Data were recorded only from those trials for which there was no conflicting traffic at the time of the decision (so that we were not measuring time that the driver was spending just waiting for traffic to clear). To some limited degree this was under the control of the experimenter, seated in the rear of the vehicle. He was able to monitor approaching traffic on a video monitor, as recorded by the array of three video cameras mounted on the roof of the vehicle. He could therefore delay turning on the "ready" signal until traffic appeared clear. The timing of the PRT was based on off-line analysis of the video record. An LED mounted just in front of one of the camera lenses illuminated when the subject pressed the acknowledgement button (initiating search); this permitted identification of the video frame in which the response occurred. The analyst could also judge from the video record the frame during which the vehicle initiated forward movement. Thus the number of frames connecting these two events provided the basis for timing, with a resolution of about 33 ms.

RESULTS

Data collection and analysis focused on daytime driving, since preliminary data found daytime PRTs to be about 0.3 s longer than those at night. Figure 1 shows cumulative probability plots of PRT for each of the three age groups. The vertical bars in the figure show the 50th percentile points and the 85th percentile points for each age group. As the figure suggests, older drivers did not require more time than younger drivers to search and proceed at intersections. The younger group actually took about 0.2 s longer to initiate movement (based on group means). A two-factor (age and sex) analysis of variance found significant main effects for both factors, as well as a significant interaction. The interaction was due to the fact that females in the oldest group were significantly slower than males, whereas there was no meaningful sex difference at the other age groups. Although older subjects did not take longer than younger subjects to initiate movement, it is possible that they spent more time visually evaluating the situation. Although the sight distance design equation is based on a strictly sequential model (search-decide-proceed), drivers will actually continue to scan for traffic after they have initiated the maneuver. In order to evaluate whether older and younger drivers employed different search strategies during the maneuver, video records were analyzed with respect to head turns. There was no indication of any age or gender effect upon the amount of time spent searching or upon the proximity of the last look to the completion of the maneuver. The 2.0 s PRT assumption appears to work reasonably well for all age groups. It approximates the 85th percentile value for all subjects combined (the 85th percentile is a commonly used, although arbitrary, value for highway design factors), and exceeds the 85th percentile values for older drivers in particular.

FIGURE 1

All Sites: Daytime only, by age group

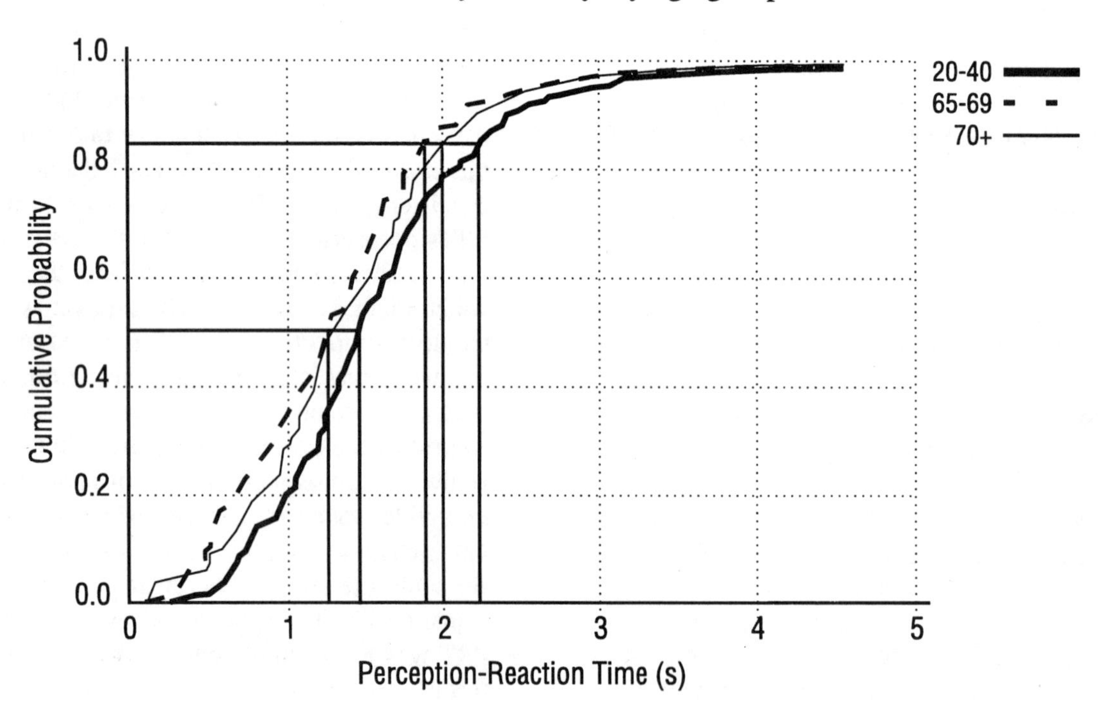

DISCUSSION

The design PRT value of 2.0 s at stop-controlled intersections appears adequate for older drivers. This should not be taken to imply that older drivers do not have difficulties at intersections, and that changes to design practice may not improve safety. However, the problem does not seem to be with PRT. Older driver intersection problems may be related to perceptual failures (looked-but-did-not-see accidents), attentional limitations (e.g., useful field of view, divided attention), dealing with visual complexity, comprehension of traffic control devices, vehicle control capabilities, or other factors. It may also be the case that summing independent PRT and maneuver components is not a fully appropriate design model. What is clear, however, is that the findings do not suggest a problem based on PRT itself. Given the assumed model, the 2.0 s PRT value appears reasonable.

Older adults are found to take longer than younger adults on many sorts of tasks. The absence of such slowing for intersection decision times may be somewhat surprising. There are a number of possible explanations. First, as in any study using volunteer subjects, there is the possibility of selection bias. The recruiting procedures used in this experiment were designed to provide a broadly representative range of older drivers, but undoubtedly the least capable extreme was less likely to be included. However, these are also the persons who will account for the fewest driving miles. While it might be possible to demonstrate different behavior for this subset, the research probably captures the characteristics of those likely to be driving to a meaningful extent. Another potential explanation of the lack of slower responding may lie in differences in the vehicles driven by older and younger drivers. However, the most likely of these differences, the use of automatic vs. standard transmissions, was examined in *post hoc* analyses, and found to have no effect. It may be the case that age differences in speeded reactions are not expressed simply because there is no urgency to initiate the response; since no one was operating near the limits of capability, the benefits of youth may not have been expressed. It may also be the case that the driving behavior dictated by the procedure (forcing the sequential process of stop-search-decide-move) may be more consistent with the normal actions of older drivers than of younger ones, and hence less disruptive. Finally, it may be the case that older drivers have come to initiate movement more quickly as a compensatory action for the longer time it may take them to actually complete the crossing or turning maneuver, once it has begun. Whatever the explanation, the empirical fact remains that older drivers in this study did not take longer to initiate movement from a stop sign; in fact, their PRTs were slightly briefer.

In another experiment conducted as part of the same project as the experiment described here, data were collected on gaps and lags in traffic that drivers judged acceptable for making various maneuvers (fully described in Lerner et al., 1993). In that experiment, subjects seated in a vehicle facing the roadway made continuous judgments about whether it would be safe to initiate a given maneuver (left or right turn, or crossing) at any particular moment. These decisions were related to the gaps (headway difference, in seconds, between two vehicles) and lags (distance, in seconds of travel time, from vehicle to observer) in the actual traffic passing by. Driver acceptance of judged safe gaps, or rejection of judged unsafe lags, may provide bases for alternative models of sight distance requirements at intersections. Because the gap/lag experiment used different sites than the intersection PRT experiment, the findings are not formally comparable. However, the general temporal aspects of driver behavior under the two experiments can be compared. Older drivers did require longer gaps than younger drivers; the gap duration accepted 50% of the time for the youngest group was 6.74 s, and for the oldest group was 7.85 s. Similarly, the mean point at which the youngest group judged an approaching vehicle to be too close for initiation of the maneuver was 5.32 s; for the oldest group, 5.86 s. Thus the PRT experiment found that older drivers did not require more time to initiate movement, nor did they require more total time from the initiation of search to the completion of the maneuver. However, the gap/lag acceptance experiment found that they demanded larger temporal intervals in order to be willing to attempt the maneuver. In general, regardless of age, the data suggested that drivers require gaps that are somewhat larger than the time durations it actually takes them to perceive, initiate, and complete a maneuver. This may reflect a margin of safety drivers allow beyond the time it normally takes to execute a maneuver. The possible advantages of design models based on gap or lag judgments are not clear, and a full discussion may be found in Lerner et al. (1993). The primary points to emphasize here are: (a) gap acceptance findings are generally consistent with, but slightly more conservative than, measured perception time-plus-maneuver time data from the PRT study; and (b) age effects are evident in gap acceptance and lag rejection, but not in measured PRT data.

In summary, no evidence was obtained of longer PRTs for older drivers at stop controlled intersections. Highway design assumptions about PRT (2.0 s) seems generally reasonable for all age groups (approximately 85th percentile), although some outlier PRTs exceed this value. Older drivers clearly suffer more problems at intersections based on other findings, and they seek longer gaps than younger drivers based on gap acceptance data. However, there is no empirical support that these problems or requirements are related to substantially longer PRTs at intersections.

REFERENCES

Birren, J., Woods, A., and Williams, M. (1980). "Behavioral Slowing with Age: Causes, Organization, and Consequences." In L. Poon (Ed.), Aging in the 1980s. Washington, DC: American Psychological Association.

Lerner, N. (1992). "Older Driver Perception-Reaction Time and Sight Distance Design Criteria." ITE 1991 Compendium of Technical Papers. Washington, DC: Institute of Transportation Engineers.

Lerner, N., Huey, R., McGee, H., and Sullivan, A. (1993). Older Driver Perception-Reaction Time for Intersection Sight Distance and Object Detection. Final Report, Contract DTFH61-90-C-00038. Washington, DC: Federal Highway Administration.

Lerner, N., Ratte', D., Huey, R. and Hussain, M. (1990). Older Driver Perception Reaction Time: Literature Review, Unpublished. Washington, D.C.: Federal Highway Administration.

McGee, H., Hooper, K., Hughes, W., and Benson, W. (1983). Highway Design and Operational Standards Affected by Driver Characteristics. FHWA-RD-78-78. Washington, DC: Federal Highway Administration.

Parasuraman, R., and Nestor, P. (1991). "Attention and Driving Skills in Aging and Alzheimer's Disease." Human Factors, 33, 539-558.

Salthouse, T. (1985). "Speed of Behavior and its Implication for Cognition." In J. Birren and K. Schaie (Ed.s), Handbook of the Psychology of Aging. New York: Van Nostrand Reinhold Company.

Shapiro, P., Upchurch, J., Loewen, J. and Siaurusaitis, V. (1986). Evaluation of MUTCD Selected Standards: Final Report, Unpublished. Washington, DC: Federal Highway Administration.

TRB (1988). Transportation in an Aging Society: Improving Mobility and Safety for Older Persons. Special Report 218. Washington, DC: Transportation Research Board, National Research Council.

FIELD MEASUREMENT OF NATURALISTIC BACKING BEHAVIOR

Jeff Harpster, Richard Huey, and Neil Lerner
COMSIS Corporation
Silver Spring, Maryland

A series of observations and measurements were made as subjects drove their own vehicles in an assortment of naturalistic backing tasks conducted on public roads in real world driving conditions. As the subjects performed eight backing tasks, the following data were collected: glance direction, hand position, velocity and acceleration, and distance to object in back of the vehicle. The results provide a set of normative data usable by automotive system designers for the design of backing warning systems and other ITS applications. The results of this study were divided into glance direction, backing speed, and time-to-collision. Glance directions were found to vary greatly between tasks. Elderly drivers demonstrated more use of their mirrors and looked over their shoulder less then the young drivers. Looking over the right shoulder was the most frequent glance location across all tasks. Except for the extended backing maneuvers, backing speeds averaged around 4.8 km/h (3 mph). The maximum backing speed for the young drivers was faster than the elderly and males backed faster than females. The time-to-collision was relatively constant over the majority of the backing sequences. Minimum time-to-collision values were generally over 2 seconds.

INTRODUCTION

In order to design an effective in-vehicle backup warning system, it is essential to understand the behavior of drivers while backing. This includes information about the sequence of events, glance direction, backing velocity, age, task and individual differences. This information is critical in determining the alarm modality, timing, location and adaptability that is required to implement a backup warning system.

The present study is the first in a sequence of experiments that will define the appropriate human factors requirements for in-vehicle back-up warning user interfaces. It addressed various key gaps in existing knowledge that were identified in the development of preliminary recommendations in the Preliminary Human Factors Guidelines for Crash Avoidance Warning Devices report (Lerner, Kotwal, Lyons, & Gardner-Bonneau, 1996). Very little empirical information exists on the nature, sequence, and timing of behaviors that occur under various vehicle backing scenarios. Information about driver behavior will be critical for addressing such issues as the location of warnings, the modality and nature of warnings, the timing of warnings, the parameters that define a hazardous situation, and the need for individually adaptive interfaces for user control. This experiment measured a range of driver behavior variables as drivers performed backing maneuvers in their own vehicles under a range of naturalistic backing scenarios. The findings provide a descriptive database that will be useful for those concerned with backing maneuvers in general, and will contribute to the development of driver warnings in particular.

This paper provides an overview of the method and major findings. Full detail may be found in Huey, Harpster, and Lerner (1995).

PROCEDURE

Data were collected in a natural setting. Participants used their own vehicles (mainly passenger cars with some minivans and utility vehicles), and drove on public roads. The ability to record drivers in the normal operation of their own vehicles allowed measurement of "normal," ecologically valid performance, such as typical velocities, distances, and mirror use. An unfamiliar instrumented vehicle might have altered this behavior, and would also have limited the findings to a specific vehicle make and model, which might be atypical. A cover story was used to minimize driver reactance to the instrumentation and experimenter during the drive. Participants were told that instrumentation for measuring vehicle performance parameters was being evaluated. Therefore, aspects of driver performance were not an explicit dependent measure and their individual performance was not at the heart of the study, and backing maneuvers were not highlighted.

An experimenter directed drivers through a course that included a wide range of backing maneuvers.

- Extended curved backing to a stop point (2 locations) [Tasks 3 and 8]
- Parallel parking against a curb with vehicles fore and aft (2 locations) [Tasks 2 and 7]
- Backing out of a perpendicular slot in a parking lot [Task 5]
- Backing out of an angle slot in a parking lot [Task 1]
- Backing to a wall [Task 4]
- Backing into a perpendicular parking slot [Task 6]

A data collection system was temporarily installed in the participant's vehicle to record measures of driver behavior and vehicle control. The variables measured included brake and accelerator use, gear changes, direction and duration of looks, vehicle velocity and acceleration, distance from objects, time-to-collision, and the sequence and timing of driver actions.

Subjects

There were 21 total subjects. Nine were elderly (mean age 73 years) and 12 were young (mean age 22.5 years). Four elderly subjects and seven young subjects were male.

Apparatus and Data Collection

The apparatus for this experiment allowed collection of both video and digital data from the vehicle and driver. All data collection equipment was controlled from a single PC located in the back seat of the vehicle with the experimenter. The PC used a custom application written in Microsoft Visual Basic to control the entire suite of equipment for data collection, reduction, and post processing procedures. The following equipment was used: 4 miniature camera systems, optical speed detector, ultrasonic distance measurement system, personal computer, computer controlled VCR, quad splitter, data acquisition system, power supply, pressure sensors (gas pedal, break pedal and shifter), and video monitor.

At the end of the data collection process, two distinct products were evident; a videotape of the entire session, and a data file representing the analog and digital sensor measurements for the various maneuvers performed by the participant. The primary purpose of the data reduction procedure was to integrate the products into single data file enriched with coded behavioral events that could be analyzed using conventional statistical analyses.

RESULTS

Glance Direction

Glance direction was determined by manually examining the video recorded during each of the backing tasks, and coding the changes into the "enriched" data file. Of particular interest was the glance direction during the backing phase of each task. Summaries of the glance direction while the vehicle was moving in reverse are shown in the following tables.

About half of the time subjects were looking over their right shoulders and almost never looked at the dash while backing. There was a very wide variation of glance location between tasks as can be seen in table 1.

There are several large differences between the young and elderly participants. Across all tasks the young participants looked over their right shoulder 59.9% of the time and the elderly participants only looked over their right shoulder 37.4% of the time while traveling in reverse. Elderly drivers were far more likely to use their mirrors than the young participants. Glances to the three mirror locations combined for about 34% of the older drivers' looking time, but only 15% (less than half as much) for younger drivers' looking. The breakdown of the glance direction by age group is shown in table 2.

Glance Direction	Total	Task 1	Task 2	Task 3	Task 4	Task 5	Task 6	Task 7	Task 8
Forward	10.6%	28.7%	16.5%	0.6%	1.8%	18.1%	3.0%	14.1%	1.8%
Dash	0.0%	0.1%	0.0%	0.0%	0.0%	0.0%	0.0%	0.1%	0.0%
Driver's mirror	8.2%	6.0%	3.6%	8.4%	7.8%	14.7%	11.6%	2.8%	11.0%
Rear mirror	4.5%	4.7%	7.4%	4.0%	1.7%	4.5%	2.5%	6.3%	5.1%
Right mirror	9.2%	7.1%	17.8%	2.7%	3.7%	13.4%	2.7%	19.3%	7.2%
Left Window	2.3%	1.5%	0.2%	0.2%	0.2%	4.9%	1.3%	0.0%	0.0%
Right Window	1.0%	3.0%	3.6%	0.1%	0.3%	5.1%	1.2%	4.6%	0.4%
Shifter	1.0%	3.2%	0.7%	0.0%	0.4%	2.1%	0.3%	1.1%	0.1%
Left shoulder	12.5%	21.6%	4.2%	27.3%	6.0%	14.0%	14.4%	5.2%	7.2%
Right shoulder	50.9%	23.6%	47.0%	56.2%	78.2%	24.7%	62.7%	47.2%	67.3%

Table 1: Glance direction for each task and overall total.

Glance Direction	Young	Elderly
Forward	9.9%	11.8%
Dash	0.0%	0.0%
Driver's mirror	4.3%	15.0% **
Rear mirror	3.3%	7.1% **
Right mirror	7.7%	12.1%
Left Window	0.7%	1.5%
Right Window	2.1%	2.1%
Shifter	0.4%	1.8% **
Left shoulder	12.8%	12.3%
Right shoulder	59.9%	37.4% **

(** significant at the 0.05 level)
Table 2 - Glance direction for Elderly and Young across all tasks.

Backing Speed

There was a wide variation of maximum backing speeds between tasks and participants. The mean maximum speed, minimum maximum speed and the maximum maximum speed while backing for each task are shown in table 3.

Task	Mean Max	Min Max	Max Max
1. Back Angle	4.8 (3.0)	2.4 (1.5)	10.1 (6.3)
2. Parallel	4.7 (2.9)	1.9 (1.2)	10.8 (6.7)
3. Extended curve	10.8 (6.7)	4.0 (2.5)	19.6 (12.2)
4. Back wall	5.6 (3.5)	3.1 (1.9)	9.5 (5.9)
5. Back out perp	5.3 (3.3)	3.2 (2.0)	11.3 (7.0)
6. Back in perp	4.7 (2.9)	2.3 (1.4)	7.4 (4.6)
7. Parallel	4.5 (2.8)	2.7 (1.7)	8.7 (5.4)
8. Extended curve	16.1 (10.0)	7.2 (4.5)	23.8 (14.8)

Table 3 - Backing speeds (KM/H -- MPH) for all participants

The maximum backing speed for each age group broken down by task is shown in table 4. On several of the tasks the younger drivers had significantly faster maximum speeds while backing.

Task	Young	Elderly
1. Back Angle	5.5 (3.4)	4.0 (2.5)
2. Parallel	5.6 (3.5)	3.4 (2.1) **
3. Extended curve	13.4 (8.3)	6.4 (4.0) **
4. Back wall	6.1 (3.8)	5.0 (3.1)
5. Back out perp	6.0 (3.7)	4.2 (2.6)
6. Back in perp	5.0 (3.1)	4.2 (2.6)
7. Parallel	4.8 (3.0)	4.0 (2.5)
8. Extended curve	18.7 (11.6)	11.3 (7.0) **

Table 4. Maximum backing speed KM/H (MPH). (** significant at the 0.05 level)

Time To Collision (TTC)

TTC at any point of the backing sequence is calculated by assuming constant velocity over the remaining distance to the target. TTC was examined across all participants and by age and gender breakdowns. Only four of the tasks had TTC values since there was no object behind the vehicle in the other four tasks.

Typically, as the driver approached an object, the vehicle decelerated at a rate that maintained a relatively constant TTC, which seldom fell below 2 seconds. Figure 1 illustrates a typical example.

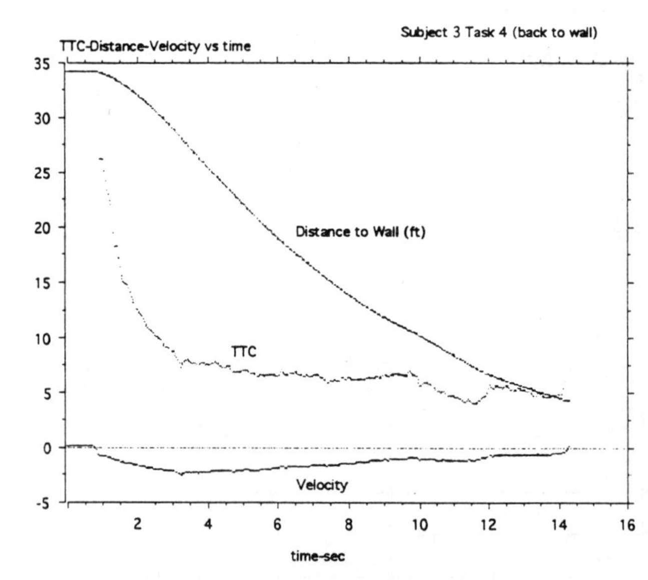

Figure 1. TTC-Distance to the Wall-Velocity vs. Time

The minimum TTC (in seconds) for each task across all participants is shown in table 5.

The age breakdown of TTC by task is shown in table 6. Minimum TTC for the young and elderly were not statistically different. It is interesting to note that there were speed differences between the elderly and young but not TTC differences.

Task	Avg Min	Min Min	10th percentile	Max Min
2. Parallel	3.4	1.0	1.3	6.3
4. Back Wall	2.4	1.1	1.5	3.9
6. Back in Perp	3.0	1.7	1.9	4.3
7. Parallel	3.7	2.0	2.1	6.3

Table 5. Minimum TTC for All Subjects

Task	**Elderly**	**Young**
2. Parallel	3.4	3.4
4. Back Wall	2.6	2.5
6. Back in Perp	3.2	3.0
7. Parallel	4.2	3.4

Table 6. Minimum TTC by Age and Task (none were significant at the 0.05 level)

Summary

This experiment provided a wealth of information about how people back up in naturalistic settings. A wide range of drivers were observed (age and gender) performing a wide variety of backing tasks. All drivers used their own vehicles and drove on public streets. Patterns of glance, hand position, velocity, and distance to targets were all measured as the driver performed eight backing tasks. The data collected in this experiment provides baseline data that can be used in the design of backing warning systems, parking lots, vehicle mirrors, vehicle windows, and special purpose devices for the elderly.

The major findings in this experiment are summarized below:

Glance direction while backing varied greatly by task

a. Glance direction while backing varied greatly by age (elderly vs young); older drivers showed more mirror use and less looking over the right shoulder

b. Glancing over the right shoulder was the most frequent location, at the initiation of backing and overall

c. Except for extended backing maneuvers, maximum backing speeds averaged around 4.8 KM/H (3 MPH), and did not exceed 11.3 KM/H (7 MPH)

d. Maximum backing speed was generally faster for the young vs. the elderly

e. Male drivers tended to back faster than female drivers

f. TTC typically dropped to an asymptotic value as the vehicle approached an object, and remained relatively stable as the vehicle slowed while approaching

g. Minimum TTC's were greater than 1.0 second, and usually exceeded 2.0 seconds

h. TTC values did not vary for males and females or for young and old

i. Drivers typically traveled less than 0.91 m (3 feet) in the first second of backing, and usually less than 2.44 m (8 feet) after two seconds.

References

Huey, H., Harpster, J., and Lerner, N. (1995). *Field Measurement of Naturalistic Backing Behavior*. (Contract number DTNH22-91-C-07004) National Highway Traffic Safety Administration.

Lerner, N. D., Kotwal, B. M., Lyons, R. D., and Gardner-Bonneau, D. J.(1996). *Preliminary Human Factors Guidelines for Crash Avoidance Warning Systems* (Contract number DTNH22-91-C-07004) National Highway Traffic Safety Administration.

RELATION OF INDIVIDUAL DIFFERENCES IN INFORMATION-PROCESSING ABILITY TO DRIVING PERFORMANCE

Thomas A. Ranney
Nathaniel H. Pulling

Liberty Mutual Research Center
Hopkinton Massachusetts

ABSTRACT

Fifty subjects ranging in age from 30 to 83 participated in a closed-course driving test and in laboratory tests of information processing. Driving tests included responding to traffic signals, selection of routes, avoidance of moving hazards, and judgment at stationary gaps. Lab tests included measures of perceptual style, selective attention, reaction time, visual acuity, perceptual speed and risk-taking propensity. Analyses were conducted to determine how well lab measures predicted driving performance. Results revealed different patterns of correlations for different age groups. For younger drivers (30-51), lab measures generally showed no association with measures of driving performance. For older drivers (74-83), measures of information-processing were associated with overall rated driving performance, while measures of reaction time showed strong correlations with objective driving measures. The results suggested that different mechanisms are utilized by drivers of different ages, and that the slowing of reaction time associated with aging has effects on driving skills related to vehicle control.

INTRODUCTION

Considerable research effort has been directed at predicting motor-vehicle accident rates with laboratory measures of information-processing skills. Barrett, Alexander and Forbes (1973), presented a conceptual analysis of pre-crash maneuvers related to accident causation and concluded that three categories of information-processing are relevant to predicting driving accidents. These categories (perceptual style, selective attention and perceptual-motor reaction time) have been the basis of a number of subsequent research studies (cf., Mihal and Barrett, 1976; Avolio, Kroeck and Panek, 1985). While much of this work has demonstrated positive results, correlations between lab measures and accident records are generally not strong. According to McKenna, Duncan and Brown (1986), this is due to the fact that accident causes may reflect chance factors as well as many different personal characteristics and psychological abilities or processes.

The stability of accident rates has been investigated by Miller and Schuster (1983), who found that past accident history is not a good predictor of future accident involvement and thus may not be a valid indicator of current or future driving competency. The need for an alternative measure of driving proficiency was asserted by McKenna et al. (1986), who concluded that detailed investigations of component skills would lead to better understanding of differences in driving performance than reliance on a single criterion such as accident rate. These authors concluded that "research should concentrate on specific skill deficiencies and their contribution to human error, rather than more immediately attempting to predict overall accident liability." Finally, practical considerations about the availability and quality of accident data, which rarely contain information in sufficient detail for research purposes, underscore the need for alternative ways of evaluating driving competency.

In response to these concerns, a battery of driving tasks was developed which focuses on the decision making and judgment involved in everyday suburban driving. The driving tasks were implemented on a half-mile closed course, which allows drivers to use their own vehicles. Use of drivers' own vehicles avoids problems of differential adaptation to unfamiliar research apparatus such as simulators and instrumented vehicles. In addition, a battery of laboratory tasks, including information-processing tasks which have been shown to be related to accident rates, was developed. The overall objective of the research program is the development of a safe driving capability profile, including both driving and laboratory tasks, for drivers of all ages. The objectives of the present paper are to: (1) describe the initial development of the driving test, and (2) evaluate the relationship between performance on the driving test and laboratory measures of information-processing.

METHOD

Subjects

Fifty subjects ranging in age from 30 to 83 participated in driving and laboratory tests. Subjects were recruited with newspaper ads and from local senior citizen activity centers. Subjects were paid $8.00 to $10.00 per hour for participation, depending on their responses to performance incentives.

Apparatus

An instrumented driving range, including a half mile of two-lane roadway, a signalized intersection, mobile hazards, and various regulatory and destination signs, was developed to enable drivers to use their own vehicles. The instrumentation and its rationale are discussed in Ranney, Pulling, Roush and Didriksen (1986). Traffic signal timing and data acquisition are controlled by a DEC PDP 11/23 computer which is housed in a van parked alongside the intersection. Spot speed data are obtained with four pairs of inductive loops buried beneath the pavement. The pairs are separated by 36.6 m (120 ft). Three pairs are before the intersection and one pair is beyond it. Time of entry into the intersection is obtained with a single loop in each approach lane located at the stopline. Traffic signal timing is related to the "temporal position", or time to the intersection of the vehicle, which is computed using approach speeds. This compensates for differences in vehicle approach speed.

Driving Test

The driving test consisted of three 30-minute "trips." Each trip, composed of up to twenty laps of the closed course, required the driver to respond to a continuous sequence of driving situations. Primary tasks included responding to traffic signals with varied timing and selection of routes using information presented on traffic signs. A gap-acceptance task required drivers to select one of two routes at a junction on the course. One route was shorter, but required drivers to drive through a variable-sized gap formed by two construction barrels. Drivers' judgments concerning the width of the gap and their willingness to attempt the gap were evaluated with this task. Secondary tasks included avoidance of unexpected moving hazards (such as a rolling ball or simulated baby stroller), responding to regulatory signs (speed limit and stop signs) and executing maneuvers created by cones and barrels.

Both subjective ratings and objective measures of drivers' responses to the driving task situations were recorded. Drivers were rated on the following ten skills:

Stop/go decision making
Gap judgment
Gap execution
Decision speed
Route selection
Speed maintenance
Vehicle control
Emergency hazard avoidance
Time to destination
Ability to follow instructions

Ratings were made on a three-point scale. Drivers were observed by two or three raters. Following each session ratings were discussed and a single consensus rating was recorded for each driver on each of the ten categories. An overall rating of driving performance was computed as the average of the ten categorical ratings.

Objective driving performance measures included the following:

Measures of intersection performance

- Stopping probability: the proportion of decision trials on which the driver stopped when faced with the yellow traffic signal (STOPPR)
- Stopping accuracy: vehicle placement relative to the stopline on stopping trials (STPACC)
- Intersection clearance margin: the mean difference between the time the vehicle exited the intersection and the onset of the red traffic signal (MARGIN)

Measures of gap performance

- Number of attempts: number of trials where driver attempted to drive through gap (NOATT)
- Number of gap judgment errors: included selection of gaps too small and avoidance of gaps of equal or greater width than the vehicle (JUDGERR)
- Number of gap execution errors: Struck barrels or excessively slow speed (EXERR)

Speed measures

- Intersection approach speed: mean over all trials (SPEED1)
- Intersection approach speed change: mean over all trials (SPDDIF)
- Mean lap time: mean over all trials in one trip (LAPTIME)
- Speed maintenance errors: instances of speeds over 56.3 km/h (35 miles/hour) (FAST35) or under 43.5 km/h (27 miles/hour) (SLOW) in the intersection approach

Measures of vehicle control consistency

Approach speed consistency: standard deviation over all trials of approach speed (SSPD1)

A measure of route selection errors was eliminated due to insufficient data.

Laboratory tasks

Visual acuity (VISION) was measured with a standard Titmus tester, similar to those used for license renewal. Perceptual style was measured with the Embedded Figures Test (EFT) (Witkin, Oltman, Raskin & Karp, 1971). Perceptual speed was measured with three tests of the Cognitive Factors Kit (Ekstrom, French, Harman & Dermen, 1976). The tests required visual search for letters (VSEARCH), matching numbers (NUMBERS), and matching figures (FIGURES). The Digit Symbol Substitution (DSS) test of the Wechsler Adult Intelligence Scale is also a measure of perceptual speed and short-term memory, and has been used widely in studies of information-processing and aging (Salthouse, 1985). Visual selective attention was measured with an analogue of the dichotic listening task (Avolio, Kroeck, and Panek, 1985). The total number of errors (VSATOT) and the number of switching errors (VSATSW) provided a measure of efficiency of switching in attention. Three measures of reaction time, included simple (SRT), simple plus movement (MRT), and movement plus (two) choice (CRT) reaction times. Risk-taking propensity (RISK) was measured by the Choice Dilemmas Questionnaire developed by Kogan and Wallach (1964).

RESULTS

Analyses were conducted to determine: (1) how well lab measures predicted driving performance; and (2) how well objective driving performance corresponded to subjective ratings. To determine how well lab measures predicted overall driving performance, correlations between overall driving performance and lab measures were examined. The correlations ranged from .21 to .61. All correlations except those for choice reaction time (CRT) and risk propensity (RISK) were significant at the .05 level, most at the .01 level. Four measures (DSS, FIGURES, VSATOT, VSATSW) exhibited correlations greater than .55 with overall driving performance. Previous research using accident rates as the criterion suggested that correlations in the range of .3 to .4 would indicate a relatively strong association. The considerably larger correlations for several of the variables, most notably the DSS, which has been used extensively in studies of aging, led us to reconsider our data. According to Salthouse (1984), spurious correlations can arise when samples are not homogeneous with regard to age. Because our sample included a wide range of ages, and the overall driving performance rating was found to be highly correlated with age, it was hypothesized that age was accounting for some of the relatively high observed correlations. Accordingly, the data were reanalyzed, separately for the two main age groups (30-51, 74-83). Results are shown in Table 1.

Table 1

Correlations of Lab Measures with Overall Driving Performance for Two Age Groups

	r	
Lab test	Young (n=21)	Old (n=20)
DSS	.07	.40
EFT	.03	-.41
VSEARCH	.01	.42
NUMBERS	-.11	.26
FIGURES	-.11	.56
SRT	.21	-.57
MRT	.28	-.31
CRT	.17	-.21
VSATOT	-.04	-.24
VSATSW	-.02	-.09
VISION	.02	-.14
RISK	-.17	-.20

The statistical power has been reduced considerably with the separation into two age groups, so that statistical significance is not comparable between the two analyses. For current purposes, correlations of .4 or greater are considered as different from zero. It is apparent that patterns of correlations differ for the two age groups. For the younger drivers, performance on the driving test was not related to any of the lab measures. However, for the older drivers, correlations greater than .4 were evident for several lab measures. Strongest correlations were associated with simple reaction time (SRT) and the figure matching test of perceptual speed (FIGURES). The DSS, EFT, and letter search task (VSEARCH) also exhibited relatively strong correlations with overall driving performance. Attention switching, as measured by the VSAT and risk propensity (RISK) were not related to overall driving performance for either group.

The results of two multiple regression analyses revealed that the lab measures accounted for 52% of the variance in overall driving scores for the younger drivers, and 84% for the older drivers.

Next, analyses were conducted to determine the relation between selected measures of driving performance and lab measures. These results are shown in Table 2.

Table 2

Correlations Among Driving Performance and Laboratory Measures for Two Age Groups

Driving Performance Measure	Lab Measures with significant correlations (p <.05)+ Younger Drivers (30-51)	Older Drivers (74-83)
STOPPR	MRT, CRT	
STPACC	RISK*	(EFT)
MARGIN		MRT*, CRT*, EFT*
NOATT		
JUDGERR	(FIGURES)	DSS, VSATOT
EXERR	VISION*	SRT, MRT, CRT
SPEED 1		SRT*, MRT*, CRT*, RISK
SPDDIF		
LAPTIME	SRT, MRT, CRT* (RISK)	SRT*, MRT,CRT
FAST35	DSS*	RISK
SLOW	(VISION)	SRT, MRT, CRT
SSPD1		SRT*, MRT*, CRT*,FIGURES, RISK*

+Correlations greater than or equal to .40, but not statistically significant at p <.05 are indicated in parenthesis.
*Correlations greater than or equal to .50

As with overall driving performance, the correlations were generally stronger for the older group. Measures of reaction time, especially simple reaction time (SRT), were correlated with a number of driving measures, including measures of speed (SPEED1, LAPTIME, SLOW) and speed consistency (SSPD1). In addition, for the older drivers, the reaction time measures were highly correlated with stopping margin (MARGIN) and execution errors on the gap task (EXERR). The fact that all three reaction time measures behaved similarly is consistent with the generally high correlations observed among the three measures for both age groups. Of interest is that the choice reaction time measure (CRT), which involved movement, was essentially identical (r=.97) to the movement reaction time (MRT) without choice for the older drivers, but not for the younger drivers (r=.87).

With several exceptions, measures of information processing speed were not strongly correlated with driving performance for either age group. Gap task judgment (JUDGERR) exhibited significant correlations with DSS and total VSAT errors for the older group only. Gap judgment for the younger group was related to the FIGURES identification task, although not strongly. For the younger group DSS scores were related to speed exceedances over 35 mph (FAST35).

The embedded figures test (EFT) was related to stopping accuracy (STPACC) and to the clearance margin (MARGIN) at the signalized intersection for the older group only; however, neither correlation was very large. The risk propensity questionnaire (RISK) was related to stopping accuracy (STPACC) and mean lap time (LAPTIME) for the younger drivers, and to two measures of speed (SPEED1, FAST35) for the older drivers.

The results of two regression analyses revealed that the measures of driving performance accounted for 69% of the variance in overall driving scores for the younger drivers and 83% for the older drivers.

DISCUSSION

Perhaps the most significant finding of the current analysis is that correlation patterns were different for different age groups. This is consistent with results of Mihal and Barrett (1976), who used accident rates as the criterion. Their explanation focused on differences between the accident rate distributions for the two age groups. Although such differences do exist for some of the current measures, it is also possible, as suggested by Salthouse (1985), that different patterns of correlations reflect use of different information processes by different-aged drivers.

For the younger drivers, none of the lab measures correlated with overall driving performance. While some of the driving measures exhibited significant correlations with lab measures, the pattern was generally not consistent with predictions of the information processing model of Barrett et al. (1973). It can thus be concluded that for drivers in our younger age group, closed-course driving involved skills different from those measured with lab tasks.

Correlations were generally stronger for the older drivers. Several information processing measures correlated with overall rated driving performance, while measures of reaction time correlated with both overall rated performance and with individual driving measures. The slowing of response time with age is well documented (Salthouse, 1985). The results of the present study indicate that for drivers over age 74, slowing of reaction time has a strong association both with overall driving performance and with specific driving measures, especially those related to vehicle control. Driving measures reflecting judgment and decision making skills were not correlated with reaction time measures.

The results suggest that the driving test was considerably more challenging for the older drivers than for the younger drivers. The absence of correlations for the younger group is consistent with the hypothesis that younger drivers were able to perform driving tasks more automatically, without complex information processing of the types required for the laboratory tasks. For the older drivers, more effortful processing, most notably visual search and identification, in addition to quick responding, were apparently required for the driving test.

Several specific hypotheses were addressed in the analysis. First, it was hypothesized that risk-taking, as measured by the Choice Dilemmas Questionnaire, would be related to drivers' willingness to attempt to negotiate different sized gaps and their decisions to stop or go when confronted with a yellow traffic signal. No such associations were found, however risk propensity scores were associated with several measures of speed, primarily for the older drivers. In this context, the Choice Dilemmas Questionnaire was not a useful predictor of risk-taking behavior.

Because accident causation may reflect different performance failures, it was hypothesized that laboratory tasks would be more highly correlated with individual driving measures than had been found in previous studies which used accident rates as the criterion. Of special interest was the Visual Selective Attention Test (VSAT), since the ability to rapidly switch attention is required in many driving situations. For the older drivers VSAT performance was related to their ability to judge whether their vehicle would fit through the stationary gap. None of the other driving measures was significantly correlated with VSAT performance, leading to the conclusion that rapid switching of attention was not generally required for the current driving test.

The results of regression analyses indicate that neither the lab measures nor the driving measures correspond to all aspects of rated driving performance. With regard to driving measures, they reflect the fact that speed of decision-making in the route selection task, avoidance of emergency hazards, and ability to follow instructions were not measured directly. With regard to the lab tasks, the above-cited correlations between information processing measures and accident rates, have in the past led to the conclusion that complex information processing is involved in accident causation. The current results are not consistent with this reasoning. Whether this reflects inadequacies of the current driving test or whether previous reliance on past accident rates is not valid cannot be determined at this time. Prediction of culpability in future accident involvement would be required for this purpose.

Finally, the analyses support the conclusion that different information processes are utilized by different age groups. They also indicate that lab tests may be more useful predictors of driving performance for older drivers than for younger drivers. If the current driving test represents the requirements of real-world driving, the results suggest that different causes are responsible for accidents of drivers of different ages.

REFERENCES

Avolio, B.J., Kroeck, K.G., & Panek, P.E. (1985). Individual differences in information processing ability as a predictor of motor vehicle accidents. Human Factors, Vol. 27, No. 5, 577-587.

Barrett, G.V., Alexander, R.A., & Forbes, B. (1973). Analysis and performance requirements for driving decision making in emergency situations. Washington D.C.: U.S. Department of Transportation, Report No. DOT HS-800 867.

Ekstrom, R.B., French, J.W., Harman, H.H., and Dermen, D. (1976). Manual for kit of factor-referenced cognitive tests. Princeton, NJ: Educational Testing Service.

Kogan, N. and Wallach, M.A. (1964). Risk taking: a study in cognition and personality. New York: Holt, Rinehart and Winston.

McKenna, F.P., Duncan, J. and Brown, I.D. (1986). Cognitive abilities and safety on the road: a re-examination of individual differences in dichotic listening and search for embedded figures. Ergonomics, Vol. 29, No. 5, 649-663.

Mihal, W.L. & Barrett, G.V. (1976). Individual differences in perceptual information processing and their relation to automobile accident involvement. Journal of Applied Psychology, Vol. 61, No. 2, 229-233.

Miller, T.M., and Schuster, D.H. (1983). Long-term predictability of driver behavior. Accident Analysis and Prevention. Vol. 15, No. 1, 11-22.

Ranney, T.A., Pulling, N.H., Roush, M.D., and Didriksen, T.D. (1986). Nonintrusive measurement of driving performance in selected decision making situations. Transportation Research Record 1059.

Salthouse, T.A. (1985). A theory of cognitive aging. Amsterdam: North Holland.

Witkin, H.A., Oltman, P.K., Raskin, E., and Karp, S.A. (1971). A manual for the embedded figures tests. Palo Alto, CA: Consulting Psychologists Press.

THE RELATIONSHIP OF AGE AND COGNITIVE CHARACTERISTICS OF DRIVERS TO PERFORMANCE OF DRIVING TASKS ON AN INTERACTIVE DRIVING SIMULATOR

José H. Guerrier, P. Manivannan, Anna Pacheco
Stein Gerontological Institute
Miami, Florida

Frances L. Wilkie
University of Miami Dept. of Psychiatry
Miami, Florida

Older adults depend highly on the automobile to satisfy their mobility needs. They use the private car for the majority of their trips. However, driving is not without risks for older drivers and those who share the road with them. Drivers 65 and older contribute to more accidents per mile driven than younger drivers except those 18-24 years old. Furthermore, they are more likely to be injured or die as a result of such accidents than their younger counterparts. Current thinking suggests that the cognitive abilities of older drivers may be the best explanation for these accidents. This study investigated the contribution of age and specific cognitive, psychomotor, and perceptual dimensions upon the performance of driving tasks on an interactive simulator. The results suggest that age as such does not explain performance of driving tasks. Rather, age-sensitive cognitive characteristics of drivers provide a better understanding of performance of specific driving tasks.

INTRODUCTION

Older adults 65 and older represent about 12% of the U.S. population and their number is expected to rise to 22% by the year 2030. They depend upon the personal automobile to accomplish most of their mobility dependent activities (Rosenbloom, 1988). However, age related cognitive and sensory changes threaten their ability to continue driving.

Older drivers are involved in a high rate of accidents per mile driven. In addition, they are more susceptible to injury and death as a result of automobile accidents than younger drivers. The accident involvement of older drivers has been attributed to: their failure to yield the right of way; failure to obey traffic signs and signals; improperly negotiating intersections; making improper turns, especially left turns; careless or inaccurate lane changes; careless backing; and driving the wrong way on one-way streets (McKnight, 1988; Yaksich, 1985; Waller, 1967). Research is therefore needed to identify the causes of these accidents in order that interventions be developed to increase the safety and mobility of older drivers.

This study's major objectives are: a) to assess the contribution of age and specific cognitive and perceptual characteristics of drivers to their performance of specific driving tasks on an interactive driving simulator, b) to assess the relationship of drivers' characteristics and driving tasks to accidents on an interactive driving simulator, and c) to assess which driving tasks can be supported on an interactive simulator.

Investigation of Driving Tasks as Opposed to Accident Involvement

Accidents are most often the yardstick used to infer about a driver's proficiency. However, they are discrete and rare events and consequently do not yield sufficient information to determine the critical cognitive and behavioral precursors of these accidents. An accident might be the result of inadequate performance of one or several of the driving tasks that a specific maneuver entails. Data on the performance of relevant driving tasks should furnish the researcher with the information necessary for explaining accidents. Simulation offers this opportunity by enabling the researcher to manipulate the driver's environment and to collect information on drivers' characteristics and their performance of various driving tasks under specific conditions without any danger to these drivers.

METHOD

Sample: Ninety-five persons with valid driver's licenses participated in the study. They ranged in age from 18 to 81 years. Information on various demographic, cognitive, psychomotor, and sensory characteristics of the participants were collected. The Demographics Questionnaire solicited information on the participant's gender, date of birth, educational level, driving experience, difficulties encountered in performing specific driving tasks, accidents in the last five years, number of days a week participant used a car, time of day car is used, reasons for time restrictions if any, and reasons for using their car and the respondents' health status.

Cognitive Battery: The cognitive battery used in this study measured a wide range of processes in the major cognitive domains and have been found to be sensitive to the effects of age. As noted by Salthouse (1991), this approach will assist in evaluating the independence and interrelationships between cognitive constructs and processing resources which may be useful in explaining cognitive processes required in the performance of specific tasks. The cognitive skills measured were: (1) Information Processing Speeds, (2) Attention, (3) Memory, (4) Visuoconstructive/Visuospatial processes, and (5) Abstraction. A brief description of the components of these measures follows.

Information Processing Speed

Posner Letter Matching Task: (Posner, 1967) measures the speed (in msec) in retrieving highly overlearned name codes from long-term memory.

Sternberg Short-Term Memory Search Task: (Sternberg, 1966; 1975) measures the speed (in msec) required to scan information in immediate memory.

Figural Visual Scanning and Discrimination task: (Ekstrom et al, 1976) involves scanning five figures to identify the one figure (clown face, truck etc) that matches a prototype.

Two-Choice Visual Reaction Time Task: (Wilkie et al, 1982) measures psychomotor response speed.

Attention

Digit Span: (Wechsler, 1981) measures memory and attention.

Variable Interstimulus Task: (Posner, 1971) measures alertness.

Continuous Paired Associates Task: (Lansman and Hunt, 1982) measures memory and attentional resources (capacity).

Memory Processes

California Verbal Learning Test (CVLT): measures verbal learning and memory

Continuous Paired Associates Learning, Secondary Task Paradigm: (Lansman and Hunt, 1982). See above

Abstraction

Trail Making Test A & B (TMTA and TMTB) (Lezak, 1983) measures visuoconceptual and visuomotor tracking (involves motor speed and attention).

Visuoconstructive/Visuospatial Skills

Digit Symbol Substitution (WAIS-R): (Wechsler, 1981) measures visuoperceptual and motor processes.

Driving Performance Measures: Driving performance was measured using an Atari Games Corporation interactive driving simulator. In order to assess driving performance on the interactive simulator, two scenarios were developed: a practice scenario to permit the driver to familiarize himself/herself with the simulator and its controls, and a test scenario which was only administered when the driver was observed to understand the operation of the simulator and when he/she reported being ready. The test scenario was developed to address various situations with which older drivers have been reported to have difficulties (e.g. making left turns, stop signs, stop lights, yielding the right of way). The simulator was set on automatic transmission to control for variations in driving due to skill in shifting. Participants were reminded that they could end their participation in the study at any point without penalty. The performance variables developed and for which results are presented in this report are described below.

The driving performance variables developed were the following: 1) collisions (with moving or stationary objects) 2) lane keeping, which was measured as the deviations from the midpoint of the lane, 3) number of times over the median, 4) number of times off the road, 5) reaction to a vehicle which stops suddenly on a hill crest, 6) reaction to stop lights, 7) reaction to stop signs, and 8) left turn performance which included whether the driver made the left turn, and the time elapsed from arrival at the intersection to the completion of the left turn. In addition to the above, the subjects' knowledge of traffic signs was assessed. This task included twelve of the most frequently encountered road sign symbols (e.g., stop sign, do not enter, railroad crossing, one way etc.,) without the text that is usually on these symbols, where applicable. The total administration time for the tests varied from 3 to 4 hours.

RESULTS

The subjects who participated in this study consisted of 48 (50.5%) females and 47 (49.5) males. Twenty-one percent (19) of the subjects were employed full time, 20% (18) worked part-time, and 60% (55) were unemployed. Eighty-one percent (77) of these subjects reported owning a car. About 91% of the subjects drove four days or more during the week with most persons (62%) driving seven days a week. There were no significant differences in the number of days per week young middle-age and older drivers drove (i.e., approximately six days per week for each group).

There was no significant difference in the educational level of the three age groups. Each averaged about fourteen years of education. While older drivers drove as many days per week as younger ones, they drove fewer miles per day than the other age groups (Young: X=35.98, SD=46.75; Middle-Age: X=39.47, SD=39.30; Old: X=21.34, SD=13.29). This difference was statistically significant between the young and old groups (F=3.78 (2,85), p=.03). This suggests that older persons generally take shorter trips.

Most (73.1%) of the subjects reported having had no accidents in the past five years, whereas 26.3% reported having had at least one accident. Older persons were less likely to have been involved in an accident than younger persons (r=-.27, n=93, p=.008). Most of the subjects (63%) considered themselves better drivers than others on the road and most (73%) considered themselves better than drivers in their own age category. The older the respondents, the more likely they were to rate their driving as better than that of others their own age r=-.28, n=94, p=.006.

Although ninety-five subjects participated in this study, only sixty-four completed all aspects of the simulator driving. The data on driving reflects their performance only.

Relationship of Age to Cognitive Characteristics

Correlational analyses of age by the cognitive characteristics mentioned above as well as visual acuity showed significantly poorer performance on all the variables except three: Digit Span and number of errors on Trail Making Tasks A&B. This supports the sensitivity of these measures to age-related effects (see Table 1).

Relationship of Age and Cognitive Characteristics of Drivers to Driving Performance

Bivariate correlation analyses of age, visual acuity, and the components of the cognitive battery and age with various aspects of driving performance were conducted. Given the large number of significant relationships found, only relationships with magnitudes of .30 or more are reported. In order to facilitate presentation, the results of the analyses are presented along the components of driving performance listed above.

Collisions: Twelve of the drivers had at least one collision on the simulator. No significant relationship was found between driver characteristics and collisions. However, collisions were significantly related to poor lane keeping (r=.33, n=63, p=.004) and going over the median (r=.40, n=64, p=.001). A multiple regression analysis showed going over the median to be the best predictor of collisions (Beta= .40 (1,61), T=3.49, p=.009).

Lane Keeping: Poor performance on the lane keeping task was related to slower speed on the TMTB task (a measure of motor speed and attention) (r=.39, n=64, p=.001) and increasing age (r=-.31, n=63, p=.006). A regression analysis to determine the contribution of these two variables to lane keeping showed speed on the TMTB task to be the best single predictor of lane keeping (R=.33 (1,61), F=7.28, p=.009). The combined contribution of speed on the TMTB and age accounted for 24% of the variance on lane keeping (R=.49 (2,60), F=9.24, p=.0003).

Table 1. Relationship of Age to Cognitive Measures and Visual Acuity.

	Age		
	Coeff.	n	Sig.
Calif. Verbal	-0.30	94	P=.002
Digit Span	-0.15	94	P=.081
Near Acuity	0.37	92	P=.000
Reaction Time	0.30	92	P=.002
Far Acuity	0.43	93	P=.000
TMTA Error	-0.02	94	P=.406
TMTA Time	0.38	94	P=.000
TMTB Error	0.01	94	P=.450
TMTB Time	0.24	94	P=.012
Alertness (VIT)	0.22	92	P=.016

Number of Times off Median: Speed on the Two Choice Reaction Time task (a measure of psychomotor response speed) was significantly related to the number of times the subject went over the median on straight roads (r=.35, n=74, p=.002). The slower the reaction time on the Two Choice task, the more often the subjects went over the median.

Number of Times off Road: The number of times the subjects went off the road increased significantly in relation to: 1) poor far visual acuity (r=.58, n=63, p<.0001), 2) slow speed on the Two-Choice Reaction Time (r=.41, n=64, p=001), 3) lower speed on the TMTA (a measure of motor speed and attention), (r=.57, n=64, p<.0001), 4) errors on the TMTB (r=.50, n=64, p<.0001), 5) slow speed on the TMTB task (r=.65, n=64, $p<.0001$).

Given the number of variables that are significantly related to the number of times a driver went off the road, a stepwise multiple regression analysis was conducted to determine which of these variables contributed most to the variance on driving off the road. Far visual acuity, the number of errors on the TMTA, the time to complete the TMTA, the time to complete the TMTB, and performance on the Two-Choice Reaction Time were entered into a Stepwise Regression. Speed on the TMTB was found to be the single best predictor of the number of times the driver went off the road (R=.67, (1,61), F=48.41, $p<.0001$). Far visual acuity was the second best predictor of that performance. These two variables combined account for 53% of the variance on the number of times the driver went off the road (R=.73, (2,60), F=34.18, $p<.0001$).

Reaction to a Vehicle's Sudden Action: Part of the scenario included the appearance of a green van ahead of the driver, stopped at the foot of a hill, then suddenly climbing up the hill and, upon reaching the crest of the hill, suddenly stopping. A significant relationship was found between performance on the Variable Interstimulus Task (VIT) (a measure of alertness) and stopping at the appearance of the van (r= -.64, n=69, $p<.0001$). The greater the percentage of errors committed by drivers on the VIT the less likely they were to stop at any point in the presence of the green van.

Reaction to a Stop Light: The only variable related to stopping at a Stop light was the score on the Digit Span Backward (a measure of attention) (r=.42, n=70, $p<.0001$). The greater their alertness/attention as indicated by higher score on the Digit Span, the more likely the driver was to stop at the stop light.

Reaction to Stop Sign: No significant relationship was found between reaction to Stop Sign and any of the drivers' characteristics.

Knowledge of Traffic Signs: Age was significantly related to the number of traffic signs identified correctly (r=-.46, n=93, $p<.0001$). The older the participant the fewer signs he/she correctly identified. The free recall score on the California Verbal Learning test (a measure of verbal learning and memory) was significantly related to the number of traffic signs correctly identified (r=.42, n=93, $p<.0001$). The greater their memory/verbal learning skills as shown by higher scores, the greater the number of signs correctly identified.

Left Turn Performance: Most of the drivers who completed the scenario (n=64) also completed the left turn (n=51) and did so without colliding with any of the oncoming cars or other objects in that environment. The amount of time the subject waited before turning increased significantly in relation to the following: age (r=.43, n=51 p= .002), poor near visual acuity (Jaeger) (r=.42, n=51 p=.002), and slow Reaction Time (r=.42, n=51, p=.003).

A stepwise regression analysis was conducted to determine which of the three variables significantly related to time to complete the left turn contributed the most to that variable. The single best predictor of time taken to make the left turn was near visual acuity (R=.42 (1,48), F=10.18 p=.003) followed by reaction time (R=.49, (2,47), F=7.60 p=.001). Age did not add significantly to the variance accounted for by the two variables mentioned.

CONCLUSION

These results show clearly that while age is significantly related to the cognitive characteristics measured in this study, in most cases it does not contribute significantly to understanding the performance of driving tasks or their outcome as measured on an interactive simulator. Moreover, as shown above, none of the cognitive characteristics of drivers were related to collisions. However, they were related to driving tasks and driving outcomes one of which (i.e., going over the median) contributed strongly to collisions. This would suggest that while collisions are important to consider, understanding how the performance of critical driving tasks is related to them is as important.

The results highlight the importance of age sensitive cognitive and perceptual characteristics of drivers in the performance of driving tasks in general and more specifically the importance of distinct cognitive characteristics to particular tasks. For example, knowing that tracking tasks (e.g., lane keeping) and their outcomes when performed inappropriately (e.g., going over the median, going off the road) are related to attention, motor speed, and psychomotor speed or that a driver's reaction to the sudden actions of a vehicle is related to a measure of attention and alertness has implications for assessment of driving skills as well as training.

If the relationships found using the simulator are replicated on the road, they will validate the interactive simulator as a useful tool for assessment of driving skills. These results would permit researchers and policy makers to develop effective ways of identifying those characteristics of persons that impact negatively upon their driving performance and, subsequently include them in drivers' evaluation.

Validation of these data would also enable the development of more effective training methods. Much of current training addresses the execution of driving tasks. While such an approach is important, the cognitive characteristics that impact upon the performance of driving tasks should also be considered and those that are sensitive to training should be identified. Otherwise, training will have addressed only a narrow aspect of driving performance.

ACKNOWLEDGMENT

This study was supported by the National Institute on Aging Grant No. 1PSO AG11748-01, through the Miami Center on Human Factors and Aging Research, Edward R. Roybal Centers for Applied Gerontological Research.

REFERENCES

Delis, D.C., Kramer, J.H., Kaplan, E., Ober, B.A (1987). California Verbal Learning Test (CVLT). The Psychological Corporation.

Ekstrom, R.B., French, J.W., Harman, H.H., and Dermen, D. (1976). Manual for kit factor-referenced cognitive tests. Princeton, N.J.: Educational Testing Service.

Lansman, M., Hunt,, E. (1982). Individual differences in secondary task performance. Memory & Cognition, 10: 10-24

Lezak, M. (1983). Neurological Assessment. New York. Oxford University Press

McKnight, J.A. (1988). Driver and pedestrian training. Transportation in an aging society. Improving mobility and safety for older persons. Vol. 2 Technical Papers. Transportation Research Board National Research Council. Washington, D.C.

Posner, M.I., Mitchell, R.F. (1967). Chronometric analysis of classification. Psychological Review, 74: 392-409.

Posner, M.I., Boies, S. (1971). Components of attention. Psychological Review, 171: 701-703.

Rosenbloom, S. (1988). The mobility needs of the elderly. In Special Report 218: Transportation in an Aging Society. Vol 2, pp. 21-71.

Sternberg, S. (1966). High speed scanning in human memory. Science, 153: 652-654.

Sternberg, S. (1975). Memory scanning: New findings and current controversies. Q.J. Experimental Psychology, 27: 1-.

Waller, J.A. (1967). Cardiovascular disease, aging, and traffic accidents. Journal of Chronic Disease, 20: pp. 615-620.

Wechsler, D. (1981). Wechsler Adult Intelligence Scale-Revised. New York. The Psychological Corporation.

Wilkie, F.L., Eisdorfer, C., Morgan, R., Lowenstein, D.A., Szapocznik, J. (1990). Cognition in early HIV-1 infection. Archives of Neurology, 47: 433-440.

Yaksich, S. (1985). Interaction of older drivers with pedestrians in traffic in: J.L. Malfetti (ed). Drivers 55+ needs and problems of older drivers: Survey results and recommendations. Proceedings of the older driver colloquium, Orlando, Florida, February 4-7, 1985.

ISOLATING RISK FACTORS FOR CRASH FREQUENCY AMONG OLDER DRIVERS

Karlene Ball*, Cynthia Owsley, Daniel Roenker*, & Michael Sloane*****

*Department of Psychology, Western Kentucky University, **Department of Ophthalmology, School of Medicine/Eye Foundation Hospital, University of Alabama at Birmingham, ***Department of Psychology, University of Alabama at Birmingham.

By the year 2024, 25% of drivers in the U.S. will be over the age of 65. Older drivers have more crashes and fatalities per mile driven than any other adult age group. Although driving is a highly visual task and vision impairment is more prevalent in the elderly, previous research has failed to identify visual factors which are strongly associated with increased crashes in the elderly. Using a comprehensive approach to assess several aspects of visual processing in a large sample of older drivers, this study has identified a measure of visual attention that had high sensitivity (89%) and specificity (85%) in predicting which older drivers had a history of crash problems, a level of predictability unprecedented in research on crash risk in older drivers. The "useful field of view", as it is called, measures the spatial area within which an individual can be rapidly alerted to visual stimuli. Older adults with substantial shrinkage in the useful field of view were six times more likely to have incurred one or more crashes in the previous five year period. By comparison, visual sensory function, cognitive status, and chronological age were poor predictors of crash involvement. This study suggests that policies which restrict driving privileges based solely on age or on stereotypes of age-related declines in vision and cognition are scientifically unfounded. With the identification of a visual attention measure highly predictive of crash problems in the elderly, decisions on the suitability of licensure in the older adult population can be based on objective, visual-performance-based criteria.

INTRODUCTION

Many older individuals are subject to age-related declines in the abilities needed to allow them to live independently. In particular, sensory and cognitive functions may deteriorate in later adulthood, and it is widely believed that these deficits contribute to a decline in the ability to carry out everyday activities. The cost of loss of independence places financial burdens on these older individuals themselves, their families, and society as a whole. One such ability, which is crucial to maintaining independence, and which is affected by multiple sensory and cognitive factors, is the continuing ability to drive. Mobility is critical in maintaining social contacts, independent functioning, and a satisfying quality of life. Since some older adults will experience behavioral and biological changes which may make driving more difficult in later life, the focus of the present paper is to evaluate screening and intervention strategies designed to enhance and/or maintain the underlying sensory/cognitive functions required for safe driving.

Recently we developed a regression model for predicting crash frequency in elderly drivers on the basis of a study which assessed visual and cognitive skills in a small sample of older adults (Owsley, et al., 1991). The most prominent feature of the model is that visual attention and mental status are the only variables that significantly predict crash frequency. Although the model acknowledges that eye health is related to visual sensory function, and visual sensory function is related to visual attention, neither eye health nor visual sensory function have a direct effect on the prediction of crashes. The test ofvisual attention employed in this study consisted of a central target identification task coupled with a peripheral target localization task, which together provided a measure of "the useful field of view" (UFOV). The size of the UFOV is a function of (at least) three variables which are varied during the test -- the duration of target presentation, the competing attentional demands of central and peripheral tasks, and the salience of the peripheral target. Thus, the UFOV incorporates stimulus and task features which seemingly reflect key components of driving.

We have subsequently evaluated a larger sample of older drivers to test this model (Ball et al., in press), assessing various aspects of visual information processing including health status of the visual system, visual sensory function, visual attentional skills, and cognitive skills. A subset of these drivers has also been prospectively evaluated over the past two years. This ongoing study will now be summarized as it relates to the concurrent evaluation of multiple risk factors among older drivers.

METHOD

The recruitment population consisted of all licensed drivers aged 55 years and older who lived in Jefferson County, Alabama (N=118,553). In order to avoid a restriction of range problem on the dependent variable, we obtained a sample which was balanced with respect to two variables, previous five year crash history and age.

In the original wave of data collection, a total of 302 subjects were tested. Six of these participants were later excluded from the sample because they had not driven in the previous five years. Two additional participants were excluded because they did not finish the test protocol. The final sample had 294 participants, with 33% of subjects having 0 crashes, 49% with 1 - 3 crashes, and 18% with 4+ crashes. Within each crash frequency category, age was evenly distributed. The mean age of the entire sample was 71 years (range 56 - 90 years); 136 were male and 158 were female. All participants lived independently in the community.

There were five parts to the protocol: visual sensory function, mental status, UFOV, driving habits questionnaire, and eye health. The order of the five parts was counterbalanced across subjects, except for the eye health exam which was always last. All subjects received a detailed eye health examination by an ophthalmologist. The visual sensory function tests consisted of visual acuity, contrast sensitivity, disability glare, stereopsis, color discrimination, and visual field sensitivity.

Mental status was assessed by the Mattis Organic Mental Status Syndrome Examination (MOMSSE), specifically designed to assess cognitive status in the elderly. Additional cognitive tests were carried out to evaluate visuospatial abilities, and included the Rey-Osterreith test, the Trailmaking test, and the block design of the Wechsler Adult Intelligence Scale (Revised).

The size of the UFOV was assessed using the Visual Attention Analyzer. This instrument uses three subtests which provide a reliable measure of UFOV size, expressed in terms of the percentage reduction (0 - 90%) of a maximum 35 degree radius field (see Ball, Roenker & Bruni, 1990 for a detailed description).

A questionnaire was administered which assessed the subject's driving habits, such as: (i) driving exposure, (ii) avoidance of potentially challenging driving situations (e.g., left-hand turns across traffic, driving alone), (iii) number of crashes incurred during the previous five year period where the police came to the scene. In addition to this "self-report" crash information, crash frequency during the previous five year period was obtained for each subject from the state computer of the Alabama Department of Public Safety (DPS).

The dependent variable utilized in this study was the total at-fault crashes recorded by the state during the five year period prior to testing. Three raters independently studied each accident report to determine whether our driver was at fault. Detailed examination of the 559 accident reports revealed 195 crashes where our research participant was clearly not at fault (e.g., subject's unoccupied parked vehicle was hit). Concordance among the three raters was perfect in identifying these cases. In the remaining 364 accident reports our research participant was judged to be solely or partially at fault.

RESULTS

The goal of this study was to test a model designed to predict crash frequency in older drivers on the basis of visual, cognitive, and attentional measures. Using representative measures from each of the different aspects of the visual information processing system, we tested our original model using the LISREL VII structural modelling program (Joreskog & Sorbom, 1989). As shown in Figure 1, our model as formulated assumes that eye health, central vision, and peripheral vision have only indirect effects on crash frequency but direct effects on visual attention (UFOV). It further asserts that mental status has a direct effect on crash frequency, as well as an indirect effect on crash frequency mediated through UFOV.

Only two variables, UFOV and mental status, had direct effects on crash frequency, jointly accounting for 28% of its variance. Even when we re-specified the LISREL model so that central and peripheral vision were forced to have direct effects on crash frequency (in addition to their indirect effect through UFOV) there was still no increase in the crash variance accounted for. The main role of central and peripheral vision in the model is their significant direct effect on the UFOV; together central and peripheral vision accounted for 30% of the UFOV variance. **Not surprisingly, visual attentional skills crucially depend on the integrity of information entering through the sensory channel.** With respect to eye health, while eye health by itself did not significantly impact UFOV, it may have exerted an indirect effect on UFOV through its association with central and peripheral visual function (indicated by the curved lines on the left side of Figure 1). These results support our hypothesis that UFOV is a mediating variable between crash frequency on the one hand, and eye health, visual function, and mental status on the other. If UFOV is removed from the overall model, the remaining visual variables jointly account for only 5% of the crash frequency variance, and the introduction of the mental status variable only accounts for an additional 11% of the variance. Therefore the model presented in Figure 1 clearly maximizes the prediction of crash frequency.

Figure 1
LISREL Model for Predicting Crash Frequency

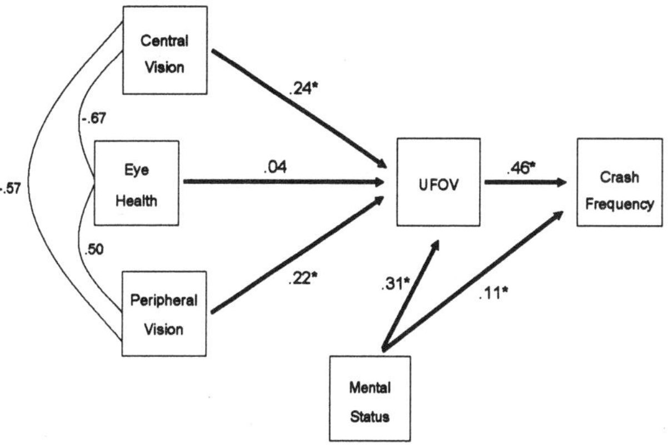

One useful way of characterizing any predictor variable is in terms of its ability to discriminate between crash involved and non-involved drivers. To determine whether an independent variable can adequately make this discrimination, we varied our definitions of "good" and "bad" performance for each independent variable, and then sorted drivers into the four categories obtained from the combination of high/low risk (good/poor performance on the independent variable) and crash history (yes/no). Figure 2 displays ROC curves for several of the variables included in our study: acuity, contrast sensitivity, peripheral vision loss, mental status, UFOV, and chronological age and clearly indicates that the UFOV was much better at identifying crash-involved older drivers than were the other independent variables evaluated.

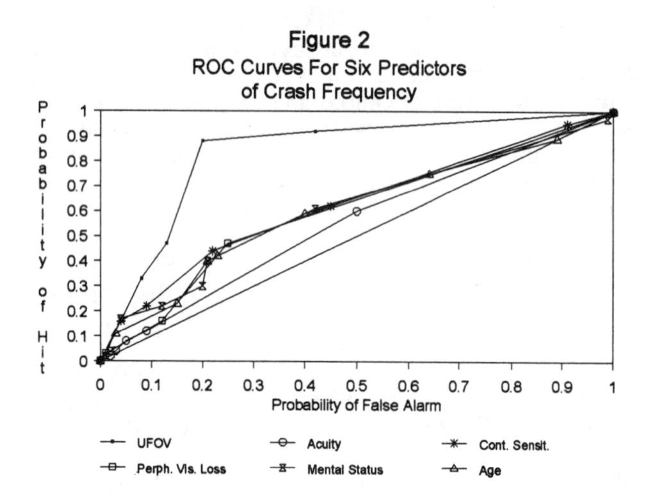

Figure 2
ROC Curves For Six Predictors of Crash Frequency

This study permits us to evaluate some of the popular hypotheses and stereotypes about which factors place an older driver at risk for crash involvement. Visual sensory impairment in later life is often suggested as the primary cause of older adults' higher crash rate. Our model indicates that although visual deficits in central and peripheral vision are significantly correlated with increased crash frequency, that the effect of visual impairment in the elderly on crash frequency is indirect. Furthermore, as Figure 2 implies, no cutoff criterion in acuity, contrast sensitivity, or peripheral vision could be adopted that would place individuals in a high risk category without including a significant number of crash-free drivers in this category as well.

Another common hypothesis is that older adults' crash problems are primarily due to cognitive confusion associated with dementing disease. Indeed mental status did have a significant, but small, direct effect on crash frequency in our study. However, mental status is also strongly related to performance in the UFOV task, which itself has the strongest relationship to crashes in our model. Figure 3 illustrates the relationship between UFOV and crash frequency for older drivers with good vs. poor mental status. The overall crash rate is indeed slightly higher for the poor mental status group. However, the association between UFOV reduction and increased crash frequency was observed in both the good mental status group and the poor mental status group. This pattern of results is reflected in the ROC curve in Figure 2 which illustrates that mental status does not successfully identify drivers at-risk for crash involvement.

Figure 3
Crash Frequency as a Function of Mental Status and UFOV

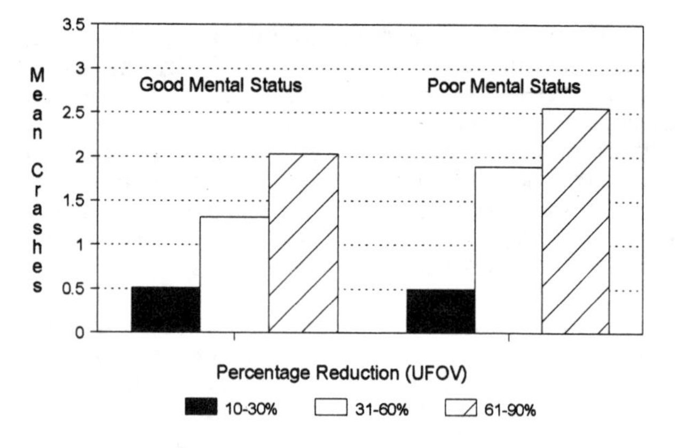

Yet another popular reason cited for increased vehicle crashes in the elderly is "old age" itself. We evaluated the relationship between chronological age and crash frequency, by stratifying our sample into three age groups. Figure 4 illustrates that the association between UFOV and crashes is similarly strong within each of three age groups evaluated ($F<1.0$). Although both UFOV reduction and crashes are more prevalent with increasing age, the ROC analysis indicates that UFOV reduction is substantially better than chronological age at differentiating drivers who are at risk for crashes from those who are not.

Figure 4
Crash Frequency as a Function of Age and UFOV

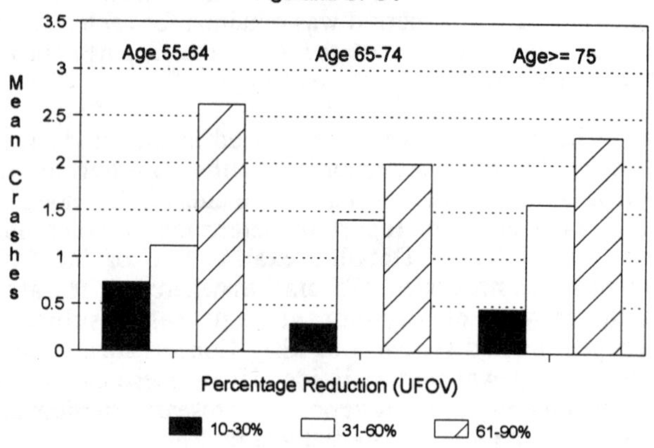

Figure 5 illustrates that the average number of crashes increases with increasing severity of UFOV reduction. We also examined the utility of UFOV using varying cutpoint criteria. The cutpoint of 40% reduction appeared to provide the best discrimination, as indicated in Figure 2. The UFOV test had both high sensitivity (89%) and high specificity of (85%) with respect to driver classification. Furthermore, the information was used to calculate an odds ratio, which indicated that individuals with UFOV reduction greater than 40% were six times more likely to be at least partially responsible for a crash than are those with minimal or no UFOV reduction. It should also be pointed out that of the 25 false-positive predictions, 19 were subjects who reported that they avoided driving in general, avoided driving alone, and/or avoided left-hand turns, which thus minimized their driving exposure. In fact, if we exclude those individuals who specified on the driving habits questionnaire that they avoided these particular aspects of driving, the correlation between UFOV and crash frequency increased from r=.52 to r=.62. It appears that driving avoidance works against the isolation of correlates of crash involvement.

Figure 5
Crash Frequency as a Function of UFOV

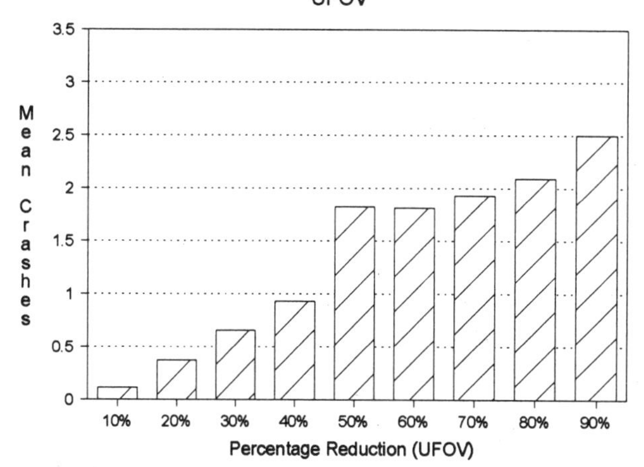

DISCUSSION

These results indicate that a measure of visual attention -- the size of the useful field of view -- has high sensitivity and specificity in predicting crash risk for older drivers. Our data also imply that current visual screening techniques are not adequate in identifying which older drivers are likely to be involved in crashes. Tests of acuity and peripheral vision may have other benefits (e.g., referral for eye care), but our analysis indicates they do not successfully screen out older drivers who pose a safety risk. Although visual attentional problems are more prevalent in the older adult population, chronological age itself did not successfully predict crash history. Thus, this study indicates that any policy to restrict driving privileges based solely on age is not scientifically well-founded. Decisions on the suitability of licensure in the older adult population are more appropriately based on an objective performance measure. A test of visual attention, such as the useful field of view, may be such a measure, given its high specificity and sensitivity for identifying drivers with a history of crash problems. The mental status test had comparatively weaker sensitivity and specificity than did UFOV, nor did it significantly improve the predictions based on the UFOV test. This finding lends little support to the idea that tests of cognitive status should be widely useful in making decisions about driving licensure for the elderly. Finally, the interrelationships among different aspects of the visual information processing system would not have been revealed (see Figure 1) if we had resorted to the conventional approach of studying visual sensory variables or cognitive variables in isolation.

Follow-up work has addressed a number of important questions which remain. The ability of UFOV to predict future crash problems in our sample of older drivers has been evaluated following one year from initial testing, and data has also being collected on a two year follow-up. An analysis of new accidents since original testing showed that those drivers in the high risk group experienced 6.5 times as many new at fault accidents in the subsequent year than the drivers in the low risk group. These data are encouraging relative to the utility of this approach in driver screening.

ACKNOWLEDGMENTS

This research was supported by the National Institute on Aging (AG04212, AG05739), the National Eye Institute (EY06390, EY03039), the AARP Andrus Foundation, the Rich Retinal Research Foundation, and a development grant from Research to Prevent Blindness to the UAB Department of Ophthalmology.

REFERENCES

Ball, K., Owsley, C., Sloane, M., Roenker, D., & Bruni, J. (In press) Visual Attention Problems as a Predictor of Vehicle Accidents among Older Drivers. Investigative Ophthalmology & Visual Science.

Ball, K., Roenker, D. & Bruni, J. (1990) Developmental changes in attention and visual search throughout adulthood. In Enns, J. (Ed.), The Development of Attention: Research and Theory. North Holland: Elsevier Science Publishers, 1990, 489-508.

Joreskog, K.G. & Sorbom, D. (1989) LISREL VII: A Guide to the Program and Application, Second Edition (SPSS, Inc, Chicago).

Owsley, C., Ball, K., Sloane, M., Roenker, D.L. & Bruni, J.R. (1991) Visual Perceptual/Cognitive Correlates of Vehicle Accidents in Older Drivers. Psychology and Aging, 6, 403-415.

OLDER COMMERCIAL VEHICLE DRIVERS: ABILITIES, AGE, AND DRIVING PERFORMANCE

Robert E. Llaneras,
Robert W. Swezey, and
John F. Brock

InterScience America
Leesburg, Virginia

This research reports upon a series of related studies which involved investigations of fifteen human abilities, their changes which occur with aging, and their effects upon commercial truck driving performance. One-hundred and seven commercially licensed truck drivers, divided into five age cohort groups, participated in the research study. An ability testbed was used to measure 15 driving related perceptual, cognitive, and psychomotor abilities. Driving performance was assessed using an interactive commercial truck driving simulator. Relationships between perceptual, cognitive, and psychomotor abilities and measures of driving performance indicated that functional levels on these tasks and driving performance were significantly related; however, age, in and of itself, was not predictive of driving performance.

INTRODUCTION

Although sensory, cognitive, and psychomotor functions tend to deteriorate with increasing age, the onset, amount, and rate, of these deteriorations vary widely among individuals. As a result, the level of competence in basic driving skills that determines overall performance, and the underlying abilities which contribute to performance may not be reliably indexed by chronological age alone. Although there is a large body of research on the abilities of people who, presumably, represent the general population, there has been relatively little work done with commercial driver performance issues, particularly in the post 50 year old age group. Further, differences in vehicles and conditions under which conventional and commercial driving occur pose difficulties in generalizing most of the research from the driving literature. Drivers in the general population tend to change their driving habits as they age; they drive fewer miles, and avoid driving at night, at high speeds, and in bad weather. Commercial vehicle drivers, on the other hand, are less apt to have the flexibility to choose the circumstances under which they drive. Thus, it is conceivable that ability deficits may play a more significant role in commercial driving performance than in conventional driving.

The primary objective of this research program (fully documented in Llaneras, Swezey, Brock, Van Cott, and Rogers, 1995) was to identify and reduce risks associated with older commercial drivers, thereby enabling this growing segment of the population to continue to effectively contribute to the workforce, while ensuring safe and productive driving. In order to address this objective, a series of empirical research studies were conducted to:

(1) Examine the effects of increasing age on perceptual, cognitive, and motor abilities by collecting data on ability measures that have been shown to be related to driving, and

(2) Investigate the effects of diminished abilities on direct measures of driving performance in order to determine how age-related ability deterioration affects performance in commercial vehicle driving.

METHOD

This effort involved investigations among fifteen human abilities, their changes which occur with aging, and their effects upon commercial truck driving performance. An ability test battery, consisting of ten unique assessment devices, was assembled and used to measure the 15 abilities listed below:

Perceptual Abilities	static visual acuity dynamic visual acuity contrast sensitivity useful field of view field dependence depth perception
Cognitive Abilities	decision-making selective attention attention sharing information processing
Psychomotor Abilities	reaction time multilimb coordination control precision tracking range of motion

Driving Performance Assessments

The influence of perceptual, cognitive, and psychomotor abilities on driving performance was assessed using a commercial truck driving simulator: the North American Van Lines' TT150 Professional Truck Driving Simulator. Using the simulator, participants drove a standardized ten mile course which included a variety of road conditions and driving tasks including: turns at intersections, straight driving along interstate highways and two-lane rural roads, lane changing, and responses to hazards (i.e., loss of air brake pressure). The scenario included critical as well as frequent commercial vehicle driving maneuvers, provided multiple opportunities to demonstrate and thus reliably measure these activities, and involved task conditions and procedures representative of commercial vehicle driving.

Two classes of driving performance measures were gathered including structured observations and automated simulator-based performance data.

Structured observations. Structured observations required that trained observers document driving related behaviors (e.g., mirror checks, search, speed control, lane position, etc.) at different junctions and various sections of the course. Observers noted the occurrence or absence of pre-specified targeted behaviors, such as checking the mirrors during a turn. This observation system allowed a composite driving performance score to be computed, as well as subsidiary scores for each driving maneuver (e.g., turning, merging, lane changing), and individual driver behaviors (e.g., search mirror checks, etc.). In addition, observers also rated each driver's performance across seven distinct categories, including an overall index of driving performance, using a five point rating scale which ranged from poor to excellent.

Simulator-based measures. Simulator-based measures collected automatically by the TT150 Professional Truck Driving Simulator summarized performance over the entire driving course in a variety of areas including speed management, space management, and fuel economy. Specific measures generated by the TT150 simulator included: time to complete the course, average miles per gallon, following distance, lane position, number of gear grinds, average brake temperature, and number of fatal and non-fatal collisions.

Participants

Ability and driving performance assessments were conducted on a sample of 107 commercially licensed truck drivers in five age cohorts: under 50, 50-54, 55-59, 60-64, and 65 and older. Each volunteer was selected from the American Trucking Association (ATA) and National Private Truck Council (NPTC) membership lists and monetarily reimbursed for their participation; all drivers volunteered to participate and each held a current Commercial Drivers License (CDL). With one noticeable exception, a fifty five year old female, all subjects in the sample were male. Drivers ranged in age from 31 to 76 years of age, with an average of 27 years driving experience. The sample encompassed both long-haul (cross-country) and local drivers; the typical driver averaged 61,000 miles per year. Data were collected in three geographic locations (Indiana, Ohio, and Maryland) with drivers representing over a dozen companies, including independent owner-operators. The sample also represented CDL's issued in 16 U.S. states.

Study participants underwent a series of tests designed to assess key driving-related abilities as well as simulation-based measures of driving performance (each of these measurement devices was described in detail previously). All testing apparatus (e.g., laboratory-based ability tests, and interactive driving simulators) were housed in the simulator trailer, allowing the lab to be transported to each of the data collection test sites.

Via use of a map, or through verbal instructions, drivers navigated through a ten mile course through an array of city, country, and mountain terrains, and featured realistic environments. Driver performance variables (e.g., lane position, shifting performance, speed maintenance, time to complete course, etc.) were recorded by the simulator system itself, and via structured observation forms. All participants received a 15 minute practice run prior to actual testing. During this time, subjects acquired the necessary information needed to operate the simulator controls, familiarized themselves with the location and layout of the cab, and experienced the feel and dynamics of the simulator input and response systems.

RESULTS

Relationships Among Abilities and Age

This section presents statistical results relating each perceptual ability to age. Age relationships are indexed via bivariate Pearson correlations, and group contrasts.

Perceptual Abilities and Age Correlations. Pearson correlations among the six perceptual abilities and age ranged from -.17 to -.51. All relationships were in the expected direction and were significant, suggesting that each of the six measured abilities tends to deteriorate with advancing age. Useful field of View, measured by the Visual Attention Analyzer, yielded the highest bivariate correlation (r= -.51); age accounted for 26 percent of the variance associated with UFOV scores.

Ability test scores across age groups were also investigated in order to identify age-related abilities. Results of ANOVA's performed on each of the six perceptual abilities indicated that drivers 65 and older had significantly poorer static visual acuity, dynamic acuity, contrast sensitivity, UFOV, field dependence, and depth perception compared to drivers below age 50.

Cognitive Abilities and Age Correlations. Pearson correlations among the four cognitive abilities ranged from .09 to -.44. With the exception of decision making accuracy (percent correct), all relationships were in the expected direction and were significant, suggesting that each of the measured abilities tends to deteriorate with advancing age. Two abilities, decision making (measured by the APT choice reaction time) and information processing (measured by the Trail Making Test), yielded the highest bivariate correlations (r= -.44).

Ability test scores across each age group were also investigated in order to identify age-related abilities. Results of ANOVA's performed on each of the four cognitive abilities indicated that drivers 65 and older showed significant deterioration in decision making, selective attention, information processing, and attention sharing abilities compared to drivers below age 50.

Psychomotor Abilities and Age Correlations. Pearson correlations of the five psychomotor abilities and age ranged from -.13 to -.50. All ability relationships were in the expected direction, and with two exceptions, all were statistically significant, suggesting that each of the measured abilities tends to deteriorate with advancing age. Two abilities, reaction time (measured by the APT simple reaction time test) and multilimb coordination (measured by the TT150 simulator) were not statistically significant. Control precision, measured by the Visuospatial II software package, yielded the highest bivariate correlation (r= -.50), followed closely by tracking (r= -.47), also measured via this software.

Average ability test scores across each age group were also investigated in order to identify age-related abilities. Significant cohort group differences occurred for all but one of the psychomotor abilities (reaction time); surprisingly, the 65 and over group had the best performance of any of the cohorts on multilimb coordination. This was the only ability measured by a driving activity, so experience may have compensated for any age induced decrements.

Relationships Among Age, Abilities, and Driving Performance

The previous analyses examined relationships among age and functional abilities. This section explores relationships among age, abilities, and driving performance. The purpose of the analyses reported here, therefore, is to identify how age is associated with driving performance, and establish which perceptual, cognitive, and psychomotor abilities are related to driving, and thereby determine how degradations in these abilities impact driving performance.

Age and driving performance. Of the 24 driving performance measures examined, 10 showed significant age relationships; two correlations indicated that driving performance actually improved with increasing age. Specifically, older drivers tended to have better lane position and fuel mileage than younger drivers. Age was inversely related to driver's ability to: negotiate curves, set-up and execute turns, monitor and extract information from the external environment as measured by mirror checks and adherence to traffic signs and signals, control the vehicle's speed, and smoothly operate the braking and steering system.

As indicated by the above results, age appears to be associated with some aspects of driving performance. These results do not, however, indicate that advancing age *causes* poor driving performance, nor do they reveal the mechanisms underlying such relationships. Given that chronological age is also strongly related to perceptual, cognitive, and psychomotor abilities (refer to the analysis section describing relationships among abilities and age), it would be misleading to suggest that age, in and of itself, is directly contributing to poor driving performance. The following section attempts to address this issue by regressing age and ability scores on driving performance.

Abilities and Driving Performance. Stepwise regressions were computed for each of 24 driving performance measures using age and ability scores as predictor variables. The goal of each analysis was to predict driving performance by identifying an optimal linear combination of predictor variables. Results indicated that as a group the perceptual, cognitive, and psychomotor abilities were able to account for a significant proportion of variance in 18 out of the 24 available indices of driving performance. Variability in overall driving was predicted by scores on five unique abilities: range of motion, attention sharing, depth perception, UFOV, and field dependence. Together, these five predictors in combination accounted for 54 percent of the variability in overall driving scores. Age only contributed to two out of the 24 driving performance measures; fuel management and shifting performance (as indexed by clutch use), suggesting that functional abilities (such as range of motion, UFOV, decision making, attention sharing, etc.), not chronological age, are generally predictive of driving performance.

DISCUSSION

Results of the laboratory tests revealed significant age-bound deteriorations occurred for all but two of the fifteen measured abilities; simple reaction time and multilimb coordination were not found to have a significant age-related effect. The ability which showed the most substantial age-related change in performance was Useful Field of View, measured by the Visual Attention Analyzer. The correlation between UFOV test scores and age was .51, suggesting that older individual's tended to have greater reductions in the size of their useful field of view. This is a particularly important and relevant finding since shrinking UFOV has been shown to be associated with an increased incidence of automobile crashes (Ball, Owsley, Sloane, Roenker, and Bruni, 1993).

Changes in perceptual, cognitive, and psychomotor abilities were accompanied by corresponding variations in driving performance. Multiple correlations between performance on ability tests and aspects of driving performance ranged from .33 to .74. Stepwise regression results provided strong support indicating that driving-related abilities, not chronological age, are critical determinants of driving performance. With the exception of fuel management and shifting performance, age did not significantly contribute to driving performance. Five abilities (3 perceptual, 1 cognitive, and 1 psychomotor) were shown to predict overall driving performance as measured by subject matter expert ratings. Of the five abilities which predicted overall driving performance, four (range of motion, attention sharing, depth perception, and UFOV) were shown to degrade with age. Based on the available evidence, it would appear that different types of driving tasks and behaviors require quite different types of abilities.

CONCLUSIONS AND RECOMMENDATIONS

Data generated as a result of this study lead to the following conclusions and recommendations:

(1) Useful field of view, control precision, tracking, decision making, and information processing are among the top functional abilities which tend to degrade with age. Dynamic acuity and field dependence demonstrated a tendency to degrade earliest, while multilimb coordination and simple reaction time were most resistant to age effects.

(2) Although age alone is not a significant predictor of driving performance, older drivers are more likely to demonstrate age related perceptual, cognitive, and psychomotor impairments which negatively influence driving performance. Age appears, therefore, to operate as a moderator variable which acts to influence driving performance indirectly through intervening variables such as perceptual, cognitive, and psychomotor abilities.

(3) Results of stepwise regression suggests that older drivers' range of motion, attention sharing, depth perception, UFOV, and field dependence significantly contribute to overall driving performance. Thus, interventions designed to compensate for degradations associated with these abilities would likely enhance driving performance.

ACKNOWLEDGMENTS

Dr. Robert E. Llaneras is now with the Intelligent Transportation Society of America (ITS America) in Washington, D.C.

REFERENCES

Ball, K., Owsley, C., Sloane, M.E., Roenker, D.L., and Bruni, J.R. (1993). Visual attention problems as a predictor of vehicle crashes in older drivers. Investigative Ophthamology & Visual Science, 34(11), 3110-3123.

Llaneras R.E., Swezey, R.W., Brock, J.F., Van Cott, H.P., and Rogers, W.C. (1995). Research to enhance the safe driving performance of older commercial vehicle drivers. Leesburg, VA, Interscience America. Prepared for the Federal Highway Administration, Office of Motor Carriers, U.S. Department of Transportation, Washington, D.C.

THE OLDER DRIVER - A CHALLENGE TO THE DESIGN OF AUTOMOTIVE ELECTRONIC DISPLAYS

T.H. Rockwell, Arol Augsburger, Stanley W. Smith, Scott Freeman

R & R Research, Inc.
Columbus, Ohio

ABSTRACT

Older drivers present unique challenges to the display designer. Approximately 30 percent of all drivers in the U.S. are over 50 years of age. Visual impairment, e.g., presbyopia, begins after 40. After age 55, approximately 91 percent of the population use bifocals. Unfortunately, bifocals with significant add power create zones of decreased acuity in the critical instrument panel viewing distances of 500-800 mm.

In this paper, the demands for vision in driving are related to the special visual disabilities associated with the older driver, such as increased sensitivity to glare, high contrast ratio blurring of electronic displays and increased time for target recognition. A computer legibility model is presented to relate the principal factors in design, namely character height and width, viewing distance, contrast ratio and background luminance with legibility impairment associated with various age groups. Implications of model predictions to display design are discussed.

THE VISUALLY IMPAIRED DRIVER

Age related vision disorders such as a reduction in focusing ability for close working distances are universal and world wide. In technological societies where the life expectancy is long enough for us to mature, this near visual disability will eventually affect the whole population, including automobile drivers. This age related visual disability called presbyopia results from the inability of one's eye to vary its optical characteristics in order to focus objects at different distances. Thus, presbyopia has been described as an irreversible optical failure, and an unexplained age related evolutionary blunder that comes as a psychological shock (Michaels, 1985).

As seen in Table 1, difficulty with close vision associated with presbyopia becomes progressively apparent after the mid-forties and peaks between 60 and 70. Consequently, most of the individuals find their visual performance is significantly better if they use reading glasses or bifocals for close work. Drivers of automobiles who use reading glasses do not wear these glasses while driving since their distance vision would be blurred. Since they do not wear glasses to drive, the presbyopic near vision disability is present during driving conditions. As these same individuals get closer to the age of sixty, this vision disability is most pronounced for near vision needs. Bifocal wearers who do not choose to use bifocals when driving are in this same category.

Bifocal wearers, on the other hand, frequently wear their glasses for driving, especially if they require power in the upper portion of the bifocal lens to see clearly at distance. Most bifocal lenses correct the presbyopic vision disability by adding a stronger plus lens power in the bifocal portion. The lens power generally used is one which gives best vision at a normal reading position of about 400 mm. At longer distance (i.e., 750 mm., which is commonly experienced as a distance from driver to instrument panel), the bifocal lenses are too powerful for clear focus at this longer distance and the vision may be less clear through the bifocal at this 750 mm distance than if through the upper (distance vision) portion of the bifocal lens.

When presbyopia is first noticeable to an individual, a fatigue problem associated with sustained near work is often noticed. Frequently, the person will push near visual material farther away in order to see clearly and comfortably. This can be done because at the onset of presbyopia a limited amount of close focusing ability (accommodation) remains. Even if a bifocal lens is prescribed, a lower power bifocal is recommended so the individual can use the limited amount of accommodation to extend the range of their clear near vision.

However, as we get older, we also have less and less accommodation. This requires a higher bifocal power in order to see clearly at close reading distances. Since there is also less accommodation remaining, there is a smaller range of clear vision through the higher power bifocal lens required as we get older (Alpern, 1969). As long as objects are within this close range, they are clear enough through the bifocal. The farther outside this range, however, the blurrier is the vision (see Figure 1). Some objects like instrument panels are not easily moved and the person consequently must move closer to see clearly through the bifocal. This can be both annoying and potentially dangerous.

Another problem facing the aged driver is blur. The blurring problem can be especially acute for CRT type instrument displays. Blur can be caused by a number of factors. Since the older driver has more problems with glare and/or dark adaptation, they may turn up the brightness of the display to a level where the CRT beam starts to defocus, causing blur. This effect, combined with the increasing opaqueness of the crystalline lens and vitreous humour may make for an intolerable or uncomfortable situation. The only means of correcting blur at this time would be to decrease background luminance or increase letter size (Shurtleff, 1980).

The problems of presbyopia may be exacerbated by the new technology introduced into the vehicle. Trip computers, diagnostic aids, trip navigation and high feature stereos are just a few of the latest applications. Electronic display characteristics are also in a state of constant evolution as the technology advances, particularly in the area of LCD's and vacuum fluorescent (VF) displays. VF technology with appropriate filters can provide contrast ratios in excess of 400:1 (Society of Automotive Engineers, 1984). A critical issue is whether the designer can design the displays to allow the older driver to have comfortable legibility with the new electronic features.

VISUAL DEMANDS IN DRIVING

Despite the visual demands in driving, most drivers can effectively use the visual system to meet most situations. Search patterns of drivers using eye movement techniques show that the driver makes 3-4 fixations/second as she/he moves through the dynamic ever changing visual scene (Rockwell, 1972). He relies on dynamic peripheral detection to sense lane deviations, presence of other cars adjacent to him and signs and signals. Once detected peripherally, the driver then makes foveal confirmation.

In the case of in-cab sampling, the driver must accommodate to near vision, 30 inches, and then back to far vision. With few exceptions, most drivers need little time for this accommodation process.

The problem becomes one of sampling allocation. How long can the driver afford to take his eyes off the road for instrument panel sampling? This, of course, is depenuent upon outside visual demands, weather, maneuver, etc. Research by Rockwell (1972) reveals an average time of 1¼ second off the roadway for inside glance duration.

DEVELOPMENT OF A COMPUTER LEGIBILITY MODEL

To ascertain the precise effect of age on display design, a computer model based on the empirical work of Blackwell (1980) which relates threshold legibility to target size (visual angle subtended), contrast and background luminance, was constructed (see Figure 2). The model includes:

Age

Age is reflected in 5-year increments from 20 to 65 years.

The Percent of an Age Group to be Accommodated

The model allows the user to designate the percent of the population that will be included in a given run through the model. For example, if 90 percent is chosen, then all output values will be based on the assumption that 90 percent of the population of each age group will be accommodated by the model output. Values ranging from 50 to 95 percent are available as inputs.

Background Luminance

Background luminance refers to the light reflected or emitted to the eye from the display area immediately adjacent to the lit target or character. It is measured in footlamberts (fL).

Target Luminance

Target luminance refers to the light reaching the eye that is emitted from an electronic display character or symbol. The light is measured in footlamberts (fL).

Contrast Ratio

Contrast ratio is defined as the simple ratio of target luminance to background luminance. It is the result of comparing the measurements obtained for background and target luminance. Contrast ratio is critical in determining the legibility of a given character or symbol.

Character Height

Character height (measured in mm) represents the overall height of an individual character that is used as an input into the program or can be a calculated output based on the values of input variables.

Character Height to Stroke Width Ratio

This term refers to the input value that determines the overall height of a given character as related to the stroke width. Input values range from 5:1 to 12:1.

Character Width

Character width is an output value and is determined from its percentage of the overall character height. For example, if 60 percent is chosen and the overall height is 4.5 mm, then the character width would be .60 x 4.5 or 2.7 mm. A default value of 70 percent was chosen for this input value based on the literature review.

Viewing Distance

Viewing distance refers to the distance from the observer's eye to the instrument display measured in millimeters. Inputting different values of this parameter will cause different values of character height. The calculation is strictly geometrical and uses the visual angle subtended for the calculation. The model can be used to predict character height, contrast ratio or background luminance (e.g., Table 2).

Model Outputs and Adjustments

The model can be used to predict character height, contrast ratio or background luminance. Table 2 is an illustration of character height output. Two fine tune adjustments were made in this model. The first involved a contrast multiplier adjustment such that the predicted required visual acuity for near vision Snellen chart performance matched the acuity characteristics of the US population as noted from a HEW study (see Table 1).

An adjustment of 5 minutes of visual arc of the stroke width was made for each model prediction to move from acuity (just legible performance) to comfortable, short time viewing performance. This brings the computer outputs in line with standards reported in the human factors literature (Haubner and Kokoschka, 1983) and complies with the optometric rule of thumb to add two lines on a viewing chart to achieve comfortable vision in comparison to threshold acuity. This adjustment also verifies the minimum character height of 2-3/4 mm found legible for all older subjects (55-65 years of age) tested in the laboratory for high contrast night conditions at 30" viewing distance.

Laboratory verification of the model using 19 subjects in the 65-year-old age bracket showed good correspondence with model predictions.

SENSITIVITY ANALYSIS OF MODEL PARAMETERS

One major advantage of a computer model is the ability to conduct sensitivity analysis. In this section, the sensitivity of some model outputs and inputs will be presented to illustrate model flexibility. Plots of required character height against contrast ratio for various background luminance are shown in Figure 3. Note the asymptotic effect of contrast ratio. It is clear that either contrast ratio or background luminance changes can be used to affect character heights. For viewing conditions of low contrast ratios, increased character heights are required. Similarly for a given contrast ratio, reduction in background luminance will allow smaller character heights. Figure 4 shows the effect of age and percent accommodated on character height for a given background luminance and contrast ratio.

The sensitivity analysis shows the nonlinear effects of age and percent accommodated. This finding should be of significant interest to the design of instrument panels for mature drivers.

CONCLUSIONS

This research has demonstrated that realistic legibility design trade-offs can be made for various aged subjects using a computer model based on past empirical research on visual acuity.

It should be noted that the model is intended to provide design guidelines. Designer judgement in the case of special technologies is usually needed. The model can start with a candidate display and work backward to required background luminance for the percent accommodated within an age bracket. It can also more typically specify the required character height given luminance conditions, the age bracket and percent accommodated.

To be conservative, the designer can use the 95th percentile of older drivers (60-65 years of age). Clearly, if high background luminance is used, display heights can become very large because good contrast ratios cannot be obtained. The use of recessed displays or filters to lower background luminance and its effects on character heights can also be investigated. In any event, the dramatic effect of age and percent accommodated on character height should prompt the designer to seek alternate mechanisms to achieve the desired legibility, such as increased contrast ratios.

This model should be considered a prototype version of a very complex human phenomenon. As the model is tested on actual instrument panels, its utility can be increased by fine tuning of model parameters.

REFERENCES

Alpern, M. Accommodation: Evaluation of Theories of Presbyopia: "The Eye," ed. H. Davson (first edition), Academic Press, Vol. 3. New York, San Francisco, 1969.

Blackwell, H.R. A Comprehensive Quantitative Method for Prediction of Visual Performance Potential as a Function of Reference Luminance. In Compendium of Technical Data in Support of CIE Publication No. 19/2.

Illuminating Engineering Research Institute. New York, NY, 1980.

Haubner, P. and Kokoschka, S. Visual Display Units: Characteristics of Performance. Survey Lecture: CIE Conference. Amsterdam, 1983.

Michaels, D. Visual Optics and Refraction, Third Edition. St. Louis: Mosby Company, 1985, 419.

Rockwell, T.H. Eye Movement Analysis of Visual Information Acquisition in Driving: An Overview (Paper No. 948). Proceedings of the Sixth Conference of the Australian Road Research Board, Vol. 6, Part 3, 1972, 316-31.

Shurtleff, D.A. How to Make Displays Legible. Human Interface Design, 1980.

Society of Automotive Engineers, Inc. SP-576 Ergonomic Aspects of Electronic Instrumentation: A Guide for Designers, Feb. 1984.

U.S. Department of Health, Education and Welfare. Number of Adults Reaching Specified Acuity Levels by Age and Sex, 1962.

Age (Years)

ACUITY	35-44	45-54	55-64	65-74
20/20	82	44	40	26
20/25	85	50	46	34
20/30	94	73	72	66
20/35	95	78	77	72
20/40	97	84	84	80
20/45	97	86	86	82
20/50	98	89	89	86
20/55	98	91	90	87
20/60	99	93	92	89

TABLE 1

PERCENT DISTRIBUTION OF POPULATION

REACHING ACUITY LEVELS FOR CORRECTED NEAR VISION

Contrast Ratio = 4.0:1
Background Luminance = 5.70 fL
Target Luminance = 22.80 fL
Viewing Distance = 750.0 mm
Character Height to Stroke Width Ratio = 5:1
Percent Resolve = 90.0%

Age	Character Height (mm)	Character Width (mm)	Stroke Width
20	2.1878	1.53	1.01
25	2.2166	1.55	1.04
30	2.2933	1.61	1.11
35	2.4467	1.71	1.25
40	2.6384	1.85	1.42
45	2.8301	1.98	1.60
50	3.0698	2.15	1.82
55	3.2424	2.27	1.98
60	3.8272	2.68	2.51
65	4.5941	3.22	3.21

TABLE 2 Calculation of Character Height and Width for Various Age Groups

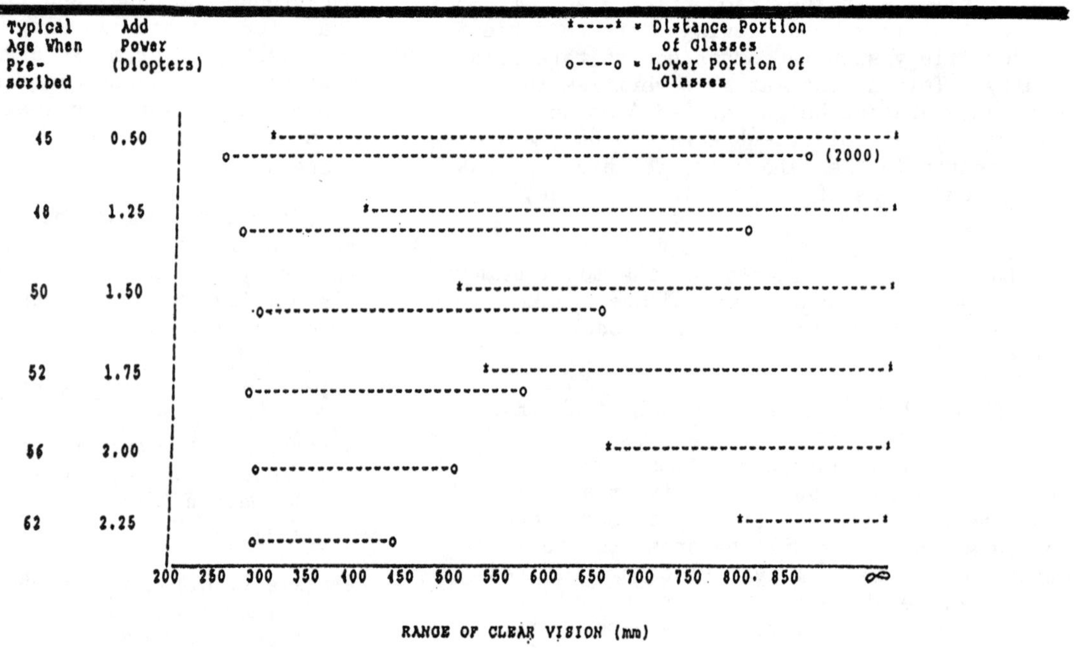

Typical Age When Prescribed | Add Power (Diopters)

---- = Distance Portion of Glasses
o----o = Lower Portion of Glasses

45 0.50
(2000)
48 1.25
50 1.50
52 1.75
56 2.00
62 2.25

200 250 300 350 400 450 500 550 600 650 700 750 800 850 ∞

RANGE OF CLEAR VISION (mm)

FIGURE 1 RANGE OF CLEAR VISION VS. ADD POWER

BLACKWELL MODEL

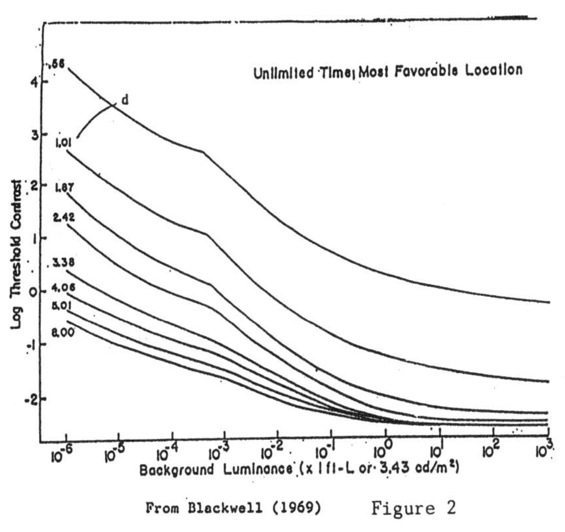

From Blackwell (1969) Figure 2

CONTRAST = $\frac{Lt-Lb}{Lb}$

Lt = Luminance of the target in cd/m^2

Lb = Luminance of the task background in cd/m^2

d = Target size in minutes of arc (corrections for time, age, % of population

VD = Viewing distance

h = stroke width = VD tan (d)

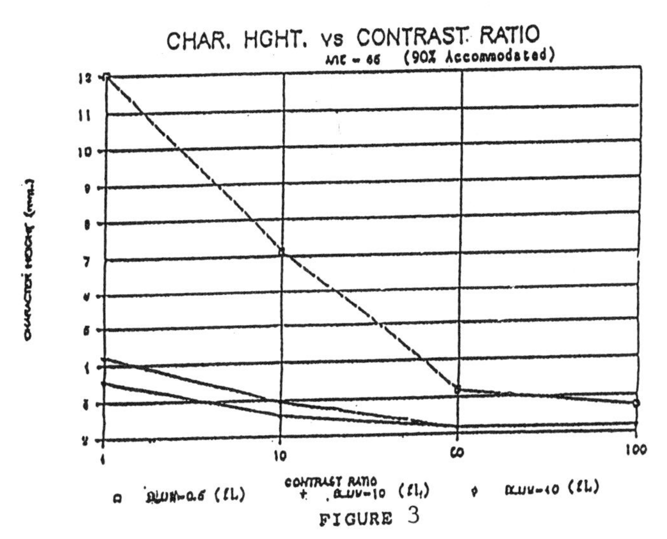

FIGURE 3

Sensitivity Analysis - Character Height vs. Contrast Ratio

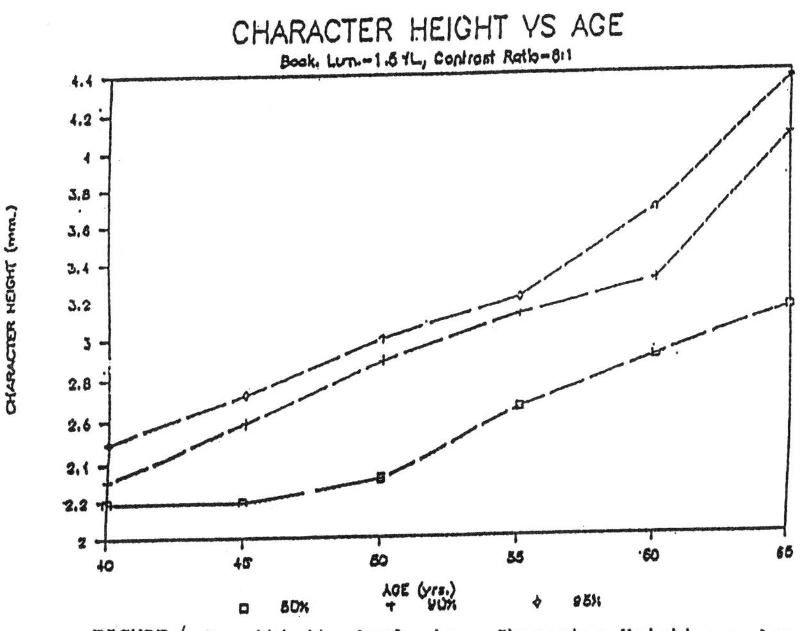

FIGURE 4 Sensitivity Analysis - Character Height vs. Age

FACTORS TO CONSIDER WHEN DESIGNING VEHICLES FOR OLDER DRIVERS

Anthony J. Yanik
General Motors Corporation
Warren, Michigan

As the Baby Boom generation gradually moves into its later years, that movement will become a Senior Boom that will have a dramatic effect upon the design of products entering the marketplace. To respond to this market, engineers and designers will require a good understanding and awareness of the changes that take place in vision and cognition as a result of the aging process, and how these changes affect the interaction of older adults with their vehicle systems such as controls and displays, mirrors, entry and exit, and lighting. This paper is an attempt to bring that understanding to the designer and engineer, as based upon current research.

When designing vehicle controls and displays that will be used by older people, one first must eliminate two stereotypes that could influence their design process; that is, 1) older people are all alike, and 2) most of them are poor, unteachable, sick and incompetent.

The fact is, unlike young adults, older adults represent quite a heterogeneous population with differing interests and capabilities. It would be more accurate to classify them as the young-old beginning at the age of about 55 to 64, the middle-old from 65 to 74, the old-old from 75 to 84, and the very old (1). Moreover, about eight of ten of older adults are active, happy, able-bodied, and competent in conducting their own affairs (2).

However, the aging process does bring a variety of changes to their visual functions and their cognition which, while gradual, may affect their interaction with the vehicle environment. This paper will examine the nature of these changes, so that engineers and designers may have a better appreciation of the needs of older drivers, and how these needs should be addressed within the vehicle.

Display Legibility: The two most obvious changes that take place in the older adult's visual process are loss of accommodation and a reduction in light transmission.

Loss of accommodation or Presbyopia means that a person cannot see things up close very clearly. The lens of the eye becomes so stiff from continual growth that the ciliary muscles cannot bend it into the required fattened shape. This is the time of life when older adults begin to wear reading glasses (3,4).

Unfortunately, reading glasses restrict near vision to a fixed near distance of 40 cm or 16 inches, but most controls and displays within a vehicle are about 50 to 80 cm away (5). As a result, older adults find themselves moving their heads back and forth to maintain this distance when reading displays and control labels - especially if those items are viewed in low light and the letter sizes are not very large.

Night time acuity also becomes a problem. Even for a healthy 20 year old with 20/20 vision, that acuity will change to 20/40 at night. For older people, it will change from 20/20 corrected vision (with glasses) to 20/70 or 20/80 in the dark. Moreover, when older adults reach the age of 60, they require least three times as much light on an object to see it as clearly as they did at 20 (6). It is no wonder, then, that they have difficulty reading road signs at night, or finding door latches, window lift buttons and other unlit controls in the dark.

Color: One of the issues that comes up frequently relative to the legibility of panel displays to older drivers is the color they should be for good discrimination.

Color discrimination is a function of the rods and cones that make up the structure of the retina upon which a visual image is formed. The cones are the tools for converting wavelengths into color sensations. They contain three photopigments: red, green and blue. Red pigment is in 64 percent of the cones, green in 32 percent, and blue in only 4 percent (which gives the first indication of it being a troublesome color) (7).

Further, the lens absorbs almost twice as much blue energy as it does of longer wavelength colors like yellow or red (7). Thus the eye is not as sensitive to display and control items colored in blue.

Finally, the lens begins to yellow after a person reaches the age of 50, so that eventually the world is seen through a yellow filter; that is, whites become yellow, yellows and oranges become brighter, but deep blues, greens, and purples become duller or muddier (8).

This being the case, what kind of colors would be best for displays for older drivers?

The most often cited work looking for such an answer is that of Galer and Simmonds. They tested the effects of five different colored instrument panel displays on 80 subjects, ages 17 to over 50, and found that both men and women over 50 favored yellow as a first choice. As a second choice men chose orange but women blue-green. Both had problems with red and green. (9)

Osaka, in his experiments, found that saturated reds and blues created eye fatigue, especially the blues, indicating the eye was struggling to focus on either color (10).

Imbeau and Wierwille, in testing 24 subjects aged 20 to 73, reported in their tests that light blue was considered to be the least attractive color among a family of colors that included orange, reddish orange, white, and amber. Older adults also tended to increase the brightness levels of the red-orange, probably because they could not bring it into focus at lower brightness levels (11).

Some have advocated red displays imitative of the display color used in aircraft. However, pilots have the need to use a color that does not interfere with their night vision; that is, does not force them to go again through the process of dark adaptation when they look away from the instruments into the night sky. Drivers have no such need since they drive with headlights on and for the most part on streets that are lit (12). At the same time, red instruments could be extremely troublesome to a color blind person.

Thus, while color researchers do not always agree, there is a thread of commonality to the extent that older drivers seem to prefer yellow, orange and white against contrasting backgrounds, or white on black. Also, a saturated blue is a poor choice for labels, thin lines and small shapes....but is excellent as a background color.

Letter Height: Another way to counter the effects of presbyopia and poor night vision is to increase the letter heights of labels featured on controls and within displays. For example, Mourant and Langolf, have worked out a table of letter heights based upon the low light reading capability of 20 different older subjects. The table indicates that when letters are viewed at a luminance level of 1.7 cd/m2 (which is typical of the car interior at night) their heights should be about 6.4 mm, and their contrast ratio 25:1 (13).

Imbeau and Wierwille, on the other hand, have recommended that letter heights be about 5.5 mm at high brightness levels, according to their own experiments (11).

Other researchers generally have advocated letter heights in the 5.0 mm to 7.0 mm range for older people in low light, so one would not be remiss to advocate a letter height of at least 5.0 mm for use in displays and on controls of vehicles intended for mature drivers.

Display Types: One question that often arises relative to displays and older drivers involves is which type would they prefer, digital or analog? Green's research indicates that drivers are more likely to understand what is signified by a moving pointer than a number (14).

Poynter calls digital displays "objects" and analog displays "features". His research indicates that "objects" are not seen too well in peripheral vision, and require high levels of attention and large display sizes. "Features", on the other hand, communicate information easily and quickly (15).

Boyce says that with numbers, one must read the entire number to understand what it represents, whereas a word can be understood quickly since often it is not necessary to read the entire word to get at its meaning (16).

Give the need that older drivers have to spend as much time as possible viewing the road for changes in traffic, the shorter glance/discrimination times of analog type displays might in various instances be preferable.

Control: Vehicle controls can present difficult challenges to older drivers because of their size, shape, color or location, or if they lack compatibility.

Reaching for a control first involves determining its general location in space, then deciding upon where to direct the hand. A final look and correction is made just before the control is reached (17).

If the control has compatability; that is, if it is where it is expected to be, finding a control is a quick and easy task. If it is not where it is expected, has a different shape or size than customary, or cannot be spotted because of its lack of contrast against its background, a recoding must take place within the thought process that may delay control operation (2). This may be a source of irritation to older drivers because it forces them to look away from the road longer than customary to find a specific control and use it.

This is not to say that new controls or new control locations should be avoided. However, the designer must keep in mind that older drivers process complex visual information slower than younger adults, and have more difficulty separating the relevant from the irrelevant when the control systems are busy or require several steps in memory to complete an operation.

If control changes are desirable, they should be accompanied by good visual cues such as obvious differences in control sizes, shapes, and colors, and accompanied with lables of a readable size and color.

Controls also should operate as expected, such as clockwise for increase, or UP for ON, and should provide good visual feedback in the ON mode.

Since control search is initiated in one's peripheral vision, it is all the more important that controls off to the side be conspicuous as to their size, shape and color to get the hand started in the right direction so that the final correction can be made quickly and correctly (16). Otherwise, the control will be ignored

Also, although functional reach is unaffected by aging, designers must be sensitive to the fact that arthritis and rheumatism are common ailments of seniors. These ailments can affect finger functions, especially as they relate to fine motor adjustments.

It would be advisable to design controls and switches for older populations that do not need to be pinched or grasped tightly to operate, and do not require large wrist rotations or input forces.

For example, a push button console shifter with high spring release pressure could be very difficult for an older person to operate relative to a column-mounted shift lever, the operation of which can benefit from full hand and arm leverage.

Another attribute of aging eyes is that the older adult's sensitivity to glare begins to accelerate once he or she begins wearing glasses. In time, glare can become one of the major reasons why older persons give up on night driving because it becomes such a painful experience for them.

While not much can be done within the vehicle to counteract direct glare, the indirect glare from rear view mirrors can be dealt with to some degree.

According to Olson and Sivak, about 70 percent of rear glare comes from the inside RV mirror. A day/night mirror can reduce that glare to 5 percent (18). The electrochromic mirror offered on several models can produce this change automatically, therefore is a distinct advantage.

The outside left mirror on vehicles creates 30 percent of total reflected glare. At present, there is no relief for it other than to turn the mirror face away which older drivers may prefer not to do because of the difficulty in repositioning the face. A day/night feature for this mirror would be a distinct advantage, and may well be the next breakthrough in glare reduction.

Paradoxically, while the removal of headlight glare is a need for older drivers, so is the need to place more light on the road at night to enhance forward vision. Some have advocated that older drivers use their high beam whenever possible out of the belief that low beam seeing distance on some vehicles may be insufficient for older drivers because of their restricted low light vision.

Olson and Sivak have demonstrated that, at 35 mph, the combined recognition/response time to avoid an object in the road is about 62 feet which is within the low beam pattern for an average driver. But, given the longer response times and lesser low light acuity of older drivers, they too speculate whether current low beams meet all of the older driver's nighttime needs (19). On the other hand, it is possible that halogen head lamps extend that nighttime field of view sufficiently for older drivers to compensate somewhat.

Entry and Exit: There is very little in the literature that addresses the subject of vehicle entry and exit parameters for older persons. Probably the most definitive study of this kind was conducted by England's Institute for Consumer Ergonamics (20). In this study, the researchers selected six passenger cars that presented a wide variety of critical dimensions bearing upon entry and exit, and had them evaluated by 64 elderly and disable subjects. From this evaluation, a set of critical dimensions for entry and exit were determined which then were programmed into an adjustable seating buck. Sixty subjects then were asked to begin with these dimensions, but adjust them accordingly to their needs. Out of these adjustments, the research team developed an idealized set of entry and exit criteria that, they believed, would satisfy over 90 percent of the population.

Following are some of the idealized entry and exit dimensions that were compiled:

Door Opening Angle - 75 degrees

Seat Height - 510 mm above ground

Sill Height - 240 mm above ground

Foot Well Depth - <50 mm

Front Pillar to Seat Edge - 450mm

Seat Edge to Outer Sill Edge - 140 mm

Summary: This paper has addressed the basic characteristics of the interior design and engineering process that the literature has signalled as being beneficial to older drivers. These are that:

-Display colors in yellows, oranges, yellow-greens, and whites on contrasting backgrounds seem to have the best legibility, and would be preferred to blues and reds which tend to create difficulty;

-Analog display may be preferred in most instances because of their more rapid recognition times;

-Letter heights of 5.5 mm to 6.4 mm appear to be a good compromise, at least for important legends;

-Control search efforts can benefit from good tactile cues via color, size, and shape, and readable identification labels, especially if the control systems are new and/or more complex;

-Specific door entry and seat dimensions are critical to ease with which elderly and disabled adults enter and leave their vehicles.

The bottom line is that older drivers have the need to spend as much time as possible examining the traffic scene for potential trouble spots. Their longer response times and greater visual difficulty in spotting signs, traffic lights, or abrupt changes in traffic make this necessary. The challenge to the automotive designer and engineer is to avoid designing controls and displays that might borrow from that available time and leave the older driver in an uncomfortable position.

References

(1) Koucelik, J. A. in Small, A. M. "Design for Older People" (See above).

(2) Small, A.M. "Design for Older People". In Salvendy, G. (Ed.) Handbook of Human Factors. John Wiley & Sons, New York, 1987, 495-504.

(3) Adler-Grinberg, D. "Questioning our Classical Understanding of Accommodation and Presbyopia". American Journal of Optometry & Psychological Optics, 63, 571-580, 1986.

(4) Kline, D. W. and Schieber, F. "Vision and Aging", Handbook of the Psychology of Aging, 2nd Edition. Von Nostrand Reinhold Co., New York, 1985, 296-330.

(5 Ophthalmology Study Guide. American Academy of Ophthalmology and Otolaryngology, Inc. 1978.

(6) Pitts, D. G. "The Effects of Aging on Selected Visual Functions: Dark Adaptation, Visual Acuity, Stereopsis, and Brightness Contrast." Aging and Human Visual Function. New York. Alan R. Liss, Inc., 1982, 131-159.

(7) Murch, G. M. "Colour Graphics - Blessing or Ballyhoo?" Computer Graphics Forum, 4, 127-135, 1985.

(8) Carter, J. H. "The Effects of Aging on Selected Visual Functions: Color Vision, Glare Sensitivity, Field of Vision, and Accommodation". Aging and Human Visual Function. New York. Alan R. Liss, Inc., 1982, 121-130.

(9) Galer, M. and Simmonds, G. R. "The Lighting of Car Instrument Panels - Drivers' Response to Five Colours". Society of Automotive Engineers, SAE 850328, 1985.

(10) Osaka, N. "The Effect of VDU Colour on Visual Fatigue in the Fovea and Periphery of the Visual Field". Displays, 138-140, July 1985.

(11) Imbeau, D. and Wierwille, W. W. "Instrument Panel Luminance and Hue". (still to be published)

(12) Connolly, P. L. "Part II", Vehicle Considerations of Man, the Vehicle and the Highway, SP-279. Society of Automotive Engineers, Inc. New York, 1966, 27-86.

(13) Mourant, R. R. and Langolf, G. D. "Luminance Specifications for Automobile Instrument Panels". Human Factors, 18, 1, 71-84, February 1976.

(14) Green, P. "Driver Understanding of Fuel and Engine Gauges". Society of Automotive Engineers, SAE 840314, 1984.

(15) Poynter, W. D. and Czarnomski, A. J. "Perceptions of 'Features' and 'Objects': Applications to the Design of Instrument Panel Displays". General Motors Research Laboratories, OS-64, October 15, 1987.

(16) Boyce, P. R. Human Factors in Lighting. Macmillan Publishing Co., Inc. New York, 1981.

(17) Paillard, J. J. (1982) in Georgopoulos, A. P. "On Reaching". Annual Reviews in Neuroscience. 9, 147-170, 1986.

(19) Olson, P. L. and Sivak, M. "Glare from Automobile Rear-Vision Mirrors". Human Factors, 26, 3, 269-282, June 1984.

(20) Olson, P. L. and Sivak, M. "Improved Low Beam Photometrics". DOT-HS-9-02304, National Highway Traffic Safety Administration, U. S. Department of Transportation, March 1983.

(21) Page, M., Spicer, J., McClelland, I., Mitchell, K., Feeney, R. and James, J. "Access to Standard Production Cars by Disabled and Elderly People". Third International Conference on Mobility and Transport of Elderly and Handicapped Persons, October 29 to 31, 1984 in Orlando, Florida.

AGE-RELATED DECREMENTS IN AUTOMOBILE INSTRUMENT PANEL TASK PERFORMANCE

Brian C. Hayes, Ko Kurokawa, and Walter W. Wierwille
Vehicle Analysis and Simulation Laboratory
Virginia Polytechnic Institute and State University
Blacksburg, Virginia

ABSTRACT

This research was undertaken, in part, to determine the magnitudes of performance decrements associated with automotive instrument panel tasks as a function of driver age. Driver eye scanning and dwell time measures and task completion measures were collected while 24 drivers aged 18 to 72 performed a variety of instrument panel tasks as each drove an instrumented vehicle along preselected routes. The results indicated a monotonically increasing relationship between driver age and task completion time and the number of glances to the instrument panel. Mean glance dwell times, either to the roadway or the instrument, were not significantly different among the various age groups. The nature of these differences for the various task categories used in the present study was examined.

INTRODUCTION

A marked increase in the proportion of elderly drivers has been observed in recent years. This trend is expected to continue. As a consequence, it has become important to understand the abilities of older drivers to cope with the displays and controls within their automobiles. While age-related decrements in visual performance are generally accepted, the exact nature and magnitudes of these decrements while driving have not been well-defined in the research literature. The aim of the present study was to determine empirically the magnitudes of age-related decrements in drivers' performance of automobile instrument panel tasks while they drove an instrumented research vehicle along a variety of road types and under different driving conditions.

METHOD

Subjects

A total of 24 subjects participated in the study, 12 males and 12 females. In addition to the 2 genders, 3 age groups were represented, including subjects between the ages of 18 and 25, between 26 and 48, and between 49 and 72. Each age group was represented by 8 subjects, which yielded 4 subjects in each Age by Gender group. Subjects were required to have a valid driver's license, and to have corrected or uncorrected near and far visual acuity of at least 20/40, as measured by a Titmus II vision tester. In addition, an informal hearing test was administered to ascertain each subject's ability to hear the task commands.

Apparatus

Experimental vehicle. The study was conducted in a 1985 Cadillac Sedan deVille that included power mirrors, a fuel data center, a digital dashboard, cruise control, and electronic climate control. The steering wheel was modified to accept a Pontiac 6000 center hub push-button radio control panel which had been removed from an experimental sample 1988 H-Car steering column. The test vehicle was further modified to accept an IBM personal computer for on-line data collection, certain safety equipment, and two video cameras and recorders. One camera was mounted on the hood of the vehicle and focused on the face of the driver to record eye movements during task execution. The second camera was mounted on the roof of the vehicle above the driver and was angled downward at 10 deg to provide a forward view of the road and the position of the vehicle within the lane.

Auxiliary instrument panel. An auxiliary instrument panel measuring approximately 9.5 in high by 12 in wide was used in the vehicle to provide a variety of control configurations in addition to those provided in the vehicle itself. The panel included a 12-button telephone keypad, a 9-position discrete rotary knob, a 4-position discrete rotary knob, a 3 x 5 push-button matrix, a Kraco KGE-601 AM/FM stereo radio/cassette player with a 3-band graphic equalizer, a Sparkomatic SR350 AM/FM stereo radio/cassette player, a Pontiac 6000 push-button radio control panel, and a 16-segment LED bar display custom designed to allow discrete or continuous adjustment. An illustration of the auxiliary panel is provided in Figure 1.

The labeling strategies on the radio control systems and the telephone keypad are standard. Two labeling strategies were used on each of the rotary selector knobs: random letters and sequential numbers (1 through 9 and 1 through 4). The labeling for the 3 x 5 push-button matrix consisted of random letters.

The location of the auxiliary panel within the vehicle was manipulated among four positions through the use of aluminum supports mounted to the center hump on the floor of the front seat compartment. The panel was located in either a low position (panel center was 13 3/4 in above the transmission hump), or high position (panel center was 22 3/4 in above the transmission hump). In addition, it was either parallel to the existing dashboard or angled toward the driver (68 deg between the vehicle longitudinal center

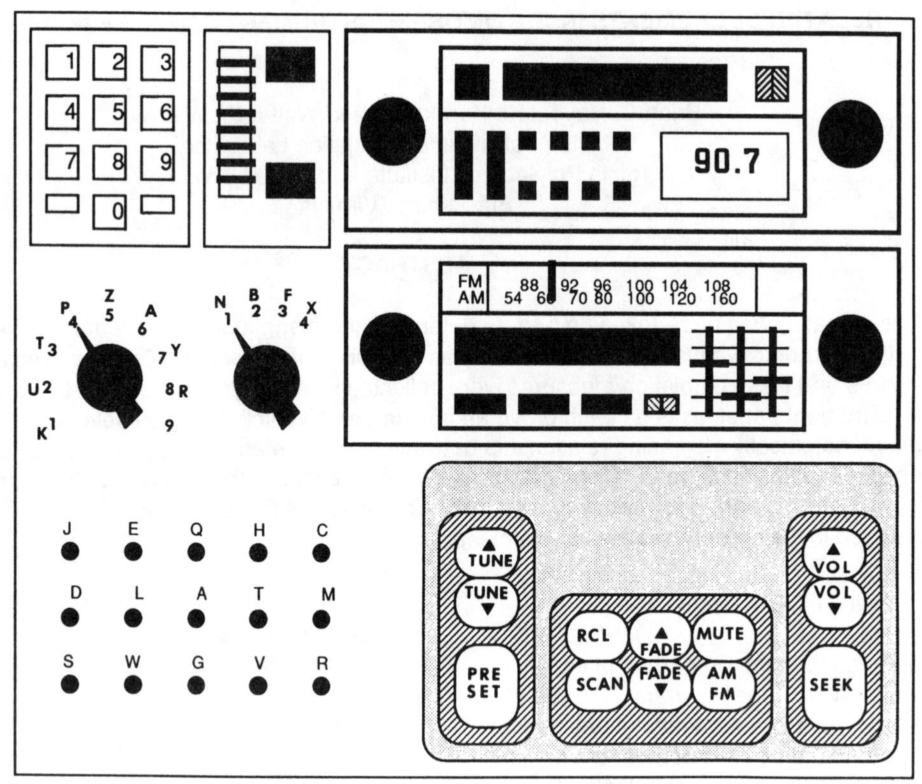

Figure 1. Layout of the auxiliary instrument panel.

line and panel). Each subject performed 50 tasks in each of the 4 positions.

Instrument panel tasks. The subjects were required to perform 200 tasks which involved the auxiliary instrument panel and the existing automotive controls and displays. The use of the auxiliary panel allowed for the separate investigation of a number of other variables, such as control clutter, task difficulty, task complexity, and random vs. sequential search tasks. The 200 tasks were divided into 4 groups of 50 equivalent but not identical tasks, one task group per panel location. Equivalent task commands were of the form: "In the top row of the red pushbuttons, press J" and "In the top row of the red pushbuttons, press C", or "On the bottom radio, press AM/FM" and "On the bottom radio, press MUTE".

Dependent variables. The dependent variables included eye glance measures and task completion measures. Eye glance measures consisted of the number of glances to the display, the mean length of a single glance to the display, the total glance time to the display, the number of glances to the roadway, the mean length of a single glance to the roadway, the total glance time to the roadway, the number of transitions between the roadway and the display, and the mean transition length. Eye glance measures were determined after data collection through a frame-by-frame analysis of the subject videotapes at a resolution of 1/30th of a second.

Task completion measures included total task completion time and hand-off-the-wheel time during task execution. The recording of task completion time began when the front-seat experimenter completed a task instruction and ended when the subject either returned the used hand to the steering wheel after successful performance in a control manipulation task or completed a verbal response to a non-manipulative task (e.g., "read the present speed"). Hand-off-the wheel time was defined as the cumulative time the subject's hand was not touching the steering wheel during task execution.

Procedure

Training. Detailed training in the arrangement and use of the controls and displays was provided by the front-seat experimenter while the vehicle was parked. Once the subject was familiarized with the use of each control, the subject practiced with the instruments while driving until error-free performance was demonstrated. Subject training required approximately 1 hour.

Data collection. Upon completion of training, 4 data runs were conducted, each lasting approximately 15 minutes. During each run, subjects were directed along a preselected route and asked to perform 50 instrumentation tasks. After each run, a short rest period was provided, during which the location of the auxiliary panel was changed and the collected data were down-loaded to disk. The presentation order of the tasks, road type, and panel location were counterbalanced according to orthogonal latin squares and randomly assigned to each subject. Upon completion of the fourth data collection run, the subject drove back to the laboratory facility and was then debriefed, paid, and dismissed. The data collection process required approximately 1.5 hours.

Two experimenters participated in the experiment. The front seat experimenter was responsible for determining the instantaneous driving task workload immediately prior to task execution, issuing the task commands, determining successful task performance, timing task completion, and monitoring safe operation of the vehicle. A second experimenter, situated in the back seat, was responsible for operating and monitoring the computer and eye scanning equipment, and recording the hand-off-the-wheel time during task execution with a momentary push-button interfaced to the computer.

RESULTS

An overall Analysis of Variance, collapsed over tasks indicated a significant effect due to driver age on task completion time [F(2,18) = 13.54, p = 0.0003], hand-off-the-wheel time [F(2,18) = 7.69, p = 0.0039], number of glances to the display [F(2,18) = 10.41, p = 0.0010], total glance time to the display [F(2,18) = 13.43, p = 0.0003], mean single glance time to the display [F(2,18) = 3.57, p = 0.0493], number of glances to the roadway [F(2,18) = 4.47, p = 0.0265], number of transitions between the roadway and display [F(2,18) = 9.12, p = 0.0018], total transition time [F(2,18) = 12.17, p = 0.0005], and mean transition length [F(2,18) = 4.48, p = 0.0264]. The magnitudes of the age-related age differences, the significance of these differences according to Newman-Keuls multiple comparison tests, and the dependencies among the dependent variables are illustrated in Figures 2 and 3 (points on the same line are significantly different at $\alpha = 0.05$ when designated with dissimilar letters). It is interesting to note that while the overall ANOVA indicated a significant effect of age on mean single glance time to the display, the Newman-Keuls test indicated no statistical differences among the age groups, as illustrated in Figure 4. The data did not reveal a significant effect of age on mean single glance time to the roadway [F(2,18) = 0.43, p = 0.6541], nor on total glance time to the roadway [F(2,18) = 3.41, p = 0.0553]. As illustrated in Figure 5 (by way of explanation), older subjects glanced at the roadway for a larger, but not significantly longer, amount of time. In addition, the mean length of a transition between the roadway and the display was significantly longer for older subjects, but did not differ significantly between the younger and middle-aged subject groups, as illustrated in Figure 6.

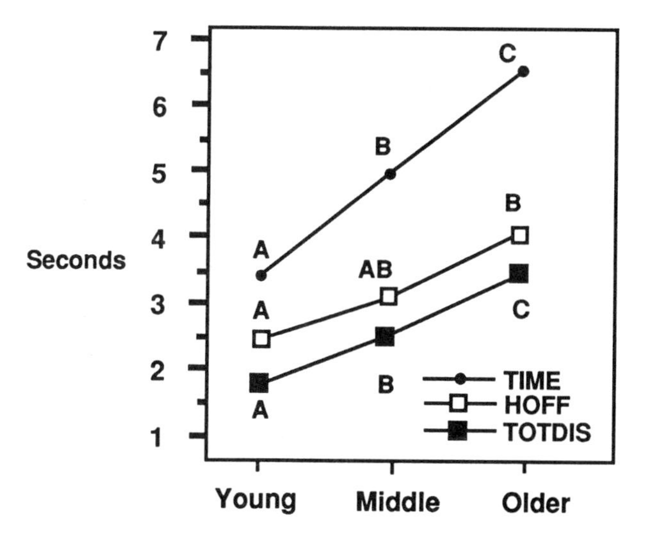

Figure 2. Mean values for hand-off-wheel (HOFF), task completion (TIME) and total display glance (DISTOT) times.

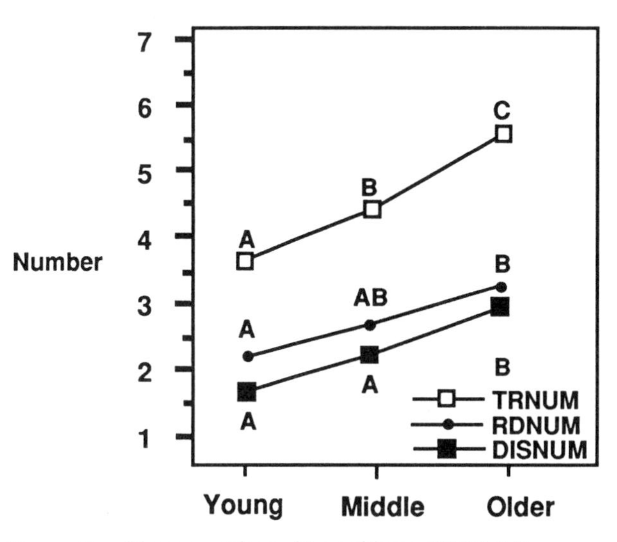

Figure 3. Mean number of transitions (TRNUM), glances to the roadway (RDNUM), and glances to the display (DISNUM).

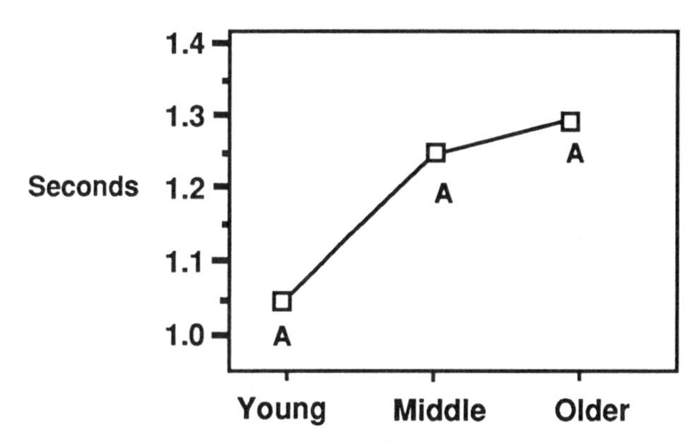

Figure 4. Mean single glance time to the display.

In addition to these results, the overall ANOVA indicated a significant effect of gender on the related variables of number of glances to the display [F(1,18) = 10.24, p = 0.0050], number of glances to the roadway [F(1,18) = 6.91, p = 0.0170], number of transitions between the roadway and the display [F(1,18) = 9.65, p = 0.0061], and total transition time [F(1,18) = 4.71, p = 0.0437]. The interaction of age and gender did not significantly effect the dependent variables.

The magnitudes of age-related decrements in instrument panel task performance may best be illustrated by examining several commonly-executed task types. A task analysis of the 200 commands used in the study yielded 58 distinct task categories, based on such

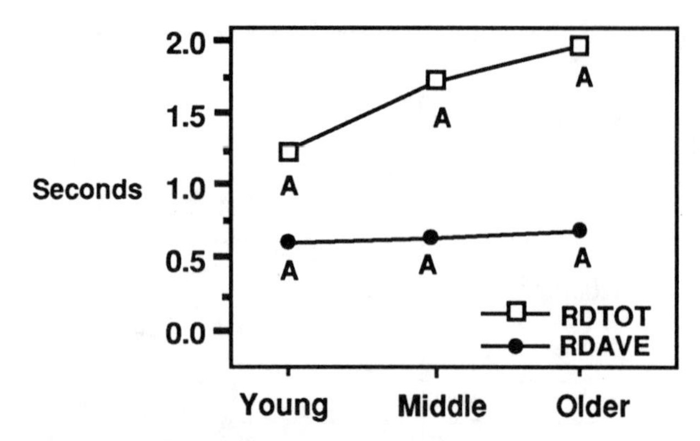

Figure 5. Mean single glance time to the roadway (RDAVE) and total glance time to the roadway (RDTOT).

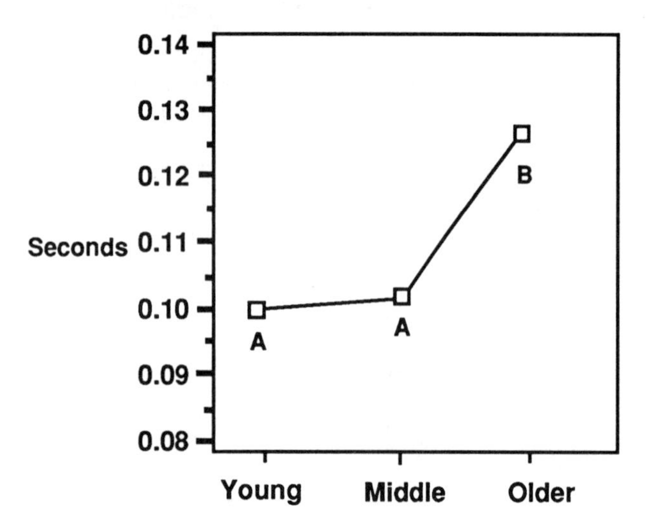

Figure 6. Mean transition time between the roadway and display.

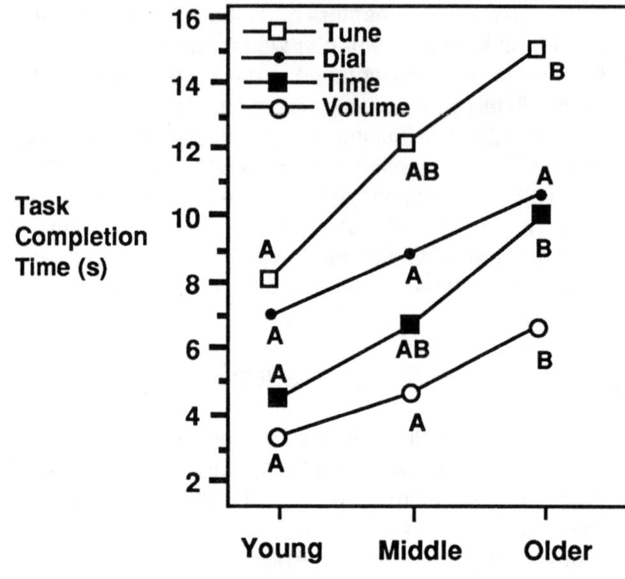

Figure 7. Task completion time as a function of age for selected tasks.

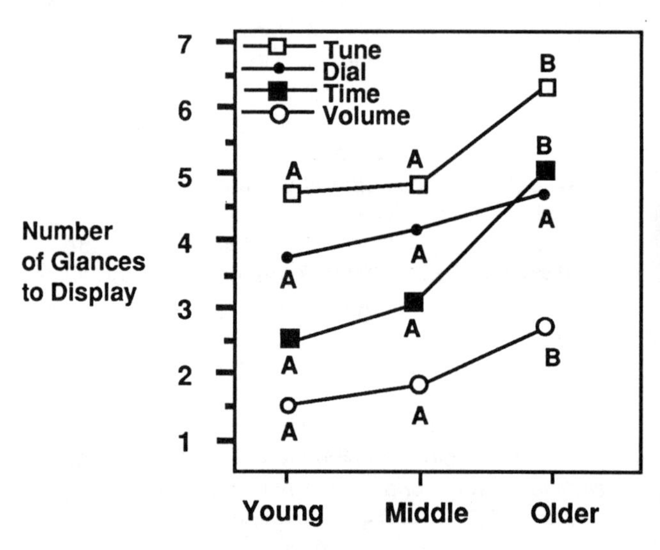

Figure 8. Number of glances to the display as a function of age for selected tasks.

parameters as the type of control used, its location, task difficulty, task complexity, and labeling. Task completion times for 4 representative tasks as a function of driver age are presented in Figure 7. These tasks were chosen as encompassing a wide range of task requirements and include: tuning an analog radio (Tune), entering a 7-digit number on the telephone keypad (Dial), pressing a button then reading the time on the radio (Time), and adjusting the volume of a radio (Volume). The mean number of glances to the display as a function of age for these same tasks is presented in Figure 8. As is evident in the figures, the number of glances to the display and the task completion time increase as a direct function of driver age. Specifically, older drivers required more glances to the display to retrieve the desired information and therefore required more time to complete the task. Similar trends were observed for most of the other tasks.

DISCUSSION

Disregarding the redundancies among the dependent variables, the results of the present study yield three salient conclusions. The older drivers tested required more glances to the instrument panel in order to retrieve the necessary information for successful task completion, required more time to complete the instrumentation tasks, and required more time to move their eyes between the roadway and the display. This latter observation may be due to the decreased motor functioning at later ages, a finding supported by Marsh (1960) and McFarland, Tune, and Welford (1964), whose studies describe a reduction in central processing capabilities due to aging.

The increased number of glances to the display and the resultant increase in task completion time may also be attributable to decreased central capacities. Older subjects may require more glances to the display in order to interpret it, and further, may retain less information while time sharing between the roadway and the display. Consequently, the time to complete the task increases.

Presbyopia further complicates the problem. The physiological changes in the visual system that naturally occur with aging include significant reductions in visual acuity, contrast sensitivity, light transfer capability, glare sensitivity, and chromatic sensitivity. Sekuler, Kline, and Dismukes (1982) and the National Research Council (1987) provide reviews of these changes.

There is evidence that the performance decrements observed in the present study may be somewhat relieved by increasing the character size of the labels used on dashboard instrumentation. Imbeau and Wierwille (1989) found that older subjects demonstrated performance degradation in simulated driving when character sizes were smaller than 17 minutes of arc. They also found, as did Poynter (1988), that low contrast ratio may cause degradation. Fowkes (1984) recommended character sizes for automotive instrument panels of 20 minutes of arc. The character sizes of the control labels used in the present study were between 8 and 16 minutes of arc for the existing instrumentation, and 14 minutes of arc for the auxiliary panel, except for certain radio legends which were as small as 7 to 9 minutes of arc.

The performance decrements observed for older subjects in the present study may therefore be viewed as resulting from some combination of reduced central processing capabilities, reduced visual capabilities, and the small character sizes used on the instrumentation. In light of these results, it is clear that further research is necessary to design appropriate automotive instrument panels for use by older drivers.

ACKNOWLEDGEMENTS

This research was sponsored by GM Project Trilby. The opinions expressed in this paper are those of the authors and do not necessarily represent those of General Motors or its employees. Special thanks are due to Dr. Brian Repa, Dr. Linda Angell, Mr. Gary Bertollini, Dr. Lenora Hardee, Dr. Ray Kiefer, Mr. Neil Schilke, and Mr. William Thomas.

REFERENCES

Fowkes, M. (1984). Presenting information to the driver. Display Technology, 5, 224-228.

Imbeau, D. and Wierwille (1989). Effects of instrument panel luminance and chromaticity on reading performance and preferences in simulated driving. Human Factors, 31 (2). (In press, April).

Marsh, B. W. (1960). Aging and driving. Traffic Engineering, 31.

McFarland, R. A., Tune, G. S., and Welford, A. T. (1964). On the driving of automobiles by older people. Journal of Gerontology, 19.

National Research Council, Committee on Vision. (1987). Work, aging, and vision. Washington, DC: National Academy Press.

Poynter, D. (1988). The effects of aging on perception of visual displays. Warrendale, PA: Society of Automotive Engineers (SAE), Technical Paper 881754.

Sekular, R., Kline, D., and Dismukes, K, (1982). Aging and human visual function. New York: Alan R., Liss, Inc.

CAR PHONE USABILITY: A HUMAN FACTORS LABORATORY TEST

Colleen Serafin, Cathy Wen, Gretchen Paelke, and Paul Green
University of Michigan Transportation Research Institute (UMTRI)
Human Factors Division
Ann Arbor, MI 48109-2150

This paper describes an experiment that examined the effect of car phone design on simulated driving and dialing performance. The results were used to help develop an easy to use car phone interface and to provide task times as input for a human performance model. Twelve drivers (six under 35 years, six over 60 years) participated in a laboratory experiment in which they operated a simple driving simulator and used a car phone. The phone was either manually dialed or voice-operated and the associated display was either mounted on the instrument panel (IP) or a simulated head-up display (HUD). The phone numbers dialed were either local (7 digits) or long distance (11 digits), and could be familiar (memorized before the experiment) or unfamiliar to the subject. Four tasks were performed after dialing a phone number; two of the tasks were fairly ordinary (listening, talking) and two required some mental processing (loose ends, listing). In terms of driving performance, dialing while driving resulted in greater lane deviation (16.8 cm) than performing a task while driving (13.2 cm). In addition, the voice-operated phone resulted in better driving performance (14.5 cm) than the manual phone (15.5 cm) using either the IP display or HUD. In terms of dialing performance, older drivers dialed 11-digit numbers faster using the voice phone (12.8 seconds) than the manual phone (19.6 seconds). Dialing performance was also affected by the familiarity of numbers. Dialing unfamiliar numbers using the voice phone was faster (9.7 seconds) than using the manual phone (13.0 seconds) and 7-digit unfamiliar numbers were dialed faster (8.2 seconds) than 11-digit unfamiliar numbers (14.5 seconds). Thus, the voice-operated design appears to be an effective way of improving the safety and performance of car phone use, but the location of the display is not important.

PREFACE

The University of Michigan Transportation Research Institute (UMTRI) under contract to the United States Department of Transportation (DOT), has undertaken a project to help develop driver information systems for cars of the future. Initially, the extent to which various driver information systems might reduce accidents, improve traffic operations, and satisfy driver needs and wants, was analyzed (Green, Serafin, Williams, and Paelke, 1991; Green, Williams, Serafin, and Paelke, 1991). That analysis led to the selection of car phones for detailed examination. This paper summarizes one of three experiments performed to improve car phone design and retrieve benchmark data to develop a driver model. For a complete description, see Serafin, Wen, Paelke, and Green (1993).

INTRODUCTION

Using a car phone while driving has become a somewhat common occurrence. The convenience of doing so attracts many people. However, its use may distract the driver's attention from the road for a longer period of time than does traditional in-vehicle equipment (speedometer, radio, etc.). Demands may be visual (needing to look at displays) or cognitive (conversing with passengers). To safely maneuver a vehicle on a highway, it is necessary to design auxiliary systems, such as car phones, that minimize the time a driver's attention is diverted from the road.

Few studies, however, directly relate to the effect of using a phone on driving performance. Of the studies that do, some include tasks other than using a phone (Brookhuis, de Vries, and de Ward, 1991; Brown, Tickner, and Simmonds, 1969). Others do not provide a primary tracking task. McKnight and McKnight (1990) had participants watch a videotape of roads, while Hanson and Bronell (1979) simply had subjects perform various tasks on a phone. Stein, Parseghian, and Allen (1987), Alm and Nilsson (1990), and Nilsson and Alm (1991) have investigated the dual tasks of driving a simulator and using a car phone. Kames (1978) has examined on-road use.

In the present study, two types of phones (manual and voice-operated) and two types of displays (instrument panel (IP) and head-up display (HUD)) were tested to determine the best design for a car phone. Tasks, simulating the range of mental and conversational activities a driver may engage in while on a car phone, were performed in place of conversations. Results were based on performance on the phone, as well as driving performance using a simulator. The following questions were addressed:

1. Which car phone interface (phone type/display type) is least distracting (as measured by steering error)?
2. Are there differences due to sex, age, or both?
3. How long does dialing take?
4. Which car phone interface do drivers prefer?

METHOD

Experimental Design

The study examined two types of phones (manual and voice-operated) combined with two types of displays (IP and HUD). Other factors that varied included the type of phone number (familiar--memorized before coming to the experiment and unfamiliar), length of phone number (7-digit and 11-digit), task combination (driving alone, dialing while driving, and performing a task while driving), type of task while driving (loose ends, listing, talking, and listening), and driver age and gender.

The design was mixed, with two between-subject variables (age and gender) and six within-subject variables (phone type, display type, type of phone number, length of phone number, task combination, and task).

The dependent variables were standard deviation of the lane position for the driving task and dialing times for the phone task. For the loose ends and listing tasks (described below), dependent measures were response time and number of items, respectively. Data from the talking and listening tasks were not analyzed.

Test Participants

Twelve licensed drivers (six men and six women), ranging in age from 20-76 years, participated in this study. They were divided into two subgroups, younger (20-35 years old, mean = 24 years) and older (over 60 years old, mean = 70 years). A requirement for participating in the experiment was that they had never used a car phone. Each participant was paid $20 for the 1-1/2 hour session.

Test Materials and Equipment

The design of the manual phone was based on several small scale human factors tests. A drawing of the manual phone design and layout is shown in Figure 1. The voice phone commands were the word equivalents of the manual phone button labels. When using the voice phone, the manual phone was in the car and the participant could refer to it.

Located next to the manual phone was a notebook with phone numbers. The IP display was located in the center console of the dashboard and the HUD was positioned to the left of the driver's view. The IP display and HUD were 68.6 cm and 119.4 cm away, respectively, from an average driver. The display indicated that power was on and showed the digits entered using the phone.

Test participants sat in a 1985 Chrysler Laser mockup. The driving scene, a single-lane road at night, was projected onto a screen 3.6 m in front of the driver. The difficulty of the driving task approximated typical nighttime driving on a single lane, slightly curved road.

A Macintosh IIcx running HyperCard was used to control the phone display and to present call processing sounds.

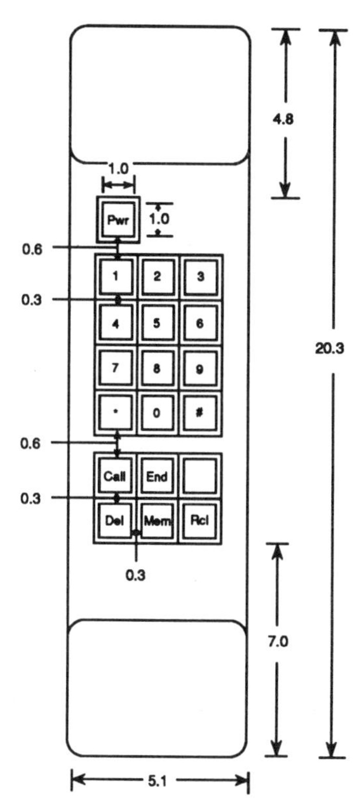

Note: All units are in cm.

Figure 1. Manual Phone Design and Layout

Test Activities and Sequence

After completing consent and biographical forms and a vision test, the participant was led to the car mockup. He/she practiced driving using the simulator for at least 2 minutes or until feeling comfortable.

Next, the dialing procedures were explained. The participant practiced dialing each type of phone number (familiar and unfamiliar 7- and 11-digit numbers) once. If he/she wanted more practice or if the experimenter thought it was necessary, the participant dialed a few more numbers. Afterwards, the participant practiced driving and using the phone at the same time. It was stressed that he/she should concentrate on driving and dial the phone numbers only when comfortable doing so.

Finally, tasks to perform in place of phone conversations were explained and practiced in the following order: loose ends, listing, talking, and listening. The loose ends task required the participant to determine how many loose ends (ends not connected to other lines) there are in a capital letter. (For example, the letter *A* has 2 loose ends at the bottom. This task was performed 3 times; a new letter was presented every 10.0 seconds.) The listing task required the participant to name as many items as he/she could in a subject category. (Some examples of the categories were a kind of fruit or a type of furniture.) The talking task required the participant to answer a question such as "What did you do last weekend?" The listening task required the participant to listen to a description of a hypothetical situation and to answer a multiple choice question about it. All of the tasks were presented aurally after the participant dialed a phone number. Each task lasted 30 seconds, and all participants were given the same questions for each task.

In the test blocks, the participant drove, used the phone, and performed the tasks together. The order of phone-display combinations was counterbalanced across participants. Within each phone-display combination, the participant dialed each of the four types of numbers (familiar and unfamiliar 7- and 11-digit numbers) and performed each of the tasks. The phone numbers and tasks were performed in the same order for all participants. After the test blocks, participants were asked to rank order the phone-display combinations (1 = best, 4 = worst).

RESULTS

Driving Performance

Driving performance was determined from the lane position on the simulator. Driving data were classified into three tasks: dialing the phone while driving, performing a task while driving, and baseline driving (no phone usage). A repeated measures analysis of variance (ANOVA) was conducted to examine the effects of the independent variables. Significant main effects resulted due to the input method (F [1, 493] = 8.34, p = 0.004), task (F [2, 493] = 11.18, $p < 0.001$), and age (F [1, 493] = 9.07, p = 0.003).

The main disturbance in driving performance was found during periods of dialing while driving (16.8 cm versus 13.2 cm for performing a task while driving and 14.2 cm for driving). When driving and dialing, voice input led to better driving performance (14.5 cm) than the manual handset (15.5 cm). Also, the driving performance of older drivers was worse (15.2 cm) than younger drivers (14.2 cm).

Dialing Performance

Dialing time was computed as the time it took to enter the digits of the phone number. A repeated measures ANOVA indicated that dialing time was significantly affected by all independent variables except display type (see Table 1). This is not surprising, since, overall, participants did not look at the displays a great deal while dialing. There were five significant two-way interactions and one significant three-way interaction.

Table 1. Significant effects from analysis of dialing performance

Significant Effects	F	p
Phone Type	12.16	0.0007
Length of Phone Number	109.02	0.0001
Type of Phone Number	29.64	0.0001
Age	119.87	0.0001
Phone Type x Phone Number Type	12.67	0.0005
Phone Number Length x Phone Number Type	4.06	0.0500
Phone Type x Age	26.19	0.0001
Phone Number Length x Age	10.24	0.0020
Gender x Age	19.19	0.0001
Phone Type x Phone Number Length x Age	8.32	0.0050

Note: Degrees of freedom for all effects are 1, 130.

The three-way interaction is shown in Figure 2. Older drivers dialed 11-digit numbers faster using the voice phone (12.8 seconds) than the manual phone (19.6 seconds). From the significant two-way interactions it was found that participants dialed unfamiliar numbers faster using the voice phone (9.7 seconds) than the manual phone (13.0 seconds), but there was no difference for familiar numbers (8.6 seconds). Also, participants dialed unfamiliar 7-digit numbers faster (8.2 seconds) than 11-digit numbers (14.5 seconds), but there was no difference between familiar numbers (8.7 seconds). Finally, older men dialed faster (11.8 seconds) than older women (13.6 seconds), while younger women dialed faster (6.2 seconds) than younger men (8.5 seconds).

Tasks

A repeated measures ANOVA on response times for the loose ends task indicated a significant effect of letters ($F[11, 93] = 5.417, p = 0.0001$) and an interaction between gender and age ($F[1, 93] = 9.417, p = 0.0028$). The response time for letter *G* was higher than every letter except *K*. While older women had longer response times (1.9 seconds) than younger women (1.6 seconds), older and younger men did not differ (1.7 seconds). For the listing task, a repeated measures ANOVA indicated no significant differences for the independent variables of interest (category, gender, age, and input method).

Driver Preferences

A survey of driver preferences revealed that the voice-HUD combination was ranked "best" by 10 of the 12 drivers; the other two preferred the voice-IP combination. The least preferred combination was the manual IP with 8 of the 12 participants ranking it "worst."

CONCLUSIONS

When compared with a manual handset, voice input with a HUD or IP display resulted in less lane position deviation for all drivers and faster dialing times for older drivers dialing unfamiliar numbers. Drivers also preferred the voice-operated phone. Thus, the voice-operated design appears to be an effective way of improving the safety and performance of car phone use. Age influenced both driving performance and dialing times, indicating that the older driver should be taken into account in the design of car phones. Finally, there may actually be location effects between a HUD and handset display location, but the study of this location was restricted by hardware limitations.

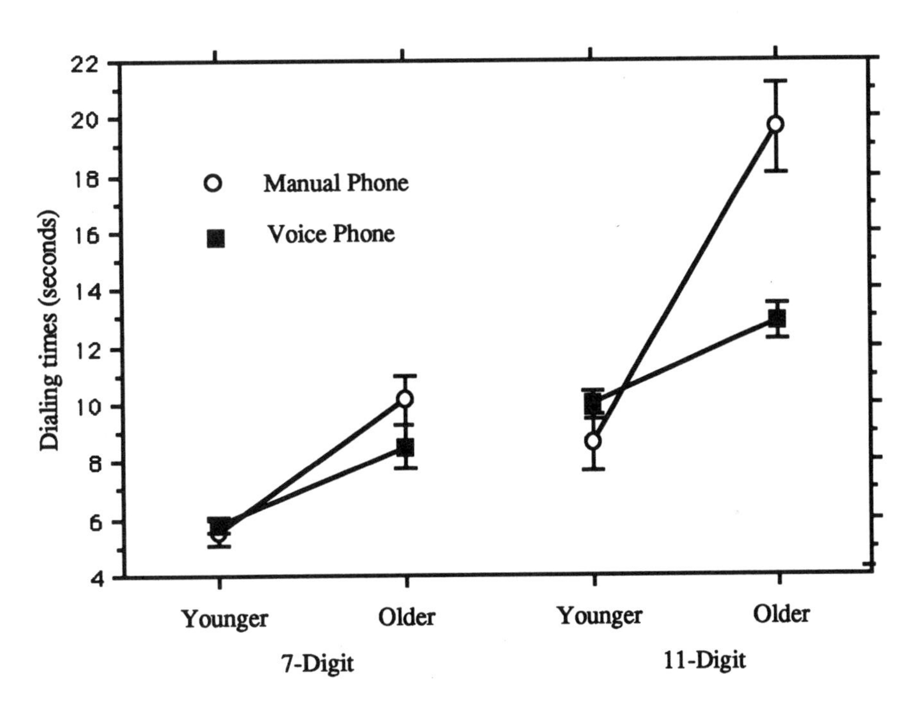

Figure 2. Dialing times using the manual and voice phones for 7- and 11-digit numbers and older and younger participants

OFFICIAL GOVERNMENT NOTICE

REFERENCES

Alm, H., and Nilsson, L. (1990, October). Changes in driver behaviour as a function of handsfree mobile telephones. (DRIVE Project V1017 (BERTIE) Report 47). Linkoping, Sweden: Swedish Road and Traffic Safety Institute.

Billheimer, J.W., Lave, R.E., Stein, A.C., Parseghian, Z., and Allen, R.W. (1986). Mobile telephone safety study (Draft Technical Report). Los Altos, CA: SYSTAN.

Brookhuis, K.A., de Vries, G., and de Waard, D. (1991). The effects of mobile telephoning on driving performance. Accident Analysis & Prevention, 23(4), 309-316.

Brown, I.D., Tickner, A.H., and Simmonds, D.C.V. (1969). Interference between concurrent tasks of driving and telephoning. Journal of Applied Psychology, 53(5), 419-424.

Green, P., Serafin, C., Williams, G., Paelke, G. (1991). What functions and features should be in driver information systems of the year 2000? In Proceedings of the Vehicle Navigation and Information Systems Conference (VNIS '91) (SAE paper 912792) (pp. 483-498). Warrendale, PA: Society of Automotive Engineers.

Green, P., Williams, M., Serafin, C., and Paelke, G. (1991). Human Factors Research on Future Automotive Instrumentation: A Progress Report. In Proceedings of the 35th Annual Meeting of the Human Factors Society. (pp. 1120-1124). Santa Monica, CA: Human Factors Society.

Hanson, B.L., and Bronell, C.E. (1979, May). Human factors evaluation of calling procedures for the advanced mobile phone system (AMPS). IEEE Transactions on Vehicular Technology, VT-28(2), 126-131.

Kames, A.J. (1978, November). A study of the effects of mobile telephone use and control unit design on driving performance. IEEE Transactions on Vehicular Technology, VT-27(4), 282-287.

McKnight, A.J., and McKnight, A.S. (1991). The effect of cellular phone use upon driver attention. Washington, D.C.: AAA Foundation for Traffic Safety.

Nilsson, L., and Alm, H. (1991, March). Effects of mobile telephone use on elderly drivers' behaviour-including comparisons to young drivers' behaviour (DRIVE Project V1017 (BERTIE) Report 53). Linkoping, Sweden: Swedish Road and Traffic Safety Institute.

Serafin, C., Wen, C., Paelke, G. and Green, P. (1993). Development and human factors tests of car telephones (Technical Report 93-17). Ann Arbor, MI: The University of Michigan Transportation Research Institute.

Smith, V.J. (1978, March). What about the customer? A survey of mobile telephone users. Presented at the 28th IEEE Vehicular Technology Conference, Denver, Colorado.

Stein, A.C., Parseghian, Z., and Allen, R.W. (1987). A simulator study of the safety implications of cellular mobile phone use. In Proceedings of the 31st Annual American Association for Automotive Medicine (pp. 181-200). New Orleans, Louisiana.

ESCAPE WORTHINESS OF VEHICLES WITH PASSIVE BELT RESTRAINT SYSTEMS

Edmundo Rodarte
Jerry L. Purswell
Robert Schlegel
University of Oklahoma
Norman, OK 73072

Richard F. Krenek
Krenek and Associates, Inc.
Norman, OK 73072

ABSTRACT

There are a variety of conditions that can exist in the post-crash environment which make rapid escape necessary for survival or to avoid further injury. These include a post-crash fire, the vehicle going into the water, or avoiding being struck in a secondary collision. The National Highway Traffic Safety Administration (NHTSA) has defined this parameter vehicle escapeworthiness. It has been estimated in past research performed by the author for NHTSA that escapeworthiness becomes important in up to 7% of all vehicle crashes. Since escapeworthiness research was performed in the early 1970's, the advent of passive shoulder belt systems has made it necessary to again review the impact of this development on escapeworthiness. In particular, the inability of the occupants to release the passive restraint because the door cannot be opened after the crash, coupled with the inability to release the passive restraint due to its design or a lack of experience, or knowledge of how to release the passive restraint while the door is closed, creates a serious problem. Thus, the present study was performed to investigate the impact of passive restraint systems on the time required to escape from the vehicle under various conditions of available escape routes, and physical condition of the occupants. The experimental design included the variables of age, gender, escape route, level of incapacitation and type of passive restraint system. The times to effect an escape as well as the method of escaping were determined through videographic analysis of all escape trials. The findings demonstrated that the use of passive restraint systems increased the time to escape significantly, ranging from 37 to 65 percent for the respective conditions. This difference may determine whether a person survives or not after some post-crash conditions. The results have significance for the design of passive restraint systems for easy release, while at the same time not creating an incentive for some users to routinely leave the passive restraint unfastened.

INTRODUCTION

The following true example illustrates the need for post-crash escapeworthiness of vehicles. A tourist couple from Europe were driving in Miami when their rented Ford Escort was struck from behind as they stopped in front of the gates which were coming down at a railway crossing. The Escort was propelled onto the tracks as it was also spun around from the collision at the right rear. The driver of the vehicle was able to exit the vehicle and his passive shoulder belt retracted as expected. Seeing the approaching train, he quickly tried to open the passenger door to help his dazed wife exit. The door was jammed from the collision. Going back into the driver's side door, he attempted to free his wife by releasing her shoulder restraint, but could not find the emergency release near the center console. He then ran to flag the train which was rapidly approaching, but the train was unable to stop in time to avoid the collision. His wife was fatally injured when the Escort was struck by the train.

There are other situations where a vehicle may be involved in a collision and come to rest on a roadway where it is subject to secondary collisions or where a fire ensues after the crash. In such circumstances there is a need for the passengers to make a rapid escape or for rescuers to quickly remove injured passengers. With the advent of passive shoulder belts, there are more actions required to release both the lap belt and passive restraint shoulder belt in an escape situation. There is also the issue of whether the driver or passenger knows how and where to release the passive shoulder belt when the in-frequent need arises. From the stand-point of overall safety of the system, however, the argument can be made that perhaps the passive shoulder belt will be released and not used if the driver or passenger of the vehicle knows that it can be readily released.

The purpose of this study was to examine this issue of whether a passive shoulder belt system would add significantly to the time to escape from a vehicle in an emergency condition or to the time required to rescue an injured passenger with such a restraint.

METHOD

Experimental Design

Five independent variables were investigated in this study. They were as follows:

1. Seat Belt Type:
   (a) non-detachable passive shoulder belt
   (b) detachable passive shoulder belt
   (c) standard three-point manual belt

2. Escape Routes:
   (a) two doors
   (b) driver's side door
   (c) driver's side window
   (d) two windows

3. Occupant Condition:
   (a) no injuries
   (b) one injury

4. Age Group:
   (a) 19-29 (younger)
   (b) 39-50 (older)

5. Gender:
   (a) two males in front seats
   (b) two females in front seats

The injured passenger condition was simulated by using an adult size anthropometric dummy minus the legs for the passenger. The legs were removed to reduce the weight of the dummy from 70 kg to 54 kg to better correspond with the average female adult weight. The Institutional Review Board of the University of Oklahoma considered that extracting a person from the vehicle in a simulated emergency condition presented too great a risk of injury.

Experimental Equipment

A 1993 Ford Escort with four doors was used for the passive restraint vehicle. A 1985 Ford Escort was used for the three-point, continuous-loop belt system. The body styles of the two vehicles were similar, except that the 1993 Escort had a slightly larger front door opening (8597 sq cm vs. 7721 sq cm). The Ford Escort with a passive restraint system originally had the release for the shoulder belt in the center console. In later models the release was placed at the point where the belt attaches to the track over the door.

The experimental vehicles were modified so that the experimenter could cause any door or window to be inoperable by either the driver or passenger for a given trial. Thus, there was the element of uncertainty for the driver or passenger after the trial started as to which escape routes would be available. This added an element of realism of the actual crash situation where the available escape exits are not likely to be known until the passengers try to escape.

The signal to start the escape trial was provided by an 80 dbA fire alarm which sounded for three seconds. The possibility of injury while escaping from the windows was minimized by the use of eight inches of foam mattresses under each window. A video camera was used to record the motions of the subjects both inside and outside the vehicle.

Subjects

A total of 12 females and 12 males participated as subjects. They were classified into two binary dimensions, the older and younger groups and the male and female groups. The older group had an age range of 39-59, while the younger group had an age range of 19-29. The height of the female subjects ranged from 1.47 m to 1.82 m ($\bar{x}$=1.63 m), while the male heights ranged from 1.68 m to 1.96 m ($\bar{x}$=1.81 m). Weights of female subjects ranged from 45.5 kg to 72.7 kg (x=61.8 kg), while that of males ranged from 63.6 kg to 97.7 kg ($\bar{x}$=79.5 kg). All were screened for any medical conditions that would have affected their safety in performing the escape trials. The subjects were assigned to the experimental conditions in such a manner as to minimize learning. Each subject participated in only three trials, and none represented the same experimental condition. While it would have been ideal to have only one group of subjects for each condition, the assignment used represented a reasonable compromise.

Procedure

The experimental trials were conducted in the University of Oklahoma armory where the interior lighting could be controlled to a level adequate for subjects to observe the interior of the Escort while effecting an escape and sufficient for the video camera to record movements both inside and outside the vehicle.

The seat position of the Escort was inclined at 95 degrees and the seat distance from the dash was adjusted for the mid position. Subjects were instructed to buckle the manual lap belt or the three point belt after entering the vehicle. They were told that the passive shoulder belt would deploy around them after they entered. They were told that one or more of the two front doors and two front windows would be available for escape. They were not to anticipate the start signal and they must be watching the video camera at the start of the trial. They were also told that the dummy represented an injured person and that it should be treated as such. In the case of the female groups, they were told that they should release the lap and shoulder belts, but not attempt to remove the dummy as done by the male groups. In the condition where the trial was for a shoulder belt that would not release from the track, the release button was taped before the trial.

Subjects were not permitted to observe other trials in order to minimize learning effects.

Criterion Measures

The two criterion measures were the time for the first person to escape and the time for both persons to exit or for the dummy to be removed from the vehicle in the case of the male groups with an injured passenger.

RESULTS

The data was analyzed using a factorial ANOVA. Results of this analysis included the following:

* Driver escape time was significantly (α=.05) affected by seat belt type, escape route, occupant condition and gender, but not by the age groups included in this study.

* Passenger escape time was significantly (α=.05) affected by escape route only.

* Total escape time was significantly (α=.05) affected by seat belt type, escape route, occupant condition and gender, but not by the age groups included in this study.

A Neuman-Keuls range test on driver, passenger and total mean escape times indicated the following:

* The three point manual belt permitted significantly (α=.05) faster escape times than either the passive detachable or passive non-detachable restraint systems (See Figure 1).

* As expected, window escape routes produce significantly longer escape times than door escape routes (See Figure 2).

* An injured person significantly (α=.05) increased total escape time (by over 100%) over the situation where neither occupant was injured. Driver escape times were significantly (α=.05) higher where there was an injury to the passenger (See Figure 3).

* Driver and total escape times for male subjects was significantly (α=.05) shorter than for female subjects (See Figure 4).

DISCUSSION

This study has demonstrated that the added safety advantage of a passive shoulder belt restraint system can add significantly to the time to escape from a vehicle under certain emergency conditions. Even when the belt was detachable, escape times did not differ significantly from non-detachable shoulder belts because of the difficulty that subjects had in understanding the method and finding the location to release the shoulder belt. Subjects were observed attempting to find the release button near the point that they had released the lap belt, rather than near the track for the belt.

This observation may indicate the need for manufacturers to carefully consider the needs for instructions and labels about how to operate the release mechanism for the shoulder belt. At a more basic level, however, standardization of both passive restraint release location and release mechanism operation among automobile manufacturers should be considered. This would be important for drivers and passengers who are not familiar with the vehicle as is the case when renting an automobile on a short term basis.

One aspect to consider in designing the passive belt release mechanism is that a passive restraint that is easy to release may encourage some users to release the passive restraint and not use it.

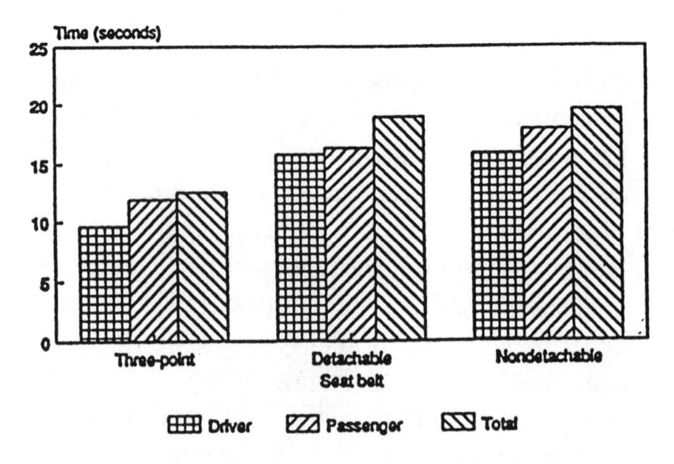

Figure 1. Driver, passenger and total escape time as a function of seat belt type (male and female).

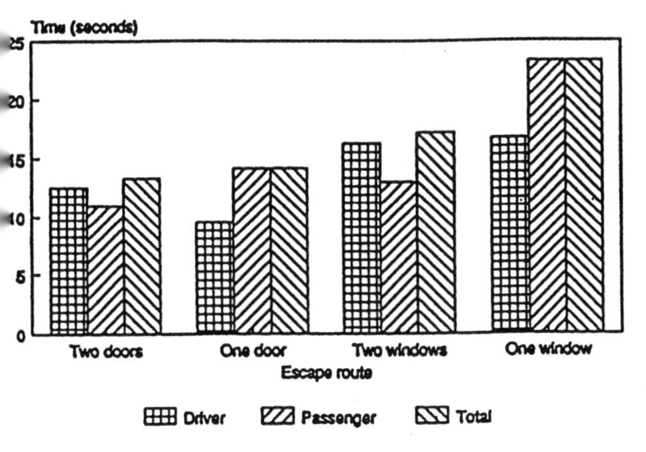

Figure 2. Driver, passenger and total escape time as a function of escape route (male and female).

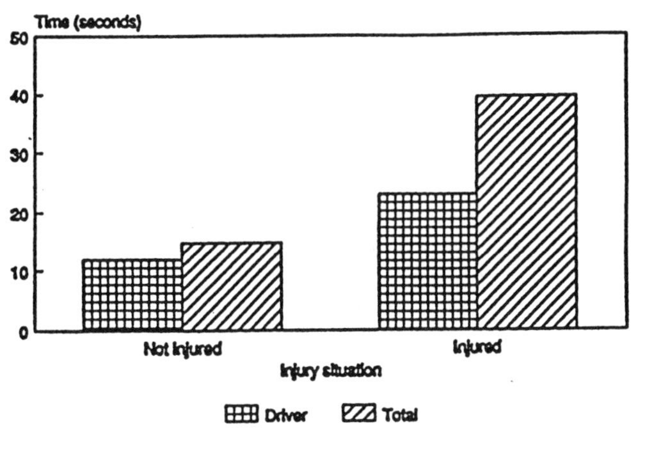

Figure 3. Driver and total escape time as a function of injury situation (male subjects).

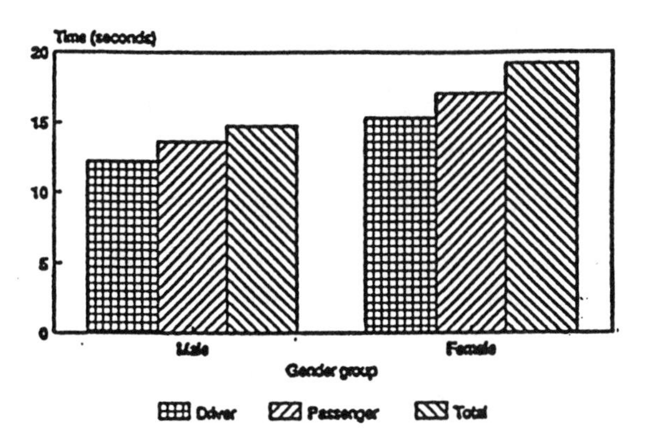

Figure 4. Driver, passenger and total escape time as a function of gender group (male and female).

REFERENCES

Congress of the United States, Technology and Handicapped People. Background Paper #1: Mandatory Passive Restraint Systems in Automobiles: Issues and Evidence, November 1982.

Department of Transportation, Passive Restraints for Automobile Occupants - A Closer Look, July 1979.

Kurylko, D., Escort Passenger Decaptitated by Motorized Shoulder belt. Automotive News, June 1991.

Miami Police, Accident report for Miko accident, 1990.

University of Oklahoma Research Institute, Escapeworthiness of Vehicles for Occupancy Survivals, Report No. 1770-FR-1-1, July 1972.

Williams A., J. Wells, A. Lund , N. Teed , Seat Belt Use in Cars with Automatic Belts. Insurance Institute for Highway Safety, August, 1990.

DESIGNING AND OPERATING SAFER HIGHWAYS FOR OLDER DRIVERS: PRESENT AND FUTURE RESEARCH ISSUES

Truman Mast
Federal Highway Administration
Turner-Fairbank Highway Research Center
McLean, Virginia 22101

A major shift is occurring in the population age distribution resulting in a growing number of older persons with an increasing demand for mobility. For a variety of reasons, meeting the present and future mobility needs of older persons will depend primarilary on the automobile. This paper will discuss key issues related to ongoing and planned research sponsored by the Federal Highway Administration (FHWA) to improve the mobility and safety of older drivers and pedestrians. The scope of FHWA research activities concerns aspects of highway design and operations that involve direct user interface. For example, geometric design and traffic control device standards explicitly consider quantifiable driver characteristics such as perception-reaction time or visual acuity. Ongoing research is addressing the extent to which existing traffic control device elements accommodate drivers with age-related diminished performance capabilities. Current work is testing older driver responses to brighter and larger signs with varying legend spacing and font characteristics in order to make recommended changes to existing highway signing standards. Older driver perception-reaction time is being evaluated in a variety of intersection, stopping, and decision sight distance situations. Studies are also underway to improve pavement markings and delineation systems to enhance their utility for older drivers. Driving simulation and field methods are being used to investigate vehicle maneuvers which cause difficulties for older drivers so that improvements can be made to highway design and operations. Future studies will determine the capabilities and limitations of older pedestrians, address the visibility problems of symbol signs and changeable message signs, and investigate all aspects of intersection design and operations in light of older driver and pedestrian capabilities.

INTRODUCTION

A major shift is taking place in the population age distribution resulting in a greater number of older persons with an increasing demand for mobility. Around age 65, older drivers exhibit a significant increase in accident rates and right-of-way violations. These facts are well documented in the Transportation Research Board's (TRB) 1988 Special Report, Transportation in an Aging Society: Improving Mobility and Safety for Older Persons.

Late in 1989 the Federal Highway Administration (FHWA) enacted a high priority national research program to improve highway travel for older drivers in the U.S. Its objectives are to identify, develop, and evaluate engineering enhancements to the highway system to meet the needs of older road users. This paper will review key issues related to ongoing and planned research under this program.

TRAFFIC CONTROL DEVICES

Intersection Control

In general, older drivers and pedestrians are over-represented in fatal and serious injury accidents at intersections. In addition, intersection design and control operations were targeted as a critical research area in the TRB (1988) report. It is clear from previous research that the older user is often confused at complex intersections and misunderstands the meaning of certain traffic signal control phases. Driver errors are most frequently associated with displays that involve flashing operations, multiple colors on the left turn signal, and different colors on the turn and through signals. When confused, many older drivers exhibit conservative behavior, such as stopping and/or waiting. Such behaviors cause traffic delays, and hesitant actions or unwarranted stops are often associated with rear-end collisions.

An FHWA study will begin in 1991 to evaluate alternate intersection traffic control devices to accommodate the perceptual, cognitive, and psychomotor capacities of older drivers and pedestrians. Urban, suburban, and rural intersections will be included. Traffic control device features to be investigated include traffic signal display type, signal placement, supplemental signing, signal phasing, novel displays, flashing displays, and left turn arrows. These traffic control device features will be studied in combination with other intersection factors -- geometry, traffic volumes, and environmental visual complexity.

Changeable Message Signs

Changeable message signs (CMS) are a primary element in real-time information systems, widely used in work zones and in many heavily congested freeway corridors. The number of CMS is expected to increase significantly during the next decade as a part of the implementation of new advanced highway technology. With the advent of Intelligent Vehicle-Highway Systems (IVHS) more real-time information will be available for motorists. CMS will be part of the highway information system to advise and re-direct drivers to less congested routes. Use of these devices to provide advance warning of incidents, adverse weather conditions, or other potentially hazardous situations on the roadway will also increase.

Drivers of all ages are having visibility problems with changeable message signs. Moreover, focus group and research studies indicate that older drivers experience major difficulties in the freeway environment in general and with CMS in particular. Unlike fixed highway signing, the Manual on Uniform Traffic Control Devices, (1988) (MUTCD) has no specific, detailed standards for CMS.

The FHWA will initiate work in 1991 to develop design guidelines and operational recommendations that will ensure adequate conspicuity and legibility of CMS. The guidelines and recommendations will be applicable to the full range of CMS hardware types currently or soon-to-be available, including light-emitting and light-reflecting technologies. The research will consider driver perceptual and cognitive processes involved in the utilization of CMS. Although drivers from all age groups will be tested, emphasis in this study will be on older drivers, age 65 and above. The design guidelines will incorporate the performance capabilities of older drivers and will consider the effects of variabilities such as sign size, letter height, stroke width, spacing, font, text and background color, daytime and nighttime luminance/contrast, glare, and complexity of visual surround.

Symbol Signing

Since the 1970's there has been an increasing trend toward the use of symbol signs. A well-designed symbol sign has advantages over standard text signs including greater legibility distance and shorter comprehension time. However, a number of symbol signs cause legibility and comprehension difficulties for drivers. Given the diminished visual capability and cognitive slowing that accompany aging, it is apparent that these problematic symbol signs are even more troublesome for the older driver. In fact, symbol signs have been identified in focus group analyses as causing difficulties for older drivers.

Research began this year to determine, through analytical and laboratory studies, which symbol signs are problematic for older drivers. Alternative symbol designs will be created and tested where problems exist, and recommendations will be made to improve these devices. Individual elements of symbol signs will be investigated to determine which graphic components, and in what combinations, are critical to good sign design. Symbol sign design guidelines will then be developed to accommodate the capabilities of all drivers, with particular emphasis on the needs of those over 65. These guidelines will consider perceptual and cognitive abilities involved in processing highway sign information. Sign attributes such as stroke width, pictorial elements, separation of elements, level of complexity, and use of color will be addressed.

Hazard Markers

Hazard markers are an essential element in highway safety, serving to warn drivers about fixed objects adjacent to the roadway, such as underpass piers, bridge abutments, narrow shoulder drop-offs, or hidden culverts. The Manual on Traffic Control Devices (1988) provides design guidelines for object markers, but offers minimal guidance on where they should be located. Because of the differing State and local practices in marking hazards, motorists have become confused regarding the meaning of these devices. In particular, confusion over the deployment and meaning of post mounted delineators versus hazard markers exists. In some cases post-mounted delineators, intended as guidance devices to mark the alignment of the roadway, are incorrectly utilized to identify severe roadside hazards such as roadside culverts or even protruding bridge abutments.

Older drivers share in this confusion over exactly what these devices indicate, furthermore, hazard markers may not be adequately designed to meet the conspicuity and legibility requirements of older drivers. Problems of conspicuity and comprehension associated with these devices can have serious consequences. Up to 14 percent of occupant fatalities are the result of collision with a fixed object. Furthermore, the severity index for collisions with fixed objects such as bridge piers is extremely high. Research will begin in 1992 to determine appropriate guidelines and standards for hazard markers to accommodate the comprehension and visibility needs of older drivers. Recommended warrants will be developed to clearly define when and how these devices should be deployed.

NIGHT DRIVING

Pavement Markings and Delineation

The average older driver is at a great disadvantage in night-time driving situations due to age-related decrements in visual capacity, especially contrast sensitivity. As a result, many older drivers limit their night driving. Despite this, there are older drivers who desire or need to be on the roadway at night, and their numbers will steadily increase in the next 30 years. Demographic changes already underway suggest that there may be a growing tendency to delay retirement. This could change current driving patterns and increase the amount of night driving by the over 65 age group.

Highway pavement markings and delineation primarily serve the night driver. They are consequently candidates for highway improvements that may help serve the needs of older motorists. Accident analyses have failed to find a relationship between enhanced highway delineation and reduced accident rate, particularly for the older driver. However, improved mobility is also an issue with older motorists and it is important to consider other measures of effectiveness for highway improvements. The Transportation Research Board report (TRB, 1988) calls special attention to the highway delineation needs of older drivers. In addition, focus group analyses and previous research have found that older drivers have difficulty seeing pavement delineation at night.

In September 1990 FHWA research began to identify the highway delineation needs of older drivers, and to evaluate situations where older driver performance may be improved by enhanced delineation and pavement markings. The work focuses on rural two-lane roadways and unlighted freeways under night-time conditions using both low and high beam headlights. Various types of delineation and pavement marking treatments are being investigated. These treatments include changes in standard width for pavement markings; utilization of raised or recessed pavement markers; addition of textured pavement associated with markings; installation of post-mounted delineators; and, use of chevron alignment signs. The use of these various treatments, individually and in combination, are being studied in conjunction with differing roadway geometries. A cost/benefit analysis will be used in the assessment of the alternative improvements to determine their potential effects on highway safety and traffic operations.

Sign Visibility

The need to increase highway signing visibility at night is well-documented in the TRB report (1988). There are two methods of increasing nighttime sign visibility. One alternative involves the use of brighter and more highly reflective sign materials. This would significantly increase sign conspicuity and give drivers more time to read and respond appropriately. The other alternative is to increase the letter height of the sign to make the sign legible from a longer distance.

Both alternatives present potential problems. Increasing letter size by one-third can result in a 60-70 percent increase in sign area. Highly reflective materials can cause irradiation, the spreading of light from very bright areas onto nearby dark ones. Irradiation has been shown to cause a reduction in letter edge contrast, making it necessary to increase the stroke width of the letters in the legend. Any change in the stroke width required for night legibility could adversely affect the daytime legibility of the sign.

Research begun in 1990 will determine the relative conspicuity and legibility for increases in sign legend size and the use of highly retroreflective materials. Letter style series and stroke width will be addressed. The study will investigate four categories of text signs: regulatory, warning, guide, and construction. Particular emphasis will be given to meeting the nighttime visibility needs of older drivers.

GEOMETRICS

Intersection Sight Distance

Sight distance is a critical factor in highway design and must allow drivers sufficient preview of the roadway to maintain safe and efficient vehicle control. Sight distance equations are a critical factor in highway geometric design. Sight distance determinations involve an equation which assumes a specific value, or range of values, for driver perception-reaction time. Sight distance equations currently used in intersection design, for example, contain a 2 to 3 second perception-reaction time, depending on the level of control. This time period includes the time needed for the driver to (1) perceive and recognize an approaching vehicle, (2) decide what evasive action is necessary, and (3) initiate the evasive action.

Numerous studies show that older drivers are over-represented in accidents and right-of-way violations at intersections. This pattern of accident involvement may be due, in part, to use of perception-reaction time values which are inadequate for older drivers. Research is now underway to determine the appropriate values needed to adequately represent older drivers in sight distance determinations. Based on measures of older driver perception-reaction time, potential changes required to sight distance equations to enhance the safety of older drivers will be identified. The effects of such changes will also be assessed from a practical, operational standpoint. Because of the possibility that sight distance equation models are flawed, the feasibility of alternate models, such as a gap acceptance model, will be investigated for use in highway and intersection design.

Intersection Geometric Design

Sight distance is only one aspect of intersection geometric design. Because intersections are the most hazardous portion of the highway system for all users, particularly for older ones, they are the subject of extensive research by FHWA. Certain geometric configurations such as acute angles, staggered intersections, and designs used for divided highways, may be difficult for older drivers and pedestrians to negotiate. Other intersection geometric features, such as large turning radii, may make intersections difficult for older pedestrians.

Research will begin in 1992 to investigate the needs of older drivers and pedestrians in intersection geometric design. Based on the research findings, recommendations will be made for new and revised intersection standards to account for these needs. Cost/benefit analyses will provide data on the practical feasibility of the recommendations. Where geometric changes are not possible, recommendations will be made for other operational or traffic control device improvements.

PERFORMANCE IN TRAFFIC

Traffic Maneuvers

Certain traffic maneuvers appear to pose particular difficulties for the older driver. In these instances, the older driver may be endangering himself as well as causing safety and operational problems for all other highway users. Driving on freeways, merging, changing lanes, and exiting are difficult for some older drivers. On roadways where access is not limited, older drivers often experience problems making left turns against oncoming traffic, judging gap acceptability, and performing two-lane road maneuvers such as overtaking and passing.

FHWA funded research has been underway since 1990 to investigate the difficulties older drivers have with certain traffic maneuvers and to propose recommended countermeasures. The recommendations will emphasize potential changes in highway design and traffic operations, and will include modifications to traffic control devices and alterations in geometric design. The research involves an accident analysis to determine the types of traffic-maneuver related accidents that older drivers are experiencing. Also included is an investigation of certain basic abilities among older drivers such as motion detection, motion perception and estimation of time to collision.

Some of the accident analysis work has been completed, and the initial findings based on police-reported accidents in Michigan and Pennsylvania have been reported (Staplin and Lyles, 1991). Some of the preliminary findings are as follows:

Left turns against traffic. Clearly the most problematic maneuvers for older drivers were left turns against oncoming traffic. This seems to be related to problems with judging time-to-collision and gap acceptability. Not only do older drivers have problems judging an approaching vehicle's motion characteristics, but their physical responses are slow.

Merging/weaving maneuvers. The age group 76+ is over-represented in merging and weaving accidents. Older drivers are more likely to fail to yield the right-of-way and to use lanes improperly (e.g. straddle the lane) compared to younger drivers. Older drivers in comparison to younger ones appear to have more problems with trucks than with automobiles in making these maneuvers.

Freeway Design Needs

Traffic engineers and others have observed that a substantial number of older drivers either avoid urban freeways, or experience difficulty using them. This raises several mobility and safety issues concerning older drivers. First of all, avoidance of freeways restricts mobility. Moreover, accident rates indicate that freeways are the safest portion of the roadway network. Older driver safety would therefore seemingly be enhanced through greater utilization of freeways. Traffic engineers from the sunbelt areas of the country observe that the slow and erratic driving behavior of large numbers of older drivers can significantly reduce freeway capacity in urban areas.

These and other issues suggest research is needed to gain a better understanding of older driver problems with freeway driving. In 1992 FHWA will fund a study to investigate the design of the freeway system in light of the needs and capabilities of older drivers. Those aspects of freeway driving that cause the greatest difficulties for older drivers will be identified. The study is intended to be a problem definition effort to be used in guiding future research activities in this area. It will be limited to an extensive literature review, freeway accident analysis, and extended focus group sessions with older drivers. There will also be observational data gathered on older driver behavior regarding freeway usage.

CONCLUDING REMARKS

Ongoing and planned research sponsored by the FHWA under a high priority national program has been briefly described, and key issues discussed. The program is in the beginning stages and is expected to continue beyond 1996. Longer term work, not discussed here, will address problems such as traffic control in construction and maintenance zones, highway lighting, older pedestrian characteristics, and advanced traffic management technologies. The program will culminate in the development of computer based older driver performance models, data handbooks, highway engineering guidelines, and recommended national warrants and standards.

The program emphasizes the improvement of older driver and pedestrian mobility as well as safety. It supports and responds to the TRB recommendation to improve highway design to accommodate older users and is guided by the knowledge that the older driver population is heterogeneous in its performance capabilities. Consequently, instead of restricting drivers' licenses at a certain maximum age, highway mobility and safety will be improved by reducing the demands of the driving task to better serve drivers of all ages.

REFERENCES

Federal Highway Administration, Manual on Uniform Traffic Control Devices, U.S. Government Printing Office, Washington, D.C., 1988 edition.

Staplin, L. and Lyles, R. "Specific Traffic Maneuver Problems for Older Drivers," Paper presented at Transportation Research Board Annual Meeting, January 14, 1991, Washington, D.C.

Transportation Research Board, Transportation in an Aging Society: Improving Mobility and Safety for Older Persons, Special Report No. 218, Transportation Research Board, Washington, D.C., 1988.

Author Index

Subject Index

Human Factors and Ergonomics Society Membership Benefits

What Is HFES?

The Human Factors and Ergonomics Society (formerly the Human Factors Society) is the principal professional association in the United States that is concerned with the study of human characteristics and capabilities and with the application of that knowledge to the design of the products, systems, and environments that people use.

Since its formation in 1957, HFES has promoted the discovery and exchange of human factors and ergonomics knowledge, as well as education and training for students and practitioners.

Members Have Diverse Backgrounds

HFES has more than 5000 members located throughout the United States and in 43 other countries. They are employed in industry, universities and colleges, government, consulting, military, public utilities, and other settings.

Members have academic specialties in psychology (39%), engineering (22%), human factors/ergonomics (9%), industrial design (2%), medicine (4%), and many other fields.

HFES Publications

All members receive four regular publications as a benefit of membership.

Human Factors. This quarterly peer-reviewed journal presents reports of basic and applied research, advances in methods and applications, and reviews of the state of the art. *Human Factors* is an invaluable source of information for those who work in the human factors and ergonomics field and a service to researchers who wish to disseminate their findings.

Ergonomics in Design. The Society's quarterly magazine contains articles, case studies, debates, commentary, and book and product reviews. The focus of *Ergonomics in Design* is the application of human factors/ergonomics research to the design, development, test, and maintenance of human-machine systems and environments.

HFES Bulletin. This monthly newsletter covers news of Society events and committee activities, reviews of meetings and courses, job opportunities, ads for products and services, calls for papers, and issues of concern to human factors/ergonomics researchers and practitioners.

Directory and Yearbook. Each year the Society updates its directory of members. Included are descriptions of the previous year's activities within HFES committees, chapters, and technical groups; alphabetical and geographical member listings; the HFES Code of Ethics; and the Society's Bylaws.

Members also receive discounts on other publications.

Standards Development Activities

Members represent the field in the development of national and international ergonomics standards on computer workstation and software design, medical devices, safety, and a number of other areas.

Technical Areas

There are 20 technical interest groups within HFES, each organized to promote information exchange: Aerospace Systems, Aging, Cognitive Engineering and Decision Making, Communications, Computer Systems, Consumer Products, Educators' Professional, Environmental Design, Forensics Professional, Individual Differences in Performance, Industrial Ergonomics, Macroergonomics, Medical Systems and Rehabilitation, Safety, Surface Transportation, System Development, Test and Evaluation, Training, Virtual Environments, and Visual Performance.

Technical groups contribute to the HFES annual meeting program, distribute newsletters, and conduct periodic meetings.

Local Chapters

Chapters offer events featuring noted speakers, tours of local facilities, symposia on developments in human factors, and social activities. For the location of a chapter in your area, call HFES at the number below.

Technical Meetings

The five-day annual meeting of the Human Factors and Ergonomics Society is held each fall and includes an extensive program featuring the latest research discoveries; methods for research, design, and training; panel discussions and debates on important issues in the field and in the practice of human factors; hands-on workshops by technical specialists; tours of technical and research facilities in the host city; and technical group meetings.

More than 90 lecture, panel, debate, symposium, demonstration, and special sessions are offered. Published proceedings are available at the meeting, representing the work of more than 300 member and nonmember contributors.

HFES members receive substantial discounts on meeting registration. The proceedings are included in the registration fee.

Placement Service

A year-round job-matching database service assists companies and job seekers. One-time searches and subscriptions are available.

To receive information about joining HFES, contact the Society at:

Human Factors and Ergonomics Society
P.O. Box 1369
Santa Monica, CA 90406-1369 USA
310/394-1811 Fax 310/394-2410
hfes@compuserve.com
http://hfes.org